Surveys in Differential Geometry

Vol. 1: Lectures given in 1990
edited by S.-T. Yau and H. Blaine Lawson

Vol. 2: Lectures given in 1993
edited by C.C. Hsiung and S.-T. Yau

Vol. 3: Lectures given in 1996
edited by C.C. Hsiung and S.-T. Yau

Vol. 4: Integrable systems
edited by Chuu Lian Terng and Karen Uhlenbeck

Vol. 5: Differential geometry inspired by string theory
edited by S.-T. Yau

Vol. 6: Essays on Einstein manifolds
edited by Claude LeBrun and McKenzie Wang

Vol. 7: Papers dedicated to Atiyah, Bott, Hirzebruch, and Singer
edited by S.-T. Yau

Vol. 8: Papers in honor of Calabi, Lawson, Siu, and Uhlenbeck
edited by S.-T. Yau

Vol. 9: Eigenvalues of Laplacians and other geometric operators
edited by A. Grigor'yan and S-T. Yau

Vol. 10: Essays in geometry in memory of S.-S. Chern
edited by S.-T. Yau

Vol. 11: Metric and comparison geometry
edited by Jeffrey Cheeger and Karsten Grove

Vol. 12: Geometric flows
edited by Huai-Dong Cao and S.-T. Yau

Vol. 13: Geometry, analysis, and algebraic geometry
edited by Huai-Dong Cao and S.-T.Yau

Vol. 14: Geometry of Riemann surfaces and their moduli spaces
edited by Lizhen Ji, Scott A. Wolpert, and S.-T. Yau

Volume V

Surveys in Differential Geometry

Differential geometry inspired by string theory

edited by

Shing-Tung Yau

International Press
www.intlpress.com

Surveys in Differential Geometry, Vol. 5
Differential geometry inspired by string theory

Editor:
Shing-Tung Yau, Harvard University

ISBN 978-1-57146-216-9

Paperback reissue 2010. Previously published in 1999 under ISBN 1-57146-070-5 (clothbound).

Printed in the United States of America.

Surveys in Differential Geometry, vol. 5

Introduction

In this volume of Surveys in Differential Geometry, we collect papers on the part of geometry that is related to the modern development of string theory. Recall that in the first volume of Surveys in Differential Geometry, we have the paper by Witten on strings with gravity which was applied to calculate chern numbers for certain line bundles over moduli space of curves. The geometric ideas initiated by consideration of string theory have been tremendous successful, and we believe that it is time to collect articles that can survey the subject in a reasonable way. The papers in this volume only represent some part of the interaction between string theory and geometry that is important and rigorous.

We have the paper by Bismuit and Labourie on the Verlinde formula, which is an accumulation of ideas from physicists and mathematicians. We have the paper of Aspinwall on K3 surface and string duality, which is a simplest nontrivial space to test some very nontrivial duality theory in string theory. We have the paper by Bryan and Leung on counting curves on surfaces, which is also inspired by works on branes in string theory. We have the paper of Kefeng Liu on how to apply ideas of localization through heat kernel to various geometric model, which in particular inspired the previous mentioned work of Bismuit and Labourie. Finally, we have collected a few papers related to mirror symmetry among Calabi-Yau manifolds. While there have been tremendous activities in this subject in the past ten years, the final rigorous treatment of the formula by Candelas and others on toric varities is only accomplished in the paper of Lian-Liu-Yau, which we reproduce here. Li and Tian's paper surveys both quantum cohomology and the vitual cycles which are part of the whole theory. In 1996, Strominger-Yau-Zaslow come up with a geometric construction of mirror manifolds, which needs a lot more rigorous mathematical treatment. Mark Gross's paper represents a good survey in the topological and some geometrical side of this.

We hope these papers will provide a good starting point for geometers who may want to get into the subject of the interaction of geometric and string theory.

Table of Contents

K3 Surfaces and String Duality
Paul S. Aspinwall 1

Symplectic Geometry and the Verlinde Formulas
Jean-Michel Bismut and François Labourie 97

Counting curves on irrational surfaces
Jim Bryan and Naichung Conan Leung 313

Special Lagrangian Fibrations II: Geometry. A Survey of Techniques in the Study of Special Lagrangian Fibrations
Mark Gross 341

Mirror Principle I
Bong H. Lian, Kefeng Liu, and S.-T. Yau 405

Mirror Principle II
Bong H. Lian, Kefeng Liu, and S.-T. Yau, ... 455

Differential Equations from Mirror Symmetry
Bong H. Lian and S.-T. Yau 510

Heat Kernels, Symplectic Geometry, Moduli Spaces and Finite Groups
Kefeng Liu 527

A Brief Tour of GW Invariants
Gang Tian and Jun Li 543

Surveys in Differential Geometry

Vol. 1: Lectures given in 1990, edited by S.-T. Yau and H. Blaine Lawson

Vol. 2: Lectures given in 1993, edited by C. C. Hsiung and S.-T. Yau

Vol. 3: Lectures given in 1996, edited by C.C. Hsiung and S.-T. Yau

Vol. 4: Integrable Systems, edited by Chuu Lian Terng and Karen Uhlenbeck

Vol. 5: Differential Geometry inspired by String Theory, edited by S.-T. Yau

Surveys in Differential Geometry, vol. 5

K3 Surfaces and String Duality

Paul S. Aspinwall

Dept. of Physics and Astronomy,
Rutgers University,
Piscataway, NJ 08855

ABSTRACT

The primary purpose of these lecture notes is to explore the moduli space of type IIA, type IIB, and heterotic string compactified on a K3 surface. The main tool which is invoked is that of string duality. K3 surfaces provide a fascinating arena for string compactification as they are not trivial spaces but are sufficiently simple for one to be able to analyze most of their properties in detail. They also make an almost ubiquitous appearance in the common statements concerning string duality. We review the necessary facts concerning the classical geometry of K3 surfaces that will be needed and then we review "old string theory" on K3 surfaces in terms of conformal field theory. The type IIA string, the type IIB string, the $E_8 \times E_8$ heterotic string, and $\mathrm{Spin}(32)/\mathbb{Z}_2$ heterotic string on a K3 surface are then each analyzed in turn. The discussion is biased in favour of purely geometric notions concerning the K3 surface itself.

CONTENTS

1. Introduction 2
2. Classical Geometry 4
3. The World-Sheet Perspective 23
4. Type II String Theory 33
5. Four-Dimensional Theories 44
6. The Heterotic String 67
Acknowledgements 89
References 89

1. Introduction

The notion of "duality" has led to something of a revolution in string theory in the past year or two. Two theories are considered dual to each other if they ultimately describe exactly the same physics. In order for this to be a useful property, of course, it is best if the two theories appear, at first sight, to be completely unrelated. Different notions of duality abound depending on how the two theories differ. The canonical example is that of "S-duality" where the coupling of one theory is inversely related to that of the other so that one theory would be weakly coupled when the other is strongly coupled. "T-duality" can be defined by similarly considering length scales rather than coupling constants. Thus a theory at small distances can be "T-dual" to another theory at large distances.

In the quest for understanding duality many examples of dual pairs have been postulated. The general scenario is that one takes a string theory (or perhaps M-theory) and compactifies it on some space and then finds a dual partner in the form of another (or perhaps the same) string theory compactified on some other space. In this form duality has become a subject dominated by geometrical ideas since most of the work involved lies in analyzing the spaces on which the string theory is compactified.

One of the spaces which has become almost omnipresent in the study of string duality is that of the K3 surface. We will introduce the K3 surface in section 2 but let us make a few comments here. Mathematicians have been studying the geometry of the K3 surface as a real surface or a complex surface for over one hundred years **[1]**. In these lectures we will always be working with complex numbers and so "surface" will mean a space of complex dimension two, or real dimension four. A curve will be complex dimension one, etc. Physicists' interest in the K3 surface (for an early paper see, for example, **[2]**) was not sparked until Yau's proof **[3]** of Calabi's conjecture in 1977. Since then the K3 surface has become a commonly-used "toy model" for compactifications (see, for example, **[4]**) as it provides the second simplest example of a Ricci-flat compact manifold after the torus.

The study of duality is best started with toy models and so the K3 surface and the torus are bound to appear often. Another reason for the appearance of the K3 surface, as we shall see in these lectures, is that the mathematics of the heterotic string appears to be intrinsically bound to the geometry of the K3 surface. Thus, whenever the heterotic string appears on one side of a pair of dual theories, the K3 surface is likely to make an appearance in the analysis.

The original purpose of these lectures was to give a fairly complete account of the way K3 surfaces appear in the subject of string duality. For reasons outlined above, however, this is almost tantamount to covering the entire subject of string duality. In order to make the task manageable therefore we will have to omit some current areas of active research. Let us first then discuss what will *not* be covered in these lectures. Note that each of these subjects are covered excellently by the other lecture series anyway.

Firstly we are going to largely ignore M-theory. M-theory may well turn out to be an excellent model for understanding string theory or perhaps even replacing string theory. It also provides a simple interpretation for many of the effects we will discussing.

Secondly we are going to ignore open strings and D-branes. There is no doubt that D-branes offer a very good intuitive approach to many of the phenomena we are going to study. It may well also be that D-branes are absolutely necessarily for a complete understanding of the foundations of string theory.

Having said that M-theory and D-branes are very important we will now do our best to not mention them. One reason for this is to able to finish giving these lectures on time but another, perhaps more important reason is to avoid introducing unnecessary assumptions. We want to take a kind of "Occam's Razor" approach and only introduce constructions as necessary. To many people's tastes our arguments will become fairly cumbersome, especially when dealing with the heterotic string, and it will certainly be true that a simpler picture could be formulated using M-theory or D-branes in some instances. What is important however is that we produce a self-consistent framework in which one may analyze the questions we wish to pose in these lectures.

Thirdly we are going to try to avoid explicit references to solitons. Since non-perturbative physics is absolutely central to most of the later portions of these lectures one may view our attitude as perverse. Indeed, one really cannot claim to understand much of the physics in these lectures without considering the soliton effects. What we will be focusing on, however, is the structure of moduli spaces and we will be able to get away with ignoring solitons in this context quite effectively. The only time that solitons will be of interest is when they become massless.

Our main goal is to understand the type II string and the heterotic string compactified on a K3 surface and what such models are dual to. Of central importance will be the notion of the *moduli space* of a given theory.

In section 2 we will introduce the K3 surface itself and describe its geometry. The facts we require from both differential geometry and algebraic geometry are introduced. In section 3 we will review the "old" approach to a string theory on a K3 surface in terms of the world-sheet conformal field theory.

In section 4 we begin our discussion of full string theory on a K3 surface in terms of the type IIA and type IIB string. The start of this section includes some basic facts about target-space supergravity which are then exploited.

The heterotic string is studied in section 6 but before that we need to take a long detour into the study of string theories compactified down to four dimensions. This detour comprises section 5 and builds the techniques required for section 6. The heterotic string on a K3 surface is a very rich and complicated subject. The analysis is far from complete and section 6 is technically more difficult than the preceding sections.

Note that blocks of text beginning with a "☞" are rather technical and may be omitted on first reading.

2. Classical Geometry

In the mid 19th century the Karakorum range of mountains in Northern Kashmir acted as a natural protection to India, then under British rule, from the Chinese and Russians to the north. Accordingly, in 1856, Captain T. G. Montgomerie was sent out to make some attempt to map the region. From a distance of 128 miles he measured the peaks across the horizon and gave them the names K1, K2, K3, ..., where the "K" stood simply for "Karakorum" [**5**]. While it later transpired that most of these mountain peaks already had names known to the Survey of India, the second peak retained the name Montgomerie had assigned it.

It was not until almost a century later in 1954 that K2, the world's second highest peak, was climbed by Achille Compagnoni and Lino Lacedelli in an Italian expedition. This event led shortly afterwards to the naming of an object of a quite different character. The first occurrence of the name "K3" referring to an algebraic variety occurs in print in [**6**]. It is explained in [**7**] that the naming is after Kummer, Kähler and Kodaira and is inspired by K2. It was Kummer who did much of the earliest work to explore the geometry of the space in question.[1]

2.1. Definition. So what exactly is a K3 surface? Let us first introduce a few definitions. For a general guide to some of the basic principles used in these lectures we refer the reader to chapter 0 of [**8**]. First we define the Hodge numbers of a space X as the dimensions of the Dolbeault cohomology groups

$$h^{p,q}(X) = \dim(H^{p,q}(X)). \tag{1}$$

Next consider the *canonical class*, K, which we may be taken to be defined as

$$K = -c_1(T_X), \tag{2}$$

where T_X is the holomorphic tangent bundle of X.

A K3 surface, S, is defined as a compact complex Kähler manifold of complex dimension two, i.e., a surface, such that

$$\begin{aligned} h^{1,0}(S) &= 0 \\ K &= 0. \end{aligned} \tag{3}$$

Note that we will sometimes relax the requirement that S be a manifold.

The remarkable fact that makes K3 surfaces so special is the following

THEOREM 1. *Any two K3 surfaces are diffeomorphic to each other.*

[1]This may explain the erroneous notion in some of the physics literature that the naming is K_3 after "Kummer's third surface" (whatever his first two surfaces may have been). To confuse the issue slightly there is a special kind of K3 surface known as a "Kummer surface" which is introduced in section 2.6.

Thus, if we can find one example of a K3 surface we may deduce all of the topological invariants. The simplest realization is to find a simple example as a complex surface embedded in a complex projective space, i.e., as an *algebraic variety*. The obvious way to do this is to consider the hypersurface defined by the equation

$$f = x_0^n + x_1^n + x_2^n + x_3^n = 0 \tag{4}$$

in the projective space $\mathbb{P}^3$ with homogeneous coordinates $[x_0, x_1, x_2, x_3]$.

It follows from the Lefschetz hyperplane theorem [8] that $h^{1,0}$ of such a hypersurface will be zero. Next we need to find if we can determine n such that $K = 0$. Associated to the canonical class is the *canonical line bundle*. This is simply the holomorphic line bundle, L, such that $c_1(L) = K$. From our definition of K it follows that the canonical line bundle for a manifold of dimension d can be regarded as the dth exterior power of the holomorphic cotangent bundle. Thus a section of the canonical line bundle can be regarded as a holomorphic d-form.

The fact that $K = 0$ for a K3 surface tells us that the canonical line bundle is trivial and thus has a holomorphic section which is nowhere zero. Consider two such sections, s_1 and s_2. The ratio s_1/s_2 is therefore a holomorphic function defined globally over the compact K3 surface. From basic complex analysis it follows that s_1/s_2 is a constant. We see that the K3 surface admits a globally defined, nowhere vanishing, holomorphic 2-form, Ω, which is unique up to a constant. It also follows that $h^{2,0}(S) = 1$.

Let us try to build Ω for our hypersurface of degree n in $\mathbb{P}^3$. First define affine coordinates in the patch $x_0 \neq 0$:

$$y_1 = \frac{x_1}{x_0}, \quad y_2 = \frac{x_2}{x_0}, \quad y_3 = \frac{x_3}{x_0}. \tag{5}$$

An obvious symmetric choice for Ω is then

$$\Omega = \frac{dy_1 \wedge dy_2}{\partial f/\partial y_3} = \frac{dy_2 \wedge dy_3}{\partial f/\partial y_1} = \frac{dy_3 \wedge dy_1}{\partial f/\partial y_2}. \tag{6}$$

This is clearly nonzero and holomorphic in our patch $x_0 \neq 0$. We can now consider another patch such as $x_1 \neq 0$. A straight forward but rather tedious calculation then shows that Ω will only extend into a holomorphic nonzero 2-form over this next patch if $n = 4$.

Our first example of a K3 surface is called the *quartic surface*, given by a hypersurface of degree 4 in $\mathbb{P}^3$. We could have arrived at this same conclusion in a somewhat more abstract way by using the *adjunction formula*. Consider the tangent bundle of S, which we denote T_S, together with the normal bundle N_S for the embedding $S \subset \mathbb{P}^3$. One can then see that

$$T_S \oplus N_S = T_{\mathbb{P}^3|S}, \tag{7}$$

where $T_{\mathbb{P}^3|S}$ is the restriction of the tangent bundle of the embedding $\mathbb{P}^3$ to the hypersurface S.

Introducing the formal sum of Chern classes of a bundle E (see, for example, [9])

$$c(E) = 1 + c_1(E) + c_2(E) + \dots, \tag{8}$$

we see that

$$c(T_{\mathbb{P}^3|S}) = c(T_S) \wedge c(N_S). \tag{9}$$

Treating a wedge product as usual multiplication from now on, it is known that [8]

$$c(T_{\mathbb{P}^k}) = (1+x)^{k+1}, \tag{10}$$

where x is the fundamental generator of $H^2(\mathbb{P}^k,\mathbb{Z})$. Since H^2 is dual to $H^{2(k-1)}$ which is dual to $H_{2(k-1)}$, we may also regard x as the homology class of a *hyperplane* $\mathbb{P}^{k-1} \subset \mathbb{P}^k$ embedded in the obvious way by setting one of the homogeneous coordinates to zero.[2] Stated in this way one can see that $c_1(N_S)$ is given by nx.

Thus we have that

$$\begin{aligned} c(T_S) &= \frac{(1+x)^4}{1+nx} \\ &= 1 + (4-n)x + (6-4n+n^2)x^2. \end{aligned} \tag{11}$$

Again we see that $n = 4$ is required to obtain $K = 0$. Now we also have c_2 which enables us to work out the Euler characteristic, $\chi(S)$, of a K3 surface:

$$\begin{aligned} \chi(S) &= \int_S c_2(T_S) \\ &= \int_{\mathbf{P}^3} l_S \wedge c_2(T_S) \\ &= \int_{\mathbf{P}^3} 4x.6x^2 \\ &= 24, \end{aligned} \tag{12}$$

where l_S is the 2-form which is the dual of the dual of the hypersurface S in the sense explained above. One may also show using the Lefschetz hyperplane theorem that

$$\pi_1(S) = 0. \tag{13}$$

We now have enough information to compute all the Hodge numbers, $h^{p,q}$. Since $\pi_1(S) = 0$, we have that the first Betti number $b_1(S) = \dim H^1(S) = h^{1,0} + h^{0,1}$ must be zero. The Euler characteristic then fixes $b_2(S)$ which then determines $h^{1,1}$ since we already know $h^{2,0} = 1$ from above. The result is

$$\begin{matrix} & & h^{0,0} & & \\ & h^{1,0} & & h^{0,1} & \\ h^{2,0} & & h^{1,1} & & h^{0,2} \\ & h^{2,1} & & h^{1,2} & \\ & & h^{2,2} & & \end{matrix} = \begin{matrix} & & 1 & & \\ & 0 & & 0 & \\ 1 & & 20 & & 1. \\ & 0 & & 0 & \\ & & 1 & & \end{matrix} \tag{14}$$

[2]Throughout these lectures we will often use the same notation for the 2-form and the associated $2(k-1)$-cycle.

2.2. Holonomy. Before continuing our discussion of K3 surfaces, we will take a detour and discuss the subject of holonomy which will be of considerable use at many points in these lectures.

Holonomy is a natural concept in the differential geometry of a manifold, M, with a vector bundle, $\pi : E \to M$, with a connection. Consider taking a point, $p \in M$, and a vector in the fibre, $e_1 \in \pi^{-1}(p)$. Now, following a closed path, Γ, starting and ending at p, parallel transport this vector according to the connection. When you are done, you will have another vector $e_2 \in \pi^{-1}(p)$. Write $e_2 = g_\Gamma(e_1)$, where $g_\Gamma(e_1)$ is an element of the structure group of the bundle. The (global) *holonomy* group of $E \to M$ is defined as the group generated by all such $g_\Gamma(e_1)$ for all closed paths Γ. The holonomy, $\mathfrak{H}_M$, of a Riemannian manifold, M, is defined as the holonomy of the tangent bundle equipped with the Levi-Civita connection from the metric.

The holonomy of a Riemannian manifold of real dimension d is contained in $\mathrm{O}(d)$. If it is orientable this becomes $\mathrm{SO}(d)$. The study of which other holonomy groups are possible is a very interesting question and will be of some importance to us. We refer the reader to [**10, 11**] for a full discussion of the results and derivations. We require the following:

1. $\mathfrak{H}_M \subseteq \mathrm{U}(\frac{d}{2})$ if and only if M is a Kähler manifold.
2. $\mathfrak{H}_M \subseteq \mathrm{SU}(\frac{d}{2})$ if and only if M is a Ricci-flat Kähler manifold.
3. $\mathfrak{H}_M \subseteq \mathrm{Sp}(\frac{d}{4})$ if and only if M is a hyperkähler manifold.
4. $\mathfrak{H}_M \subseteq \mathrm{Sp}(\frac{d}{4}).\mathrm{Sp}(1)$ if and only if M is a quaternionic Kähler manifold.[3]
5. A "symmetric space" of the form G/H where G and H are Lie groups has holonomy H.

Actually in each case the specific representation of the group in which the fibre transforms is also fixed. A celebrated theorem due to Berger, with contributions from Simons [**11**], then states that *the only other possibilities not yet mentioned are $\mathfrak{H}_M \cong G_2$ where the fibre transforms as a* **7** *or $\mathfrak{H}_M \cong \mathrm{Spin}(7)$ where the fibre transforms as an* **8**.

There is a fairly clear relationship between the holonomy groups and the invariant forms of the natural metric on M in the first cases. For the most general case we have that the form

$$\sum_{i=1}^{d} dx^i \otimes dx^i \tag{15}$$

on $\mathbb{R}^d$ admits $\mathrm{O}(d)$ as the group of invariances. In the complex Kähler case we consider the Hermitian form

$$\sum_{i=1}^{d/2} d\bar{z}^i \otimes dz^i \tag{16}$$

on $\mathbb{C}^{d/2}$ which admits $\mathrm{U}(\frac{d}{2})$ as the invariance group.

---

[3]The "." denotes that we take the direct product except that the $\mathbf{Z}_2$ centers of each group are identified.

For the next case we consider the quaternionic numbers, $\mathbb{H}$. In this case the natural form

$$\sum_{i=1}^{d/4} d\zeta^i \otimes d\bar{\zeta}^i \tag{17}$$

on $\mathbb{H}^{d/4}$ is preserved by $\mathrm{Sp}(\frac{d}{4})$. Note that writing quaternions as 2×2 matrices in the usual way gives an embedding $\mathrm{Sp}(\frac{d}{4}) \subset \mathrm{SU}(\frac{d}{2})$. Thus, a hyperkähler manifold is always a Ricci-flat Kähler manifold. In fact, one is free to choose one of a family of complex structures. Let us denote a quaternion by $q = a + bI + cJ + cK$, where $a, b, c, d \in \mathbb{R}$, $I^2 = J^2 = K^2 = -1$ and $IJ = K$, $JI = -K$, etc. Given a hyperkähler structure we may choose a complex structure given by q, where $q^2 = -1$. This implies $a = 0$ and

$$b^2 + c^2 + d^2 = 1. \tag{18}$$

Thus for a given hyperkähler structure we have a whole S^2 of possible complex structures. We will see that the K3 surface is hyperkähler when equipped with a Ricci-flat metric.

Because of the fact that quaternionic numbers are not commutative, we also have the notion of a quaternionic Kähler manifold in addition to that of the hyperkähler manifold. The space $\mathbb{H}^n$ admits an action of $\mathrm{Sp}(n).\mathrm{Sp}(1)$ by multiplication on the right by $n \times n$ quaternionic matrices in $\mathrm{Sp}(n)$ and by a quaternion of unit norm on the left. This also leads to the notion of a manifold with a kind of quaternionic structure — this time the "quaternionic Kähler manifold". The main difference between this and the hyperkähler manifold is that the extra $\mathrm{Sp}(1)$ can act on the S^2 of complex structures between patches and so destroy any global complex structure. All that remains is an S^2 bundle of almost complex structures which need have no global section. Thus a generic quaternionic Kähler manifold will *not* admit a complex structure. When this bundle is trivial the situation reduces to the hyperkähler case. As an example, the space $\mathbb{HP}^n$ is quaternionic Kähler. Note that the case $n = 1$ is somewhat redundant as this reduces to $\mathrm{Sp}(1).\mathrm{Sp}(1) \cong SO(4)$, which gives a generic orientable Riemannian manifold.

2.3. Moduli space of complex structures. We now want to construct the moduli space of all K3 surfaces. In order to determine the moduli space it is very important to specify exactly what data defines a particular K3 surface. By considering various data we will construct several different moduli spaces throughout these lectures. To begin with we want to consider the K3 surface purely as an object in algebraic geometry and, as such, we will find the moduli space of complex structures for K3 surfaces.

To "measure" the complex structure we need some relatively simple quantity which depends on the complex structure. This will be provided by "periods" which are simply integrals of the holomorphic 2-form, Ω, over integral 2-cycles within S. To analyze periods we first then require an understanding of the integral 2-cycles $H_2(S, \mathbb{Z})$.

Since $b_2(S) = 22$ from the previous section, we see that $H_2(S,\mathbb{Z})$ is isomorphic to $\mathbb{Z}^{22}$ as a group.[4] In addition to this group structure we may specify an inner product between any two elements, $\alpha_i \in H_2(S,\mathbb{Z})$, given by

$$\alpha_1.\alpha_2 = \#(\alpha_1 \cap \alpha_2), \tag{19}$$

where the notation on the right refers to the oriented intersection number, which is a natural operation on homology cycles [**9**]. This abelian group structure with an inner product gives $H_2(S,\mathbb{Z})$ the structure of a lattice.

The signature of this lattice can be determined from the index theorem for the *signature complex* [**9**]

$$\tau = \int_S \tfrac{1}{3}(c_1^2 - 2c_2) = -\tfrac{2}{3}\chi(S) = -16. \tag{20}$$

Thus the 22-dimensional lattice has signature $(3,19)$.

Poincaré duality tells us that given a basis $\{e_i\}$ of 2-cycles for $H_2(S,\mathbb{Z})$, for each e_i we may find an e_j^* such that

$$e_i.e_j^* = \delta_{ij}, \tag{21}$$

where the set $\{e_j^*\}$ also forms a basis for $H_2(S,\mathbb{Z})$. Thus $H_2(S,\mathbb{Z})$ is a *self-dual* (or *unimodular*) lattice. Note that this also means that the lattice of integral cohomology, $H^2(S,\mathbb{Z})$, is isomorphic to the lattice of integral homology, $H_2(S,\mathbb{Z})$.

The next fact we require is that the lattice $H_2(S,\mathbb{Z})$ is *even*. That is,

$$e.e \in 2\mathbb{Z}, \quad \forall e \in H_2(S,\mathbb{Z}). \tag{22}$$

This is a basic topology fact for any spin manifold, i.e., for $c_1(T_X) = 0 \pmod 2$. We will not attempt a proof of this as this is rather difficult (see, for example, Wu's formula in [**13**]).

The classification of even self-dual lattices is extremely restrictive. We will use the notation $\Gamma_{m,n}$ to refer to an even self-dual lattice of signature (m,n). It is known that m and n must satisfy

$$m - n = 0 \pmod 8 \tag{23}$$

and that if $m > 0$ and $n > 0$ then $\Gamma_{m,n}$ is unique up to isometries [**14, 15**]. An isometry is an automorphism of the lattice which preserves the inner product.

[4] Actually we need the result that $H_2(S,\mathbb{Z})$ is *torsion-free* to make this statement to avoid any finite subgroups appearing. This follows from $\pi_1(S) = 0$ and the various relations between homotopy and torsion in homology and cohomology [**12**].

In our case, one may chose a basis such that the inner product on the basis elements forms the matrix

$$\left(\begin{array}{c|c|c|c|c} -E_8 & & & & \\ \hline & -E_8 & & & \\ \hline & & U & & \\ \hline & & & U & \\ \hline & & & & U \end{array}\right), \tag{24}$$

where $-E_8$ denotes the 8×8 matrix given by minus the Cartan matrix of the Lie algebra E_8 and U represents the "hyperbolic plane"

$$U \cong \begin{pmatrix} 0 & 1 \\ 1 & 0 \end{pmatrix}. \tag{25}$$

Now we may consider periods

$$\varpi_i = \int_{e_i} \Omega. \tag{26}$$

We wish to phrase these periods in terms of the lattice $\Gamma_{3,19}$ we have just discussed. First we will fix a specific embedding of a basis, $\{e_i\}$, of 2-cycles into the lattice $\Gamma_{3,19}$. That is, we make a specific choice of which periods we will determine. Such a choice is called a "marking" and a K3 surface, together with such a marking, is called a "marked K3 surface".

There is the natural embedding

$$\Gamma_{3,19} \cong H^2(S,\mathbb{Z}) \subset H^2(S,\mathbb{R}) \cong \mathbb{R}^{3,19}. \tag{27}$$

We may now divide $\Omega \in H^2(S,\mathbb{C})$ as

$$\Omega = x + iy, \tag{28}$$

where $x, y \in H^2(S,\mathbb{R})$. Now a (p,q)-form integrated over S is only nonzero if $p = q = 2$ [8] and so

$$\begin{aligned} \int_S \Omega \wedge \Omega &= (x+iy).(x+iy) = (x.x - y.y) + 2ix.y \\ &= 0. \end{aligned} \tag{29}$$

Thus, $x.x = y.y$ and $x.y = 0$. We also have

$$\begin{aligned} \int_S \Omega \wedge \overline{\Omega} &= (x+iy).(x-iy) = (x.x + y.y) \\ &= \int_S \|\Omega\|^2 > 0. \end{aligned} \tag{30}$$

The vectors x and y must be linearly independent over $H^2(S,\mathbb{R})$ and so span a 2-plane in $H^2(S,\mathbb{R})$ which we will also give the name Ω. The relations (29) and (30) determine that this 2-plane must be space-like, i.e., any vector, v, within it satisfies $v.v > 0$.

Note that the 2-plane is equipped with a natural orientation but that under complex conjugation one induces $(x, y) \to (x, -y)$ and this orientation is reversed.

We therefore have the following picture. The choice of a complex structure on a K3 surface determines a vector space $\mathbb{R}^{3,19}$ which contains an even self-dual lattice $\Gamma_{3,19}$ and an oriented 2-plane Ω. If we change the complex structure on the K3 surface we expect the periods to change and so the plane Ω will rotate with respect to the lattice $\Gamma_{3,19}$.

We almost have enough technology now to build our moduli space of complex structures on a K3 surface. Before we can give the result however we need to worry about special things that can happen within the moduli space. A K3 surface which gives a 2-plane, Ω, which very nearly contains a light-like direction, will have periods which are only just acceptable and so this K3 surface will be near the boundary of our moduli space. As we approach the boundary we expect the K3 surfaces to degenerate in some way. Aside from this obvious behaviour we need to worry that some points away from this natural boundary may also correspond to K3 surfaces which have degenerated in some way. It turns out that there are such points in the moduli space and these will be of particular interest to us in these lectures. They will correspond to *orbifolds*, as we will explain in detail in section 2.6. For now, however, we need to include the orbifolds in our moduli space to be able to state the form of the moduli space.

The last result we require is the following

THEOREM 2 (Torelli). *The moduli space of complex structures on a marked K3 surface (including orbifold points) is given by the space of possible periods.*

For an account of the origin of this theorem we refer to [**16**]. Thus, the moduli space of complex structures on a marked K3 surface is given by the space of all possible oriented 2-planes in $\mathbb{R}^{3,19}$ with respect to a fixed lattice $\Gamma_{3,19}$.

Consider this space of oriented 2-planes in $\mathbb{R}^{3,19}$. Such a space is called a Grassmannian, which we denote $\mathrm{Gr}^+(\Omega, \mathbb{R}^{3,19})$, and we have

$$\mathrm{Gr}^+(\Omega, \mathbb{R}^{3,19}) \cong \frac{\mathrm{O}^+(3,19)}{(\mathrm{O}(2)\times \mathrm{O}(1,19))^+}. \tag{31}$$

This may be deduced as follows. In order to build the Grassmannian of 2-planes in $\mathbb{R}^{3,19}$, first consider all rotations, $\mathrm{O}(3,19)$, of $\mathbb{R}^{3,19}$. Of these, we do not care about internal rotations within the 2-plane, $\mathrm{O}(2)$, or rotations normal to it, $\mathrm{O}(1,19)$. For oriented 2-planes we consider only the index 2 subgroup which preserves orientation on the space-like directions. We use the "+" superscripts to denote this.

This Grassmannian builds the moduli space of marked K3 surfaces. We now want to remove the effects of the marking. There are diffeomorphisms of the K3 surface, which we want to regard as trivial as far as our moduli space is concerned, but which have a nontrivial action on the lattice $H^2(S,\mathbb{Z})$. Clearly any diffeomorphism induces an isometry of $H^2(S,\mathbb{Z})$, preserving the inner product. We denote the full group of such isometries as $\mathrm{O}(\Gamma_{3,19})$.[5] Our moduli space of marked K3 surfaces can

[5]Sometimes the less precise notation $\mathrm{O}(3,19;\mathbb{Z})$ is used.

be viewed as a kind of Teichmüller space, and the image of the diffeomorphisms in $O(\Gamma_{3,19})$ can be viewed as the modular group. The moduli space is the quotient of the Teichmüller space by the modular group.

What is this modular group? It was shown in [**17, 18**] that any element of $O^+(\Gamma_{3,19})$ can be induced from a diffeomorphism of the K3 surface. It was shown further in [**19**] that any element of $O(\Gamma_{3,19})$ which is not in $O^+(\Gamma_{3,19})$ *cannot* be induced by a diffeomorphism. Thus our modular group is precisely $O^+(\Gamma_{3,19})$.

Treating (31) as a right coset we will act on the left for the action of the modular group. The result is that the moduli space of complex structures on a K3 surface (including orbifold points) is

$$\mathscr{M}_c \cong O^+(\Gamma_{3,19})\backslash O^+(3,19)/(O(2)\times O(1,19))^+. \tag{32}$$

When dealing with $\mathscr{M}_c$ it is important to realize that $O^+(\Gamma_{3,19})$ has an ergodic action on the Teichmüller space and thus $\mathscr{M}_c$ is actually not Hausdorff. Such unpleasant behaviour is sometimes seen in string theory in fairly pathological circumstances [**20**] but it seems reasonable to expect that under reasonable conditions we should see a fairly well-behaved moduli space. As we shall see, the moduli space $\mathscr{M}_c$ does not appear to make any natural appearance in string theory and the related moduli spaces which do appear will actually be Hausdorff.

2.4. Einstein metrics. The first modification we will consider is that of considering the moduli space of Einstein metrics on a K3 surface. We will always assume that the metric is Kähler.

An Einstein metric is a (pseudo-)Riemannian metric on a (pseudo-)Riemannian manifold whose Ricci curvature is proportional to the metric. Actually, for a K3 surface, this condition implies that the metric is Ricci-flat [**21**]. We may thus use the terms "Einstein" and "Ricci-flat" interchangeably in our discussion of K3 surfaces.

The Hodge star will play an essential rôle in the discussion of the desired moduli space. Recall that [**9, 10**]

$$\alpha\wedge *\beta = (\alpha,\beta)\omega_g, \tag{33}$$

where α and β are p-forms, ω_g is the volume form and (α,β) is given by

$$(\alpha,\beta) = p!\int \sum_{i_1 i_2 j_1 j_2\ldots} g^{i_1 i_2} g^{j_1 j_2}\alpha_{i_1 j_1\ldots}\beta_{i_2 j_2\ldots}\, dx_1 dx_2\ldots, \tag{34}$$

in local coordinates. In particular, if α is self-dual, in the sense $\alpha = *\alpha$, then $\alpha.\alpha > 0$ in the notation of section 2.3. Similarly an anti-self-dual 2-form will obey $\alpha.\alpha < 0$. On our K3 surface S we may decompose

$$H^2(S,\mathbb{R}) = \mathscr{H}^+ \oplus \mathscr{H}^-, \tag{35}$$

where $\mathscr{H}^\pm$ represents the cohomology of the space of (anti-)self-dual 2-forms. We then see that

$$\dim \mathscr{H}^+ = 3, \quad \dim \mathscr{H}^- = 19, \tag{36}$$

from section 2.3.

The curvature acts naturally on the bundle of (anti)-self-dual 2-forms. By standard methods (see, for example, [**10**]) one may show that the curvature of the bundle of self-dual 2-forms is actually zero when the manifold in question is a K3 surface. This is one way of seeing directly the action of the SU(2) holonomy of section 2.2. Since a K3 is simply-connected, this shows that the bundle $\mathscr{H}^+$ is trivial and thus has 3 linearly independent sections.

Consider a local orthonormal frame of the cotangent bundle $\{e_1, e_2, e_3, e_4\}$. We may write the three sections of $\mathscr{H}^+$ as

$$\begin{aligned} s_1 &= e_1 \wedge e_2 + e_3 \wedge e_4 \\ s_2 &= e_1 \wedge e_3 + e_4 \wedge e_2 \\ s_3 &= e_1 \wedge e_4 + e_2 \wedge e_3. \end{aligned} \tag{37}$$

Clearly an SO(4) rotation of the cotangent directions produces an SO(3) rotation of $\mathscr{H}^+$. That is, a rotation within $\mathscr{H}^+$ is induced by a reparametrization of the underlying K3 surface. One should note that the orientation of $\mathscr{H}^+$ is fixed.

Let us denote by Σ the space $\mathscr{H}^+$ viewed as a subspace of $H^2(S, \mathbb{R})$. Putting $dz_1 = e_1 + ie_2$ and $dz_2 = e_3 + ie_4$ we obtain a Kähler form equal to s_1, and the holomorphic 2-form $dz_1 \wedge dz_2$ is given by $s_2 + is_3$. This shows that Σ is spanned by the 2-plane Ω of section 2.3 together with the direction in $H^2(S, \mathbb{R})$ given by the Kähler form.

This fits in very nicely with Yau's theorem **[3]** which states that for any manifold M with $K = 0$, and a fixed complex structure, given a cohomology class of the Kähler form, there exists a unique Ricci-flat metric. Thus, we fix the complex structure by specifying the 2-plane, Ω, and then choose a Kähler form, J, by specifying another direction in $H^2(S, \mathbb{R})$. Clearly this third direction is space-like, since

$$\mathrm{Vol}(S) = \int_S J \wedge J > 0, \tag{38}$$

and it is perpendicular to Ω as the Kähler form is of type $(1, 1)$. Thus Σ, spanned by Ω and J, is space-like.

The beauty of Yau's theorem is that we need never concern ourselves with the explicit form of the Einstein metric on the K3 surface. Once we have fixed Ω and J, we know that a unique metric exists. Traditionalists may find it rather unsatisfactory that we do not write the metric down — indeed no explicit metric on a K3 surface has ever been determined to date — but one of the lessons we appear to have learnt from the analysis of Calabi–Yau manifolds in string theory is that knowledge of the metric is relatively unimportant.

As far as our moduli space is concerned, one aspect of the above analysis which is important is that rotations within the 3-plane, Σ, may affect what we consider to be the Kähler form and complex structure but they do not affect the underlying Riemannian metric. We see that a K3 surface viewed as a Riemannian manifold may admit a whole family of complex structures. Actually this family is parametrized by the sphere, S^2, of ways in which Σ is divided into Ω and J.

This property comes from the fact that a K3 surface admits a hyperkähler structure. This is obvious from section 2.2 as a K3 is Ricci-flat and Kähler and thus has holonomy SU(2), and SU(2) $\cong$ Sp(1). The sphere (18) of possible complex structures is exactly the S^2 degree of freedom of rotating within the 3-plane Σ we found above.

Some care should be taken to show that the maps involved are surjective [**22, 23**] but we end up with the following [**24**]

THEOREM 3. *The moduli space of Einstein metrics, $\mathscr{M}_E$, for a K3 surface (including orbifold points) is given by the Grassmannian of oriented 3-planes within the space $\mathbb{R}^{3,19}$ modulo the effects of diffeomorphisms acting on the lattice $H^2(S,\mathbb{Z})$.*

In other words, we have a relation similar[6] to (32):

$$\mathscr{M}_E \cong \mathrm{O}^+(\Gamma_{3,19})\backslash \mathrm{O}^+(3,19)/(\mathrm{SO}(3)\times \mathrm{O}(19))\times \mathbb{R}_+, \tag{39}$$

where the $\mathbb{R}_+$ factor denotes the volume of the K3 surface given by (38).

This is actually isomorphic to the space

$$\mathscr{M}_E \cong \mathrm{O}(\Gamma_{3,19})\backslash \mathrm{O}(3,19)/(\mathrm{O}(3)\times \mathrm{O}(19))\times \mathbb{R}_+, \tag{40}$$

since the extra generator required, $-1\in \mathrm{O}(3,19)$, to elevate $\mathrm{O}^+(3,19)$ to $\mathrm{O}(3,19)$, is in the center and so taking the left-right coset makes no difference.

Note that (40) is actually a Hausdorff space as the discrete group $\mathrm{O}(\Gamma_{3,19})$ has a properly discontinuous action [**25**]. Thus we see that the global description of this space of Einstein metrics on a K3 surface is much better behaved than the moduli space of complex structures discussed earlier.

2.5. Algebraic K3 Surfaces. In section 2.4 we enlarged the moduli space of complex structures of section 2.3 and we found a space with nice properties. In this section we are going to find another nice moduli space by going in the opposite direction. That is, we will restrict the moduli space of complex structures by putting constraints on the K3 surface. We are going to consider *algebraic* K3 surfaces, i.e., K3 surfaces that may be embedded by algebraic (holomorphic) equations in homogeneous coordinates into $\mathbb{P}^N$ for some N.

The analysis is done in terms of algebraic curves, that is, Riemann surfaces which have been holomorphically embedded into our K3 surface S. Clearly such a curve, C, may be regarded as a homology cycle and thus an element of $H_2(S,\mathbb{Z})$. By the "dual of the dual" construction of section 2.1 we may also regard it as an element $C\in H^2(S,\mathbb{Z})$. The fact that C is holomorphically embedded may also be used to show that $C\in H^{1,1}(S)$ [**8**]. We then have $C\in \mathrm{Pic}(S)$, where we define

$$\mathrm{Pic}(S) = H^2(S,\mathbb{Z})\cap H^{1,1}(S), \tag{41}$$

which is called the "Picard group", or "Picard lattice", of S. We also define the "Picard number", $\rho(S)$, as the rank of the Picard lattice. Any element of the Picard group, $e\in \mathrm{Pic}(S)$, corresponds to a line bundle, L, such that $c_1(L)=e$ [**8**]. Thus the Picard group may be regarded as the group of line bundles on S, where the group composition is the Whitney product.

As the complex structure of S is varied, the Picard group changes. This is because an element of $H^2(S,\mathbb{Z})$ that was regarded as having type purely $(1,1)$ may pick up parts of type $(2,0)$ or $(0,2)$ as we vary the complex structure. A completely generic K3 surface will have completely trivial Picard group, i.e., $\rho=0$.

The fact that S contains a curve C is therefore a restriction on the complex structure of S. An algebraic K3 surface similarly has its deformations restricted as

[6]Note that the orientation problem makes this look more unlike (32) than it needs to! We encourage the reader to not concern themselves with these orientation issues, at least on first reading.

the embedding in $\mathbb{P}^N$ will imply the existence of one or more curves. As an example let us return to the case where S is a quartic surface

$$f = x_0^4 + x_1^4 + x_2^4 + x_3^4 = 0, \tag{42}$$

in $\mathbb{P}^3$. A hyperplane $\mathbb{P}^2 \subset \mathbb{P}^3$ will cut $f = 0$ along a curve and so shows the existence of an algebraic curve C. Taking various hyperplanes will produce various curves but they are all homologous as 2-cycles and thus define a unique element of the Picard lattice. Thus the Picard number of this quartic surface is at least one. Actually for the "Fermat" form of f given in (42) it turns out[7] that $\rho = 20$ (which is clearly the maximum value we may have since $\dim H^{1,1}(S) = 20$). A quartic surface need not be in the special form (42). We may consider a more general

$$f = \sum_{i+j+k+l=4} a_{ijkl}\, x_0^i x_1^j x_2^k x_3^l, \tag{43}$$

for arbitrary $a_{ijkl} \in \mathbb{C}$. Generically we expect no elements of the Picard lattice other than those generated by C and so $\rho(S) = 1$.

Now consider the moduli space of complex structures on a quartic surface. Since C is of type $(1,1)$, the 2-plane Ω of section 2.3 must remain orthogonal to this direction.

We may determine $C.C$ by taking two hyperplane sections and finding the number of points of intersection. The intersection of two hyperplanes in $\mathbb{P}^3$ is clearly $\mathbb{P}^1$ and so the intersection $C \cap C$ is given by a quartic in $\mathbb{P}^1$, which is 4 points. Thus $C.C = 4$ and C spans a space-like direction.

Our moduli space will be similar to that of (32) except that we may remove the direction generated by C from consideration. Thus our 2-plane is now embedded in $\mathbb{R}^{2,19}$ and the discrete group is generated by the lattice $\Lambda_C = \Gamma_{3,19} \cap C^\perp$. Note that we do not denote Λ_C as $\Gamma_{2,19}$ as it is *not* even-self-dual. The moduli space in question is then

$$\mathcal{M}_{\text{Quartic}} \cong \mathrm{O}(\Lambda_C)\backslash \mathrm{O}(2,19)/(\mathrm{O}(2) \times \mathrm{O}(19)). \tag{44}$$

This is Hausdorff. Note also that it is a space of complex dimension 19 and that a simple analysis of (43) shows that $f = 0$ has 19 deformations of a_{ijkl} modulo reparametrizations of the embedding $\mathbb{P}^3$. Thus embedding this K3 surface in $\mathbb{P}^3$ has brought about a better-behaved moduli space of complex structures but we have "lost" one deformation as (32) has complex dimension 20.

One may consider a more elaborate embedding such as a hypersurface in $\mathbb{P}^2 \times \mathbb{P}^1$ given by an equation of bidegree $(3,2)$, i.e.,

$$f = \sum_{\substack{a_0+a_1+a_2=3\\ b_0+b_1=2}} A_{a_0a_1a_2,b_0b_1}\, x_0^{a_0} x_1^{a_1} x_2^{a_2} y_0^{b_0} y_1^{b_1}. \tag{45}$$

This is an algebraic K3 since $\mathbb{P}^2 \times \mathbb{P}^1$ itself may be embedded in $\mathbb{P}^5$ (see, for example, [8]). Taking a hyperplane $\mathbb{P}^1 \times \mathbb{P}^1 \subset \mathbb{P}^2 \times \mathbb{P}^1$ one cuts out a curve C_1. Taking the hyperplane $\mathbb{P}^2 \times \{p\}$ for some $p \in \mathbb{P}^1$ cuts out C_2. By the same method we used for the quartic above we find the intersection matrix

$$\begin{pmatrix} 2 & 3 \\ 3 & 0 \end{pmatrix}. \tag{46}$$

[7] I thank M. Gross for explaining this to me.

Thus, denoting $\Lambda_{C_1C_2}$ by the sublattice of $\Gamma_{3,19}$ orthogonal to C_1 and C_2 we have

$$\mathscr{M} \cong \mathrm{O}(\Lambda_{C_1C_2})\backslash \mathrm{O}(2,18)/(\mathrm{O}(2)\times \mathrm{O}(18)). \tag{47}$$

This moduli space has dimension 18 and this algebraic K3 surface has $\rho(S) = 2$ generically. In general it is easy to see that the dimension of the moduli space plus the generic Picard number will equal 20. Note that the Picard lattice will have signature $(1, \rho - 1)$. This follows as it is orthogonal to Ω inside $\mathbb{R}^{3,19}$ and thus has at most one space-like direction but the natural Kähler form inherited from the ambient $\mathbb{P}^N$ is itself in the Picard lattice and so there must be at least one space-like direction. Thus the moduli space of complex structures on an algebraic K3 surface will always be of the form

$$\mathscr{M}_{\mathrm{Alg}} \cong \mathrm{O}(\Lambda)\backslash \mathrm{O}(2,20-\rho)/(\mathrm{O}(2)\times \mathrm{O}(20-\rho)), \tag{48}$$

where Λ is the sublattice of $H^2(S,\mathbb{Z})$ orthogonal to the Picard lattice. This lattice is often referred to as the *transcendental lattice* of S. Note that this lattice is rarely self-dual.

2.6. Orbifolds and blow-ups. When discussing the moduli spaces above we have had to be careful to note that we may be including K3 surfaces which are not manifolds but, rather, *orbifolds*. The term "orbifold" was introduced many years ago by W. Thurston after their first appearance in the mathematics literature in **[26]** (where they were referred to as "V-manifolds"). The general idea is to slightly enlarge the concept of a manifold to objects which contain singularities produced by quotients. They have subsequently played a celebrated rôle in string theory after they made their entry into the subject in **[27]**. It is probably worthwhile noting that the definition of an orbifold is slightly different in mathematics and physics. We will adopt the mathematics definition which, for our purposes, we take to be defined as follows:

> *An orbifold is a space which admits an open covering, $\{U_i\}$, such that each patch is diffeomorphic to $\mathbb{R}^n/G_i$.*

The G_i's are discrete groups (which may be trivial) which can be taken to fix the origin of $\mathbb{R}^n$. The physics definition however is more global and defines an orbifold to be a space of the form M/G where M is a manifold and G is a discrete group. A physicist's orbifold is a special case of the orbifolds we consider here. Which definition is applicable to string theory is arguable. Most of the vast amount of analysis that has been done over the last 10 years on orbifolds has relied on the global form M/G. Having said that, one of the appeals of orbifolds is that string theory is (generically) well-behaved on such an object, and this behaviour only appears to require the local condition.

The definition of an orbifold can be extended to the notion of a complex orbifold where each patch is biholomorphic to $\mathbb{C}^n/G_i$ and the induced transition functions are holomorphic. Since we may define a metric on $\mathbb{C}^n/G_i$ by the natural inherited metric on $\mathbb{C}^n$, we have a notion of a metric on an orbifold. In fact, it is not hard to extend the notion of a Kähler-Einstein metric to include orbifolds **[28]**. Similarly, a complex orbifold may be embeddable in $\mathbb{P}^N$ and can thus be viewed as an algebraic variety. In this case the notion of canonical class is still valid. Thus, there is nothing to stop the definition of the K3 surface being extended to include orbifolds. As we will see in this section, such K3 surfaces lie naturally at the boundary of the moduli space of K3 manifolds.

An example of such a K3 orbifold is the following, which is often the first K3 surface that string theorists encounter. Take the 4-torus defined as a complex manifold of dimension two by dividing the complex plane $\mathbb{C}^2$ with affine coordinates (z_1, z_2) by the group $\mathbb{Z}^4$ generated by

$$z_k \mapsto z_k + 1, \quad z_k \mapsto z_k + i, \quad k = 1, 2. \tag{49}$$

Then consider the $\mathbb{Z}_2$ group of isometries generated by $(z_1, z_2) \mapsto (-z_1, -z_2)$. It is not hard to see that this $\mathbb{Z}_2$ generator fixes 16 points: $(0,0)$, $(0,\frac{1}{2})$, $(0,\frac{1}{2}i)$, $(0,\frac{1}{2}+\frac{1}{2}i)$, $(\frac{1}{2},0)$, $\ldots$, $(\frac{1}{2}+\frac{1}{2}i, \frac{1}{2}+\frac{1}{2}i)$. Thus we have an orbifold, which we will denote S_0.

Since the $\mathbb{Z}_2$-action respects the complex structure and leaves the Kähler form invariant we expect S_0 to be a complex Kähler orbifold. Also, a moment's thought shows that any of the non-contractable loops of the 4-torus may be shrunk to a point after the $\mathbb{Z}_2$-identification is made. Thus $\pi_1(S_0) = 0$. Also, the holomorphic 2-form $dz_1 \wedge dz_2$ is invariant. We thus expect $K = 0$ for S_0. All said, the orbifold S_0 has every right to be called a K3 surface.

We now want to see what the relation of this orbifold S_0 might be to the general class of K3 manifolds. To do this, we are going to modify S_0 to make it smooth. This process is known as "blowing-up". This procedure is completely local and so we may restrict attention to a patch within S_0. Clearly the patch of interest is $\mathbb{C}^2/\mathbb{Z}_2$.

The space $\mathbb{C}^2/\mathbb{Z}_2$ can be written algebraically by embedding it in $\mathbb{C}^3$ as follows. Let (x_0, x_1, x_2) denote the coordinates of $\mathbb{C}^3$ and consider the hypersurface A given by

$$f = x_0 x_1 - x_2^2 = 0. \tag{50}$$

A hypersurface is smooth if and only if $\partial f/\partial x_0 = \ldots = \partial f/\partial x_2 = f = 0$ has no solution. Thus $f = 0$ is smooth everywhere except at the origin where it is singular. We can parameterize $f = 0$ by putting $x_0 = \xi^2$, $x_1 = \eta^2$, and $x_2 = \xi\eta$. Clearly then (ξ, η) and $(-\xi, -\eta)$ denote the same point. This is the only identification and so $f = 0$ in $\mathbb{C}^3$ really is the orbifold $\mathbb{C}^2/\mathbb{Z}_2$ we require.

Consider now the following subspace of $\mathbb{C}^3 \times \mathbb{P}^2$:

$$\{(x_0, x_1, x_2), [s_0, s_1, s_2] \in \mathbb{C}^3 \times \mathbb{P}^2;\ x_i s_j = x_j s_i,\ \forall i, j\}. \tag{51}$$

This space may be viewed in two ways — either by trying to project it onto the $\mathbb{C}^3$ or the $\mathbb{P}^2$. Fixing a point in $\mathbb{P}^2$ determines a line (i.e., $\mathbb{C}^1$) in $\mathbb{C}^3$. Thus, (51) determines a line bundle on $\mathbb{P}^2$. One may determine $c_1 = -H$ for this bundle, where H is the hyperplane class. We may thus denote this bundle $\mathcal{O}_{\mathbb{P}^2}(-1)$. Alternatively one may fix a point in $\mathbb{C}^3$. If this is not the point $(0,0,0)$, this determines a point in $\mathbb{P}^2$. At $(0,0,0)$ however we have the entire $\mathbb{P}^2$. Thus $\mathcal{O}_{\mathbb{P}^2}(-1)$ can be identified pointwise with $\mathbb{C}^3$ except that the origin in $\mathbb{C}^3$ has been replaced by $\mathbb{P}^2$. The space (51) is thus referred to as a *blow-up of* $\mathbb{C}^3$ *at the origin*. The fact the $\mathcal{O}_{\mathbb{P}^2}(-1)$ and $\mathbb{C}^3$ are generically isomorphic as complex spaces in this way away from some subset means that these spaces are *birationally equivalent* [8]. A space X blown-up at a point will be denoted $\widetilde{X}$ and the birational map between them will usually be

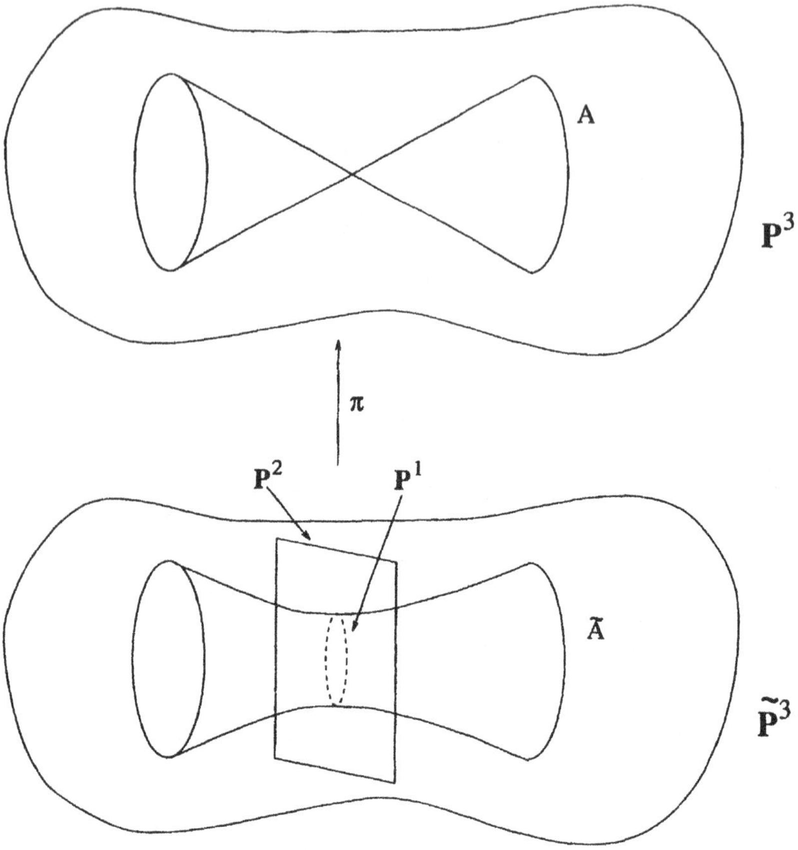

FIGURE 1. Blowing up $A \cong \mathbb{C}^2/\mathbb{Z}_2$.

written

$$\pi : \widetilde{X} \to X. \tag{52}$$

That is, π represents the *blow-down* of $\widetilde{X}$. The $\mathbb{P}^2$ which has grown out of the origin is called the *exceptional divisor.*

Now let us consider what happens to the hypersurface A given by (50) in $\mathbb{C}^3$ as we blow up the origin. We will consider the *proper transform,* $\widetilde{A} \subset \widetilde{X}$. If X is blown-up at the point $p \in X$ then $\widetilde{A}$ is defined as the closure of the point set $\pi^{-1}(A \setminus p)$ in $\widetilde{X}$.

Consider following a path in A towards the origin. In the blow-up, the point we land on in the exceptional $\mathbb{P}^2$ in the blow-up depends on the angle at which we approached the origin. Clearly the line given by (x_0t, x_1t, x_2t), $t \in \mathbb{C}$, $x_0x_1 - x_2^2 = 0$, will land on the point $[s_0, s_1, s_2] \in \mathbb{P}^2$ where again $s_0s_1 - s_2^2 = 0$. Thus the point set that provides the closure away from the origin is a quadric $s_0s_1 - s_2^2 = 0$ in $\mathbb{P}^2$. It is easy to show that this curve has $\chi = 2$ and is thus a sphere, or $\mathbb{P}^1$.

We have thus shown that when the origin is blown-up for $A \subset \mathbb{C}^3$, the proper transform of A replaces the old origin, i.e., the singularity, by a $\mathbb{P}^1$. Within the context of blowing up A, this $\mathbb{P}^1$ is viewed as the exceptional divisor and we denote it E. What is more, this resulting space, $\widetilde{A}$, is now smooth. We show this process in figure 1.

Carefully looking at the coordinate patches in $\widetilde{A}$ around E, we can work out the normal bundle for $E \subset \widetilde{A}$. The result is that this line bundle is equal to $\mathcal{O}_{\mathbb{P}^1}(-2)$. We will refer to such a rational (i.e., genus zero) curve in a complex surface as a "(-2)-curve".

Let us move our attention for a second to the general subject of complex surfaces with $K = 0$ and consider algebraic curves within them. Consider a curve C of genus g. The self-intersection of a curve may be found by deforming the curve to another one, homologically equivalent, and counting the numbers of points of intersection (with orientation giving signs). In other words, we count the number of points which remain fixed under the deformation. Suppose we may deform and keep the curve algebraic. Then an infinitesimal such deformation may be considered as a section of the normal bundle of C and the self-intersection is the number of zeros, i.e., the value of c_1 of the normal bundle integrated over C. Thus $c_1(N)$ gives the self-intersection, where N is the normal bundle. Note that two algebraic curves which intersect transversely always have positive intersection since the complex structure fixes the orientation. Thus this can only be carried out when $c_1(N) > 0$. We may extend the concept however when $c_1(N) < 0$ to the idea of negative self-intersection. In this case we see that C cannot be deformed to a nearby algebraic curve.

The adjunction formula tells us the sum $c_1(N) + c_1(T)$, where T is the tangent bundle of C, must give the first Chern class of the embedding surface restricted to C. Thus, if $K = 0$, we have

$$C.C = 2(g-1). \tag{53}$$

That is, any rational curve in a K3 surface must be a (-2)-curve. Note that (53) provides a proof of our assertion in section 2.3 that the self-intersection of a cycle is always an even number — at least in the case that the cycle is a smooth algebraic curve.

Actually, if we blow-up all 16 fixed points of our original orbifold S_0 we obtain a smooth K3 surface. To see this we need only show that the blow-up we have done does not affect $K = 0$. One can show that this is indeed the case so long as the exceptional divisor satisfies (53), i.e., it is a (-2)-curve [8]. A smooth K3 surface obtained as the blow-up of $T^4/\mathbb{Z}_2$ at all 16 fixed points is called a *Kummer surface*. Clearly the Picard number of a Kummer surface is at least 16. A Kummer surface need not be algebraic, just as the original T^4 need not be algebraic.

Now we have enough information to find how the orbifolds, such as S_0, fit into the moduli space of Einstein metrics on a K3 surface. If we blow down a (-2)-curve in a K3 surface we obtain, locally, a $\mathbb{C}^2/\mathbb{Z}_2$ quotient singularity. The above description appeared somewhat discontinuous but we may consider doing such a process gradually as follows. Denote the (-2)-curve as E. The size of E is given by the integral of the Kähler form over E, that is, $J.E$. Keeping the complex structure of the K3 surface fixed we may maintain E in the Picard lattice but we may move

J so that it becomes orthogonal to E. Thus E has shrunk down to a point — we have done the blow-down.

We have shown that any rational curve in a K3 surface is an element of the Picard lattice with $C.C = -2$. Actually the converse is true [16]. That is, given an element of the Picard lattice, e, such that $e.e = -2$, then either e or $-e$ gives the class of a rational curve in the K3 surface. This will help us prove the following. Let us define the *roots* of $\Gamma_{3,19}$ as $\{\alpha \in \Gamma_{3,19};\ \alpha.\alpha = -2\}$.

THEOREM 4. *A point in the moduli space of Einstein metrics on a K3 surface corresponds to an orbifold if and only if the 3-plane, Σ, is orthogonal to a root of $\Gamma_{3,19}$.*

If we take a root which is perpendicular to Σ, then it must be perpendicular to Ω and thus in the Picard lattice. It follows that this root (or minus the root, which is also perpendicular to Σ) gives a rational curve in the K3. Then, since J is also perpendicular to this root, the rational curve has zero size and the K3 surface must be singular. Note also that any higher genus curve cannot be shrunk down in this way as it would be a space-like or light-like direction in the Picard lattice and could thus not be orthogonal to Σ. What remains to be shown is that the resulting singular K3 surface is always an orbifold.

For a given point in the moduli space of Einstein metrics consider the set of roots orthogonal to Σ. Suppose this set can be divided into two mutually orthogonal sets. These would correspond to two sets of curves which did not intersect and thus would be blown down to two (or more) separate isolated points. Since the orbifold condition is local we may confine our attention to the case when this doesn't happen. The term "root" is borrowed from Lie group theory and we may analyze our situation in the corresponding way. We may choose the "simple roots" in our set which will span the root lattice in the usual way (see, for example, [29]). Now consider the intersection matrix of the simple roots. It must have -2's down the diagonal and be negative definite (as it is orthogonal to Σ). This is entirely analogous to the classification of simply-laced Lie algebras and immediately tells us that there is an *A-D-E* classification of such events.

We have already considered the A_1 case above. This was the case of a single isolated (-2)-curve which shrinks down to a point giving locally a $\mathbb{C}^2/\mathbb{Z}_2$ quotient singularity. To proceed further let us try another example. The next simplest situation is that of a $\mathbb{C}^2/\mathbb{Z}_3$ quotient singularity given by $(\xi, \eta) \mapsto (\omega\xi, \omega^2\eta)$, where ω is a cube root of unity. As before we may rewrite this as a subspace of $\mathbb{C}^3$ as

$$f = x_0x_1 - x_2^3 = 0. \tag{54}$$

The argument is very similar to the $\mathbb{C}^2/\mathbb{Z}_2$ case. The difference is that now as we follow a line into the blown-up singularity and consider the path (x_0t, x_1t, x_2t) within $x_0x_1 - x_2^3$, the x_2^3 term becomes irrelevant and the closure of the point set becomes $s_0s_1 = 0$ within $\mathbb{P}^2$. This consists of two rational curves ($s_0 = 0$ and $s_1 = 0$) intersecting transversely (at $[s_0, s_1, s_2] = [0, 0, 1]$). Thus, when we blow-up $\mathbb{C}^2/\mathbb{Z}_3$, we obtain as an exceptional divisor two (-2)-curves crossing at one point. Clearly this is the A_2 case.

Now consider the general case of a cyclic quotient $\mathbb{C}^2/\mathbb{Z}_n$. This is given by $f = x_0x_1 - x_2^n = 0$. At first sight the discussion above for the case $n = 3$ would appear to be exactly the same for any value of $n > 3$ but actually we need to be careful that after the blow-up we really have completely resolved the singularity. Consider the coordinate patch in $\mathcal{O}_{\mathbb{P}^2}(-1)$ written as (51) where $s_2 \neq 0$. We may use $y_0 = s_0/s_2$, $y_1 = s_1/s_2$ as good affine coordinates on the base $\mathbb{P}^2$ and $y_2 = x_2$ as a good coordinate in the fibre. Since $x_0 = y_0y_2$ and $x_1 = y_1y_2$, our hypersurface becomes

$$y_2^2(y_0y_1 - y_2^{n-2}) = 0. \tag{55}$$

Now $y_2 = 0$ is the equation for the exceptional divisor $\mathbb{P}^2 \subset \mathbb{P}^3$ in our patch. We are interested in the proper transform of our surface and thus we do not want to include the full $\mathbb{P}^2$ in our solution — just the intersection with our surface. Thus we throw this solution away and are left with

$$y_0y_1 - y_2^{n-2} = 0. \tag{56}$$

If $n = 2$ or 3 this is smooth. If $n > 3$ however we have a singularity at $y_0 = y_1 = y_2 = 0$. This point is at $s_0 = s_1 = 0$ which is precisely where the two $\mathbb{P}^1$'s produced by the blow-up intersect. One may check that the other patches contain no singularities.

What we have shown then is that starting with the space $\mathbb{C}^2/\mathbb{Z}_n$, $n > 2$, the blow-up replaces the singularity at the origin by two $\mathbb{P}^1$'s which intersect at a point but, in the case $n > 3$, this point of intersection is locally of the form $\mathbb{C}^2/\mathbb{Z}_{n-2}$. To resolve the space completely, the procedure is clear. We simply repeat the process until we are done. Note that when we blow up the point of intersection of two $\mathbb{P}^1$'s intersecting transversely, the fact that the $\mathbb{P}^1$'s approach the point of intersection at a different "angle" means that after the blow-up they pass through different points of the exceptional divisor and thus become disjoint. We show the process of blowing-up a $\mathbb{C}^2/\mathbb{Z}_6$ singularity in figure 2. In this process we produce a chain of 5 $\mathbb{P}^1$'s, $E_1, \dots, E_5$, when completely resolving the singularity. We show the $\mathbb{P}^1$'s as lines.

Clearly we see that resolving the $\mathbb{C}^2/\mathbb{Z}_n$ singularity produces the A_{n-1} intersection matrix for the (-2)-curves. Thus we have deduced the form of the A-series. We now ponder the D- and E-series.

Consider the general form of the quotient $\mathbb{C}^2/G$. We are interested in the cases which occur locally in a K3 surface in which $K = 0$. This requires that G leaves the holomorphic 2-form $dz_1 \wedge dz_2$ invariant. This implies that G must be a discrete subgroup of SU(2). One may also obtain this result by noting that the holonomy of the orbifold near the quotient singularity can be viewed as being isomorphic to G. The subgroups of SU(2) are best understood from the well-known exact sequence

$$1 \to \mathbb{Z}_2 \to \mathrm{SU}(2) \to \mathrm{SO}(3) \to 1. \tag{57}$$

Thus any subgroup of SU(2) can be projected into a subgroup of SO(3) and considered as a symmetry of a 3-dimensional solid. The cyclic groups $\mathbb{Z}_n$ may be thought of as, for example, the symmetries of cones over regular polygons. The other possibilities are the dihedral groups which are the symmetries of a prism over a regular polygon, and the symmetries of the tetrahedron, the octahedron (or cube), and the icosahedron (or dodecahedron). Each of these latter groups are nonabelian and correspond to a subgroup of SU(2) with twice as many elements as the subgroup of SO(3). They are thus called the binary dihedral, binary tetrahedral, binary octahedral, and binary icosahedral groups respectively. In each case the quotient $\mathbb{C}^2/G$

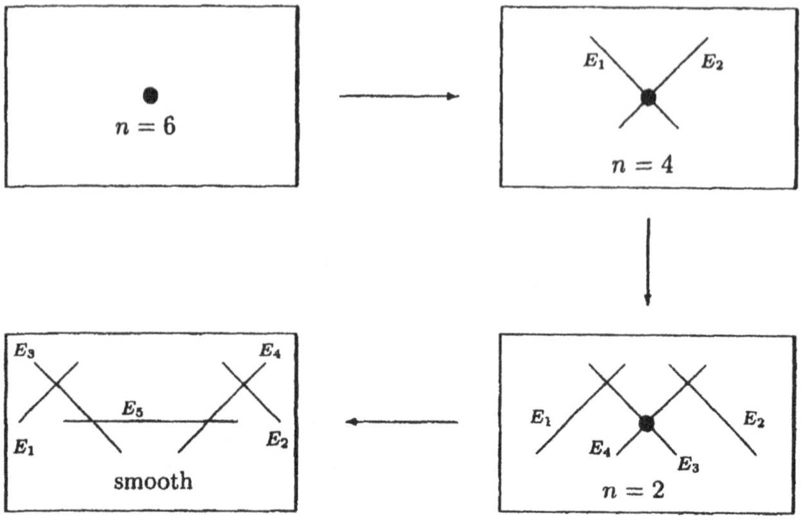

FIGURE 2. Blowing up $\mathbb{C}^2/\mathbb{Z}_6$.

can be embedded as a hypersurface in $\mathbb{C}^3$. This work was completed by Du Val (see **[30]** and references therein), after whom the singularities are sometimes named. The case of an icosahedron was done by Felix Klein last century **[31]** and so they are also often referred to as "Kleinian singularities".

Once we have a hypersurface in $\mathbb{C}^3$ we may blow-up as before until we have a smooth manifold. The intersection matrices of the resulting (-2)-curves can then be shown to be (minus) the Cartan matrix of D_n, E_6, E_7, or E_8 respectively. This process is laborious and is best approached using slightly more technology than we have introduced here. We refer the reader to **[16]** or **[32]** for more details. The results are summarized in table 1(see **[33]** for some of the details). We have thus shown that any degeneration of a K3 surface that may be achieved by blowing down (-2)-curves leads to an orbifold singularity.

One might also mention that in the case of the A_n blow-ups, explicit metrics are known which are asymptotically flat **[34, 35, 36]**. Unfortunately, since the blow-up inside a K3 surface is not actually flat asymptotically, such metrics represent only an approximation to the situation we desire. As we said earlier however, lack of an explicit metric will not represent much of a problem.

This miraculous correspondence between the *A*-*D*-*E* classification of discrete subgroups of SU(2) and Dynkin diagrams for simply-laced simple Lie groups must count as one of the most curious interrelations in mathematics. We refer to **[37]** or **[38]** for the flavour of this subject. We will see later in section 4.3 that string theory will provide another striking connection.

We have considered the resolution process from the point of view of blowing-up by changing the Kähler form. In terms of the moduli space of Einstein metrics on a K3 surface this is viewed as a rotation of the 3-plane Σ so that there are no longer

Group	Generators	Hypersurface	Resolution
Cyclic	$\begin{pmatrix} \alpha & 0 \\ 0 & \alpha^{-1} \end{pmatrix}, \alpha = e^{\frac{2\pi i}{n}}$	$x^2 + y^2 + z^n = 0$	A_{n-1}
Binary Dihedral	$\begin{pmatrix} \beta & 0 \\ 0 & \beta^{-1} \end{pmatrix}, \begin{pmatrix} 0 & 1 \\ -1 & 0 \end{pmatrix}, \beta = e^{\frac{\pi i}{n}}$	$x^2 + y^2 z + z^{n+1} = 0$	D_{n+2}
Binary Tetrahedral	$D_4, \frac{1}{\sqrt{2}} \begin{pmatrix} \varepsilon^7 & \varepsilon^7 \\ \varepsilon^5 & \varepsilon \end{pmatrix}, \varepsilon = e^{\frac{2\pi i}{8}}$	$x^2 + y^3 + z^4 = 0$	E_6
Binary Octahedral	$E_6, \begin{pmatrix} \varepsilon & 0 \\ 0 & \varepsilon^7 \end{pmatrix}, \varepsilon = e^{\frac{2\pi i}{8}}$	$x^2 + y^3 + yz^3 = 0$	E_7
Binary Icosahedral	$-\begin{pmatrix} \eta^3 & 0 \\ 0 & \eta^2 \end{pmatrix}, \frac{1}{\eta^2 - \eta^3} \begin{pmatrix} \eta + \eta^4 & 1 \\ 1 & -\eta - \eta^4 \end{pmatrix}, \eta = e^{\frac{2\pi i}{5}}$	$x^2 + y^3 + z^5 = 0$	E_8

TABLE 1. *A-D-E* Quotient Singularities.

any roots in the orthogonal complement. Since Σ is spanned by the Kähler form and Ω, which measures the complex structure, we may equally view this process in terms of changing the complex structure, rather than the Kähler form. In this language, the quotient singularity is *deformed* rather than blown-up. The process of resolving is now seen, not as giving a non-zero size to a shrunken rational curve, but rather changing the complex structure so that the rational curve no longer exists. This deformation process is actually very easy to understand in terms of the singularity as a hypersurface in $\mathbb{C}^3$. Consider the A_{n-1} singularity in table 1. A deformation of this to

$$x^2 + y^2 + z^n + a_{n-2}z^{n-2} + a_{n-3}z^{n-3} + \ldots + a_0 = 0 \tag{58}$$

will produce a smooth hypersurface for generic values of the a_i's. (Note that the $a_{n-1}z^{n-1}$ term can be transformed away by a reparametrization.)

It is worth emphasizing that in general, when considering any algebraic variety, blow-ups and deformations are quite different things. We will discuss later how the difference between blowing up a singularity and deforming it away can lead to topology changing processes in complex dimension three, for example. It is the peculiar way in which the complex structure moduli and Kähler form get mixed up in the moduli space of Einstein metrics on a K3 surface that makes them amount to much the same thing in this context. The relationship between blowing up and deformations will be deepened shortly when we discuss mirror symmetry.

3. The World-Sheet Perspective

In this section we are going to embark on an analysis of string theory on K3 surfaces from what might be considered a rather old-fashioned point of view. That is, we are going to look at physics on the world-sheet. One point of view that was common more than a couple of years ago was that string theory could solve difficult problems by "pulling back" physics in the target space, which has a large number of dimensions and is hence difficult, to the world-sheet, which is two-dimensional and hence simple. Thus an understanding of two-dimensional physics on the world-sheet would suffice for understanding the universe.

More recently it has been realized that the world-sheet approach is probably inadequate as it misses aspects of the string theory which are nonperturbative in the string coupling expansion. Thus, attention has switched somewhat away from the world-sheet and back to the target space. One cannot forget the world-sheet however and, as we will see later in section 5, in some examples it would appear that the target space point of view appears on an equal footing with the world-sheet point of view. We must therefore first extract from the world-sheet as much as we can.

3.1. The Nonlinear Sigma Model. "Old" string theory is defined as a two-dimensional theory given by maps from a Riemann surface, Σ, into a target manifold X:

$$x : \Sigma \to X. \tag{59}$$

In the conformal gauge, the action is given by (see, for example, [**39**] for the basic ideas and [**40**] for conventions on normalizations)

$$S = \frac{i}{8\pi\alpha'} \int_\Sigma (g_{ij} - B_{ij})\, \partial x^i \bar{\partial} x^j \, d^2z - 2\pi \int \Phi R^{(2)}\, d^2z + \dots, \tag{60}$$

where we have ignored any terms which contain fermions. The terms are identified as follows. g_{ij} is a Riemannian metric on X and B_{ij} are the components of a real 2-form, B, on X. Φ is the "dilaton" and is a real number (which might depend on x) and $R^{(2)}$ is the scalar curvature of Σ. This two-dimensional theory is known as the "non-linear σ-model".

In order to obtain a valid string theory, we require that the resulting two-dimensional theory is conformally invariant with a specific value of the "central charge". (See [**41**] for basic notions in conformal field theory.) Conformal invariance puts constraints on the various parameters above [**42, 43**]. In general the result is in terms of a perturbation theory in the quantity α'/R^2, where R is some characteristic "radius" (coming from the metric) of X, assuming X to be compact.

The simplest way of demanding conformal invariance to leading order is to set the dilaton, Φ, to be a constant, and let B be closed and g_{ij} be Ricci-flat. There are other solutions, such as the one proposed in [**44**] and these do play a rôle in string duality as solitons (see, for example, [**45**]). It is probably safe to say however that the solution we will analyze, with the constant dilaton, is by far the best understood.

In many ways one may regard this conformal invariance calculation to be the string "derivation" of general relativity. To leading order in α'/R^2 we obtain Einstein's field equation and then perturbation theory "corrects" this to higher orders. Nonperturbative effects, i.e., "world-sheet instantons", should also modify notions in general relativity. Anyway, we see that in this simple case, a vacuum solution for string theory is the same as that for general relativity — namely a Ricci-flat manifold.

There is a simple and beautiful relationship between supersymmetry in this non-linear σ-model and the Kähler structure of the target space manifold X. We have neglected to include any fermions in the action (60) but they are of the form ψ^i and transform, as far as the target space is concerned, as sections of the cotangent bundle. A supersymmetry will be roughly of the form

$$\delta_\epsilon X^i = \bar{\epsilon}\, l^i_j \psi^j. \tag{61}$$

The object of interest here is l^i_j. With one supersymmetry ($N = 1$) on the world-sheet one may simply reparameterize to make it equal to δ^i_j. When the $N > 1$

however we have more structure. It was shown [46] that for $N = 2$ the second l^i_j acts as an almost complex structure and gives X the structure of a Kähler manifold. In the case $N = 4$, we have 3 almost complex structures, as in section 2.4, and this leads to a hyperkähler manifold. The converse also applies.

Note that this relationship between world-sheet supersymmetry and the complex differential geometry of the target space required no reference to conformal invariance. In the case that we have conformal invariance one may also divide the analysis into holomorphic and anti-holomorphic parts and study them separately. In this case we have separate supersymmetries in the left-moving (holomorphic) and right-moving (anti-holomorphic) sectors.

The case of interest to us, of course, is when X is a smooth K3 manifold. From what we have said, this will lead to an $N = (4,4)$ superconformal field theory, at least to leading order in α'/R^2. Actually for $N = 4$ non-linear σ-models, the perturbation theory becomes much simpler and it can be shown [47] that there are no further corrections to the Ricci-flat metric after the leading term. Additionally, one may present arguments that this is even true nonperturbatively [48]. Thus our Ricci-flat metric on the K3 surface is an exact solution. One must contrast this to the $N = 2$ case where there are both perturbative corrections and nonperturbative effects in general.

3.2. The Teichmüller space. The goal of this section is to find the moduli space of conformally invariant non-linear σ-models with K3 target space. This may be considered as an intermediate stage to that of the last section, where we considered classical geometry, and that of the following sections where we consider supposedly fully-fledged string theory. This will prove to be a very important step however.

Firstly note that there are three sets of parameters in (60) which may be varied to span the moduli space required. In each case we need to know which deformations will be effective in the sense that they really change the underlying conformal field theory. Here we will have to make some assumptions since a complete analysis of these conformal field theories has yet to be completed.

First consider the metric g_{ij}. We have seen that this must be Ricci-flat to obtain conformal invariance. We will assume that any generic deformation of this Ricci-flat metric to another inequivalent Ricci-flat metric will lead to an inequivalent conformal field theory. Since the dimension of the moduli space of Einstein metrics on a K3 surface given in (40) is 58, we see that the metric accounts for 58 parameters.

Next we have the 2-form, B. This appears in the action in the form

$$\int_{x(\Sigma)} B. \tag{62}$$

Thus, since the image of the world-sheet in X under the map x is a closed 2-cycle, any exact part of B is irrelevant. All we see of B is its cohomology class. As b_2 of a K3 surface is 22, this suggests we have 22 parameters from the B-field.

Lastly we have the dilaton, Φ. This plays a very peculiar rôle in our conformal field theory. Since Φ is a constant over X, by assumption, we may pull it outside the integral leaving a contribution of $2\Phi(g-1)$ to the action, where g is the genus of Σ. In this section we really only care about the conformal field theory for a fixed Σ and so this quantity remains constant. To be more complete we should sum over the genera of Σ. Taking the limit of $\alpha' \to \infty$, the world-sheet image in the target space will degenerate to a Feynman diagram and then g will count the number of loops. Thus we have an effective target space coupling of

$$\lambda = e^{\Phi}. \tag{63}$$

Anyway, since we want to ignore this summation of Σ for the time being we will ignore the dilaton. See, for example, **[49]** for a further discussion of world-sheet properties of the dilaton from a string field theory point of view.

All said then we have a moduli space of $58 + 22 = 80$ real dimensions. To proceed further we need to know some aspects about the holonomy of the moduli space. The local form of the moduli space was first presented in **[50]** but we follow more closely here the method of **[51]**.

Let us return to the moduli space of Einstein metrics on a K3 surface. Part of the holonomy algebra of the symmetric space factor in (40) is $\mathfrak{so}(3) \cong \mathfrak{su}(2)$. This rotation in $\mathbb{R}^3$ comes from the choice of complex structures given a quaternionic structure as discussed in section 2.2. Thus, this part of the holonomy can be understood as arising from the symmetry produced by the S^2 of complex structures.

This $\mathfrak{su}(2)$ symmetry must therefore be present in the non-linear σ-model. In the case where we have a conformal field theory however, we may divide the analysis into separate left- and right-moving parts. Thus each sector must have an independent $\mathfrak{su}(2)$ symmetry. Indeed, it is known from conformal field theory that an $N = 4$ superconformal field theory contains an affine $\mathfrak{su}(2)$ algebra and so an $N = (4,4)$ superconformal field theory has an $\mathfrak{su}(2) \oplus \mathfrak{su}(2) \cong \mathfrak{so}(4)$ symmetry.

This symmetry acts on the tangent directions to a point in the moduli space (i.e., the "marginal operators") and so will be a subgroup of the holonomy. Thus the $\mathfrak{so}(3)$ appearing in the holonomy algebra of the moduli space of Einstein metrics is promoted to $\mathfrak{so}(4)$ for our moduli space of conformal field theories.

Now we are almost done. We need to find a space whose holonomy contains SO(4) as a factor and has dimension 80. One could suggest spaces such as $A \times B$, where A is a Riemannian manifold of dimension 4 and B is a Riemannian manifold of dimension 76. Such a factorization is incompatible with what we know about the conformal field theory, however. Analyzing the marginal operators in terms of superfields shows that each one is acted upon non-trivially by at least part of the SO(4). Given the work of Berger and Simons therefore leaves us with only one possibility.[8]

THEOREM 5. *Given the assumptions about the effectiveness of deformations on the underlying conformal field theory, any smooth neighbourhood of the moduli space of conformally invariant non-linear σ-models with a K3 target space is isomorphic to an open subset of*

$$\mathcal{T}_\sigma = \frac{\mathrm{O}(4,20)}{\mathrm{O}(4) \times \mathrm{O}(20)}. \tag{64}$$

We now want to know about the global form of the moduli space. Here we are forced to make assumptions about how reasonable our conformal field theories

[8]Actually we should rule out the compact symmetric space possibility. This is done by our completeness assumption, as we know we may make a K3 surface arbitrarily large.

can be. We will assume that from theorem 6 it follows that the moduli space of conformal field theories is given by $\mathscr{M}_\sigma \cong G_\sigma \backslash \mathscr{T}_\sigma$, where G_σ is some discrete group. That is, $\mathscr{T}_\sigma$ is the Teichmüller space. All we are doing here is assuming that our moduli space is "complete" in the sense that there are no pathological limit points in it possibly bounding some bizarre new region. While this assumption seems extremely reasonable I am not aware of any rigorous proof that this is the case.

3.3. The geometric symmetries. All that remains then is the determination of the modular group, G_σ. To begin with we should relate (64) to the Teichmüller spaces we are familiar with from section 2. This will allow us the incorporate the modular groups we have already encountered. This is a review of the work that first appeared in [**52, 53**].

The space $\mathscr{T}_\sigma$ is the Grassmannian of space-like 4-planes in $\mathbb{R}^{4,20}$. We saw that the moduli space of Einstein metrics on a K3 surface is given by the Grassmannian of space-like 3-planes in $\mathbb{R}^{3,19}$ and this must be a subspace of $\mathscr{T}_\sigma$ since the Einstein metric appears in the action (60). This gives us a clear way to proceed.

Let us introduce the even self-dual lattice $\Gamma_{4,20} \subset \mathbb{R}^{4,20}$. It would be nice if we could show that this played the same rôle as $\Gamma_{3,19}$ played in the moduli space of Einstein metrics. That is, we would like to show $G_\sigma \cong \mathrm{O}(\Gamma_{4,20})$. We will see this is indeed true.

First we want a natural way of choosing $\Gamma_{3,19} \subset \Gamma_{4,20}$. To do this fix a primitive element $w \in \Gamma_{4,20}$ such that $w.w = 0$. Now consider the space, $w^\perp \subset \mathbb{R}^{4,20}$, of all vectors x such that $x.w = 0$. Clearly w is itself contained in this space. Now project onto the codimension one subspace $w^\perp/w$ by modding out by the w direction. We now embed $w^\perp/w$ back into $\mathbb{R}^{4,20}$ such that

$$\frac{w^\perp}{w} \cap \Gamma_{4,20} \cong \Gamma_{3,19}. \tag{65}$$

It is important to be aware of the fact that all statements about $\Gamma_{3,19}$ are dictated by a choice of w. The embedding $w^\perp/w \subset \mathbb{R}^{4,20}$ can be regarded as a choice of a second lattice vector, w^*, such that w^* is orthogonal to $w^\perp/w$, $w^*.w^* = 0$ and $w^*.w = 1$. As we shall see, the choice of w^* is not as significant as the choice of w.

Now perform the same operation on the space-like 4-plane. We will denote this plane $\Pi \subset \mathbb{R}^{4,20}$. First define $\Sigma' = \Pi \cap w^\perp$ and then project this 3-plane into the space $w^\perp/w$ and embed back into $\mathbb{R}^{4,20}$ to give Σ. This Σ may now be identified with that of section 2.4 to give the Einstein metric on the K3 surface of some fixed volume.

Fixing Σ we may look at how we may vary Π to fill out the other deformations. Let Π be given by the span of Σ' and B', where B' is a vector, orthogonal to Σ', normalized by $B'.w = 1$. We may project B' into $w^\perp/w$ to give $B \in \mathbb{R}^{3,19}$. Note that $\mathbb{R}^{3,19}$ is the space $H^2(X, \mathbb{R})$ and so B is a 2-form as desired.

Lastly we require the volume of the K3 surface. Let us decompose B' as

$$B' = \alpha w + w^* + B, \tag{66}$$

We claim that α is the volume of the K3 surface. To see this we need to analyze the explicit form of the moduli space further.

We have effectively decomposed the Teichmüller space of conformal field theories as

$$\frac{\mathrm{O}(4,20)}{\mathrm{O}(4)\times\mathrm{O}(20)} \cong \frac{\mathrm{O}(3,19)}{\mathrm{O}(3)\times\mathrm{O}(19)} \times \mathbb{R}^{22} \times \mathbb{R}_{+}, \tag{67}$$

where the three factors on the right are identified as the Teichmüller spaces for the metric on the K3 surface (given by Σ), the B-field, and the volume respectively. Each of the spaces in the equation (67) has a natural metric, given by the invariant metric for the group action in the case of the symmetric spaces. This can be shown to coincide with the natural metric from conformal field theory — the Zamolodchikov metric — given the holonomy arguments above. The isomorphism (67) will respect this metric if "warping" factors are introduced as explained in [**54**]. It is these warping factors which determine the identification of the volume of the K3 surface as α in (66). This will be explained further in [**53**].

The part of G_σ we can understand directly is the part which fixes w but acts on $w^\perp/w$. This will affect the metric on the K3 surface of fixed volume and the B-field. We know that the modular group coming from global diffeomorphisms of the K3 surface is $\mathrm{O}^+(\Gamma_{3,19})$, which should be viewed as $\mathrm{O}^+(H^2(X,\mathbb{Z}))$. The discrete symmetries for the B-field meanwhile can be written as $B \cong B + e$, where $e \in H^2(X,\mathbb{Z})$. To see this note that shifting B by a integer element will shift the action (60) by a multiple of $2\pi i$ and hence will not effect the path integral. Thanks to our normalization of B', a shift of B by an element of $\Gamma_{3,19}$ amounts to a rotation of Π which is equivalent to an element of $\mathrm{O}(\Gamma_{4,20})$. Note that this can also be interpreted as a redefinition of w^*. That is, the freedom of choice of w^* is irrelevant once we take into account the B-field shifts.

We may also consider taking the complex conjugate of the non-linear σ-model action. This has the effect of reversing the sign of B while providing the extra element required to elevate $\mathrm{O}^+(\Gamma_{3,19})$ to $\mathrm{O}(\Gamma_{3,19})$.

The result is that the subgroup of G_σ that we see directly from the non-linear σ-model action is a subgroup of $\mathrm{O}(\Gamma_{4,20})$ consisting of rotations *and* translations of $\Gamma_{3,19} \subset \Gamma_{4,20}$. This may be viewed as the *space group* of $\Gamma_{3,19}$, or equivalently, the semi-direct product

$$G_\sigma \supset \mathrm{O}(\Gamma_{3,19}) \ltimes \Gamma_{3,19}. \tag{68}$$

This is as much as we can determine from $w^\perp/w$.

3.4. Mirror symmetry. To proceed any further in our analysis of G_σ we need to know about elements which do not correspond to a manifest symmetry of the non-linear σ-model action (60). This knowledge will be provided by *mirror symmetry*. Mirror symmetry is a much-studied phenomenon in Calabi–Yau threefolds (see, for example, [**55**] for a review). The basic idea in the subject of threefolds is that the notion of deformation of complex structure is exchanged with deformation of complexified Kähler form. The character of mirror symmetry in K3 surfaces is somewhat different since, as we have seen, the notion of what constitutes a deformation of complex structure and what constitutes a deformation of the Kähler form can be somewhat ambiguous. Also, we have yet to mention the possibility of complexifying the Kähler form, as that too is a somewhat ambiguous process.

Indeed mirror symmetry itself is somewhat meaningless when viewed in terms of the intrinsic geometry of a K3 surface and does not begin to make much sense until viewed in terms of *algebraic* K3 surfaces. The results for algebraic K3 surfaces were first explored by Martinec [**56**], whose analysis actually predates the discovery (and naming) of mirror symmetry in the Calabi–Yau threefold context. See also [**57, 58**] for a discussion of some of the issues we cover below.

Recall that the Picard lattice of section 2.5 was defined as the lattice of integral 2-cycles in $H^2(X,\mathbb{Z})$ which were of type $(1,1)$. The transcendental lattice, Λ, was defined as the orthogonal complement of the Picard lattice in $H^2(X,\mathbb{Z})$. The signature of Λ is $(2, 20-\rho)$, where ρ is the Picard number of X.

It is clear that $\Gamma_{4,20} \cong \Gamma_{3,19} \oplus U$, where U is the hyperbolic plane of (25). We will extend the Picard lattice to the "quantum Picard lattice", Υ, by defining

$$\Upsilon = \mathrm{Pic}(X) \oplus U, \tag{69}$$

as the orthogonal complement of Λ within $\Gamma_{4,20}$.

Given a K3 surface with an Einstein metric specified by Σ and a given algebraic structure, we know that the complex structure 2-plane, Ω, is given by $\Sigma \cap (\Lambda \otimes_{\mathbf{Z}} \mathbb{R})$.[9] The Kähler form direction, J is then given by the orthogonal complement of $\Omega \subset \Sigma$. Accordingly, J lies in $\mathrm{Pic}(X) \otimes_{\mathbf{Z}} \mathbb{R}$.

We want to extend this to the non-linear σ-model picture. Keep Ω defined as above, or equivalently,

$$\Omega = \Pi \cap (\Lambda \otimes_{\mathbf{Z}} \mathbb{R}), \tag{70}$$

and introduce a new space-like 2-plane, $\mho$, as the orthogonal complement of $\Omega \subset \Pi$. Clearly we have

$$\mho = \Pi \cap (\Upsilon \otimes_{\mathbf{Z}} \mathbb{R}). \tag{71}$$

Our notion of a mirror map will be to exchange

$$\mu : (\Lambda, \Omega) \leftrightarrow (\Upsilon, \mho). \tag{72}$$

Note that as $\mho$ encodes the Kähler form and the value of the B-field we have the notion of exchange of complex structure data with that of Kähler form + B-field as befits a mirror map.

The moduli space of non-linear σ-models on an algebraic K3 surface will be the subspace of (64) which respects the division of Π into Ω and $\mho$. This is given by

$$\mathcal{T}_{\sigma,\mathrm{alg}} \cong \frac{O(2,20-\rho)}{O(2)\times O(20-\rho)} \times \frac{O(2,\rho)}{O(2)\times O(\rho)}, \tag{73}$$

where the first factor is the moduli space of complex structures and the second factor is moduli space of the Kähler form + B-field.

Given an algebraic K3 surface X with quantum Picard lattice $\Upsilon(X)$ we have a mirror K3 surface, Y, with quantum Picard lattice $\Upsilon(Y)$ such that $\Upsilon(Y)$ is the orthogonal complement of $\Upsilon(X) \subset \Gamma_{4,20}$. Translating this back into classical ideas, $\mathrm{Pic}(Y) \oplus U$ will be the orthogonal complement of $\mathrm{Pic}(X) \subset \Gamma_{3,19}$. If X is such that the orthogonal complement of $\mathrm{Pic}(X) \subset \Gamma_{3,19}$ has no U sublattice then X has no classical mirror.

---

[9]The notation $\Lambda \otimes_{\mathbf{Z}} \mathbb{R}$ denotes the real vector space generated by the generators of Λ.

Such mirror pairs of K3 surfaces were first noticed some time ago by Arnold (see for example, [**59**]). The non-linear σ-model moduli space gives a nice setting for this pairing to appear.

Now we also have the mirror construction of Greene and Plesser [**60**] which takes an algebraic variety X as a hypersurface in a weighted projective space and produces a "mirror" Y as an quotient of X by some discrete group such that X and Y as target spaces produce completely identical conformal field theories.

An interesting case is when X is the hypersurface

$$f = x_0^2 + x_1^3 + x_2^7 + x_3^{42} = 0 \tag{74}$$

in $\mathbb{P}^3_{\{21,14,6,1\}}$. X is a K3 surface which happens to be an orbifold due to the quotient singularities in the weighted projective space. Blowing this up provides a K3 surface with Picard lattice isomorphic to the even self-dual lattice $\Gamma_{1,9}$. Thus in this case $\Lambda \cong \Upsilon \cong \Gamma_{2,10}$. What is interesting is that in this case, the Greene-Plesser construction says that the group by which X should be divided to obtain the mirror is trivial. This mirror map does not act trivially on the marginal operators however but one may show actually exchanges the factors in (73). This is little delicate and we refer the reader to [**53**] for details. Anyway, what this shows is that the Greene-Plesser mirror map, which is an honest symmetry in the sense that it produces an identical conformal field theory, really is given by the exchange, μ, in (72).

This implicitly provides us with another element of G_σ, namely an element of $\mathrm{O}(\Gamma_{4,20})$ which exchanges two orthogonal $\Gamma_{2,10}$ sublattices. It is now possible to show that this new element, together with the subgroup given in (68), is enough to generate $\mathrm{O}(\Gamma_{4,20})$. Thus G_σ at least contains $\mathrm{O}(\Gamma_{4,20})$.

To prove the required result that $G_\sigma \cong \mathrm{O}(\Gamma_{4,20})$ we use the result of [**25**] which states that any further generator will destroy the Hausdorff nature of the moduli space contrary to our assumption.

We thus have

THEOREM 6. *Given theorem 5 and the assumption that the moduli space is Hausdorff, we have that the moduli space of conformally invariant non-linear σ-models on a K3 surface is*

$$\mathcal{M}_\sigma = \mathrm{O}(\Gamma_{4,20})\backslash\, \mathrm{O}(4,20)/(\mathrm{O}(4)\times \mathrm{O}(20)) \tag{75}$$

Lastly we should discuss the meaning of $\Gamma_{4,20}$. The lattice $\Gamma_{3,19}$ was associated to the lattice $H^2(X,\mathbb{Z})$. Extend the notion of our inner product (19) to that of any p-cycle (or equivalently p-form) by saying that the product of two cycles is zero unless their dimensions add up to 4. Now one can see that the lattice of total cohomology, $H^*(X,\mathbb{Z})$, is isomorphic to $\Gamma_{4,20}$. This gives a tempting interpretation. We will see later that this really is the right one.

This tells us how to view w. A point in the moduli space $\mathcal{M}_\sigma$ determines a conformal field theory uniquely but we are required to make a choice of w before we can determine the geometry of the K3 surface in terms of metrics and B-fields. This amounts to deciding which direction in $\Gamma_{4,20}$ will be $H^0(X,\mathbb{Z})$. This choice is arbitrary and different choices will lead to potentially very different looking K3 surfaces which give the same conformal field theory. This may be viewed as a kind of T-duality.

3.5. Conformal field theory on a torus. Although our main interest in these lectures are K3 surfaces, it turns out that the idea of compactification of strings on a torus will be intimately related. We are thus required to be familiar with the situation when the σ-model has a torus as a target space.

The problem of the moduli space in this case was solved in [**61, 62**]. This subject is also covered by H. Ooguri's lectures in this volume and we refer the reader there for further information (see also [**63**]). Here we will quickly review the result in a language appropriate for the context in which we wish to use it.

To allow for the heterotic string case, we are going to allow for the possibility that the left-moving sector of the σ-model may live on a different target space to that of the right-moving sector. We thus consider the left sector to have a torus of n_L real dimensions as its target space and let the right-moving sector live on a n_R-torus. Of course, for the simple picture of a string propagating on a torus, we require $n_L = n_R$.

The moduli space of conformal field theories is then given by

$$\mathrm{O}(\Gamma_{n_L,n_R})\backslash \mathrm{O}(n_L,n_R)/(\mathrm{O}(n_L)\times \mathrm{O}(n_R)). \tag{76}$$

The even self-dual condition, required for world-sheet modular invariance [**61**], on the lattice Γ_{n_L,n_R} enforces $n_L - n_R \in 8\mathbb{Z}$. Note that whether we impose any supersymmetry requirements or not makes little difference. The only Ricci-flat metric on a torus is the flat one, which guarantees conformal invariance to all orders. The trivial holonomy then means that a torus may be regarded as Kähler, hyperkähler or whatever so long as the dimensions are right.

Let us assume without loss of generality that $n_L \leq n_R$.[10] Our aim here is to interpret the moduli space (76) in terms of the Grassmannian of space-like n_L-planes in $\mathbb{R}^{n_L,n_R}$. Let us use Π to denote a space-like n_L-plane. Any vector in $\mathbb{R}^{n_L,n_R}$ may be written as a sum of two vectors, $p_L + p_R$, where

$$\begin{aligned} p_L &\in \Pi \\ p_R &\in \Pi^\perp, \end{aligned} \tag{77}$$

and $\Pi^\perp$ is the orthogonal complement of Π. Thus $p_L.p_L \geq 0$ and $p_R.p_R \leq 0$.

The winding and momenta modes of the string on the torus are now given by points in the lattice Γ_{n_L,n_R} according to Narain's construction [**61**]. The left and right conformal weights of such states are then given by $\frac{1}{2}p_L.p_L$ and $-\frac{1}{2}p_R.p_R$, thus allowing the mass to be determined in the usual way.

It is easy to see how the moduli space (76) arises from this point of view. The position of Π determines p_L and p_R for each mode. The modular group $\mathrm{O}(\Gamma_{n_L,n_R})$ merely rearranges the winding and momenta modes.

Let us try to make contact with the σ-model description of the string. We choose a null (i.e., light-like) n_L-plane, W, in $\Gamma_{n_L,n_R}\otimes_{\mathbb{Z}}\mathbb{R}\cong\mathbb{R}^{n_L,n_R}$ which is "aligned" along the lattice Γ_{n_L,n_R}. By this we mean that W is spanned by a subset of the generators of Γ_{n_L,n_R}. We then define W^* as the null n_L-plane dual to W,

[10] This means that the heterotic string will be chosen to be a superstring in the left-moving sector and a bosonic string in the right-moving sector. Unfortunately this differs from the usual convention. This is imposed on us, however, when we consider duality to a type IIA string and the natural orientation of a K3 surface.

and the time-like space $V \cong \mathbb{R}^{n_R - n_L}$ so that

$$\Gamma_{n_L,n_R} \otimes_{\mathbb{Z}} \mathbb{R} \cong W \oplus W^* \oplus V. \tag{78}$$

This decomposition is also aligned with the lattice so that $\Gamma_{n_L,n_R} \cap W$, $\Gamma_{n_L,n_R} \cap W^*$, and $\Gamma_{n_L,n_R} \cap V$ generate Γ_{n_L,n_R}.

We may write Π as given by

$$\Pi = \{x + \psi(x) + A(x);\ \forall x \in W\}, \tag{79}$$

where

$$\begin{aligned} \psi &: W \to W^* \\ A &: W \to V \end{aligned} \tag{80}$$

Viewing ψ as

$$\psi : W \times W \to \mathbb{R}, \tag{81}$$

we may divide ψ into symmetric, G, and anti-symmetric parts, B, such that $\psi = B + G$.

Physically W represents, under the metric G, the target space in which the string lives. To be precise, the target space n_L-torus is

$$\frac{W}{\Gamma_{n_L,n_R} \cap W}. \tag{82}$$

W^* is then the dual space in which the momenta live. Clearly B is the B-field. Lastly V represents the gauge group generated by the extra right-moving directions. To be precise, the group is given by

$$U(1)^{n_R - n_L} \cong \frac{V}{\Gamma_{n_L,n_R} \cap V}. \tag{83}$$

We will discuss in section 4.3 how this group can be enhanced to a nonabelian group for particular Π's. The quantity A then represents $\mathfrak{u}(1)$-valued 1-forms. We will follow convention and refer to these degrees of freedom within A as "Wilson Lines".

One may write G, B and A in terms of matrices to make contact with the expressions in [**62, 63**]. This is a somewhat cumbersome process but the reader should check it if they are unsure of this construction. Here we illustrate the procedure in the simplest case, namely that of a circle as target space. Here we have $n_L = n_R = 1$ and $\Gamma_{1,1} \cong U$ as in (25). An element of $\mathrm{O}(1,1)$ preserving the form U may be written in one of the forms

$$\begin{pmatrix} t & 0 \\ 0 & t^{-1} \end{pmatrix}, \begin{pmatrix} -t & 0 \\ 0 & -t^{-1} \end{pmatrix}, \begin{pmatrix} 0 & t \\ t^{-1} & 0 \end{pmatrix}, \begin{pmatrix} 0 & -t \\ -t^{-1} & 0 \end{pmatrix}, \tag{84}$$

where t is real and positive.

We divide this space by $\mathrm{O}(1) \times \mathrm{O}(1)$ from the right and by $\mathrm{O}(\Gamma_{1,1})$ from the left. Both these groups are isomorphic to $\mathbb{Z}_2 \times \mathbb{Z}_2$ and are given by (84) with $t = 1$. The result is that the moduli space is given by the real positive line mod $\mathbb{Z}_2$ represented by

$$\begin{pmatrix} t & 0 \\ 0 & t^{-1} \end{pmatrix} \cong \begin{pmatrix} t^{-1} & 0 \\ 0 & t \end{pmatrix}, \tag{85}$$

where t is real and positive.

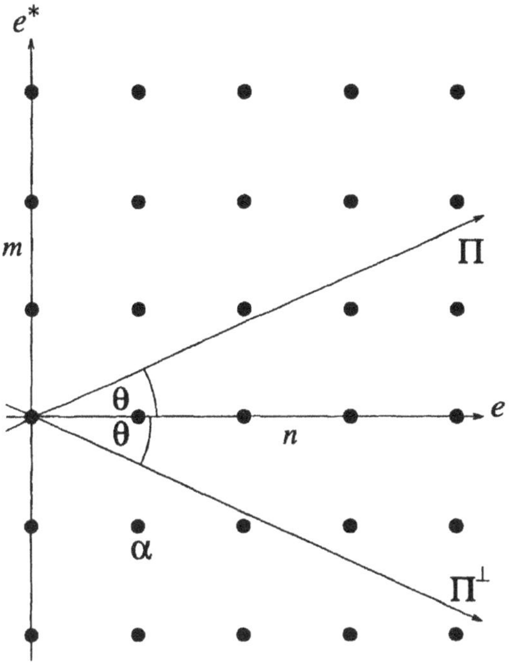

FIGURE 3. The moduli space picture for a string on a circle.

Let us chose a basis, $\{e, e^*\}$, for $\Gamma_{1,1}$ so that $e.e = e^*.e^* = 0$ and $e.e^* = 1$. Now let the space-like direction, Π, be given, as in figure 3, by

$$\Pi = \{xe + xe^* \tan\theta;\ x \in \mathbb{R}\}, \tag{86}$$

where $t = \tan\theta$. (Note that Π and $\Pi^\perp$ may not look particularly orthogonal in figure 3 but don't forget we are using the hyperbolic metric given by U!) Denoting a state in $\Gamma_{1,1}$ by $ne + me^*$, for $m, n \in \mathbb{Z}$, one can show

$$\begin{aligned} p_L{\cdot}p_L &= \tfrac{1}{2}n^2 \tan\theta + nm + \frac{m^2}{2\tan\theta} \\ p_R{\cdot}p_R &= -\tfrac{1}{2}n^2 \tan\theta + nm - \frac{m^2}{2\tan\theta}. \end{aligned} \tag{87}$$

Since G is given by $\tan\theta$, the radius of the circle is proportional to $\tan\theta$. The identification in (85) then gives the familiar "$R \leftrightarrow 1/R$" T-duality relation.

4. Type II String Theory

Now we have the required knowledge from classical geometry and "old" string theory on the world-sheet to tackle full string theory on a K3 surface. We begin with the most supersymmetry to make life simple. This is the type IIA or IIB superstring which has 2 supersymmetries when viewed as a ten-dimensional field theory. The heterotic string on a K3 surface constitutes a much more difficult problem and we won't be ready to tackle it until section 6.

d	N	Rep
11	1	$\mathbb{R}^{32}$
10	2	$\mathbb{R}^{16}$
9	2	$\mathbb{R}^{16}$
8	2	$\mathbb{C}^{8}$
7	2	$\mathbb{H}^{8}$
6	4	$\mathbb{H}^{4}$
5	4	$\mathbb{H}^{4}$
4	8	$\mathbb{C}^{2}$
3	16	$\mathbb{R}^{2}$

TABLE 2. Maximum numbers of supersymmetries.

4.1. Target space supergravity and compactification. We begin by switching our attention from the quantum field theory that lives in the world-sheet to the effective quantum field theory that lives in the target space in the limit that $\alpha'/R^2 \to 0$. Of particular interest will be theories with N supersymmetries in d dimensions (one of which is time).

A spinor is an irreducible representation of the algebra $\mathfrak{so}(1,d-1)$ and has dimension $2^{[\frac{d+1}{2}]-1}$, where the bracket means round down to the nearest integer. This is not the only information about the representation we require however. A spinor may be real ($\mathbb{R}$), complex ($\mathbb{C}$), or quaternionic[11] ($\mathbb{H}$) depending on d. The rule is

$$\begin{array}{l} \mathbb{R} \text{ if } d = 1,2,3 \pmod 8 \\ \mathbb{C} \text{ if } d = 0 \pmod 4 \\ \mathbb{H} \text{ if } d = 5,6,7 \pmod 8. \end{array}$$

A complex representation has twice as many degrees of freedom as a real representation. A quaternionic representation has the same number of degrees of freedom as a complex representation (as, even though one may define the representation naturally over the quaternions, one must subject it to constraints [29]). In terms of the language often used in the physics literature: in even numbers of dimensions we use Weyl spinors; real spinors are "Majorana" and quaternionic spinors are "symplectic Majorana".[12]

We may now list the maximal number of supersymmetries in each dimension subject to the constraint that no particle with spin > 2 appears. It was determined in [64] that $N = 1$ in $d = 11$ was maximal in this regard. To get the other values we simply maintain the number of degrees of freedom of the spinors. For example, in reducing to ten dimensions we go from 32 real degrees of freedom per spinor to 16 real degrees of freedom per spinor. We thus need two spinors in ten dimensions. The result is shown in table 2.

When $d \in 4\mathbb{Z}+2$, the left and right spinors in the field theory are independent and the supersymmetries may be separated. As we did in section 3.1, we will denote a theory with N_L left supersymmetries and N_R right supersymmetries as having

[11] Sometimes the terminology "pseudo-real" is used.

[12] Note that two symplectic Majorana spinors make up one quaternionic spinor.

$N = (N_L, N_R)$ supersymmetry. For purposes of counting total supersymmetries, $N = N_L + N_R$.

Now we want to see what happens when we compactify such a theory down to a lower number of dimensions. That is, we replace the space $\mathbb{R}^{1,d_0-1}$ by $\mathbb{R}^{1,d_1-1} \times X$, for some compact manifold X of dimension $d_0 - d_1$. To see what happens to the supersymmetries we need to consider how a spinor of $\mathfrak{so}(1, d_0 - 1)$ decomposes under the maximal subalgebra $\mathfrak{so}(1, d_0 - 1) \supset \mathfrak{so}(1, d_1 - 1) \oplus \mathfrak{so}(d_0 - d_1)$. The holonomy of X will act upon representations of $\mathfrak{so}(d_0 - d_1)$. Any representation of $\mathfrak{so}(1, d_1 - 1) \oplus \mathfrak{so}(d_0 - d_1)$ which is invariant under this action will lead to representations of $\mathfrak{so}(1, d_1 - 1)$ in our new compactified target space. This tells us how to count supersymmetries.

As the first example, consider toroidal compactification. In this case the torus is flat and so the holonomy is trivial. Thus every representation survives. Compactifying $N = 1$ supergravity in eleven dimensions down to d dimensions results in reproducing the maximal supersymmetries in table 2. Note that for $d = 10$ the supersymmetry is $N = (1,1)$ and for $d = 6$ the supersymmetry is $N = (2,2)$. One simple way of deducing this immediately is from the general result [**65**] that compactification of eleven dimensional supergravity on any manifold results in a non-chiral theory.[13]

Naturally our next example will be compactification on a K3 surface. In this case the holonomy is SU(2). Let us consider the case of an $N = 1$ theory in ten dimensions compactified down to six dimensions on a smooth K3 surface. The required decomposition is

$$\mathfrak{so}(1,9) \supset \mathfrak{so}(1,5) \oplus \mathfrak{so}(4) \cong \mathfrak{so}(1,5) \oplus \mathfrak{su}(2) \oplus \mathfrak{su}(2). \tag{88}$$

The spinor decomposes accordingly as $\mathbf{16} \to (\mathbf{4}, \mathbf{2}, \mathbf{1}) \oplus (\mathbf{4}, \mathbf{1}, \mathbf{2})$. We may take the last $\mathfrak{su}(2)$ in (88) as the holonomy. Thus the $(\mathbf{4}, \mathbf{2}, \mathbf{1})$ part is preserved. It may look like we have 2 spinors in 6 dimensions at this point but remember that our spinor in ten dimensions was real and spinors in six dimensions are quaternionic. Thus the degrees of freedom give only a single spinor in six dimensions. That is, $N = 1$ supergravity in ten dimensions compactified on a smooth K3 surface will give an $N = 1$ theory in six dimensions. The general rule is that compactification on a smooth K3 surface will preserve half as many supersymmetries as compactifying on a torus.

In the case that the number the supersymmetries or the number of dimensions is large, the form of the moduli space of possible supergravity theories becomes quite constrained. Holonomy is again the agent responsible for this. Let us write the notion of extended supersymmetry very roughly in the form

$$\{\overline{Q}^i, Q^j\} = \delta^{ij} P, \tag{89}$$

where Q are the supersymmetry generators and P is translation (we may ignore any central charge for purposes of our argument). Now such a relationship must clearly be preserved as we go around a loop in the moduli space. However, the supersymmetries may transform among themselves as we do this. This gives us a restriction on the holonomy of the bundle of supersymmetries over the moduli

[13]Fortunately recent progress in M-theory appears to tell us that compactification on spaces which are not manifolds can circumvent this statement (see, for example, [**66**]).

space. Comparing (89) to the analysis of invariant forms in section 2.2 tells us immediately what this restriction is. In the case that the spinor is of type $\mathbb{R}$, $\mathbb{C}$, or $\mathbb{H}$, the holonomy algebra will be (or contain) $\mathfrak{so}(N)$, $\mathfrak{u}(N)$, or $\mathfrak{sp}(N)$ respectively. Note that when we have chiral spinors in $4\mathbb{Z}+2$ dimensions we may factor the holonomy into separate left and right parts since these sectors will not mix.

Now the tangent directions in the moduli space are given by massless scalar fields which lie in supermultiplets. These multiplets have a definite transformation property under the holonomy group above. For a thorough account of this process we refer to [**67**]. This relates the holonomy of the bundle of supersymmetries to the holonomy of the tangent bundle of the moduli space. This knowledge can tell us a great deal about the form of the moduli space.

As an example consider $N = 4$ supergravity in five dimensions from table 2. We see immediately that the holonomy algebra of the moduli space is $\mathfrak{sp}(4)$. An analysis of the representation theory of the supergravity multiplet shows that the massless scalars transforms in a **42** of $\mathfrak{sp}(4)$. The only possibility from Berger and Simons result, and Cartan's classification of noncompact symmetric spaces, is that the moduli space is locally $E_{6(6)}/\,\mathrm{Sp}(4)_{\sim}$, where the tilde subscript denotes a quotient by the central $\mathbb{Z}_2$. The moduli spaces for all the entries in table 2 are given in [**68**].

4.2. The IIA string. The type IIA superstring in ten dimensions yields, in the low-energy limit, a theory of ten-dimensional supergravity with $N = (1,1)$ supersymmetry. If we compactify this theory on a K3 surface then each of these supersymmetries gives a supersymmetry in six dimensions and so the result is an $N = (1,1)$ theory in six dimensions.

The local holonomy algebra from above must therefore be $\mathfrak{sp}(1) \oplus \mathfrak{sp}(1) \cong \mathfrak{su}(2) \oplus \mathfrak{su}(2) \cong \mathfrak{so}(4)$. There are two types of supermultiplet in six dimensions which contain moduli fields:

1. The supergravity multiplet contains the dilaton which is a real scalar.
2. Matter multiplets each contain 4 real massless scalars which transform as a **4** of $\mathfrak{so}(4)$.

Thus the moduli space must factorize (at least locally) into a product of a real line for the dilaton times the space parametrized by the moduli coming from the matter multiplets. Thus, given the assumption concerning completeness, the Teichmüller space is of the form

$$\frac{\mathrm{O}(4,m)}{\mathrm{O}(4)\times\mathrm{O}(m)} \times \mathbb{R}, \tag{90}$$

where m is the number of matter multiplets.

As well as the metric, B-field and dilaton, the type II string also contains "Ramond-Ramond" states (see, for example, [**39**]). For the type IIA string these may be regarded as a 1-form and a 3-form.[14] A p-form field can produce massless fields upon compactification by integrating it over nontrivial p-cycles in the compact manifold. As a K3 surface has no odd cycles, no moduli come from the R-R sector. Thus, comparing (90) to (75) we see that $m = 20$.

[14] In a sense all odd forms may be included [**69**].

The moduli space of conformal field theories may be considered as living at the boundary of the moduli space of string theories in the limit that $\Phi \to -\infty$, i.e., weak string-coupling. Thus $O(\Gamma_{4,20})$ acts on this boundary. Given the factorization of the holonomy between the dilaton and the matter fields (as they transform in different representations of $\mathfrak{so}(4)$) the action of $O(\Gamma_{4,20})$ also acts away from the boundary in a trivial way on the dilaton. It is believed that moduli space is given simply by

$$\mathscr{M}_{\text{IIA}} \cong \mathscr{M}_\sigma \times \mathbb{R}. \tag{91}$$

That is, there are no identifications which incorporate the dilaton. Certainly any duality which mixed the factors would not respect the holonomy. Thus the only possibility remaining would be an action of the form $\Phi \to -\Phi$ which would be a strong-weak coupling duality (acting trivially on all other moduli). Instead of such an S-duality, a far more curious type of duality was suggested in **[70, 71]** as we now discuss.

The form of the moduli space of toroidal conformal field theories (76) in section 3.5 bares an uncanny resemblance to that of the moduli space of K3 conformal field theories, $\mathscr{M}_\sigma$. Indeed they are identical if $n_L = 4$ and $n_R = 20$. This is precisely the moduli space of conformal field theories associated to a heterotic string compactified on a 4-torus. Recall that the heterotic string consists of a superstring in the left-moving sector and a bosonic string in the right-moving sector. As such there are 16 extra dimensions in the right-moving sector which are compactified on a 16-torus and contribute towards the gauge group. In ten dimensions the heterotic string is an $N = (1, 0)$ theory and thus yields an $N = (1, 1)$ theory in six dimensions when compactified on a 4-torus.

The suggestion, which at first appears outrageous, is that the type IIA string compactified on a K3 surface is the same thing, physically, as the heterotic string compactified on a 4-torus. Although the moduli space of conformal field theories is identical for these models, the world-sheet formulation is so different that any notion that the two conformal field theories could be shown to be equivalent is doomed from the start. The claim however is not that the conformal field theories are equivalent, but that the full string theories are equivalent. The heterotic string has a dilaton, just like the type IIA string, and so the moduli space of heterotic string theories is also $\mathscr{M}_\sigma \times \mathbb{R}$. The only way we can map these two moduli spaces into each other without identifying the conformal field theories is thus to map Φ from one theory to $-\Phi$ from the other.

This then, is the alternative to S-duality for the type IIA string.

PROPOSITION 1. *The type IIA string compactified on a K3 surface is equivalent to the heterotic string compactified on a 4-torus. The moduli spaces are mapped to each other in the obvious way except that the strongly-coupled type IIA string maps to the weakly-coupled heterotic string and vica versa.*

It would be nice at this point if we could prove proposition 1. Unfortunately it appears that string theory is simply not sufficiently defined to allow this. We should check first that this proposition is consistent with what we do know about string theory.

The fact that the moduli spaces of the type IIA string on a K3 surface and the heterotic string on a 4-torus are identical is a good start. Next we may check that the effective six-dimensional field theory given by each is the same. This analysis was done in [**71**]. The result is affirmative but there is some new information. To achieve complete agreement the flat six-dimensional spaces given by the two string theories are not identical but instead the metrics are scaled with respect to each other. That is, let g_6 denote the six-dimensional space-time metric and Φ the dilaton, then

$$\begin{aligned} \Phi_{\text{Het}} &= -\Phi_{\text{IIA}} \\ g_{6,\text{Het}} &= e^{2\Phi_{\text{Het}}} g_{6,\text{IIA}}. \end{aligned} \tag{92}$$

Other checks may be performed too. Since the strings are related by a strong-weak coupling relations, the fundamental particles in one theory should map to solitons of the other theory. This has been analyzed to a great extent but we refer to J. Harvey's lectures for an account of this.

To date nothing has been discovered in known string theory to disprove proposition 1. Assuming there are no such inconsistencies one may wish to boldly assert that proposition 1 is true *by definition*. That is, whatever string theory may turn out to be, we will demand that it satisfies proposition 1. This is the point of view we will take from now on.

4.3. Enhanced gauge symmetries. One of the interesting questions we are now fully equipped to address is that of the gauge symmetry group of the six-dimensional theory resulting from either a compactification of a type IIA string on a K3 surface, or a heterotic string on a 4-torus.

Firstly we may consider this question from the point of view of conformal field theory. The type IIA string produces gauge fields from the R-R sector. The 1-form gives a U(1). The 3-form may be compactified down to a 1-form over 2-cycles in $b_2(S) = 22$ ways. Lastly, writing the 3-form as C_3, we have a dual field C_3^* given by

$$dC_3^* = *dC_3, \tag{93}$$

where C_3^* is a one form and also gives a U(1). Thus, all told, we have a gauge group of $\mathrm{U}(1)^{24}$. Note that it is not possible to obtain a nonabelian gauge theory, as far as the conformal field theory is concerned, because the R-R fields are so reluctant to couple to any other fields (see, for example, [**72**]).

Now consider the heterotic string picture. As explained in section 3.5, generically we obtain a $\mathrm{U}(1)^{16}$ contribution to the gauge group from the extra 16 right-moving degrees of freedom for the heterotic string. We also obtain 4 "Kaluza-Klein" U(1) factors from the metric from the 4 isometries of the torus. Lastly, the B-field is a 2-form and so contributes $b_1(T^4) = 4$ more U(1)'s. All told we have $\mathrm{U}(1)^{24}$ again as befits proposition 1.

Things become more interesting however when we realize that the heterotic string can exhibit larger gauge groups at particular points in the moduli space. What happens is that some winding/momentum modes of $\Gamma_{4,20}$ may happen to give physical massless vectors for special values of the moduli. Such states will be charged with respect to the generic $\mathrm{U}(1)^{24}$ and so a nonabelian group results.

The heterotic string is subject to a GSO projection to yield a supersymmetric field theory and this effectively prevents any new left-moving physical states from becoming massless. This may be thought of as a similar statement to the assertion that the type IIA string could never yield extra massless vectors. The right-moving sector of the heterotic string is not subject to such constraints however, and we may use our knowledge from section 3.5 to determine exactly when this gauge group enhancement occurs.

The vertex operator for one of our new massless states will be as follows. For the left-moving part we want simply ∂X^{μ} to give a vector index. For the right-moving part we require another operator with conformal weight one. This results in a requirement that we desire a state with $p_L = 0$ and $p_R.p_R = -2$. Thus we require an element $\alpha \in \Gamma_{4,20}$ which is orthogonal to Π and is such that $\alpha.\alpha = -2$.

The charge of such a state, α, with respect to the $\mathrm{U}(1)^{24}$ gauge group is simply given by the coordinates of α. This follows from the conformal field theory of free bosons and we refer to [41] for details. Comparing this to the standard way of building Lie algebras, we see that α looks like a root of a Lie algebra, whose Cartan subalgebra is $\mathfrak{u}(1)^{\oplus 24}$ (except that the Killing form is negative definite rather than positive definite).

In the simplest case, there will be only a single pair $\pm\alpha$ which satisfy the required condition. This then will add two generators to the gauge group charged with respect to one of the 24 U(1)'s. Thus the gauge group will be enhanced to $\mathrm{SU}(2) \times \mathrm{U}(1)^{23}$. Note that this enhancement to an SU(2) gauge group can also be seen in the simple case of a string on a circle as was pictured in figured 3. If $\theta = 45°$ then the lattice element marked α in the figure (together with $-\alpha$) will generate SU(2).

The general rule then should be that the set of all vectors given by

$$\mathscr{A} = \{\alpha \in \Gamma_{4,20} \cap \Pi^{\perp};\ \alpha.\alpha = -2\}, \tag{94}$$

will form the *roots* of the nonabelian part of the gauge group. Note that the rank of the gauge group always remains 24. Note also that the roots all have the same length, i.e., the gauge group is always *simply-laced* and falls into the *A-D-E* classification.

This allows us to build large gauge groups. One thing one might do, for example, is to split $\Gamma_{4,20} \cong \Gamma_{4,4} \oplus \Gamma_8 \oplus \Gamma_8$, where Γ_8 is the Cartan matrix of E_8 (with a negative-definite signature). We may then consider the case where $\Pi \subset \Gamma_{4,4} \otimes_{\mathbb{Z}} \mathbb{R}$. This would mean that any element in $\Gamma_8 \oplus \Gamma_8$ is orthogonal to Π and thus the gauge group is at least $E_8 \times E_8$. Looking at (79) we see that this is equivalent to putting $A = 0$. Thus we reproduce the simple result that the $E_8 \times E_8$ heterotic string compactified on a torus has gauge group containing $E_8 \times E_8$ if no Wilson lines are switched on.

Proposition 1 now tells us something interesting. Despite the fact that the conformal field theory approach to the type IIA string insisted that it could never have any gauge group other than $\mathrm{U}(1)^{24}$, the dual picture in the heterotic string dictates otherwise. There must be some points in the moduli space of a type IIA string compactified on a K3 surface where the conformal field theory misses part of the story and we really do get an enhanced gauge group.

Since we know exactly where the enhanced gauge groups appear in the moduli space of the heterotic string and we know exactly how to map this to the moduli space of K3 surfaces, we should be able to see exactly when the conformal field theory goes awry.

We will determine just which K3 surfaces give rise to this behaviour for the type IIA string. To do this we are required to choose w as in section 3.3 so that we can find a geometric description. Let us first assume that we may choose w so that

$$\alpha \in \frac{w^{\perp}}{w}, \quad \forall \alpha \in \mathscr{A}. \tag{95}$$

Let w^* be the same vector as was introduced in equation (66). Any vector in Π can be written as a sum $x + bw^* + cw$, where $b, c \in \mathbb{R}$ and $x \in \Sigma$. Thus, the statement that α is orthogonal to Π implies that α is orthogonal to Σ. Now we use the results of section 2.6 which tell us that this implies that the K3 surface is an orbifold. To be precise, the set $\mathscr{A}$ corresponds to the root diagram of the A-D-E singularity given in table 1.

What we have just shown is a remarkable fact (and was first shown by Witten in [**71**]).

PROPOSITION 2. *If we have a K3 surface with an orbifold singularity then a type IIA string compactified on this surface can exhibit a nonabelian gauge group such that the A-D-E classification of orbifold singularities coincides perfectly with the A-D-E classification of simply-laced Lie groups.*

This proposition rests on proposition 1 and the assumptions that went into building the moduli spaces. As an example, we see that an $\mathrm{SU}(n)$ gauge group corresponds to a singularity locally of the form $\mathbb{C}^2/\mathbb{Z}_n$.

Note that the orbifold singularity is not a sufficient condition for an enhanced gauge symmetry. For Π to be orthogonal to α we also require that B', and hence B, is also perpendicular to α. One way of stating this is to say that *the component of B along the direction dual to α is zero.* Note that the volume of the K3 surface does not matter in this context.

It is probably worth emphasizing here that this statement about the B-field can be important [**73**]. Orbifolds are well-known as "good" target spaces for string theory in that they lead to finite conformal field theories. It might first appear then that we are saying that the conformal field theory picture is breaking down at a point in the moduli space where an enhanced gauge group appears but when the conformal field theory appears to be perfectly reasonable. This is not actually the case. The enhanced gauge symmetry appears when $B = 0$ along the relevant direction. Conformal field theory orbifolds however tend to give the value $B = \frac{1}{2}$. Thus, the point in the moduli space corresponding to the happy conformal field theory orbifold and the point where the enhanced gauge group appears are not the same. There is good reason to believe the conformal field theory at the point where the enhanced gauge group appears is not well-behaved [**74**].

What happens if we relax our condition on w given in (95)? Now the situation is not so clear. Σ need not be perpendicular to any α and so the K3 surface may be smooth. If this is the case, the volume of the K3 surface cannot be arbitrarily large. This is because in the large radius limit, Π is roughly the span of Σ and w, but α is not perpendicular to w. In fact, the volume of the K3 surface must be of order one in units of $(\alpha')^2$. Thus we see, assuming the B-field is tuned to the right value, that an enhanced gauge group arises when the K3 surface has orbifold singularities and is any size, or if the K3 surface is very small and is given just the right (possibly smooth) shape.

Note that to get a very large gauge group we probably require the K3 surface be singular *and* it to be very small. This is necessary, for example, if we want a gauge group $E_8 \times E_8 \times \mathrm{SU}(2)^4 \times \mathrm{U}(1)^4$. Note that this latter group is "maximal" in the sense that the nonabelian part is of rank 20, which is the maximal rank sublattice that can be orthogonal to Π.

We have proven the appearance of nonabelian gauge groups by assuming proposition 1. Instead one might like to attempt some direct justification. This is probably best seen by using Strominger's notion of "wrapping p-branes" [**75**]. The general idea is that the R-R solitons are associated with cycles in the target space and the mass of these states is given by the area (or volume) of these cycles. Thus, as the K3 surface acquires a quotient singularity an S^2 shrinks down and thus a soliton becomes massless. We will not pursue the details of this construction as solitons lie somewhat outside our intended focus of these lectures. We also refer the reader to [**76, 77**] for further discussion.

4.4. The IIB string. The type IIB superstring in ten dimensions yields, in the low-energy limit, a theory of ten-dimensional supergravity with $N = (2, 0)$ supersymmetry. If we compactify this theory on a K3 surface then each of these supersymmetries gives a supersymmetry in six dimensions and so the result is an $N = (2, 0)$ theory in six dimensions.

The local holonomy algebra from above must therefore be $\mathfrak{sp}(2) \cong \mathfrak{so}(5)$. There is only one type of supermultiplet in six dimensions which contains massless scalars and that is the matter supermultiplet. Each such multiplet contains five scalars transforming as a **5** of $\mathfrak{so}(5)$. Thus, given the assumption concerning completeness, the Teichmüller space is of the form [**78**]

$$\frac{\mathrm{O}(5, m)}{\mathrm{O}(5) \times \mathrm{O}(m)}, \tag{96}$$

where m is the number of matter multiplets.

In addition to the metric, B-field and dilaton, moduli may also arise from the R-R fields. The type IIB string in ten dimensions has a 0-form, a 2-form and a self-dual 4-form. The 0-form gives $b_0(S) = 1$ modulus. The 2-form gives $b_2(S) = 22$ moduli. The 4-form gives $b_4(S) = 1$ modulus. One might also try to take the dual of the 4-form to give another modulus. This would be over-counting the degrees of freedom however, as the 4-form is self-dual. Thus, the number of moduli are given by

Metric	58
B-field	22
Dilaton	1
0-form	1
2-form	22
4-form	1
Total	105

To get the dimension of the Teichmüller space correct we require $m = 21$.

To find the discrete group acting on the Teichmüller space we go through a procedure remarkably similar to that of section 3.4 where we found the moduli space of conformal field theories. Firstly we may consider the Teichmüller space as the Grassmannian of space-like 5-planes in $\mathbb{R}^{5,21}$. Denote the 5-plane by Θ. Then we choose a primitive null element, u, of $\Gamma_{5,21}$. Now we have $u^\perp/u \cong \mathbb{R}^{4,20}$. We may define $\Pi' = \Theta \cap u^\perp$. Define the vector R' by demanding that Π' and R' span Θ, R' be orthogonal to Π' and $u.R' = 1$. Then project Π' and R' into $u^\perp/u$ to obtain the space-like 4-plane, Π, and the vector, R, respectively.

Clearly we now interpret Π in terms of the underlying conformal field theory on the K3 surface. R represents the degrees of freedom coming from the R-R sector. The six-dimensional dilaton may be deduced from $R' - R$ just as the volume of the K3 surface was deduced from $B' - B$ in section 3.4.

It is important to note that this construction provides considerable evidence for our identification of $\mathbb{R}^{4,20}$ with $H^*(S,\mathbb{Z})$ in section 3.4. This is because R, which represents the R-R degrees of freedom, is a vector in $\mathbb{R}^{4,20}$ — the same space in which Π lives. We also saw how these moduli arise from $H^0 \oplus H^2 \oplus H^4 \cong H^*$.

Generating the discrete modular group, G_{IIB} (which, as the reader will have guessed, will turn out to be $\mathrm{O}(\Gamma_{5,21})$) is slightly different to the way we built $\mathrm{O}(\Gamma_{4,20})$ in section 3.4, but we show here that we can reduce it to the same problem. First note that we have $\mathrm{O}(\Gamma_{4,20}) \subset G_{\text{IIB}}$. That is, any symmetry of the conformal field theory will be a symmetry of the string theory (just as any symmetry of the classical geometry is a symmetry of the conformal field theory).

The next ingredient we use will be that of S-duality of the type IIB string in ten dimensions [**70, 71**]. This asserts that there is an $\mathrm{SL}(2,\mathbb{Z})$ symmetry acting on the ten-dimensional dilaton and the axion (i.e., the R-R 0-form). This group is generated by a strong-weak coupling interchange of the form $\Phi_{10,\text{IIB}} \to -\Phi_{10,\text{IIB}}$, and a translation of the axion by one. While one might assert this S-duality statement as a distinct conjecture, it is certainly intimately related to other duality statements. One simple way of "deriving" it is to see that it is almost an inevitable consequence of M-theory [**79, 80**] (see also J. Schwarz's lectures). We will also see in section 5.1 that it follows from proposition 1 and mirror symmetry.

To embed $\mathrm{SL}(2,\mathbb{Z})$ into $\Gamma_{5,21}$ we note that $\mathrm{SL}(2,\mathbb{Z})$ is the group of automorphisms of $\mathbb{Z}^2$, which means that it is the group of isometries of a null lattice of rank two. We may split $\Gamma_{5,21} \cong \Gamma_{3,19} \oplus \Gamma_{2,2}$ and then let the $\mathrm{SL}(2,\mathbb{Z})$ act on a null 2-plane in the $\Gamma_{2,2}$ part. If we identify $\Gamma_{3,19}$ with $H^2(S,\mathbb{Z})$ then we see that the group of classical symmetries of the K3 surface (i.e., $\mathrm{O}(\Gamma_{3,19})$) commutes with S-duality. This is exactly what we desire from the effective target space theory.

Now the shift of the axion is a shift in the $H^0(S,\mathbb{Z})$ direction. Since $\mathrm{O}(\Gamma_{4,20})$ acts transitively on primitive null vectors, we immediately see that any shift by an element of $\Gamma_{4,20}$ is a symmetry of the string theory. This is an analogue of the "integral B-field shift". That is, *shifting any R-R modulus by an element of $H^*(S,\mathbb{Z})$ is a symmetry of the string theory.*

The other generator of $\mathrm{SL}(2,\mathbb{Z})$ may be taken as one which exchanges the two null vectors generating the null 2-plane in $\Gamma_{2,2}$. That is, we swap the axion direction, $H^0(S,\mathbb{Z})$, with u, a null vector outside $\mathbb{R}^{4,20}$. This is the exact analogue of mirror symmetry (which exchanged an element of $H^2(S)$ with $H^0(S)$). We have reduced the problem to one completely analogous to finding the modular group for conformal field theories on a K3 surface. Thus we deduce that we can generate all of $\mathrm{O}(\Gamma_{5,21})$ (as asserted first in [**71**]). Assuming the moduli space is Hausdorff we have

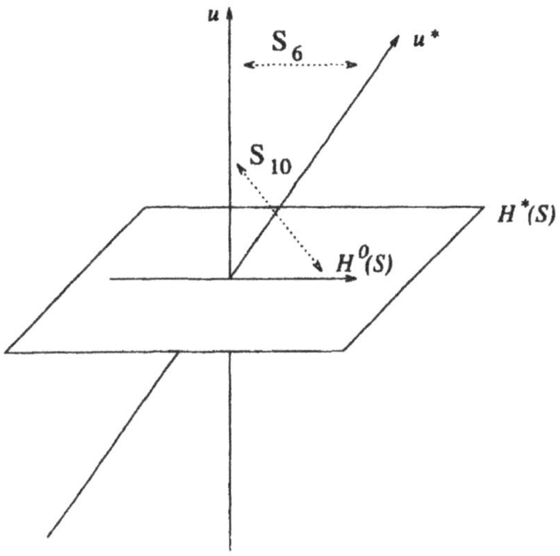

FIGURE 4. The S-dualities of the type IIB string.

PROPOSITION 3. *The moduli space of type IIB string theories compactified on a K3 surface is*

$$\mathcal{M}_{\text{IIB}} = \mathrm{O}(\Gamma_{5,21})\backslash \mathrm{O}(5,21)/(\mathrm{O}(5)\times \mathrm{O}(21)). \tag{97}$$

This proposition depends on the S-duality conjecture for the type IIB string in ten dimensions, theorem 6, and the completeness and Hausdorff constraints.

We should mention that there is a strong-weak coupling duality in the resulting six-dimensional theory. Let u^* be a null vector in $\Gamma_{5,21}$ dual to u such that u and u^* span the $\Gamma_{1,1}$ sublattice orthogonal to $H^*(S,\mathbb{Z}) \cong \Gamma_{4,20}$. One of the elements in $\mathrm{O}(\Gamma_{5,21})$ exchanges u and u^* and has the effect of reversing the sign of the six-dimensional dilaton. Note that this is not at all the same as the element of $\mathrm{O}(\Gamma_{5,21})$ which changed the sign of the *ten*-dimensional dilaton as the latter exchanged u with $H^0(S,\mathbb{Z})$ as shown in figure 4. In general an S-duality in a given number of dimensions will not give rise to an S-duality in a lower number of dimensions upon compactification. In the type IIB string however, we see S-duality in both six and ten dimensions.

The behaviour of the type IIA and type IIB string can be contrasted. The strongly-coupled type IIA string on a K3 surface is dual to a different string theory (the heterotic string) which is weakly-coupled. The type IIB strongly-coupled string on a K3 surface is dual to a weakly-coupled version of itself.

Finally in this section let us mention some strange properties of the type IIB string on a K3 surface. We know that when the type IIA string is compactified on an orbifold with $B=0$ then an enhanced gauge symmetry appears. This indicated some divergence within the underlying conformal field theory. It should be true therefore that a type IIB string compactified on the same space must have some interesting nonperturbative physics since it is associated with the same divergent conformal field theory. It was explained in [74] that these new theories are associated with massless string-like solitons which appear at these points in the moduli space.

5. Four-Dimensional Theories

Now let us explore what happens when we compactify string theories down to four dimensions. This process need not involve a K3 surface in general but we will find that in all the easy cases K3 surfaces will be present in abundance!

5.1. $N = 4$ theories. Supersymmetries are not chiral in four dimensions and so, in contrast to the six-dimensional case, there is only one kind of $N = 4$ theory.

The local holonomy algebra must be $\mathfrak{u}(4) \cong \mathfrak{u}(1) \oplus \mathfrak{su}(4) \cong \mathfrak{u}(1) \oplus \mathfrak{so}(6)$. There are two types of $N = 4$ supermultiplet in four dimensions which contain moduli fields:

1. The supergravity multiplet contains the dilaton-axion field which is a complex object under the $\mathfrak{u}(1)$ holonomy.
2. Matter multiplets each contain 6 real massless scalars which transform as a **6** of $\mathfrak{so}(6)$.

Thus the moduli space must factorize into a product of a complex plane, for the dilaton-axion, times the space parametrized by the moduli coming from the matter multiplets. Thus, given the assumption concerning completeness, the Teichmüller space may be written in the form

$$\frac{\mathrm{O}(6,m)}{\mathrm{O}(6)\times \mathrm{O}(m)} \times \frac{\mathrm{SL}(2)}{\mathrm{U}(1)}, \tag{98}$$

where m is the number of matter multiplets.

A very simple way to arrive at this theory is to compactify a heterotic string on a 6-torus. The first factor of (98) is then clearly the moduli space of the conformal field theories on the torus with 6 left-moving dimensions and 22 right-moving dimensions — that is, $m = 22$. The $\mathrm{SL}(2)/\mathrm{U}(1)$ term then comes from the dilaton-axion system: the dilaton being the string dilaton as usual and the axion from dualizing the B-field to obtain a scalar.

The type IIA string may be compactified on $\mathrm{K3}\times T^2$ to obtain an $N{=}4$ theory too. The conformal field theory on a K3 has 80 real deformations and for T^2 it has 4 deformations. The 1-form R-R field gives $b_1(\mathrm{K3}\times T^2) = 2$ moduli and the 1-form R-R field gives $b_3(\mathrm{K3}\times T^2) = 44$ moduli. The 3-form may also be compactified down to a 2-form in $b_1(\mathrm{K3}\times T^2) = 2$ ways and then dualized to give 2 more scalars. Adding the dilaton-axion we have $80 + 4 + 2 + 44 + 2 + 2 = 134$. This implies $m = 22$ again. Actually proposition 1 tells us that this must be the same theory as the heterotic string on a 6-torus.

Let us examine the moduli space of conformal field theories on a 2-torus. As far as the Teichmüller space is concerned we have

$$\frac{\mathrm{O}(2,2)}{\mathrm{O}(2)\times \mathrm{O}(2)} \cong \frac{\mathrm{SL}(2)}{\mathrm{U}(1)} \times \frac{\mathrm{SL}(2)}{\mathrm{U}(1)}, \tag{99}$$

up to $\mathbb{Z}_2$ identifications. One of the $\mathrm{SL}(2)/\mathrm{U}(1)$ factors may be regarded as the complex structure of the torus and the other $\mathrm{SL}(2)/\mathrm{U}(1)$ factor represents the Kähler form and B-field on the torus. We refer to [63] for a review of this. What does the second factor in (98) represent? There are many ways of approaching this problem. Here we use a trick following [81] that will come in use later on.

Begin with a six-dimensional field theory given by the heterotic string compactified on a 4-torus. The effective field theory in six dimensions will be roughly of the form

$$S = \int d^6X\,\sqrt{g_6}e^{-2\Phi_{6,\mathrm{Het}}}(R+\dots) \tag{100}$$

Now compactify over a 2-torus of area A_{Het} (as measured by the heterotic string).

$$\begin{aligned} S &= \int d^4X\,\sqrt{g_4}\,e^{-2\Phi_{6,\mathrm{Het}}}A_{\mathrm{Het}}(R+\dots) \\ &= \int d^4X\,\sqrt{g_4}\,e^{-2\Phi_{4,\mathrm{Het}}}(R+\dots), \\ &= \int d^4X\,\sqrt{g_4}\,A_{\mathrm{IIA}}(R+\dots), \end{aligned} \tag{101}$$

where A_{IIA} is the area of the 2-torus when we consider building the same theory by compactifying the type IIA string on K3 $\times\, T^2$. We have made use of (92) to derive this. Looking at the second line of (101) we see that the area of the T^2 is actually playing the rôle of the coupling constant. Thus, in going from the heterotic string description of the situation to the type IIA description of the same situation, the dilaton of the heterotic string has been replaced by the area of the T^2 of the type IIA string. Thus, *the* SL(2)/ U(1) *factor in (98) must represent the Kähler form and B-field of the 2-torus in the type IIA picture.*

We know from conformal field theory that $\mathrm{SL}(2,\mathbb{Z})$ acts on the SL(2)/ U(1) part of the Teichmüller space giving the Kähler form and B-field of the 2-torus. Combining this knowledge with what we found from the heterotic string, we see that $\mathrm{O}(\Gamma_{6,22}) \times \mathrm{SL}(2,\mathbb{Z})$ acts as the modular group for our $N = 4$ theory.

PROPOSITION 4. *The type IIA string compactified on K3 × T^2 is equivalent to the heterotic string compactified on a 6-torus and they form the moduli space*

$$\mathscr{M}_{N=4} = (\mathrm{O}(\Gamma_{6,22})\backslash \mathrm{O}(6,22)/(\mathrm{O}(6)\times\mathrm{O}(22))) \times (\mathrm{SL}(2,\mathbb{Z})\backslash \mathrm{SL}(2)/\mathrm{U}(1))\,. \tag{102}$$

This rests on the same assumptions as proposition 3.[15]

The $\mathrm{SL}(2,\mathbb{Z})$ factor of the modular group acts as an S-duality in the effective four-dimensional theory. Thus we have *derived*, from proposition 1 the existence of Montonen-Olive S-duality [**83, 84, 85**] for $N = 4$ theories in four dimensions. (Again we are going to neglect to discuss solitons — see [**82**] for such analysis.)

Lastly we may consider the type IIB string compactified on K3 $\times\, T^2$. Again there are a multitude of ways of arriving at the desired result. One of the easiest ways is to take following proposition from [**86, 87**]:

[15] Depending on one's tastes, in the case of the moduli space of the heterotic string on a 6-torus one may wish to consider this statement as more fundamental than proposition 1 as it may be analyzed directly in terms of solitons [**82**].

PROPOSITION 5. *The type IIA superstring and type IIB superstring compactified down to nine dimensions on a circle are equivalent except that the radii of the circles are inversely related.*

One also needs to shift the dilaton of one theory relative to the other to achieve the same target space effective field theory for the two string theories.

Thus, since the T^2 of K3 $\times$ T^2 contains a circle, the type IIB theory compactified on K3 $\times$ T^2 can also be bundled into proposition 4. Note that an $R \leftrightarrow 1/R$ transformation on one of the circles in the T^2 is a mirror map in the sense that the notions of deformation of complex structure and complexified Kähler form are interchanged. Thus, the $\mathrm{SL}(2,\mathbb{Z})\backslash\mathrm{SL}(2)/\mathrm{U}(1)$ factor in the moduli space in (102) represents the complex structure moduli space of the 2-torus in the case of the type IIB string.

Thus the $\mathrm{SL}(2,\mathbb{Z})\backslash\mathrm{SL}(2)/\mathrm{U}(1)$ factor in the moduli space in (102) can play 3 rôles [**88, 89**]:

1. The dilaton-axion variable in the case of the heterotic string.
2. The area and B-field of the 2-torus in the case of the type IIA string.
3. The complex structure of the 2-torus in the case of the type IIB string.

Comparing the heterotic string to the type IIA string in this setup may be regarded as fairly profound. The dilaton of the heterotic string, i.e., the coupling of the space-time field theory, is mapped to an area in the type IIA theory, i.e., the coupling of the world-sheet field theory. Thus, in a sense *we are mapping the space-time field theory associated to one string theory to the world-sheet field theory associated to another.* One might take this as evidence that neither the target space point of view nor the world-sheet point of view of string theory may be regarded as more fundamental than the other since they may be exchanged.

The $\mathrm{SL}(2,\mathbb{Z})$ S-duality of the type IIB string in ten dimensions is now sitting in the group $\mathrm{O}(\Gamma_{6,22})$. In fact, the above analysis can be used to "prove" the existence of this S-duality group. This can be viewed as an analogue of the deduction of this same S-duality group from M-theory as was done in [**79, 80**].

Note that we can play the same game as in section 4.3 to find the enhanced gauge groups. In this case we have a space-like 6-plane in $\Gamma_{6,22} \otimes_{\mathbb{Z}} \mathbb{R}$ and we look for roots perpendicular to this plane. The gauge group is always of rank 28.

5.2. More $N = 4$ theories. Does the moduli space (102) represent all possible $N = 4$ theories in four dimensions? It seems unlikely as one expects to be able to build theories with a gauge group of rank < 28. Consider compactification of the type II string. All we demand to obtain the desired theory in four dimensions is that the manifold on which we compactify have SU(2) holonomy. The only complex surface with SU(2) holonomy is a K3 surface. We have more possibilities in complex dimension three however. Thus we expect that the moduli spaces we discussed in section 4 to give the complete story for $N = 2$ theories in six dimensions but we are not done yet for $N = 4$ theories in four dimensions.

Note that this is a similar statement to the one that Seiberg gave using anomalies [**50**]. When using a conformal field theory to compactify a ten-dimensional theory to six dimensions one may consider the case of a type IIB string compactified to a chiral $N = (2,0)$ six-dimensional theory and analyze the anomalies. The Hodge numbers of a K3 surface are found to be necessary for a consistent theory.

We should add that one can find further theories in six dimensions if one is willing to drop the requirement that the compactification has some conformal field theory description (and switching to something like M-theory instead). See **[90]** for an example.

Let us consider how to build a complex threefold with holonomy SU(2). First we note the existence of a covariantly-constant holomorphic 2-form and thus $h^{2,0} = 1$. The Dolbeault index **[9]** may then be used to establish $h^{1,0} = 1$. Thus our manifold cannot be simply-connected. Now we may use the Cheeger–Gromoll theorem **[91]** which tells us that the universal cover of the manifold is isometric to $M \times \mathbb{R}^n$ for some compact simply-connected manifold, M. It is clear that we require that the universal cover to be $\mathrm{K3} \times \mathbb{R}^2$. In other words, *any complex threefold with holonomy* SU(2) *is isometric to K3 × T^2 or some quotient thereof.*

To build more $N = 4$ theories we will consider compactifying type II strings on a quotient of $\mathrm{K3} \times T^2$. This quotient must of course preserve the global SU(2) holonomy and thus any element of the quotienting group must preserve the holomorphic 2-form on the K3 surface. Any such action has fixed points on the K3 surface. Thus, to avoid getting a quotient singularity, any such action on the K3 surface must be accompanied by a translation on the T^2. That is, the quotienting group must have translations in $\mathbb{R}^2$ as a faithful representation. An immediate consequence is that the quotienting group must be abelian.

The classification of such groups, G, has been done by Nikulin **[92]** and we list the results in table 3. M is the rank of the maximal sublattice of $H^2(\mathrm{K3}, \mathbb{Z})$ that transforms nontrivially under G.

This action of G on $\Gamma_{6,22} \cong \Gamma_{4,20} \oplus \Gamma_{2,2}$ is now determined. Firstly, the action on K3 is a geometric symmetry and so must preserve w and w^*. The K3 part of the action is then determined by the action on $H^2(\mathrm{K3}, \mathbb{Z}) \cong \Gamma_{3,19} \subset \Gamma_{4,20}$ For the explicit form of the action on the lattice $H^2(\mathrm{K3}, \mathbb{Z})$ we refer the reader to **[93]**. Lastly we need the action on $\Gamma_{2,2}$. This encodes the action of G on T^2 which we require to be a translation. We also want this action to be geometric and therefore left-right symmetric. This forces the shift to be a *null* direction. This is sufficient to determine the shift up to isomorphism.

Now that we know the action of G on $\Gamma_{6,22}$ we may copy the description of the quotienting procedure over into the heterotic string picture. The result is that we are now describing an *asymmetric orbifold* of a heterotic string on T^6. Such objects were first analyzed in **[94]**. An asymmetric orbifold is an string-theory orbifold in which the left-movers and right-movers of the conformal field theory are not treated identically. Because of this the geometric description of the quotienting process in terms of target space geometry is obscure. Also the chiral nature of quotienting can produce anomalies. One manifestation of this can be lack of modular invariance of the resulting conformal field theory.

Let us consider an asymmetric orbifold of a toroidal theory built on a lattice Λ. Consider an element of the quotienting group, $g \in G$, and represent it as a rotation

G	$\mathbb{Z}_2$	$\mathbb{Z}_2 \times \mathbb{Z}_2$	$\mathbb{Z}_2 \times \mathbb{Z}_4$	$\mathbb{Z}_2 \times \mathbb{Z}_6$	$\mathbb{Z}_3$	$\mathbb{Z}_3 \times \mathbb{Z}_3$	$\mathbb{Z}_4$	$\mathbb{Z}_4 \times \mathbb{Z}_4$	$\mathbb{Z}_5$	$\mathbb{Z}_6$	$\mathbb{Z}_7$	$\mathbb{Z}_8$
M	8	12	16	18	12	16	14	18	16	16	18	18

TABLE 3. Nikulin's K3 quotienting groups.

of the lattice followed by a shift, $\delta \in \Lambda \otimes_{\mathbb{Z}} \mathbb{R}$. Let the eigenvalues of the rotation be of the form $\exp(2\pi i.r_j)$. A necessary condition for modular invariance is that [**94, 95**]

$$|g|\left(\tfrac{1}{4}\sum_j r_j(1-r_j)+\tfrac{1}{2}\delta.\delta\right)\in\mathbb{Z};\quad \forall g\in G, \tag{103}$$

where $|g|$ is the order of g. One may check the groups in table 3 and show that this condition is indeed satisfied in every case.

Did we have the right to expect that (103) should be satisfied for all the groups in table 3? To this author the result seems a little mysterious. Indeed, it is the case that if one considers quotients which destroy the $N = 4$ supersymmetry then one need not be so lucky [**96, 97**]. For now though, since we are concerned with $N = 4$ theories at this point, we may content ourselves with the knowledge that the anomalies appear to be looking after themselves and press on.

Now let us determine the moduli space. Firstly any deformation of the original theory of the type II string on $\mathrm{K3}\times T^2$ or the heterotic string on T^6 which is invariant under G will be a deformation of the resulting quotient. Secondly we need to worry in string theory that we may introduce some "twisted marginal operators" — that is, massless modes associated with fixed points. Since there are no fixed points of G (at least in the type II picture) we may ignore the latter possibility. To obtain the first type of deformation we may simply restrict attention to the invariant sublattice $\Lambda_G \subset \Gamma_{6,22}$ under (the rotation part of) G. Note that all of the space-like directions of $\Gamma_{6,22}$ are not rotated by G and so the resulting invariant sublattice, Λ_G, will have signature $(6, m)$, where $m = 22 - M$ from table 3.

What we have done is to build a moduli space of the form

$$\mathrm{O}(\Lambda_G)\backslash\,\mathrm{O}(6,m)/(\mathrm{O}(6)\times\mathrm{O}(m)), \tag{104}$$

of an $N = 4$ theory in four dimensions which we viewed either as a type IIA string on a freely-acting quotient of $\mathrm{K3}\times T^2$ or an asymmetric orbifold of a heterotic string on T^6.

Are there any further possibilities beyond those listed in table 3? We restricted ourselves to classical symmetries of the K3 surface. There should certainly be more symmetries from the stringy geometry of the K3 surface when it is Planck-sized. One may also look at M-theory to provide more possibilities. We refer the reader to [**93, 98**] for further discussion.

Although (104) looks suspiciously like a Narain moduli space for a heterotic string on a torus, it is important to notice that Λ_G is, in general, not self-dual. Any attempt to describe this theory in a straight forward way as a toroidal compactification is doomed as it would imply that the theory is not modular invariant.

On a similar point we have to be careful when considering which enhanced gauge groups can appear. One can view the moduli space as a Grassmannian of space-like 6-planes in $\mathbb{R}^{6,m}$ but it is no longer the case that only elements, $v \in \Lambda_G$, perpendicular to this 6-plane with $v.v = -2$ will give massless vector fields. The simplest way to approach the question of enhanced gauge groups is as follows (see [**99, 100**] for the original analysis in terms of heterotic strings). Consider the original theory before we divide by G. In this case, we know what the roots of the enhanced gauge group are, given the space-like 6-plane. As G has a nontrivial

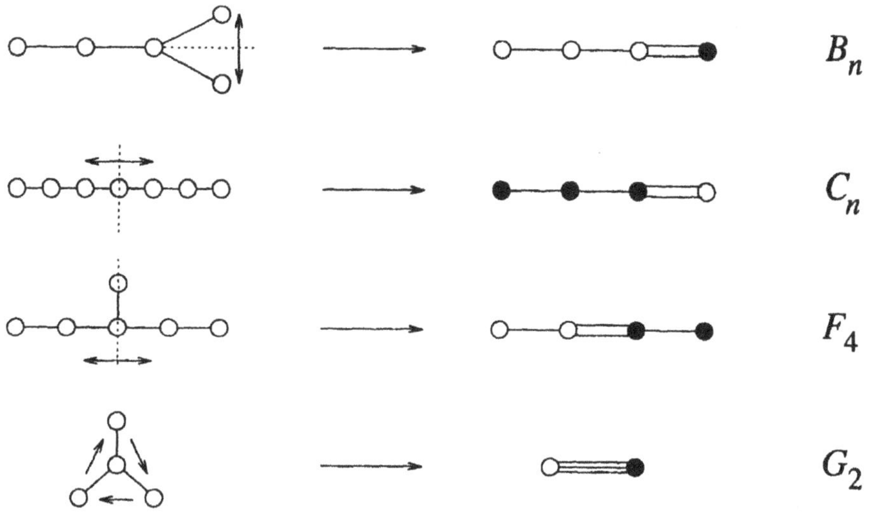

FIGURE 5. Quotients of simply-laced groups.

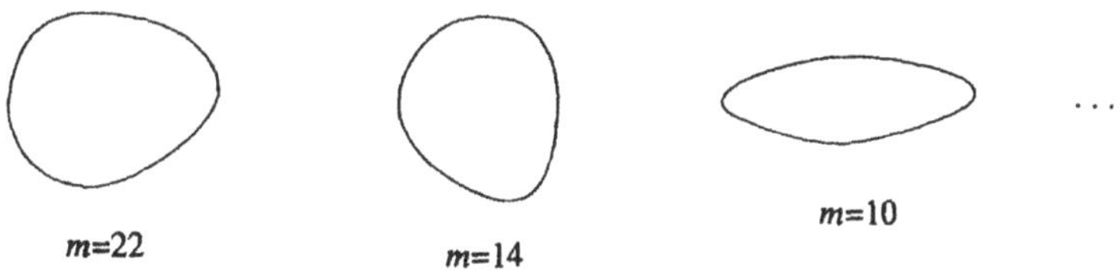

FIGURE 6. The moduli space of $N = 4$ theories.

action on the lattice $\Gamma_{6,22}$, it may also act on the roots of the gauge group. Since our desired theory is the invariant part of the original theory under the quotient by G, the resulting gauge group will be the invariant part of the original gauge group under the action of the discrete group G.

The problem we have therefore is as follows. Given a simply-laced gauge group and an action of a discrete group, G, on the roots of this gauge group, find the subgroup of the gauge group which is invariant under this action. This will be the enhanced gauge group of the desired quotient theory. Fortunately this is a well-known problem in Lie group theory (see, for example, exercise 22.24 in [**29**]). The outer automorphism of the group, given by an action on the roots can be written as a symmetry of the Dynkin diagram in the obvious way.[16] The results are shown in figure 5 and show that non-simply-laced Lie groups can result. In particular one may show that *any* Lie group (of sufficiently small rank) can appear as an enhanced gauge symmetry.

Lastly let us note an important point about the $N = 4$ moduli spaces. This is that they are *disconnected* from each other as shown in figure 6. If components of the moduli space with different values of m were to touch each other then, at such

[16]Except for the case of $\mathfrak{su}(2n+1)$ in which case the outer automorphism yields $\mathfrak{so}(2n+1)$ as the invariant subalgebra.

a point of contact, we would have a theory with very special properties. As one approached such a theory from within the interior of one of the regions, extra states would become massless to furnish the deformations into the other region. This does not happen according to the conformal field description of either the heterotic string or the type II strings. Thus it would appear unreasonable to expect it to occur in the full string theory. This is to be contrasted with the behaviour of $N = 2$ theories in four dimensions, as we discuss in section 5.7.

5.3. Generalities for $N = 2$ theories. Now that we have understood the main features of the moduli space of $N = 4$ theories in four dimensions we are ready to embark on a study of the much richer field of $N = 2$ theories. Much of the recent interest in duality was sparked by Seiberg and Witten's work on $N = 2$ Yang-Mills field theory [**101, 102**]. Here we are hoping to analyze full string theory in the same context. Thus we expect the subject to be at least as rich as Seiberg-Witten theory. In the short period that $N = 2$ theories have been studied in the context of string duality, the subject is already vast and it will be difficult to do justice to it here. As in the rest of these lectures, we will attempt to confine our attention to matters related to the moduli space of theories.

How can we obtain an $N = 2$ theory in four dimensions from string theory? Two answers immediately appear given the usual holonomy argument. Firstly one may take a heterotic string theory in ten dimensions and compactify it on a complex threefold with SU(2) holonomy. We have already discussed such manifolds in section 5.2 and found that they are of the form K3 $\times$ T^2, or some free quotient thereof. Secondly one might take a type II string and compactify it on a complex threefold of SU(3) holonomy, i.e., a Calabi–Yau manifold. Given the story for $N = 4$ theories above it is tempting to conjecture that there may be dual pairs of such theories. That is, we wish that a heterotic string when compactified in a specific way on K3 $\times$ T^2 be physically equivalent to a type II string compactified on a Calabi–Yau threefold. This story began with the papers of [**103, 96**] and, as we shall see, the full picture is still to be uncovered.

There is one immediately apparent curiosity which is associated with such a conjecture. This is that there are a very large number of topological classes of Calabi–Yau threefolds. The exact number is not known since a classification remains elusive. Indeed, one cannot rule out the possibility that the number is infinite. Contrasted to this are the few manifolds of K3 $\times$ T^2 and its quotients. At first it might appear that only a tiny fraction of the type II compactifications can have heterotic partners. This argument is flawed, however, as the heterotic string requires more data to specify its class than just the topology of the space on which it is compactified.

The heterotic string in ten dimensions has a gauge group which is either $E_8 \times E_8$ or Spin(32)/$\mathbb{Z}_2$. Let us consider the $E_8 \times E_8$ string for purposes of discussion. This "primordial" gauge group must be compactified in addition to the extra dimensions. The generally accepted way to do this is to take a vector bundle $E \to X$ over the compactification manifold, X, with a structure group contained in $E_8 \times E_8$. The embedding of this structure group into the heterotic string's gauge group then gives a recipe for compactifying the heterotic string on X including the gauge degrees of freedom.

This is exactly what we were doing in section 3.5. In the case of a heterotic string we considered a rank 16 principal $U(1)^{16}$-bundle over a torus. The structure group was embedded as the Cartan subgroup of $E_8 \times E_8$ and the connection on the bundle was specified by the parameters of the matrix A. The equations of motion demand that A be a constant and so the bundle is flat. The Narain moduli space then gives the full moduli space of such flat vector bundles on a torus.

In the case of compactification over a more general manifold, one way of solving the equations of motion **[104]** is to demand that the vector bundle be *holomorphic* and that the curvature satisfies

$$g^{i\bar{\jmath}}F_{i\bar{\jmath}} = 0. \tag{105}$$

One also requires $c_1(E) \in H^2(X, 2\mathbb{Z})$ and that

$$c_2(E) = c_2(T_X), \tag{106}$$

for anomaly cancelation. Clearly the torus fits into this picture if we replace the principal $U(1)^{16}$-bundle by the associated sum of holomorphic line bundles.

The analysis of this bundle for the case of a heterotic string on a K3 surface is going to be much harder than the toroidal case because now the bundle *cannot* be flat as it must satisfy $c_2(E) = 24$.[17] We can see hope then that the large number of choices of possible Calabi–Yau manifolds for compactification of the type II string might be matched by the large number of choices of suitable bundles over K3 $\times\, T^2$ for the heterotic string compactification.

It will take a fairly long argument before we are able to give an explicit example of such a pair so we will discuss the situation in general first. Let us start in our usual way by thinking about the holonomy of the moduli space. For an $N = 2$ theory in four dimensions the holonomy algebra is $\mathfrak{u}(2) \cong \mathfrak{u}(1) \oplus \mathfrak{sp}(1)$. There are two types of $N = 2$ supermultiplet which contain massless scalars:

1. The vector multiplets each contain 2 real fields which form a complex object under the $\mathfrak{u}(1)$ holonomy.
2. The hypermultiplets each contain 4 real massless scalars which transform as a quaternionic object under the $\mathfrak{sp}(1)$ holonomy.

Thus, at least away from points where the manifold structure may break down, we expect the moduli space to be in the form of a product

$$\mathcal{M}_{N=2} = \mathcal{M}_V \times \mathcal{M}_H, \tag{107}$$

where $\mathcal{M}_V$ is a Kähler manifold spanned by moduli in vector supermultiplets and $\mathcal{M}_H$ is a quaternionic Kähler manifold spanned by moduli in hypermultiplets. It can be shown **[105]** that $\mathcal{M}_H$ is not hyperkähler.

The effective four-dimensional theory always contains a dilaton-axion system. This will govern the string-coupling. As it plays such an important rôle we will look first at whether the dilaton-axion lives in a hypermultiplet or a vector multiplet. One way to do this is simply to count the dimensions of the moduli space and use the fact that $\mathcal{M}_V$ has an even number of real dimensions and $\mathcal{M}_H$ has a multiple of four dimensions. For a more direct way of justifying which kind of supermultiplet the dilaton-axion lives in see **[97]**.

[17] As is common, we assume integration over the base K3 surface in this notation.

Consider first the type IIA superstring on a Calabi–Yau manifold X. First consider the moduli space of underlying conformal field theories (see, for example, [**40**] for a full discussion). We have $h^{1,1}(X)$ complex dimensions of moduli space coming from the deformations of Kähler form and B-field and we have $h^{2,1}(X)$ complex dimensions coming from deformations of complex structure. The R-R sector moduli come from a 3-form giving $b_3(X) = 2(h^{2,1}(X)+1)$ real deformations. Finally we have 2 real deformations given by the dilaton-axion. Given that $h^{1,1}(X)$ and $h^{2,1}(X)$ can be even or odd, the only way to arrange these deformations in a way consistent with the dimensionality of the moduli space is to arrange:

- There are $h^{1,1}(X)$ vector supermultiplets.
- There are $h^{2,1}(X) + 1$ hypermultiplets, one of which contains the dilaton-axion.

Next consider the type IIB superstring compactified on a Calabi–Yau manifold, Y. The moduli space of conformal field theories is as for the type IIA string with $h^{1,1}(Y)$ complex deformations of complexified Kähler form and $h^{2,1}$ complex deformations of complex structure. The R-R moduli consist of one from the 0-form, $b_2(Y) = h^{1,1}(Y)$ from the 2-form, $b_4(Y) = h^{1,1}(Y)$ from the self-dual 4-form and one more from dualizing the 2-form. Lastly we have two more moduli from the dilaton-axion. Now we are forced to arrange as follows.

- There are $h^{2,1}(Y)$ vector supermultiplets.
- There are $h^{1,1}(Y) + 1$ hypermultiplets, one of which contains the dilaton-axion.

Note that the type IIA picture and the type IIB picture are related by an exchange

$$
\begin{aligned}
h^{1,1}(X) &\leftrightarrow h^{2,1}(Y) \\
h^{2,1}(X) &\leftrightarrow h^{1,1}(Y).
\end{aligned}
\tag{108}
$$

If we have a pair of Calabi–Yau varieties, X and Y, such that type IIA string theory compactified on X is equivalent to type IIB string theory on Y then we may use this as definition of the statement that "X and Y are a mirror pair". See [**106**] for further discussion of this point.

Lastly we consider the heterotic string compactified on a product of a K3 surface and a 2-torus. First let us assume that there are no nasty obstructions in the moduli space and we can count the number of deformations of the K3 surface, the bundle over the K3 surface, the 2-torus, and the bundle over the 2-torus and simply sum the result. There are 80 deformations of the K3 surface as far as conformal field theory is concerned and the resulting moduli space is a quaternionic Kähler manifold. It was shown by Mukai [**107**] that the moduli space of holomorphic vector bundles over the K3 will be hyperkähler. Thus it appears very reasonable to expect that the complete moduli space coming from the K3 surface will be a quaternionic Kähler manifold. Certainly its dimension is a multiple of four assuming unobstructedness. The moduli space coming from the torus will be locally of the form $\mathrm{O}(2,m)/(\mathrm{O}(2)\times\mathrm{O}(m))$ and has an even number of dimensions. Thus we assemble the supermultiplets as follows

- The vector multiplets come from the 2-torus, together with its bundle, and the dilaton-axion.

- The hypermultiplets are associated to the K3 surface, together with its bundle.

Note that we are assuming we can give simple geometric interpretations to all the moduli. We will see later that there may be other vector or hypermultiplet moduli we have not included in the above lists.

5.4. K3 fibrations. Suppose we are able to find a pair of theories, one a type II string on a Calabi–Yau manifold and the other a heterotic string theory compactified over some bundle on K3 $\times T^2$. The first thing one would do would be to line up the moduli spaces of the two theories so that the parameters of one theory could be understood in terms of the other. We begin by analyzing what would happen in the case of the vector multiplet moduli space. The is the first step required before we can actually propose a dual pair of such theories. We now follow an argument first presented in [**108**].

For definiteness we choose a type IIA string rather than a type IIB. The reason for this is that we will ultimately be able to tie our analysis to proposition 1, which was also phrased in terms of the type IIA string.

The holonomy argument leads to a factorization of the moduli space, which in turn has another consequence given the fact that all the matter fields are related to the moduli by supersymmetry. This is that the couplings between fields in the vector multiplets can only depend on the moduli from the vector multiplets and the couplings between fields in the hypermultiplets can only be affected by the moduli from the hypermultiplets. This may also be deduced directly from the supergravity Lagrangian [**109**].

Let us consider just the moduli space, $\mathscr{M}_V$, coming from the scalars in the vector multiplets. We know this is a complex Kähler manifold. Further analysis of the supergravity Lagrangian puts more constraints on the geometry of the moduli space [**110**]. A manifold satisfying these extra conditions is called "special Kähler".[18] The main importance of special Kähler geometry is the fact that all the information we require about the theory is encoded in a single holomorphic function $\mathscr{F}$ on the moduli space. If we use specific complex coordinates, the "special coordinates", on the moduli space, the metric is of the form

$$\begin{aligned} K &= -\log\left(2(\mathscr{F}+\overline{\mathscr{F}}) - (q^i - \bar{q}^{\bar{\imath}})\left(\frac{\partial\mathscr{F}}{\partial q^i} - \frac{\partial\overline{\mathscr{F}}}{\partial\bar{q}^{\bar{\imath}}}\right)\right) \\ g_{i\bar{\jmath}} &= \frac{\partial K}{\partial q^i \partial\bar{q}^{\bar{\jmath}}}. \end{aligned} \tag{109}$$

When viewed from the point of view of the heterotic string, we expect the dilaton-axion to be contained in this moduli space. Let us suppose for the time being that all the other moduli can be understood from the world-sheet perspective of the heterotic string. This should mean that we have a moduli space of conformal field theories spanning all but one of the complex directions in the moduli space with the extra dimension being given by the dilaton-axion system. In the limit that the string coupling becomes very small, i.e., the dilaton approaches $-\infty$, we expect

[18]The quaternionic Kähler manifold parametrized by the hypermultiplets is also subject to extra constraints [**111**].

that the moduli space as described by the conformal field theory becomes exact. Thus, in the limit of small dilaton, the moduli space should factorize into a product of the moduli space of conformal field theories, and the extra bit spanned by the dilaton-axion. In **[112]** precisely this problem was analyzed. It was discovered that the only way a special Kähler manifold could factorize was if it became locally a product of the form

$$\frac{\mathrm{O}(2,m)}{\mathrm{O}(2)\times \mathrm{O}(m)}\times\frac{\mathrm{SL}(2)}{\mathrm{U}(1)}. \tag{110}$$

This of course is excellent news. The first term in (110) looks suspiciously like the Narain moduli space for a 2-torus and the second term looks like a dilaton-axion. This is exactly what we wanted. Note that the form (110) is only expected in the limit that the dilaton approaches $-\infty$. Away from this limit we expect the two factors to begin to interfere with each other.

Now we want to carry this information over to the type IIA compactification on the Calabi–Yau manifold, X. All of the vector multiplet moduli are expected to be associated to deformations of the Kähler form and B-field on X. Let us fix some notation. In contrast to the K3 surface, degrees of freedom of the Kähler form and the B-field for the Calabi–Yau threefold, X, can be nicely paired-up. Introduce a basis of divisors, or 4-cycles, $\{D_k\}$, spanning $H_4(X,\mathbb{Z})$, where $k = 0,\ldots,h^{1,1}(X)-1$. Dual to the dual of these divisors we have a basis of 2-forms, $\{e_k\}$, generating $H^2(X,\mathbb{Z})$.[19] Expand out the Kähler form and B-field as

$$B+iJ=\sum_{k=0}^{h^{1,1}-1}(B_k+iJ_k)e_k, \tag{111}$$

for real numbers B_k, J_k. We will take e_0 to correspond to the generator associated to the direction in moduli space given by the heterotic dilaton-axion.

Our information for the heterotic side is in terms of the local form of the moduli space. Thanks to special Kähler geometry we can translate this into information concerning couplings between certain fields. The fields we are interested in are the superpartners of the moduli of the vector superfields — i.e., the gauginos, the vector bosons and the moduli themselves. One may consider couplings in the effective action of the form of "Yukawa couplings", i.e., $\kappa_{ijk} = \langle a_i\psi_j\psi_k\rangle$, or other terms equivalent by supersymmetry. It was shown in **[113]** that, to leading order in the non-linear σ-model, the coupling between three fields is given by

$$\kappa_{ijk}=\#(D_i\cap D_j\cap D_k), \tag{112}$$

where the D's are the divisors in X associated to the fields. The reader is also referred to **[114]** for an account of this.[20] It is also known from special Kähler geometry that

$$\kappa_{ijk}=\frac{\partial\mathscr{F}}{\partial q_i\partial q_j\partial q_k}. \tag{113}$$

We now have some approximate knowledge about both the heterotic string and the type IIA string. In the case of the heterotic string we know that, in the small

[19] For simplicity let us assume all cohomology is torsion-free.

[20] Note that **[114]** explicitly refers to a heterotic string compactified on a Calabi–Yau manifold whereas we are considering a type IIA string. Most of the calculations are unaffected however.

dilaton limit, the moduli space factorizes in the form (110) and in the case of the type IIA string, we know that, in the small α'/R^2 limit, the couplings are of the form (112). We can make a useful statement about a heterotic-type II dual pair if both of these approximations happen to be simultaneously true. Note that, for the heterotic string, the dilaton lies in a vector multiplet and that, in the type IIA string, the size parameters lie in vector multiplets. Thus we wish to assert that the moduli spaces of the theories are aligned in the right way so that as the dilaton in the heterotic string approaches $-\infty$, some size in the Calabi–Yau space on which the type IIA string is compactified is becoming very large.

To picture this presumed aligning of the moduli spaces it is best to picture exactly what makes corrections to the approximations we are considering. In the case of the heterotic string, corrections arise from instantons in the Seiberg-Witten theory [**101**]. The action of such an instanton becomes very large, and hence the contribution to any physical quantity becomes very small, when the dilaton becomes close to $-\infty$. In the case of the non-linear σ-model, corrections come from world-sheet instantons [**115**]. A world-sheet instanton takes the form of a holomorphic map from the world-sheet into the target space. We will be interested only in tree-level effects and so as far as we are considered *world-sheet instantons are rational curves*. The action for such an instanton is simply the area of the curve. Thus the effect of an instanton gets weaker as the Kähler form is varied so as to make the rational curve bigger.

The picture one should have mind therefore is that as the heterotic dilaton is decreased down to $-\infty$, some rational curve (or some set or family of rational curves) is getting bigger. The important thing is that no curve should shrink down during this process.

Let us assume we are now at the edge of our moduli space where, in the type IIA interpretation, all of the rational curves are very large compared to α'. Thus, by assumption, the heterotic string's dilaton is close to $-\infty$. We can take the moduli space given by (110) and deduce the form of $\mathscr{F}$. We may then translate this into a statement about κ_{ijk} and thus about the topology of X from (112). The result is that [**112**]

$$\begin{aligned}
&\#(D_0 \cap D_0 \cap D_0) = 0 \\
&\#(D_0 \cap D_0 \cap D_i) = 0, \quad i = 1, \ldots, h^{1,1} - 1 \\
&\#(D_0 \cap D_i \cap D_j) = \eta_{ij}, \quad i,j = 1 \ldots, h^{1,1} - 1,
\end{aligned} \tag{114}$$

where η_{ij} is a matrix of nonzero determinant and signature $(+,-,-,\ldots,-)$.

Suppose that X is such that there is some smooth complex surface embedded in X whose class, as a divisor, is D_0. From (114) we see that $D_0 \cap D_0 = 0$ and so the normal bundle for this surface is trivial. It then follows from the adjunction formula, and the fact that X is a Calabi–Yau space, that the tangent bundle for this surface has trivial c_1. Thus, the surface representing D_0 must either be a K3 surface or an algebraic 4-torus (also known as an "abelian surface"). The fact that the normal bundle is trivial also suggests that the K3 or abelian surface can be "moved" parallel to itself to sweep out the entire space X. That is, X is a fibration where the generic fibre is either a K3 surface or an abelian surface.

We can make the above more rigorous by appealing to a theorem by Oguiso [**116**] which states that

THEOREM 7. *Let X be a minimal Calabi–Yau threefold. Let D be a nef divisor on X. If the numerical D-dimension of D equals one then there is a fibration $\Phi : X \to W$, where W is $\mathbb{P}^1$ and the generic fibre is either a K3 surface or an abelian surface.*

The numerical D-dimension of a divisor is the maximal number of times it may be intersected with itself to produce something nontrivial. We see above that D_0 has D-dimension equal to one. The statement that D is "nef" is the assertion that for any algebraic curve, $C \subset X$, we have that

$$\#(D \cap C) \geq 0, \quad \forall C \subset X. \tag{115}$$

We can come very close to proving that D_0 is nef following our assumption about the way that moduli spaces are aligned. The special coordinates on a our moduli space of type IIA string theories may be written in the form

$$q_k = e^{2\pi i(B_k + iJ_k)}. \tag{116}$$

We are in an area of moduli space where all $q_k \ll 1$. The contribution of a curve, C, to the instanton sum will then scale roughly as

$$\prod_{k=1}^{h^{1,1}} q_k^{\#(D_k \cap C)}. \tag{117}$$

Clearly, if C is not nef then negative powers of q_k appear and the instanton sum will fail to converge, in contradiction to expectation.

The above argument that C is nef is not actually complete. When one computes the instanton sum, one also has to compute the coefficient in front of the monomial of the form (117). This can be done using the methods discussed in [**117, 118**]. In the simple case of an isolated curve, the coefficient is simply one (although extra contributions arise from multiple covers). C may not always be isolated however. One may be able to deform C into a whole family of rational curves. In this case the coefficient might be zero. One cannot then rule out that $\#(D \cap C) < 0$ as the field theory is simply unaffected by C.

An example of a case where rational curves don't count in this way is that of $N = 4$ theories. Compactifying the type IIA string on K3 $\times T^2$, there are no instanton corrections but the K3 surface may contain rational curves. Each rational curve will clearly be in the form of a family $C \times T^2$ inside K3 $\times T^2$. An algebraic surface of the form $\mathbb{P}^1 \times T^2$ is known as an "elliptic scroll".[21] Since the calculation in [**117, 118**] is essentially local we may generalize this result to any $N = 2$ Calabi–Yau compactification. That is we claim that *rational curves in an elliptic scroll do not contribute to the instanton sum.* Note that rational curves in a K3 surface are unstable in the sense that a generic deformation of complex structure of the K3 surface will kill them. The work of Wilson [**119**] showed that essentially the same is true for elliptic scrolls in Calabi–Yau manifolds. Thus, if we choose a generic complex structure on the Calabi–Yau manifold then there will be no curves lying in an elliptic scroll.

If we assume that the only curves contributing zero to the instanton sum lie in an elliptic scroll then we may complete our proof that D_0 is nef by choosing a generic complex structure. We do not know if this assumption is valid but it seems reasonable.

Note that the fibration $\Phi : X \to W$ is allowed to "degenerate" at a finite number of points in W. Indeed, if this were not the case then X could not be a Calabi–Yau variety. Note that the degenerate fibre need not be some degenerate

[21] An "elliptic curve" is the algebraic geometer's name for an algebraic curve of genus 1.

limit of a K3 surface. It could be a perfectly smooth manifold (which is neither a K3 or abelian surface).

Finally we would like to know if the generic fibre is a K3 surface or an abelian surface. We can determine this by finding c_2, and hence the Euler characteristic, of the fibre. This may be determined by using the *holomorphic anomaly* of **[120]**. The result is that the Euler characteristic of the generic fibre is 24 and so the fibration is of the K3 type. We refer the reader to **[108]** for details of this calculation.

We thus arrive at the following:

PROPOSITION 6. *Given a dual pair of theories, one of which is a heterotic string compactified on K3* $\times T^2$ *(or some free quotient thereof) and the other is a type IIA string compactified on a Calabi–Yau manifold* X*, then if there is a region of moduli space in which both the respective perturbation theories converge, then* X *must be of the form of a K3-fibration over* $\mathbb{P}^1$.

This proposition depends upon our statement about zero contributions from rational curves only arising from elliptic scrolls. The fact that the base of the fibration is a $\mathbb{P}^1$ may be deduced from the fact that H^1 of the base injects[22] into the total cohomology of X and that X has $b_1 = 0$.

Not only do we know that X is a K3 fibration, but we also know that the divisor D_0 is the generic K3 fibre. This tells us immediately which deformation of X corresponds to a deformation of the heterotic dilaton — it is the component of the Kähler form that affects the area of the curve dual to D_0.

Let us first assume that the K3-fibration of X has a global section. This means that as well as the fibration map:

$$\Phi : X \to W, \tag{118}$$

we also have a holomorphic embedding

$$\gamma : W \to X. \tag{119}$$

The image of this embedding is a rational curve in X. Clearly this curve is dual to D_0. Thus the value of the dilaton in the heterotic string is given by the area of this rational curve. The instanton sums converge as the dilaton approaches $-\infty$ and as this rational curve gets very large. Thus the weakly-coupled limit of the heterotic string is given by the limit of X in which the base $\mathbb{P}^1$ swells up to infinite size.

Note that if the K3 fibration does not have a global section, but rather a multi-valued section, we may play a similar game. In this case the multi-section will not define an embedding of the base $\mathbb{P}^1$ into X but rather an embedding of a multiple cover of X. Thus, the algebraic curve within X whose area gives the heterotic dilaton will be of genus greater than zero.

One should note that what we have discussed here for the $N = 2$ dual pairs is actually very similar to the $N = 4$ case discussed above. In that case we had a type IIA string compactified on K3 $\times T^2$ or a quotient thereof. This space may be viewed as a smooth K3-fibration over T^2 (that is, an elliptic curve). This fibration is trivial in the case that the target space is K3 $\times T^2$ but becomes nontrivial when we take a free quotient. We saw in section 5.1 that the dilaton for the heterotic string is given by the area of the T^2 — i.e., the size of the base of the fibration.

[22]This follows from the Leray spectral sequence for a fibration.

The general picture we see is the following. If we have a heterotic-type IIA dual pair in four dimensions, we may expect that the type IIA string is compactified on

- a K3-fibration over a 2-torus in the case of $N = 4$ supersymmetry, or,
- a K3-fibration over a $\mathbb{P}^1$ in the case of $N = 2$ supersymmetry.

In either case, the area of the base of the fibration gives the value of the heterotic dilaton.

The link to proposition 1 is clear in the $N = 4$ case. For the $N = 2$ case we should note that the K3 surface on which the heterotic string is compactified might be viewable as an elliptic fibration itself — that is, a fibration over $\mathbb{P}^1$ with generic fibre given by a T^2. Thus the K3 $\times T^2$ (or quotient thereof) upon which the heterotic string is compactified may be viewed as an "abelian fibration" over $\mathbb{P}^1$ — i.e., a fibration over $\mathbb{P}^1$ with generic fibre given by a T^4. When viewed in this way the relationship between the $N = 2$ theories in four dimensions given by the heterotic string and the type IIA string may be viewed as a *fibre-wise* application of the duality of proposition 1.

Such "fibre-wise duality", which was first suggested in [**97**], is a potentially very powerful tool. It has been extended to fibre-wise mirror symmetry in [**121**] and has recently been applied to the problem of mirror symmetry itself in [**122**]. Both of these developments deserve to be covered here in some detail since they both have direct relevance to K3 surfaces, but we do not have time to do so. We refer the reader to [**123, 124**] for more details of the latter in the context of K3 surfaces.

Let us note that the general appearance of K3-fibrations in the area of heterotic-type II duality was first noted in [**125**] where some naturalness arguments based on the work of [**126**] were presented. It is interesting to note that in this context it was really the type IIB string that being studied rather than the type IIA. This raises the question as to whether the mirror of a K3 fibration is another K3 fibration, which again raises the possibility of some fibre-wise duality argument.

5.5. More enhanced gauge symmetries. So far we have identified one of the moduli lying in a vector multiplet. This is the dilaton-axion in the heterotic string, or the component of the complexified Kähler form associated to the base of the K3-fibration on which the type IIA string is compactified. Now we wish to turn our attention to some of the other vector multiplet moduli.

Let us first look at things from the type IIA perspective. We know that our Calabi–Yau space, X, is a K3-fibration and we wish to analyze elements of $H^2(X)$, or equivalently, $H_4(X)$. In general, given a fibration $X \to W$, the cohomology of X may be determined from the cohomology of the base together with the cohomology of the fibre. The mechanism by which this happens is called a "spectral sequence". We do not wish to get involved with the technicalities of spectral sequences here and refer the reader to [**127**] for the general idea.

The result is that the contributions to $H_4(X)$ are as follows (see figure 7):

1. The generic fibre, D_0, will generate an element of $H_4(X)$.
2. Take an algebraic 2-cycle in the fibre, i.e., an element of $H_2(D_0)$ and use it to "sweep out" a 4-cycle in X by transporting it around the base, W. Note that this 2-cycle needs to be monodromy invariant for this to make sense. Thus the 2-cycle might be an irreducible curve in D_0 which is monodromy

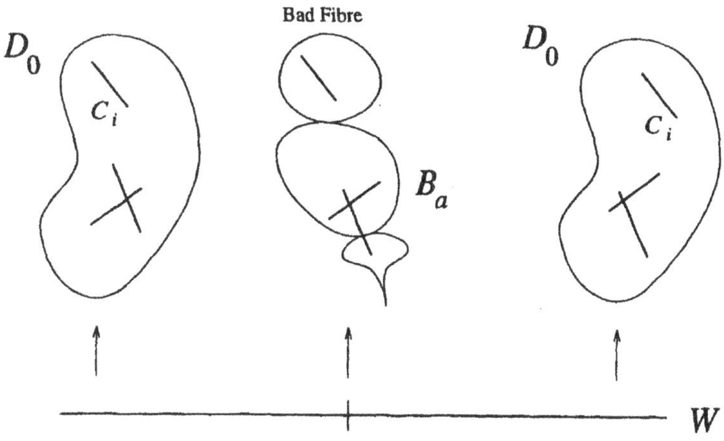

FIGURE 7. X as a K3 fibration over W.

invariant or it may be the sum of two curves which are interchanged under monodromy, etc.

3. When we have a bad fibre which is a reducible divisor in X, we may vary the volumes of the components of this bad fibre independently. Thus such fibres will contribute extra pieces to $H_4(X)$.

The second class above is clearly generated by elements of the Picard group of the generic fibre. A monodromy-invariant element of the group C_i will sweep out a divisor D_i, where $C_i = D_0 \cap D_i$. Thus

$$\begin{aligned}\#(D_0 \cap D_i \cap D_j)_X &= \#((D_0 \cap D_i) \cap (D_0 \cap D_j))_{D_0} \\ &= \#(C_i \cap C_j)_{D_0},\end{aligned} \tag{120}$$

where the subscript denotes the space within which we are considering the intersection theory. This agrees very nicely with our earlier result of (114). We see that η_{ij} is simply the natural inner product of the monodromy-invariant part of the Picard lattice of the generic fibre. As we mentioned in section 2.5, the Picard lattice has signature $(+,-,-,\ldots,-)$. When we take the monodromy-invariant part we retain the positive eigenvalue as the Kähler form restricted to the generic K3 fibre must be monodromy invariant and the generic fibre has positive volume. Thus η_{ij} has the correct signature.

Now consider the third class. We denote such an element by B_a. Since this class is supported away from the generic fibre, we have $D_0 \cap B_a = 0$. Now comparing to (114) we are in trouble. The moduli coming from such vector supermultiplets cannot live in the space (110) we expected from the heterotic string. What can have gone wrong? The only place our argument was flawed was in the *perturbative* analysis of the heterotic string. It turns out that the classes B_a cannot be understood perturbatively from the perspective of the heterotic string. We will encounter such objects later in section 6.2 but in this section we will restrict our attention to perturbative questions.

We now know exactly how to interpret the Teichmüller space

$$\frac{\mathrm{O}(2,m)}{\mathrm{O}(2)\times\mathrm{O}(m)}\times\frac{\mathrm{SL}(2)}{\mathrm{U}(1)}, \tag{121}$$

both in terms of the heterotic string compactified on $\mathrm{K3}\times T^2$ and its supposed type IIA dual compactified on X. The second term comes from the heterotic dilaton-axion and the complexified Kähler form associated to the class $D_0 \in H_4(X)$. The first term is considered to be associated to the Narain moduli space of the heterotic string on T^2, or, from what we have just said, the complexified Kähler form on the (monodromy-invariant part of) the K3 fibre. We computed the stringy moduli space of the Kähler-form and B-field on an algebraic K3 surface earlier and obtained the second term of (73). Fortunately it turns out to have just the right form!

Note that we have reduced our problem once again to a comparison of a heterotic string on a torus (this time a 2-torus) with a type IIA string on a K3 surface (this time an algebraic fibre in X). We should be able to use the same old arguments we used in sections 4.3 and 5.2 to obtain the enhanced gauge group.

Let us first assume there is no monodromy acting on the Picard group of the K3 fibre for X. We expect the slice of the moduli space coming from varying the complexified Kähler form on the generic K3 fibre to be of the form

$$\mathrm{O}(\Upsilon)\backslash\,\mathrm{O}(2,\rho)/(\mathrm{O}(2)\times\mathrm{O}(\rho)), \tag{122}$$

where Υ is the quantum Picard lattice introduced in (69). Identifying this with the Narain moduli space of T^2 that we expect to see in the weak-coupling limit of the heterotic string, we see that the Narain lattice for the T^2 is isomorphic to Υ.[23] That is, we have a gauge bundle of rank $\rho-2$ compactified over T^2.

To obtain the enhanced gauge groups we take (122) to be the Grassmannian of space-like 2-planes in $\mathbb{R}^{2,\rho}$ and look for points where the 2-plane becomes orthogonal to roots in the lattice Υ. The roots will give the root diagram of the enhanced gauge group in the usual way.

Let us clarify all this general discussion with an example taken from **[103]**. Let us take X_0 to be the hypersurface

$$z_1^2+z_2^3+z_3^{12}+z_4^{24}+z_5^{24}=0, \tag{123}$$

in the weighted projective space $\mathbb{P}^4_{\{12,8,2,1,1\}}$. The weighted projective space contains various quotient singularities. We will blow these up and take the proper transform of X_0 to be X. There is a $\mathbb{Z}_2$-quotient singularity along the locus $[z_1,z_2,z_3,0,0]$. We may blow this up, replacing each point in the locus by $\mathbb{P}^1$. The projection of X onto to this $\mathbb{P}^1$ will be the K3-fibration. That is, roughly speaking, we view $[z_4,z_5]$ as the homogeneous coordinates of the base $W\cong\mathbb{P}^1$. To find the fibre fix a point in the base by fixing z_4/z_5. This projects $\mathbb{P}^4_{\{12,8,2,1,1\}}$ onto the subspace $\mathbb{P}^3_{\{12,8,2,1\}}$. Now $\mathbb{P}^3_{\{12,8,2,1\}}$ may be viewed as a $\mathbb{Z}_2$-quotient of $\mathbb{P}^3_{\{6,4,1,1\}}$ by taking $z_4\mapsto -z_4$. Such

[23]Note that there is no reason to expect that Υ should be self-dual. In many proposed dual pairs (e.g., some of those of **[103]**) it is not self-dual. On the heterotic side this says that the heterotic string on T^2 is not modular invariant. We shall assume that modular invariance is satisfied once the K3 factor is taken into account too. In general one might worry that strange effects such as those encountered in **[96]** might cause problems with this. We will assume here that modular invariance looks after itself.

codimension quotients are equivalent to reparametrizations in complex geometry and so the fibre may be taken to be the hypersurface

$$z_1^2 + z_2^3 + z_3^{12} + z_4^{12} = 0, \tag{124}$$

in the weighted projective space $\mathbb{P}^3_{\{6,4,1,1\}}$. This is a K3 surface as expected. Note that we still have a $\mathbb{Z}_2$-quotient singularity in the fibre along $[z_1, z_2, 0, 0]$. This intersects the generic K3 fibre at a single point. Thus for a smooth X we blow-up again to introduce a single (-2)-curve into each generic fibre.

Now let us work out the Picard lattice of the generic fibre. A generic hyperplane in $\mathbb{P}^3_{\{6,4,1,1\}}$ may be written as $az_3 + bz_4 = 0$ for some $a, b \in \mathbb{C}$. Any two such hyperplanes will intersect at the point $[z_1, z_2, 0, 0]$ but this is exactly where we are blowing up. Thus the hyperplane doesn't intersect itself at all but will intersect the (-2)-curve (i.e., the exceptional divisor) once. The Picard lattice, for a generic value of complex structure of X, will be generated by the hyperplane and the single (-2)-curve and has intersection matrix

$$\begin{pmatrix} 0 & 1 \\ 1 & -2 \end{pmatrix}. \tag{125}$$

A simple change of basis shows that this is $\Gamma_{1,1} \cong U$. Note that neither the hyperplane nor the exceptional (-2)-curve are effected by monodromy around W and so we needn't worry about monodromy invariance in this case.

The bad fibres occur when we fix a point on the base W such that $z_4^{24} + z_5^{24} = 0$. Thus, at 24 points, we have fibres

$$z_1^2 + z_2^3 + z_3^{12} = 0, \tag{126}$$

in $\mathbb{P}^3_{\{6,4,1,1\}}$. Although singular, this equation does not factorize and so the bad fibres are irreducible. Thus there are no contributions to $h^{1,1}$ from bad fibres. Thus, $h^{1,1} = 1 + 2 = 3$ since we have the generic fibre itself together with $\rho = 2$.

Now $\Upsilon = U \oplus U \cong \Gamma_{2,2}$ and the moduli space of vector multiplets coming from the quantum Picard lattice will be

$$\mathrm{O}(\Gamma_{2,2}) \backslash \mathrm{O}(2,2)/(\mathrm{O}(2) \times \mathrm{O}(2)). \tag{127}$$

This is exactly the moduli space for a string on T^2. Thus if the type IIA string compactified on X is dual to a heterotic string on K3 $\times T^2$ then the T^2 part of the latter has none of the gauge group from the ten-dimensional string wound around it. All must be wound around the K3 factor.

The lattice $\Gamma_{2,2}$ contains root diagrams for $\mathfrak{su}(2) \oplus \mathfrak{su}(2)$ and $\mathfrak{su}(3)$ and so we should be able to obtain these gauge symmetries for suitable choices of vector moduli. Clearly the (-2)-curve in the generic K3 fibre may be shrunk down to a point to obtain SU(2). To obtain more gauge symmetry one must shrink the K3 fibre itself down to a size of order $(\alpha')^2$ as discussed in section 4.3.

There is an important point we should note in general about enhanced gauge groups. It is known that in $N = 2$ theories quantum effects may break the gauge group down to a Cartan subgroup of the classical gauge group. Exactly how this happens depends on whether any extra hypermultiplets are becoming massive when the point of classical enhanced gauge symmetry occurs. We don't want to discuss hypermultiplets yet but, at least in simple cases, there are usually a small number, if any, of such massless particles and the theory will be "asymptotically free". In this case the gauge group will be broken.

Thus we only really expect the nonabelian gauge group to appear in the heterotic string for the case that the coupling tends to zero. In the type IIA picture this corresponds to the base $\mathbb{P}^1$ blowing up to infinite size. This means that we are "decompactifying" the type IIA picture so that it is compactified, not on X, but on the generic fibre — a K3 surface. In this respect we really are saying little more than we already said in section 4.3. We can obtain enhanced gauge symmetries when we have a theory in six dimensions from a type IIA string compactified on a certain K3 surface. This is approximately true in four dimensions so long as the base $\mathbb{P}^1$ is very large.

For the heterotic string, the gauge group is broken in the quantum theory by Yang-Mills instantons. For the type IIA string we note that world-sheet instantons wrapping around the base $\mathbb{P}^1$ presumably play an analogous rôle. This is another example of one string's target space field theory being another string's world-sheet field theory as we saw in section 5.1. It would be interesting to see if these instantons can be explicitly mapped to each other.

To determine the gauge group we should say what happens when there is monodromy in the Picard lattice of the generic K3 fibre as we move about W. There really is no difference between this and the $N = 4$ analogue we discussed in section 5.2. The monodromy of the Picard lattice should be translated into an action on the heterotic string on T^2 and divided out. Thus we expect an *asymmetric orbifold* of T^2 for the heterotic dual. When finding the enhanced gauge group we should take the monodromy-invariant subdiagram of the root diagram. This may lead to non-simply-laced gauge groups again.

One should also worry about the global form of the gauge group. That is, one may have the simply connected form of the group or one may have to mod out by part of the center, e.g., the gauge group might be SU(2) or SO(3). All we have said above is only really enough to determine the *algebra* of the gauge symmetry. We will evade this issue where possible in these lectures by only specifying gauge algebras rather than gauge groups. There will be times later on in these lectures when we have to confront this problem, however.

Let us mention here that compactifying the type IIB string, rather than the type IIA string, on a K3 fibration, can lead to a very direct link between the geometry of the Calabi–Yau threefold and Seiberg-Witten theory as explored in [**128, 129**]. Therefore an understanding of mirror symmetry within K3 fibrations may shed considerable light on some of the details of string duality.

Finally let us note that further analysis may be done on the moduli space of vector multiplets to check that string duality is working as expected. We refer the reader to [**130, 131**] for examples and especially to [**132**] where further direct links to geometry were established.

5.6. Heterotic-heterotic duality. We have seen how a type IIA string compactified on a Calabi–Yau space with a K3 fibration may have as a dual partner a heterotic string compactified on $\mathrm{K3} \times T^2$. An obvious question springs immediately to mind. What happens if X can admit more than one K3 fibration? One might expect it may be dual to more than one heterotic string. This implies that we can find pairs of heterotic strings that are dual to each other. We will follow this line of

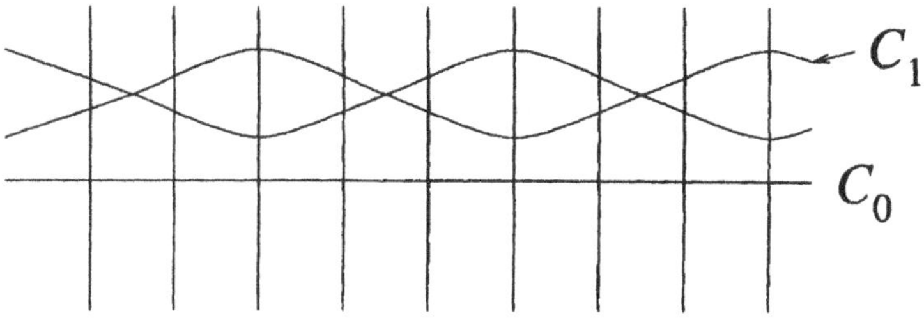

FIGURE 8. The Hirzebruch surface $\mathbf{F}_n$.

logic to analyze a case introduced in [**133**]. This geometric approach was discussed originally in [**134, 135**].

We are going to consider the example we introduced in the last section based on the hypersurface in $\mathbb{P}^4_{\{12,8,2,1,1\}}$ given by (123). In the last section we projected onto the last 2 coordinates of the $\mathbb{P}^4_{\{12,8,2,1,1\}}$ to obtain a K3 fibration over $\mathbb{P}^1$. Now let us project into the last 3 coordinates. Our base space will now be $\mathbb{P}^2_{\{2,1,1\}}$. The fibre will be an algebraic 2-torus, that is, an elliptic curve. Thus X may be considered as an elliptic fibration as well as a K3 fibration.

The space $\mathbb{P}^2_{\{2,1,1\}}$ is singular. Writing the homogeneous coordinates as $[x_0, x_1, x_2]$, there is a $\mathbb{Z}_2$ quotient singularity at the point $[1,0,0]$. This may be blown up, introducing a (-2)-curve. This exceptional curve provided the base of the K3 fibration in the last section. We may also use it to write the blow-up of $\mathbb{P}^2_{\{2,1,1\}}$ as a fibration. From the same argument as we used in the last section, it is straight forward to see that the fibre will be $\mathbb{P}^1$. Thus our blown-up $\mathbb{P}^2_{\{2,1,1\}}$ is a fibration with base space given by $\mathbb{P}^1$ and with fibre given by $\mathbb{P}^1$. Note that the fibre never degenerates. Such complex surfaces are called "Hirzebruch surfaces". These objects will turn out to be important for the analysis of the heterotic string on a K3 surface so we discuss the geometry here in some detail.

A $\mathbb{P}^1$ bundle over $\mathbb{P}^1$ may be regarded as the compactification of a complex line bundle over $\mathbb{P}^1$ by adding a point to each fibre "at infinity". Such line bundles are classified by an integer — the first Chern class of the bundle integrated over the base $\mathbb{P}^1$. We use the notation $\mathbf{F}_n$, to denote the Hirzebruch surface built from the line bundle with $c_1 = -n$. Assume first that $n \geq 0$. Denote the base rational curve by C_0. The line bundle with $c_1 = -n$ represents the normal bundle to C_0 and so the self-intersection of C_0 equals $-n$. Thus C_0 is isolated (assuming $n > 0$). Denote the fibre by f. Clearly $\#(f \cap f) = 0$ and $\#(f \cap C_0) = 1$. We may introduce a class $C_1 = C_0 + nf$. This intersects f once and C_0 not at all. C_1 is a section of the bundle, and hence a rational curve, away from the isolated section C_0. The self intersection of C_1 is $+n$. Note that C_1 can be deformed. Thus we view $\mathbf{F}_n$ as a $\mathbb{P}^1$ bundle over $\mathbb{P}^1$ with one isolated section, of self-intersection $-n$, and a family of sections, disjoint from the isolated section of self-intersection $+n$. This picture of a Hirzebruch surface, with two sections in the class C_1 shown, is drawn in figure 8. If $n = 0$ this picture degenerates and we simply have $\mathbf{F}_0 \cong \mathbb{P}^1 \times \mathbb{P}^1$.

Note that if $n < 0$ we may exchange the rôles of C_0 and C_1 and recover the surface $\mathbf{F}_{-n}$. Thus we may assume $n \geq 0$.

For the case we are concerned with in this section C_0 is the exceptional divisor with self-intersection -2 and so we have $\mathbf{F}_2$ as the base of X as an elliptic fibration. It turns out however that $\mathbf{F}_2$ is somewhat "unstable" in this context. Note that one may embed $\mathbb{P}^2_{\{2,1,1\}}$ into $\mathbb{P}^3$ as follows. Denote the homogeneous coordinates by $[x_0, x_1, x_2]$ and $[y_0, y_1, y_2, y_3]$ respectively. By putting

$$\begin{aligned} y_0 &= x_0 \\ y_1 &= x_1^2 \\ y_2 &= x_2^2 \\ y_3 &= x_1 x_2, \end{aligned} \tag{128}$$

we embed $\mathbb{P}^2_{\{2,1,1\}}$ as the hypersurface

$$y_1 y_2 - y_3^2 = 0. \tag{129}$$

This is singular, as expected, but we may deform this hypersurface to a generic quadric. It is well-known that a generic quadric in $\mathbb{P}^3$ is isomorphic to $\mathbb{P}^1 \times \mathbb{P}^1 \cong \mathbf{F}_0$ (see, for example, [8]). Thus $\mathbb{P}^2_{\{2,1,1\}}$ may be blown-up to give $\mathbf{F}_2$ but deformed to give $\mathbf{F}_0$. A generic point in the moduli space of X is actually an elliptic fibration over $\mathbf{F}_0$ rather than $\mathbf{F}_2$.[24]

We have arrived at the result that X is an elliptic fibration over $\mathbf{F}_0$. This may be viewed as a two-stage fibration. X may be projected onto a $\mathbb{P}^1$ to produce a K3 fibration. The fibre of this map may be projected onto the other $\mathbb{P}^1$ to write the K3 as an elliptic fibration.

Since the base space of the elliptic fibration is $\mathbb{P}^1 \times \mathbb{P}^1$, if one of the $\mathbb{P}^1$'s in the base may be viewed as the base for X as a K3 fibration, then so may the other $\mathbb{P}^1$. Thus X can be written as a K3 fibration in two different ways. This suggests that there should be two heterotic strings dual to the type IIA theory on X and hence dual to each other. What is the relationship between these two heterotic strings?

One of the heterotic strings will be compactified on $S_1 \times T_1^2$ and the other on $S_2 \times T_2^2$, where S_1 and S_2 are K3 surfaces. We examined the moduli space coming from the vector multiplets in section 5.5 and saw that the 3 moduli described the dilaton-axion and the Narain moduli of the T^2 (without any Wilson lines switched on). The dilaton is given by the area of the base $\mathbb{P}^1$ in X. Let us determine the area of the heterotic T^2 in terms of the moduli of the K3 fibre within X.

The generic K3 fibre within X has Picard number 2 and has a moduli space of Kähler form and B-field given by (127). Let $\Gamma_{2,2}$ be generated by the null vectors w and v together with their duals w^* and v^*. For simplicity we will avoid switching on any B-fields. Thus we consider a point in the moduli space (127) to be given by a space-like 2-plane, $\mho$, spanned by $w^* + \alpha w$ and $v^* + \beta v$, for $\alpha, \beta > 0$. Following our construction to determine the Kähler form, we have that $\mho \cap w^\perp$ is spanned

[24]To see this note that when we blow-up the fibration of X over $\mathbb{P}^2_{\{2,1,1\}}$ to a fibration over $\mathbf{F}_2$ we introduce an elliptic scroll just as in section 5.4. The results of [119] tell us that the (-2)-curve will therefore vanish for a generic complex structure.

by $v^* + \beta v$ which is contained in $w^\perp/w$. Thus, as promised, B is zero. From our analysis in section 3.3 we see that the Kähler form determines the volume to be

$$J.J = \alpha. \tag{130}$$

We also have that the direction of J is given by $v^* + \beta v$. Thus the Kähler form is

$$J = \sqrt{\frac{\alpha}{2\beta}}v^* + \sqrt{\frac{\alpha\beta}{2}}v. \tag{131}$$

We have seen that the generic K3 fibre is itself an elliptic fibration over $\mathbb{P}^1$. Knowing the intersection numbers of the curves within this fibration together with the positivity of the Kähler class determines the class of the base $\mathbb{P}^1$ to be $v^* - v$ and that of the elliptic fibre to be v. Thus, the area of the $\mathbb{P}^1$ within the K3 is given by

$$J.C = \sqrt{\frac{\alpha\beta}{2}} - \sqrt{\frac{\alpha}{2\beta}}. \tag{132}$$

Now let us reinterpret $\mho$ in terms of the moduli of the T^2. From section 3.5 we obtain a map from W into W^* given by[25]

$$\psi = \begin{pmatrix} \alpha & 0 \\ 0 & \beta \end{pmatrix}. \tag{133}$$

This is symmetric and thus gives the metric. Therefore, the area of the T^2 is given by $\sqrt{\det(\psi)} = \sqrt{\alpha\beta}$.

We may now obtain an interesting result confirming the analysis of [**133**] by going to the limit where α and β are taken to be very large. In this case, our heterotic string is compactified on K3 $\times T^2$ where the 2-torus is very large. In the type IIA string, the generic K3 fibre contains a rational curve which becomes very large. From the equation (132), the area of this rational curve is proportional to the heterotic string's 2-torus in this limit.

This limit as $\alpha, \beta \to \infty$ may be viewed as a decompactification of the model to a six-dimensional theory given by the heterotic string compactified on a K3 surface. Let us consider the coupling of this six-dimensional theory, λ_6. This is given by the four-dimensional coupling prior to decompactification and the area of the T^2. The former is given by the size of the base $\mathbb{P}^1$ of the K3 fibration, which we refer to as $\mathbb{P}^1_1$. The latter is given by the area of the base of the K3 fibre itself written as an elliptic fibration, which we will refer to as $\mathbb{P}^1_2$. We have that

$$\lambda_6^2 = \lambda_4^2.\,\mathrm{Area}(T^2) \sim \frac{\mathrm{Area}(\mathbb{P}^1_2)}{\mathrm{Area}(\mathbb{P}^1_1)}. \tag{134}$$

The other heterotic string is obtained by exchanging the rôles of $\mathbb{P}^1_1$ and $\mathbb{P}^1_2$. Thus we have

PROPOSITION 7. *Let X be the Calabi–Yau manifold given by a resolution of the degree 24 hypersurface in $\mathbb{P}^4_{\{12,8,2,1,1\}}$. The two heterotic string theories dual to the type IIA string theory on X decompactify to two heterotic string theories compactified on K3 surfaces. From (134) these two six-dimensional theories are*

[25] Note that W is spanned by w^* and v^*; and W^* is spanned by w and v. There is no simple choice of conventions which would have circumvented this notational irritation!

S-dual in the sense that the coupling of one theory is inversely proportional to the coupling of the other theory.

The main assumption underlying this proposal is that these heterotic string theories actually exist. We will get closer to identifying these theories in section 6.3.

We may vary the complex structure of X and this should correspond to deformations of the K3 together with its vector bundle on the heterotic side. Note that the K3 of one of the heterotic strings will, in general, have quite different moduli than the K3 of the other heterotic string since exchanging the rôles of $\mathbb{P}^1_1$ and $\mathbb{P}^1_2$ need not be a geometrical symmetry of X — only a topological symmetry. Before we can explicitly give the map between these K3 surfaces and their vector bundles we need to map out the moduli space of the moduli from the hypermultiplets. This remains to be done.

There are many examples of Calabi–Yau manifolds which admit more than one K3 fibration. This will lead to many examples of heterotic-heterotic duality. In most cases however the result will be a rather tortuous mapping between four-dimensional theories and will not be as simple as the above example.

5.7. Extremal transitions and phase transitions. We will take our first tentative steps into the moduli space of hypermultiplets in this section. This will deal with the simplest aspects of the bundle over K3 — namely when this bundle, or part of this bundle, becomes trivial.

Our heterotic string theory is compactified on a bundle over K3 $\times$ T^2 which we view as the product of a bundle over K3 and a bundle over T^2. Generically one would expect the deformation space of this bundle structure to be smooth. Where this can break down however is when part of the bundle becomes trivial. To deform away from such a bundle we may wrap the trivial part around either the K3 surface or the T^2. Thus we obtain branches in the moduli space. A deformation in the K3 part will be a hypermultiplet modulus while a deformation in the T^2 part will be a vector modulus. Our picture therefore for a transition across this branch will consist of moving in the moduli space of hypermultiplets until suddenly a vector scalar becomes massless and can be used as a marginal operator to move off into a new branch of the moduli space, at which point some of the moduli in the hypermultiplets in the original theory may acquire mass.

When we compactify the $E_8 \times E_8$ or Spin(32)/$\mathbb{Z}_2$ heterotic string on a vector bundle, the original gauge group is broken by the holonomy of the bundle. Thus, when we move to a transition point where part of the vector bundle becomes trivial we may well expect the holonomy group to shrink and thus the observed gauge group is enhanced. Since we have only $N = 2$ supersymmetry this observed gauge group enhancement may get killed by quantum effects but should be present in the zero-coupling limit.

The picture is the geometric version of the "Higgs" transitions explored in, for example, [**102**]. We wish to see how this transition appears from the type IIA picture.

We have already seen what the type IIA picture is for moving within the moduli space of vector scalars to a point at which an enhanced gauge symmetry appears. This is where we vary the Kähler form on the generic K3 fibre to shrink down some rational curves (or go to Planck scale effects). Thus, every K3 fibre becomes singular. That is, we have a curve of singularities in X. Now given such a singular Calabi–Yau manifold, it may be possible to deform this space by a deformation of complex structure to obtain a new smooth Calabi–Yau manifold. This would correspond to a deformation of each singular K3 fibre to obtain a smooth K3 fibre. This process will decrease the Picard number of the generic fibre (as we have lost the class of rational curves we shrank down) and will change the topology of the underlying Calabi–Yau threefold.

In the type IIA language then, this Higgs transition consists of deforming the Kähler form on X to obtain a singular space and then smoothing by a deformation to another smooth manifold. Such a topology-changing process is called an "extremal transition".

One example of an extremal transition is the "conifold" transition of [**136**]. A conifold transition consists of shrinking down isolated rational curves and then deforming away the singularities. These were explored in the context of full string theory in [**75, 137**] (see also B. Greene's lectures). In our case however we are not shrinking down isolated curves but whole curves of curves and so we are not discussing a conifold transition.

There has been speculation [**138, 137**] that the moduli space of all Calabi–Yau threefolds is connected because of extremal transitions (based on an older, much weaker, statement by Reid [**139**]). See [**140, 141**] for recent results in this direction. Certainly no counter example is yet known to this hypothesis. The heterotic picture of these extremal transitions is simple to understand in the case of singularities developing within the generic fibre. Such specific extremal transitions are certainly not sufficient to connect the moduli space and we will require an understanding of nonperturbative heterotic string theory to complete the picture. One obvious example to worry about is when the Calabi–Yau threefold on the type IIA side goes through an extremal transition from something that is a K3 fibration to something that is not. It is not difficult to see that such a transition must involve shrinking down the base, W, of the fibration and thus going to a strongly-coupled heterotic string. It is not surprising therefore that we cannot understand such a transition perturbatively in the heterotic picture.

Given a dual pair of a type IIA string compactified on a Calabi–Yau manifold and a heterotic string compactified on $\mathrm{K3} \times T^2$ we may generate more dual pairs by following each through these phase transitions we do understand perturbatively. Such "chains" of dual pairs were first identified in [**142**] and many examples have been given in [**143**]. In order to understand where these chains come from, and indeed the original Kachru–Vafa examples of dual pairs in [**103**], we need to confront the issue of compactifying the heterotic string on a K3 surface, a subject we have done our best to avoid up to this point!

6. The Heterotic String

This section will be concerned with the heterotic string compactified on a K3 surface. In particular we would like to find a string theory dual to this. An obvious answer is another heterotic string compactified on another K3 surface, as we saw in section 5.6. We will endeavor to find a type II dual. It turns out the we will not be able to find a type II dual directly but will have to go via the construction of section 5. This process (in its various manifestations) is often called "F-theory". A great deal of the following analysis is based on work by Morrison and Vafa [**144, 135, 145**].

6.1. $N = 1$ **theories.** The heterotic string compactified on a K3 surface gives a theory of $N = 1$ supergravity in six dimensions. From section 4.1 the holonomy algebra of the moduli space coming from supersymmetry will be $\mathfrak{sp}(1)$. There are two types of supermultiplet in six dimensions which contain moduli fields:

1. The hypermultiplets each contain 4 real massless scalars which transform as quaternionic objects under the $\mathfrak{sp}(1)$ holonomy.
2. The "tensor multiplets" each contain one real massless scalar.

Thus, at least away from points where the manifold structure may break down, we expect the moduli space to be in the form of a product

$$\mathcal{M}_{N=1} = \mathcal{M}_T \times \mathcal{M}_H, \tag{135}$$

(possibly divided by a discrete group) where $\mathcal{M}_T$ is a generic Riemannian manifold spanned by moduli in tensor supermultiplets and $\mathcal{M}_H$ is a quaternionic Kähler manifold spanned by moduli in hypermultiplets.

For a review of some of the aspects of these theories we refer to [**146**]. The six-dimensional dilaton lives in a tensor multiplet. An interesting feature of these theories is that it appears to be impossible to write down an action for these theories unless there is exactly one tensor multiplet.[26] Thus moduli in tensor multiplets other than the dilaton should be regarded as fairly peculiar objects.

It is useful to compare $N = 1$ theories in six dimensions with the resulting $N = 2$ theory in four dimensions obtained by compactifying the six-dimensional theory on a 2-torus. To explain the moduli which appear in four dimensions we need to consider another supermultiplet of the $N = 1$ theory in six dimensions. This is the vector multiplet which contains a real vector degree of freedom but no scalars. Each supermultiplet in six dimensions produces moduli in four dimensions as follows:

1. The 4 real scalars of a hypermultiplet in six dimensions simply produce the 4 real scalars of a hypermultiplet in four dimensions.
2. The 1 real scalar of a tensor multiplet together with the anti-self-dual two-form compactified on T^2 produce the two real scalars of a vector multiplet in four dimensions.
3. The vector field of a vector multiplet compactified on the two 1-cycles of T^2 of produces the two real scalars of a vector multiplet in four dimensions.

That is, hypermultiplets in six dimensions map to hypermultiplets in four dimension but both tensor multiplets and vector multiplets in six dimensions map to vector multiplets in four dimensions.

We should emphasize that quantum field theories in four and six dimensions are quite different. In particular, conventional arguments imply that six-dimensional quantum field theories should always be infra-red free and therefore rather boring.

[26]This is because the gravity supermultiplet (of which there is always exactly one) contains a self-dual two-form and the tensor multiplets contain anti-self-dual two-forms. It is problematic to write down Lorentz-invariant actions for theories with net (anti-)self-dual degrees of freedom [**147**].

This notion has been revised recently in light of many of the results coming from string duality [**146, 148**] where it is now believed that nontrivial theories can occur in six dimensions as a result of "tensionless string-like solitons" appearing. Such theories potentially appear when one goes through extremal transitions between tensor multiplet moduli and hypermultiplet moduli.

Despite the strange properties of these exotic six-dimensional field theories we will be able to avoid having to explain them here. This is because all of our discussion will really happen in four dimensions — the six-dimensional picture is only considered as a large T^2 limit. It is important to realize however that some of the things we will say, based on four-dimensional physics, may well be rather subtle in six dimensions. An example of this will be when certain enhanced gauge symmetries are said to appear when the hypermultiplet moduli are tuned to a certain value. If massless tensor moduli also happen to appear at the same time (which will happen for E_8 as we shall see) then any conventional description of the resulting six-dimensional field theory is troublesome. Declaring what the massless spectrum of such a theory is not an entirely well-defined question and one should move the theory slightly by perturbing either a massless hypermultiplet modulus or a tensor modulus before asking such a question.

6.2. Elliptic fibrations. Our method of approach will be to consider a type IIA string compactified on a Calabi–Yau threefold, X, dual to a heterotic string compactified on K3 $\times T^2$ and then take the volume of the 2-torus to infinity thus decompactifying our theory to a heterotic string on a K3 surface. Actually, decompactification is a rather delicate process and perhaps we should be more pragmatic and say that we will consider a heterotic string on K3 $\times T^2$ and try to systematically ignore all aspects of the type IIA string on X coming from the T^2 part of the heterotic string.

Much of the analysis we require we have already covered in section 5.6. There, for a specific example, we did precisely the decompactification we require. We need to consider how general we can make this process. Clearly we should insist that our moduli space of vector multiplets in the four-dimensional theory corresponding to the heterotic string on K3 $\times T^2$ contain the space

$$\mathrm{O}(\Gamma_{2,2})\backslash \mathrm{O}(2,2)/(\mathrm{O}(2)\times \mathrm{O}(2)). \tag{136}$$

That is, we have the moduli space of the string on T^2. If this were not the case, we could not claim that we had really compactified on a true product of K3$\times T^2$. From the analysis of section 5.4 we expect X to be a K3 fibration. Given the appearance of $\Gamma_{2,2}$ in (136) we expect from section 5.5 that the Picard lattice of the generic fibre of X must contain the lattice $\Gamma_{1,1}$.

A K3 surface with a Picard lattice containing $\Gamma_{1,1}$ is an elliptic fibration. To see this, let v and v^* be a basis for $\Gamma_{1,1}$. The class $v - v^*$ is a primitive element of self-intersection -2 and thus (see, for example, [**16**]) either this class, or minus this class, corresponds to a rational curve within the K3. Now either v or v^* is a nef curve of zero self-intersection. This will be the class of the elliptic fibre. Note that we have a rational curve in the K3 surface intersecting each fibre once — our elliptic fibration has a global section. What's more, it is easy to see that this section is unique as there is no other (-2)-curve.[27]

We thus claim that X is a K3 fibration and that each K3 fibre can be written as an elliptic fibration with a unique section. Together these give [**149**]

PROPOSITION 8. *If a heterotic string compactified on $K3 \times T^2$ is dual to a type IIA string on X and there are no obstructions in the moduli space to taking the size of T^2 to infinity and thereby ignoring the effects associated to it, then X is an elliptic fibration over a complex surface with a section.*

This proposition is subject to the same conditions as proposition 6 and to the caveat in the footnote above.

Let us denote this elliptic fibration as $p: X \to \Theta$. Θ itself is a $\mathbb{P}^1$ fibration over $\mathbb{P}^1$, $\Theta \to W$. The simplest possibility is that Θ is the Hirzebruch surface $\mathbf{F}_n$. This need not be the case however — there may be some bad fibres over some points in W.

The subject of elliptic fibrations has been studied intensively by algebraic geometers, thanks largely to Kodaira [**150**]. This can be contrasted to the subject of K3 fibrations, about which relatively little is known.

Let us fix our notation. Let W, the base of X as a K3 fibration, be $\mathbb{P}^1$ with homogeneous coordinates $[t_0, t_1]$. We will also use the affine coordinate $t = t_1/t_0$. Θ is a $\mathbb{P}^1$ fibration over W. Over a generic point in W, the $\mathbb{P}^1$ fibre will have homogeneous coordinates $[s_0, s_1]$ and affine coordinate $s = s_1/s_0$. Now we want to write down the fibre of X as an elliptic fibration over Θ. Any elliptic curve may be written as a cubic hypersurface in $\mathbb{P}^2$. Writing this in affine coordinates we may put the fibration in "Weierstrass form" for a generic point $(s,t) \in \Theta$:

$$y^2 = x^3 + a(s,t)x + b(s,t), \tag{137}$$

where x and y are affine coordinates in a patch of $\mathbb{P}^2$ and a and b are arbitrary polynomials. If Θ itself has bad fibres — that is, it is not $\mathbf{F}_n$ — then more coordinate patches need to be introduced to give a global description of the elliptic fibration.

Note that not any elliptic fibration can be written in Weierstrass form. Homogenizing the coordinates of $\mathbb{P}^2$ putting $x = x_1/x_0$ and $y = x_2/x_0$, we see that $[0,0,1]$ always lies in (137), which gives the fibration a global section. Thus only elliptic fibrations with a section can be written in this form. Luckily this is the case we are interested in.

The j-invariant of the elliptic fibre (137) is $4a^3/\delta$, where δ is the discriminant of (137) given by

$$\delta = 4a^3 + 27b^2. \tag{138}$$

If $\delta = 0$ then the elliptic fibre is singular. The hypersurface, or divisor, $\delta(w,z) = 0$ in Θ thus gives the locus of bad fibres. It is commonly called the discriminant locus.

[27] Actually we have cheated here. We have assumed that the bundle over K3 may be chosen so that it breaks the $E_8 \times E_8$ or $\mathrm{Spin}(32)/\mathbf{Z}_2$ gauge group. Then the Narain moduli space for the T^2 really is given by (136). There can be times however, as we shall see soon, when this is not possible and then the Narain moduli space for the 2-torus becomes larger. This allows for more than one section of the elliptic fibration. We claim this is not important for the examples we discuss in these lectures, however, as we may begin in a case where the gauge group is broken by the K3 bundle and then proceed via extremal transitions to the case we desire, preserving the section of the elliptic fibration if X.

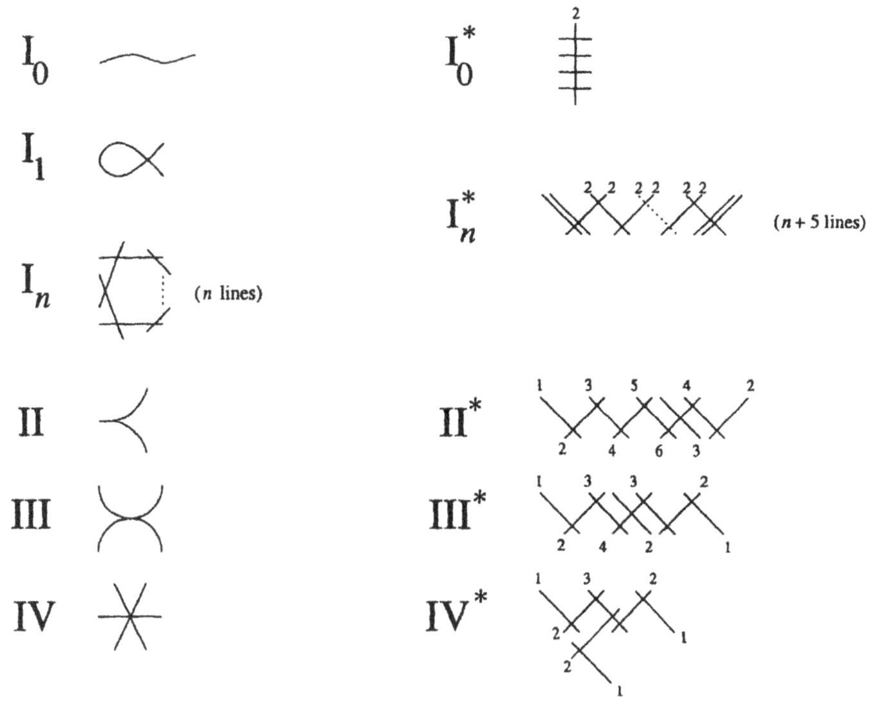

FIGURE 9. Classification of elliptic fibres.

There are only so many things that can happen at a generic bad fibre of an elliptic K3 surface and these have been classified (see, for example, [**16**]). Let us take a small disc $D \subset \mathbb{C}$, embedded in Θ, with coordinate z which cuts a generic point on the discriminant locus transversely. Let $z = 0$ be the location of the discriminant locus. Let us analyze the restriction of the fibration of X to the part which is fibred over D, We are thus considering an open set of a fibration of a complex *surface*. Assuming that the total space of the fibration is smooth and there are no (-1)-curves, which will certainly be true if we are talking about a Calabi–Yau manifold, the possibilities for what happens to the fibre when $z = 0$ is shown in figure 9. The case I_0 is when there is no zero of δ and the elliptic fibre is smooth. In all other cases the lines and curves in figure 9 represent rational curves. Case I_1 is a rational curve with a "double point", i.e., it looks locally like $y^2 = x^2$ at one point and case II is a rational curve with a "cusp", i.e., it looks locally like $y^2 = x^3$ at one point. All the other cases consist of multiple rational curves. All of the singular fibres should be homologous to the smooth fibre. To achieve this, some of the rational curves in the bad fibre must be counted more than once to obtain the correct homology. The multiplicity of the curves are shown as the small numbers in figure 9. If omitted, multiplicity one is assumed.[28]

[28]There is also a possibility that I_n fibres can appear with multiplicity greater than one. We ignore this as the canonical class of such a fibration cannot be trivial.

L	K	N	Fibre	Υ'
≥ 0	≥ 0	0	I_0	
0	0	> 0	I_N	A_{N-1}
≥ 1	1	2	II	
1	≥ 2	3	III	A_1
≥ 2	2	4	IV	A_2
≥ 2	≥ 3	6	I_0^*	D_4
2	3	≥ 7	I_{N-6}^*	D_{N-2}
≥ 3	4	8	IV^*	E_6
3	≥ 5	9	III^*	E_7
≥ 4	5	10	II^*	E_8

TABLE 4. Weierstrass classification of fibres.

The possibilities listed in figure 9 can also be classified according to the Weierstrass form. On our disc, D, the polynomials, a and b, in (137) will be polynomials in z. Let us define the non-negative integers (L, K, N) by

$$\begin{aligned} a(z) &= z^L a_0(z) \\ b(z) &= z^K b_0(z) \\ \delta(z) &= z^N \delta_0(z), \end{aligned} \tag{139}$$

where $a_0(z)$, $b_0(z)$, and $\delta_0(z)$ are all nonzero at $z = 0$. The triple (L, K, N) then determines which fibre we have according to table 4. See **[32]** for an explanation of this. To be precise, the Weierstrass form of the fibration in x, y and z will produce a surface singularity which, when blown-up, will have the fibres in figure 9. This is closely linked to the results in table 1.

It is important to note that this classification procedure only applies to a smooth point on the discriminant locus. Only in this case can we characterize the bad fibre in terms of the family of elliptic curves over our small disc, D. When the discriminant is singular the nature of the bad fibre need not be expressible in terms of the geometry of a complex surface — it will be higher-dimensional in character. In general, for Calabi–Yau threefolds, we should expect to encounter some singular fibres not listed above. Such exotic fibres are important in string theory but we will try to avoid such examples here as it makes the analysis somewhat harder.

For our elliptic fibration, $p : X \to \Theta$, a knowledge of the explicit Weierstrass form is enough to calculate the canonical class, K_X. This may be done as follows. In homogeneous coordinates, the Weierstrass form is

$$x_0 x_2^2 = x_1^3 + a x_0^2 x_1 + b x_0^3, \tag{140}$$

giving a cubic curve in $\mathbb{P}^2$. Now since the elliptic fibration is not trivial, this $\mathbb{P}^2$ will vary nontrivially as we move over Θ. We may describe such a $\mathbb{P}^2$ as the projectivization of a sum of three line bundles over Θ. We are free to declare that x_0 is a section of a trivial line bundle. We may then find a line bundle, $\mathscr{L}$, such that x_1 is a section of $\mathscr{L}^2$ and x_2 is a section of $\mathscr{L}^3$ in order to be compatible with (140). It also follows that a is a section of $\mathscr{L}^4$ and b is a section of $\mathscr{L}^6$.

Now let us consider the normal bundle of the section, σ, given by $[x_0, x_1, x_2] = [0,0,1]$, embedded in X. We may use affine coordinates $\xi_1 = x_1/x_2$ and $\xi_2 = x_0/x_2$ whose origin gives the section σ. Note that ξ_1 is a section of $\mathscr{L}^{-1}$ and ξ_2 is a section of $\mathscr{L}^{-3}$. Near σ, (140) becomes

$$\xi_2 = \xi_1^3. \tag{141}$$

Thus, ξ_1 is a good coordinate to describe the fibre of the normal bundle of σ in X. This implies that this normal bundle is given by $\mathscr{L}^{-1}$.

We may now use the adjunction formula for $\sigma \subset X$ to give

$$K_X|_\sigma = K_\sigma + \mathscr{L}. \tag{142}$$

Actually, this is the only contribution towards the canonical class of X. That is,

$$K_X = p^*(K_\Theta + \mathscr{L}). \tag{143}$$

For the case we will be interested in, we want $K_X = 0$ and so $\mathscr{L} = -K_\Theta$. We will give examples of such constructions later.

Note that if there is a divisor within Θ over which a vanishes to order ≥ 4 and b vanishes to order ≥ 6, we may redefine $\mathscr{L}$ to "absorb" this divisor and lower the degrees of a and b accordingly. This is why no such fibres appear in table 4. As this will change K_X, such occurrences cannot happen in a Calabi–Yau variety.

Let us consider a K3 surface written as an elliptic fibration with a section. The Picard number of the K3 is at least two — we have the section and a generic fibre as algebraic curves. If we have any of the fibres I_n, for $n > 2$, or I_n^*, III, IV, II*, III*, or IV* we will also have a contribution to the Picard group from reducible fibres. Each of these fibres contains rational curves in the form of a root lattice of a simply-laced group. Let us denote this lattice Υ'. The possibilities are listed in table 4. Thus, shrinking down these rational curves will induce the corresponding gauge group for a type IIA string.

We know that for a Calabi–Yau manifold compactified on a K3 fibration, the moduli coming from varying the Kähler form on the K3 fibre map to the T^2 part of the heterotic string compactified on K3 $\times$ T^2. In particular, the act of blowing-up rational curves in the K3 to resolve singularities, and hence break potential gauge groups, is identified with switching on Wilson lines on T^2. Thus, to ignore Wilson lines, these rational curves must all be blown down and held at zero area. That is, *any of the fibres* I_n, *for* $n > 2$, *or* I_n^*, III, IV, II*, III*, *or* IV* *appearing in the elliptic fibration will produce an enhanced gauge symmetry in the theory.*

From section 5.6 and, in particular (131), the size of an elliptic fibre within this K3 will be fixed to some constant $\sqrt{\alpha/2\beta}$ as $\alpha, \beta \to \infty$ to make the T^2 infinite area. Thus this size is "frozen out" as a degree of freedom. To ignore the 2-torus degrees of freedom for the type IIA string compactified on X we should take the K3 fibre within X, consider it as an elliptic fibration with an elliptic fibre of frozen area and blow down any rational curves which may take the Picard number of the K3 fibre beyond 2. In summary, any degrees of freedom coming from sizes within the elliptic fibre structure are ignored.

Consider the base, Θ, as a $\mathbb{P}^1$ bundle over W. Suppose we have bad fibres in this case. These must correspond to reducible fibres. Now, when we build X as a K3 fibration over W, these reducible fibres in Θ will build reducible fibres in X. This is exactly case 3 listed at the beginning of section 5.5. That is, varying the size of irreducible parts of these reducible fibres will give moduli in vector supermultiplets which cannot be understood perturbatively on the heterotic side. The important point to note here is that these degrees of freedom will not go away when we unwind any Wilson lines around T^2 and take its area to infinity. Thus, these degrees of freedom must be associated to the six-dimensional theory of the heterotic string compactified on a K3. From section 6.1 these moduli must therefore come from tensor multiplets in the six-dimensional theory. It is the peculiar nature of tensor moduli which prevented us from having a perturbative understanding of these moduli in section 5.5.

We saw in section 5.6 how the size of W and the size of the fibre of Θ were used to produce the area of the heterotic string's T^2 and the size of its dilaton. The area of the T^2 is lost as a degree of freedom in our six-dimensional theory. We see then *that the number of tensor multiplets in our theory will be the Picard number of* Θ *minus one.* In the case that Θ is $\mathbf{F}_n$, that is, there are no bad fibres, the number of tensor multiplets will be one and this single multiplet contains the heterotic dilaton as a modulus.

Questions concerning hypermultiplets between the four-dimensional theory and the six-dimensional theory are unchanged. In particular, we retain the relationship from section 5.3 that the number of hypermultiplets is given by $h^{2,1}(X)+1$.

Now we have counted massless tensor multiplets and hypermultiplets, let us count massless vector multiplets. We know that any vector multiplet in the four-dimensional theory must have its origin in either a vector multiplet or a tensor multiplet in six dimensions. Thus we can count the number of six-dimensional vectors by subtracting $h^{1,1}(\Theta)-1$ from the number of four-dimensional vectors.

☞ Many of these vectors can be seen directly in terms of the enhanced nonabelian gauge symmetry but there is an additional contribution. In section 5.5 we saw how $H^2(X)$, where X is a K3 fibration, could be built from elements from the base, from the generic fibre and from the bad fibres in a fairly obvious way. Now we want to consider if the same thing is true for an elliptic fibration. Analyzing the spectral sequence one can see that $H^2(\Theta,\mathbb{Z})$ contributes, for the base, to give the tensor multiplets and $H^0(\Theta,R^2p_*\mathbb{Z})$ contributes for the fibres. This latter piece accounts for the enhanced nonabelian symmetry discussed above. The object of interest will be the term $H^1(\Theta,R^1p_*\mathbb{Z})$. In the case of K3 fibrations, this is trivial since $H^1(\mathrm{K3})=0$. This ceases to be true for an elliptic fibration however. This contribution may be associated to the group of sections of the bundle as seen in **[151]**. Later on, in section 6.5, we will consider a case where this is a finite group, but here we note that if this group is infinite, then its rank will contribute to the dimension of $H^2(X)$.

We see then that if the elliptic fibration has an infinite number of sections, there will be massless vector fields beyond those accounted for from the nonabelian gauge symmetry from bad fibres. These may contribute extra $\mathfrak{u}(1)$ terms to the gauge symmetry **[145]** and, conceivably, more nonabelian parts. As this part of the gauge group is rather difficult to analyze we will restrict ourselves to examples in these lectures where there is no such contribution.

6.3. Small instantons. Our goal in this section will be to find the map from at least part of the moduli space of hypermultiplets for a type IIA string on a Calabi–Yau manifold to the moduli space of hypermultiplets of the heterotic string compactified on a K3×T^2. As we have seen, the T^2 factor is irrelevant for the hypermultiplet moduli space and we are free to consider the latter as a heterotic string on a K3 surface so long as the Calabi–Yau is an elliptic fibration with a section. We will be able to go some way to determining "which" heterotic string a given elliptic threefold is dual to.

The best policy when finding a map between two moduli spaces is to start with a particularly special point in the moduli space of one theory with hopefully unique properties which allow it to be identified with a correspondingly special point in the other theory's moduli space. This special point will usually be very symmetric in some sense. The trick, invented in [**135**] and inspired by the work of [**152**], is to look for very large gauge symmetries resulting from collapsed instantons in the heterotic string.

Recall that the K3 surface, S, on which the heterotic string is compactified comes equipped with a bundle, $E \to S$, with $c_2(E) = 24$. Fixing S, the bundle E will have moduli. A useful trick when visualizing the moduli space of bundles is to try to flatten out as much of the connection on the bundle as possible. Of course, the fact that $c_2(E) = 24$ makes it impossible to completely flatten out E but we can concentrate the parts of the bundle with significant curvature into small regions isolated from each other. In this picture, an approximate point of view of at least part of the moduli space of the bundle can be viewed as 24 "instantons" localized in small regions over S each of which contributes one to $c_2(E)$.

As well as its position, each instanton will have a degree of freedom associated to its size — that is, the characteristic length away from the centre of the instanton where the curvature becomes small. At least from a classical point of view, there is nothing to stop one shrinking this length scale down to zero.[29] Such a process naturally takes one to the boundary of the moduli space of instantons.

Let us consider the $E_8 \times E_8$ heterotic string. The observed gauge group of a heterotic string theory compactified on S will be the part of the original $E_8 \times E_8$ which is not killed by the holonomy of E. It is the "centralizer" of the embedding of the holonomy in $E_8 \times E_8$ — i.e., all the elements of $E_8 \times E_8$ which commute with the holonomy. What is the holonomy around a collapsed instanton? In general the global holonomy is generated by contractable loops due to the curvature of the bundle, and from non-contractable loops in the base. The curvature is zero everywhere when the instanton has become a point. Also, if S is smooth we may look at a 3-sphere surrounding the instanton to look for holonomy effects. Since $\pi_1(S^3) = 0$ we cannot generate non-contractable loops. Thus, the holonomy of a point-like instanton is trivial.[30]

One possibility is to shrink all 24 instantons down to zero size. When we do this, E will have no holonomy whatsoever and the resulting heterotic string theory will retain its full $E_8 \times E_8$ gauge symmetry. As this is such a big gauge group it is a good place to start analyzing our duality map.

We thus want to find a Calabi–Yau space on which we may compactify the type IIA string to give an $E_8 \times E_8$ gauge symmetry. That is, we want an elliptic fibration over $\mathbf{F}_n$ such that when viewed as a K3 fibration, the generic K3 fibre has two E_8 singularities. Let us discuss what this implies about the discriminant locus within $\Theta \cong \mathbf{F}_n$.

First, let us be a little sloppy with notation and not distinguish between curves and their divisor class (roughly speaking, homology class) in Θ. Thus we use C_0 to

[29] Although there is no reason to suppose that one may do this independently for all 24 instantons.

[30] Note that this need not be the case if S has an orbifold singularity and the instanton sits at this point. Then we can only surround the instanton by a lens space which is not simply connected. One should note that we have conveniently ignored those loops which happen to go through the point where the point-like instanton sits.

denote the class of base, i.e., the $(-n)$-curve within Θ, and f to denote the class of the generic $\mathbb{P}^1$ fibre. Let us determine K_Θ in terms of these classes. Consider the adjunction formula for a curve $C \in \Theta$. Integrating this over C we obtain its Euler characteristic

$$\chi(C) = -C.(C + K_\Theta). \tag{144}$$

Knowing that C_0 and f are spheres is enough to determine

$$K_{\mathbf{F}_n} = -2C_0 - (2+n)f. \tag{145}$$

Let us use the letters A, B, and Δ to denote the divisors associated to the equations $a = 0$, $b = 0$, and $\delta = 0$ respectively. From (143), the classes of these divisors will be

$$\begin{aligned} A &= 8C_0 + (8+4n)f \\ B &= 12C_0 + (12+6n)f \\ \Delta &= 24C_0 + (24+12n)f, \end{aligned} \tag{146}$$

in order that X be Calabi–Yau.

The locus of E_8 singularities in X will map to curves in Θ. As these E_8's are independent, we want to make these curves disjoint sections of Θ, that is, one curve will be in the class C_0 (the isolated zero section of the Hirzebruch surface as a $\mathbb{P}^1$ bundle over $\mathbb{P}^1$) and the other in the class $C_\infty = C_0 + nf$ (a section in the non-isolated class which we view as a section "at infinity" of the Hirzebruch surface). From table 4 we see that we want II* fibres over these curves.

These two curves of II* fibres will account for a large portion of the A, B, and Δ divisors. Let us write A', B', and Δ' for the remaining parts of the divisors not contained in the curves C_0 and C_∞. From table 4 we have

$$\begin{aligned} A' &= A - 4(C_0) - 4(C_0 + nf) = 8f \\ B' &= B - 5(C_0) - 5(C_0 + nf) = 2C_0 + (12+n)f \\ \Delta' &= \Delta - 10(C_0) - 10(C_0 + nf) = 4C_0 + (24+2n)f. \end{aligned} \tag{147}$$

This means that what is left of the discriminant, Δ', will collide with the curve C_0 a total number of $C_0.\Delta' = 2(12-n)$ times and with the curve C_∞ a total number of $2(12+n)$ times. These 48 points of intersection are not independent. The reason that Δ' collides with the curves of II* fibres is precisely because B' also collides with these curves. Each time B' hits these curves, the degree of the discriminant will rise by 2 and hence Δ' hits them twice in the same place. To see exactly what shape this intersection is one may explicitly write out the equations. The result is that Δ' crosses itself transversely at these points as well as hitting C_0 or C_∞. We see then that, generically, Δ' collides with C_0 at $12 - n$ points and with C_∞ at $12 + n$ points. Within Δ', away from these collisions, we expect the discriminant to behave reasonably nicely.[31] The result is shown in the upper part of figure 10.

We know how to deal with all of the points on the discriminant from Kodaira's list in figure 9 except for the 24 points of collision of Δ' with the two lines of II* fibres. Here we have to work harder to obtain a smooth model for X.

[31] Although it will have cusps.

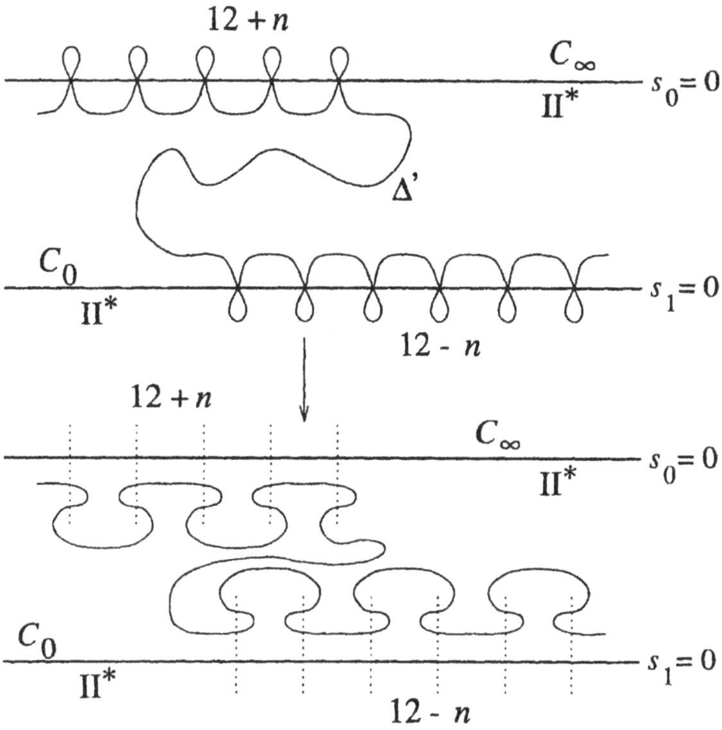

FIGURE 10. 24 small instantons in the $E_8 \times E_8$ string.

Let us focus on one of these points where either C_0 or C_∞ hits Δ' twice within Θ. Blow up this point of intersection, $\pi : \widetilde{\Theta} \to \Theta$. This will introduce a new rational curve, E, in the blown-up surface $\widetilde{\Theta}$ such that $K_{\widetilde{\Theta}} = \pi^* K_\Theta + E$. Pulling back $\mathscr{L}$ onto $\widetilde{\Theta}$ will show that a now vanishes to degree 4 on E and b vanishes to degree 6 on E. Introduce $\mathscr{L}' = \pi^* \mathscr{L} - E$ and write a and b in terms of $\mathscr{L}'$ instead of $\mathscr{L}$. Now a and b will not vanish at all on a generic point on E. Note that the effect on K_X of blowing up the base, Θ, and then subtracting E from $\mathscr{L}$ nicely cancels out and so X is still Calabi–Yau.

Thus, to obtain a smooth X we need to blow up all 24 points of collision. This process is shown in the lower part of figure 10. The dotted lines represent the 24 new $\mathbb{P}^1$'s introduced into the base. Note that they are not part of the discriminant.

From our interpretation of moduli in section 6.2 we see that in terms of the underlying six-dimensional field theory we have 24 new moduli from tensor multiplets as soon as we try to enhance the gauge group to $E_8 \times E_8$ in this way. Remember that in the language of the heterotic string this large gauge group was meant to be the result of shrinking down 24 instantons to zero size. What we have therefore shown here is that shrinking down 24 instantons to zero size results in the appearance of 24 new tensor moduli.

Given that the appearance of tensor multiplets should be a nonperturbative phenomenon in the heterotic string, it would seem unreasonable to expect them to appear when the target space and vector bundle is smooth. At least in the case that the underlying K3 surface is large, it is then clear that each tensor modulus can be tied to each shrunken instanton in this picture. That is, one point-like instanton will result in one tensor modulus appearing.

Suppose we try to give size to some of the instantons. This should correspond to a deformation of the complex structure of the Calabi–Yau threefold. We know that this should result in the disappearance of (at least) one of the tensor multiplets and thus (at least) one of the blow-ups in Θ. Thus the deformation has to disturb one of the curves of II^* and thus lower the size of the effective gauge group. What is important to notice is that $12+n$ of the small instantons are thus embedded in the E_8 factor associated to C_∞ and $12-n$ of the instantons are embedded in the E_8 associated to C_0. To put this another way, when we smooth everything out to obtain a theory with no extra tensor moduli we expect to have an $(E_8\times E_8)$-vector bundle which is a sum of two E_8-bundles, one of which has $c_2 = 12+n$ and the other with $c_2 = 12-n$.

This pretty well specifies exactly which heterotic string our type IIA string is dual to. For a G-bundle on a K3 surface, where G is semi-simple and simply-connected, the topological class of this bundle is specified by a map $H_4(\mathrm{K3})\to\pi_3(G)$ (for a clear explanation of this see **[153]**). This may be viewed as given by the second Chern class of each sub-bundle associated to each factor of G. In our case we fix the total second Chern class and so the only freedom remaining is specified by how c_2 of the bundle is split between the factors of G. Thus n determines the class of our $E_8\times E_8$ bundle. We have arrived at the following:

PROPOSITION 9. *Let E be a sum, $E_1\oplus E_2$, of two E_8-bundles on a smooth K3 surface such that $c_2(E_1) = 12+n$ and $c_2(E_2) = 12-n$. Then a heterotic string compactified on this bundle on K3 is dual to (a limit of) a type IIA string compactified on a Calabi–Yau threefold which is an elliptic fibration with a section over the Hirzebruch surface $\mathbf{F}_n$.*

Our main assumption here is that the type IIA string on the Calabi–Yau manifold really is dual to a heterotic string. If it is, then we have certainly identified the correct one subject to the provisos of proposition 6.

This proposition first appeared in **[135]**. Although virtually all of the mathematics above has been copied from that paper our presentation has been slightly different. Rather than take the limit of decompactifying a T^2 in a heterotic string on $\mathrm{K3}\times T^2$, the line of attack in **[135]** was effectively to decompactify the dual type IIA string (or its mirror partner, the type IIB string) to a twelve-dimensional theory compactified on a Calabi–Yau manifold. It is not clear whether this twelve-dimensional "F-theory" exists in the usual sense of ten-dimensional string theory or eleven-dimensional M-theory or whether it serves simply as useful mnemonic for the above analysis. Another point of view of F-theory is to think of it as the type IIB string compactified down to six dimensions on Θ (see, for example, **[154]** for some nice results along these lines). The fact that Θ is not a Calabi–Yau space is corrected for by placing fixed D-branes within it. Again this is essentially equivalent to the above. The key ingredient to associate to the term "F-theory" is the elliptic fibration structure. Whether one wishes to think of this in terms of a mysterious twelve-dimensional theory or a type IIA string or a type IIB string is up to the reader.

The above proposition establishing the link between how the 24 of the second Chern class is divided between the two E_8's, and over which Hirzebruch surface X is elliptically fibred, is in agreement with all the relevant conjectured dual pairs of [**103**] and [**142**] for example. The case of $n = 12$ was established in [**132**].

The appearance of tensor moduli for small instantons was first noted in [**146**]. Since we are not allowing ourselves to appeal to M-theory or D-branes in these lectures we will not reproduce the argument here but just note that all necessary information appears to be contained in the type IIA approach we use here.

6.4. Aspects of the $E_8 \times E_8$ string. Let us now follow the analysis of [**135, 145**] and continue to explore the duality between the type IIA string on X and the $E_8 \times E_8$ heterotic string.

In the previous section we looked at the case of fixing hypermultiplet moduli in order to break none of the $E_8 \times E_8$ gauge group. We should ask the opposite question of what the gauge group is at a generic point in the hypermultiplet moduli space. We shall do this as follows. If any of the $E_8 \times E_8$ gauge group remains unbroken we expect either of our curves C_0 or C_∞ to contain part of the discriminant locus, Δ. Let us focus on C_0. Split off the part of the discriminant not contained in C_0 by putting

$$\Delta = NC_0 + \Delta', \tag{148}$$

where $N \geq 0$ and Δ' does not contain C_0. Since the only way that the intersection number of two algebraic curves in an algebraic surface can be negative is if one of the curves contains the other, we have

$$\Delta'.C_0 \geq 0. \tag{149}$$

Following (146) we have, for $n \geq 0$,

$$N \geq 12 - \frac{24}{n}. \tag{150}$$

Similarly we may analyze the divisors A and B to obtain the respective orders, L and K, to which a and b vanish on C_0. This gives

$$\begin{aligned} L &\geq 4 - \frac{8}{n} \\ K &\geq 6 - \frac{12}{n}. \end{aligned} \tag{151}$$

We may now use table 4 to determine the fibre over a generic point in C_0. Repeating this procedure for the other "primordial" E_8 along C_∞ shows that no singular fibres are required there for $n \geq 0$.

We see that in the case $n > 2$, we will have singular fibres over C_0 generating a curve of singularities within X. Thus we expect an enhanced gauge group. Loosely speaking, the gauge group can be read from the last column of table 4. The only thing we have to worry about is the monodromy of section 5.2 — it may be that there is monodromy on the singular fibres as we move about C_0. If $\Delta'.C_0 = 0$ then there can be no monodromy since the fibre is the same over every point of C_0. This occurs for $n = 2, 3, 4, 6, 8, 12$. When $n = 7, 9, 10, 11$, the fibre admits no symmetry

n	L	K	N	Fibre	Mon.	G	H_0
≤ 2	0	0	0	I_0			E_8
3	2	2	4	IV		$\mathfrak{su}(3)$	E_6
4	2	3	6	I_0^*		$\mathfrak{so}(8)$	$\mathfrak{so}(8)$
5	3	4	8	IV*	$\mathbb{Z}_2$	F_4	G_2
6	3	4	8	IV*		E_6	$\mathfrak{su}(3)$
7	3	5	9	III*		E_7	$\mathfrak{su}(2)$
8	3	5	9	III*		E_7	$\mathfrak{su}(2)$
≥ 9	4	5	10	II*		E_8	

TABLE 5. Generic gauge symmetries, G.

and thus there cannot be any monodromy. Therefore, the only time we have to worry about monodromy is for the IV* fibre in the case $n = 5$. There is indeed monodromy in this case [**149**]. (See [**155**] for an account of this in terms of "Tate's algorithm" or [**156**] for an alternative approach.) Thus, whereas one associates E_6 with a type IV* fibre, this becomes F_4 from figure 5 when $n = 5$. The gauge algebras for generic moduli are summarized in table 5.

This agrees nicely with the heterotic picture. For a given value of c_2 of a bundle, find the largest possible structure group, H, of a vector bundle. Then the desired gauge group, G, will be the centralizer of H within $E_8 \times E_8$. One "rough and ready" approach to this question is as follows. Consider an H-bundle E with fibre in an irreducible representation, R, of the structure group. The Dolbeault index theorem on the K3 surface then gives [**9**]

$$\begin{aligned}\dim H^0(E) - \dim H^1(E) + \dim H^2(E) &= \int_S td(T) \wedge ch(E) \\ &= 2\,\mathrm{rank}(R) - l(R)c_2(E),\end{aligned} \tag{152}$$

where $l(R)$ is the index of R using the conventions of [**157**]. If E really is a strict H-bundle with fibre R, it should have no nonzero global sections, otherwise the structure group would be a strict subgroup of H. Thus $H^0(E)$ is trivial. Similarly, by Serre duality, we expect the same for $H^2(E)$. Thus, the right hand side of (152) must be non-positive. That is,

$$c_2(E) \geq \frac{2\,\mathrm{rank}(R)}{l(R)}, \tag{153}$$

for any irreducible representation, R. The bundle with $c_2 = 12 + n$ may have the full E_8 as structure group so one E_8 of the $E_8 \times E_8$ will be broken generically for any $n \geq 0$. Table 5 then shows how the other E_8 is broken down to G by a bundle with structure group H_0 and $c_2 = 12 - n$. For example, the $\mathbf{3}$ of $\mathfrak{su}(3)$ has $l(\mathbf{3}) = 1$ and so c_2 of a generic $\mathfrak{su}(3)$-bundle with no global sections is at least 6. This is why it appears on the row for $n = 6$ in the table.

Note that there need not exist a bundle that saturates the bound in (153) and so we cannot reproduce all of the rows in table 5. For example, in the case $n = 3$ one sees that (153) does not rule out a bundle with the full E_8 structure group but we see that only an E_6-bundle is expected. See [**133**] for a discussion of this. It is interesting to note that proposition 9 implies that $\mathfrak{su}(3)$ *must* appear as the gauge symmetry in the case $n = 3$ and so the E_8-bundle must not exist. If a smooth E_8-bundle on a K3 surface with $c_2 = 9$ is discovered it will violate proposition 9.

The cases $n = 9, 10, 11$ are somewhat peculiar since we appear to be suggesting that we have a bundle with trivial structure group and yet $c_2 > 0$. Clearly this is not possible classically. The fact that classical reasoning is breaking down somewhat can be seen from the fact that the reasoning of section 6.3 applies to this case and we have $12 - n$ tensor multiplets. That is, we have $12 - n$ point-like instantons which cannot be given nonzero size.

The fact that $n \leq 12$ can be understood from both the type IIA side and the heterotic side. In the case of our elliptic fibration over $\mathbf{F}_n$, if $n > 12$ then (L, K, N) as determined from (150) and (151) is at least $(4, 6, 12)$. As discussed in section 6.2, this means that we may redefine $\mathscr{L}$ to absorb C_0 to reduce the fibre to something in the list in figure 9. This kills $K_X = 0$ however and so we do not have a Calabi–Yau space. On the heterotic side, $c_2 < 0$ would be a clear violation of (153). One might worry about "point-like anti-instantons" but such objects would break supersymmetry and as such do not solve the equations of motion.

In section 5.7 we discussed extremal transitions between Calabi–Yau manifolds which, in the heterotic language, corresponded to unwrapping part of the gauge bundle around the K3 surface and rewrapping it around the T^2. Such transitions are not of much interest to us in this section as we are concerned only with the K3 part of the story. There are other possible extremal transitions which will effect us though. One kind which is of interest are ones which will take us from an elliptic fibration over $\mathbf{F}_n$ to another elliptic fibration over $\mathbf{F}_{n-1}$. In the heterotic string this will correspond to a transition from splitting the $E_8 \times E_8$ bundle into two bundles with c_2 equal to $12 + n$ and $12 - n$, to a splitting of $12 + n - 1$ and $12 - n + 1$ respectively. In this way we may "join up" all the theories considered so far into one connected moduli space. This phenomenon was first observed in **[146]** but we will again follow the argument as presented in **[135, 145]**.

As explained in section 6.3, when the $E_8 \times E_8$ heterotic string has a point-like instanton, we expect the dual Calabi–Yau space for the type IIA string to admit a blow-up in the base, Θ, of the elliptic fibration. When all 24 instantons are point-like we saw this as a collision between a curve of II^* fibres and other parts of the discriminant locus as shown in figure 10. Let us concentrate on what happens when one of these points is blown up.

As in section 6.3, we use Δ' to denote the part of discriminant left over after we subtract the contribution from the two curves of II^* fibres. In the first diagram in figure 11 we show locally how Δ' loops around a collision between it and a line, C_∞, of II^* fibres together with the class, f, that passes though this point. Now when we do the blow-up by switching on a scalar in a tensor multiplet we go to the middle diagram. The exceptional divisor is a line of self-intersection -1. The line that was in the class f also becomes a (-1)-line. The middle diagram of figure 11 is obviously symmetric and we may blow-down the latter (-1)-curve to push the loop onto the bottom line of II^* fibres as shown in the last diagram.

The effect of this is to change C_∞ into a line of self-intersection $n - 1$ and C_0 into a line of self-intersection $-n + 1$. Thus we have turned $\mathbf{F}_n$ into $\mathbf{F}_{n-1}$. It is also clear that we have moved the small instanton from one of the E_8's into the other

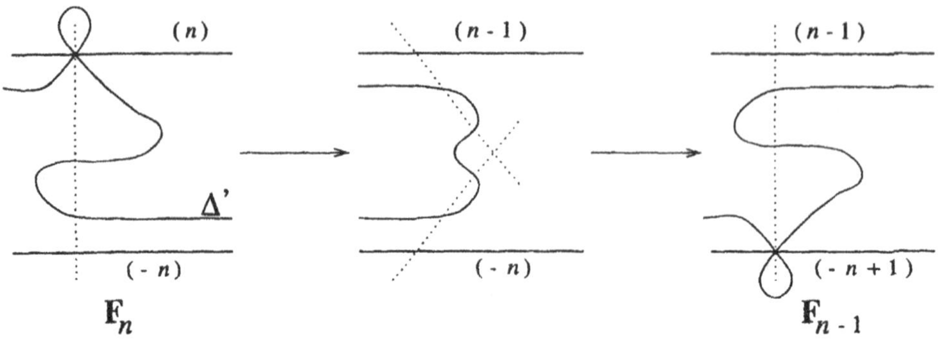

FIGURE 11. The transition $n \to n-1$.

E_8. Thus we may connect up all our theories which are elliptic fibrations over $\mathbf{F}_n$. Note that the use of tensor moduli means that we do not expect a perturbative interpretation of this process in the heterotic string language.

6.5. The Spin(32)/$\mathbb{Z}_2$ heterotic string. Now that we can content ourselves with the knowledge that we know how to build a type IIA dual to the generic $E_8 \times E_8$ heterotic string on a K3 surface we turn our attention to the Spin(32)/$\mathbb{Z}_2$ heterotic string. Whereas the topology of the $E_8 \times E_8$ bundle required specifying how the 24 instantons were divided between the two E_8's, the Spin(32)/$\mathbb{Z}_2$ heterotic string is quite different. In this case the gauge group is not simply-connected and the topology of the bundle is not simply specified by c_2.

It will be important to recall some fundamentals of the construction of the Spin(32)/$\mathbb{Z}_2$ heterotic string from [**158**]. The 16 extra right-movers of the heterotic string are compactified on an even self-dual lattice of definite signature. There are two such lattices, which we denote $\Gamma_8 \oplus \Gamma_8$ and Γ_{16}. The former is two copies of the root lattice of $E_8 \times E_8$ with which we have become well-acquainted in these talks. The second lattice is the "Barnes-Wall" lattice [**15**]. This may be constructed by supplementing the root lattice of $\mathfrak{so}(32)$ by the weights of one of its spinors. Such spinor weights are never of length squared 2 are so do not give massless states. Thus, as far as massless states are concerned, the lattice is the root lattice of SO(32) and the string states fill out the adjoint representation. Massive representations may fill out spinor representations for one of the spinors and but we never have representation in the vector representation or the other spinor representation. The gauge group can be viewed as a $\mathbb{Z}_2$ quotient of Spin(32) which does not admit vector representations. Hence π_1 of our gauge group is $\mathbb{Z}_2$. When we try to build Spin(32)/$\mathbb{Z}_2$-bundles the situation is very similar to real vector bundles over M where the fact that $\pi_1(\mathrm{SO}(d)) \cong \mathbb{Z}_2$ leads to the notion of the second "Stiefel-Whitney" class, w_2, of a bundle as an element of $H^2(M,\mathbb{Z}_2)$. Here we have a similar object characterizing the topology of the bundle which we denote $\tilde{w}_2$. If $\tilde{w}_2 \neq 0$ then bundles with fibre in the vector representation are obstructed just as spinor representations are obstructed for non-spin bundles with $w_2 \neq 0$. See [**159**] for a detailed account of this.

Thus, rather than being classified by how 24 is split between second Chern class, which was the case for the $E_8 \times E_8$ string, the topological class of an Spin(32)/$\mathbb{Z}_2$

heterotic string compactified over a K3 surface is characterized by $\tilde{w}_2 \in H^2(S, \mathbb{Z}_2)$. Whereas elements $\tilde{w}_2$ are in one-to-one correspondence with the homotopy classes of $\mathrm{Spin}(32)/\mathbb{Z}_2$-bundles on a fixed (marked) K3 surface, in our moduli space we are also allowed to vary the moduli of the K3 surface. Thus two elements of $H^2(S, \mathbb{Z}_2)$ should be considered to be equivalent if they can be mapped to each other by a diffeomorphism of the K3 surface. That is,

$$\tilde{w}_2 \in \frac{\Gamma_{3,19}/2\Gamma_{3,19}}{\mathrm{O}^+(\Gamma_{3,19})}. \tag{154}$$

It was shown in **[159]** that there are only three possibilities:

1. $\tilde{w}_2 = 0$,
2. $\tilde{w}_2 \neq 0$ and $\tilde{w}_2.\tilde{w}_2 = 0 \pmod 4$,
3. $\tilde{w}_2 \neq 0$ and $\tilde{w}_2.\tilde{w}_2 = 2 \pmod 4$.

We will focus on the case $\tilde{w}_2 = 0$ in this section. This will allow us to use the same arguments as in the last section about shrinking instantons down to retrieve the entire primordial gauge group. If $\tilde{w}_2$ were not zero, the topology of the bundle would obstruct the existence of arbitrary point-like instantons at smooth points in the K3 surface. See **[159]** for an account of this.

We have seen already that if a perturbatively understood heterotic string is dual to a limit of a type IIA string on Calabi–Yau threefold, X, then X must be an elliptic fibration with a section over $\mathbf{F}_n$, where $0 \leq n \leq 12$. Thus, if our duality picture is going to continue working for any $\mathrm{Spin}(32)/\mathbb{Z}_2$ heterotic string then we must have already encountered it in disguise as the $E_8 \times E_8$ heterotic string for a particular n.

The statement that the $\mathrm{Spin}(32)/\mathbb{Z}_2$ heterotic string compactified on a K3 surface is the same thing as an $E_8 \times E_8$ heterotic string compactified on a K3 surface should not come as a surprise as the same thing has been known to be true for toroidal compactifications for some time **[61, 160]**. The identification of the dilaton as the size of the base of X as a K3 fibration has nothing to do with whether we deal with the $E_8 \times E_8$ or the $\mathrm{Spin}(32)/\mathbb{Z}_2$ string and so the duality between these theories cannot effect the string coupling. Thus there must be some T-duality statement that connects these two theories. This may be highly nontrivial however, as it may mix up the notion of what constitutes the base of the bundle over K3 and what constitutes the fibre. A construction of such a T-duality at a special point in moduli space was given in **[159]**.

Now let us return to the issue concerning the extra states in the Γ_{16} lattice not contained in the root lattice of $\mathfrak{so}(32)$. What is the dual analogue of these extra massive states coming from the spinor of $\mathfrak{so}(32)$? Let us think in terms of the type IIA string compactified on a K3 surface versus the $\mathrm{Spin}(32)/\mathbb{Z}_2$ heterotic string compactified on a 4-torus as in section 4.3. We may switch off all the Wilson lines of the heterotic string compactification by rotating the space-like 4-plane, Π, so that $\Gamma_{4,20} \cap \Pi^\perp \cong \Gamma_{16}$. Now view this as the "fibre" of the type IIA string compactified on X versus the heterotic string compactified on $\mathrm{K3} \times T^2$. This restores the full $\mathfrak{so}(32)$ gauge symmetry and so can be thought of as shrinking down all 24 instantons. As discussed in section 5.5 we may now vary the T^2 and the Wilson lines by varying the 2-plane, $\mho$, in $\Upsilon \otimes_{\mathbb{Z}} \mathbb{R} \cong \mathbb{R}^{2,\rho}$, where Υ is the quantum Picard lattice of the generic fibre of X as a K3 fibration. Thus we see that $\Upsilon \cong \Gamma_{2,2} \oplus \Gamma_{16}$. In other words, *the Picard lattice of the generic K3 fibre of X is $\Gamma_{1,1} \oplus \Gamma_{16} \cong \Gamma_{1,17}$ and is therefore self-dual.*

Let the limit of a type IIA string compactified on X_0 be dual to the Spin(32)/$\mathbb{Z}_2$ heterotic string compactified on a K3 surface (times a 2-torus of large area) with $\tilde{w}_2 = 0$ and all the instantons shrunk down and let X be the blow-up of X_0. This theory will have an $\mathfrak{so}(32)$ gauge symmetry. We know the following:

1. X is a K3 fibration and an elliptic fibration with a section over a Hirzebruch surface.
2. X_0 contains a curve of singularities of type D_{16}.
3. The generic K3 fibre of X has a self-dual Picard lattice (of rank 18).

Let us construct X.

Table 4 tells is that the base, Θ, of X as an elliptic fibration will contain a curve of I_{12}^* fibres. We may put this curve along C_0, the isolated section of $\mathbf{F}_n$. A generic fibre of X as a K3 fibration will be a K3 surface built as an elliptic fibration with a I_{12}^* fibre (and, generically, 6 I_1 fibres). Let us denote this K3 surface as S_t. What is Pic(S_t)?

Let σ denote the section of S_t as an elliptic fibration "at infinity" guaranteed by the Weierstrass form. Let R be the sublattice of Pic(S_t) generated by the irreducible curves within the fibres not intersecting σ. Let Φ be the set of sections. One may show **[161]**

$$\operatorname{disc}(R) = |\Phi|^2 \operatorname{disc}(\operatorname{Pic}(S_t)), \tag{155}$$

where disc denotes the "discriminant" of a lattice, i.e., the determinant of the inner product on the generators.

In our case, R is generated by a set of rational curves forming the Dynkin diagram for D_{16}. Thus, disc(R) is the determinant of the Cartan matrix of D_{16} which is 4. In order for Pic(S_t) to be unimodular we see that we require exactly two sections. Writing an elliptic fibration with two sections in Weierstrass form is easy. We are guaranteed one section at infinity. Put the other section along $(y = 0, x = p(s,t))$. Thus the general Weierstrass form with two sections is

$$y^2 = (x - p(s,t))(x^2 + p(s,t)x + q(s,t)), \tag{156}$$

where

$$\begin{aligned} a(s,t) &= q(s,t) - p(s,t)^2 \\ b(s,t) &= -p(s,t)q(s,t). \end{aligned} \tag{157}$$

From (146) the divisors P and Q, given by the zeros of p and q, are in the class $4C_0 + (4+2n)f$ and $8C_0 + (8+4n)f$ respectively. The discriminant is

$$\begin{aligned} \delta &= 4a^3 + 27b^2 \\ &= (q + 2p^2)^2(4q - p^2). \end{aligned} \tag{158}$$

The fact that the discriminant factorizes will have some profound consequences. In terms of divisor classes let us write $\Delta = 2M_1 + M_2$, where M_1 is the divisor given by $q + 2p^2$ and M_2 corresponds to $4q - p^2$.

We know that $2M_1 + M_2$ contains $18C_0$ from the I_{12}^* fibres. Using the fact that a and b vanish to an order no greater than 2 and 3 respectively along C_0 and the fact that M_1 and M_2 must contain a nonnegative number of C_0 and f, one may show that M_1 contains $8C_0$ and M_2 contains $2C_0$ as their contributions towards the I_{12}^* fibres. What remains is

$$\begin{aligned} M_1' &= (8+4n)f \\ M_2' &= 6C_0 + (8+4n)f. \end{aligned} \tag{159}$$

Putting $M_2'.C_0 \geq 0$ fixes $n \leq 4$. There are no other constraints.

We have not yet achieved our goal in determining what n is for the $\mathrm{Spin}(32)/\mathbb{Z}_2$ heterotic string. All we know is that $n \leq 4$. Note however that along the $8+4n$ zeros of M_1 we have a double zero of Δ. Thus, generically we have $8+4n$ parallel lines along the f direction of I_2 fibres. Thus means the gauge algebra acquires an extra $\mathfrak{su}(2)^{8+4n}$ factor. The necessary appearance of an extra gauge symmetry can ultimately be traced to the way the determinant factorized. We will denote this as $\mathfrak{sp}(1)^{8+4n}$ to fit in with later analysis. It is very striking how different this behaviour is from the $E_8 \times E_8$ analysis of the last section. In the latter case we acquired 24 massless tensors when the instantons were shrunk down to zero size, whereas for the $\mathrm{Spin}(32)/\mathbb{Z}_2$ heterotic string we acquire new massless gauge fields.

The natural thing to do would be to identify each $\mathfrak{sp}(1)$ with a small instanton. To do this fixes $n=4$ as one can have 24 small instantons, at least in the case that the underlying K3 surface is smooth and $\tilde{w}_2 = 0$. Thus we arrive at the following

PROPOSITION 10. *The* $\mathrm{Spin}(32)/\mathbb{Z}_2$ *heterotic string compactified smoothly on a K3 surface with* $\tilde{w}_2 = 0$ *is dual to (a limit of) the type IIA string compactified on an elliptic fibration over* $\mathbf{F}_4$.

This proposition depends on the assumptions governing proposition 6 and the somewhat schematic way we associated each of the $\mathfrak{sp}(1)$ factors to a small instanton to show $n=4$.

Another way [**135**] to argue $n=4$ is that the generic point in moduli space of $n=4$ theories has gauge symmetry $\mathfrak{so}(8)$ and then compare to the analysis in section 6.4. We can try to break as much of $\mathfrak{so}(32)$ as possible by finding the largest subalgebra of $\mathfrak{so}(32)$ consistent with (153). Since $\tilde{w}_2 = 0$ we may consider vector representations in which case $\mathfrak{so}(24)$ is the largest such algebra with $c_2 = 24$ and this breaks $\mathfrak{so}(32)$ down to $\mathfrak{so}(8)$. A glance at table 5 then confirms that $n=4$.

Actually, the analysis of the $\mathrm{Spin}(32)/\mathbb{Z}_2$ heterotic string was first done by Witten [**152**], where the appearance of the $\mathfrak{sp}(1)$ factors from small instantons was argued directly. Unfortunately this involves methods beyond the scope of these talks. It is worth contrasting Witten's method with the above. It is a remarkable achievement of string duality that the same answer results.

We may consider a specialization of the moduli to bring together some of the components of M_1' along f. Suppose we bring k of the 24 lines together. This will produce an I_{2k} line of fibres. As usual we have to worry about monodromy to get the full gauge group. In this case the transverse collision between the I_{2k} line and the I_{12}^* line produces monodromy acting on the I_{2k} fibres to give an $\mathfrak{sp}(k)$ gauge symmetry.[32] This corresponds to k of the 24 point-like instantons *coalescing* [**152**].

[32] Actually, since the f curves are topologically spheres, there had better be more than one point within them around which there is nontrivial monodromy. There are three other points within each f curve where there is monodromy provided by non-transverse collisions of f with what is

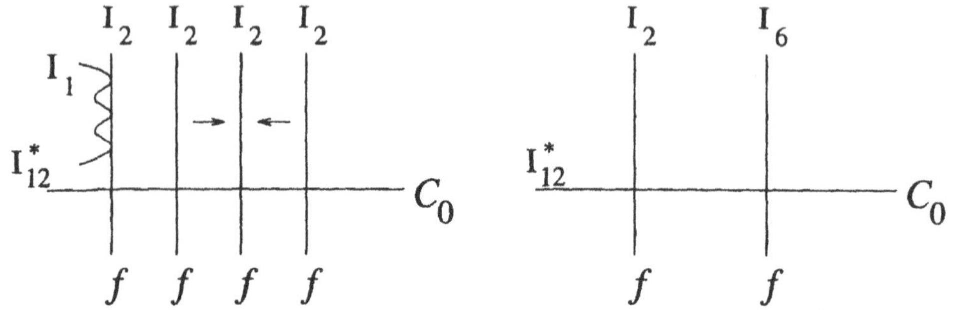

FIGURE 12. Three $\mathfrak{so}(32)$ point-like instantons coalescing to form $\mathfrak{sp}(3)$.

We show an example of this in figure 12. As an extreme case, all 24 instantons may be pushed to the same point. This results in a gauge algebra of $\mathfrak{so}(32) \oplus \mathfrak{sp}(24)$.

As we said above, it is interesting to contrast the behaviour of the point-like E_8 instantons with the point-like $\mathfrak{so}(32)$ instantons. The former produced massless tensor multiplets whereas the latter produce extra massless vector multiplets to enhance the gauge group. Of course, since both theories are meant to live in the same moduli space for $n = 4$, it should be possible to continuously deform one type of point-like instanton into another. This may well require a deformation of the base K3 surface as well as the bundle. The T-duality analysis of **[159]** should provide a good starting point for such a description.

6.6. Discussion of the heterotic string. We have analyzed both the $E_8 \times E_8$ and $\mathrm{Spin}(32)/\mathbb{Z}_2$ heterotic string compactified on a K3 surface in terms of a dual type IIA string compactified on an elliptically fibred Calabi–Yau threefold. In both cases we discovered curious nonperturbative behaviour associated to point-like instantons.

It is worth noting that in the case of the $\mathrm{Spin}(32)/\mathbb{Z}_2$ string we appear to have acquired a free lunch concerning the analysis of the gauge group. To see this note the following. In old perturbative string theory a type IIA string compactified on a Calabi–Yau manifold never has a nonabelian gauge group. Similarly a heterotic string may have a subgroup of the original $E_8 \times E_8$ or $\mathrm{Spin}(32)/\mathbb{Z}_2$ as its gauge group together with perhaps a little more from massless strings. This latter can only have rank up to a certain value limited by the central charge of the corresponding conformal field theory (see, for example, **[108]** for a discussion of this). In the case of point-like instantons we have claimed above to have found gauge algebras as large as $\mathfrak{so}(32) \oplus \mathfrak{sp}(24)$. Such a gauge group must be nonperturbative from both the type IIA *and* heterotic point of view.

When discussing duality one would normally claim that its power lies in its ability to relate nonperturbative aspects of one theory to perturbative aspects of another. This then allows the nonperturbative quantities to be calculated. In the above we appear to have discovered things about a subject that was nonperturbative in both theories at the same time! How did we do this? The answer is that we first

---

left of the discriminant. We show this on one of the f curves in figure 12. We omit this on the other f curves to make the figure more readable.

related perturbative nonabelian gauge groups in the heterotic string to curves of orbifold singularities in the type IIA picture. To do this the curves of singularities were always formed by each K3 fibre of X, as a K3 fibration, acquiring a singular point. There is no reason why a curve of singularities need be in this form — it may be completely contained in a bad fibre and not seen by the generic fibre at all. Why should the type IIA string intrinsically care about the fibration structure? Assuming it does not, this latter kind of curve should have just as much right to produce an enhanced gauge group as those we could understand perturbatively from the heterotic string. This is exactly the type of curve that gives the $\mathfrak{sp}(24)$ gauge symmetry which is now nonperturbative in terms of the heterotic string.

The example of heterotic-heterotic duality studied in section 5.6 gives a clear picture of how nonperturbative effects are viewed in this way. Begin with a perturbative gauge group in the heterotic string. This corresponds to a curve of singularities in the type IIA's K3 fibration passing through each generic fibre. Heterotic-heterotic duality corresponds to exchanging the two $\mathbb{P}^1$'s in the base of the Calabi–Yau viewed as an elliptic fibration. This gives another K3 fibration but now the curve of singularities lies totally within a bad fibre. Thus it has become a nonperturbative gauge group for the dual heterotic string.

An issue which is very interesting but we do not have time to discuss here concerns the appearance of extra massless hypermultiplets at special points in the moduli space. This question appears to be rather straight forward in the case of hypermultiplets in the adjoint representation, as shown in [**162**]. The more difficult question of other representations has been analyzed in [**96, 145, 149, 155, 163**]. It is essential to do this analysis to complete the picture of possible phase transitions in terms of Higgs transitions in the heterotic string. As usual, in discussions about duality, it also hints at previously unsuspected relationships in algebraic geometry.

Another very important issue we have not mentioned so far concerns anomalies. The heterotic string compactified on a K3 surface produces a chiral theory with potential anomalies. This puts constraints on the numbers of allowed massless supermultiplets (see [**164**] for a discussion of this). One constraint may be reduced to the condition [**165, 166, 167**]

$$273 - 29n_T - n_H + n_V = 0, \tag{160}$$

where n_T, n_H and n_V count the number of massless tensors, hypermultiplets, and vectors respectively. The reason we have been able to ignore this seemingly important constraint is that, assuming one does the geometry of elliptic fibrations correctly, it always appears to be obeyed. At present this appears to be another string miracle!

Ex As an example consider the following. Compactify the $E_8 \times E_8$ heterotic string, with c_2 split as $12+n$ and $12-n$ between the two E_8 factors, on a K3 surface so that the unbroken gauge symmetry is precisely E_8. This means that the corresponding elliptic threefold, X, has a curve of II* fibres which we may assume lies along C_0 in the Hirzebruch surface $\mathbf{F}_n$. We will calculate n_T, n_H, and n_V. What is left of the divisors A, B, and Δ after subtracting the contribution from the curve of II* fibres is given by

$$\begin{aligned} A' &= 4C_0 + (8+4n)f \\ B' &= 7C_0 + (12+6n)f \\ \Delta' &= 14C_0 + (24+12n)f. \end{aligned} \tag{161}$$

B' collides with C_0 generically $12-n$ times. Each of these points corresponds to a point-like instanton required to produce the E_8 gauge group. Each such point must be blown up within the base to produce a Calabi–Yau threefold. Thus we have $12-n$ massless tensor multiplets in addition to the one from the six-dimensional dilaton. Since the gauge group is E_8, we have 248 massless vectors furnishing the adjoint representation.

To count the massless hypermultiplets we require $h^{2,1}(X)$. It is relatively simple to compute $h^{1,1}(X)$ (assuming there is a finite number of sections) as this is given by 2, from the Hirzebruch surface, plus $12-n$ from the blow-ups within the base, plus 1 from the generic fibre, plus 8 from II* fibres. That is, $h^{1,1}(X) = 23-n$. Now $h^{2,1}$ may be determined from the Euler characteristic of X, $\chi(X) = 2(h^{1,1} - h^{2,1})$. To find this, recall that the Euler characteristic of a smooth bundle is given by the product of the Euler characteristic of the base multiplied by the Euler characteristic of the fibre. Thanks to the nice way Euler characteristics behave under surgery, we may thus apply this rule to each part of fibration separately. The Euler characteristic of any fibre is given by N in table 4.

Over most of the base, the fibre is a smooth elliptic curve which has Euler characteristic zero. Thus only the degenerate fibres contribute to our calculation. We have a curve of II* fibres over C_0 which contributes 10×2. The rest of the contribution comes from Δ'. Let us determine the geometry of Δ'. Firstly we know that, prior to blowing up the base, Δ' has $12-n$ double points as seen in the upper part of figure 10. Secondly, whenever A' and B' collide, which happens at $A'.B' = 24n+104$ points, Δ' will have a cusp (assuming everything is generic). Naïvely, the Euler characteristic of Δ' can be given by the adjun

$$-\Delta'.(\Delta'+K) = -596-130n. \tag{162}$$

However, each cusp will increase this by value by 2 and each double point by 1. In addition, when we do the blow-up of the base, the double points will be resolved and an additional 1 must be added for each double point. Thus

$$\begin{aligned}\chi(\Delta') &= -596-130n+2(24n+104)+2(12-n)\\ &= -364-84n.\end{aligned} \tag{163}$$

Over most of the points in Δ' the fibre will be type I_1 but there will be type II fibres over each cusp. Thus the Euler characteristic of X is given by

$$\begin{aligned}\chi(X) &= 10.2+1.(-364-84n-(24n+104))+2.(24n+104)\\ &= -240-60n,\end{aligned} \tag{164}$$

and so $h^{2,1}(X) = 143+29n$. To obtain the number of hypermultiplets we add one to this figure from the four-dimensional dilaton in the type IIA string. In summary, we have

$$\begin{aligned}n_T &= 13-n\\ n_H &= 144+29n\\ n_V &= 248.\end{aligned} \tag{165}$$

It can be seen that this satisfies (160). See [**164, 145**] for more examples.

In section 6.5 we analyzed the Spin(32)/$\mathbb{Z}_2$ heterotic string compactified on a K3 surface with $\tilde{w}_2 = 0$. What about the case $\tilde{w}_2 \neq 0$? An example of this was studied in [**159**] based on the construction of Gimon and Polchinski [**168**]. This uses open string models which we have not discussed here. In this case, $\tilde{w}_2$ is dual to $\frac{1}{2}\sum_{i=1}^{16} C_i$, where $\{C_i\}$ are sixteen disjoint (-2)-curves. The fact that such a class lies in $H^2(S,\mathbb{Z})$ can be seen in [**16**]. Such a $\tilde{w}_2$ is conjectured to be dual to the case of the $E_8\times E_8$ heterotic string with $n=0$. It would be interesting to analyze this from the elliptic fibration point if view presented here. One indication that things will work is that the algebra $\mathfrak{sp}(8)$ appears naturally in the nonperturbative group in the $n=0$ case from the analysis following (159). This agrees with the model of Gimon and Polchinski. The problem is that we have the full $\mathfrak{so}(32)$ present as a gauge symmetry which does not appear in Gimon and Polchinski's model. One can also show that massless tensor multiplets appear from the type IIA approach. These issues are resolved in [**156**].

An interesting issue we had no time to pursue is that concerning what happens as we take the coupling of the heterotic string to be strong. All of the above analysis was done for a weakly-coupled heterotic string. It has been noted that one can expect some kind of phase transition to occur as the coupling reaches a particular value [**133**]. This can be analyzed in the context of elliptic fibrations [**145**].

As concluding remarks to the discussion concerning the heterotic string on a K3 surface we should note that this analysis is far from complete. In the case of the type IIA and type IIB string we were able to give a fairly complete picture of the entire moduli space. For the heterotic string we have been able to study a few components of the moduli space and a few points at extremal transitions. A first requirement to study the full moduli space will be a classification of elliptic fibrations. A more immediate shortcoming in our discussion is that we do not yet have an explicit map between the moduli of the elliptically fibered Calabi–Yau manifold and the K3 surface and bundle on which the heterotic string is compactified. This must be understood before we can really answer questions such as what happens when point-like instantons collide with singularities of the K3 surface.

Acknowledgements

I would like to thank my collaborators M. Gross and D. Morrison for their contribution to much of the work discussed above. I would also like to thank them for answering a huge number of questions concerning algebraic geometry many of which I should have looked up in a book first. I thank J. Louis, with whom I have also collaborated on some of the work covered above. It is a pleasure to thank O. Aharony, S. Kachru, E. Silverstein, and H. Tye for useful conversations and of course B. Greene and K. T. Mahanthappa for organizing TASI 96. The author was supported by an NSF grant while at Cornell University, where part of these lecture notes were written, and is supported in part by DOE grant DE-FG02-96ER40959 at Rutgers University.

References

[1] R. W. H. T. Hudson, *Kummer's Quartic Surface*, Cambridge, 1905.
[2] S. W. Hawking and C. N. Pope, *Symmetry Breaking by Instantons in Supergravity*, Nucl. Phys. **B146** (1978) 381–392.
[3] S.-T. Yau, *Calabi's Conjecture and Some New Results in Algebraic Geometry*, Proc. Natl. Acad. Sci. **74** (1977) 1798–1799.
[4] M. J. Duff, B. E. W. Nilsson, and C. N. Pope, *Compactification of $d = 11$ Supergravity on $K3 \times T^3$*, Phys. Lett. **129B** (1983) 39–42.
[5] R. Messner and A. Gogna, *K2 Mountain of Mountains*, Oxford University Press, 1981.
[6] A. Weil, *Final Report on Contract AF 18(603)–57*, 1958.
[7] A. Weil, *Collected Works*, volume 2, Springer-Verlag, 1979.
[8] P. Griffiths and J. Harris, *Principles of Algebraic Geometry*, Wiley-Interscience, 1978.
[9] T. Eguchi, P. Gilkey, and A. Hansen, *Gravitation, Gauge Theories and Differential Geometry*, Phys. Rept. **66** (1980) 213–393.
[10] A. L. Besse, *Einstein Manifolds*, Springer-Verlag, Berlin, 1987.

[11] S. Salamon, *Riemannian Geometry and Holonomy Groups*, Pitman Research Notes in Math. **201**, Longman, 1989.

[12] R. Bott and L. W. Tu, *Differential Forms in Algebraic Topology*, Springer-Verlag, New York, 1982.

[13] E. H. Spanier, *Algebraic Topology*, Springer-Verlag, 1966.

[14] J. P. Serre, *A Course in Arithmetic*, Springer-Verlag, 1973.

[15] J. H. Conway and N. J. A. Sloane, *Sphere Packings, Lattices, and Groups*, Springer-Verlag, 1993.

[16] W. Barth, C. Peters, and A. van de Ven, *Compact Complex Surfaces*, Springer, 1984.

[17] C. Borcea, *Diffeomorphisms of a K3 surface*, Math. Ann. **275** (1986) 1–4.

[18] T. Matumoto, *On Diffeomorphisms of a K3 Surface*, in M. Nagata et al, editor, "Algebraic and Topological Theories — to the memory of Dr. Takehiko Miyaka", pages 616–621, Kinukuniya, Tokyo, 1985.

[19] S. K. Donaldson, *Polynomial Invariants for Smooth Four-Manifolds*, Topology **29** (1990) 257–315.

[20] G. Moore, *Finite in All Directions*, hep-th/9305139.

[21] N. Hitchin, *On Compact Four-Dimensional Einstien Manifolds*, J. Diff Geom. **9** (1974) 435–442.

[22] A. Todorov, *Applications of Kähler–Einstein–Calabi–Yau Metric to Moduli of K3 Surfaces*, Inv. Math. **61** (1980) 251–265.

[23] D. R. Morrison, *Some Remarks on the Moduli of K3 Surfaces*, in K. Ueno, editor, "Classification of Algebraic and Analytic Manifolds", Progress in Math. **39**, pages 303–332, Birkhäuser, 1983.

[24] R. Kobayashi and A. N. Todorov, *Polarized Period Map for Generalized K3 Surfaces and the Moduli of Einstein Metrics*, Tôhoku Math. J. (2) **39** (1987) 341–363.

[25] N. D. Allan, *Maximality of Some Arithmetic Groups; A Note of the Arithmetic of the Orthogonal Group*, An. Acad. Brasil. Ci. **38** (1966) 223–227 and 243–244.

[26] I. Satake, *On a Generalization of the Notion of Manifold*, Proc. Nat. Acad. Sci. USA **42** (1956) 359–363.

[27] L. Dixon, J. A. Harvey, C. Vafa, and E. Witten, *Strings on Orbifolds*, Nucl. Phys. **B261** (1985) 678–686, and **B274** (1986) 285–314.

[28] R. Kobayashi, *Einstein-Kähler V-Metrics on Open Satake V-Surfaces with Isolated Quotient Singularities*, Math. Ann. **272** (1985) 385–398.

[29] W. Fulton and J. Harris, *Representation Theory: A First Course*, Springer-Verlag, 1991.

[30] P. Du Val, *Homographies, Quaternions and Rotations*, Oxford, 1964.

[31] F. Klein, *Vorlesen über das Ikosaeder und die Auflösung der Gleichungen vom fünften Grade*, Teubner, Leipzig 1884.

[32] R. Miranda, *Smooth Models for Elliptic Threefolds*, in R. Friedman and D. R. Morrison, editors, "The Birational Geometry of Degenerations", Birkhauser, 1983.

[33] P. Slodowy, *Simple Singularities and Simple Algebraic Groups*, Lecture Notes in Mathematics **815**, Springer-Verlag, 1980.

[34] T. Eguchi and A. J. Hanson, *Asymptotically Flat Self-Dual Solutions to Euclidean Gravity*, Phys. Lett. **74B** (1978) 249–251.

[35] G. W. Gibbons and S. W. Hawking, *Gravitational Multi-Instantons*, Phys. Lett. **78B** (1978) 430–432.

[36] N. Hitchin, *Polygons and Gravitons*, Math. Proc. Cam. Phil. Soc. **85** (1979) 465–476.

[37] A. H. Durfee, *Fifteen Characterizations of Rational Double Points and Simple Critical Points*, l'Enseign. Math. **25** (1979) 131–163.

[38] A. Dimca, *Singularities and Topology of Hypersurfaces*, Springer-Verlag, 1992.

[39] M. Green, J. Schwarz, and E. Witten, *Superstring Theory*, Cambridge University Press, 1987, 2 volumes.

[40] P. S. Aspinwall, *The Moduli Space of $N = 2$ Superconformal Field Theories*, in E. Gava et al., editors, "1994 Summer School in High Energy Physics and Cosmology", pages 352–401, World Scientific, 1995, hep-th/9412115.

[41] P. Ginsparg, *Applied Conformal Field Theory*, in E. Brésin and J. Zinn-Justin, editors, "Fields, Strings, and Critical Phenomena", pages 1–168, Elsevier Science Publishers B.V., 1989.

[42] D. Friedan, *Nonlinear Models in $2+\varepsilon$ Dimensions*, Ann. Phys. **163** (1985) 318–419.

[43] C. G. Callan, D. Friedan, E. J. Martinec, and M. J. Perry, *Strings in Background Fields*, Nucl. Phys. **B262** (1985) 593–609.

[44] C. G. Callan, J. A. Harvey, and A. Strominger, *World Sheet Approach to Heterotic Instantons and Solitons*, Nucl. Phys. **B359** (1991) 611–634.

[45] J. A. Harvey and A. Strominger, *The Heterotic String is a Soliton*, Nucl. Phys. **B449** (1995) 535–552, hep-th/9504047.

[46] L. Alvarez-Gaumé and D. Z. Freedman, *Geometric Structure and Ultraviolet Finiteness in the Supersymmetric σ-Model*, Commun. Math. Phys. **80** (1981) 443–451.

[47] L. Alvarez-Gaumé and P. Ginsparg, *Finiteness of Ricci-flat Supersymmetric Nonlinear σ-Models*, Commun. Math. Phys. **102** (1985) 311–326.

[48] T. Banks and N. Seiberg, *Nonperturbative Infinities*, Nucl. Phys. **B273** (1986) 157–164.

[49] A. Belopolsky and B. Zwiebach, *Who Changes the String Coupling?*, Nucl. Phys. **B472** (1996) 109–138, hep-th/9511077.

[50] N. Seiberg, *Observations on the Moduli Space of Superconformal Field Theories*, Nucl. Phys. **B303** (1988) 286–304.

[51] S. Cecotti, *$N=2$ Landau-Ginzburg vs. Calabi–Yau Sigma Model: Nonperturbative Aspects*, Int. J. Mod. Phys. **A6** (1991) 1749–1814.

[52] P. S. Aspinwall and D. R. Morrison, *String Theory on K3 Surfaces*, in B. Greene and S.-T. Yau, editors, "Mirror Symmetry II", pages 703–716, International Press, 1996, hep-th/9404151.

[53] P. S. Aspinwall and D. R. Morrison, *Mirror Symmetry and the Moduli Space of K3 Surfaces*, in preparation.

[54] A. Borel, *Some Metric Properties of Arithmetic Quotients of Symmetric Spaces and an Extention Theorem*, J. Diff. Geom. **6** (1972) 543–560.

[55] S.-T. Yau, editor, *Essays on Mirror Manifolds*, International Press, 1992.

[56] E. Martinec, *Criticality, Catastrophes and Compactifications*, in L. Brink et al., editors, "Physics and Mathematics of Strings", World Scientific, 1990.

[57] I. V. Dolgachev, *Mirror symmetry for lattice polarized K3-surfaces*, J. Math. Sciences **81** (1996) 2599–2630, alg-geom/9502005.

[58] C. Borcea, *K3 Surfaces with Involution and Mirror Pairs of Calabi–Yau Manifolds*, in B. Greene and S.-T. Yau, editors, "Mirror Symmetry II", pages 717–743, International Press, 1996.

[59] H. Pinkham, *Singularités exceptionnelles, la dualité étrange d'Arnold et les surfaces K3*, C. R. Acad. Sci. Paris **284A** (1977) 615–618.

[60] B. R. Greene and M. R. Plesser, *Duality in Calabi–Yau Moduli Space*, Nucl. Phys. **B338** (1990) 15–37.

[61] K. S. Narain, *New Heterotic String Theories in Uncompactified Dimensions < 10*, Phys. Lett. **169B** (1986) 41–46.

[62] K. S. Narain, M. H. Samadi, and E. Witten, *A Note on the Toroidal Compactification of Heterotic String Theory*, Nucl. Phys. **B279** (1987) 369–379.

[63] A. Giveon, M. Porrati, and E. Rabinovici, *Target Space Duality in String Theory*, Phys. Rept. **244** (1994) 77–202, hep-th/9401139.

[64] E. Cremmer and B. Julia, *The $SO(8)$ Supergravity*, Nucl. Phys. **B159** (1979) 141–212.

[65] E. Witten, *Search For a Realistic Kaluza-Klein Theory*, Nucl. Phys. **B186** (1981) 412–428.

[66] P. Hořava and E. Witten, *Heterotic and Type I String Dynamics from Eleven Dimensions*, Nucl. Phys. **B460** (1996) 506–524, hep-th/9510209.

[67] J. Strathdee, *Extended Poincaré Supersymmetry*, Int. J. Mod. Phys. **A1** (1987) 273–300.

[68] B. Julia, *Group Disintegrations*, in S. Hawking and M. Roček, editors, "Superspace and Supergravity", pages 331–350, Cambridge University Press, 1981.

[69] J. Polchinski, *Dirichlet-Branes and Ramond-Ramond Charges*, Phys. Rev. Lett. **75** (1995) 4724–4727, hep-th/9510017.

[70] C. Hull and P. Townsend, *Unity of Superstring Dualities*, Nucl. Phys. **B438** (1995) 109–137, hep-th/9410167.

[71] E. Witten, *String Theory Dynamics in Various Dimensions*, Nucl. Phys. **B443** (1995) 85–126, hep-th/9503124.

[72] L. J. Dixon, V. Kaplunovsky, and C. Vafa, *On Four-Dimensional Gauge Theories from Type-II Superstrings*, Nucl. Phys. **B294** (1987) 43–82.

[73] P. S. Aspinwall, *Enhanced Gauge Symmetries and K3 Surfaces*, Phys. Lett. **B357** (1995) 329–334, hep-th/9507012.

[74] E. Witten, *Some Comments on String Dynamics*, hep-th/9507121.

[75] A. Strominger, *Massless Black Holes and Conifolds in String Theory*, Nucl. Phys. **B451** (1995) 96–108, hep-th/9504090.

[76] M. Bershadsky, V. Sadov, and C. Vafa, *D-Strings on D-Manifolds*, Nucl. Phys. **B463** (1996) 398–414, hep-th/9510225.

[77] M. R. Douglas and G. Moore, *D-branes, Quivers, and ALE Instantons*, hep-th/9603167.

[78] L. Romans, *Self-Duality for Interacting Fields: Covariant Field Equations for Six-Dimensional Chiral Supergravities*, Nucl. Phys. **B276** (1986) 71–92.

[79] P. S. Aspinwall, *Some Relationships Between Dualities in String Theory*, in "S-Duality and Mirror Symmetry", Nucl. Phys. (Proc. Suppl.) **B46**, pages 30–38, North Holland, 1996, hep-th/9508154.

[80] J. H. Schwarz, *An SL(2,Z) Multiplet of Type IIB Superstrings*, Phys. Lett. **B360** (1995) 13–18, hep-th/9508143.

[81] M. Duff, *Strong/Weak Coupling Duality from the Dual String*, Nucl. Phys. **B442** (1995) 47–63, hep-th/9501030.

[82] A. Sen, *Dyon-Monopole Bound States, Selfdual Harmonic Forms on the Multi-Monopole Moduli Space, and SL(2,Z) Invariance in String Theory*, Phys. Lett. **B329** (1994) 217–221, hep-th/9402032.

[83] C. Montonen and D. Olive, *Magnetic Monopoles as Gauge Particles?*, Phys. Lett. **72B** (1977) 117–120.

[84] H. Osborn, *Topological Charges for $N = 4$ Supersymmetric Gauge Theories and Monopoles of Spin 1*, Phys. Lett. **83B** (1979) 321–326.

[85] A. Sen, *Strong-Weak Coupling Duality in Four-Dimensional String Theory*, Int. J. Mod. Phys. **A9** (1994) 3707–3750, hep-th/9402002.

[86] J. Dai, R. Leigh, and J. Polchinski, *New Connections Between String Theories*, Mod. Phys. Lett. **A4** (1989) 2073–2083.

[87] M. Dine, P. Huet, and N. Seiberg, *Large and Small Radius in String Theory*, Nucl. Phys. **B322** (1989) 301–316.

[88] P. S. Aspinwall and D. R. Morrison, *U-Duality and Integral Structures*, Phys. Lett. **355B** (1995) 141–149, hep-th/9505025.

[89] M. J. Duff, J. T. Liu, and J. Rahmfeld, *Four-Dimensional String/String/String Triality*, Nucl. Phys. **B459** (1996) 125–159, hep-th/9508094.

[90] J. H. Schwarz and A. Sen, *Type IIA Dual of the Six-dimensional CHL Compactification*, Phys. Lett. **B357** (1995) 323–328, hep-th/9507027.

[91] G. Cheeger and M. Gromoll, *The Splitting Theorem for Manifolds of Non-Negative Ricci Curvature*, J. Diff. Geom. **6** (1971) 119–128.

[92] V. V. Nikulin, *Finite Automorphism Groups of Kähler K3 Surfaces*, Trans. Moscow Math. Soc. **38** (1979) 71–135.

[93] S. Chaudhuri and D. A. Lowe, *Type IIA-Heterotic Duals with Maximal Supersymmetry*, Nucl. Phys. **B459** (1996) 113–124, hep-th/9508144.

[94] K. S. Narain, M. H. Sarmadi, and C. Vafa, *Asymmetric Orbifolds*, Nucl. Phys. **B288** (1987) 551–577.

[95] D. Freed and C. Vafa, *Global Anomalies on Orbifolds*, Commun. Math. Phys. **110** (1987) 349–389.

[96] S. Ferrara, J. Harvey, A. Strominger, and C. Vafa, *Second Quantized Mirror Symmetry*, Phys. Lett. **361B** (1995) 59–65, hep-th/9505162.

[97] C. Vafa and E. Witten, *Dual String Pairs With $N = 1$ and $N = 2$ Supersymmetry in Four Dimensions*, in "S-Duality and Mirror Symmetry", Nucl. Phys. (Proc. Suppl.) **B46**, pages 225–247, North Holland, 1996, hep-th/9507050.

[98] S. Chaudhuri and D. A. Lowe, *Monstrous String-String Duality*, Nucl. Phys. **B469** (1996) 21–36, hep-th/9512226.

[99] S. Chaudhuri, G. Hockney, and J. D. Lykken, *Maximally Supersymmetric String Theories in $D < 10$*, Phys. Rev. Lett. **75** (1995) 2264–2267, hep-th/9505054.

[100] S. Chaudhuri and J. Polchinski, *Moduli Space of CHL Strings*, Phys. Rev. **D52** (1995) 7168–7173, hep-th/9506048.
[101] N. Seiberg and E. Witten, *Electric - Magnetic Duality, Monopole Condensation, and Confinement in N=2 Supersymmetric Yang-Mills Theory*, Nucl. Phys. **B426** (1994) 19–52, hep-th/9407087.
[102] N. Seiberg and E. Witten, *Monopoles, Duality and Chiral Symmetry Breaking in N=2 Supersymmetric QCD*, Nucl. Phys. **B431** (1994) 484–550, hep-th/9408099.
[103] S. Kachru and C. Vafa, *Exact Results For N=2 Compactifications of Heterotic Strings*, Nucl. Phys. **B450** (1995) 69–89, hep-th/9505105.
[104] E. Witten, *New Issues in Manifolds of SU(3) Holonomy*, Nucl. Phys. **B268** (1986) 79–112.
[105] J. Bagger and E. Witten, *Matter Couplings in N=2 Supergravity*, Nucl. Phys. **B222** (1983) 1–10.
[106] D. R. Morrison, *Mirror Symmetry and the Type II String*, in "S-Duality and Mirror Symmetry", Nucl. Phys. (Proc. Suppl.) **B46**, pages 146–153, North Holland, 1996, hep-th/9512016.
[107] S. Mukai, *Symplectic Structure of the Moduli Space of Sheaves on an Abelian or K3 Surface*, Invent. Math. **77** (1984) 101–116.
[108] P. S. Aspinwall and J. Louis, *On the Ubiquity of K3 Fibrations in String Duality*, Phys. Lett. **369B** (1996) 233–242, hep-th/9510234.
[109] B. de Wit, P. Lauwers, and A. Van Proeyen, *Lagrangians of N = 2 Supergravity-Matter Systems*, Nucl. Phys. **B255** (1985) 569–608.
[110] B. de Wit and A. Van Proeyen, *Potentials and Symmetries of General Gauged N=2 Supergravity–Yang-Mills Models*, Nucl. Phys. **B245** (1984) 89–117.
[111] S. Cecotti, S. Ferrara, and L. Girardello, *Geometry of Type II Superstrings and the Moduli of Superconformal Field Theories*, Int. J. Mod. Phys. **A4** (1989) 2475–2529.
[112] S. Ferrara and A. Van Proeyen, *A Theorem on N=2 Special Kähler Product Manifolds*, Class. Quant. Grav. **6** (1989) L243–L247.
[113] A. Strominger and E. Witten, *New Manifolds for Superstring Compactification*, Commun. Math. Phys. **101** (1985) 341–361.
[114] J. Distler and B. Greene, *Some Exact Results on the Superpotential from Calabi–Yau compactifications*, Nucl. Phys. **B309** (1988) 295–316.
[115] M. Dine, N. Seiberg, X. G. Wen, and E. Witten, *Nonperturbative Effects on the String World-Sheet*, Nucl. Phys. **B278** (1986) 769–789, and Nucl. Phys. **B289** (1987) 319–363.
[116] K. Oguiso, *On Algebraic Fiber Space Structures on a Calabi-Yau 3-fold*, Int. J. of Math. **4** (1993) 439–465.
[117] E. Witten, *The N-Matrix Model and Gauged WZW Models*, Nucl. Phys. **B371** (1992) 191–245.
[118] P. S. Aspinwall and D. R. Morrison, *Topological Field Theory and Rational Curves*, Commun. Math. Phys. **151** (1993) 245–262, hep-th/9110048.
[119] P. M. H. Wilson, *The Kähler Cone on Calabi–Yau Threefolds*, Invent. Math. **107** (1992) 561–583.
[120] M. Bershadsky, S. Cecotti, H. Ooguri, and C. Vafa, *Holomorphic Anomalies in Topological Field Theories*, Nucl. Phys. **B405** (1993) 279–304, hep-th/9302103.
[121] H. Ooguri and C. Vafa, *Two-Dimensional Black Hole and Singularities of CY Manifolds*, Nucl. Phys. **B463** (1996) 55–72, hep-th/9511164.
[122] A. Strominger, S.-T. Yau, and E. Zaslow, *Mirror Symmetry is T-Duality*, hep-th/9606040.
[123] M. Gross and P. M. H. Wilson, *Mirror Symmetry via 3-Tori for a Class of Calabi-Yau Threefolds*, alg-geom/9608004.
[124] D. Morrison, *The Geometry Underlying Mirror Symmetry*, alg-geom/9608006.
[125] A. Klemm, W. Lerche, and P. Mayr, *K3-Fibrations and Heterotic-Type II String Duality*, Phys. Lett. **357B** (1995) 313–322, hep-th/9506112.
[126] B. H. Lian and S.-T. Yau, *Mirror Maps, Modular Relations and Hypergeometric Series I,II*, in "S-Duality and Mirror Symmetry", Nucl. Phys. (Proc. Suppl.) **B46**, pages 248–262, North Holland, 1996, hep-th/9507151, hep-th/9507153.
[127] J. McCleary, *User's Guide to Spectral Sequence*, Publish or Perish, 1985.
[128] S. Kachru et al., *Nonperturbative Results on the Point Particle Limit of N=2 Heterotic String Compactifications*, Nucl. Phys. **B459** (1996) 537–558, hep-th/9508155.

[129] A. Klemm et al., *Self-Dual Strings and N=2 Supersymmetric Field Theory*, Nucl. Phys. **B477** (1996) 746–766, hep-th/9604034.

[130] V. Kaplunovsky, J. Louis, and S. Theisen, *Aspects of Duality in N=2 String Vacua*, Phys. Lett. **357B** (1995) 71–75, hep-th/9506110.

[131] I. Antoniadis, E. Gava, K. S. Narain, and T. R. Taylor, *N=2 Type II Heterotic Duality and Higher Derivative F Terms*, Nucl. Phys. **B455** (1995) 109–130, hep-th/9507115.

[132] J. A. Harvey and G. Moore, *Algebras, BPS States, and Strings*, Nucl. Phys. **B463** (1996) 315–368, hep-th/9510182.

[133] M. J. Duff, R. Minasian, and E. Witten, *Evidence for Heterotic/Heterotic Duality*, Nucl. Phys. **B465** (1996) 413–438, hep-th/9601036.

[134] P. S. Aspinwall and M. Gross, *Heterotic-Heterotic String Duality and Multiple K3 Fibrations*, Phys. Lett. **B382** (1996) 81–88, hep-th/9602118.

[135] D. R. Morrison and C. Vafa, *Compactifications of F-Theory on Calabi–Yau Threefolds — I*, Nucl. Phys. **B473** (1996) 74–92, hep-th/9602114.

[136] P. Green and T. Hübsch, *Possible Phase Transitions Among Calabi–Yau Compactifications*, Phys. Rev. Lett. **61** (1988) 1163–1166.

[137] B. R. Greene, D. R. Morrison, and A. Strominger, *Black Hole Condensation and the Unification of String Vacua*, Nucl. Phys. **B451** (1995) 109–120, hep-th/9504145.

[138] P. Candelas, P. Green, and T. Hübsch, *Rolling Among Calabi–Yau Vacua*, Nucl. Phys. **B330** (1990) 49–102.

[139] M. Reid, *The Moduli Space of 3-folds with* $K = 0$ *May Nevertheless be Irreducible*, Math Ann. **278** (1987) 329–334.

[140] T.-M. Chiang, B. R. Greene, M. Gross, and Y. Kanter, *Black Hole Condensation and the Web of Calabi-Yau Manifolds*, in "S-Duality and Mirror Symmetry", Nucl. Phys. (Proc. Suppl.) **B46**, pages 248–262, North Holland, 1996, hep-th/9511204.

[141] A. C. Avram, P. Candelas, D. Jančic, and M. Mandelberg, *On the Connectedness of Moduli Spaces of Calabi-Yau Manifolds*, Nucl. Phys. **B465** (1996) 458–472, hep-th/9511230.

[142] G. Aldazabal, A. Font, L. E. Ibáñez, and F. Quevedo, *Chains of N=2, D=4 Heterotic/Type II Duals*, Nucl. Phys. **B461** (1996) 85–100, hep-th/9510093.

[143] P. Candelas and A. Font, *Duality Between Webs of Heterotic and Type II Vacua*, hep-th/9603170.

[144] C. Vafa, *Evidence for F-Theory*, Nucl. Phys. **B469** (1996) 403–418, hep-th/9602022.

[145] D. R. Morrison and C. Vafa, *Compactifications of F-Theory on Calabi–Yau Threefolds — II*, Nucl. Phys. **B476** (1996) 437–469, hep-th/9603161.

[146] N. Seiberg and E. Witten, *Comments on String Dynamics in Six Dimensions*, Nucl. Phys. **B471** (1996) 121–134, hep-th/9603003.

[147] N. Marcus and J. H. Schwarz, *Field Theories That Have No Manifestly Lorentz-Invariant Formulation*, Phys. Lett. **115B** (1982) 111–114.

[148] N. Seiberg, *Non-trivial Fixed Points of The Renormalization Group in Six Dimensions*, hep-th/9609161.

[149] P. S. Aspinwall and M. Gross, *The SO(32) Heterotic String on a K3 Surface*, Phys. Lett. **B387** (1996) 735–742, hep-th/9605131.

[150] K. Kodaira, *On Compact Analytic Surfaces II*, Ann. Math. **77** (1963) 563–626.

[151] D. A. Cox and S. Zucker, *Intersection Numbers of Sections of Elliptic Surfaces*, Invent. Math. **53** (1979) 1–44.

[152] E. Witten, *Small Instantons in String Theory*, Nucl. Phys. **B460** (1996) 541–559, hep-th/9511030.

[153] E. Witten, *Topological Tools in 10-Dimensional Physics*, Int. J. Mod. Phys. **A1** (1986) 39–64.

[154] A. Sen, *F-theory and Orientifolds*, Nucl. Phys. **B475** (1996) 562–578, hep-th/9605150.

[155] M. Bershadsky et al., *Geometric Singularities and Enhanced Gauge Symmetries.*, Nucl. Phys. **B481** (1996) 215–252, hep-th/9605200.

[156] P. S. Aspinwall, *Point-like Instantons and the* $\mathrm{Spin}(32)/\mathbb{Z}_2$ *Heterotic String*, hep-th/9612108.

[157] R. Slansky, *Group Theory for Unified Model Building*, Phys. Rept. **79** (1981) 1–129.

[158] D. J. Gross, J. A. Harvey, E. Martinec, and R. Rohm, *The Heterotic String*, Phys. Rev. Lett. **54** (1985) 502–505.

[159] M. Berkooz et al., *Anomalies, Dualities, and Topology of $D = 6$ $N = 1$ Superstring Vacua*, Nucl. Phys. **B475** (1996) 115–148, hep-th/9605184.

[160] P. Ginsparg, *On Toroidal Compactification of Heterotic Strings*, Phys. Rev. **D35** (1987) 648–654.

[161] R. Miranda and U. Persson, *On Extremal Rational Elliptic Surfaces*, Math. Z. **193** (1986) 537–558.

[162] S. Katz, D. R. Morrison, and M. R. Plesser, *Enhanced Gauge Symmetry in Type II String Theory*, Nucl. Phys. **B477** (1996) 105–140, hep-th/9601108.

[163] S. Katz and C. Vafa, *Matter From Geometry*, hep-th/9606086.

[164] J. H. Schwarz, *Anomaly-Free Supersymmetric Models in Six Dimensions*, Phys. Lett. **B371** (1996) 223–230, hep-th/9512053.

[165] L. Alvarez-Gaumé and E. Witten, *Gravitational Anomalies*, Nucl. Phys. **B234** (1984) 269–330.

[166] M. B. Green, J. H. Schwarz, and P. C. West, *Anomaly Free Chiral Theories in Six-Dimensions*, Nucl. Phys. **B254** (1985) 327–348.

[167] A. Sagnotti, *A Note on the Green-Schwarz mechanism in open string theories*, Phys. Lett. **B294** (1992) 196–203, hep-th/9210127.

[168] E. G. Gimon and J. Polchinski, *Consistency Conditions for Orientifolds and D-Manifolds*, Phys. Rev. **D54** (1996) 1667–1676, hep-th/9601038.

Surveys in Differential Geometry, vol. 6

Symplectic Geometry and the Verlinde formulas

Jean-Michel Bismut and François Labourie

ABSTRACT. The purpose of this paper is to give a proof of the Verlinde formulas by applying the Riemann-Roch-Kawasaki theorem to the moduli space of flat G-bundles on a Riemann surface Σ with marked points, when G is a connected simply connected compact Lie group G. Conditions are given for the moduli space to be an orbifold, and the strata are described as moduli spaces for semisimple centralizers in G. The contribution of the strata are evaluated using the formulas of Witten for the symplectic volume, methods of symplectic geometry, including formulas of Witten-Jeffrey-Kirwan, and residue formulas. Our paper extends prior work by Szenes on SU(3) and Jeffrey-Kirwan for SU(n) to general groups G.

CONTENTS

Introduction 98
1. Simple Lie groups and their centralizers 103
2. Fourier analysis on the centralizers of semisimple Lie groups 126
3. Symplectic manifolds and moment maps 153
4. The affine space of connections 170
5. The moduli space of flat bundles on a Riemann surface 194
6. The Riemann-Roch-Kawasaki formula on the moduli space of flat bundles 248
7. Residues and the Verlinde formula 277
8. The Verlinde formulas 302
References 308

Supported by Institut Universitaire de France (I.U.F.).
Supported by Institut Universitaire de France (I.U.F.).

Introduction

The Verlinde formula **[62]**, **[3]** computes the dimension of spaces of holomorphic sections of canonical line bundles over the moduli space $\mathcal{M}$ of semistable $G_{\mathbb{C}}$-bundles on a Riemann surface Σ with marked points, with G a connected and simply connected compact Lie group. For a given "level" p, the Verlinde formula is a sum over the finite collection of weights parametrizing the representations of the central extension of the loop group LG at level p. This formula, discovered by Verlinde in the context of quantum field theory, has received a number of rigorous proofs first for $G = \mathrm{SU}(2)$ by Thaddeus **[57]**, Bertram and Szenes **[7]** and Szenes **[52]** (see also Donaldson **[17]** and Jeffrey-Weitsman **[31]** for related questions), and for more general groups by Tsuchiya-Ueno-Yamada **[59]**, Beauville-Laszlo **[4]** (for $G = \mathrm{SU}(n)$), Faltings **[20]**, Kumar, Narasimhan, Ramanathan **[38]** for general groups G. A common feature of many of these proofs is that establishing the fusion rules for the Verlinde numbers is an essential step in the proof, a second step being the description of the fusion algebra.

The purpose of this paper is to give a proof of the Verlinde formula for connected simply connected compact Lie groups, by methods of symplectic geometry. This program has been already carried out by Szenes **[53]** for SU(3), and Jeffrey-Kirwan **[30]** in the case of $\mathrm{SU}(n)$. The main point of the paper is to extend their approach a to general groups. More precisely, we will obtain the Verlinde formula by an application of Riemann-Roch.

The theorem of Narasimhan-Sheshadri **[46]** asserts that $\mathcal{M}$ can be identified with the set of representations of $\pi_1(\Sigma)$ with values in G, with given conjugacy classes of holonomies at the marked points. For generic choices of holonomies, this last space is a symplectic orbifold $(M/G, \omega)$, which carries an orbifold Hermitian line bundle with connection $(\lambda^p, \nabla^{\lambda^p})$, such that $c_1(\lambda^p, \nabla^{\lambda^p}) = p\omega$. In particular the orbifold M/G is complex. This orbifold carries a canonical Dirac operator $D_{p,+}$, unique up to homotopy. Its index $\mathrm{Ind}(D_{p,+})$ is the Euler characteristic of λ^p. We will compute the index $\mathrm{Ind}(D_{p,+})$ using the formula of Riemann-Roch-Kawasaki **[32, 33]**. We show that, for p large, it is given by the Verlinde formula. The fact that p has to be large may be related to the fact that a priori, higher cohomology may well not vanish for small p. In fact, the Verlinde formula computes $\dim H^0(\mathcal{M}, \lambda^p)$, while $\mathrm{Ind}(D_{p,+})$ is the corresponding Euler characteristic (we will come back to this point at the end of the introduction in connection with results by Teleman **[55, 56]**). For non generic holonomies, we also show that small perturbations of the holonomies still produce an orbifold moduli space. For suitable perturbations, we show that for any p, the index of the corresponding Dirac operator $D_{p,+}$ is still given by the Verlinde formula. The typical case where such a perturbation is needed is when Σ does not have marked points.

Our proof contains various interrelated steps.

- A first step is the description of the strata of the orbifold moduli space M/G. These strata are in fact moduli spaces for the semisimple centralizers in G. Up to conjugacy, there is only a finite family of such centralizers. In general, they are non simply connected. The strata split into a union of substrata indexed by the fundamental group of the centralizers. The description of the geometry of M/G involves results contained in Sections 1, 4, 5, 6. Observe that if $G = \mathrm{SU}(n)$, there are no non trivial semisimple centralizers, which explains the smoothness of the moduli space M/G (this case which was already considered by Jeffrey-Kirwan [**30**]).

- A second step consists in reproving Witten's formula [**63**] for the symplectic volume of the moduli spaces.
- Another step is the detailed construction of the orbifold line bundle λ^p. Also we have to compute the action of the finite stabilizers of elements of M on λ^p. This is done in Sections 4 and 5.
- In Section 5, we show that the formalism of the moment map can be applied to each stratum of the moduli space. In Section 6, we use a formula of Witten [**64**] , Jeffrey-Kirwan [**28**] (see Vergne [**61**]) and Liu [**39, 40**] to express the contribution of each stratum as the action of a differential operator on a locally polynomial function on a maximal torus T. This locally polynomial function is just the symplectic volume of a deformation of the moduli space M/G. Our treatment of these formulas is very close in spirit to Liu [**39, 40**].
- Witten's formula [**63**] for the deformed symplectic volume of each stratum is a Fourier series on T. In order to calculate the contribution of each stratum explicitly, it is of critical importance to express Witten's formula using residues techniques, which will make obvious the fact that the given Fourier series is indeed a local polynomial on T. These residue techniques are developed in Section 2.
- In Section 7, we give a residue formula for the index of the Dirac operator on M/G, by putting together the contribution of all strata.
- Another step is to express the Verlinde sums as residues. This step, which is carried out of Section 7, has many formal similarities with what is done in Section 2 for the Fourier series on T.
- In Section 8, a comparison of the results of Sections 6 and 7 leads us to our main result.

We now review our techniques in more detail.

1. The residue techniques

Residue techniques play an important role in the whole paper, in order to convert the Witten Fourier series [**63**] for the symplectic volumes into expressions which make them local polynomials in an "obvious" way, so that differential operators can be applied to these polynomials. Similar residue techniques are also applied to the Verlinde sums.

Szenes [**53, 54**] initiated the use of residue techniques to treat Verlinde formulas. In [**53**], Szenes applied such residue techniques to the case of SU(3) and obtained the corresponding Verlinde formulas. In [**54**], Szenes developed a cohomological approach in terms of arrangement of hyperplanes to treat the Witten sums for any group G.

In [**30**], Jeffrey and Kirwan gave a proof of the Verlinde formulas for $\mathrm{SU}(n)$ using residue techniques from a point of view which is very different from ours. In fact they use non abelian localization formulas [**64**], [**28**] applied to an extended moduli space. In the spirit of the formula of Duistermaat-Heckman [**19**], they use residue techniques to evaluate the Fourier transform of the contribution of the fixed points. In particular they reobtain Witten's formulas [**63**] for the symplectic volume of the moduli space of $G = \mathrm{SU}(n)$.

The strategy used in the present paper goes in some sense in an opposite direction. First, as in Liu [**39, 40**], our computations are local, and only use the action of differential operators on the symplectic volumes of the moduli spaces. Then, we express the symplectic volumes as residues in order to evaluate explicitly the action of certain differential operators on the symplectic volumes .

In the present paper, multidimensional residues are used in a rather "naive" way. The Witten sums are expressed as sums over a lattice identified to $\mathbf{Z}^r$. We compute the given sums by summing in succession in the variables $k_1, \ldots, k_r \in \mathbf{Z}$, and by applying standard residue techniques to these one dimensional sums. Handling the recursion requires the development of a trivial, but heavy linear algebra. We believe that Szenes's techniques [**54**] can put put to fruitful use to give a more conceptual approach to this part of our work.

2. A combinatorial description of the moduli spaces

Let $\mathcal{O}_1, \ldots, \mathcal{O}_s$ be s adjoint orbits in G. Put $X = G^{2g} \times \prod_{j=1}^s \mathcal{O}_j$. Let $\phi : X \to G$ be given by

$$\phi(u_1, v_1, \ldots, u_g, v_g, w_1, \ldots, w_s) = \prod_{i=1}^{g} [u_i, v_i] \prod_{j=1}^{s} w_j. \tag{0.1}$$

Put $M = \phi^{-1}(1)$. Under a genericity assumption on the $\mathcal{O}_j, 1 \leq j \leq s$, the condition (A) of Definition 5.17, which requires that $s \geq 1$, in Theorem 5.18, we show that M is a smooth manifold on which G acts locally freely. The moduli space is the orbifold M/G.

To prove Witten's formula [**63**], we show in Theorem 5.45 that the image by ϕ of the Haar measure on X has a density with respect to the Haar measure on G, which is essentially given by the symplectic volume of the quotient fibres of ϕ, which are themselves moduli spaces with an extra marked point. This approach was initiated by Liu [**39, 40**], who showed in particular that the differential of ϕ can be expressed in terms of the combinatorial complexes which compute the cohomology of the flat adjoint vector bundle E. Liu then obtains the intersection numbers of the moduli spaces by applying certain differential operators to the symplectic volumes. Inspired by Witten [**63, 64**], Liu gave a special role to the heat kernel on the group G to establish Witten's formula, while in our approach, we do not use any heat kernel. Needless to say, the heat kernel on G remains crucial in understanding connections with 2-dimensional Yang-Mills theory, and also with Witten's non Abelian localization [**64**].

3.Moment maps and the quantization conjecture

In Section 5.10, under genericity assumptions, we show that a G-invariant neighborhood $\widehat{X}$ of M in X can be equipped with a symplectic form, that G acts on $\widehat{X}$ with a moment map, and that the standard symplectic structure on the quotients on the fibres of ϕ included in $\widehat{X}$ come from the symplectic structure on $\widehat{X}$. We can then use directly the formulas of Witten [**64**] and Jeffrey-Kirwan [**28**] to express the integrals of certain characteristic classes in terms of differential operators acting on the symplectic volume of the fibres.

When the genericity assumptions are not verified, we replace the given moduli space by a generic perturbation, which still carries a Dirac operator to which the

Riemann-Roch-Kawasaki theorem [**32, 33**] can be applied. We then show that the above index results still hold.

We will now put our results in perspective from the point of view of geometric quantization, especially in connection with the Guillemin-Sternberg conjecture [**23**]. By Atiyah-Bott [**2**, Section 9], we know that when there is one marked point with central holonomy, M/G is a symplectic reduction of the affine space $\mathcal{A}$ of G-connections with respect to the action of the gauge group ΣG, which acts on $\mathcal{A}$ with a moment map μ, which is the curvature, so that $M/G = \mu^{-1}(0)/\Sigma G$. Similarly the line bundle λ^p is itself the reduction of a universal line bundle L^p on $\mathcal{A}$. This theory can be extended to the case with marked points (this we do in part in Sections 4 and 5). If $\mathcal{A}$ was instead a compact manifold, and ΣG a compact connected Lie group, the Guillemin-Sternberg conjecture [**23**] asserts that the Riemann-Roch number of $(M/G, \lambda^p)$ is equal to the multiplicity of the trivial representation in the action of ΣG on the cohomology of L^p. The Guillemin-Sternberg conjecture has been proved in various stages, the most general result being given by Meinrenken [**41**],[**42**] (for a more analytic proof, see Tian and Zhang [**58**]). In Meinrenken's formalism, when the considered group does not act locally freely on $\mu^{-1}(0)$, one replaces 0 by any regular value of μ close to 0 , and one still gets a corresponding version of Guillemin-Sternberg's conjecture.

Chang [**16**] initiated the study of Verlinde formulas in the context of geometric quantization. A new twist was introduced to the story of the proof of the Verlinde formula in work by Meinrenken and Woodward [**43**], [**44**], Alekseev, Malkin, Meinrenken [**1**], and later work by these authors. In [**43, 44**], Meinrenken and Woodward gave a symplectic proof of the fusion rules for the Verlinde numbers. In [**1**], the authors develop a theory of group actions with moment maps taking their values in the given Lie group G. This theory is in fact a theory of the standard moment map for an action of a central extension of the loop group LG. The space X is the prototype of such a manifold, the moment map being just ϕ. These authors then develop a localization formula in equivariant cohomology, which is an analogue of the formula of Duistermaat-Heckman [**19**], Berline-Vergne [**5**]. By using the theory of symplectic cuts and the previous results by Meinrenken [**41, 42**] on the Guillemin-Sternberg conjecture, they announce a proof of the Verlinde formulas.

In some way, our paper represents a direct attempt to prove Verlinde formulas directly, by a method which resembles the proof given by Jeffrey-Kirwan [**29**] of the Guillemin-Sternberg conjecture. The proof of [**29**] consists in extracting the Riemann-Roch number of $\mu^{-1}(0)/\Sigma G$ from the Lefschetz formulas.

In [**55, 56**], Teleman gave a proof of the vanishing of the higher cohomology groups of λ^p for a small perturbation of the moduli space M/G. As a consequence, one should always have $\mathrm{Ind}(D_{p,+}) = \dim H^0(\mathcal{M}, \lambda^p)$. As explained before, we only prove such an equality for large p, and otherwise, we have to perturb the moduli space by a perturbation which can be 'large' for small p, so that there could be a wall-crossing discrepancy between $\mathrm{Ind}(D_{p,+})$ and $\dim H^0(\mathcal{M})$. Inspection of the proof shows that this discrepancy also vanishes for g large enough, but Teleman's results indicate that the perturbation described above should never be needed to get the above equality.

Our paper is organized as follows. In Section 1, we establish basic simple facts on compact simply connected simple Lie groups and their semisimple centralizers. In Section 2, we develop our basic residue techniques in several variables. In particular, we express certain Fourier series on T, which are local polynomials, as residues. In Section 3, we reestablish well-known results on symplectic actions with moment maps, and we give a proof of the formula of Witten and Jeffrey-Kirwan. In Section 4, we construct the canonical line bundle L on the moduli space of G-connections on the Riemann surface Σ with fixed holonomy at the given marked points. We show that a suitable central extension of the gauge group ΣG acts on L, and we compute the action of certain stabilizers on L and on a related line bundle λ_p. In Section 5, we describe the moduli space M/G. We show that the formalism of the moment map can be applied to the action of G on M. We apply the formula of Witten **[64]** and Jeffrey-Kirwan **[28]** to the moduli spaces associated to semisimple centralizers. Also we give a formula for $c_1(TM/G)$. In Section 6, we apply the theorem of Riemann-Roch-Kawasaki to the orbifold M/G, and we give a residue formula for the index $\mathrm{Ind}(D_{p,+})$. In Section 7, we give a residue formula for the Verlinde sums. Finally in Section 8, we compare $\mathrm{Ind}(D_{p,+})$ and the Verlinde formula, and give a number of conditions under which they coincide.

The results contained in this paper were announced in **[12]**.

1. Simple Lie groups and their centralizers

Let G be a connected and simply connected compact simple Lie group. The purpose of this Section is to give the basic facts which will be needed in the description of the strata of the moduli space of flat G-bundles on a Riemann surface Σ. This involves in particular a complete description of the semisimple centralizers in G.

This Section is organized as follows. In Section 1.1, we recall elementary properties of roots and coroots. In Section 1.2, we construct the basic scalar product on the Lie algebra $\mathfrak{g}$ of G. In Section 1.3, we compute the volume of a maximal torus T. In Section 1.4, we relate the quadratic form attached to a representation to the basic quadratic form. In Section 1.5, we introduce the dual Coxeter number. In Section 1.6, we construct an embedding of the center $Z(G)$ in the Weyl group W. In Section 1.7, we give simple properties of the element $\rho/c \in T$, in particular in its relations with the center $Z(G)$. In Section 1.8, we review elementary properties of the characters of G. In Sections 1.9-1.12, we describe the semisimple centralizers, and give some of their properties. In Section 1.13, we consider the intersection of an adjoint orbit with such a centralizer. Finally in Section 1.14, we recall various properties of the stabilizers of elements of $\mathfrak{g}$, and we construct the symplectic structure on the coadjoint orbits, and the corresponding line bundles.

1.1. Roots and coroots. Let G be a connected simply connected simple compact Lie group of rank r. Let $\mathfrak{g}$ be its Lie algebra, let $\mathfrak{g}^*$ be its dual. Let T be a maximal torus in G, let $\mathfrak{t}$ be its Lie algebra, let $\mathfrak{t}^*$ be its dual. Let W be the corresponding Weyl group.

We will denote the composition law multiplicatively in G, but often additively in T.

Let $\Gamma \subset \mathfrak{t}$ be the lattice of integral elements in $\mathfrak{t}$, i.e.

$$\Gamma = \{t \in \mathfrak{t}, \exp(t) = 1 \text{ in } T\}. \tag{1.1}$$

Let $\Lambda = \Gamma^* \subset \mathfrak{t}^*$ be the lattice of weights in $\mathfrak{t}^*$, so that if $h \in \Gamma$, $\lambda \in \Lambda$,

$$\langle \lambda, h \rangle \in \mathbf{Z}. \tag{1.2}$$

Let $R = \{\alpha\} \subset \Lambda$ be the root system of G. Then R is a finite family of elements of Λ, which span $\mathfrak{t}^*$. Let $\overline{R} \subset \Lambda$ be the lattice generated by R, let $\overline{R}^* \supset \Gamma$ be the lattice dual to $\overline{R}$.

Let $CR = \{h_\alpha\} \subset \mathfrak{t}$ be the family of coroots attached to R. Let $\overline{CR} \subset \mathfrak{t}$ be the lattice generated by CR, let $\overline{CR}^* \subset \mathfrak{t}^*$ be the corresponding dual lattice. Since G is simply connected, by [**15**, Theorem V.7.1],

$$\begin{aligned} \Gamma &= \overline{CR}, \\ \Lambda &= \overline{CR}^*. \end{aligned} \tag{1.3}$$

Let $Z(G)$ be the center of G. By [**15**, Proposition V.7.16],

$$\overline{R}^* / \overline{CR} = Z(G). \tag{1.4}$$

Let R_s, R_ℓ and CR_s, CR_ℓ be the short, long roots and coroots. Note that R_s corresponds to CR_ℓ and R_ℓ to CR_s. Recall that G is said to be simply laced if all the roots (or coroots) have the same length. In this case, all the roots will be considered as long, and the coroots as short, so that

$$R_s = \emptyset, \qquad CR_\ell = \emptyset. \tag{1.5}$$

In the sequel, when G is simply laced, any statement concerning R_s or CR_ℓ should be disregarded.

Let $\overline{R_\ell}, \overline{R_s}, \overline{CR_\ell}, \overline{CR_s}$ be the lattices generated by R_ℓ, R_s, CR_ℓ, CR_s. It follows from the classification of Lie groups that

$$\overline{CR_s} = \overline{CR}, \qquad \overline{R_s} = \overline{R}. \tag{1.6}$$

Note that when G is simply laced, the second equality in (1.6) is empty.

1.2. The basic scalar product on the Lie algebra.

DEFINITION 1.1. Let $\langle\ ,\ \rangle$ be the G-invariant scalar product on $\mathfrak{g}$ such that if $\|.\|$ is the corresponding norm, if $h_\alpha \in CR_s$,

$$\|h_\alpha\|^2 = 2. \tag{1.7}$$

If G is not simply laced, there is one $m \in \mathbf{N}$ (equal to 2 or 3) such that if $h_\alpha \in CR_\ell$,

$$\|h_\alpha\|^2 = 2m. \tag{1.8}$$

By [**15**], if $f, f' \in \overline{CR}$,

$$\langle f, f'\rangle \in \mathbf{Z}. \tag{1.9}$$

Using $\langle\ ,\ \rangle$, we may and we will identify $\mathfrak{t}$ and $\mathfrak{t}^*$. By [**15**, V, eq. (2.14)], under this identification, if $\alpha \in R$, if $h_\alpha \in CR$ corresponds to α, then

$$\alpha = \frac{2}{\|h_\alpha\|^2} h_\alpha. \tag{1.10}$$

By (1.10) , we find that

$$\overline{CR} \subset \overline{R} \cap \overline{R}^*. \tag{1.11}$$

Also, by (1.6), (1.8), (1.10),

$$\overline{CR} = \overline{R_\ell}, \qquad \overline{CR_\ell} = m\overline{R}. \tag{1.12}$$

By (1.12),

$$m\overline{R} \subset \overline{R_\ell} \subset \overline{R}, \qquad m\overline{CR} \subset \overline{CR_\ell} \subset \overline{CR}. \tag{1.13}$$

From (1.12),(1.13),

$$\begin{aligned} m\overline{CR} &= m\overline{R_\ell} \subset \overline{CR_\ell} = m\overline{R} \subset \overline{CR} \\ &= \overline{R_\ell} \subset \frac{\overline{CR_\ell}}{m} = \overline{R} \subset \frac{\overline{CR}}{m} = \frac{\overline{R_\ell}}{m}. \end{aligned} \tag{1.14}$$

Also

$$\Gamma = \overline{CR} \subset \overline{R} \cap \overline{R}^* \subset \overline{CR}^* = \Lambda. \tag{1.15}$$

Observe that the Weyl group W preserves all the objects which have been constructed above. Also note that

$$T = \mathfrak{t}/\overline{CR}. \tag{1.16}$$

Moreover $T' = \mathfrak{t}/\overline{R}^*$ is a maximal torus in the adjoint group $G' = G/Z(G)$.

1.3. The volume of T. Let $\mathrm{Vol}(T)$ be the volume of T for the metric $\langle , \rangle$.

PROPOSITION 1.2. *The following identity holds*

$$\mathrm{Vol}(T)^2 = |\overline{CR}^*/\overline{CR}|. \tag{1.17}$$

PROOF. Consider the exact sequence of lattices

$$0 \to \overline{CR} \to \overline{CR}^* \to \overline{CR}^*/\overline{CR} \to 0, \tag{1.18}$$

which induces the exact sequence

$$0 \to \overline{CR}^*/\overline{CR} \to \mathfrak{t}/\overline{CR} \to \mathfrak{t}/\overline{CR}^* \to 0. \tag{1.19}$$

From (1.19), we obtain

$$\mathrm{Vol}(\mathfrak{t}/\overline{CR}) = |\overline{CR}^*/\overline{CR}| \mathrm{Vol}(\mathfrak{t}/\overline{CR}^*). \tag{1.20}$$

Now $\mathfrak{t}/\overline{CR}$ and $\mathfrak{t}^*/\overline{CR}^* \simeq \mathfrak{t}/\overline{CR}^*$ are dual tori. Therefore

$$\mathrm{Vol}(\mathfrak{t}/\overline{CR})\mathrm{Vol}(\mathfrak{t}/\overline{CR}^*) = 1. \tag{1.21}$$

By (1.20), (1.21), we obtain (1.17). The proof of our Proposition is completed. □

PROPOSITION 1.3. *The following identity holds*

$$\mathrm{Vol}(T)^2 = |Z(G)|\ |\overline{R}/\overline{R}_\ell|. \tag{1.22}$$

In particular, if G is simply laced,

$$\mathrm{Vol}(T)^2 = |Z(G)|. \tag{1.23}$$

PROOF. Clearly

$$|\overline{CR}^*/\overline{CR}| = |\overline{CR}^*/\overline{R}|\ |\overline{R}/\overline{CR}|. \tag{1.24}$$

Also by (1.4),

$$(\overline{CR}^*/\overline{R})^* = \overline{R}^*/\overline{CR} = Z(G). \tag{1.25}$$

From (1.12), (1.17), (1.24), (1.25), we get (1.22) . The identity (1.23) follows. □

1.4. The quadratic form attached to a representation. Assume temporarily that G is a compact connected semisimple Lie group, which is not necessarily simply connected. Otherwise, we use the notation of Sections 1.1-1.3.

Let $\sigma : G \to \mathrm{Aut}(V)$ be a finite dimensional representation of G.

DEFINITION 1.4. If $A, B \in \mathfrak{g}$, put

$$\langle A, B\rangle^\sigma = \frac{1}{4\pi^2}\mathrm{Tr}^V[\sigma(A)\sigma(B)]. \tag{1.26}$$

Then $\langle\ ,\ \rangle$ is an ad-invariant symmetric bilinear form on $\mathfrak{g}$.

Let $x \in \mathbf{R} \mapsto [x] \in [0,1[$ be the periodic function of period 1 such that for $x \in [0,1[$, $[x] = x$.

PROPOSITION 1.5. *If $u \in \overline{R}^*$, $h \in \Gamma$, then*

$$\langle u, h\rangle^\sigma \in \mathbf{Z}. \tag{1.27}$$

PROOF. Let $M \subset \Lambda$ be the set of T-weights in the representation σ. Then

$$\langle u, h\rangle^\sigma = -\sum_{\lambda \in M} \langle \lambda, u\rangle\langle \lambda, h\rangle. \tag{1.28}$$

Also for $\lambda \in M, \langle \lambda, h\rangle \in \mathbf{Z}$. Therefore, mod($\mathbf{Z}$),

$$\begin{aligned}\langle u, h\rangle^\sigma &= -\sum_{\lambda \in M} [\langle \lambda, u\rangle]\langle \lambda, h\rangle \\ &= -\sum_{s\in[0,1[} \langle \sum_{\substack{\lambda \in M \\ [\langle\lambda,u\rangle]=s}} \lambda, h\rangle.\end{aligned} \tag{1.29}$$

Also the image of u in $T = \mathfrak{t}/\Gamma$ lies in $Z(G)$.Therefore the representation σ splits into a sum of representations on which u acts like $e^{2i\pi s}, 0 \leq s < 1$. The corresponding T-weights are given by $\{\lambda \in M, [\langle\lambda, u\rangle] = s\}$. Since G is semisimple

$$\sum_{\substack{\lambda \in M \\ [\langle\lambda,u\rangle]=s}} \lambda = 0 \tag{1.30}$$

From (1.29), (1.30), we get (1.27) . The proof of our Proposition is completed. □

1.5. The dual Coxeter number. Again we assume that G is a connected and simply connected simple compact Lie group. Also we use the assumptions and notation of Sections 1.1-1.3.

Let $K \subset \mathfrak{t}$ be a Weyl chamber. Let R_+ be the corresponding system of positive roots, so that

$$R = R_+ \cup R_-. \tag{1.31}$$

Then

$$K = \{t \in \mathfrak{t}, \text{ for any } \alpha \in R_+, \langle \alpha, t\rangle > 0\}. \tag{1.32}$$

Let $P \subset \mathfrak{t}$ be the alcove in K whose closure contains 0. Then

$$P = \{t \in \mathfrak{t}, \text{ for any } \alpha \in R_+,\ 0 < \langle \alpha, t\rangle < 1\}. \tag{1.33}$$

Since G is simple, the adjoint representation of G' on $\mathfrak{g}$ is irreducible. The corresponding T-weights are given by $\{0\} \cup R$. Let $\alpha_0 \in R_+ \cap R_\ell$ be the corresponding highest root. Then

$$P = \{t \in K, \langle \alpha_0, t\rangle < 1\}. \tag{1.34}$$

DEFINITION 1.6. Let $\rho \in \mathfrak{t}^*$ be given by

$$\rho = \frac{1}{2}\sum_{\alpha \in R_+} \alpha. \tag{1.35}$$

By [15, Proposition V.4.12], if $\alpha \in R_+$ is a simple root, $\langle \rho, h_\alpha\rangle = 1$. In particular $\rho \in \overline{CR}^*$, and $\rho \in K$.

DEFINITION 1.7. Let $c \in \mathbf{N}$ be the dual Coxeter number, given by

$$c = \langle \rho, h_{\alpha_0}\rangle + 1. \tag{1.36}$$

Let $\tau : G' \to \mathrm{Aut}(\mathfrak{g})$ be the adjoint representation. Then by [**47**, p 285], [**22**, eq. (1.6.45)],

$$\langle,\rangle^{\tau} = -2c\langle,\rangle. \tag{1.37}$$

PROPOSITION 1.8. *If* $t \in \mathfrak{t}$,

$$ct = \sum_{\alpha \in R_+} \langle \alpha, t\rangle \alpha. \tag{1.38}$$

In particular

$$c\overline{R}^* \subset \overline{R}. \tag{1.39}$$

PROOF. Since $\{0\} \cup R$ are the T-weights of τ, if $t, t' \in \mathfrak{t}$,

$$\langle t, t'\rangle^{\tau} = -2 \sum_{\alpha \in R_+} \langle \alpha, t\rangle\langle \alpha, t'\rangle. \tag{1.40}$$

Using (1.37), (1.40), we get (1.38). From (1.38), (1.39) follows. □

Recall that we have identified $\mathfrak{t}$ and $\mathfrak{t}^*$ by $\langle,\rangle$. Then $\alpha_0 = h_{\alpha_0}$. In particular, for any $\alpha \in R_+$,

$$0 < \langle \alpha, \rho/c\rangle < 1, \tag{1.41}$$

i.e. $\rho/c \in P$.

DEFINITION 1.9. For $t \in \mathfrak{t}$, put

$$\sigma(t) = \prod_{\alpha \in R_+} (e^{i\pi\langle \alpha, t\rangle} - e^{-i\pi\langle \alpha, t\rangle}). \tag{1.42}$$

Equivalently,

$$\sigma(t) = e^{2i\pi\langle \rho, t\rangle} \prod_{\alpha \in R_+} (1 - e^{-2i\pi\langle \alpha, t\rangle}). \tag{1.43}$$

By (1.43), we find that $\sigma(t)$ descends to a function defined on $T = \mathfrak{t}/\overline{CR}$.
Recall that W is the Weyl group of G . If $w \in W$, set

$$\epsilon_w = \det(w|_{\mathfrak{t}}). \tag{1.44}$$

Then by [**15**, Theorem VI.1.7], if $w \in W, t \in \mathfrak{t}$,

$$\sigma(wt) = \epsilon_w \sigma(t). \tag{1.45}$$

Put

$$\ell = |R_+|\,. \tag{1.46}$$

Now we recall a result stated in [**3**, Lemma 9.17].

PROPOSITION 1.10. *The following identity holds*

$$(i^{\ell}\sigma(e^{\rho/c}))^2 = |\overline{CR}^*/c\overline{CR}|. \tag{1.47}$$

1.6. An embedding of the center in the Weyl group. By [**15**, Theorem V.4.1], $wK \cap K \neq \emptyset$ if and only if $w = 1$.

Take $q \in \mathbf{Z}^*$. Then $\bigcup_{w \in W} w(qP \cap \overline{CR}^*)$ is a disjoint union of finite sets.

PROPOSITION 1.11. *The set* $\bigcup_{w \in W} w(qP \cap \overline{CR}^*)$ *embeds naturally as a subset of* $\overline{CR}^*/q\overline{CR}$. *More precisely,*

$$\bigcup_{w \in W} w(qP \cap \overline{CR}^*) = \{\lambda \in \overline{CR}^*/q\overline{CR}, \sigma^2(\lambda/q) \neq 0\}. \tag{1.48}$$

Also

$$cP \cap \overline{CR}^* = \{\rho\}. \tag{1.49}$$

PROOF. Let $W_{\text{aff}} = W \ltimes \overline{CR}$ be the affine Weyl group. By [**15**, Proposition V.7.10], W_{aff} acts freely and effectively on the set of alcoves in $\mathfrak{t}$. Thus we get (1.48). If $\lambda \in cP \cap \overline{CR}^*$, then $\lambda \in \overline{CR}^* \cap K$. By [**15**, Note V.4.14],

$$\lambda - \rho \in \overline{CR}^*_+ = \overline{CR}^* \cap \overline{K}. \tag{1.50}$$

Also, by (1.36),

$$\langle \rho, h_{\alpha_0} \rangle = c - 1. \tag{1.51}$$

Since $\alpha_0 = h_{\alpha_0}$, by (1.34), since $\lambda \in cP$,

$$\langle \lambda, h_{\alpha_0} \rangle < c. \tag{1.52}$$

By (1.50)-(1.52), we obtain

$$0 \leq \langle \lambda - \rho, h_{\alpha_0} \rangle < 1. \tag{1.53}$$

Since $\langle \lambda - \rho, h_{\alpha_0} \rangle \in \mathbf{N}$, from (1.53), we obtain

$$\lambda = \rho. \tag{1.54}$$

The proof of our Proposition is completed. □

PROPOSITION 1.12. *Let f be a* W*-invariant function on* $\overline{CR}^*/q\overline{CR}$. *Then*

$$\sum_{\substack{\lambda \in \overline{CR}^*/q\overline{CR} \\ \sigma^2(\lambda/q) \neq 0}} f(\lambda) = |W| \sum_{\lambda \in qP \cap \overline{CR}^*} f(\lambda) \tag{1.55}$$

PROOF. This is a trivial consequence of Proposition 1.11. □

PROPOSITION 1.13. *The set* $qP \cap \overline{CR}^*$ *embeds naturally into* $\{\lambda \in \overline{CR}^*/q\overline{R}^*, \sigma^2(\lambda/q) \neq 0\}$.

PROOF. By (1.33), (1.42), if $\lambda \in qP$, then $\sigma^2(\lambda/q) \neq 0$. Let $\lambda, \lambda' \in qP \cap \overline{CR}^*$, and assume there is $\mu \in \overline{R}^*$ such that

$$\lambda - \lambda' = q\mu. \tag{1.56}$$

By (1.33), for $\alpha \in R_+$,

$$-q < \langle \alpha, \lambda - \lambda' \rangle < q, \tag{1.57}$$

and so

$$-1 < \langle \alpha, \mu \rangle < 1. \tag{1.58}$$

Since $\mu \in \overline{R}^*$, for $\alpha \in R_+, \langle \alpha, \mu \rangle \in \mathbf{Z}$. By (1.58), we get

$$\mu = 0. \tag{1.59}$$

The proof of our Proposition is completed. □

PROPOSITION 1.14. *If* $\lambda \in \overline{CR}^*$, $\sigma^2(\lambda/q) \neq 0$, *there is* $w \in W$, $\lambda' \in qP \cap \overline{CR}^*$ *such that* $w\lambda - \lambda' \in q\overline{R}^*$.

PROOF. Take $\lambda \in \overline{CR}^*$. Then by [**15**, Proposition V.7.10], there exists $w \in W$, $h \in \overline{P}$, $f \in \overline{CR}$, such that

$$\frac{\lambda}{q} = wh + f\,. \tag{1.60}$$

Clearly $qf \in q\overline{CR}$. Also by (1.45),

$$\sigma^2(\lambda/q) = \sigma^2(h)\,. \tag{1.61}$$

So if $\sigma^2(\lambda/q) \neq 0$, then $\sigma^2(h) \neq 0$, so that $h \in P$. The proof of our Proposition is completed. □

Recall that $Z(G) = \overline{R}^*/\overline{CR}$. Also W acts trivially on $Z(G) \subset G$. Equivalently if $f \in \overline{R}^*, w \in W$

$$wf - f \in \overline{CR}\,. \tag{1.62}$$

If $g \in G$, let $Z(g) \subset G$ be the centralizer of g, let $\mathfrak{z}(g)$ be its Lie algebra. By [**14**, Corollaire 5.3.1] , since G is simply connected, $Z(g)$ is a connected Lie subgroup of G.

An element $t \in T$ is said to be regular (resp. very regular) if $\mathfrak{z}(t) = \mathfrak{t}$ (resp. $Z(t) = T$). By the above, $t \in T$ is regular if and only if t is very regular. Let $T_{\mathrm{reg}} \subset T$ be the set of regular elements in T. By [**15**, Proposition V.7.10] , P embeds into T_{reg}. More precisely,

$$T_{\mathrm{reg}} = \bigcup_{w \in W} wP \tag{1.63}$$

and the union in (1.63) is disjoint.

Let $u \in Z(G)$. Then $u + P \subset T$ is another alcove in T_{reg}. Therefore there is a well-defined $w_u \in W$ such that

$$u + P = w_u P\,. \tag{1.64}$$

PROPOSITION 1.15. *The map* $u \in Z(G) \mapsto w_u \in W$ *embeds* $Z(G)$ *as a commutative subgroup of* W. *In particular* $|Z(G)|$ *divides* $|W|$.

PROOF. If $u \in Z(G)$, $w_u = 1$, then $u + P = P$ in T. Therefore there is $v \in \overline{R}^*$ mapping into $u \in \overline{R}^*/\overline{CR}$ such that $v + P = P$ in T. By proceeding as in (1.57), (1.58), we find that $v = 0$, i.e. $u = 1$.

If $u, u' \in Z(G)$, then

$$u + u' + P = u' + w_u P = w_u(u' + P) = w_u w_{u'} P \text{ in } T\,. \tag{1.65}$$

From (1.65) , we get

$$w_{u+u'} = w_u w_{u'}\,. \tag{1.66}$$

The proof of our Theorem is completed. □

By Proposition 1.13 , we can view $qP \cap \overline{CR}^*$ as a subset of $\overline{CR}^*/q\overline{R}^*$, which itself is stable by W. In particular if $\lambda \in qP \cap \overline{CR}^*$, $w \in W$, $w\lambda$ will be viewed as an element of $\overline{CR}^*/q\overline{R}^*$. So the equality $w\lambda = \lambda$ says here that $w\lambda - \lambda \in q\overline{R}^*$.

PROPOSITION 1.16. *If $\lambda \in qP \cap \overline{CR}^* \subset \overline{CR}^*/q\overline{R}^*$, $w \in W$, then $w\lambda \in qP \cap \overline{CR}^* \subset \overline{CR}^*/q\overline{R}^*$ if and only if there is $u \in Z(G)$ such that $w = w_u$.*

PROOF. Take $\lambda \in qP \cap \overline{CR}^*$, $u \in \overline{R}^*$ representing an element of $Z(G) = \overline{R}^*/\overline{CR}$. Since $\lambda/q \in P$, by (1.64) , there is $\mu \in P$ such that

$$w_u\lambda/q - \mu - u \in \overline{CR}. \tag{1.67}$$

Therefore

$$w_u\lambda - q\mu \in q\overline{R}^*. \tag{1.68}$$

Then $\lambda' = q\mu \in qP \cap \overline{CR}^*$, and $w_u\lambda - \lambda' \in q\overline{R}^*$.

Conversely, if $\lambda, \lambda' \in qP \cap \overline{CR}^*$, $u \in \overline{R}^*$ are such that

$$w\lambda - \lambda' = qu, \tag{1.69}$$

then

$$w\lambda/q = \lambda'/q + u. \tag{1.70}$$

From (1.70) , since $\lambda/q, \lambda'/q \in P$, we get $w = w_u$. The proof of our Proposition is completed. □

Put

$$h = \frac{|W|}{|Z(G)|}. \tag{1.71}$$

Let $w^1, \ldots, w^s \in W$ be distinct representatives in W of the classes of $W/Z(G)$.

THEOREM 1.17. *The set $\bigcup_1^n w^i(qP \cap \overline{CR}^*)$ is a disjoint union, which embeds into $\overline{CR}^*/q\overline{R}^*$. More precisely*

$$\bigcup_1^n w_i(qP \cap \overline{CR}^*) = \{\lambda \in \overline{CR}^*/q\overline{R}^*\ ,\ \sigma^2(\lambda/q) \neq 0\}. \tag{1.72}$$

PROOF. Our Theorem follows from Propositions 1.14 and 1.16. □

THEOREM 1.18. *Let f a W-invariant function on $\overline{CR}^*/q\overline{R}^*$. Then*

$$\sum_{\substack{\lambda \in \overline{CR}^*/q\overline{R}^* \\ \sigma^2(\lambda/q) \neq 0}} f(\lambda) = \frac{|W|}{|Z(G)|} \sum_{\lambda \in qP \cap \overline{CR}^*} f(\lambda). \tag{1.73}$$

PROOF. Our identity follows from Theorem 1.17. □

REMARK 1.19. Our Theorem is also a consequence of Proposition 1.12. In fact, if f is a W-invariant function on $\overline{CR}^*/q\overline{R}^*$,

$$\sum_{\substack{\lambda \in \overline{CR}^*/q\overline{CR} \\ \sigma^2(\lambda/q) \neq 0}} f(\lambda) = \left|\frac{\overline{R}^*}{\overline{CR}}\right| \sum_{\substack{\lambda \in \overline{CR}^*/q\overline{R}^* \\ \sigma^2(\lambda/q) \neq 0}} f(\lambda). \tag{1.74}$$

Now $\overline{R}^*/\overline{CR} \simeq Z(G)$, and so using Proposition 1.12 , we finally obtain (1.73).

1.7. The element ρ/c.

PROPOSITION 1.20. *If $s \in \mathfrak{t}$, $t \in P$, then*

$$\sum_{\alpha\in R}\langle\alpha,s\rangle[\langle\alpha,t\rangle] = 2\langle ct-\rho,s\rangle\,. \tag{1.75}$$

PROOF. By (1.35),(1.38),

$$\begin{aligned}\sum_{\alpha\in R}\langle\alpha,s\rangle[\langle\alpha,t\rangle] &= \sum_{\alpha\in R_+}\left(\langle\alpha,s\rangle\langle\alpha,t\rangle-\langle\alpha,s\rangle(1-\langle\alpha,t\rangle)\right)\\ &= \sum_{\alpha\in R}\langle\alpha,s\rangle\langle\alpha,t\rangle-\langle 2\rho,s\rangle\\ &= 2\langle ct-\rho,s\rangle\,.\end{aligned} \tag{1.76}$$

The proof of our Proposition is completed. □

Let $t \in T_{\mathrm{reg}}$. Then t determines a Weyl chamber K and an alcove P of the above type. In the sequel, we consider t as an element of $P \subset T$. The element ρ is still given by (1.35). Of course ρ depends implicitly on $t \in T_{\mathrm{reg}}$.

THEOREM 1.21. *The following identity holds*

$$ct-\rho = \frac{1}{2}\sum_{\alpha\in R}[\langle\alpha,t\rangle]\alpha\,. \tag{1.77}$$

In particular the map $t \in T_{\mathrm{reg}} \mapsto ct-\rho \in \mathfrak{t}$ descends to a map $T'_{\mathrm{reg}} = T_{\mathrm{reg}}/Z(G) \to \mathfrak{t}$.

PROOF. By Proposition 1.20 , we get (1.77) . Also if $u \in Z(G)$, if $t \in T_{\mathrm{reg}}$ is replaced by $t+u \in T_{\mathrm{reg}}$, $[\langle\alpha,t\rangle]$ is unchanged. By (1.77) , we find that $t \in T_{\mathrm{reg}} \mapsto ct-\rho \in \mathfrak{t}$ descends to a map from T'_{reg} into $\mathfrak{t}$. The proof of our Theorem is completed. □

Observe that since $\rho \in \overline{CR}^*$ and $2\rho \in \overline{R}$, then $u \in Z(G) = \overline{R}^*/\overline{CR} \mapsto \exp(2i\pi\langle\rho,u\rangle) = \pm 1$ is a character of $Z(G)$.

Recall that K is a Weyl chamber, and that $Z(G)$ has been embedded in W, by an embedding which depends explicitly on K.

THEOREM 1.22. *If $w \in W$, then $w\rho/c-\rho/c \in \overline{CR}$ if and only if $w=1$. Also if $w \in W$, then $w\rho/c-\rho/c = u \in \overline{R}^*$ if and only if $w = w_u$, so that $w \in Z(G)$. The map $w \in Z(G) \mapsto w\rho/c-\rho/c \in \overline{R}^*/\overline{CR} = Z(G)$ is a group isomorphism. Also if $u \in Z(G)$,*

$$\exp(2i\pi\langle\rho,u\rangle) = \epsilon_{w_u}\,. \tag{1.78}$$

PROOF. By (1.41), $\rho/c \in P$. Using [15, Proposition V.7.10] , we find that if $w\rho/c-\rho/c \in \overline{CR}$, then $w=1$. By Theorem 1.21 , if $u \in Z(G) = \overline{R}^*/\overline{CR}$,

$$\rho/c+u = w_u(\rho/c) \text{ in } \mathfrak{t}/\overline{CR}\,. \tag{1.79}$$

By (1.79) , we find that $w_u(\rho/c)-\rho/c \in \overline{R}^*$. Conversely let $w \in W$ be such that

$$u = w(\rho/c)-\rho/c \in \overline{R}^*\,. \tag{1.80}$$

By (1.79), (1.80), we get

$$w_u(\rho/c) = w(\rho/c) \;\text{ in }\; T = \mathfrak{t}/\overline{CR}, \tag{1.81}$$

so that $w = w_u$. Clearly if $u \in Z(G)$, by (1.45),

$$\sigma(w_u e^{\rho/c}) = \varepsilon_{w_u} \sigma(e^{\rho/c}). \tag{1.82}$$

Also by (1.79) and by the above,

$$\sigma(w_u e^{\rho/c}) = \sigma(e^{\rho/c+u}) = e^{2i\pi\langle\rho,u\rangle}\sigma(e^{\rho/c}). \tag{1.83}$$

Since $\sigma(e^{\rho/c}) \neq 0$, from (1.82), (1.83), we get (1.78) . The proof of our Theorem is completed. □

Let $\overline{P}$ be the closure of the alcove P. By (1.33),

$$\overline{P} = \{t \in \mathfrak{t}, \text{ for any } \alpha \in R_+,\ 0 \leq \langle\alpha, t\rangle \leq 1\}. \tag{1.84}$$

PROPOSITION 1.23. *For $h \in \overline{CR}$, then $\overline{P} \cap (\overline{P} + h) \neq \emptyset$ if and only if $h = 0$.*

PROOF. Recall that $W_{\text{aff}} = W \ltimes \overline{CR}$. By [**15**, Lemma V.4.3], if $f \in \overline{P}$, $v \in W_{\text{aff}}$, then $vf \in \overline{P}$ if and only if $vf = f$. In particular, if $f \in \overline{P}$, $h \in \overline{CR}$, then $f + h \in \overline{P}$ if and only if $h = 0$. The proof of our Proposition is completed. □

PROPOSITION 1.24. *The center $Z(G) = \overline{R}^* / \overline{CR}$ embeds as a finite subset of $\overline{P}$.*

PROOF. If $u \in Z(G) = \overline{R}^* / \overline{CR}$, there is an alcove Q containing 0 such that u is represented in $\mathfrak{t}$ by an element $v \in \overline{Q}$. By [**15**, Theorem V.4.1], there is $w \in W$ such that $wQ = P$. Also since $u \in \overline{R}^* / \overline{CR}$, then $wv - v \in \overline{CR}$. Therefore $wv \in \overline{P}$ still represents u. So any element $u \in Z(G)$ has a representative in $\overline{P}$. By Proposition 1.23, this representative is unique. The proof of our Proposition is completed. □

REMARK 1.25. If $u \in Z(G)$, we still denote by u the corresponding representative in $\overline{P}$. Then if $w \in W$, $wu \in w\overline{P}$ is the unique representative of u in $w\overline{P}$. Of course, if $wu \in \overline{P}$, the above implies that $wu = u$. However this also follows from [**15**, Theorem V.4.1].

Let $w_o \in W$ be the unique element of the Weyl group such that $w_o K = -K$.

THEOREM 1.26. *Let $u \in \overline{P}$ be the unique representative in $\overline{P}$ of an element of $Z(G)$. Then if $t \in \overline{P}$, if $[t + u] \in w_u \overline{P}$ represents $t + u \in T$, then*

$$[t + u] = t + w_u u. \tag{1.85}$$

Also

$$w_u u = w_o u. \tag{1.86}$$

In particular

$$w_u \overline{P} = w_u u + \overline{P}. \tag{1.87}$$

Moreover

$$u = \rho/c - w_u \rho/c \text{ in } \mathfrak{t}. \tag{1.88}$$

In particular, if $t \in \overline{P}$,

$$w_u \rho - c[t + u] = \rho - ct \text{ in } \mathfrak{t}. \tag{1.89}$$

PROOF. Clearly $w_o(-u)$ lies in $\overline{P}$, and so it is the unique representative in $\overline{P}$ of $u^{-1} \in T$. If $t \in P$, by (1.84), it is clear that $t + w_o u$ lies in some alcove containing 0. Also $t + w_o u$ represents $t + u \in T$. From the above, we get

$$[t + u] = t + w_o u. \tag{1.90}$$

The alcove which contains $t + w_o u$ is necessarily equal to $w_u P$. Therefore $w_u u$ and $w_o u$ both lie in $w_u \overline{P}$, and represent $u \in Z(G)$. By Proposition 1.24, we get (1.86). By (1.86) and (1.90), we obtain (1.85) and (1.87).

Using Theorem 1.21 and (1.85), if $t \in P$,

$$c(t + w_u u) - w_u \rho = ct - \rho \text{ in } \mathfrak{t}. \tag{1.91}$$

From (1.91), we get (1.88) and (1.89). The proof of our Theorem is completed. □

1.8. Some properties of the characters of G. Put

$$\overline{CR}^*_+ = \overline{CR}^* \cap \overline{K}. \tag{1.92}$$

If $\lambda \in \overline{CR}^*_+$, let χ_λ be the character of G which is the character of the irreducible representation of G with highest weight λ. By the Weyl character formula, if $t \in T_{\mathrm{reg}}$,

$$\chi_\lambda(t) = \sum_{w \in W} \frac{e^{2i\pi\langle w\lambda, t\rangle}}{\prod_{\alpha \in R_+} (1 - e^{2i\pi\langle w\alpha, t\rangle})}. \tag{1.93}$$

Equivalently

$$\chi_\lambda(t) = \frac{1}{\sigma(t)} \sum_{w \in W} \varepsilon_w e^{2i\pi\langle w(\rho+\lambda), t\rangle}. \tag{1.94}$$

Recall that by [**15**, LemmaVI.1.2], if $\mu \in \overline{CR}^*$ is not included in a Weyl chamber,

$$\sum_{w \in W} \varepsilon_w e^{2i\pi\langle w\mu, t\rangle} = 0. \tag{1.95}$$

If $\mu \in \overline{CR}^*$ lies in K, then by [**15**, Note V.4.14], there is $\lambda \in \overline{CR}^*_+$ such that $\mu = \rho + \lambda$.

If $\lambda \in \overline{CR}^*$, there exists a Weyl chamber K such that $\lambda \in \overline{K}$. In general K is not unique.

PROPOSITION 1.27. *The character χ_λ does not depend on K.*

PROOF. Assume that $\lambda \in \overline{K}$, $\lambda \in \overline{K}'$. Then by [**15**, Lemma V.4.2], there is $w' \in W$ such that $w'K = K'$. Let R_+, R'_+ be the system of positive roots attached to K, K'. Clearly

$$R'_+ = w' R_+. \tag{1.96}$$

Also since $\lambda \in \overline{K} \cap \overline{K}'$, and $w'\overline{K} = \overline{K}'$, then $\lambda \in \overline{K}, w'^{-1}\lambda \in \overline{K}$. By [**15**, Theorem V.4.1], it follows that

$$w'^{-1}\lambda = \lambda. \tag{1.97}$$

Let $\chi_\lambda^K(t)$, $\chi_\lambda^{K'}(t)$ be the characters attached to K, K', of highest weight λ. Using (1.93), (1.97) , we get

$$\chi_\lambda^K(t) = \chi_\lambda^{K'}(t)\,. \tag{1.98}$$

Our Proposition follows. □

For $\lambda \in \overline{CR}^*$, we can then write $\chi_\lambda(t)$ without explicitly mentioning a Weyl chamber K.

PROPOSITION 1.28. *If* $\lambda \in \overline{CR}^*$, $w \in W$,

$$\chi_{w\lambda} = \chi_\lambda\,. \tag{1.99}$$

PROOF. We may and we will assume that $\lambda \in \overline{CR}^*_+ = \overline{CR}^* \cap \overline{K}$. Then $w\lambda \in \overline{CR}^* \cap wK$. Equation (1.99) now follows from (1.93). □

Recall that if $\lambda \in \overline{CR}^*$, $w \in W$, then $w\lambda - \lambda \in \overline{R}$. It follows that if $\theta_1, \dots, \theta_s \in \overline{CR}^*$, then $\sum_{j=1}^{s} \theta_j \in \overline{R}$ if and only if, given $(w^1, \dots, w^s) \in W^s$, then $\sum_{j=1}^{s} w^j\theta_j \in \overline{R}$.

PROPOSITION 1.29. *If* $\sum_{j=1}^{s} \theta_j \in \overline{R}$, *then* $\prod_{j=1}^{s} \chi_{\theta_j}$ *descends to a function on the adjoint group* $G' = G/Z'(G)$.

PROOF. By Proposition 1.27 , we may and will assume that the θ_j's lie in $\overline{CR}^*_+ = \overline{CR}^* \cap K$. By (1.93) , for $t \in T_{\mathrm{reg}}$,

$$\prod_{j=1}^{s} \chi_{\theta_j}(t) = \sum_{(w^1,\dots,w^s)\in W^s} \frac{e^{2i\pi\langle \sum_{k=1}^{s} w^k\theta_k, t\rangle}}{\prod_{j=1}^{s}\prod_{\alpha\in R_+}\left(1 - e^{-2i\pi\langle w^j\alpha, t\rangle}\right)}\,. \tag{1.100}$$

If $\sum_{j=1}^{s} \theta_j \in \overline{R}$, then $\sum_{j=1}^{s} w^j\theta_j \in \overline{R}$, and so by (1.100), $\prod_{j=1}^{s} \chi_{\theta_j}(t)$ descends to a function on the adjoint torus $T' = \mathfrak{t}/\overline{R}^*$. The proof of our Proposition is completed. □

PROPOSITION 1.30. *If* $\sum_{j=1}^{s} \theta_j \notin \overline{R}$, *then*

$$\sum_{\mu\in\overline{R}^*/\overline{CR}} \prod_{j=1}^{s} \chi_{\theta_j}(e^{t+\mu}) = 0\,. \tag{1.101}$$

PROOF. This follows from (1.100). □

PROPOSITION 1.31. *If* $\lambda \in \overline{CR}^*$, $\lambda \notin \overline{R}$, *then*

$$\chi_\lambda(e^{\rho/c}) = 0\,. \tag{1.102}$$

PROOF. We may and we will assume that $\lambda \in \overline{CR}^*_+$. If $w \in W$, then

$$\chi_\lambda(e^{\rho/c}) = \chi_\lambda(e^{w\rho/c}). \tag{1.103}$$

Take $u \in \overline{R}^*$. Then by Theorem 1.22,

$$w_u\rho/c - \rho/c = u \text{ in } T. \tag{1.104}$$

By (1.103), (1.104),

$$\chi_\lambda(e^{\rho/c}) = \chi_\lambda(e^{\rho/c+u}). \tag{1.105}$$

Now since the representation associated to λ is irreducible, the central element e^u acts in the λ representation like $e^{2i\pi\langle\lambda,u\rangle}$. From (1.105) , we get

$$\chi_\lambda(e^{\rho/c}) = e^{2i\pi\langle\lambda,u\rangle}\chi_\lambda(e^{\rho/c}). \tag{1.106}$$

If $\lambda \notin \overline{R}$, it is then clear that (1.102) holds. The proof of our Proposition is completed. □

Now we have the result of Kostant [**36**].

THEOREM 1.32. *If* $\lambda \in \overline{R}$, *then*

$$\chi_\lambda(e^{\rho/c}) = 0, \ +1 \ \textit{or} \ -1. \tag{1.107}$$

Take now $p \in \mathbf{N}^*$ and $\lambda \in p\overline{P}$, so that $\lambda \in \overline{CR}^*_+$. Take $u \in Z(G) = \overline{R}^*/\overline{CR}$. Then $\lambda/p+u \in T$ is represented uniquely by an element $[\lambda/p+u] \in w_u\overline{P}$. Observe that by ((1.85),

$$[\lambda/p+u] = \lambda/p + w_u u \text{ in } \mathfrak{t}. \tag{1.108}$$

Also

$$p[\lambda/p+u] - (\lambda + pu) \in p\overline{CR}. \tag{1.109}$$

THEOREM 1.33. *If* $\mu \in \overline{CR}^*/(p+c)\overline{CR}$, $\sigma(\mu/(p+c)) \neq 0$, *then*

$$\chi_{p[\lambda/p+u]}(e^{\mu/(p+c)}) = \varepsilon_{w_u} e^{2i\pi\langle u,\mu\rangle}\chi_\lambda(e^{\mu/(p+c)}). \tag{1.110}$$

In particular

$$\chi_{pu}(e^{\mu/(p+c)}) = \epsilon_{w_u} e^{2i\pi\langle u,\mu\rangle}. \tag{1.111}$$

PROOF. By (1.94),

$$\begin{aligned} \chi_\lambda(e^{\mu/(p+c)}) &= \frac{1}{\sigma(e^{\mu/(p+c)})}\sum_{w\in W}\varepsilon_w e^{2i\pi\langle w(\lambda+\rho),\frac{\mu}{p+c}\rangle} \\ &= \frac{1}{\sigma(e^{\mu/(p+c)})}\sum_{w\in W}\varepsilon_w e^{2i\pi\langle w\lambda/p,\mu\rangle} \\ &\quad e^{2i\pi\langle w(\rho-c\lambda/p),\frac{\mu}{p+c}\rangle}. \end{aligned} \tag{1.112}$$

Also $p[\lambda/p+u] \in w_u\overline{P} \cap \overline{CR}^*$. When replacing $\overline{P}$ by $w_u\overline{P}$, ρ is replaced by $w_u\rho$. When replacing λ by $p[\lambda/p+u]$, using (1.109), we find that $e^{2i\pi\langle w\lambda/p,\mu\rangle}$ is replaced by $e^{2i\pi\langle w\lambda/p,\mu\rangle}e^{2i\pi\langle wu,\mu\rangle}$. Also by equation (1.89) in Theorem 1.26, we find that when replacing λ by $p[\lambda/p+u], \rho - c\lambda/p \in \mathfrak{t}$ is unchanged. Finally $\sigma(e^{\mu/(p+c)})$ is changed into $\varepsilon_{w_u}\sigma(e^{\mu/(p+c)})$. Equation (1.110) follows from the above arguments. Equation (1.111) is a consequence of (1.110). □

REMARK 1.34. If we use (1.110) with $p = 0$, and $\mu = \rho$, we get

$$1 = \epsilon_{w_u} e^{2i\pi\langle\rho,u\rangle}, \tag{1.113}$$

which is precisely equation (1.78).

Theorem 1.33 will be used in Remark 8.4 as a consistency check on our index theoretic computations.

1.9. The elements whose centralizer is semisimple. We still assume that G is a connected and simply connected simple compact Lie group.

DEFINITION 1.35. Let $C \subset \mathfrak{t}$ be the set of $u \in \mathfrak{t}$ such that $\{\alpha \in R; \langle\alpha, u\rangle \in \mathbf{Z}\}$ spans $\mathfrak{t}^*$.

Clearly

$$\overline{R}^* \subset C. \tag{1.114}$$

Since by (1.6), (1.12), $\overline{CR} = \overline{CR}_s = \overline{R}_\ell$, then

$$\overline{R}^* \subset \overline{CR}^* \subset C. \tag{1.115}$$

In particular $\overline{R}^* \subset C$ acts by translations on C. Also the Weyl group W acts on C.

PROPOSITION 1.36. *The set $C/\overline{R}^*$ is a finite subset of the adjoint torus $T' = \mathfrak{t}/\overline{R}^*$, which contains $\overline{CR}^*/\overline{R}^*$. Also W acts on $C/\overline{R}^*$.*

PROOF. Let $R' \subset R$ be such that the elements of R' span $\mathfrak{t}^*$, and let $\overline{R}'$ be the associated lattice. Clearly $\overline{R}/\overline{R}'$ is a finite set. Then

$$C/\overline{R}^* = \bigcup_{R'} \overline{R}'^*/\overline{R}^*. \tag{1.116}$$

From (1.116) , its follows that $C/\overline{R}^*$ is finite. The proof of our Proposition is completed. □

Let now K be a Weyl chamber in $\mathfrak{t}$. Let R_+ be the corresponding system of positive roots so that

$$R = R_+ \cup (-R_+). \tag{1.117}$$

DEFINITION 1.37. If $u \in G' = G/Z(G)$, let $Z(u) \subset G$ be the centralizer of u.

Recall that by [**14**, Corollaire 5.3.1], since G is connected and simply connected, $Z(u)$ is a connected Lie subgroup of G. Clearly if $u \in T' = \mathfrak{t}/\overline{R}^*$, T is a maximal torus in $Z(u)$.

Also W acts like the identity on $Z(G) = \overline{R}^*/\overline{CR}$. Therefore, if $u \in T' = \mathfrak{t}/\overline{R}^*, w \in W$, then $wu - u$ is well-defined in $T = \mathfrak{t}/\overline{CR}$. Put

$$W_u = \{w \in W; wu - u = 0 \text{ in } T = \mathfrak{t}/\overline{CR}^*\}. \tag{1.118}$$

THEOREM 1.38. *If $u \in T' = \mathfrak{t}/\overline{R}^*$, the root system R_u of $Z(u)$ is given by*

$$R_u = \{\alpha \in R, \langle\alpha, u\rangle \in \mathbf{Z}\}. \tag{1.119}$$

Also $R_{u,+} = R_u \cap R_+$ is a system of positive roots for $Z(u)$. If $Z(Z(u)) \subset Z(u)$ is the center of $Z(u)$, its Lie algebra $\mathfrak{z}(Z(u))$ is given by

$$\mathfrak{z}(Z(u)) = \{f \in \mathfrak{t}, \text{ for any } \alpha \in R_u, \langle\alpha, f\rangle = 0\}. \tag{1.120}$$

Also the Weyl group $W_{Z(u)}$ of $Z(u)$ is given by

$$W_{Z(u)} = W_u. \tag{1.121}$$

Finally

$$C/\overline{R}^* = \{u \in T', Z(u) \text{ is semisimple}\}. \tag{1.122}$$

PROOF. Let $\mathfrak{z}(u) \subset \mathfrak{g}$ be the Lie algebra of $Z(u)$. Then

$$\mathfrak{z}(u) = \{f \in \mathfrak{g}, u \cdot f = f\}. \tag{1.123}$$

As a T-space, $\mathfrak{g} \otimes_{\mathbf{R}} \mathbf{C}$ splits as

$$\mathfrak{g} \otimes_{\mathbf{R}} \mathbf{C} = (\mathfrak{t} \otimes_{\mathbf{R}} \mathbf{C}) \oplus (\bigoplus_{\alpha \in R} \mathfrak{g}_\alpha). \tag{1.124}$$

From (1.124), we get

$$\mathfrak{z}(u) \otimes_{\mathbf{R}} \mathbf{C} = (\mathfrak{t} \otimes_{\mathbf{R}} \mathbf{C}) \oplus (\bigoplus_{\substack{\alpha \in R \\ (\alpha, u) \in \mathbf{Z}}} \mathfrak{g}_\alpha). \tag{1.125}$$

From (1.125), it is clear that (1.119) holds.

Since the forms α in R do not vanish on K, the same is true for elements in R_u, i.e. K is included in a $Z(u)$ Weyl chamber K_u. If $R_{u,+}$ is the corresponding system of positive roots, then

$$R_{u,+} = R_u \cap R_+. \tag{1.126}$$

The identity (1.120) is trivial. Let $N(T) \subset G$ be the normalizer of T. Then

$$W = N(T)/T. \tag{1.127}$$

Similarly let $N_u(T)$ be the normalizer of T in $Z(u)$. Then

$$W_{Z(u)} = N_u(T)/T. \tag{1.128}$$

Clearly

$$N_u(T) = N(T) \cap Z(u). \tag{1.129}$$

Therefore $W_{Z(u)}$ is a subgroup of W. Since $u \in Z(Z(u))/Z(G)$, if $w \in W_{Z(u)}$,

$$wu - u = 0 \text{ in } T. \tag{1.130}$$

Conversely if $w \in W$, let $w' \in N(T)$ represent w. If $wu - u = 0$ in T, then

$$w'uw'^{-1} = u \text{ in } T, \tag{1.131}$$

i.e. $w' \in Z(u)$. Therefore $w' \in N_u(T)$, and w lies in $W_{Z(u)}$.

By definition, $Z(u)$ is semisimple if and only if $\mathfrak{z}(Z(u)) = 0$. From (1.120), we get (1.122).

The proof of our Theorem is completed. □

REMARK 1.39. Clearly, if we identify u to a corresponding element in $\mathfrak{t}$, then

$$W_u = \{w \in W, wu - u \in \overline{CR}\}. \tag{1.132}$$

By Theorem 1.38, $W_u = W_{Z(u)}$. Let $\overline{CR}_u$ be the lattice generated by the coroots of $Z(u)$. Since $u \in Z(Z(u))$, if $w \in W_{Z(u)}$, $wu - u \in \overline{CR}_u$. Therefore

$$W_u = \{w \in W, wu - u \in \overline{CR}_u\}. \tag{1.133}$$

PROPOSITION 1.40. *If $n \geq 2$ and $G = \mathrm{SU}(n)$, then*

$$C = \overline{R}^{*}. \tag{1.134}$$

PROOF. By [15, Proposition V.6.3], it is clear that if $\alpha_1, \dots, \alpha_r$ is a basis of $\mathfrak{t}^*$ over $\mathbf{R}$, then $\alpha_1, \dots, \alpha_r$ spans $\overline{R}$. Equation (1.134) follows. □

1.10. Some properties of semisimple centralizers. We still assume that G is a connected simply connected simple compact Lie group of rank r. Also we use the notation of Sections 1.1-1.3 and 1.9.

Take $u \in C/\overline{R}^{*}$. To the Lie group $Z(u)$, we can associate the objects we considered in Section 1.1 for G, with the reservation that since the action of W_u on $\mathfrak{t}$ may be reducible, $Z(u)$ is semisimple and in general not simple.

However, we equip $\mathfrak{z}(u)$ with the scalar product induced from the scalar product of $\mathfrak{g}, \langle\ ,\ \rangle$. Therefore $\mathfrak{t}$ is equipped with the scalar product $\langle\ ,\ \rangle$, and the identification $\mathfrak{t} \simeq \mathfrak{t}^*$ will still be the one we used for G.

The objects we considered before which are attached to $Z(u)$ will be denoted with the index u. The lattices $\Gamma_u \subset \mathfrak{t}, \Lambda_u = \Gamma_u^* \subset \mathfrak{t}^*$ are given by

$$\Gamma_u = \overline{CR},\ \Lambda_u = \overline{CR}^{*}. \tag{1.135}$$

The roots R_u have already been described in Theorem 1.38 . Clearly

$$CR_u = \{h_\alpha, \alpha \in R_u\}. \tag{1.136}$$

Note that in general

$$\pi_1(Z(u)) = \overline{CR}/\overline{CR_u}, \tag{1.137}$$

and so $Z(u)$ is in general not simply connected.

If $u \in C/\overline{R}^{*}$, put as in (1.35),

$$\rho_u = \frac{1}{2} \sum_{\alpha \in R_{u,+}} \alpha. \tag{1.138}$$

Then $\rho_u \in \overline{CR}_u^{*}$.

THEOREM 1.41. *For any $u \in C$,*

$$\begin{aligned} 2cu &\in \overline{R}, \\ 2\rho_u &\in \overline{R}. \end{aligned} \tag{1.139}$$

If $h \in \overline{R}^{*}/\overline{CR}_u$, then $2c\langle u, h\rangle \in \mathbf{Z}$, $2\langle \rho - \rho_u, h\rangle \in \mathbf{Z}$, and moreover

$$c\langle u, h\rangle = \langle \rho - \rho_u, h\rangle \bmod(\mathbf{Z}). \tag{1.140}$$

In particular, if $h \in \pi_1(Z(u)) = \overline{CR}/\overline{CR}_u$, then $2c\langle u, h\rangle \in \mathbf{Z}$, $2\langle \rho_u, h\rangle \in \mathbf{Z}$, and

$$c\langle u, h\rangle = \langle \rho_u, h\rangle \bmod(\mathbf{Z}). \tag{1.141}$$

PROOF. Let $\tau : G' \to \mathrm{End}(\mathfrak{g})$ be the adjoint representation. Then by (1.37), if $A, B \in \mathfrak{g}$

$$\langle A, B\rangle^{\tau} = -2c\langle A, B\rangle. \tag{1.142}$$

Also τ induces a representation $Z(u)/Z(G) \to \mathrm{End}(\mathfrak{g})$. Then $\overline{R}^{*} \subset \mathfrak{t}$ is exactly the lattice of integral elements in $\mathfrak{t}$ with respect to $Z(u)/Z(G)$. By Proposition 1.5 and by (1.142), we then find that $2cu \in \overline{R}$. Also $2\rho_u \in \overline{R}_u \subset \overline{R}$.

If $h \in \overline{R}^*$, it follows that $2c\langle u,h\rangle \in \mathbf{Z}$, $2\langle \rho_u, h\rangle \in \mathbf{Z}$. Also since $\overline{CR}_u \subset \overline{R}_u$, and $u \in \overline{R}_u^*$, if $h \in \overline{CR}_u$, $\langle u,h\rangle \in \mathbf{Z}$, and, since $\rho_u \in \overline{CR}_u^*$, then $\langle \rho_u, h\rangle \in \mathbf{Z}$. Therefore mod $(\mathbf{Z})$, if $h \in \overline{R}^*$, $c\langle u,h\rangle$ and $\langle \rho_u, h\rangle$ only depend on the class of h in $\overline{R}^*/\overline{CR}_u$.

Now we establish (1.141). By [**15**, Proposition V.7.10], we may suppose that $u \in C$ is such that for $\alpha \in R$

$$|\langle \alpha, u\rangle| \leq 1. \tag{1.143}$$

Take $h \in \overline{R}^*$. Since $Z(u)$ is semisimple, and since the $\alpha \in R$ are the weights of the restriction of τ to $Z(u)$, then by proceeding as in (1.29),

$$\sum_{\alpha \in R} [\langle \alpha, u\rangle]\langle \alpha, h\rangle = \sum_{s\in[0,1[} s\langle \sum_{\substack{\alpha\in R \\ [\langle\alpha,u\rangle]=s}} \alpha, h\rangle = 0. \tag{1.144}$$

Also,

$$\begin{aligned} \sum_{\alpha\in R}[\langle \alpha, u\rangle]\langle \alpha, h\rangle &= \sum_{\alpha \in R\setminus R_u} [\langle \alpha, u\rangle]\langle \alpha, h\rangle \\ &= \sum_{\alpha\in R_+\setminus R_{u,+}} (\langle \alpha, u\rangle\langle \alpha, h\rangle - (1-\langle \alpha, u\rangle)\langle \alpha, h\rangle) \\ &= \sum_{\alpha\in R\setminus R_u} \langle \alpha, u\rangle\langle \alpha, h\rangle - \langle \sum_{\alpha\in R_+\setminus R_{u,+}} \alpha, h\rangle . \end{aligned} \tag{1.145}$$

If $\alpha \in R_u$, $\langle \alpha, u\rangle \in \mathbf{Z}$, and $\langle \alpha, h\rangle \in \mathbf{Z}$. Since roots in R_u come by pairs,by (1.40),(1.142), (1.144), (1.145),

$$2c\langle u,h\rangle = \langle \sum_{\alpha\in R_+\setminus R_{u,+}} \alpha, h\rangle \mod (2\mathbf{Z}), \tag{1.146}$$

which is equivalent to (1.140).

If $h \in \overline{CR}$, then $\langle \rho, h\rangle \in \mathbf{Z}$. From (1.140), we get (1.141). The proof of our Theorem is completed. □

REMARK 1.42. Needless to say, the class of ρ_u in $\mathfrak{t}/\overline{R}$ does not depend on the choice of R_+. This fact fits with (1.141).

1.11. The first homotopy group in a semisimple centralizer. Take $u \in C/\overline{R}^*$. Then $Z(u)$ is a connected semisimple subgroup of G. Also by (1.135), (1.137) and by [**15**, Theorem V.7.1],

$$\begin{aligned} \pi_1(Z(u)) &= \overline{CR}/\overline{CR}_u, \\ Z(Z(u)) &= \overline{R}_u^*/\overline{CR}. \end{aligned} \tag{1.147}$$

By (1.147),

$$\frac{Z(Z(u))}{Z(G)} = \overline{R}_u^*/\overline{R}^* . \tag{1.148}$$

Let $\pi_u : \widetilde{Z}(u) \to Z(u)$ be the universal cover of $Z(u)$. Then by [**15**, Proposition V.7.16],

$$Z(\widetilde{Z}(u)) = \overline{R}_u^*/\overline{CR}_u, \tag{1.149}$$

and $\pi_1(Z(u))$ is a subgroup of $Z(\widetilde{Z}(u))$.

Clearly

$$\frac{Z(Z(u))}{Z(G)} = \overline{R}^*_u/\overline{R}^* \subset C/\overline{R}^* . \tag{1.150}$$

Take $v \in Z(Z(u))/Z(G)$. Then $Z(v) \supset Z(u)$, $CR_v \supset CR_u$. Therefore $\pi_1(Z(u)) = \overline{CR}/\overline{CR}_u$ surjects on $\pi_1(Z(v)) = \overline{CR}/\overline{CR}_v$. Let $\tau_{u,v} : \overline{CR}/\overline{CR}_u \to \overline{CR}/\overline{CR}_v$ be this surjection. Clearly $\overline{R}^*_u/\overline{R}^*$ maps into $\overline{CR}^*_u/\overline{R}^*$.

Let $h \in \pi_1(Z(u)) = \overline{CR}/\overline{CR}_u$. Then h defines a character δ_h of $Z(Z(u))/Z(G)$ given by

$$\delta_h : v \in Z(Z(u))/Z(G) \mapsto \exp(2i\pi\langle h, v\rangle) . \tag{1.151}$$

Recall that by Proposition 1.36, $\overline{CR}^*/\overline{R}^* \subset C/\overline{R}^*$.

PROPOSITION 1.43. *The element $u \in C/\overline{R}^*$ is such that for any $h \in \overline{CR}/\overline{CR}_u$, $\delta_h(u) = 1$ if and only if $u \in \overline{CR}^*/\overline{R}^*$.*

PROOF. This is trivial. □

PROPOSITION 1.44. *For any $u \in C/\overline{R}^*$, $v \in \dfrac{Z(Z(u))}{Z(G)} = \overline{R}^*_u/\overline{R}^*$, $h \in \pi_1(Z(u)) = \overline{CR}/\overline{CR}_u$,*

$$\exp(2i\pi\langle h, v\rangle) = \exp(2i\pi\langle \tau_{u,v}h, v\rangle) . \tag{1.152}$$

PROOF. We have the exact sequence

$$0 \to \overline{CR}_v/\overline{CR}_u \to \overline{CR}/\overline{CR}_u \overset{\tau_{u,v}}{\to} \overline{CR}/\overline{CR}_v \to 0 . \tag{1.153}$$

Also if $h' \in \overline{CR}_v/\overline{CR}_u$, $\exp(2i\pi\langle h', v\rangle) = 1$. Our Proposition follows. □

REMARK 1.45. Proposition 1.44 will be used in Remark 4.40.

1.12. Centralizers in a connected and non simply connected semisimple Lie group. Let G be a compact connected semisimple Lie group. We use otherwise the same notation as in Sections 1.1-1.3 and 1.11.

Let $\pi_1(G)$ be the first homotopy group of G. Let $\widetilde{G}$ be the universal cover of G. Then $\widetilde{G}$ is a compact connected and simply connected semisimple Lie group, $\pi_1(G)$ is a subgroup of of $Z(\widetilde{G})$ and

$$G = \widetilde{G}/\pi_1(G) \tag{1.154}$$

Let T be a maximal torus in G. Let $u \in T' = \mathfrak{t}/\overline{R}^*$. Let $Z(u) \subset G$ be the centralizer of u.By [**14**, Corollaire 5.3.1], if G is simply connected, then $Z(u)$ is connected. Let $Z(u)_o$ be the connected component of the identity in $Z(u)$. Then $Z(u)_o$ is a normal subgroup of $Z(u)$, and so $Z(u)_o\backslash Z(u)$ is a finite group. Moreover $T \subset Z(u)_o$, so that $u \in Z(u)_o$.

If $g \in Z(u)$, let $\tilde{g} \in \widetilde{G}$ be a lift of g, let $\tilde{u} \in \widetilde{G}$ be a lift of u. Then $[\tilde{g}, \tilde{u}] \in \pi_1(G)$ does not depend on $\tilde{g}, \tilde{u}$.

PROPOSITION 1.46. *The map $g \in Z(u) \mapsto [\tilde{g}, \tilde{u}] \in \pi_1(G)$ induces an embedding of $Z(u)_o\backslash Z(u)$ into $\pi_1(G)$. In particular $Z(u)_o\backslash Z(u)$ is commutative.*

PROOF. The Lie group $Z(\tilde{u}) \subset \widetilde{G}$ is connected. Therefore $Z(\tilde{u})/\pi_1(G)$ is a connected subgoup of $Z(u)$.Since $Z(\tilde{u})/\pi_1(G)$ and $Z(u)_o$ have the same Lie algebra, it follows that

$$Z(u)_o = Z(\tilde{u})/\pi_1(G). \tag{1.155}$$

Let $g, g' \in Z(u)$, let $\tilde{g}, \tilde{g'} \in \widetilde{G}$ lift g, g'. Let $h, h' \in \pi_1(G)$ be such that

$$\begin{aligned} \tilde{g}\tilde{u}\tilde{g}^{-1} &= \tilde{u}\tilde{h}, \\ \tilde{g'}\tilde{u}\tilde{g'}^{-1} &= \tilde{u}\tilde{h'}. \end{aligned} \tag{1.156}$$

Then

$$\tilde{g'}\tilde{g}\tilde{u}\tilde{g}^{-1}\tilde{g'}^{-1} = \tilde{g'}\tilde{u}h\tilde{g}^{-1} = \tilde{g'}\tilde{u}\tilde{g'}^{-1}h = \tilde{u}h'h. \tag{1.157}$$

If $g \in Z(u)_o$, if $\tilde{g} \in Z(\tilde{u})$ lifts g, then

$$[\tilde{g}, \tilde{u}] = 1. \tag{1.158}$$

Conversely if (1.158) holds, then $\tilde{g} \in Z(\tilde{u})$ and $g \in Z(u)_o$. The proof of our Proposition is completed. □

Let $N(T) \subset G$ be the normalizer of T in G. Then $W = N(T)/T$ is the Weyl group. Put

$$W_u = \{w \in W, wu - u = 0 \text{ in } T = \mathfrak{t}/\Gamma\}. \tag{1.159}$$

PROPOSITION 1.47. *The following identity holds*

$$W_u = (N(T) \cap Z(u))/T. \tag{1.160}$$

PROOF. The proof of our Proposition is the same as in (1.128)-(1.131). □

Let $W_{Z(u)_o}$ be the Weyl group of $Z(u)_o$. Then $W_{Z(u)_o} = W_{Z(\tilde{u})}$. Also $W_{Z(u)_o}$ is a normal subgroup of W_u. Then $W_{Z(u)_o} \backslash W_u$ is a finite group.

Let $\overline{CR} \subset \mathfrak{t}$ be the lattice in $\mathfrak{t}$ spanned by the coroots of $\widetilde{G}$. Let $\Gamma \subset \mathfrak{t}$ be the lattice of integrals elements in $\mathfrak{t}$, i.e. whose exponential in $T \subset G$ is equal to 1. Then $\overline{CR} \subset \Gamma$, and

$$\Gamma/\overline{CR} = \pi_1(G) \tag{1.161}$$

We define $\overline{R}, \overline{R}^*$ as in Section 1.1 . Let $\overline{u} \in \mathfrak{t}$ represent $u \in T' = \mathfrak{t}/\overline{R}^*$.Then $w \in W \mapsto w\overline{u} - \overline{u} \in \Gamma/\overline{CR} = \pi_1(G)$ is a well-defined map.

Recall that $Z(\widetilde{G}) = \overline{R}^*/\overline{CR}$, and that $\pi_1(G) = \Gamma/\overline{CR}$ is a subgroup of $Z(\widetilde{G})$. In particular, by Proposition 1.15, $\pi_1(G)$ embeds as a commutative subgroup of W.

THEOREM 1.48. *The map $w \in W_u \mapsto w\overline{u} - \overline{u} \in \Gamma/\overline{CR} = \pi_1(G)$ induces an embedding of $W_{Z(u)_o} \backslash W_u$ into $\pi_1(G)$. In particular $W_{Z(u)_o} \backslash W_u$ is commutative. If $w \in W_{Z(u)_o} \backslash W_u$ is identified to the corresponding element $w' \in \pi_1(G) \subset Z(\widetilde{G})$, the image of w' in W lies in W_u and represents w in $W_{Z(u)_o} \backslash W_u$.*

PROOF. Let $w, w' \in W_u$. Let $h, h' \in \Gamma$ be such that

$$w\overline{u} - \overline{u} = h, w'\overline{u} - \overline{u} = h'. \tag{1.162}$$

Then

$$w'w\overline{u} - \overline{u} = w'h + h'. \tag{1.163}$$

Now $\Gamma \subset \overline{R}^*$, and so

$$w'h - h \in \overline{CR}. \tag{1.164}$$

Therefore by (1.163), (1.164)

$$w'w\overline{u} - \overline{u} = h + h' \text{ in } \Gamma/\overline{CR}. \tag{1.165}$$

If $w \in W_{Z(u)_o}$, since $u \in Z(u)_o$ is in the center of $Z(u)_o$, then

$$w\overline{u} - \overline{u} = 0 \text{ in } \Gamma/\overline{CR}. \tag{1.166}$$

Conversely let $w \in W$ be such that

$$w\overline{u} - \overline{u} \in \overline{CR}. \tag{1.167}$$

Let $\tilde{g} \in \widetilde{G}$ represent w.By (1.167),

$$[\tilde{g}, \tilde{u}] = 1, \tag{1.168}$$

so that $\tilde{g}$ maps into an element of $Z(u)_o$. In particular $w \in W_{Z(u)_o}$. The proof of our Proposition is completed. □

Observe that

$$W_{Z(u)_o}\backslash W_u = (N(T) \cap Z(u)_o)\backslash(N(T) \cap Z(u)). \tag{1.169}$$

Therefore $W_{Z(u)_o}\backslash W_u$ embeds naturally into $Z(u)_o\backslash Z(u)$.

THEOREM 1.49. *We have the identity*

$$W_{Z(u)_o}\backslash W_u \simeq Z(u)_o\backslash Z(u). \tag{1.170}$$

This identity is compatible with the embeddings of both groups into $\pi_1(G) = \Gamma/\overline{CR}$.

PROOF. Let $g \in Z(u)$. Then gTg^{-1} is a maximal torus in $Z(u)_o$.Therefore there is $h \in Z(u)_o$ such that

$$(hg)T(hg)^{-1} = T, \tag{1.171}$$

so that $hg \in N(T) \cap Z(u)$. Therefore the embedding $W_{Z(u)_o}\backslash W_u \to Z(u)_o\backslash Z(u)$ is in fact one to one. It is trivial to verify that the above identification is compatible with the given embeddings into $\pi_1(G)$. The proof of our Theorem is completed. □

1.13. The intersection of an adjoint orbit with a centralizer. We make the same assumptions as in Section 1.12.

Let $t \in T$. Let $\mathcal{O}_t$ be the adjoint orbit of t in G. By [**15**, Lemma IV.2.5],

$$\mathcal{O}_t \cap T = \{wt\}_{w \in W} \quad . \tag{1.172}$$

More generally, if H is a Lie subgroup of G, and $t \in H$, let $\mathcal{O}_H(t)$ be the adjoint orbit of t in H. In particular $\mathcal{O}_t = \mathcal{O}_G(t)$.

THEOREM 1.50. *If G is simply connected, if $u \in T'$, then*

$$\mathcal{O}_t \cap Z(u) = \bigcup_{w \in W_u\backslash W} \mathcal{O}_{Z(u)}(wt). \tag{1.173}$$

If t is regular, the above union is disjoint.

If G is not necessarily simply connected, if $t \in T$ is very regular,

$$(1.174) \qquad \mathcal{O}_t \cap Z(u) = \bigcup_{w \in W_{Z(u)_o} \backslash W} \mathcal{O}_{Z(u)_o}(wt),$$

$$\mathcal{O}_t \cap Z(u) = \bigcup_{w \in W_u \backslash W} \mathcal{O}_{Z(u)}(wt),$$

and the above unions are disjoint.

PROOF. If G is simply connected, then $Z(u)$ is connected, and T is a maximal torus in $Z(u)$. Let $g \in \mathcal{O}_t \cap Z(u)$. There is $g' \in Z(u)$ such that $g'gg'^{-1} \in \mathcal{O}_t \cap T$. By [**15**, Lemma IV.2.5], there is $w \in W$ such that

$$(1.175) \qquad g'gg'^{-1} = wt,$$

so that

$$(1.176) \qquad g \in \mathcal{O}_{Z(u)}(wt).$$

Therefore (1.173) holds. If $t \in T$ is regular, since G is simply connected, t is very regular. Therefore the $\{wt\}$ are distinct in T. Moreover two elements in T lie in the same $Z(u)$-orbit if and only if they lie in the same W_u-orbit. Using Theorem 1.38, it follows that when $t \in T$ is regular, the union in (1.173) is disjoint.

If G is non necessarily simply connected, if $g \in \mathcal{O}_t \cap Z(u)$, then $u \in Z(g)$. If t is very regular, $Z(g)$ is a maximal torus, which is included in $Z(u)_o$. Therefore $g \in Z(u)_o$. The above argument shows that there is $g' \in Z(u)_o$, and $w \in W$ such that

$$(1.177) \qquad g'gg'^{-1} = wt,$$

which is equivalent to

$$(1.178) \qquad g \in \mathcal{O}_{Z(u)_o}(wt).$$

So we have proved the first identity in (1.174). Since t is very regular in G, it is very regular in $Z(u)_o$. So the union in the first identity of (1.174) is disjoint.

Clearly

$$(1.179) \qquad \bigcup_{w \in W_u \backslash W} \mathcal{O}_{Z(u)}(wt) \subset \mathcal{O}_t \cap Z(u),$$

and also

$$(1.180) \qquad \bigcup_{w \in W_{Z(u)_o} \backslash W} \mathcal{O}_{Z(u)_o}(wt) \subset \bigcup_{w \in W_u \backslash W} \mathcal{O}_{Z(u)}(wt).$$

Therefore the second identity in (1.174) holds.

If $g \in Z(u), w \in W$ are such that $gtg^{-1} = wt$, if $g' \in N(T)$ represents w, then $g'^{-1}gt(g'^{-1}g)^{-1} = t$. Since t is very regular, $g'^{-1}g \in T$, so that $g \in Z(u) \cap N(T)$.It follows that $w \in W_u$. Therefore the second union in (1.174) is also disjoint.

The proof of our Theorem is completed. □

1.14. The stabilizer of an element of the Lie algebra, and coadjoint orbits. Let G be a compact connected semisimple Lie group. We use otherwise the notation of Section 1.1. Recall that $\tau : G \to \mathrm{Aut}(\mathfrak{g})$ is the adjoint representation.

DEFINITION 1.51. If $p \in \mathfrak{t}$, put

$$Z(p) = \{g \in G \ ; \ \tau(g) \cdot p = p\} . \tag{1.181}$$

By [**15**, Theorem IV.2.3], $Z(p)$ is a connected Lie subgroup of G. Then T is a maximal torus in $Z(p)$. Let $\mathfrak{z}(p)$ be the Lie algebra of $Z(p)$.

THEOREM 1.52. *If $p \in \mathfrak{t}$, then*

$$\mathfrak{z}(p) \otimes_{\mathbf{R}} \mathbf{C} = \mathfrak{t} \oplus \bigoplus_{\substack{\alpha \in R \\ \langle \alpha, p \rangle = 0}} \mathfrak{g}_\alpha \, . \tag{1.182}$$

Also the root system R_p of $Z(p)$ is given by

$$R_p = \{\alpha \in R \ ; \ \langle \alpha, p \rangle = 0\} . \tag{1.183}$$

If $Z(Z(p)) \subset Z(p)$ is the center of $Z(p)$, its Lie algebra $\mathfrak{z}(Z(p))$ is given by

$$\mathfrak{z}(Z(p)) = \{f \in \mathfrak{t}, \quad \text{for any } \alpha \in R_p, \langle \alpha, f \rangle = 0\} . \tag{1.184}$$

PROOF. The proof of these results is left to the reader. It is essentially the same as the proof of Theorem 1.38. □

DEFINITION 1.53. Let $\pi : \mathfrak{t} \otimes_{\mathbf{R}} \mathbf{C} \to \mathbf{C}$ be the monomial

$$\pi(t) = \prod_{\alpha \in R_+} \langle 2i\pi\alpha, t \rangle . \tag{1.185}$$

By [**15**, Corollary V.4.6 and Lemma V.4.10], if $w \in W$

$$\pi(wt) = \varepsilon_w \pi(t) . \tag{1.186}$$

Also if $t \in \mathfrak{t}$, one has the obvious

$$\det \mathrm{Ad}(t)_{|\mathfrak{g}/\mathfrak{t}} = \pi^2(t/i) . \tag{1.187}$$

By (1.187), we find that $\pi^2(t/i)$ does not depend on K, and lifts to a G-invariant function on $\mathfrak{g}$.

DEFINITION 1.54. Set

$$\begin{aligned} \mathfrak{g}_{\mathrm{reg}} &= \{p \in \mathfrak{g} \ , \ Z(p) \text{ is a maximal torus}\} , \\ \mathfrak{t}_{\mathrm{reg}} &= \{t \in \mathfrak{t} \ ; \ Z(t) = T\} . \end{aligned} \tag{1.188}$$

Clearly

$$\mathfrak{t}_{\mathrm{reg}} = \mathfrak{g}_{\mathrm{reg}} \cap \mathfrak{t} . \tag{1.189}$$

PROPOSITION 1.55. *The following identity holds*

$$\begin{aligned} \mathfrak{g}_{\mathrm{reg}} &= \{p \in \mathfrak{g} \ , \ \pi^2(p/i) \neq 0\} , \\ \mathfrak{t}_{\mathrm{reg}} &= \{t \in \mathfrak{t} \ ; \ \pi^2(t/i) \neq 0\} . \end{aligned} \tag{1.190}$$

PROOF. Take $t \in \mathfrak{t}$. Since $Z(t)$ is connected, $t \in \mathfrak{t}_{\mathrm{reg}}$ if and only if

$$\mathfrak{z}(t) = \mathfrak{t} . \tag{1.191}$$

Using Theorem 1.52, we get the second identity in (1.190). Also by [**15**, Theorem IV.1.6], any G-orbit in $\mathfrak{g}$ intersects $\mathfrak{t}$. Our Proposition follows. □

By (1.190), $t \in \mathfrak{t}_{\text{reg}}$ if and only if t lies in a Weyl chamber. The G-orbit of t in $\mathfrak{g}$ intersects $\mathfrak{t}$ at $|W|$ distinct elements, which form the W-orbit of t.

If $p \in \mathfrak{g}$, let $\mathcal{O}_p$ be the G-orbit of p. Clearly,

$$\pi_p : \ g \in G/Z(p) \mapsto gpg^{-1} \in \mathcal{O}_p \tag{1.192}$$

is a one to one map. Also by [**6**, Lemma 7.22], $\mathcal{O}_p$ is equipped with a canonical symplectic form $\sigma_{\mathcal{O}_p}$. In fact G acts on the left on $\mathcal{O}_p$. If $X \in \mathfrak{g}$, let $X^{\mathcal{O}_p}$ be the corresponding vector field on $\mathcal{O}_p$. Then if $X, Y \in \mathfrak{g}$, $q \in \mathcal{O}_p$,

$$\sigma_{\mathcal{O}_p,q}(X^{\mathcal{O}_p}, Y^{\mathcal{O}_p}) = \langle q, [X,Y] \rangle. \tag{1.193}$$

Let f_p be the left invariant 1-form on G

$$f_p = \langle p, g^{-1}dg \rangle \,. \tag{1.194}$$

Let π_p be the projection $G \to G/Z(p) \simeq \mathcal{O}_p$.

PROPOSITION 1.56. *The following identity holds*

$$df_p = -\pi_p^* \sigma_p \,. \tag{1.195}$$

In particular the restriction of f_p to $Z(p)$ is a closed 1-form.

PROOF. Clearly

$$df_p = \langle p, -\frac{1}{2}[g^{-1}dg, g^{-1}dg] \rangle \tag{1.196}$$

from which (1.195) follows. Since π_p maps $Z(p)$ to a constant in $G/Z(p)$, our Proposition follows. □

PROPOSITION 1.57. *If $p \in \overline{CR}^*$, f_p is an integral closed 1-form on $Z(p)$.*

PROOF. Clearly T is a maximal torus in $Z(p)$. Then by [**15**, Proposition V.7.6], $Z(p)/T$ is simply connected. Therefore $\pi_1(T)$ surjects on $\pi_1(Z(p))$. To verify that f_p is an integral 1-form, we only need to check that if $s \in S_1 \mapsto g_s \in T$ is smooth, the integral of f_p on this loop lies in $\mathbf{Z}$. Since $p \in \overline{CR}^*$, this is obvious. The proof of our Proposition is completed. □

Now we assume that $p \in \overline{CR}^*$. Take $g \in Z(p)$. Let $s \in [0,1] \mapsto g_s \in Z(p)$ be a smooth path such that $g_0 = 1, g_1 = g$.

PROPOSITION 1.58. *The map $g \mapsto \exp\left(2i\pi \int_0^1 g_s^* f_p ds\right) \in S_1$ defines a representation ρ_p of $Z(p)$.*

PROOF. This follows from Proposition 1.57. □

DEFINITION 1.59. Let L_p be the Hermitian line bundle on $\mathcal{O}_p$

$$L_p = G \times_{Z(p)} \mathbf{C} \,. \tag{1.197}$$

Clearly the connection

$$d + 2i\pi f_p \tag{1.198}$$

descends to a connection ∇^{L_p} on L_p.

PROPOSITION 1.60. *The following identity holds*

$$c_1(L_p, \nabla^{L_p}) = \sigma_p \,. \tag{1.199}$$

PROOF. This is obvious by (1.195). □

2. Fourier analysis on the centralizers of semisimple Lie groups

The purpose of this Section is to express certain Fourier series on T (which will later turn out to be the symplectic volumes of the stratas of the moduli space of flat G-bundles on the Riemann surface Σ, first computed by Witten **[63, 64]**) as residues of certain holomorphic functions in several complex variables. The main point is that it is then possible to compute explicitly the action of certain differential operators on these Fourier series. The results of this Section will be used in Sections 5, 6 and 7.

This Section is organized as follows. In Section 2.1, we make elementary constructions in linear algebra. In Section 2.2, we apply these constructions to the root system of G. In Section 2.3, given $u \in C/\overline{R}^*$ and the corresponding semisimple centralizer $Z(u)$, we consider an associated Fourier series $Q_u(t,x)$, which we express as a simple integral along the fibre of a torus fibration. In Section 2.4, we express $Q_u(t,x)$ in terms of iterated residues. In Sections 2.5-2.8, we introduce other related Fourier series, which are related in particular to the universal cover of semisimple centralizers. In Sections 2.9-2.11, we introduce our symplectic volume Fourier series, which are local polynomials on T. We express these Fourier series as residues. Finally, in Section 2.12, we compute the action of any power series of differential operators on these local polynomials.

As explained in the Introduction, residue techniques have been developed by Szenes **[53]**, **[54]** to handle the Witten Fourier series **[63, 64]**. The methods of Szenes are more conceptual than ours, which only usesimple linear algebra. It is probable that the results of this Section can be rephrased using Szenes's formalism.

2.1. Some linear algebra. Let V be a real vector space of dimension r. Let $e_1, \dots, e_r$ be a basis of V, let $e^1, \dots, e^r$ be the dual basis of V^*.

For $1 \le i \le r$, if $u_1, \dots, u_i \in V$, put

$$\langle u_1, \dots, u_i \rangle = \langle u_1 \wedge \dots \wedge u_i, e^1 \wedge \dots \wedge e^i \rangle . \tag{2.1}$$

Equivalently

$$\langle u_1, \dots, u_i \rangle = \frac{u_1 \wedge \dots u_i \wedge e_{i+1} \wedge \dots \wedge e_r}{e_1 \wedge \dots \wedge e_r} . \tag{2.2}$$

Let $f_1, \dots, f_r$ be another basis of V, let $f^1, \dots, f^r$ be the corresponding dual basis of V^*.

Let E, E', F, F' be the flags in V,

$$\begin{aligned} E &: \; 0 \subset \{e_1\} \subset \{e_1, e_2\} \dots \subset \{e_1, \dots, e_r\} = V, \\ E' &: \; V = \{e_1, \dots, e_r\} \supset \{e_2, \dots, e_r\} \dots \supset \{e_r\} \supset 0, \\ F &: \; 0 \subset \{f_1\} \subset \{f_1, f_2\} \dots \subset \{f_1, \dots, f_r\} = V, \\ F' &: \; \{f_1 \dots, f_r\} \supset \{f_2, \dots, f_r\} \supset \dots \{f_r\} \supset 0 . \end{aligned} \tag{2.3}$$

We assume that F lies in the orbit of E. Equivalently for i, $1 \le i \le r$, $f_1, \dots, f_i$, $e_{i+1}, \dots, e_r$ is a basis of E, i.e.

$$\langle f_1, \dots, f_i \rangle \neq 0 . \tag{2.4}$$

Let A be the (r,r) matrix expressing $f_1, \dots, f_r$ on the basis $e_1, \dots, e_r$. Then (2.4) says that the principal minors of A do not vanish.

On V^*, we can define the flags E^*, E'^*, F^*, F'^* associated to the basis $e^1, \dots, e^r$ and $f^1, \dots, f^r$. Now E^* lies in the orbit of F^*, i.e. for any i, $e^1, \dots, e^i, f^{i+1}, \dots, f^r$ is a basis of V^*.

For $0 \leq i \leq r-1$, let p_i be the projection from V on $\{e_{i+1}, \dots, e_r\}$ with kernel $\{f_1, \dots, f_i\}$. Clearly

$$p_{i+1} = p_{i+1} p_i \,. \tag{2.5}$$

Also for $0 \leq i \leq r-1$, $p_i f_{i+1}, \dots, p_i f_r$ is a basis of $\{e_{i+1}, \dots, e_r\}$. More precisely for $0 \leq j \leq r-i$, $p_i f_{i+1}, \dots, p_i f_{i+j}, e_{i+j+1}, \dots, e_r$ is a basis of $\{e_{i+1}, \dots, e_r\}$.

Clearly, if $t \in V$,

$$p_1 t = t - \frac{\langle t, e^1 \rangle}{\langle f_1, e^1 \rangle} f_1 \,. \tag{2.6}$$

By the above, there are similar formulas for $p_2, \dots, p_r$. In particular, by (2.5), (2.6), for $1 \leq i \leq r$, and $t \in V$,

$$p_i t = p_{i-1} t - \frac{\langle p_{i-1} t, e^i \rangle}{\langle p_{i-1} f_i, e^i \rangle} p_{i-1} f_i. \tag{2.7}$$

From (2.7), we get

$$p_i t = t - \sum_{j=1}^{i} \frac{\langle p_{j-1} t, e^j \rangle}{\langle p_{j-1} f_j, e^j \rangle} p_{j-1} f_j. \tag{2.8}$$

Using (2.8) with $i = r$, we obtain

$$t = \sum_{j=1}^{r} \frac{\langle p_{j-1} t, e^j \rangle}{\langle p_{j-1} f_j, e^j \rangle} p_{j-1} f_j. \tag{2.9}$$

For $0 \leq i \leq r-1$, let q_i be the projection from V^* on $\{f^{i+1}, \dots, f^r\}$ with kernel $\{e^1, \dots, e^i\}$. Then q_i is the transpose of p_i, i.e.

$$q_i = \tilde{p}_i. \tag{2.10}$$

Also, as in (2.6), if $x \in V^*$,

$$q_1 x = x - \frac{\langle x, f_1 \rangle}{\langle e^1, f_1 \rangle} e^1 \,. \tag{2.11}$$

Moreover the results which hold for the p_i's also hold for the q_i's.

THEOREM 2.1. *For* $1 \leq i \leq r$,

$$\sum_{j=1}^{i} \frac{\langle p_{j-1} t, e^j \rangle \langle f_j, q_{j-1} x \rangle}{\langle p_{j-1} f_j, q_{j-1} e^j \rangle} = \langle t, x \rangle - \langle p_i t, q_i x \rangle \,. \tag{2.12}$$

In particular,

$$\sum_{i=1}^{r} \frac{\langle p_{i-1} t, e_i \rangle \langle q_{i-1} x, f_i \rangle}{\langle p_{i-1} f_i, q_{i-1} e^i \rangle} = \langle t, x \rangle \,. \tag{2.13}$$

PROOF. Equation (2.12) follows from (2.8), and equation (2.13) from (2.9). □

PROPOSITION 2.2. *If $t \in V$, $x \in V^*$, for $0 \leq i \leq r$,*

$$
\begin{aligned}
(1-p_i)t &= \sum_{j=1}^{i} \frac{\langle f_1 \wedge \ldots \wedge f_{j-1} \wedge t \wedge f_{j+1} \wedge \ldots \wedge f_i, e^1 \wedge \ldots \wedge e^i \rangle}{\langle f_1 \wedge \ldots \wedge f_i, e^1 \wedge \ldots \wedge e^i \rangle} f_j, \\
p_i t &= \sum_{j=i+1}^{r} \frac{\langle f_1 \wedge \ldots \wedge f_i \wedge t, e^1 \wedge \ldots \wedge e^i \wedge e^j \rangle}{\langle f_1 \wedge \ldots \wedge f_i, e^1 \wedge \ldots \wedge e^i \rangle} e_j \\
&= \sum_{j=i+1}^{r} \Bigg(\langle t, e^j \rangle - \sum_{k=1}^{i} \\
&\quad \frac{\langle f_1 \wedge \ldots \wedge f_i, e^1 \wedge \ldots \wedge e^{k-1} \wedge e^j \wedge e^{k+1} \wedge \ldots \wedge e^i \rangle \langle t, e^k \rangle}{\langle f_1 \wedge \ldots \wedge f_i, e^1 \wedge \ldots \wedge e^i \rangle} \Bigg) e_j,
\end{aligned}
\tag{2.14}
$$

$$
\begin{aligned}
(1-q_i)x &= \sum_{j=1}^{i} \frac{\langle f_1 \wedge \ldots \wedge f_i, e^1 \wedge \ldots \wedge e^{j-1} \wedge x \wedge e^{j+1} \wedge \ldots \wedge e^i \rangle}{\langle f_1 \wedge \ldots \wedge f_i, e^1 \wedge \ldots \wedge e^i \rangle} e^j, \\
q_i x &= \sum_{j=i+1}^{r} \frac{\langle f_1 \wedge \ldots \wedge f_i \wedge f_j, e^1 \wedge \ldots \wedge e^i \wedge x \rangle}{\langle f_1 \wedge \ldots \wedge f_i, e^1 \wedge \ldots \wedge e^i \rangle} f^j \\
&= \sum_{j=i+1}^{r} \Bigg(\langle f_j, x \rangle - \sum_{k=1}^{i} \\
&\quad \frac{\langle f_1 \wedge \ldots \wedge f_{k-1} \wedge f_j \wedge f_{k+1} \wedge \ldots \wedge f_i, e^1 \wedge \ldots \wedge e^i \rangle \langle f_k, x \rangle}{\langle f_1 \wedge \ldots \wedge f_i, e^1 \wedge \ldots \wedge e^i \rangle} \Bigg) f^j.
\end{aligned}
$$

PROOF. Clearly we only need to prove the first two series of identities in (2.14). The first identity and the first part of the second identity are standard linear algebra. Also for $j \geq i+1$,

$$
\begin{aligned}
&\langle f_1 \wedge \ldots \wedge f_i \wedge t, e^1 \wedge \ldots \wedge e^i \wedge e^j \rangle = \langle f_1 \wedge \ldots \wedge f_i, e^1 \wedge \ldots \wedge e^i \rangle \langle t, e^j \rangle \\
&- \sum_{k=1}^{i} (-1)^{k-i} \langle f_1 \wedge \ldots \wedge f_i, e^1 \wedge \ldots \wedge e^{k-1} \wedge e^{k+1} \wedge \ldots \wedge e^i \wedge e^j \rangle \langle t, e^k \rangle \\
&= \langle f_1 \wedge \ldots \wedge f_i, e^1 \wedge \ldots \wedge e^i \rangle \langle t, e^j \rangle \\
&- \sum_{k=1}^{i} \langle f_1 \wedge \ldots \wedge f_i, e^1 \wedge \ldots \wedge e^{k-1} \wedge e^j \wedge e^{k+1} \wedge \ldots \wedge e^i \rangle \langle t, e^k \rangle .
\end{aligned}
\tag{2.15}
$$

The proof of our Proposition is completed. □

PROPOSITION 2.3. *For any j, $1 \leq j \leq r$*

$$
\begin{aligned}
&\langle f_j, q_{j-1} x \rangle = \\
&\langle f_j, x \rangle - \sum_{k=1}^{j-1} \frac{\langle f_1 \wedge \ldots \wedge f_{k-1} \wedge f_j \wedge f_{k+1} \wedge \ldots \wedge f_{j-1}, e^1 \wedge \ldots \wedge e^{j-1} \rangle}{\langle f_1 \wedge \ldots \wedge f_{j-1}, e^1 \wedge \ldots \wedge e^{j-1} \rangle} \langle f_k, x \rangle .
\end{aligned}
\tag{2.16}
$$

PROOF. This is a consequence of the last equality in (2.14). □

REMARK 2.4. From (2.16), we find that $\langle f_j, q_{j-1} x \rangle = \langle p_{j-1} f_j, x \rangle$ depends only on the $\langle f_k, x \rangle$, $1 \leq k \leq j$.

2.2. The linear algebra of the basis of a root system. Let G be a compact connected and simply connected simple Lie group of rank r. Let T be a maximal torus in G. Let $\mathfrak{t}$ be the Lie algebra of T. Let K be a Weyl chamber in $\mathfrak{t}$. Otherwise, we use the notation of Section 1.1.

We will also use the notation of Section 2.1, with $V = \mathfrak{t}^*$. We identify $\mathfrak{t}$ and $\mathfrak{t}^*$ by the scalar product $\langle \, , \, \rangle$ of Section 1.2. Then

$$r = \dim \mathfrak{t} . \tag{2.17}$$

Let $e_1, \dots, e_r \subset R_+$ be the simple basis of $\mathfrak{t}^*$ associated to K [**15**, Proposition V.4.5]. Any $\alpha \in R_+$ is a linear combination with non negative integral coefficients of $e_1, \dots, e_r$. Then $e_1, \dots, e_r$ generate $\overline{R}$. Let $e^1, \dots, e^r$ be the corresponding dual basis of $\overline{R}^* \subset \mathfrak{t}$.

Let $\{\alpha_1, \dots, \alpha_\ell\}$ be an ordering of R_+, such that

$$\alpha_i = e_i \text{ for } 1 \leq i \leq r . \tag{2.18}$$

Recall that we use the notation of Section 2.1 . Clearly all the $\langle \alpha_{i_1}, \dots, \alpha_{i_j} \rangle$ lie in $\mathbf{Z}$.

If G is not simply laced, m was defined in (1.8) and is equal to 2 or to 3. By convention, if G is simply laced, we take $m = 1$. By (1.12),

$$m\overline{R} \subset \overline{CR}. \tag{2.19}$$

DEFINITION 2.5. Let $d \in \mathbf{N}^*$ be a common multiple of the $m|\langle \alpha_{i_1} \wedge \dots \wedge \alpha_{i_r} \rangle|$.

Observe that since $e_1, \dots, e_r \in R_+$, for any j, $1 \leq j \leq r$, d is a multiple of $m\langle \alpha_{i_1}, \dots, \alpha_{i_j} \rangle \in \mathbf{Z}$. Also by (2.19),

$$d\overline{R} \subset \overline{CR}. \tag{2.20}$$

Recall that $C \subset \mathfrak{t}$ was defined in Definition 1.35.

PROPOSITION 2.6. *The following identity holds*

$$dC \subset \overline{R}^* . \tag{2.21}$$

PROOF. Let $u \in C$, let $\alpha_{i_1}, \dots, \alpha_{i_r} \in R_+$ be a basis of $\mathfrak{t}^*$ such that for $1 \leq j \leq r$, $\langle \alpha_{i_j}, u \rangle \in \mathbf{Z}$. Since $e_1, \dots, e_r$ is a basis of $\overline{R}$, we find that if H is the lattice generated by $\alpha_{i_1}, \dots, \alpha_{i_r}$,

$$d\overline{R} \subset H , \tag{2.22}$$

which is equivalent to

$$dH^* \subset \overline{R}^* . \tag{2.23}$$

Since $u \in H^*$, from (2.23) , we get (2.21). The proof of our Proposition is completed. □

REMARK 2.7. Take $n \geq 2$. Since SU(n) is simply laced, $m = 1$. Also by [**15**, Proposition V.6.3], all the $|\langle \alpha_{i_1}, \dots, \alpha_{i_j} \rangle|$ are equal to 1. Therefore, for $G = \mathrm{SU}(n)$, we can take $d = 1$. Using (1.114) and (2.21), we recover Proposition 1.40.

DEFINITION 2.8. A family $I = (i_1, \dots, i_r)$ of distinct indices in $\{1, \dots, \ell\}$ is said to be generic if

$$\langle \alpha_{i_1}, \dots, \alpha_{i_j} \rangle \neq 0 \text{ for } 1 \leq j \leq r . \tag{2.24}$$

Given a generic family $I = (i_1, \dots, i_r)$, we now use the notation in Section 2.1 associated to the given basis $e_1, \dots, e_r$ and $\alpha_{i_1}, \dots, \alpha_{i_r}$ of $\mathfrak{t}^* \simeq \mathfrak{t}$. In particular the operators which appear in Section 2.1 will be denoted with the superscript I, to mark their dependence on I.

Let $\alpha^{i_1}, \dots, \alpha^{i_r}$ be the basis of $\mathfrak{t}^* \simeq \mathfrak{t}$ which is dual to $\alpha_{i_1}, \dots, \alpha_{i_r}$.

Recall that (2.18) holds.

DEFINITION 2.9. A family $I = (i_1, \dots, i_{j-1})$ of $j-1$ distinct elements of $\{1, \dots, \ell\}$ is said to be generic if $I_{j-1} = \{i_1, \dots, i_{j-1}, j, j+1, \dots, r\}$ is generic.

If $I = (i_1, \dots, i_r)$ is generic, $I_{j-1} = (i_1, \dots, i_{j-1})$ is also generic, and by construction (or by (2.14)),

$$p^I_{j-1} = p^{I_{j-1}}_{j-1}, \qquad q^I_{j-1} = q^{I_{j-1}}_{j-1}. \tag{2.25}$$

DEFINITION 2.10. If $x \in \mathbf{C}^\ell$, and if $I = (i_1, \dots, i_r)$ is generic, put

$$x^I = \sum_1^r x_{i_j} \alpha^{i_j} \in \mathfrak{t} \simeq \mathfrak{t}^*. \tag{2.26}$$

By (2.25),

$$q^I_{j-1} x^I = q^{I_{j-1}}_{j-1} x^I. \tag{2.27}$$

Also by Proposition 2.3,

$$\langle \alpha_{i_j}, q^I_{j-1} x^I \rangle = \langle p^I_{j-1} \alpha_{i_j}, x^I \rangle = x_{i_j} - \sum_{k=1}^{j-1} \frac{\langle \alpha_{i_1}, \dots, \alpha_{i_{k-1}}, \alpha_{i_j}, \alpha_{i_{k+1}}, \dots, \alpha_{i_{j-1}} \rangle}{\langle \alpha_{i_1}, \dots, \alpha_{i_{j-1}} \rangle} x_{i_k}. \tag{2.28}$$

Note that the right hand side of (2.28) only depends on $x_{i_1}, \dots, x_{i_j}$.

Assume that $I = (i_1, \dots, i_r)$ is generic. Then by Proposition 2.2, if $t \in \mathfrak{t}$, $y \in \mathfrak{t}^*$

$$\begin{aligned} \langle p^I_{j-1} \alpha_{i_j}, e^j \rangle &= \frac{\langle \alpha_{i_1}, \dots, \alpha_{i_j} \rangle}{\langle \alpha_{i_1}, \dots, \alpha_{i_{j-1}} \rangle}, \\ \langle p^I_{j-1} \alpha_{i_j}, t \rangle &= \frac{\langle \alpha_{i_1} \wedge \dots \wedge \alpha_{i_j}, e^1 \wedge \dots \wedge e^{j-1} \wedge t \rangle}{\langle \alpha_{i_1}, \dots, \alpha_{i_{j-1}} \rangle}, \\ \langle p^I_{j-1} y, e^j \rangle &= \frac{\langle \alpha_{i_1}, \dots, \alpha_{i_{j-1}}, y \rangle}{\langle \alpha_{i_1}, \dots, \alpha_{i_{j-1}} \rangle}. \end{aligned} \tag{2.29}$$

In particular, by (2.29), we find that

$$\frac{d}{\langle p^I_{j-1} \alpha_{i_j}, e^j \rangle} \in \mathbf{Z}. \tag{2.30}$$

Also by (2.28), (2.29), if $x \in \mathbf{Z}^\ell$,

$$d \frac{\langle p^I_{j-1} \alpha_{i_j}, x^I \rangle}{\langle p^I_{j-1} \alpha_{i_j}, e^j \rangle} \in \mathbf{Z}. \tag{2.31}$$

DEFINITION 2.11. For $u \in C/\overline{R}^*$, let $I_u \subset \{1, \dots, \ell\}$ be given by

$$I_u = \{i, 1 \le i \le \ell, \alpha_i \in R_{u,+}\}. \tag{2.32}$$

Put

(2.33) $$\ell_u = |I_u|.$$

2.3. Fourier series and integration along the fibre.

DEFINITION 2.12. For $u \in C/\overline{R}^*$, let $\mathbf{R}^{I_u} \subset \mathbf{R}^\ell$ be given by

(2.34) $$\mathbf{R}^{I_u} = \{s = (s^1, \dots, s^\ell) \in \mathbf{R}^\ell \,,\ s^i = 0 \ \text{ for } i \notin I_u\}\,.$$

Let $a_u : \mathbf{R}^{I_u} \to \mathfrak{t}^* \simeq \mathfrak{t}$ be given by

(2.35) $$a_u(s) = \sum_{i \in I_u} s^i \alpha_i\,.$$

The transpose $\widetilde{a}_u : \mathfrak{t} \simeq \mathfrak{t}^* \to \mathbf{R}^{I_u}$ is given by

(2.36) $$\widetilde{a}_u(t) = (\langle \alpha_i, t \rangle)_{i \in I_u}\,.$$

Set

(2.37) $$V_u = \ker a_u.$$

Then we have the exact sequence

(2.38) $$0 \to V_u \to \mathbf{R}^{I_u} \xrightarrow[a_u]{} \mathfrak{t} \to 0\ .$$

We define $\mathbf{Z}^{I_u} \subset \mathbf{Z}^\ell$, $(\mathbf{R}/\mathbf{Z})^{I_u} \subset (\mathbf{R}/\mathbf{Z})^\ell$ as before. Then

(2.39) $$a_u(\mathbf{Z}^{I_u}) = \overline{R}_u\,.$$

Also a_u induces a surjection $(\mathbf{R}/\mathbf{Z})^{I_u} \to \mathfrak{t}/\overline{R}$. Set

(2.40) $$K_u = \ker a_u \subset (\mathbf{R}/\mathbf{Z})^{I_u}.$$

We have the exact sequence

(2.41) $$0 \to K_u \to (\mathbf{R}/\mathbf{Z})^{I_u} \xrightarrow[a_u]{} \mathfrak{t}/\overline{R} \to 0\ .$$

Set

(2.42) $$\gamma_u = \mathbf{Z}^{I_u} \cap V_u\,.$$

Then K_u is a union of $|\overline{R}/\overline{R}_u|$ tori V_u/γ_U .

DEFINITION 2.13. For $u \in C/\overline{R}^*$, let $H_u \subset \mathfrak{t}/\overline{R}$ be given by

(2.43) $$H_u = \{t \in \mathfrak{t}/\overline{R}, t = \sum_{j \in \mathcal{J} \subset I_u} t^j \alpha_j, \ \text{ and } \{\alpha_j, j \in \mathcal{J}\} \ \text{ does not span } \mathfrak{t}^* \simeq \mathfrak{t}\}\,.$$

Then H_u is a finite union of hypertori in T. Put

(2.44) $$H = H_0\,.$$

Clearly,

(2.45) $$H = \{t \in \mathfrak{t}/\overline{R}, t = \sum_{j \in \mathcal{J} \subset \{1, \dots, \ell\}} t^j \alpha_j, \ \text{ and } \{\alpha_j, j \in \mathcal{J}\} \ \text{ does not span } \mathfrak{t}^* \simeq \mathfrak{t}\}\,.$$

For any $u \in C/\overline{R}^*$,

(2.46) $$H_u \subset H\,.$$

PROPOSITION 2.14. *Let $I = (i_1, \dots, i_r) \subset I_u$ be generic. Then if $t \in (\mathfrak{t}/\overline{R}) \setminus H$ is represented by $\bar{t} \in \mathfrak{t}$,*

$$\langle p^I_{j-1}\bar{t}, e^j\rangle \notin \mathbf{Z}. \tag{2.47}$$

PROOF. By (2.29),

$$\langle p^I_{j-1}\bar{t}, e^j\rangle = \frac{\langle \alpha_{i_1}, \dots, \alpha_{i_{j-1}}, \bar{t}\rangle}{\langle \alpha_{i_1}, \dots, \alpha_{i_{j-1}}\rangle}. \tag{2.48}$$

Now for $\bar{t} = e_j$, the expression (2.48) is equal to 1. Also $\langle \alpha_{i_1}, \dots, \alpha_{i_{j-1}}, \bar{t}\rangle$ vanishes if and only if $\bar{t}$ is a linear combination of $\alpha_{i_1}, \dots, \alpha_{i_{j-1}}, e_{j+1}, \dots, e_r$. Therefore the condition

$$\langle p^I_{j-1}\bar{t}, e^j\rangle \in \mathbf{Z} \tag{2.49}$$

is equivalent to

$$\bar{t} = \sum_{k=1}^{j-1} a^k \alpha_{i_k} + b e_j + \sum_{k=j+1}^{r} c^k e_k \ , \ a^k, c^k \in \mathbf{R}, b \in \mathbf{Z}. \tag{2.50}$$

Then,

$$\bar{t} = \sum_{k=1}^{j-1} a^k \alpha_{i_k} + \sum_{k=j+1}^{r} c^k e_k \ \text{ in } \ \mathfrak{t}/\overline{R}, \tag{2.51}$$

so that $t \in H$.

The proof of our Proposition is completed. □

Let dt be the Lebesgue measure on $\mathfrak{t}$ associated with $\langle\ ,\ \rangle$. We still denote by dt the Lebesgue measure on $\mathfrak{t}/\overline{R}$. We map $L^1(\mathfrak{t}/\overline{R})$ into $\mathcal{D}'(\mathfrak{t}/\overline{R})$ by the map $f \mapsto \dfrac{f\,dt}{\mathrm{Vol}(\mathfrak{t}/\overline{R})}$. We use the same convention for other tori.

Clearly a_u induces a map a_{u*} from $\mathcal{D}'((\mathbf{R}/\mathbf{Z})^{I_u})$ into $\mathcal{D}'(\mathfrak{t}/\overline{R})$.

PROPOSITION 2.15. *If $g \in L^1((\mathbf{R}/\mathbf{Z})^{I_u})$, then $a_{u*}g \in L^1(\mathfrak{t}/\overline{R})$. More precisely,*

$$a_{u*}g(t) = \int_{K_u} g(t+s)\frac{ds}{\mathrm{Vol}(K_u)}. \tag{2.52}$$

Also if $k \in \mathbf{Z}^{I_u}$, then

$$\begin{aligned} a_{u*}[e^{2i\pi\langle k,s\rangle}] &= e^{2i\pi\langle\lambda,t\rangle} \ \textit{if there is} \ \lambda \in \overline{R}^* \ \textit{such that} \ k_i = \langle\lambda,\alpha_i\rangle, i \in I_u, \\ &= 0 \qquad \textit{otherwise}. \end{aligned} \tag{2.53}$$

PROOF. The proof of this result is trivial. □

DEFINITION 2.16. For $u \in C/\overline{R}^*$, $x \in (\mathbf{C} \setminus 2i\pi\mathbf{Z})^\ell$, $t \in \mathfrak{t}/\overline{R}$, set

$$Q_u(t,x) = \sum_{\lambda\in\overline{R}^*} \frac{\exp(2i\pi\langle\lambda,t\rangle)}{\prod_{i\in I_u}(2i\pi\langle\alpha_i,\lambda\rangle - x_i)}. \tag{2.54}$$

Clearly as a function of t, $Q_u(t,x)$ is a well-defined distribution on $\mathfrak{t}/\overline{R}$.

For $M \in \mathbf{N}$, we also consider the partial sums

$$Q^M_u(t,x) = \sum_{\substack{\lambda\in\overline{R}^* \\ |\lambda|\le M}} \frac{\exp(2i\pi\langle\lambda,t\rangle)}{\prod_{i\in I_u}(2i\pi\langle\alpha_i,\lambda\rangle - x_i)}. \tag{2.55}$$

Note that $Q_u(t,x)$ depends only on the projection of x on $\mathbf{C}^{I_u}$.

THEOREM 2.17. *For any* $n \in \mathbf{N}^*$, $u \in C/\overline{R}^*$, $v \in \frac{\overline{R}^*}{n}$, $t \in \mathfrak{t}/\overline{R}$, *the following identity holds*

$$n^{\ell_u - r} \sum_{h \in \overline{R}/n\overline{R}} \exp(2i\pi\langle v, t+h\rangle) Q_u\Big(\frac{t+h}{n}, nx\Big) = Q_u(t, x + 2i\pi\widetilde{a}_u v). \tag{2.56}$$

PROOF. Clearly, if $\lambda \in \overline{R}^*$, $v \in \overline{R}^*/n$,

$$\frac{1}{n^r} \sum_{h \in \overline{R}/n\overline{R}} \exp(2i\pi\langle\frac{\lambda}{n} + v, t+h\rangle) = \exp(2i\pi\langle\frac{\lambda}{n} + v, t\rangle) \ \text{ if } \ \frac{\lambda}{n} + v \in \overline{R}^*, \tag{2.57}$$
$$= 0 \quad \text{otherwise}.$$

From (2.54),(2.57) , we get

$$\begin{aligned} n^{\ell_u - r} \sum_{h \in \overline{R}/n\overline{R}} \exp(2i\pi\langle v, t+h\rangle) Q_u(\frac{t+h}{n}, nx) \\ = \sum_{\lambda \in \overline{R}^*} \frac{\exp(2i\pi\langle\lambda, t\rangle)}{\prod_{i \in I_u} (2i\pi\langle\alpha_i, \lambda\rangle - (x_i + 2i\pi\langle\alpha_i, v\rangle))} \\ = Q_u(t, x + 2i\pi\widetilde{a}_u v). \end{aligned} \tag{2.58}$$

The proof of our Theorem is completed. □

THEOREM 2.18. *The partial sums* $Q_u^M(t,x)$ *converge uniformly together with their derivatives to* $Q_u(t,x)$ *on compact subsets of* $(\mathfrak{t}/\overline{R}) \setminus H_u \times (\mathbf{C} \setminus 2i\pi\mathbf{Z})^\ell$. *The following identity of distributions holds*

$$Q_u(t,x) = (-1)^{\ell_u} a_{u*} \left[\frac{1}{\prod_{i \in I_u} (\exp(x_i) - 1)} \exp(\langle x, s\rangle) \right] \quad \text{in } \mathcal{D}'(\mathfrak{t}/\overline{R}). \tag{2.59}$$

In particular for $x \in (\mathbf{C} \setminus 2i\pi\mathbf{Z})^\ell$, $Q_u(t,x)$ *is a distribution in* $L^\infty(\mathfrak{t}/\overline{R})$. *Also (2.59) is an identity of smooth functions on* $(\mathfrak{t}/\overline{R}) \setminus H_u \times (\mathbf{C} \setminus 2i\pi\mathbf{Z})^\ell$.

PROOF. For $x \in \mathbf{C} \setminus 2i\pi\mathbf{Z}$, the Fourier series for e^{xs} ($s \in [0,1]$) is given by

$$e^{xs} = (e^x - 1) \sum_{k \in \mathbf{Z}} \frac{e^{2i\pi ks}}{-2i\pi k + x}. \tag{2.60}$$

Since e^{xs} is smooth on $[0,1]$, the partial sums in (2.60) converge uniformly together with their derivatives to e^{xs} on compact subsets of $\mathbf{R}/\mathbf{Z} \setminus \{0\} \times \mathbf{C} \setminus 2i\pi\mathbf{Z}$.

From Proposition 2.15 and from (2.60) , we get

$$a_{u*}[e^{\langle x,s\rangle}] = \prod_{i \in I_u} (e^{x_i} - 1) \sum_{\lambda \in \overline{R}^*} \frac{e^{2i\pi\langle\lambda,t\rangle}}{\prod_{i \in I_u} (-2i\pi\langle\alpha_i, \lambda\rangle + x_i)}, \tag{2.61}$$

which coincides with (2.59).

Also for $x \neq 0$, the wave front set of the distribution e^{xs} on S^1 is just $\{0\} \times \mathbf{R}^*$. By [**26**, Theorem 8.2.13], we see that $(t,p) \in \mathfrak{t}/\overline{R} \times \mathbf{R}^{I_u} \setminus \{0\}$ lies in the wave front set of $Q_u(t,x)$ only if $t = \sum_{i \in I_u} t^i \alpha_i$, and $\langle p, \alpha_i\rangle = 0$ when $t^i \notin \mathbf{Z}$. Therefore $Q_u(t,x)$ is smooth on $(\mathfrak{t}/\overline{R}) \setminus H_u$.

The proof of our Theorem if completed. □

2.4. Iterated residues and the series $Q_u(t,x)$. Recall that by Theorem 2.18, for generic values of x, $Q_u(t,x)$ is smooth on $(\mathfrak{t}/\overline{R}) \setminus H_u$. Therefore by (2.46), $Q_u(t,x)$ is smooth on $(\mathfrak{t}/\overline{R}) \setminus H$.

Let $x \in \mathbf{R} \mapsto [x] \in [0,1[$ be the periodic function of period 1, such that $[x] = x$ on $[0,1[$.

In the sequel, if $t \in \mathfrak{t}/\overline{R}$, we represent t by a given element in $\mathfrak{t}$, which we also denote by t.

Also the map $(f^1,\dots,f^r) \in \{0,1,\dots,d-1\}^r \mapsto f = \sum_1^r f^i e_i \in \overline{R}$ defines a one to one map into $\overline{R}/d\overline{R}$. In the sequel, we will always identify $f \in \overline{R}/d\overline{R}$ to the corresponding element in $\{0,1,\dots,d-1\}^r$.

THEOREM 2.19. *For any $u \in C/\overline{R}^*$, for generic values of $x \in (\mathbf{C} \setminus 2i\pi\mathbf{Z})^\ell$, for $t \in (\mathfrak{t}/\overline{R}) \setminus H$, if we still denote by t a representative in $\mathfrak{t}$,*

$$
\begin{aligned}
Q_u(t,x) = (-1)^r & \sum_{\substack{I=(i_1,\dots,i_r)\subset I_u \\ I \text{ generic},\ f\in \frac{\overline{R}}{d\overline{R}}}} \frac{1}{\langle \alpha_{i_1},\dots,\alpha_{i_r}\rangle} \\
& \exp\left\{ d\sum_{j=1}^{r} \frac{\langle p^I_{j-1}\alpha_{i_j}, x^I\rangle}{\langle p^I_{j-1},\alpha_{i_j}, e^j\rangle}\left[\frac{1}{d}\langle p^I_{j-1}(t+f), e^j\rangle\right]\right\} \\
& \frac{1}{\prod_{i\in I_u\setminus I}\left(\langle \alpha_i, x^I\rangle - x_i\right)} \prod_{j=1}^{r} \frac{1}{\exp\left(\frac{d\langle p^I_{j-1}\alpha_{i_j}, x^I\rangle}{\langle p^I_{j-1}\alpha_{i_j}, e^j\rangle}\right) - 1}.
\end{aligned}
\tag{2.62}
$$

PROOF. Take $1 \le j \le r$, $I = (i_1,\dots,i_{j-1}) \subset I_u$ such that $\langle \alpha_{i_1}\rangle \neq 0,\dots,$ $\langle \alpha_{i_1},\dots,\alpha_{i_{j-1}}\rangle \neq 0$. For $\widehat{k} = (k_{j+1},\dots,k_r) \in \mathbf{Z}^{r-j}$, we identify $\widehat{k}$ with $\sum_{i=j+1}^{r} k_i e^i \in \mathfrak{t} \simeq \mathfrak{t}^*$. In the sequel, p^I_{j-1} denotes the projection $\mathfrak{t} \to \{e_j,\dots,e_r\}$ with kernel $\{\alpha_{i_1},\dots,\alpha_{i_{j-1}}\}$. In view of (2.28), for $i \in I_u$, $i \notin I$, we will use the abusive notation

$$
\begin{aligned}
\langle p^I_{j-1}\alpha_i, x^I\rangle = & \\
& x_i - \sum_{k=1}^{j-1} \frac{\langle \alpha_{i_1},\dots,\alpha_{i_{k-1}},\alpha_i,\alpha_{i_{k+1}},\dots,\alpha_{i_{j-1}}\rangle}{\langle \alpha_{i_1},\dots,\alpha_{i_{j-1}}\rangle} x_{i_k}.
\end{aligned}
\tag{2.63}
$$

Equation (2.63) will in fact be a definition for the left-hand side. For $s \in \mathbf{R}$, if $1 \le j \le r$, set

$$
Q_I = \sum_{k\in\mathbf{Z}} \frac{e^{2i\pi ks}}{\prod_{i\in I_u\setminus I}\left(\langle p^I_{j-1}\alpha_i, 2i\pi(ke^j + \widehat{k}) - x^I\rangle\right)}.
\tag{2.64}
$$

We claim that for at least one $i \in I_u \setminus I$,

$$
\langle p^I_{j-1}\alpha_i, e^j\rangle \neq 0.
\tag{2.65}
$$

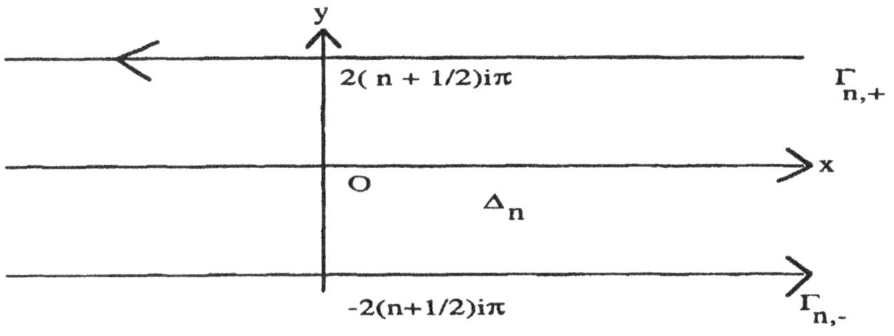

FIGURE 2.1

In fact by (2.29),

$$\langle p^I_{j-1}\alpha_i, e^j\rangle = \frac{\langle \alpha_{i_1}, \dots, \alpha_{i_{j-1}}, \alpha_i\rangle}{\langle \alpha_{i_1}, \dots, \alpha_{i_{j-1}}\rangle}. \tag{2.66}$$

Since the elements of $R_{u,+}$ span $\mathfrak{t}^* \simeq \mathfrak{t}$, for at least one $i \in I_u \setminus I$, (2.66) does not vanish.

Clearly

$$Q_I = \frac{1}{d} \sum_{\substack{k\in\mathbf{Z}\\ 0\le f<d}} \frac{\exp(2i\pi\frac{k}{d}(s+f))}{\prod_{i\in I_u\setminus I} \left(2i\pi\langle p^I_{j-1}\alpha_i, e^j\rangle \frac{k}{d} + \langle p^I_{j-1}\alpha_i, 2i\pi\widehat{k} - x^I\rangle\right)}. \tag{2.67}$$

Then for generic $x \in \mathbf{C}^\ell$, we can write (2.67) in the form

$$Q_I = \frac{1}{d} \sum_{0\le f<d} \sum_{k\in\mathbf{Z}} \operatorname{Res}_{a=2i\pi k} \left\{ \frac{e^{a[\frac{s+f}{d}]}}{\left(\prod_{i\in I_u\setminus I} \left(\frac{\langle p^I_{j-1}\alpha_i, e^j\rangle a}{d} + \langle p^I_{j-1}\alpha_i, 2i\pi\widehat{k} - x^I\rangle\right)\right)(e^a - 1)} \right\}. \tag{2.68}$$

Assume that $s \notin \mathbf{Z}$. Then for $f \in \mathbf{N}$, $0 \le f < d$,

$$0 < \left[\frac{s+f}{d}\right] < 1. \tag{2.69}$$

Also by (2.29) , if $\langle \alpha_{i_1}, \dots, \alpha_{i_{j-1}}, \alpha_i\rangle \neq 0$, then

$$\frac{d}{\langle p^I_{j-1}\alpha_i, e^j\rangle} \in \mathbf{Z}, \qquad \frac{d\langle p^I_{j-1}\alpha_i, \widehat{k}\rangle}{\langle p^I_{j-1}\alpha_i, e^j\rangle} \in \mathbf{Z}. \tag{2.70}$$

For $n \in \mathbf{N}$, consider the contour $\Gamma_n = \Gamma_{n,+} \cup \Gamma_{n,-}$ given in Figure 2.1, and its interior Δ_n. For $f \in \mathbf{N}$, $0 \le f < d$, put

$$g_f(a) = \frac{e^{a[\frac{s+f}{d}]}}{\left(\prod_{i\in I_u\setminus I}\left(\frac{\langle p^I_{j-1}\alpha_i, e^j\rangle a}{d} + \langle p^I_{j-1}\alpha_i, 2i\pi\widehat{k} - x^I\rangle\right)\right)(e^a-1)}. \tag{2.71}$$

By (2.69), as $a \in \Delta_n$, $|a| \to +\infty$,

$$g_f(a) \to 0. \tag{2.72}$$

Also the integral $\int_{\Gamma_n} g_f(a)da$ converges. We can then use the residue theorem to evaluate $\int_{\Gamma_n} g_f(a)da$. Finally by (2.65), (2.71), as $n \to +\infty$, $\int_{\Gamma_n} g_f(a)da \to 0$. Then we find that for generic $x \in \mathbf{C}^\ell$,

$$\sum_{a\in\mathbf{C}} \mathrm{Res}_a g_f(a) = 0. \tag{2.73}$$

Now for generic $x \in \mathbf{C}^\ell$, the poles of $g_f(a)$ other than $\{2i\pi k\}_{k\in\mathbf{Z}}$ are simple (this follows in particular from (2.63)), and given by

$$a = -d\frac{\langle p^I_{j-1}\alpha_{i_j}, 2i\pi\widehat{k} - x^I\rangle}{\langle p^I_{j-1}\alpha_{i_j}, e^j\rangle}, \ i_j \in I_u \setminus I,\ \langle \alpha_{i_1}, \dots, \alpha_{i_j}\rangle \neq 0. \tag{2.74}$$

In the sequel, we use the notation

$$(I, i_j) = (i_1, \dots, i_j). \tag{2.75}$$

Observe that if $(i_1, \dots, i_j)$ is generic, if $y \in \mathfrak{t}^*$ for $1 \le k \le j-1$,

$$\begin{aligned}\frac{\langle \alpha_{i_1}, \dots, \alpha_{i_{k-1}}, y, \alpha_{i_{k+1}}, \dots, \alpha_{i_j}\rangle}{\langle \alpha_{i_1}, \dots, \alpha_{i_j}\rangle} = \\ \frac{1}{\langle \alpha_{i_1}, \dots, \alpha_{i_{j-1}}\rangle}\Big(\langle \alpha_{i_1}, \dots, \alpha_{i_{k-1}}, y, \alpha_{i_{k+1}}, \dots, \alpha_{i_{j-1}}\rangle \\ - \frac{\langle \alpha_{i_1}, \dots, \alpha_{i_{k-1}}, \alpha_{i_j}, \alpha_{i_{k+1}}, \dots, \alpha_{i_{j-1}}\rangle}{\langle \alpha_{i_1}, \dots, \alpha_{i_j}\rangle}\langle \alpha_{i_1}, \dots, \alpha_{i_{j-1}}, y\rangle\Big).\end{aligned} \tag{2.76}$$

In fact if y lies the vector space spanned by $\alpha_{i_1}, \dots, \alpha_{i_{k-1}}, \alpha_{i_{k+1}}, \dots, \alpha_{i_{j-1}}, \alpha_{i_j}$, $e_{j+1}, \dots, e_r$ both sides of (2.76) vanish, and if $y = \alpha_{i_k}$ both sides are equal to 1. So by (2.6), (2.63), (2.66), (2.68), (2.70), (2.73) and (2.76) with $y = \alpha_i$, $i \notin (I, i_j)$, we get

$$\begin{aligned}Q_I = -\sum_{\substack{i_j\in I_u\setminus I\\ \langle\alpha_{i_1},\dots,\alpha_{i_j}\rangle\neq 0,\, 0\le f<d}} \frac{\langle \alpha_{i_1}, \dots, \alpha_{i_{j-1}}\rangle}{\langle \alpha_{i_1}, \dots, \alpha_{i_j}\rangle}. \\ \exp\left(-2i\pi\frac{\langle p^I_{j-1}\alpha_{i_j}, \widehat{k}\rangle}{\langle p^I_{j-1}\alpha_{i_j}, e^j\rangle}(s+f) + \frac{d\langle p^I_{j-1}\alpha_{i_j}, x^I\rangle}{\langle p^I_{j-1}\alpha_{i_j}, e^j\rangle}\left[\frac{s+f}{d}\right]\right) \\ \frac{1}{\left(\prod_{i\in I_u\setminus(I,i_j)}\left(\langle p_j^{(I,i_j)}\alpha_i, 2i\pi\widehat{k} - x^I\rangle\right)\right)\left(\exp\left(\frac{d\langle p^I_{j-1}\alpha_{i_j}, x^I\rangle}{\langle p^I_{j-1}\alpha_{i_j}, e^j\rangle}\right) - 1\right)}.\end{aligned} \tag{2.77}$$

Now by (2.7), for $j' \geq j+1$,

$$\langle p_j^{(I,i_j)} e_j, e^{j'} \rangle = -\frac{\langle p_{j-1}^I \alpha_{i_j}, e^{j'} \rangle}{\langle p_{j-1}^I \alpha_{i_j}, e^j \rangle}. \tag{2.78}$$

Using (2.77), (2.78), we obtain

$$Q_I = - \sum_{\substack{i_j \in I_u \setminus I \\ \langle \alpha_{i_1}, \ldots, \alpha_{i_j} \rangle \neq 0, 0 \leq f < d}} \frac{\langle \alpha_{i_1}, \ldots, \alpha_{i_{j-1}} \rangle}{\langle \alpha_{i_1}, \ldots, \alpha_{i_j} \rangle} \tag{2.79}$$

$$\frac{\exp\left(2i\pi \langle p_j^{(I,i_j)}(s+f)e_j, \widehat{k}\rangle + \frac{d\langle p_{j-1}^I \alpha_{i_j}, x^I\rangle}{\langle p_{j-1}^I \alpha_{i_j}, e^j\rangle}\left[\frac{s+f}{d}\right]\right)}{\left(\prod_{i \in I_u \setminus (I,i_j)} \left(\langle p_j^{(I,i_j)} \alpha_i, 2i\pi \widehat{k} - x^I\rangle\right)\right)\left(\exp\left(\frac{d\langle p_{j-1}^I \alpha_{i_j}, x^I\rangle}{\langle p_{j-1}^I \alpha_{i_j}, e^j\rangle}\right) - 1\right)}.$$

Clearly

$$Q_u(t,x) = \sum_{k=(k_1,\ldots,k_r)\in \mathbf{Z}^r} \frac{\exp(2i\pi \sum_{i=1}^r k_i \langle t, e^i\rangle)}{\prod_{i\in I_u} (2i\pi \langle \alpha_i, \sum_{i=1}^r k_i e^i\rangle - x_i)}. \tag{2.80}$$

Also with the notation in (2.63),

$$\langle p_0^{\emptyset} \alpha_i, x^I \rangle = x_i. \tag{2.81}$$

So using (2.79) with $I = \emptyset$, $s = \langle t, e^1\rangle$, we find that for $\langle t, e^1 \rangle \notin \mathbf{Z}$,

$$Q_u(t,x) = - \sum_{\substack{i_1 \in I_u \\ \langle \alpha_{i_1}\rangle \neq 0 \\ 0 \leq f^1 < d}} \frac{1}{\langle \alpha_{i_1}\rangle} \sum_{k=(k^2,\ldots,k^r)\in \mathbf{Z}^{r-1}}$$

$$\frac{\exp\left(2i\pi \langle p_1^{(i_1)}(t+f^1 e_1), k\rangle + d\frac{x_{i_1}}{\langle \alpha_{i_1}\rangle}\left[\frac{\langle t + f^1 e_1, e^1\rangle}{d}\right]\right)}{\left(\prod_{i \in I_u \setminus \{i_1\}} \left(\langle p_1^{(i_1)} \alpha_i, 2i\pi k - x^I\rangle\right)\right)\left(\exp(\frac{d x_{i_1}}{\langle \alpha_{i_1}\rangle}) - 1\right)}. \tag{2.82}$$

Clearly, if $t \in (t/\overline{R}) \setminus H$, then $\langle t, e^1\rangle \notin \mathbf{Z}$, and so (2.82) holds. More generally, if $t \in (t/\overline{R}) \setminus H$, $f \in \overline{R}$, by Proposition 2.14, $\langle p_{j-1}^I (t+f), e_j\rangle \notin \mathbf{Z}$. Therefore using (2.79), (2.82), we can iterate the procedure. Finally observe that by (2.63), if $I = (i_1, \ldots, i_r)$ is generic, for $i \notin I$,

$$\begin{aligned} \langle p_r^I \alpha_i, x^I \rangle &= x_i - \sum_{k=1}^r \frac{\langle \alpha_{i_1}, \ldots, \alpha_{i_{k-1}}, \alpha_i, \alpha_{i_{k+1}}, \ldots, \alpha_{i_r}\rangle}{\langle \alpha_{i_1}, \ldots, \alpha_{i_r}\rangle} x_{i_k} \\ &= x_i - \langle \alpha_i, x^I\rangle. \end{aligned} \tag{2.83}$$

Using (2.83), we get (2.62).

The proof of our Theorem is completed. □

REMARK 2.20. If we had assumed instead that $e_1, \ldots, e_r$ is a simple basis of $R_{u,+}$, we would have obtained a better equality in (2.62), with $t \in (\mathfrak{t}/\overline{R}) \setminus H_u$. However, it is essential here that we have use the same simple basis $e_1, \ldots, e_r$ of R_+ for all the $u \in C/\overline{R}^*$ simultaneously. Finally observe that Theorem 2.19 can possibly be reformulated using the formalism of Szenes [**54**].

REMARK 2.21. Take $n \in \mathbf{N}^*$. Then we have the exact sequence

$$0 \to \overline{R}/d\overline{R} \xrightarrow[n]{} \overline{R}/dn\overline{R} \xrightarrow[p]{} \overline{R}/n\overline{R} \to 0 \,. \tag{2.84}$$

In (2.84), the map n is just multiplication by n, and p is the obvious projection.

Now in (2.62), we may replace d by nd. We get

$$\begin{aligned} Q_u(t,x) = n^{\ell_u - r}(-1)^r & \sum_{\substack{I=(i_1,\ldots,i_r)\subset I_u, I \text{ generic} \\ f\in\overline{R}/d\overline{R}, h\in\overline{R}/n\overline{R}}} \\ & \frac{1}{\langle\alpha_{i_1},\ldots,\alpha_{i_r}\rangle}\exp\left\{d\sum_{j=1}^r \frac{\langle p^I_{j-1}\alpha_{i_j}, nx^I\rangle}{\langle p^I_{j-1}\alpha_{i_j}, e^j\rangle}\right. \\ & \left.\left[\frac{1}{d}\langle p^I_{j-1}\left(\frac{(t+h)}{n}+f\right), e^j\rangle\right]\right\}\frac{1}{\prod_{i\in I_u\setminus I}(\langle\alpha_i, nx^I\rangle - nx_i)} \\ & \frac{1}{\prod_{j=1}^r\left(\exp\left(\frac{d\langle p^I_{j-1}\alpha_{i_j}, nx^I\rangle}{\langle p^I_{j-1}\alpha_{i_j}, e^j\rangle}\right)-1\right)} . \end{aligned} \tag{2.85}$$

Comparing (2.62) and (2.85) gives the identity (2.56), with $v = 0$.

By definition, $Q_u(t,x)$ is well-defined on $\mathfrak{t}/\overline{R}$. However it is not entirely clear that the right-hand side of (2.62) is indeed well-defined on $\mathfrak{t}/\overline{R}$. We will now check this fact directly.

THEOREM 2.22. *As a function of $t \in \mathfrak{t}$, the right-hand side of (2.62) descends to a function on $\mathfrak{t}/\overline{R}$.*

PROOF. By Proposition 2.2, $\langle\alpha_{i_1},\ldots,\alpha_{i_k}\rangle p^I_k e_k$ is an integral linear combination of the $(e_\rho)_{\rho\geq k+1}$. Therefore for $k \leq j-1$, $dp^I_{j-1}e_k$ is an integral linear combination of the $(p^I_{j-1}e_\rho)_{\rho\geq k+1}$. So for $k \leq j-1$, $d\langle p^I_{j-1}e_k, e^j\rangle$ is an integral linear combination of the $\langle p^I_{j-1}e_{k'}, e^j\rangle (k+1 \leq k' \leq j)$.

To prove that the right-hand side of (2.62) is well-defined on $(\mathfrak{t}/\overline{R}) \setminus H$, we only need to show that if we add to t an element of $d\overline{R}$, the total expression does not change.

Take $f \in \overline{R}/d\overline{R}$. Then f is uniquely represented by an element we also note f,

$$f = \sum_1^r f^k e_k \quad, 0 \leq f^k < d \,. \tag{2.86}$$

Clearly

$$\langle p^I_{j-1}(t+f), e^j\rangle = \langle p^I_{j-1}(t + \sum_1^j f^k e_k), e^j\rangle \,. \tag{2.87}$$

Put

$$A_j = \left[\frac{1}{d}\langle p^I_{j-1}(t+f), e^j\rangle\right]. \tag{2.88}$$

Since $p^I_{j-1}e_j = e_j$, $\langle p^I_{j-1}e_j, e^j\rangle = 1$. Therefore if we add to f^j an integral multiple of d, A_j is unchanged.

By the above, if we add to f^r an integral multiple of d, the right-hand side of (2.62) is unchanged. If we add to f^{r-1} an integral multiple of d, then only the term A_r is possibly affected. However as we saw before, $d\langle p^I_{r-1}e_{r-1}, e^r\rangle$ is an integral multiple of $\langle p^I_{r-1}e_r, e^r\rangle = 1$, i.e.

$$d\langle p^I_{r-1}e_{r-1}, e^r\rangle = q\langle p^I_{r-1}e_r, e^r\rangle \quad , q \in \mathbf{Z}. \tag{2.89}$$

Therefore adding to f^{r-1} an integral multiple of d is equivalent to adding to f^r an integer. This show the right-hand side of (2.62) is invariant under this change.

A trivial downward recursion procedure shows that $Q_u(t,x)$ is indeed well-defined on $\mathfrak{t}/\overline{R}$.

The proof of our Theorem is completed. □

2.5. The Fourier transform on quotient of lattices. Let Δ, Δ' be lattices in $\mathfrak{t}$, with $\Delta \subset \Delta'$. Then there is a projection $\mathfrak{t}/\Delta \to \mathfrak{t}/\Delta'$. Let $f \in \mathcal{D}'(t/\Delta)$. For $\mu \in \Delta^*/\Delta'^*$, put

$$\widehat{f}_\mu(t) = \frac{1}{\left|\frac{\Delta'}{\Delta}\right|} \sum_{k\in\Delta'/\Delta} e^{-2i\pi\langle\mu,k\rangle} f(t+k). \tag{2.90}$$

Then $\widehat{f}_\mu \in \mathcal{D}'(t/\Delta)$. Moreover if $k \in \Delta'/\Delta$,

$$\widehat{f}_\mu(t+k) = \exp(2i\pi\langle\mu,k\rangle)\widehat{f}_\mu(t). \tag{2.91}$$

Also

$$f(t) = \sum_{\mu\in\Delta^*/\Delta'^*} \widehat{f}_\mu(t). \tag{2.92}$$

By (2.90)

$$f(t+k) = \sum_{\mu\in\Delta^*/\Delta'^*} \exp(2i\pi\langle\mu,k\rangle)\widehat{f}_\mu(t). \tag{2.93}$$

Equation (2.93) is just an aspect of Fourier transform. Note here that if $A \subset \mathfrak{t}/\Delta$ is such that f is C^∞ on $(\mathfrak{t}/\Delta) \setminus A$, then $\widehat{f}_\mu$ is C^∞ on $(\mathfrak{t}/\Delta) \setminus \bigcup_{k\in\Delta'/\Delta} (A+k)$. Then formula (2.93) only expresses f as a function which is smooth on $(\mathfrak{t}/\Delta)\setminus \bigcup_{k\in\Delta'/\Delta} (A+k)$, i.e. there is a loss of regularity in (2.93).

2.6. Fourier series on T. In the constructions of Section 2.3 , we may as will replace R by CR, $\overline{R}$ by $\overline{CR}$, $\overline{R}^*$ by $\overline{CR}^*$. For $u \in C/\overline{R}^*$, we still define I_u, $\mathbf{R}^{I_u}$ as in (2.32), (2.34). However in (2.36), $a_u : \mathbf{R}^{I_u} \to \mathfrak{t}^* \simeq \mathfrak{t}$ is replaced by $b_u : \mathbf{R}^{I_u} \to \mathfrak{t} \simeq \mathfrak{t}^*$ given by

$$b_u(t) = \sum_{i\in I_u} t^i h_{\alpha_i}. \tag{2.94}$$

Clearly

$$b_u(\mathbf{Z}^{I_u}) = \overline{CR_u} \subset \overline{CR}. \tag{2.95}$$

Therefore there is a a surjection $b_u : (\mathbf{R}/\mathbf{Z})^{I_u} \to T = \mathfrak{t}/\overline{CR}$. Then we have an exact sequence

$$0 \to L_u \to (\mathbf{R}/\mathbf{Z})^{I_u} \xrightarrow[b_u]{} T \to 0\,. \tag{2.96}$$

DEFINITION 2.23. For $u \in C/\overline{R}^*$, let $S_u \subset T = \mathfrak{t}/\overline{CR}$ be given by

$$S_u = \{t \in \mathfrak{t}/\overline{CR},\ t = \sum_{j \in \mathcal{J} \subset I_u} t^j h_{\alpha_j},\ \text{and } \{h_{\alpha_j}, j \in \mathcal{J}\} \text{ do not span } \mathfrak{t}\}\,. \tag{2.97}$$

Put

$$S = S_0\,. \tag{2.98}$$

Then

$$S = \{t \in T = \mathfrak{t}/\overline{CR}, t = \sum_{j \in \mathcal{J} \subset \{1,\dots,\ell\}} t^j h_{\alpha_j},\ \text{and } \{h_{\alpha_j}, j \in \mathcal{J}\} \text{ do not span } \mathfrak{t}\}\,. \tag{2.99}$$

As in (2.46) , for any $u \in C/\overline{R}^*$,

$$S_u \subset S\,. \tag{2.100}$$

By (1.12) , it is clear that if $\lambda \in \overline{CR}^*$, $\alpha \in R_\ell$, $\beta \in R$,

$$\begin{aligned} &\langle \lambda, \alpha\rangle \in \mathbf{Z}\,, \\ &m\langle \lambda, \beta\rangle \in \mathbf{Z}\,. \end{aligned} \tag{2.101}$$

In the sequel, when G is simply laced, we will make $m = 1$.

DEFINITION 2.24. For $u \in C/\overline{R}^*$, $x \in (\mathbf{C} \setminus (2i\pi\frac{\mathbf{Z}}{m})^\ell$, $t \in T = \mathfrak{t}/\overline{CR}$, put

$$R_u(t,x) = \sum_{\lambda \in \overline{CR}^*} \frac{\exp(2i\pi\langle \lambda, t\rangle)}{\prod_{i \in I_u}(2i\pi\langle \alpha_i, \lambda\rangle - x_i)}\,. \tag{2.102}$$

Clearly, by (1.8), (1.10),

$$R_u(t,x) = m^{|R_{u,+} \cap R_s|} \sum_{\lambda \in \overline{CR}^*} \frac{\exp(2i\pi\langle \lambda, t\rangle)}{\prod_{i \in I_u}(2i\pi\langle h_{\alpha_i}, \lambda\rangle - \frac{\|h_{\alpha_i}\|^2}{2} x_i)}\,. \tag{2.103}$$

Of course in (2.103) , $\dfrac{\|h_{\alpha_i}\|^2}{2} = 1$ or m. From (2.103) , it should now be clear that Theorem 2.18 can be applied to $R_u(t,x)$. In particular on compact subsets of $T \setminus S_u \times (\mathbf{C} \setminus \frac{2i\pi\mathbf{Z}}{m})^\ell$, $R_u(t,x)$ is a smooth function of (t,x).

If G is simply laced, the objects we just constructed are the ones we already obtained in Section 2.3.

Now we use the notation in (2.35). Put

$$a = a_0\,. \tag{2.104}$$

Then if $s \in \mathbf{R}^\ell$,

$$as = \sum_{i=1}^{\ell} s^i \alpha_i\,. \tag{2.105}$$

Let $\widetilde{a} : \mathfrak{t} \simeq \mathfrak{t}^* \to \mathbf{R}^\ell$ be the transpose of a. Then

$$\widetilde{a}t = (\langle\alpha_1, t\rangle, \dots, \langle\alpha_\ell, t\rangle), \tag{2.106}$$

and $\widetilde{a}$ maps $\overline{R}^*$ into $\mathbf{Z}^\ell$.

Now we will use (2.90)-(2.93), with $\Delta = \overline{CR}$, $\Delta' = \overline{R}$. In the sequel, we view $R_u(t,x)$ as an element of $\mathcal{D}'(\mathfrak{t}/\overline{CR})$. If $\mu \in \overline{CR}^*/\overline{R}^*$, we define $(\widehat{R}_u)_\mu(t,x)$ as in (2.90).

PROPOSITION 2.25. *If $\mu \in \overline{CR}^*/\overline{R}^*$, if $\lambda_\mu \in \overline{CR}^*$ represents μ, then*

$$(\widehat{R}_u)_\mu(t,x) = \exp(2i\pi\langle\lambda_\mu, t\rangle) Q_u(t, x - 2i\pi\widetilde{a}\lambda_\mu). \tag{2.107}$$

PROOF. If $\lambda \in \overline{CR}^*$, then

$$\begin{aligned}\widehat{\exp(2i\pi\langle\lambda, t\rangle)}_\mu &= \exp(2i\pi\langle\lambda, t\rangle) \text{ if } \lambda \in \overline{CR}^* \text{ maps to } \mu \in \overline{CR}^*/\overline{R}^*, \\ &= 0 \text{ otherwise}.\end{aligned} \tag{2.108}$$

Then

$$(\widehat{R}_u)_\mu(t,x) = \sum_{\lambda \in \overline{R}^*} \frac{\exp(2i\pi\langle\lambda + \lambda_\mu, t\rangle)}{\prod_{i \in I_u} (2i\pi\langle\alpha_i, \lambda\rangle - (x_i - 2i\pi\langle\alpha_i, \lambda_\mu\rangle))}. \tag{2.109}$$

From (2.109) , we get (2.107). The proof of our Proposition is completed. □

REMARK 2.26. By Proposition 2.25 , we get the otherwise obvious fact that the right-hand side of (2.107) only depends on μ and not on λ_μ.

For any $\mu \in \overline{CR}^*/\overline{R}^*$, we choose $\lambda_\mu \in \overline{CR}^*$ representing μ.

THEOREM 2.27. *The following identity holds*

$$R_u(t,x) = \sum_{\mu \in \overline{CR}^*/\overline{R}^*} \exp(2i\pi\langle\lambda_\mu, t\rangle) Q_u(t, x - 2i\pi\widetilde{a}\lambda_\mu). \tag{2.110}$$

PROOF. This follows from (2.92) and (2.107). □

2.7. Iterated residues and the series $R_u(t,x)$. Let $\tau : \mathfrak{t}/\overline{CR} \to \mathfrak{t}/\overline{R}$ be the obvious projection. For $u \in C/\overline{R}^*$, put

$$\begin{aligned}\overline{S}_u &= \tau^{-1}(H_u), \\ \overline{S} &= \tau^{-1}(H).\end{aligned} \tag{2.111}$$

Clearly τ maps S_u into H_u, S into H. Therefore

$$\begin{aligned}S_u &\subset \overline{S}_u, \\ S &\subset \overline{S}.\end{aligned} \tag{2.112}$$

If G is simply laced

$$\begin{aligned}S_u &= \overline{S}_u, \\ S &= \overline{S}.\end{aligned} \tag{2.113}$$

Recall that by (2.20), $d\overline{R} \subset \overline{CR}$.

THEOREM 2.28. *For any $u \in C/\overline{R}^*$, $t \in T \setminus \overline{S}$, then*

$$R_u(t,x) = \left|\frac{\overline{R}}{\overline{CR}}\right| (-1)^r \sum_{\substack{I=(i_1,\dots,i_r)\subset I_u \\ I \text{ generic} \\ f\in \overline{CR}/d\overline{R}}} \frac{1}{\langle \alpha_{i_1},\dots,\alpha_{i_r}\rangle} \tag{2.114}$$

$$\frac{1}{\prod_{i\in I_u\setminus I} (\langle \alpha_i, x^I\rangle - x_i)} \exp\left\{ d\sum_{j=1}^{r} \frac{\langle p^I_{j-1}\alpha_{i_j}, x^I\rangle}{\langle p^I_{j-1}\alpha_{i_j}, e^j\rangle}\left[\frac{1}{d}\langle p^I_{j-1}(t+f), e^j\rangle\right]\right\}$$

$$\prod_{j=1}^{r} \frac{1}{\exp\left(\frac{d\langle p^I_{j-1}\alpha_{i_j}, x^I\rangle}{\langle p^I_{j-1}\alpha_{i_j}, e^j\rangle}\right) - 1}.$$

PROOF. We use Theorems 2.19 and 2.27 . Also, if $h \in \mathfrak{t}$,

$$(\tilde{a}h)^I = h. \tag{2.115}$$

So we deduce from (2.115) that if $i \in I_u \setminus I$,

$$\langle \alpha_i, (\tilde{a}\lambda_\mu)^I\rangle - \langle \alpha_i, \lambda_\mu\rangle = 0. \tag{2.116}$$

By (2.19), (2.28), if $\lambda \in \overline{CR}^*$,

$$\frac{d\langle p^I_{j-1}\alpha_{i_j}, \tilde{a}\lambda\rangle}{\langle p^I_{j-1}\alpha_{i_j}, e^j\rangle} \in \mathbf{Z}. \tag{2.117}$$

From (2.13), (2.62), (2.110), (2.116), (2.117), we get (2.114). The proof of our Theorem is completed. □

2.8. Fourier series for the universal cover of a semisimple centralizer. Take $u \in C/\overline{R}^*$. Recall that by

$$\pi_1(Z(u)) = \overline{CR}/\overline{CR_u}. \tag{2.118}$$

Also

$$\overline{CR} \subset \overline{R}^*_u. \tag{2.119}$$

Let $\widetilde{Z}(u)$ be the universal cover of $Z(u)$. Then by [**15**, Theorem V.7.1],

$$Z(\widetilde{Z}(u)) = \overline{R}^*_u/\overline{CR_u}. \tag{2.120}$$

Therefore $\pi_1(Z(u)) = \overline{CR}/\overline{CR_u}$ is a subgroup of $Z(\widetilde{Z}(u))$. Also $\overline{CR}^*_u$ is the lattice of weights of $\widetilde{Z}(u)$.

DEFINITION 2.29. For $u \in C/\overline{R}^*$, $t \in \mathfrak{t}/\overline{CR_u}$, $x \in \left(\mathbf{C} \setminus \frac{2i\pi\mathbf{Z}}{m}\right)^\ell$, put

$$\widetilde{R}_u(t,x) = \sum_{\lambda\in\overline{CR}^*_u} \frac{\exp(2i\pi\langle\lambda, t\rangle)}{\prod_{i\in I_u}(2i\pi\langle\lambda,\alpha_i\rangle - x_i)}. \tag{2.121}$$

Clearly

$$\pi_1(Z(u))^* = \overline{CR}^*_u/\overline{CR}^*. \tag{2.122}$$

Also $\overline{R}^*_u/\overline{R}^*$ maps into $\overline{CR}^*_u/\overline{CR}^*$, with kernel $\dfrac{\overline{R}^*_u \cap \overline{CR}^*}{\overline{R}^*}$. In particular $u \in C/\overline{R}^*$ maps to an element of $\pi_1(Z(u))^*$.

DEFINITION 2.30. For $u \in C/\overline{R}^*$, $q \in \mathbf{Z}$, $t \in T = \mathfrak{t}/\overline{CR}$, put

$$R_{u,q}(t,x) = \frac{1}{|\frac{\overline{CR}}{\overline{CR}_u}|} \sum_{h \in \frac{\overline{CR}}{\overline{CR}_u}} \exp(2i\pi\langle qu, h\rangle)\widetilde{R}_u(t+h,x)\,. \tag{2.123}$$

PROPOSITION 2.31. *If $u \in C$, the following identity holds*

$$R_{u,q}(t,x) = R_u(t, x + 2i\pi q\widetilde{a}u)\exp(-2i\pi\langle qu,t\rangle)\,. \tag{2.124}$$

PROOF. Since $u \in \overline{CR}_u^*$, our identity follows from (2.123). □

THEOREM 2.32. *For any $u \in C/\overline{R}^*$, $q \in \mathbf{Z}$, $t \in T \setminus \overline{S}$, then*

$$\begin{aligned} R_{u,q}(t,x) = \left|\frac{\overline{R}}{\overline{CR}}\right|(-1)^r &\sum_{\substack{I=(i_1,\dots,i_r)\subset I_u \\ I \text{ generic} \\ f\in\overline{CR}/d\overline{R}}} \frac{1}{\langle\alpha_{i_1},\dots,\alpha_{i_r}\rangle}\exp(2i\pi(\langle qu,f\rangle)) \\ &\frac{1}{\prod_{i\in I_u\setminus I}(\langle\alpha_i,x^I\rangle - x_i)}\exp\left\{d\sum_{j=1}^r \frac{\langle p^I_{j-1}\alpha_{i_j},x^I\rangle}{\langle p^I_{j-1}\alpha_{i_j},e^j\rangle}\left[\frac{1}{d}\langle p^I_{j-1}(t+f),e^j\rangle\right]\right\} \\ &\prod_{j=1}^r \frac{1}{\exp(\frac{d\langle p^I_{j-1}\alpha_{i_j},x^I\rangle}{\langle p^I_{j-1}\alpha_{i_j},e^j\rangle}) - 1}\,. \end{aligned} \tag{2.125}$$

PROOF. We use Theorem 2.28 and Proposition 2.31. Since for $i \in I_u$, $\langle u,\alpha_i\rangle \in \mathbf{Z}$, by (2.13), (2.29), (2.116), we get (2.125). The proof of our Theorem is completed. □

2.9. Bernouilli polynomials and the Fourier series $P_n(t)$.

DEFINITION 2.33. For $n \in \mathbf{N}$, $t \in T = \mathfrak{t}/\overline{CR}$, put

$$P_n(t) = -\sum_{\substack{\lambda\in\overline{CR}^* \\ \pi(\lambda)\neq 0}} \frac{\exp(2i\pi\langle\lambda,t\rangle)}{[\pi(\lambda)]^n}\,. \tag{2.126}$$

For $M \in \mathbf{N}^*$, we will consider the partial sums

$$P_n^M(t) = -\sum_{\substack{\lambda\in\overline{CR}^* \\ \pi(\lambda)\neq 0 \\ |\lambda|\leq M}} \frac{\exp(2i\pi\langle\lambda,t\rangle)}{[\pi(\lambda)]^n}\,. \tag{2.127}$$

Clearly $P_n(t)$ is a well-defined distribution on $T = \mathfrak{t}/\overline{CR}$. Recall that

$$K = \{t \in \mathfrak{t}, \text{ for } \alpha \text{ in } R_+,\ \langle\alpha,t\rangle > 0\}\,. \tag{2.128}$$

Then by [**15**, Note V.4.14] and (1.186),

$$\begin{aligned} P_n(t) &= -\sum_{\substack{\lambda\in\overline{CR}^*\cap\overline{K} \\ w\in W}} \frac{\varepsilon_w\exp(2i\pi\langle w(\lambda+\rho),t\rangle)}{[\pi(\lambda+\rho)]^n} \quad \text{for } n \text{ odd}\,, \\ &= -\sum_{\substack{\lambda\in\overline{CR}^*\cap\overline{K} \\ w\in W}} \frac{\exp(2i\pi\langle w(\lambda+\rho),t\rangle)}{[\pi(\lambda+\rho)]^n} \quad n \text{ even}\,. \end{aligned} \tag{2.129}$$

By (1.186), (2.126), for $w \in W$,

$$P_n(wt) = \varepsilon_w^n P(t). \tag{2.130}$$

Since $\pi(t)$ is a polynomial, $\pi\left(\frac{\partial/\partial t}{2i\pi}\right)$ is a differential operator. Then we have the identity of distributions on $T = \mathfrak{t}/\overline{CR}$,

$$\pi\left(\frac{\partial/\partial t}{2i\pi}\right) P_n(t) = P_{n-1}(t) \text{ for } n \geq 1. \tag{2.131}$$

DEFINITION 2.34. For $t \in [0,1]$, $n \geq 0$, put

$$p_n(t) = -\sum_{k \neq 0} \frac{e^{2i\pi kt}}{(2i\pi k)^n}. \tag{2.132}$$

If $G = SU(2)$, then $T \simeq S_1 \simeq \mathbf{R}/\mathbf{Z}$. One verifies easily that the $p_n(t)$'s are exactly the P_n's associated to $G = SU(2)$. Then (2.131) is the equation of distribution on S_1

$$p_n'(t) = p_{n-1}(t), \quad n \geq 1. \tag{2.133}$$

Also

$$p_0(t) = 1 - \delta_{\{0\}}, \tag{2.134}$$

So by (2.133), (2.134), we get

$$\left(\frac{d}{dt}\right)^n p_n(t) = 1 - \delta_{\{0\}}. \tag{2.135}$$

By (2.135), it is clear that the $p_n(t)$'s are polynomials on $S_1 \setminus \{0\}$. Also for $n \geq 2$, the series in (2.132) is absolutely convergent on $[0,1]$. For $n \geq 1$, the series in (2.132) converges uniformly together with its derivatives on compact sets of S_1, not containing 0.

By [**45**, Appendix B], the p_n's are exactly the Bernouilli polynomials. In the sequel, we will consider the p_n's as polynomials on $\mathbf{R}$, whose restriction to $]0,1[$ is given by (2.132).

Recall that

$$\mathrm{Td}(x) = \frac{x}{1-e^{-x}}. \tag{2.136}$$

Then

$$\mathrm{Td}(x) - \mathrm{Td}(-x) = 1. \tag{2.137}$$

Finally by [**51**, p147], if the B_k, $k \geq 1$, are the Bernoulli numbers,

$$\mathrm{Td}(x) = 1 + \frac{x}{2} + \sum_{k=1}^{+\infty} (-1)^{k+1} B_k \frac{x^{2k}}{(2k)!}. \tag{2.138}$$

PROPOSITION 2.35. *For $n \geq 0$, $t \in \mathbf{R}$,*

$$p_n(t) = \mathrm{Res}_{a=0}\left[\frac{e^{ta}}{a^n}\frac{1}{e^a - 1}\right]. \tag{2.139}$$

PROOF. For $0 \le t \le 1$, put

$$f_{t,n}(a) = \frac{e^{ta}}{a^n} \frac{1}{e^a - 1}. \tag{2.140}$$

Clearly $f_{t,n}(a)$ has simple poles at $a = 2i\pi k, k \in \mathbf{Z}^*$, and a pole of order $n+1$ at $a = 0$. Then, for $k \in \mathbf{Z}^*$

$$\operatorname{Res}_{a=2i\pi k} f_{t,n}(a) = \frac{e^{2i\pi kt}}{(2i\pi k)^n}. \tag{2.141}$$

Now we use the Cauchy residue theorem inside a circle of centre 0 and radius $2\pi(M+1/2)$, as $M \to +\infty$. For $n \ge 2$, or for $n = 1$, $t \in]0,1[$, the integral of $f_{t,n}(a)$ on the circle tends to 0 as $M \to +\infty$. So we find that

$$\sum_{k \in \mathbf{Z}} \operatorname{Res}_{a=2i\pi k} f_{t,n}(a) = 0. \tag{2.142}$$

From (2.142) , we get (2.139) for $n \ge 2$, or $n = 1$, $0 < t < 1$. Then since $p_n(t)$ is a polynomial, (2.139) holds for any $t \in \mathbf{R}$. For $n = 0$, (2.139) is trivial. The proof of our Proposition is completed. □

PROPOSITION 2.36. *For $n \ge 0$, $t \in \mathbf{R}$,*

$$\begin{aligned} p_n(t) &= \operatorname{Td}(-\partial/\partial t)\frac{t^n}{n!}, \\ p_{n+1}(t+1) - p_{n+1}(t) &= \frac{t^n}{n!}. \end{aligned} \tag{2.143}$$

PROOF. Clearly,

$$\frac{\partial}{\partial t} e^{at} = a e^{at}. \tag{2.144}$$

By (2.144) , we find that for $|a| < 2\pi$,

$$\operatorname{Td}(-\partial/\partial t)e^{at} = \operatorname{Td}(-a)e^{at}. \tag{2.145}$$

Using (2.139), (2.145), we get

$$\begin{aligned} p_n(t) = \operatorname{Res}_{a=0}\left[\frac{e^{ta}}{a^{n+1}}\operatorname{Td}(-a)\right] &= \operatorname{Td}(-\partial/\partial t)\operatorname{Res}_{a=0}\left[\frac{e^{ta}}{a^{n+1}}\right] \\ &= \operatorname{Td}(-\partial/\partial t)\frac{t^n}{n!}. \end{aligned} \tag{2.146}$$

By (2.139)

$$p_{n+1}(t+1) - p_{n+1}(t) = \operatorname{Res}\left[\frac{e^{ta}}{a^{n+1}}\right] = \frac{t^n}{n!}. \tag{2.147}$$

The proof of our Proposition is completed. □

Recall that $R_{s,+}$ denote the set of short positive roots.

THEOREM 2.37. *For $n \ge 1$, as $M \to +\infty$, the partial sums $P_n^M(t)$ converge to $P_n(t)$ uniformly on the compact subsets of $T \setminus S$. The following identity of distributions holds on $T = \mathfrak{t}/\overline{CR}$,*

$$P_n(t) = (-1)^{\ell+1} m^{n|R_{s,+}|} b_* \left[\prod_{i=1}^{\ell} p_n(t^i)\right]. \tag{2.148}$$

In particular $P_n(t) \in L_\infty(T)$. *Also (2.148) is an identity of smooth functions on* $T \setminus S$. *Finally* $P_n(t)$ *is a polynomial on* $T \setminus S$.

PROOF. By proceeding as in the proof of Theorem 2.18 , we get (2.148). By the same argument as in Theorem 2.18 , we find that $P_n(t)$ is smooth on $T \setminus S$. Also the uniform convergence results for the $p_n^M(t)$'s and (2.148) imply the corresponding uniform convergence for $P_n^M(t)$.

Now, we will show that $P_n(t)$ is a polynomial on $T \setminus S$. In fact we will prove that for $1 \le i \le \ell$, for p large enough,

$$h_{\alpha_i}^p P_n(t) = 0 \text{ on } T \setminus S \,. \tag{2.149}$$

Clearly $b_* \partial/\partial t_i = h_{\alpha_i}$. Then using (2.135) , we obtain

$$\begin{aligned} h_{\alpha_1}^n b_*[p_n(t^1)\dots p_n(t^\ell)] &= b_*[p_n(t^2)\dots p_n(t^\ell)] \\ &\quad - b_*[\delta_{t_1=0} p_n(t^2)\dots p_n(t^\ell)] \,. \end{aligned} \tag{2.150}$$

Clearly

$$h_{\alpha_1} b_*[p_n(t^2)\dots p_n(t^\ell)] = 0 \,. \tag{2.151}$$

Let V_1 be the vector space in $\mathfrak{t}$ spanned by $h_{\alpha_2}, \dots, h_{\alpha_\ell}$. If V_1 is not equal to $\mathfrak{t}$, the support of $b_*[\delta_{t_1=0} p_n(t^2)\dots p_n(t^\ell)]$ is included in S. From (2.150) , we then get

$$h_{\alpha_1}^{n+1} b_*[p_n(t^1)\dots p_n(t^\ell)] = 0 \text{ on } T \setminus S \,. \tag{2.152}$$

If $V_1 = \mathfrak{t}$, we can express h_{α_1} in the form

$$h_{\alpha_1} = \sum_{j=2}^{\ell} a^j h_{\alpha_j} \,. \tag{2.153}$$

By (2.135),

$$\begin{aligned} h_{\alpha_2}^n b_*[\delta_{t_1=0} p_n(t^2)\dots p_n(t^\ell)] &= b_*[\delta_{t_1=0} p_n(t^3)\dots p_n(t^\ell)] \\ &\quad - b_*[\delta_{t_1=0,t_2=0} p_n(t^3)\dots p_n(t^\ell)] \,. \end{aligned} \tag{2.154}$$

Let V_2 be the vector space spanned by $h_{\alpha_2}, \dots, h_{\alpha_\ell}$. If $V_2 \neq \mathfrak{t}$, then by the analogue of (2.152),

$$h_{\alpha_2}^{n+1} q_*[\delta_{t_1=0} p_n(t^2)\dots p_n(t^\ell)] = 0 \text{ on} T \setminus S. \tag{2.155}$$

Now using (2.155) , and iterating the above argument, it should be clear that for $p \in \mathbb{N}$ large enough,

$$h_{\alpha_1}^p P_n(t) = 0 \quad \text{on } T \setminus S \,. \tag{2.156}$$

In (2.156), we may as well replace the index 1 by the index i, $1 \le i \le \ell$. Therefore we have established (2.149).

The proof of our Theorem is completed. □

2.10. The Fourier series $\tilde{P}_{u,n}(t)$. Take $u \in C/\overline{R}^*$. Recall that $\widetilde{Z}(u)$ is the universal cover of $Z(u)$. Then $\widetilde{Z}(u)$ is connected and simply connected. Therefore to $\widetilde{Z}(u)$, we can associate the objects we just constructed for G. Observe that $\mathfrak{t}/\overline{CR}_u$ is a maximal torus in $\widetilde{Z}(u)$, and $\mathfrak{t}$ is its Lie algebra.

DEFINITION 2.38. Let $\pi_u(t)$ be the function on $\mathfrak{t}$

$$\pi_u(t) = \prod_{\alpha \in R_{u,+}} \langle 2i\pi\alpha, t \rangle \,. \tag{2.157}$$

Then $\pi_u(t)$ is the analogue of $\pi(t)$.

DEFINITION 2.39. For $u \in C/\overline{R}^*$, $n \in \mathbf{N}$, $t \in \mathfrak{t}/\overline{CR_u}$, put

$$\widetilde{P}_{u,n}(t) = -\sum_{\substack{\lambda \in \overline{CR}_u^* \\ \pi_u(\lambda) \neq 0}} \frac{\exp(2i\pi\langle\lambda, t\rangle)}{[\pi_u(\lambda)]^n}. \tag{2.158}$$

Then $\widetilde{P}_{u,n}(t)$ is the analogue of $P_n(t)$ for $\widetilde{Z}(u)$. Take $w \in W$. We identify w with a representative in $N(T)/T$. Then if $u \in C/\overline{R}^*$,

$$\begin{aligned} Z(wu) &= w\, Z(u) w^{-1}, \\ \overline{CR}_{wu} &= w\overline{CR}_u, \\ R_{Z(wu)} &= wR_u, \\ R_{Z(wu),+} &= wR_u \cap R_+. \end{aligned} \tag{2.159}$$

By (2.159), we get

$$\pi_{wu}(t) = (-1)^{|R_+ \cap w(-R_{u,+})|} \pi_u(w^{-1}t). \tag{2.160}$$

PROPOSITION 2.40. *If $u \in C/\overline{R}^*$, $n \in \mathbf{N}$, $w \in W$, then*

$$\widetilde{P}_{u,n}(t) = (-1)^{n|R_+ \cap w(-R_{u,+})|} \widetilde{P}_{wu,n}(wt). \tag{2.161}$$

PROOF. Clearly by (2.159), (2.160),

$$\begin{aligned} \widetilde{P}_{u,n}(t) &= -\sum_{\substack{\lambda \in \overline{CR}_{wu}^* \\ \pi_u(w^{-1}\lambda) \neq 0}} \frac{e^{2i\pi\langle\lambda, wt\rangle}}{[\pi_u(w^{-1}\lambda)]^n} \\ &= -(-1)^{n|R_+ \cap w(-R_{u,+})|} \sum_{\substack{\lambda \in \overline{CR}_{wu}^* \\ \pi_{wu}(\lambda) \neq 0}} \frac{e^{2i\pi\langle\lambda, wt\rangle}}{[\pi_{wu}(\lambda)]^n} \\ &= (-1)^{n|R_+ \cap w(-R_{u,+})|} \widetilde{P}_{wu,n}(wt), \end{aligned} \tag{2.162}$$

which coincides with (2.161). □

Let $\widetilde{S}_u \subset \mathfrak{t}/\overline{CR_u}$ be the analogue of $S \subset \mathfrak{t}/\overline{CR}$. Of course $\widetilde{S}_u$ projects into S_u. Then by Theorem 2.37 , $\widetilde{P}_{u,n}(t)$ is polynomial on $(\mathfrak{t}/\overline{CR_u}) \setminus \widetilde{S}_u$.

Let $i_1, \ldots, i_{\ell_u}$ be the elements of $I_u = \{i \in \{1, \ldots, \ell\}, \alpha_i \in R_{u,+}\}$ arranged in increasing order, i.e. $i_1 < i_2 \ldots < i_{\ell_u}$.

If $f(x) = f(x_{i_1}, \ldots, x_{i_{\ell_u}})$ is a meromorphic function of $x \in \mathbf{R}^{I_u}$, we denote by $\mathrm{Res}_{x=0}^{I_u} f$ the expression

$$\mathrm{Res}_{x=0}^{I_u} f = \mathrm{Res}_{x_{i_{\ell_u}}=0} \ldots \mathrm{Res}_{x_{i_1}=0} f(x_{i_1}, \ldots, x_{i_{\ell_u}}). \tag{2.163}$$

It will be of fundamental importance that the order in (2.163) is fixed once and for all.

THEOREM 2.41. *For $u \in C/\overline{R}^*$, $n \geq 1$,*

$$\widetilde{P}_{u,n}(t) = -\mathrm{Res}_{x=0}^{I_u} \frac{1}{(\prod_{i \in I_u} x_i)^n} \widetilde{R}_u(t, x). \tag{2.164}$$

PROOF. Clearly

$$\operatorname{Res}_{x_i=0}\frac{1}{x_i^n}\frac{1}{2i\pi\langle\alpha_i,\lambda\rangle - x_i} = \frac{1}{(2i\pi\langle\alpha_i,\lambda\rangle)^n} \quad \text{if } \langle\alpha_i,\lambda\rangle \neq 0, \tag{2.165}$$
$$= 0 \quad \text{if } \langle\alpha_i,\lambda\rangle = 0.$$

Using (2.165) , we get (2.164). The proof of our Theorem is completed. □

REMARK 2.42. The identity (2.164) can be viewed as an identity of distributions on $\mathfrak{t}/\overline{CR_u}$, of smooth functions on $(\mathfrak{t}/\overline{CR_u}) \setminus \widetilde{S}_u$, of elements of $L^\infty(\mathfrak{t}/\overline{CR_u})$. Also for the moment, the ordering in (2.164) is still irrelevant.

2.11. The function $P_{u,n,q}(t)$. If $t \in \mathfrak{t}$, we still denote by t the corresponding element in $T = \mathfrak{t}/\overline{CR}$.

DEFINITION 2.43. For $u \in C/\overline{R}^*$, $q \in \mathbf{Z}$, $n \in \mathbf{N}$, $t \in \mathfrak{t}/\overline{CR}$, put

$$P_{u,n,q}(t) = \frac{1}{\left|\frac{\overline{CR}}{\overline{CR_u}}\right|}\sum_{h\in\frac{\overline{CR}}{\overline{CR_u}}}\exp(2i\pi\langle qu,h\rangle)\widetilde{P}_{u,n}(t+h). \tag{2.166}$$

Observe that the function $\exp(2i\pi\langle qu,t\rangle)P_{u,n,q}(t)$ descends to a well defined function on $T = \mathfrak{t}/\overline{CR}$. Equivalently, we may consider $P_{u,n,q}(t)$ as a section of the flat line bundle L_u on T, associated to $u \in \mathfrak{t}^*/\overline{CR}^*$.

PROPOSITION 2.44. *If $u \in C/\overline{R}^*$, $q \in \mathbf{Z}$, $n \in \mathbf{N}$, $w \in W$, then*

$$P_{u,n,q}(t) = (-1)^{n|R_+\cap w(-R_{u,+})|}P_{wu,q,n}(wt). \tag{2.167}$$

PROOF. Clearly $w : \frac{\overline{CR}}{\overline{CR_u}} \to \frac{\overline{CR}}{\overline{CR_{wu}}}$ is one to one. Then (2.167) follows from Proposition 2.40. □

THEOREM 2.45. *The section $P_{u,n,q}(t)$ is a polynomial on $T \setminus S_u$.*

PROOF. Since $\widetilde{P}_{u,n}(t)$ is polynomial on $(\mathfrak{t}/\overline{CR_u}) \setminus \widetilde{S}_u$, it is clear from (2.166) that $P_{u,n,q}$ is polynomial on $T \setminus S_u$. □

THEOREM 2.46. *For $u \in C/\overline{R}^*$, $n \in \mathbf{N}$, $q \in \mathbf{Z}$, $t \in \mathfrak{t}/\overline{CR}$, the following identity holds*

$$P_{u,n,q}(t) = -\operatorname{Res}^{I_u}_{x=0}\frac{1}{\left(\prod_{i\in I_u}x_i\right)^n}R_{u,q}(t,x). \tag{2.168}$$

PROOF. This follows from (2.123), (2.164), (2.166) . □

DEFINITION 2.47. For $u \in C/\overline{R}^*$, let $\mathcal{I}_u$ be the set of generic $I = (i_1,\dots,i_r) \subset I_u$ such that if $\sigma^I \in \mathcal{S}_r$ is defined by $i_{\sigma^I(1)} < i_{\sigma^I(2)} < \dots < i_{\sigma^I(r)}$, if $j \in I_u \setminus I$, either $j < \sigma^I(1)$, or if p_j is defined by the condition

$$i_{\sigma^I(1)} < \dots < i_{\sigma^I(p_j)} < j < i_{\sigma^I(p_j+1)} < \dots, \tag{2.169}$$

then

$$\alpha_{i_{\sigma^I(1)}} \wedge \dots \wedge \alpha_{i_{\sigma^I(p_j)}} \wedge \alpha_j \neq 0. \tag{2.170}$$

Condition (2.170) is equivalent to

$$\alpha_j \notin \{\alpha_{i_{\sigma^I(1)}}, \dots, \alpha_{i_{\sigma^I(p_j)}}\}. \tag{2.171}$$

If $I \in \mathcal{I}_u$, we define $\mathrm{Res}^I_{x=0}$ by the formula

$$\mathrm{Res}^I_{x=0} = \mathrm{Res}_{x_{i_{\sigma(r)}}=0} \dots \mathrm{Res}_{x_{i_{\sigma(1)}}=0}. \tag{2.172}$$

THEOREM 2.48. *For $u \in C/\overline{R}^*$, $n \geq 1$, $q \in \mathbf{N}$, if $t \in \mathfrak{t}$ represents an element of $T \setminus \overline{S}$, then*

$$\begin{aligned}(2.173)\quad P_{u,n,q}(t) = \left|\frac{\overline{R}}{\overline{CR}}\right| (-1)^{r+1} \sum_{\substack{I=(i_1,\dots,i_r)\in\mathcal{I}_u \\ f\in\overline{CR}/d\overline{R}}} \frac{1}{\langle\alpha_{i_1},\dots,\alpha_{i_r}\rangle} \exp(2i\pi(\langle qu, f\rangle))\\
\mathrm{Res}^I_{x=0}\left(\frac{1}{\prod_{\alpha\in R_{u,+}}\langle\alpha, x^I\rangle}\right)^n \exp\left\{d\sum_{j=1}^r \frac{\langle p^I_{j-1}\alpha_{i_j}, x^I\rangle}{\langle p^I_{j-1}\alpha_{i_j}, e^j\rangle}\right.\\
\left.\left[\frac{1}{d}\langle p^I_{j-1}(t+f), e^j\rangle\right]\right\} \prod_{j=1}^r \frac{1}{\exp(\frac{d\langle p^I_{j-1}\alpha_{i_j}, x^I\rangle}{\langle p^I_{j-1}\alpha_{i_j}, e^j\rangle}) - 1}.\end{aligned}$$

PROOF. By Theorems 2.32 and 2.46, for $t \in T \setminus \overline{S}$, we get

$$\begin{aligned}P_{u,n,q}(t) = \left|\frac{\overline{R}}{\overline{CR}}\right| (-1)^{r+1} \sum_{\substack{I=(i_1,\dots,i_r)\subset I_u \\ I\,\text{generic} \\ f\in\overline{CR}/d\overline{R}}} \frac{1}{\langle\alpha_{i_1},\dots,\alpha_{i_r}\rangle} \exp(2i\pi(\langle qu, f\rangle))\\
(2.174)\quad \mathrm{Res}^{I_u}_{x=0} \frac{1}{(\prod_{i\in I_u} x_i)^n} \frac{1}{\prod_{i\in I_u\setminus I}(\langle\alpha_i, x^I\rangle - x_i)} \exp\left\{d\textstyle\sum_{j=1}^r \frac{\langle p^I_{j-1}\alpha_{i_j}, x^I\rangle}{\langle p^I_{j-1}\alpha_{i_j}, e^j\rangle}\right.\\
\left.[\tfrac{1}{d}\langle p^I_{j-1}(t+f), e^j\rangle]\right\} \prod_{j=1}^r \frac{1}{\exp\left(\frac{d\langle p^I_{j-1}\alpha_{i_j}, x^I\rangle}{\langle p^I_{j-1}\alpha_{i_j}, e^j\rangle}\right) - 1}.\end{aligned}$$

Now x^I depends only on $x_{i_1}, \dots, x_{i_r}$. Therefore in (2.174), the dependence on the $(x_i)_{i\in I_u\setminus I}$ is only via the term $\dfrac{1}{\prod_{i\in I_u\setminus I} x_i^n(\langle\alpha_i, x^I\rangle - x_i)}$.

Take $\epsilon \geq 0$. Let $v \in \mathbf{C}$, $|v| \ll \epsilon$, $w \in \mathbf{C}$, $|w| > \epsilon$. Then by the theorem of residues, for $n \geq 1$,

$$\begin{aligned}(2.175)\quad &\frac{1}{2\pi i}\int_{\substack{x_j\in\mathbf{C}\\|x_j|=\epsilon}} \frac{dx_j}{x_j^n(v+w-x_j)} = \frac{1}{(v+w)^n},\\
&\frac{1}{2\pi i}\int_{\substack{x_j\in\mathbf{C}\\|x_j|=1}} \frac{dx_j}{x_j^n(v-x_j)} = 0.\end{aligned}$$

Take $j \in I_u \setminus I$. If

$$\alpha_j = \sum_{k=1}^r a^k \alpha_{i_{\sigma^I(k)}}, \tag{2.176}$$

then

$$\langle \alpha_j, x^I \rangle = \sum_{k=1}^{r} a^k x_{i_{\sigma^I(k)}} . \tag{2.177}$$

Put

$$v = \sum_{k=1}^{p_j} a^k x_{i_{\sigma^I(k)}} \quad , \quad w = \sum_{k=p_j+1}^{r} a^k x_{i_{\sigma^I(k)}} . \tag{2.178}$$

Observe that w is identically 0 if and only if $a^k = 0$ for $k \geq p_j + 1$, i.e. if $\alpha_j \in \{\alpha_{i_{\sigma^I(1)}}, \dots, \alpha_{i_{\sigma^I(p_j)}}\}$.

To evaluate (2.174), we will use (2.175). Namely take a sequence $\epsilon_1, \dots, \epsilon_{\ell_u}$ in $\mathbf{R}_+$, with $0 < \epsilon_{i_1} < \dots < \epsilon_{\ell_u}$. Then if $f(x_{i_1}, \dots, x_{i_{\ell_u}})$ is a meromorphic function, by definition,

$$\mathrm{Res}^{I_u}_{x=0} f = \int_{\substack{|x_{i_j}| = \epsilon_{i_j} \\ 1 \leq j \leq \ell_u}} f(x_{i_1}, \dots, x_{i_{\ell_u}}) \frac{dx_{i_1}}{2i\pi} \cdots \frac{dx_{i_{\ell_u}}}{2i\pi}. \tag{2.179}$$

If the sequence$\epsilon_1, \dots, \epsilon_{\ell_u}$ is enough decreasing, when taking the x_{i_j} as in (2.179), if $\alpha_j \notin \{\alpha_{i_{\sigma^I(1)}}, \dots, \alpha_{i_{\sigma^I(r)}}\}$, in (2.178), then $|w| > \epsilon_j$.

Using (2.175), (2.179), we find easily that for $j \in I_u \setminus I$,

$$\begin{aligned} \mathrm{Res}_{x_j=0} \frac{1}{x_j^n(\langle \alpha_j, x^I \rangle - x_j)} &= 0 && \text{if } \alpha_j \in \{\alpha_{i_{\sigma^I(1)}}, \dots, \alpha_{i_{\sigma^I(p_j)}}\}, \\ &= \frac{1}{\langle \alpha_j, x^I \rangle^n} && \text{if } \alpha_j \notin \{\alpha_{i_{\sigma^I(1)}}, \dots, \alpha_{i_{\sigma^I(p_j)}}\}. \end{aligned} \tag{2.180}$$

A related argument is as follows. Define v, w by (2.178). If $w = 0$, then since the $x_{i_{\sigma^I(k)}}$, $k \leq p_j$ have been made nearly equal to 0 before x_j,

$$\frac{1}{v + w - x_j} = -\sum_{k=0}^{+\infty} \frac{v^k}{x_j^{k+1}} . \tag{2.181}$$

From (2.181) , we get

$$\mathrm{Res}_{x_j=0} \left[\frac{1}{x_j^n} \frac{1}{v + w - x_j} \right] = 0 . \tag{2.182}$$

If $w \neq 0$, then

$$\frac{1}{v + w - x_j} = \sum_{k=0}^{+\infty} \frac{(x_j - v)^k}{w^{k+1}} . \tag{2.183}$$

By (2.183) , we get

$$\mathrm{Res}_{x_j=0} \frac{1}{x_j^n} \frac{1}{v + w - x_j} = \sum_{k=n-1}^{+\infty} \frac{C_k^{n-1}(-v)^{k-n+1}}{w^{k+1}} = \frac{1}{(v + w)^n} , \tag{2.184}$$

which gives another proof of (2.180). Finally observe that if $j \in I$,

$$\langle \alpha_j, x^I \rangle = x_j . \tag{2.185}$$

From (2.174), (2.180), (2.185), we get (2.173). The proof of our Theorem is completed. □

By Theorem 2.22, we already know that, as a section of a line bundle, $P_{u,n,q}(t)$ is polynomial on $T\backslash\overline{S}$. By Theorem 2.22, the right-hand side of (2.173) descends to a section of the same line bunlde . We will give a direct proof that the right-hand side of (2.173) is polynomial on $T\setminus\overline{S}$.

THEOREM 2.49. *The right-hand side of (2.173) is a polynomial on* $T\setminus\overline{S}$.

PROOF. In the right-hand side of (2.173), up to a locally constant factor on $T\setminus\overline{S}$, we may and will take off the brackets in $\left[\frac{1}{d}\langle p^I_{j-1}(t+f),e^j\rangle\right]$. By (2.13),

$$\sum_{j=1}^{r}\frac{\langle p^I_{j-1}\alpha_{i_j},x^I\rangle}{\langle p^I_{j-1}\alpha_{i_j},e^j\rangle}\langle p^I_{j-1}(t+f),e^j\rangle=\langle t+f,x^I\rangle\,. \tag{2.186}$$

Then

$$\langle t+f,x^I\rangle=\sum_{1}^{r}F^j(t)x_{\sigma^I(j)}\,, \tag{2.187}$$

and the $F^j(t)$ are affine functions of $t\in\mathfrak{t}$. Therefore

$$\exp(\langle t+f,x^I\rangle)=\prod_{j=1}^{r}\left(\sum_{k=0}^{+\infty}\frac{(F^j(t)x_{\sigma^I(j)})^k}{k!}\right)\,. \tag{2.188}$$

Moreover

$$\frac{1}{\exp\left(\frac{d\langle p^I_{j-1}\alpha_{i_j},x^I\rangle}{\langle p^I_{j-1}\alpha_{i_j},e^j\rangle}\right)-1}=\frac{\operatorname{Td}\left(\frac{-d\langle p^I_{j-1}\alpha_{i_j},x^I\rangle}{\langle p^I_{j-1}\alpha_{i_j},e^j\rangle}\right)}{\frac{d\langle p^I_{j-1}\alpha_{i_j},x^I\rangle}{\langle p^I_{j-1}\alpha_{i_j},e^j\rangle}}\,. \tag{2.189}$$

Also by (2.28), (2.29),

$$\frac{\langle p^I_{j-1}\alpha_{i_j},x^I\rangle}{\langle p^I_{j-1}\alpha_{i_j},e^j\rangle}=\frac{1}{\langle\alpha_{i_1},\dots,\alpha_{i_j}\rangle}\left(\langle\alpha_{i_1},\dots,\alpha_{i_{j-1}}\rangle x_{i_j}-\sum_{k=1}^{j-1}\langle\alpha_{i_1},\dots,\alpha_{i_{k-1}},\alpha_{i_j},\alpha_{i_{k+1}},\dots,\alpha_{i_{j-1}}\rangle x_{i_k}\right)\,. \tag{2.190}$$

Finally the $\langle\alpha_j,x^I\rangle$ are linear combinations of $x_{i_1},\dots,x_{i_r}$.

Freeze now $x_{i_{\sigma^I(2)}},\dots,x_{i_{\sigma^I(r)}}$, and consider one of the terms in (2.173) as a function of $x_{i_{\sigma^I(1)}}$. Then it is clear that this term is a meromorphic function of $x_{i_{\sigma^I(1)}}$ with a pole of finite order at $x_{i_{\sigma^I(1)}}=0$. When taking the residue at 0 in the variable $x_{i_{\sigma^I(1)}}$, by the above, it is clear this will introduce a finite power $F^1(t)^k$, i.e. a polynomial function of t.

It is now clear that the procedure can be iterated. The proof of our Theorem is completed. □

2.12. The action of a differential operator on $P_{u,n,q}(t)$. Let $s\in\mathfrak{t}^*\mapsto F(s)\in C[s]$ be a power series, which converges on a neighborhood of 0. Then $F(\partial/\partial t)$ is a formal power series of differential operators. Since $P_{u,n}(t)$ is a polynomial on $T\setminus S_u$, $F(\partial/\partial t)P_{u,n}(t)$ is a well-defined polynomial on $T\setminus S_u$.

Recall that we have identified $\mathfrak{t}$ and $\mathfrak{t}^*$. Then if $I=(i_1,\dots,i_r)$ is generic, if $x\in\mathbf{C}^\ell$, then $x^I\in\mathfrak{t}\simeq\mathfrak{t}^*$.

THEOREM 2.50. *For $u \in C/\overline{R}^*$, $n \in \mathbf{N}^*$, $q \in \mathbf{N}$, if $t \in \mathfrak{t}$ represents an element of $T \setminus \overline{S}$, then*

$$F(\partial/\partial t)P_{u,n,q}(t) = \left|\frac{\overline{R}}{\overline{CR}}\right|(-1)^{r+1} \sum_{\substack{I=(i_1,\dots,i_r)\in\mathcal{I}_u \\ f\in\overline{CR}/d\overline{R}}} \frac{1}{\langle\alpha_{i_1},\dots,\alpha_{i_r}\rangle}$$

$$\exp(2i\pi(\langle qu, f\rangle))\operatorname{Res}^I_{x=0}\Bigg(F(x^I)\frac{1}{(\prod_{j\in I_u}\langle\alpha_j, x^I\rangle)^n}$$

$$\exp\left\{d\frac{\langle p^I_{j-1}\alpha_{i_j}, x^I\rangle}{\langle p^I_{j-1}\alpha_{i_j}, e^j\rangle}\left[\frac{1}{d}\langle p^I_{j-1}(t+f), e^j\rangle\right]\right\}$$

$$\frac{1}{\prod_{j=1}^r\left(\exp\left(\frac{\langle p^I_{j-1}\alpha_{i_j}, x^I\rangle}{\langle p^I_{j-1}\alpha_{i_j}, e^j\rangle}\right)-1\right)}. \tag{2.191}$$

PROOF. For $y \neq 0$, $[y]-y$ is locally constant. Using (2.186), if $s \in \mathfrak{t}$ is identified with the corresponding vector field, then

$$\begin{aligned} & s\left\{\exp\left(d\sum_{j=1}^r \frac{\langle p^I_{j-1}\alpha_{i_j}, x^I\rangle}{\langle p^I_{j-1}\alpha_{i_j}, e^j\rangle}\left[\frac{1}{d}\langle p^I_{j-1}(t+f), e^j\rangle\right]\right)\right\} \\ & = \langle s, x^I\rangle \exp\left(d\sum_{j=1}^r \frac{\langle p^I_{j-1}\alpha_{i_j}, x^I\rangle}{\langle p^I_{j-1}\alpha_{i_j}, e^j\rangle}\left[\frac{1}{d}\langle p^I_{j-1}(t+f), e^j\rangle\right]\right). \end{aligned} \tag{2.192}$$

From (2.173), (2.192), we get (2.191).

The proof of our Theorem is completed. □

3. Symplectic manifolds and moment maps

In this Section, we recall known results on symplectic manifolds and moment maps. In particular we prove a form of the formula of Witten [64] Jeffrey-Kirwan [28], which expresses the integral of certain characteristic classes on symplectic reductions in terms of the action of differential operators on the symplectic volume. The results of this Section will be used in Section 5, where we will give a formula for the integral of certain characteristic classes on the strata of the moduli space.

This Section is organized as follows. In Section 3.1, we recall elementary facts on orbifolds. In Section 3.2-3.4, we give elementary properties of moment maps. In Section 3.5, we give a direct simple proof that the image of the symplectic volume measure by the moment map can be evaluated in terms of the symplectic volume of the symplectic reductions. In Section 3.6, when 0 is a regular value of the moment map μ, we recall Jeffrey-Kirwan's expression of the volume of the neighbouring fibres in terms of integrals of characteristic classes on the symplectic reduction of $\mu^{-1}(0)$. In Section 3.7, we prove the formula of Witten-Jeffrey-Kirwan. Finally in Section 3.8, we apply the above to the symplectic coadjoint orbits of G.

3.1. Orbifolds. Let X be a smooth compact manifold. Let G be a compact connected Lie group, and let $\mathfrak{g}$ be its Lie algebra. We assume that G acts on X on the right. If $Y \in \mathfrak{g}$, let $Y^X \in \mathrm{Vect}(X)$ be the corresponding vector field.

We assume that G acts locally freely on X, i.e. for any non zero $Y \in \mathfrak{g}$, Y^X is a non vanishing vector field on X.

Let $\mathfrak{g}^X$ be the subvector bundle of TX which is the image of $\mathfrak{g}$ by $Y \to Y^X$. Then we have an exact sequence of G vector bundles

$$0 \to \mathfrak{g}^X \to TX \to TX/\mathfrak{g}^X \to 0\,. \tag{3.1}$$

The above data define an orbifold X/G, and the G-bundle $TX/\mathcal{G}^X$ is also called the tangent bundle TX/G to X/G. If the G-bundle $TX/\mathfrak{g}^X$ is orientable (or equivalently if TX is orientable), we will say that the orbifold X/G is orientable.

If G acts freely on X, then X/G is just the standard quotient.

If $y \in X$, let $Z(y) = \{g \in G; yg = y\}$ be the stabilizer of y. Then $Z(y)$ is a finite subgroup of G. By [24, Proposition 27.4] , there are finitely many conjugacy classes of finite subgroups of G, which occur as stabilizers.

Inclusion induces a partial ordering on the set of conjugacy classes of finite subgroup of G. On each connected component of X, there is a unique minimal conjugacy class of stabilizers S, called the generic conjugacy class of stabilizers. This minimal conjugacy class then acts as the identity on the considered connected component. The order $|S|$ of a generic stabilizer is locally constant on X.

Let X_{reg} be the set of $y \in X$ such that $Z(y)$ lies in the minimal conjugacy class. Then X_{reg} is open in X, and X_{reg}/G is a smooth manifold included in the orbifold X/G.

DEFINITION 3.1. A 1-form $\theta : TX \to \mathfrak{g}$ is said to be a connection form on $\pi : X \to X/G$ if

- For $Y \in \mathfrak{g}$,

$$\theta(Y^X) = Y. \tag{3.2}$$

- For $g \in G, Y \in \mathfrak{g}$,

$$g^*\theta = \theta.g. \tag{3.3}$$

One verifies trivially that G-connections exist. Then the curvature Θ of θ is defined by

$$d\theta = -\frac{1}{2}[\theta, \theta] + \Theta. \tag{3.4}$$

Also if $Y \in \mathfrak{g}, g \in G$,

$$i_{Y^X}\Theta = 0, g^*\Theta = \Theta.g. \tag{3.5}$$

Put

$$T^H X = \{U \in TX, \theta(U) = 0\}. \tag{3.6}$$

Then $T^H X$ is a G-invariant subbundle of TX such that

$$TX = T^H X \oplus \mathfrak{g}^X. \tag{3.7}$$

A G-invariant form α on X is said to be basic if for $Y \in \mathfrak{g}$, $i_{Y^X}\alpha = 0$. From now on, we suppose that X/G is an oriented orbifold. Then X_{reg}/G is an oriented manifold. By definition

$$\int_{X/G} \alpha = \int_{X_{\text{reg}}/G} \alpha\,. \tag{3.8}$$

Let $E_1, \dots, E_n$ be a basis of $\mathfrak{g}$, let $E^1, \dots, E^n$ be the corresponding dual basis of $\mathfrak{g}^*$. We write the connection θ in the form

$$\theta = \Sigma_1^n \theta^i E_i\,. \tag{3.9}$$

Then $E^1 \wedge \dots \wedge E^n$ defines a volume form on G. Let $\text{Vol}(G)$ be the corresponding volume of G.

We equip $\mathfrak{g}^X$ with the orientation induced by the orientation of $\mathfrak{g}$. Then $TX \simeq TX/\mathfrak{g}^X \oplus \mathfrak{g}^X$ is naturally oriented. If α is a G-invariant basic form on X,

$$\int_{X/G} \alpha = \frac{|S|}{\text{Vol}(G)} \int_X \alpha \wedge \theta^1 \wedge \dots \wedge \theta^n\,. \tag{3.10}$$

If α is a G-invariant basic form on X, $d\alpha$ is also G-invariant and basic. Then since $(X/G) \setminus (X_{\text{reg}}/G)$ is a union of submanifolds of codimension ≥ 2,

$$\int_{X/G} d\alpha = 0\,. \tag{3.11}$$

In fact note that

$$\int_X d\alpha \wedge \theta^1 \wedge \dots \wedge \theta^n = (-1)^{\deg(\alpha+1)} \int_X \alpha \wedge d(\theta^1 \wedge \dots \wedge \theta^n)\,. \tag{3.12}$$

By (3.4),

(3.13)

$$\int_X \alpha \wedge d(\theta^1 \wedge \dots \wedge \theta^n) = \int_X \alpha \wedge \sum_{i=1}^n (-1)^{i-1} \theta^1 \wedge \dots \wedge \theta^{i-1} \wedge \Theta^i \wedge \dots \wedge \theta^n = 0,$$

which provides another proof of (3.11).

Let $E \to X$ be a complex G-vector bundle on E. Let ∇^E be a G-invariant horizontal connection on X. Namely suppose first that G acts freely on X. Then ∇^E is just a connection on the vector bundle $X \times_G E$ over X/G. More generally, if G only acts locally freely on X, the above construction still makes sense. Let $F^E = \nabla^{E,2}$ be the curvature of ∇^E. Then F^E is a G-invariant basic 2 form on X with values in $\mathfrak{g}$. Let P be an ad-invariant function on $\mathfrak{g}$. Then $P(E, \nabla^E) = P(\frac{-F^E}{2i\pi})$ is a G-invariant basic closed form on X. If X/G is an oriented orbifold, the integral $\int_{X/G} P(F, \nabla^E)$ is well-defined and does not depend on ∇^E.

3.2. Symplectic manifolds and moment maps. Let (X, σ) be a symplectic manifold, so that σ is a nondegenerate closed 2-form. Let $H : X \to \mathbf{R}$ be a smooth function. The corresponding hamiltonian vector field Y_H is defined by the equation

$$dH = i_{Y_H}\sigma, \tag{3.14}$$

which we rewrite in the form

$$(d - i_{Y_H})(H + \sigma) = 0. \tag{3.15}$$

From (3.15), we get

$$L_{Y_H}\sigma = 0. \tag{3.16}$$

If H, H' are two smooth functions, put

$$\{H, H'\} = -Y_{H'}.H = Y_H.H' = -\sigma(Y_H, Y_{H'}). \tag{3.17}$$

Then

$$Y_{\{H,H'\}} = [Y_H, Y_{H'}]. \tag{3.18}$$

Let G be a compact connected Lie group acting on the right on X, and preserving σ. Let $\mathfrak{g}$ be the Lie algebra of G. Let $\eta : G \to \mathrm{Aut}(\mathfrak{g}^*)$ be the coadjoint representation. If $Y \in \mathfrak{g}, p \in \mathfrak{g}^*$, let $\{Y, p\} \in \mathfrak{g}^*$ be the infinitesimal (left) action of Y on p.

We say that $\mu : X \to \mathfrak{g}^*$ is a moment map if

- For $x \in X, g \in G$,

$$\mu(xg) = \mu(x).g. \tag{3.19}$$

- If $Y \in \mathfrak{g}$, if Y^X is the corresponding vector field on X, then $\langle \mu, Y \rangle$ is a Hamiltonian for Y^X.

In particular, by (3.17),(3.19),

$$\langle \mu, [Y, Y'] \rangle = -\sigma(Y^X, Y'^X). \tag{3.20}$$

Also by (3.19),

$$Y^X \mu = -\{Y, \mu\}. \tag{3.21}$$

Assume now that G acts locally freely on X. Of course, this never occurs if X is compact. Let θ be a G-connection form on $\pi : X \to X/G$.

PROPOSITION 3.2. *There is a unique closed 2-form η on X/G such that*

$$\sigma = \pi^*\eta - d\langle \mu, \theta \rangle. \tag{3.22}$$

PROOF. By (3.3) , (3.19) , the 1-form $\langle \mu, \theta \rangle$ is G-invariant, so that if $Y \in \mathfrak{g}$,

$$L_{Y^X}\langle \mu, \theta \rangle = 0. \tag{3.23}$$

Also by (3.16),

$$L_{Y^X}\sigma = 0. \tag{3.24}$$

Then by (3.14), (3.23),

$$i_{Y^X}(\sigma + d\langle \mu, \theta\rangle) = i_{Y^X}\sigma + L_{Y^X}\langle \mu, \theta\rangle - di_{Y^X}\langle \mu, \theta\rangle = 0 \tag{3.25}$$

By (3.25), we find that the G-invariant 2-form $\sigma + d\langle \mu, \theta\rangle$ is basic. Therefore it is of the form $\pi^*\eta$. The proof of our Proposition is completed. □

Let σ^H be the restriction of σ to $T^H X$. Since σ^H is G-invariant, it descends to a 2-form on X/G. The same is true for the 2-form $\langle \mu, \Theta\rangle$.

THEOREM 3.3. *The following identity of 2-forms holds on X/G,*

$$\sigma^H = \pi^*\eta - \langle \mu, \Theta\rangle. \tag{3.26}$$

PROOF. By (3.4),

$$d\langle \mu, \theta\rangle = \langle d\mu, \theta\rangle + \langle \mu, -\frac{1}{2}[\theta, \theta] + \Theta\rangle. \tag{3.27}$$

From (3.27), we get

$$[d\langle \mu, \theta\rangle]^H = \langle \mu, \Theta\rangle. \tag{3.28}$$

From (3.22) , (3.28), we get (3.26). □

REMARK 3.4. If $X \in \mathfrak{g}$, by (3.15),

$$(d - i_{Y^X})(\sigma + \langle \mu, Y\rangle) = 0. \tag{3.29}$$

Classically [**61**], (3.29) shows that $\sigma^H + \langle \mu, \Theta\rangle$ descends to a closed 2-form on X/G. Of course this also follows from Proposition 3.2 and Theorem 3.3.

REMARK 3.5. Let $\langle , \rangle$ be a G- invariant scalar product on $\mathfrak{g}$. Then $\mathfrak{g}$ and $\mathfrak{g}^*$ can be identified. Let $H \subset G$ be a Lie subgroup of G, and let $\mathfrak{h} \subset \mathfrak{g}$ be the corresponding Lie algebra. Let $\mathfrak{h}^\perp$ be the orthogonal space to $\mathfrak{h}$ in $\mathfrak{g}$. Then $\mathfrak{g} = \mathfrak{h} \oplus \mathfrak{h}^\perp$ is a H-invariant splitting. Let $P^{\mathfrak{h}} : \mathfrak{g} \to \mathfrak{h}$ be the corresponding projection. Put

$$\theta^{\mathfrak{h}} = P^{\mathfrak{h}}\theta, \ \theta^{\mathfrak{h}^\perp} = P^{\mathfrak{h}^\perp}\theta. \tag{3.30}$$

Then $\theta^{\mathfrak{h}}$ is a connection on $X \to X/H$, whose curvature $\Theta^{\mathfrak{h}}$ is given by

$$\Theta^{\mathfrak{h}} = P^{\mathfrak{h}}\Theta - \frac{1}{2}P^{\mathfrak{h}}[\theta^{\mathfrak{h}^\perp}, \theta^{\mathfrak{h}^\perp}]. \tag{3.31}$$

Take $\mu \in \mathfrak{g}^*$, and suppose that $Z(\mu) = H$. Then

$$\langle \mu, \frac{1}{2}[\theta, \theta]\rangle = \langle \mu, \frac{1}{2}[\theta^{\mathfrak{h}^\perp}, \theta^{\mathfrak{h}^\perp}]\rangle = \langle \mu, \frac{1}{2}P^{\mathfrak{h}}[\theta^{\mathfrak{h}^\perp}, \theta^{\mathfrak{h}^\perp}]\rangle. \tag{3.32}$$

Therefore by (3.31), (3.32),

$$\langle \mu, -\frac{1}{2}[\theta, \theta] + \Theta\rangle = \langle \mu, \Theta^{\mathfrak{h}}\rangle. \tag{3.33}$$

So by (3.4), (3.22),(3.27), (3.33), if $Z(\mu) = H$,

$$\sigma = \pi^*\eta - \langle d\mu, \theta\rangle - \langle \mu, \Theta^{\mathfrak{h}}\rangle. \tag{3.34}$$

Equation (3.34) is of special interest when $\mu \in \mathfrak{t}^* \simeq \mathfrak{t}$, and $H = T$.

Put

$$E = X \times_G \mathfrak{g}^*. \tag{3.35}$$

Then E is an orbifold vector bundle over X/G. Moreover (3.19) says that μ descends to a section of E over X/G.

3.3. Symplectic reduction. We make the same assumptions as in Section 3.2. We do no longer assume that G acts locally freely on X. If $p \in \mathfrak{g}^*$, put

$$X_p = \mu^{-1}(p). \tag{3.36}$$

Let $Z(p) = \{g \in G, g.p = p\}$ be the stabilizer of p in G, and let $\mathfrak{z}(p) = \{X \in \mathfrak{g}, \{X,p\} = 0\}$ be its Lie algebra. By [**15**, Theorem IV.2.3], $Z(p)$ is a connected Lie subgroup of G. Clearly $Z(p)$ acts on X_p.

PROPOSITION 3.6. *The element $p \in \mathfrak{g}^*$ is a regular value of μ if and only if for $x \in X_p, Y \in \mathfrak{g} \mapsto Y^X(x) \in T_xX$ is injective, or equivalently if and only if $Z(p)$ acts locally freely on X_p.*

PROOF. By (3.29) if $x \in X_p$, $Y \in \mathfrak{g}$ lies in coker$(d\mu(x))$ if and only if $Y^X(x) = 0$. Also by (3.21), if $\mu(x) = p, Y^X(x) = 0$, then $\{Y,p\} = 0$, i.e. $Y \in \mathfrak{z}(p)$. The proof of our Proposition is completed. □

Assume now that p is a regular value of μ, and that $X_p \neq \emptyset$. Then X_p is a smooth submanifold of X, on which $Z(p)$ acts locally freely, so that $X_p/Z(p)$ is a $Z(p)$-orbifold.

Let i_p be the embedding $X_p \to X$. Then by (3.21), (3.29), for $Y \in \mathfrak{z}(p)$,

$$i_{Y^X} i_p^* \sigma = 0. \tag{3.37}$$

By (3.37) , we find that the $Z(p)$ -invariant closed 2-form $i_p^*\sigma$ descends to a closed 2-form σ_p on $X_p/Z(p)$. Then $(X_p/Z(p), \sigma_p)$ is a symplectic orbifold.

If $p \in \mathfrak{g}^*$, $g \in G$ maps X_p into $X_{p.g}$. Also $p \in \mathfrak{g}^*$ is regular if and only if $p.g$ is regular, and there is an obvious symplectomorphism $(X_p/Z(p), \sigma_p) \to (X_{p.g}/Z(p.g), \sigma_{p.g})$.

Let $\mathcal{O} \subset \mathfrak{g}^*$ be a coadjoint orbit. If $p \in \mathcal{O}, Y, Z \in \mathfrak{g}$, put

$$\sigma_{\mathcal{O}}(Y,Z) = \langle p, [Y,Z] \rangle. \tag{3.38}$$

By (1.193), $\sigma_{\mathcal{O}}$ is a G-invariant symplectic form on $\mathcal{O}$. Moreover $\mu_{\mathcal{O}} : p \in \mathcal{O} \mapsto -p \in \mathfrak{g}^*$ is a moment map for the right action of G on $\mathcal{O}$.

Put

$$X_{\mathcal{O}} = \mu^{-1}(\mathcal{O}). \tag{3.39}$$

Then G acts on the right on $X_{\mathcal{O}} \subset X$.

Let p_1, p_2 be the projections $X \times \mathcal{O} \to X, X \times \mathcal{O} \to \mathcal{O}$. Then $(X \times \mathcal{O}, p_1^*\sigma + p_2^*\sigma_{\mathcal{O}})$ is a symplectic manifold on which G acts symplectically on the right, with moment map $p_1^*\mu + p_2^*\mu_{\mathcal{O}}$.Then

$$(X \times \mathcal{O})_0 = \{(x,p) \in X \times \mathcal{O}, \mu(x) - p = 0\}. \tag{3.40}$$

So, if $p \in \mathcal{O}$,

$$\begin{aligned} &(X \times \mathcal{O}_p)_0 \simeq X_{\mathcal{O}_p}, \\ &(X \times \mathcal{O}_p)_0/G \simeq X_p/Z(p). \end{aligned} \tag{3.41}$$

Moreover $p \in \mathfrak{g}^*$ is a regular value of μ if and only if 0 is a regular value of $p_1^*\mu + p_2^*\mu_{\mathcal{O}}$. Then one finds easily that (3.41) identifies the symplectic forms on the corresponding orbifolds.

3.4. Symplectic reduction with respect to a fixed stabilizer. Let G be a compact connected semisimple Lie group. We use the same notation as in Section 1.14.

Let $\langle , \rangle$ be a G-invariant scalar product on $\mathfrak{g}$. So we can identify $\mathfrak{g}$ and $\mathfrak{g}^*$, $\mathfrak{t}$ and $\mathfrak{t}^*$. Let $\eta : G \to \mathrm{Aut}(\mathfrak{g}^*)$ be the coadjoint representation. Of course we can now identify the representations τ and η.

Take $p_0 \in \mathfrak{g}^*$. Put

$$\begin{aligned} Z &= Z(p_0), \\ \mathfrak{z} &= \mathfrak{z}(p_0). \end{aligned} \tag{3.42}$$

DEFINITION 3.7. Put

$$\mathfrak{R} = \{p \in \mathfrak{t}^*, Z(p) = Z\}. \tag{3.43}$$

Since the $Z(p)$ are connected,

$$\mathfrak{R} = \{p \in \mathfrak{t}^*, \mathfrak{z}(p) = \mathfrak{z}\}. \tag{3.44}$$

By (1.182), (3.44), $\mathfrak{R}$ is the complement of a finite union of hyperplanes in $\{p \in \mathfrak{t}^*, \text{if } \langle p_0, h_\alpha\rangle = 0, \langle p, h_\alpha \rangle = 0\}$.

Now we make the same assumptions as in Section 3.2. Suppose that G acts locally freely on X. Then by Proposition 3.6 , any $p \in \mathfrak{g}^*$ is a regular value of μ.

DEFINITION 3.8. Put

$$\mathfrak{S} = \mu^{-1}(\mathfrak{R}). \tag{3.45}$$

Then $\mathfrak{S}$ is a submanifold of X, on which Z acts locally freely. Let $\sigma_{\mathfrak{S}}$ be the restriction of σ to S. Then $\sigma_{\mathfrak{S}}$ is a closed 2-form on $\mathfrak{S}$, which in general is not symplectic. Let $j : \mathfrak{g}^* \to \mathfrak{z}^*$ be the obvious projection. Then $j\mu : \mathfrak{S} \to \mathfrak{z}^*$ is a moment map for the action of Z on $\mathfrak{S}$ with respect to $\sigma_{\mathfrak{S}}$. Finally $\mathfrak{S}/Z$ embeds into X/G.

Let P be the orthogonal projection $\mathfrak{g} \to \mathfrak{z}$. Let θ be a connection form on $X \to X/G$. Put

$$\tilde{\theta} = P\theta. \tag{3.46}$$

Then $\tilde{\theta}$ defines a Z-connection on $X \to X/Z$. Let $\tilde{\Theta}$ be the curvature of θ. Clearly, over $\mathfrak{S}$, since $\mu \in \mathfrak{z}$,

$$\langle \mu, \theta \rangle = \langle \mu, \tilde{\theta} \rangle. \tag{3.47}$$

By (3.34), over $\mathfrak{S}$,

$$\sigma = \pi^*\eta - (\langle d\mu, \tilde{\theta}\rangle + \langle \mu, \tilde{\Theta}\rangle). \tag{3.48}$$

In particular, if $p \in \mathfrak{R}$,

$$i_p^*\sigma = i_p^*(\pi^*\eta - \langle \mu, \tilde{\Theta}\rangle). \tag{3.49}$$

Recall that the vector bundle E was defined in (3.35). Let $\tilde{\nabla}^E$ be the connection on $E_{|\mathfrak{S}/Z}$ induced by $\tilde{\theta}$. Then over X_p/Z,

$$\tilde{\nabla}^E \mu = 0. \tag{3.50}$$

Equation (3.50) explains why over X_p, σ given by (3.49) is closed. In fact it is the pull-back of the symplectic form σ_p on $X_p/Z(p)$.

PROPOSITION 3.9. *Locally, over $\mathfrak{S}$, the cohomology class of σ_p depends linearly on $p \in \mathfrak{R}$.*

PROOF. This is obvious by (3.49). □

3.5. The image of the symplectic volume by the moment map. Let G be a compact connected semisimple Lie group. We use otherwise the notation of Section 1. Let (X, σ) be a compact symplectic manifold, on which G acts symplectically, with moment map $\mu : X \to \mathfrak{g}^*$.

We make the assumption that

$$Z(x) = 1 \text{ for a.e. } x \in X. \tag{3.51}$$

Then a.e., G acts locally freely on X. By Proposition 3.6, a.e., $d\mu(x)$ surjects on $\mathfrak{g}^*$. By Sard's theorem, a.e. $p \in \mathfrak{g}^*$ is a regular value of μ.

Let dp, dt be the Lebesgue measures on $\mathfrak{g}, \mathfrak{t}$ with respect to $\langle , \rangle$. Let dg be the Lebesgue measure on G. If $p \in \mathfrak{g}$, let dg_p be the Lebesgue measure on $Z(p)$.

Clearly

$$\{e^\sigma\}^{\max} = \frac{\sigma^{\dim X/2}}{(\dim X/2)!}. \tag{3.52}$$

Then $\frac{\sigma^{\dim X/2}}{(\dim X/2)!}$ is a $\dim X$ form on X, which does not vanish, and so defines an orientation of X. If $f : X \to \mathbf{R}$ is a bounded measurable function, put

$$\int_X f(x)|e^\sigma| = \int_X f(x) e^\sigma, \tag{3.53}$$

The notation (3.53) emphasizes the fact that the left-hand side is an integral with respect to a nonnegative measure. We will use a similar notation over $X_p/Z(p)$.

Now we will compute the disintegration of the symplectic volume on X with respect to the moment map μ. The point of this proof is that it is parallel to a corresponding for moduli spaces given in Theorem 5.45.

Recall that the monomial $\pi : \mathfrak{t} \to \mathbf{C}$ was defined in Definition 1.53.

THEOREM 3.10. *Let $f : X \to \mathbf{R}$ be a bounded measurable function. Then*

$$\begin{aligned} \int_X f(x)|e^\sigma| &= \int_{\mathfrak{t}/W} |\pi(t)| dt \int_{X_t/T} |e^{\sigma_t}| \int_G f(x.g) dg, \\ \int_X f(x)|e^\sigma| &= \int_{\mathfrak{g}} \frac{dp}{|\pi(p)|} \int_{X_p/Z(p)} |e^{\sigma_p}| \int_{Z(p)} f(x.g) dg_p. \end{aligned} \tag{3.54}$$

PROOF. Since $d\mu(x)$ is a.e. surjective, it is clear that $\mu_* \frac{\sigma^{\dim X/2}}{(\dim X/2)!}$ is absolutely continuous with respect to dp. Also for a.e. x, $d\mu(x)$ is surjective and $\mu(x) \in \mathfrak{g}_{\text{reg}}$.

Let $K \subset \mathfrak{t}$ be a Weyl chamber. Let q be the projection $\mathfrak{g}_{\text{reg}} \to \mathfrak{t}_{\text{reg}}/W \simeq K$. If $x \in X$ is such that $\mu(x) \in \mathfrak{g}_{\text{reg}}$, we have a G-equivariant complex

$$(C'_x, \partial) : 0 \to \mathfrak{g} \to T_x X \overset{dq\mu}{\to} \mathfrak{t} \to 0, \tag{3.55}$$

where the map $\mathfrak{g} \to T_x X$ is just $Y \to Y^X(x)$. If $d\mu(x)$ is surjective, by Proposition 3.6, the cohomology of the complex (C'_x, ∂) is concentrated in degree 1. More precisely, one finds easily that in this case,

$$H^1(C'_x, \partial) \simeq T_{\pi(x)} X_{q\mu(x)}/Z(q\mu(x)). \tag{3.56}$$

In particular, by [**35**] and by (3.55), (3.56), we have a canonical isomorphism of real lines

$$(\det C'_x)^{-1} \simeq \det(T_{\pi(x)} X_{q\mu(x)}/Z(q\mu(x))). \tag{3.57}$$

Now $\det(T_x X)$ is equipped with the volume form associated to σ_x.Also $\mathfrak{g}$ and $\mathfrak{t}$ are equipped with the volume forms dp and dt. Therefore $\det(C'_x)$ is equipped with a natural metric. Let $dv_{X_{q\mu(x)}/Z(q\mu(x))}$ be the corresponding volume form on $T_{\pi(x)} X_{q\mu(x)}/Z(q\mu(x))$ via the isomorphism (3.57).

By the formula of change of variables,we get

$$\int_X f(x)|e^\sigma| = \int_{\mathfrak{t}_{\mathrm{reg}}/W} dt \int_{X_t/T} dv_{X_t/T}(x) \int_G f(x.g)dg. \tag{3.58}$$

Take $t \in \mathfrak{t}_{\mathrm{reg}}, x \in X$ such that $\mu(x) = t$. Consider the double complex

$$\begin{array}{ccccccccc}
 & & 0 & & 0 & & 0 & & \\
 & & \uparrow & & \uparrow & & \uparrow & & \\
0 & \longrightarrow & \mathfrak{g} & \longrightarrow & T_x X & \xrightarrow{d(q\mu)} & \mathfrak{t} & \longrightarrow & 0 \\
 & & \uparrow & & \uparrow & & \uparrow & & \\
0 & \longrightarrow & \mathfrak{g} & \longrightarrow & T_x X \oplus \mathrm{Im\,Ad}(t) & \longrightarrow & \mathfrak{g} & \longrightarrow & 0 \\
 & & \uparrow & & \uparrow & & \uparrow & & \\
0 & \longrightarrow & 0 & \longrightarrow & \mathrm{Im\,Ad}(t) & \longrightarrow & \mathrm{Im\,Ad}(t) & \longrightarrow & 0 \\
 & & \uparrow & & \uparrow & & \uparrow & & \\
 & & 0 & & 0 & & 0 & &
\end{array} \tag{3.59}$$

In (3.59) , the map $\mathfrak{g} \to \mathrm{Im\,Ad}(t)$ is just $f \mapsto -\mathrm{Ad}(t)f = -[t,f]$, the map $T_x X \to \mathfrak{g}$ is $d\mu(x)$. Also the columns and the lowest row are acyclic. In particular the cohomology groups of the two upper rows are isomorphic. Therefore they are concentrated in degree 1 and equal to $H^1(C'_x, \partial)$.

In the second column of (3.59), we equip $\mathrm{Im\ Ad}(t) \simeq T_t \mathcal{O}_t$ with the 2-form $\sigma_{\mathcal{O}_t}$ given in (3.38). In the third column of (3.59), we equip $\mathrm{Im\ Ad}(t) \simeq T_t \mathcal{O}_t$ with the metric induced by the metric of $\mathfrak{g}$. By (3.22), (3.27), the volume induced by the second row on its determinant is just the symplectic volume $\frac{\sigma^{\dim(X_t/T)/2}}{(\dim(X_t/T)/2)!}$ on $\det H^1(C'_x, \partial)$. On the other hand, let R be the lowest row in (3.59). Since it is acyclic, $\det R \simeq \mathbf{R}$ has a canonical section η. Clearly

$$|\eta| = |\pi(t)|. \tag{3.60}$$

Therefore

$$dv_{X_t/T} = |\pi(t)| \frac{|\sigma_t^{\dim(X_t/T)/2}|}{(\dim(X_t/T)/2)!}. \tag{3.61}$$

By (3.58), (3.61), we get the first identity in (3.54) .

By [**14**, Proposition 6.3.4], if $h : \mathfrak{g} \to \mathbf{R}$ is a bounded measurable function,

$$\int_{\mathfrak{g}} h(p)dp = \frac{1}{\mathrm{Vol}(T)} \int_{\mathfrak{t}/W} |\pi(t)|^2 dt \int_G h(t.g)dg. \tag{3.62}$$

From the first identity in (3.54) and from (3.62), we obtain

$$(3.63)\quad \int_X f(x)|e^{\sigma}| = \int_{\mathfrak{t}/W} |\pi(t)|dt \int_{X_t/T} |e^{\sigma_t}| \int_{T\backslash G} d\dot{g} \int_T f(x.t'\dot{g})dt' = \int_{\mathfrak{g}} \frac{dp}{|\pi(p)|} \int_{X_p/Z(p)} |e^{\sigma_p}| \int_{Z(p)} f(x.g)dg_p,$$

which is the second identity in (3.54). The proof of our Theorem is completed. □

DEFINITION 3.11. If $t \in \mathfrak{t}$ is a regular value of μ, let $|V(t)|$ be the absolute value of the symplectic volume of $X_t/Z(t)$ with respect to σ_t.

THEOREM 3.12. *Let $f : \mathfrak{g}^* \to \mathbf{R}$ be a bounded measurable function. Then*

$$(3.64)\quad \begin{aligned} \int_X f(\mu(x))|e^{\sigma}| &= \int_{\mathfrak{t}/W} \left[\int_G f(t.g)dg\right] |\pi(t)||V(t)|dt, \\ \int_X f(\mu(x))|e^{\sigma}| &= \mathrm{Vol}(T) \int_{\mathfrak{g}^*} f(p)\frac{|V(p)|}{|\pi(p)|}dp. \end{aligned}$$

PROOF. Our Theorem follows from Theorem 3.10. □

Clearly, if 0 is a regular value of μ, $\mu_*|e^{\sigma}|$ has a smooth density with respect to dp near $p = 0$. In particular

$$(3.65)\quad \lim_{\substack{p\in\mathfrak{g}_{reg} \\ p\to 0}} \frac{|V(p)|}{|\pi(p)|}$$

exists and is the value of the above density at $p = 0$. A more precise statement is as follows.

PROPOSITION 3.13. *If $0 \in \mathfrak{g}^*$ is a regular value of μ, and if*

$$(3.66)\quad Z(x) = 1 \,\text{a.e. on}\, X_0,$$

then

$$(3.67)\quad \lim_{\substack{p\in\mathfrak{g}_{reg} \\ p\to 0}} \frac{|V(p)|}{|\pi(p)|} = \frac{\mathrm{Vol}(G)}{\mathrm{Vol}(T)}|V(0)|.$$

PROOF. By 3.12, one gets (3.67) easily. □

3.6. The symplectic volume as a polynomial near the origin. In the sequel, we will assume that 0 is a regular value of μ, and also that

$$(3.68)\quad Z(x) = 1 \,\text{a.e. on}\, X_0.$$

Let U be a W-invariant open neighborhood of 0 in $\mathfrak{t}^*$ consisting of regular values of μ. Let $\mathfrak{g}^*_U \subset \mathfrak{g}^*$ be the corresponding union of coadjoint orbits. Then $\mathfrak{g}^*_U$ is an open neighborhood of 0 in $\mathfrak{g}$.

Classically, U is small enough, there is a G-invariant open neighborhood V of X_0 in X such that we have the identification of G-spaces

$$(3.69)\quad V \simeq X_0 \times \mathfrak{g}^*_U,$$

so that in the right hand-side of (3.69), μ is just the projection $M \times \mathfrak{g}^*_U \to \mathfrak{g}^*_U$.

By (3.69), if $p \in U$,

$$(3.70)\quad X_{\mathcal{O}_p} \simeq X_0 \times \mathcal{O}_p,$$

so that

$$X_{\mathcal{O}_p}/G \simeq X_0 \times_G \mathcal{O}_p \simeq X_0/Z(p) \tag{3.71}$$

is a $\mathcal{O}_p$-fibre bundle over the orbifold X_0/G.

Now we construct an orientation over X_0/T. In fact if $t \in T_{\text{reg}}$, then $\mathcal{O}_t \simeq G/T$, and so

$$X_0/T = X_0 \times_G G/T. \tag{3.72}$$

Let $K \subset \mathfrak{t}$ be a Weyl chamber, and let $t \in K$. Then the symplectic form $\sigma_{\mathcal{O}_t}$ orients $G/T \simeq \mathcal{O}_t$, and the corresponding orientation on G/T does not depend on $t \in K$. Also X_0/G carries the symplectic form σ_0. Therefore once K is fixed, X_0/T carries a canonical orientation.

Take $t \in U$. Let $i_t : X_t \to X$ be the obvious embedding. Then by (3.37), $i_t^*\sigma$ descends to a closed 2-form on X_t/T, which we denote $\bar{\sigma}_t$. Note that if $t \in \mathfrak{t}_{\text{reg}}$, $\sigma_t = \bar{\sigma}_t$. If $t \notin \mathfrak{t}_{\text{reg}}$, $T \subset Z(t), T \neq Z(t)$, and $\bar{\sigma}_t$ is in general not symplectic.

By (3.69),

$$X_t \simeq X_0 \times \{t\}. \tag{3.73}$$

We equip $X_t/T \simeq X_0/T$ with the given fixed orientation.

DEFINITION 3.14. Put

$$P(t) = \int_{X_t/T} e^{\bar{\sigma}_t}. \tag{3.74}$$

Then $P(t)$ is a smooth function on U.

THEOREM 3.15. *One has the identity*

$$\begin{aligned} |P(t)| &= |V(t)| \text{ if } t \in \mathfrak{t}_{\text{reg}}, \\ &= 0 \text{ if } t \notin \mathfrak{t}_{\text{reg}}. \end{aligned} \tag{3.75}$$

If $w \in W, t \in U$,

$$P(wt) = \epsilon_w P(t). \tag{3.76}$$

Also, near 0, $P(t)$ *is a polynomial. More precisely, if* θ *is a connection form on* $\pi : X_0 \to X_0/G$, *and if* Θ *is its curvature,*

$$P(t) = \int_{X_0/T} \exp\left(\pi^*\sigma_0 - \langle t, \Theta\rangle + \langle t, \frac{1}{2}[\theta,\theta]\rangle\right). \tag{3.77}$$

Near 0, $P(t)$ *and* $\pi(t/i)$ *either vanish together or are nonzero, and then they have the same sign. In particular*

$$\pi(t/i)P(t) \geq 0 \text{ near } 0. \tag{3.78}$$

PROOF. By the considerations we made after (3.72), (3.75) holds. If $w \in W$, let $g \in N(T)$ represent $w \in W = N(T)/T$. Since G is connected, it acts on $X_0, \mathfrak{g}, \mathfrak{g}^*$ by orientation preserving maps. Clearly

$$TX_0 = \pi^*(TX_0/G) \oplus \mathfrak{g}^X. \tag{3.79}$$

Also

$$\mathfrak{g} \otimes_{\mathbf{R}} \mathbf{C} = \mathfrak{t} \otimes_{\mathbf{R}} \mathbf{C} \oplus \left(\bigoplus_{\alpha \in R} \mathfrak{g}_\alpha\right), \tag{3.80}$$

so that

$$(\mathfrak{g}/\mathfrak{t}) \otimes_{\mathbf{R}} \mathbf{C} = \bigoplus_{\alpha \in R} \mathfrak{g}_\alpha. \tag{3.81}$$

Then $(\mathfrak{g}/\mathfrak{t}) \otimes_{\mathbf{R}} \mathbf{C}$ descends to the complexified tangent space to the fibre G/T over X_0/G. Since $g \in N(T)$, g preserves $\mathfrak{t}$ and its orthogonal $\mathfrak{t}^\perp$ in $\mathfrak{g}$. Since g changes the orientation of $\mathfrak{t}$ by the factor ϵ_w, it changes the orientation of $\mathfrak{g}/\mathfrak{t}$ by the same factor.

So g changes the orientation of X_0/T by the factor ϵ_w. Since σ is G-invariant, and g acts on X_0/T,

$$\bar{\sigma}_{gt} = \bar{\sigma}_t. \tag{3.82}$$

From the above, we get (3.76).

By (3.22),(3.27),

$$\sigma = \pi^*\eta - \langle d\mu, \theta\rangle - \langle \mu, \Theta - \frac{1}{2}[\theta,\theta]\rangle. \tag{3.83}$$

Clearly $i_t^*\pi^*\eta$ is cohomologous to $i_0^*\pi^*\eta = \pi^*\sigma_0$. So by (3.83), we get (3.77). From (3.77), it is clear that $P(t)$ is a polynomial near 0.

If $t \in U \backslash \mathfrak{t}_{\text{reg}}$, by the considerations after (3.72), $\bar{\sigma}_t$ is an everywhere degenerate 2-form on X_t/T. Therefore

$$P(t) = 0 \,\text{on}\, U \backslash \mathfrak{t}_{\text{reg}}. \tag{3.84}$$

For $t \in U \cap \mathfrak{t}_{\text{reg}}$, $\bar{\sigma}_t$ is a symplectic 2-form on X_t/T, so that

$$P(t) \neq 0 \,\text{on}\, U \cap \mathfrak{t}_{\text{reg}}. \tag{3.85}$$

By (1.190), (3.85), $P(t)$ and $\pi(t/i)$ have the same zeroes on U.

By (3.38), on X_0/T, the form $\langle t, \frac{1}{2}[\theta,\theta]\rangle$ is just the symplectic 2-form along the fibre $G/T \simeq \mathcal{O}_t$. If $t \in K \cap U$ is close to 0, by (3.72), (3.77), $P(t) > 0$. Therefore, on $K \cap U$, $P(t)$ and $\pi(t/i)$ have the same sign. Using (1.186) and (3.76), we get the end of our Theorem, the proof of which is now completed. □

REMARK 3.16. Recall that we have a fibration $X_0/T \overset{G/T}{\to} X_0/G$. Then we can rewrite (3.77) in the form

$$P(t) = \int_{X_0/G} e^{\sigma_0} \int_{G/T} e^{-\langle t,\Theta\rangle + \langle t, \frac{1}{2}[\theta,\theta]\rangle}. \tag{3.86}$$

In (3.86), $\int_{G/T} \cdots$ is an integral along the fibre. In Remark 3.26, we will reinterpret identity (3.86).

3.7. A formula of Witten-Jeffrey-Kirwan. We make the same assumptions as in Section 3.6. Also we suppose the set U to be bounded. Let $\langle , \rangle$ be a G-invariant scalar product on $\mathfrak{g}$. If $f : \mathfrak{g}^* \to \mathbf{R}$ is a bounded measurable function, let $Mf : \mathfrak{g}^* \to \mathbf{R}$ be given by

$$Mf(p) = \frac{1}{\text{Vol}(G)} \int_G f(p.g)dg. \tag{3.87}$$

Recall that $\mathfrak{g}$ and $\mathfrak{g}^*$ have been identified by $\langle , \rangle$, so that (3.87) also makes sense when $f : \mathfrak{g} \to \mathbf{R}$ is bounded and measurable.

DEFINITION 3.17. If $f : \mathfrak{g}^* \to \mathbf{R}$ is a bounded measurable function with support in $\mathfrak{g}^*_U$, for $u \in \mathfrak{g}$, put

$$\widehat{f}_{\mathfrak{g}}(u) = \int_{\mathfrak{g}^*} f(p) e^{\langle u,p \rangle} dp. \tag{3.88}$$

Clearly

$$M\widehat{f}_{\mathfrak{g}} = \widehat{Mf}_{\mathfrak{g}}. \tag{3.89}$$

Also $\widehat{f}_{\mathfrak{g}}(u)$ is an analytic function of u.

If $f : \mathfrak{t}^* \to \mathbf{R}$ is a bounded measurable function with support in U, for $t \in \mathfrak{t}$, set

$$\widehat{f}_{\mathfrak{t}}(t) = \int_{\mathfrak{t}^*} f(p) e^{\langle t,p \rangle} dp. \tag{3.90}$$

If f is W-invariant, then $\widehat{f}_{\mathfrak{t}}$ is also W-invariant.

The maps $f \mapsto f_{\mathfrak{g}}$ and $f \mapsto f_{\mathfrak{t}}$ are injective.

Also by Rossmann's formula [**50**], [**60**], if $f : \mathfrak{g}^* \to \mathbf{R}$ is taken as before and is G-invariant, then,

$$\widehat{\pi f}_{\mathfrak{t}} = \pi\left(\frac{\cdot}{2\pi}\right) \widehat{f}_{\mathfrak{g}}. \tag{3.91}$$

In the sequel, we orient X by the symplectic form σ. Now we prove a formula related to a formula of Jeffrey-Kirwan [**28**] in a work where they prove a formula by Witten [**64**]. Our approach is closely related to Vergne [**61**] and especially to Liu [**39, 40**] who worked out similar formulas in the context of moduli spaces.

THEOREM 3.18. *Let $f : \mathfrak{g}^* \to \mathbf{R}$ be a bounded measurable function with support in $\mathfrak{g}^*_U$. Let θ be a connection form on $X_0 \to X_0/G$, and let Θ be its curvature. Then*

$$\int_X f(\mu) e^{\sigma} = \mathrm{Vol}(G) \int_{X_0/G} \widehat{Mf}_{\mathfrak{g}}(-\Theta) e^{\sigma_0}. \tag{3.92}$$

PROOF. By Theorem 3.12,

$$\int_X f(\mu) e^{\sigma} = \mathrm{Vol}(G) \int_{\mathfrak{t}/W} Mf(t) |\pi(t)| |V(t)| dt. \tag{3.93}$$

By Theorem 3.15,

$$|\pi(t)||V(t)| = \pi(t/i) P(t). \tag{3.94}$$

So (3.93) can be rewritten in the form

$$\int_X f(\mu) e^{\sigma} = \mathrm{Vol}(G) \int_{\mathfrak{t}/W} (Mf)(t) \pi(t/i) P(t) dt. \tag{3.95}$$

By (3.77),

$$P(t) = \int_{X_0/T} e^{\pi^* \sigma_0 - \langle t, -\frac{1}{2}[\theta,\theta] + \Theta \rangle}. \tag{3.96}$$

It follows from the above that we may as well assume that over $X_0 \times \mathfrak{g}^*$, σ is given by

$$\sigma = \pi^* \sigma_0 - d\langle p, \theta \rangle. \tag{3.97}$$

Equivalently,

$$\sigma = \pi^*\sigma_0 - \langle dp, \theta\rangle - \langle p, -\frac{1}{2}[\theta,\theta] + \Theta\rangle. \tag{3.98}$$

Then

$$\int_X f(\mu)e^{\sigma} = \int_{X_0\times\mathfrak{g}^*} f(p)e^{\pi^*\sigma_0 - \langle p,\Theta\rangle - \langle dp,\theta\rangle + \frac{1}{2}\langle p,[\theta,\theta]\rangle}. \tag{3.99}$$

Since only the term of top degree in $\Lambda(\mathfrak{g}^*)$ contributes to the integral in the right hand-side of (3.99), we can rewrite (3.99) in the form

$$\int_X f(\mu)e^{\sigma} = \int_{X_0\times\mathfrak{g}^*} f(p)e^{\pi^*\sigma_0 - \langle p,\Theta\rangle - \langle dp,\theta\rangle}. \tag{3.100}$$

Now recall that X is oriented by σ, and X_0/G by σ_0. In particular, near $X_0 \subset X_0 \times \mathfrak{g}^*$, $X_0 \times \mathfrak{g}^*$ is oriented by σ. Also Θ is G-equivariant. From (3.100), we get

$$\int_X f(\mu)e^{\sigma} = \int_{X_0/G} e^{\sigma_0} \int_{\mathfrak{g}^*} e^{-\langle p,\Theta\rangle} \left[\int_G f(p.g)dg\right] dp, \tag{3.101}$$

which coincides with (3.92). The proof of our Theorem is completed. □

REMARK 3.19. In [**28**], Jeffrey-Kirwan use instead the coisotropic embedding theorem [Theorem 39.2] [**24**], which asserts there is a G-equivariant identification $V \simeq X_0 \times \mathfrak{g}^*_U$, so that σ is exactly given by

$$\sigma = p_1^*\sigma_0 - d\langle p, \theta\rangle. \tag{3.102}$$

Recall that by Theorem 3.15, $P(t)$ is a polynomial near $t = 0$. Therefore $P(\frac{\partial}{\partial t})$ is a differential operator.

THEOREM 3.20. *If $f : \mathfrak{t} \to \mathbf{R}$ is a bounded measurable W-invariant function with support in U, then*

$$\int_X f(\mu)e^{\sigma} = \frac{\mathrm{Vol}(G)}{|W|}\left[P(\frac{\partial}{\partial t})\widehat{\pi(t/i)f_{|\mathfrak{t}}}(t)\right]_{|t=0}. \tag{3.103}$$

PROOF. By (3.95),

$$\int_X f(\mu)e^{\sigma} = \frac{\mathrm{Vol}(G)}{|W|}\int_{\mathfrak{t}} f(t)\pi(t/i)P(t)dt. \tag{3.104}$$

From (3.90), (3.104), we get (3.103). The proof of our Theorem is completed. □

Let θ be a G-connection form on $X_0 \to X_0/G$, and let Θ be its curvature. Let Q be a G- invariant C^∞ function defined on $\mathfrak{g}$ with values in $\mathbf{R}$. We define $Q(-\Theta)$ by its Taylor expansion, which only contains a finite number of terms. Then $Q(-\Theta)$ is a closed form on X_0/G, and $\int_{X_0/G} Q(-\Theta)e^{\sigma_0}$ does not depend on θ.

Similarly, we define the differential operator $Q(\frac{\partial}{\partial t})$ as the formal power series associated to the Taylor expansion of Q. When applying the power series $Q(\frac{\partial}{\partial t})\pi(\frac{\partial/\partial t}{2i\pi})$ to the polynomial $P(t)$, only a finite number of terms in the Taylor expansion contribute, so that $Q(\frac{\partial}{\partial t})\pi(\frac{\partial/\partial t}{2i\pi})P(t)$ is a well-defined polynomial.

THEOREM 3.21. *The following identity holds,*

$$\int_{X_0/G} Q(-\Theta)e^{\sigma_0} = \frac{1}{|W|}\left[Q(\partial/\partial t)\pi(\frac{\partial/\partial t}{2i\pi})P(t)\right]_{|t=0}. \tag{3.105}$$

PROOF. Let $f : \mathfrak{g}^* \to \mathbf{R}$ be a G- invariant bounded measurable function with support in $\mathfrak{g}^*_U$. By Theorems 3.18 and 3.20,

$$\int_{X_0/G} \widehat{f}_{\mathfrak{g}}(-\Theta)e^{\sigma_0} = \frac{1}{|W|}\left[P(\partial/\partial t)\widehat{\pi(./i)f_{|\mathfrak{t}}}\right](0). \tag{3.106}$$

By Rossmann's formula (3.91) and by (3.106), we obtain

$$\int_{X_0/G} \widehat{f}_{\mathfrak{g}}(-\Theta)e^{\sigma_0} = \frac{1}{|W|}\left[P(\partial/\partial t)\pi(\frac{t}{2i\pi})\widehat{f}_{\mathfrak{g}}(t)\right](0). \tag{3.107}$$

Equivalently

$$\int_{X_0/G} \widehat{f}_{\mathfrak{g}}(-\Theta)e^{\sigma_0} = \frac{1}{|W|}\left[\widehat{f}_{\mathfrak{g}}(\partial/\partial t)\pi(\frac{\partial/\partial t}{2i\pi})P(t)\right](0). \tag{3.108}$$

Then (3.108) is exactly (3.105) when $Q = \widehat{f}_{\mathfrak{g}}$.

Clearly, it is enough to verify (3.105) when Q is a polynomial. Then Q is the Fourier transform of a distribution whose support is $\{0\}$. Using (3.108) for f with support in $\mathfrak{g}^*_U$, and a simple limit procedure, we get (3.105) when Q is a polynomial. The proof of our Theorem is completed. □

3.8. The volume of symplectic coadjoint orbits. Let $t \in \mathfrak{t}$, and let $\mathcal{O}_t \subset \mathfrak{g} = \mathfrak{g}^*$ be the G-orbit of t. Recall that $\sigma_{\mathcal{O}_t}$ is the canonical symplectic form on $\mathcal{O}_t$ given in 1.193.

Let $p_t : G/T \to \mathcal{O}_t$ be given by

$$p_t g = g.t. \tag{3.109}$$

Then p_t is one to one if and only if $t \in \mathfrak{t}_{\mathrm{reg}}$. If $t \notin \mathfrak{t}_{\mathrm{reg}}$,

$$\dim(G/T) > \dim \mathcal{O}_t. \tag{3.110}$$

We fix a positive Weyl chamber $K \subset \mathfrak{t}$. This defines an orientation on G/T, which is fixed once and for all.

DEFINITION 3.22. If $t \in \mathfrak{t}$, $X \in \mathfrak{g}$, put

$$H(t,X) = \int_{G/T} e^{\langle p_t x, X\rangle + p_t^* \sigma_{\mathcal{O}_t}}. \tag{3.111}$$

Then since G acts on the right on $\mathcal{O}_t$ and preserves $\sigma_{\mathcal{O}_t}$, $H(t,X)$ is a G-invariant function of $X \in \mathfrak{g}^*$.

Let $K \subset \mathfrak{t}$ be a Weyl chamber, and let $\pi(t)$ be the corresponding function on $\mathfrak{t}$ defined in (1.185).

PROPOSITION 3.23. *If* $t \notin \mathfrak{t}_{\mathrm{reg}}$,

$$H(t,X) = 0. \tag{3.112}$$

If $t \in \overline{K}, X \in \mathfrak{t}_{\mathrm{reg}}$, *then*

$$H(t,X) = \frac{1}{\pi(X/2i\pi)} \sum_{w\in W} \epsilon_w e^{\langle wt, X\rangle}. \tag{3.113}$$

PROOF. Using (3.110), we get (3.112). Recall that G acts on the right on $\mathcal{O}_t$. Also if $X \in \mathfrak{g}$,

$$d\langle p, X\rangle + i_{X^{\mathcal{O}_t}}\sigma_{\mathcal{O}_t} = 0 \tag{3.114}$$

Then when $t \in K$, we get (3.113) from the localization formula in equivariant cohomology of Duistermaat-Heckman[19], Berline-Vergne [5], [6, Theorem 7.11]. This equality obviously extends to the case where $t \in \bar{K}$. The proof of our Proposition is completed. □

REMARK 3.24. From (3.112), (3.113), we recover the well-known fact [15, Lemma VI.1.2] that if $t \notin \mathfrak{t}_{\mathrm{reg}}$,

$$\sum_{w \in W} \epsilon_w e^{\langle wt, . \rangle} = 0. \tag{3.115}$$

Recall that $H(t, X)$ is a G- invariant function of $X \in \mathfrak{g}^*$. Also the function $P(t)$ was defined in Definition 3.14.

PROPOSITION 3.25. *For $t_0 \in U \cap \bar{K}$,*

$$P(t_0) = \frac{1}{|W|} \left[H(t_0, \partial/\partial t) \pi \left(\frac{\partial/\partial t}{2i\pi} \right) P(t) \right]_{|t=0}. \tag{3.116}$$

PROOF. By (3.113), if $X \in \mathfrak{t}_{\mathrm{reg}}$,

$$H(t_0, X) \pi (\frac{X}{2i\pi}) = \sum_{w \in W} \epsilon_w e^{\langle wt_0, X \rangle}. \tag{3.117}$$

Of course the identity extends to arbitrary $X \in \mathfrak{t}$. In particular, by (3.117), we have the identity of formal power series of differential operators,

$$H(t_0, \partial/\partial t) \pi (\frac{\partial/\partial t}{2i\pi}) = \sum_{w \in W} \epsilon_w e^{\langle wt_0, \partial/\partial t \rangle}. \tag{3.118}$$

Now in view of (3.76),

$$\sum_{w \in W} \epsilon_w e^{\langle wt_0, \partial/\partial t \rangle} P(t)_{|t=0} = |W| P(t_0). \tag{3.119}$$

From (3.118), (3.119), we get (3.116). The proof of our Proposition is completed. □

REMARK 3.26. In view of (3.105), (3.116), we get

$$P(t_0) = \int_{X_0/G} H(t_0, -\Theta) e^{\sigma_0} \tag{3.120}$$

One then verifies that (3.120) is just a reformulation of (3.77).

Recall that we have identified $\mathfrak{g}$ and $\mathfrak{g}^*$, $\mathfrak{t}$ and $\mathfrak{t}^*$ by the scalar product $\langle , \rangle$. For $t \in \mathfrak{t} \simeq \mathfrak{t}^*$, the orbit $\mathcal{O}_t$ is equipped with the symplectic form $\sigma_{\mathcal{O}_t}$.

THEOREM 3.27. *The following identity holds*

$$\left| \int_{G/T} e^{p_t^* \sigma_{\mathcal{O}_t}} \right| = \frac{\mathrm{Vol}(G)}{\mathrm{Vol}(T)} |\pi(t)|, \tag{3.121}$$

$$\frac{\mathrm{Vol}(G)}{\mathrm{Vol}(T)} \frac{1}{|W|} \pi (\frac{\partial/\partial t}{2i\pi}) \pi(t/i) = 1.$$

PROOF. If $t \in \mathfrak{t}_{\rm reg}$, then $\mathcal{O}_t \simeq G/T$. So the first identity in (3.121) is trivial.

Let T^*G be the cotangent bundle of G. We identify $X = T^*G$ to $G \times \mathfrak{g}^*$ via the left action of G on T^*G. If θ is the canonical left-invariant 1-form on G with values in $\mathfrak{g}$, if $p \in \mathfrak{g}^*$, $\langle p, \theta\rangle$ is the canonical real 1-form on $T^*G \simeq G \times \mathfrak{g}^*$, and $-d\langle p,\theta\rangle$ is the canonical symplectic form σ on T^*G. Also G acts on the right on T^*G and preserves σ. Then $(g,p) \in G \times \mathfrak{g}^* \mapsto p \in \mathfrak{g}^*$ is a moment map for this action.

Now we use the notation of Section 3.7. Take $t \in \mathfrak{t}$. Then

$$X_t = G \times \{t\}, \tag{3.122}$$

and $X_t/Z(t) \simeq G/Z(t)$ can be identified with $\mathcal{O}_t$ by $g \in G/Z(t) \mapsto g.(t) \in \mathcal{O}_t$. Then one verifies easily that the symplectic form σ_t on $X_t/Z(t)$ is just $\sigma_{\mathcal{O}_t}$. By the first identity in (3.121), we get

$$\int_{X_t/T} e^{\bar{\sigma}_t} = \frac{\mathrm{Vol}(G)}{\mathrm{Vol}(T)}\pi(t/i), \tag{3.123}$$

so that

$$P(t) = \frac{\mathrm{Vol}(G)}{\mathrm{Vol}(T)}\pi(t/i). \tag{3.124}$$

Finally observe that the moment map $\mu : (g,p) \in G \times \mathfrak{g}^* \mapsto p \in \mathfrak{g}^*$ is regular, that G acts freely on $X_0 \simeq G$, and $X_0/G = 1$. We apply Theorem 3.21 with $Q = 1$, use (3.124), and we get

$$\frac{\mathrm{Vol}(G)}{\mathrm{Vol}(T)}\frac{1}{|W|}\left[\pi(\frac{\partial/\partial t}{2i\pi})\pi(t/i)\right]_{|t=0} = 1. \tag{3.125}$$

Also $\pi(\frac{\partial/\partial t}{2i\pi})\pi(t/i)$ is constant. It is now clear that the second equation in 3.121) follows from (3.125). The proof of our Theorem is completed. □

REMARK 3.28. It is of some interest to verify that as should be the case, the right-hand side in the second equation in (3.121) is positive. In fact

$$\pi(\frac{\partial/\partial t}{2i\pi})\pi(t/i) = (2\pi)^\ell \sum_{\sigma\in S_\ell} \langle\alpha_1, \alpha_{\sigma(1)}\rangle \ldots \langle\alpha_\ell, \alpha_{\sigma(\ell)}\rangle. \tag{3.126}$$

Put

$$(\alpha_1 \otimes \ldots \otimes \alpha_\ell)^\sigma = \frac{1}{l!}\sum_{\sigma\in S_\ell} \alpha_{\sigma(1)} \otimes \ldots \alpha_{\sigma(\ell)} \in S^\ell \mathfrak{t}^*. \tag{3.127}$$

Then (3.126) can be written as

$$\pi(\frac{\partial/\partial t}{2i\pi})\pi(t/i) = (2\pi)^\ell \ell! \|(\alpha_1 \otimes \ldots \otimes \alpha_\ell)^\sigma\|^2_{S^\ell\mathfrak{t}^*}, \tag{3.128}$$

which is indeed positive.

Recall that $\rho \in K$, so that $\pi(\rho/i) > 0$. We now recover a well-known formula for $\mathrm{Vol}(G/T)$ [6, Corollary 7.27].

THEOREM 3.29. *The following identity holds*

$$\frac{\mathrm{Vol}(G)}{\mathrm{Vol}(T)} = \frac{1}{\pi(\rho/i)}. \tag{3.129}$$

PROOF. We proceed as in [**6**]. Let $\lambda \in \Lambda \cap \overline{K}$. Let χ_λ be the character of highest weight λ. Then by Kirillov's formula [**34**], [**5**], [**6**, Theorem 8.4], for $t \in \mathfrak{t}$, $|t|$ small enough,

$$\chi_\lambda(e^t) = \frac{\pi(t)}{\sigma(t)} \int_{\mathcal{O}_{\rho+\lambda}} e^{2i\pi\langle\mu,t\rangle + \sigma_{\mathcal{O}_{\rho+\lambda}}}. \tag{3.130}$$

In particular

$$\chi_\lambda(1) = \int_{\mathcal{O}_{\rho+\lambda}} e^{\sigma_{\mathcal{O}_{\rho+\lambda}}}. \tag{3.131}$$

By Theorem 3.27,

$$\int_{\mathcal{O}_{\rho+\lambda}} e^{\sigma_{\mathcal{O}_{\rho+\lambda}}} = \frac{\mathrm{Vol}(G)}{\mathrm{Vol}(T)} \pi((\rho+\lambda)/i). \tag{3.132}$$

Also by Weyl's dimension formula [**15**, Theorem VI.1.7],

$$\chi_\lambda(1) = \frac{\pi(\rho+\lambda)}{\pi(\rho)}. \tag{3.133}$$

From (3.131)-(3.133), we get (3.129). The proof of our Theorem is completed. □

4. The affine space of connections

In this Section, we construct a canonical line bundle L on the affine space of connections on the trivial G-bundle on a Riemann surface Σ with marked points, and an action of a central extension of the gauge group ΣG on L. We compute the action of the stabilizers of certain connections on the line bundle L. We also construct a line bundle λ_p, which, as we shall see in Section 6.3, will descend to the moduli space of flat G-bundles. The main purpose of this Section is to evaluate the angles of the action of the stabilizers on the line bundle λ_p, in order to apply the theorem of Riemann-Roch-Kawasaki [**32, 33**] to this moduli space, which we will do in Section 6.

Our Section is organized as follows. In Section 4.1, we briefly recall the construction of the central extension $\widetilde{LG}$ of the loop space LG. In Section 4.2, we consider the coadjoint orbits of $\widetilde{LG}$, their symplectic form, and the corresponding line bundles. In Section 4.3, we construct a central extension of $(LG)^s$. In Section 4.4, we give a formula for the holonomy of the canonical connection on the S^1-bundle $\widetilde{LG} \overset{S^1}{\to} LG$. In Section 4.5, we describe the symplectic affine manifold $\mathcal{A}$ of G-connections on Σ, and a symplectic action of a central extension $\widetilde{\Sigma G}$ of the gauge group ΣG. In Section 4.6, we construct the line bundle L on $\mathcal{A}$, and an action of $\widetilde{\Sigma G}$ on L. In Section 4.7, when G is not simply connected, we classify the G-bundles on Σ. In Section 4.8, we specialize the results of Section 4.7 when H is a connected subgroup of a simply connected group G. In Section 4.9, we compute the action of certain elements of $\widetilde{\Sigma G}$ on L. Finally, in Section 4.10, we define the line bundle λ_p , on which ΣG acts, and we compute the action of certain elements of ΣG on λ_p.

4.1. The central extension of the loop group LG. Let G be a compact connected and simply connected simple Lie group. We will use the notation of Section 1. In particular $\langle,\rangle$ denotes the basic scalar product on $\mathfrak{g}$.

Let T be a maximal torus in G, let $\mathfrak{t} \subset \mathfrak{g}$ be its Lie algebra. Let W be the corresponding Weyl group. Then W acts on $\overline{CR}$. Let W_{aff} be the affine Weyl group

$$W_{\text{aff}} = W \ltimes \overline{CR}. \tag{4.1}$$

Let $S^1 \simeq \mathbf{R}/\mathbf{Z}$, and let $t \in [0,1[$ be the canonical coordinate on S^1. Then $\partial/\partial t$ trivializes TS^1 and dt trivializes T^*S^1.

Let LG be the loop group of G, i.e. the group of smooth maps $s \in S^1 \mapsto g_s \in G$. Let $L\mathfrak{g}$ be the Lie algebra of LG, i.e. the set of smooth maps $s \in S^1 \mapsto f_s \in \mathfrak{g}$.

If $a, b \in L\mathfrak{g}$, put

$$\eta(\alpha,\beta) = \int_{S^1} \langle \alpha, d\beta \rangle . \tag{4.2}$$

Then by [**47**, Section 4.2], η is a cocycle on $L\mathfrak{g}$. By [**47**, Theorem 4.4.1], there is a unique central extension $\rho : \widetilde{LG} \xrightarrow{S^1} LG$ associated to η. The Lie algebra of $\widetilde{LG}$ is $\widetilde{L\mathfrak{g}} = L\mathfrak{g} \oplus \mathbf{R}$. If $(\alpha, a), (\alpha', a') \in \widetilde{L\mathfrak{g}} = L\mathfrak{g} \oplus \mathbf{R}$,

$$[(\alpha, a), (\alpha', a')] = \left([\alpha, \alpha'], \int_{S^1} \langle \alpha, d\alpha' \rangle \right) . \tag{4.3}$$

Observe that the Lie group $G \times S^1$ embeds as a Lie subgroup of $\widetilde{L}G$, and that the Lie algebra of $G \times S^1$, $\mathfrak{g} \oplus \mathbf{R}$ (equipped with its standard Lie algebra structure) embeds correspondingly as a Lie subalgebra of $L\mathfrak{g} \oplus \mathbf{R}$.

Clearly $s \in S^1$ acts on LG by $g \in LG \mapsto k_s g = g_{.+s} \in LG$. We can then form the semidirect product $\widehat{L}G = S^1 \ltimes LG$. The Lie algebra $\widehat{L}\mathfrak{g}$ of $\widehat{L}G$ is given by $\widehat{L}\mathfrak{g} = \mathbf{R} \oplus L\mathfrak{g}$, and can be identified to the Lie algebra of differential operators $a\frac{d}{dt} + \alpha$, $a \in \mathbf{R}$, $\alpha \in L\mathfrak{g}$, so that

$$\left[a\frac{d}{dt} + \alpha, a'\frac{d}{dt} + \alpha'\right] = [\alpha, \alpha'] + a\frac{d}{dt}\alpha' - a'\frac{d}{dt}\alpha . \tag{4.4}$$

By [**47**, Theorem 4.4.1], the action of S^1 on LG lifts to $\widetilde{L}G$, so that we can form the semidirect product $\widehat{\widetilde{L}}G = S^1 \ltimes \widetilde{L}G$. Its Lie algebra $\widehat{\widetilde{L}}\mathfrak{g}$ is given by

$$\begin{aligned} \widehat{\widetilde{L}}\mathfrak{g} &= \mathbf{R} \oplus L\mathfrak{g} \oplus \mathbf{R} \\ &= \mathbf{R} \oplus \widetilde{L}\mathfrak{g} \\ &= \widehat{L}\mathfrak{g} \oplus \mathbf{R} . \end{aligned} \tag{4.5}$$

The Lie bracket in $\widehat{\widetilde{L}}\mathfrak{g}$ is given by

$$[(a,\alpha,b),(a',\alpha',b')] = \left(0, [\alpha,\alpha'] + a\frac{d\alpha'}{dt} - a'\frac{d\alpha}{dt}, \int_{S^1} \langle \alpha, d\alpha' \rangle \right) . \tag{4.6}$$

Also $S^1 \times G \times S^1$ embed as a Lie subgroup of $\widehat{\widetilde{L}}\mathfrak{g}$, and $\mathbf{R} \oplus \mathfrak{g} \oplus \mathbf{R}$ (with its standard structure of Lie algebra) embeds correspondingly as a Lie subalgebra of $\mathbf{R} \oplus L\mathfrak{g} \oplus \mathbf{R}$.

Observe that we can reverse the orientation of S^1. Namely let $\psi : LG \to LG$ be the morphism $g_s \mapsto g_{-s}$. Then ψ lifts to a morphism $\widehat{\widetilde{L}}G \to \widehat{\widetilde{L}}G$, so that if $(x, y) \in S^1 \times S^1$,

$$\psi(x,y) = (x^{-1}, y^{-1}) . \tag{4.7}$$

By [**47**, Section 4.9], on $\widehat{\widetilde{L}}\mathfrak{g}$, there is a $\widehat{\widetilde{L}}G$-invariant symmetric bilinear form

$$\langle (a,\alpha,b),(a',\alpha',b') \rangle = \int_{S^1} \langle \alpha, \alpha' \rangle dt - ab' - ba' . \tag{4.8}$$

By (4.8), we have the embedding

$$\mathbf{R} \oplus L\mathfrak{g} \subset (L\mathfrak{g} \oplus \mathbf{R})^* . \tag{4.9}$$

Equivalently

$$\widehat{L}\mathfrak{g} \subset (\widetilde{L}\mathfrak{g})^* . \tag{4.10}$$

4.2. The coadjoint orbits of LG**.** The group LG has a coadjoint action on the right on $(\widetilde{L}\mathfrak{g})^*$. If we restrict this action to $\widehat{L}\mathfrak{g} \subset (\widetilde{L}\mathfrak{g})^*$, for $g \in LG$, we get

$$\left(a\frac{d}{dt} + \alpha\right) \cdot g = a\frac{d}{dt} + ag^{-1}\frac{dg}{dt} + g^{-1}\alpha g . \tag{4.11}$$

Let $P = S^1 \times G$ be the trivial G-bundle on S^1. Let $\mathcal{A}^{S^1}$ be the affine space of G-connections on S^1. A connection in $\mathcal{A}^{S^1}$ can be written as

$$d + A , \ A \in \Omega^1(S^1, \mathfrak{g}) . \tag{4.12}$$

In the sequel, we identify $d+A \in \mathcal{A}^{S^1}$ with the differential operator $\frac{d}{dt}+A(\partial/\partial t) \in \widehat{L\mathfrak{g}}$. So $\mathcal{A}^{S^1}$ is an affine subspace of $\widehat{L\mathfrak{g}}$.

Clearly LG acts on the right on $\mathcal{A}^{S^1}$, so that if $g \in LG$,

$$d+A\cdot g = g^{-1}(d+A)g \tag{4.13}$$

i.e.

$$A\cdot g = g^{-1}dg + g^{-1}Ag\,. \tag{4.14}$$

By (4.11), (4.14), $\mathcal{A}^{S^1}$ embeds as an affine suspace of $(\widetilde{L\mathfrak{g}})^* = \widehat{L\mathfrak{g}}$ and the action of LG on $(\widetilde{L\mathfrak{g}})^*$ induces the corresponding action of LG on $\mathcal{A}^{S^1}$.

Now we briefly develop the theory of coadjoint orbits for $\widetilde{LG}$.

DEFINITION 4.1. If $A \in L\mathfrak{g}$, let $w \in G$ be the holonomy of the operator $\frac{d}{dt}+A$ on S^1, i.e. if g_t is the solution of

$$\frac{dg_t}{dt}g_t^{-1} + A_t = 0\,, \tag{4.15}$$

then

$$w = g_1\,. \tag{4.16}$$

Then Floquet's theory of differential equations (Frenkel [**21**, Section 3.2], Segal [**47**, Proposition 4.3.6]) shows that the orbits of $\frac{d}{dt}+A$ in $\widehat{L\mathfrak{g}}$ can be expressed in terms of the adjoint orbits in G. Namely $\frac{d}{dt}+A$ and $\frac{d}{dt}+A'$ lie in the same LG-orbit if and only if the corresponding holonomies w and w' lie in the same G-orbit. In particular the LG orbits of the differential operator $\frac{d}{dt}+A$ always contain representatives of the form $\frac{d}{dt}+\lambda$, $\lambda \in \mathfrak{t}$, and two $\frac{d}{dt}+\lambda$ lie in the same LG-orbit if and only if λ and λ' lie in the same W_{aff}-orbit.

If $\lambda \in \mathfrak{t}$, let $Z(d/dt+\lambda) \subset LG$ be the stabilizer of $d/dt+\lambda$.

PROPOSITION 4.2. *If $\lambda \in \mathfrak{t}$,*

$$Z(d/dt+\lambda) \simeq Z(e^{-\lambda})\,. \tag{4.17}$$

Namely if $g \in Z(e^{-\lambda})$, the corresponding element of $Z(d/dt+\lambda)$ is $t \in S^1 \mapsto e^{-t\lambda}ge^{t\lambda} \in G$. If $e^{-\lambda} \in T_{\mathrm{reg}}$, then

$$Z(d/dt+\lambda) = T \subset LG\,. \tag{4.18}$$

PROOF. The proof of this simple result is left to the reader. □

Let $\mathcal{O}_{d/dt+\lambda} \subset \widehat{L\mathfrak{g}}$ be the LG orbit of $d/dt+\lambda$. Clearly the map

$$g \in LG \mapsto g(d/dt+\lambda)g^{-1} \in \widehat{L\mathfrak{g}} \tag{4.19}$$

induces the identification

$$LG/Z(d/dt+\lambda) \simeq \mathcal{O}_{d/dt+\lambda}\,. \tag{4.20}$$

In particular if $\lambda \in P$,

$$LG/T \simeq \mathcal{O}_{d/dt+\lambda}\,. \tag{4.21}$$

Also the theory of coadjoint orbits developed in Section 1.14 tells us that $\mathcal{O}_{d/dt+\lambda}$ is equipped with a symplectic form $\sigma_{\mathcal{O}_{d/dt+\lambda}}$. Namely let $d/dt+A =$

$D^A \in \mathcal{O}_{d/dt+\lambda}$. Recall that LG acts on the right on $\mathcal{O}_{d/dt+\lambda}$. If $\alpha \in L\mathfrak{g}$, let $\alpha^{\mathcal{O}_{d/dt+\lambda}}$ be the corresponding vector field on $\mathcal{O}_{d/dt+\lambda}$. Clearly

$$\alpha_A^{\mathcal{O}_{d/dt+\lambda}} = D^A \alpha . \tag{4.22}$$

Then formula (1.193) for the symplectic form $\sigma_{\mathcal{O}_{d/dt+\lambda}}$ is

$$\sigma(D^A\alpha, D^A\beta) = \langle d/dt + A, [\alpha, \beta] \rangle . \tag{4.23}$$

By (4.3), (4.8), (4.23), we get

$$\sigma(D^A\alpha, D^A\beta) = \int_{S^1} \langle D^A\alpha, \beta \rangle . \tag{4.24}$$

Finally $\widetilde{L}G$ acts symplectically on the left and on the right $\mathcal{O}_{d/dt+\lambda}$, and $d/dt + A \mapsto (d/dt + A) \in (\widetilde{L}\mathfrak{g})^*$ is a moment map for the left action of $\widetilde{L}G$ on $\mathcal{O}_{d/dt+\lambda}$.

If $p \in \mathbf{R}^*$, $\lambda \in \mathfrak{t}$, if $pd/dt + A \in \mathcal{O}_{pd/dt+A}$, then $d/dt + A/p = D^{A/p}$ is a connection, and

$$\sigma_{\mathcal{O}_{pd/dt+\lambda}}(D^A\alpha, D^A\beta) = p \int_{S^1} \langle D^{A/p}\alpha, \beta \rangle . \tag{4.25}$$

Recall that $T \times S^1 \subset \widetilde{L}G$.

DEFINITION 4.3. If $(\lambda, p) \in \overline{CR}^* \times \mathbf{Z}^*$, if $\rho_{(\lambda,-p)}$ is the one dimensional representation of $T \times S^1$ of weight $(\lambda, -p)$, let $H_{(\lambda,p)}$ be the line bundle on $\widetilde{L}G/(T \times S^1) = LG/T$,

$$H_{(\lambda,p)} = \widetilde{L}G \times_{\rho_{(\lambda,-p)}} \mathbf{C} . \tag{4.26}$$

In the same way as the Weyl group W acts on the right on G/T, the affine Weyl group $W_{\text{aff}} \simeq W \rtimes \overline{CR}$ acts on the right LG/T. In fact if $w \in W$, w acts on LG/T by $g \mapsto gw$ and $\mu \in \overline{CR}$ acts on LG/T by $g \mapsto ge^{\mu t}$. Then W_{aff} acts on the left on $\overline{CR}^* \times \mathbf{Z}^*$ by the formula

$$\begin{aligned} w(\lambda, p) &= (w\lambda, p) \\ \mu(\lambda, p) &= (\lambda + p\mu, p) . \end{aligned} \tag{4.27}$$

We can then restrict (λ, p) to vary in a fundamental domain of the action of W_{aff}. This just says that

$$p \in \mathbf{Z}^* \ , \ \lambda \in |p|\overline{P} . \tag{4.28}$$

Equivalently if $p \in \mathbf{Z}^*$,

$$\text{for } \alpha \in R_+ , \quad 0 \leq \langle \alpha, \lambda \rangle \leq \langle \alpha_0, \lambda \rangle \leq p . \tag{4.29}$$

Let now $p \in \mathbf{Z}^*$, $\lambda \in |p|P \cap \overline{CR}^*$. Then by (4.18), (4.20),

$$\mathcal{O}_{pd/dt+\lambda} \simeq LG/T . \tag{4.30}$$

Therefore, the line bundle $H_{(\lambda,p)}$ is well defined on $\mathcal{O}_{pd/dt+\lambda}$. Also by proceeding as in Section 1.14, the Hermitian line bundle $H_{(\lambda,p)}$ is equipped with a unitary connection $\nabla^{H_{(\lambda,p)}}$ and

$$c_1\left(H_{(\lambda,p)}, \nabla^{H_{(\lambda,p)}}\right) = \sigma_{\mathcal{O}_{pd/dt+\lambda}} . \tag{4.31}$$

If $p \in \mathbf{Z}^*$, $\lambda \in |p|\overline{P} \cap \overline{CR}^*$, but $\lambda \notin pP$, the theory is slightly more involved.

Now, we will just assume that $\lambda \in \overline{CR}^*$ and explain the construction of the line bundle $H_{(\lambda,p)}$ on $\mathcal{O}_{pd/dt+\lambda}$ in full generality.

By (4.17),

$$Z(pd/dt + \lambda) \simeq Z(e^{-\lambda/p}). \tag{4.32}$$

DEFINITION 4.4. Let $\widetilde{Z}(pd/dt + \lambda)$ be the stabilizer of $pd/dt + \lambda$ in $\widetilde{L}G$.

Then $\widetilde{Z}(pd/dt+\lambda)$ is a central extension of $Z(pd/dt+\lambda)$. Let $\mathfrak{z}(e^{-\lambda/p}), \widetilde{\mathfrak{z}}(pd/dt+\lambda)$ be the Lie algebras of $Z(e^{-\lambda/p})$, $\widetilde{Z}(pd/dt + \lambda)$. Then

$$\widetilde{\mathfrak{z}}(pd/dt + \lambda) \simeq \mathfrak{z}(e^{-\lambda/p}) \oplus \mathbf{R}. \tag{4.33}$$

PROPOSITION 4.5. *If* $(X,a),(Y,b) \in \widetilde{\mathfrak{z}}(pd/dt + \lambda) \simeq \mathfrak{z}(e^{-\lambda/p}) \oplus \mathbf{R}$, *then*

$$[(X,a),(Y,b)] = \left([X,Y], \frac{1}{p}\langle \lambda, [X,Y]\rangle\right). \tag{4.34}$$

PROOF. Clearly, if $X \in \mathfrak{z}(e^{-\lambda/p})$, the corresponding element in the Lie algebra $\mathfrak{z}(pd/dt + \lambda)$ of $Z(pd/dt + \lambda)$ is just $e^{-t\lambda/p}Xe^{t\lambda/p}$. In view of (4.3),

$$\begin{aligned}&\left[(e^{-t\lambda/p}Xe^{t\lambda/p},0)\ ,\ (e^{-t\lambda/p}Ye^{t\lambda/p},0)\right]\\ &\left(e^{-t\lambda/p}[X,Y]e^{t\lambda/p}\ ,\ \langle X,-[\lambda/p,Y]\rangle\right).\end{aligned} \tag{4.35}$$

Also

$$\langle X, -[\lambda, Y]\rangle = \langle \lambda, [X,Y]\rangle. \tag{4.36}$$

The proof of our Proposition is completed. □

PROPOSITION 4.6. *For any* $\lambda \in \overline{CR}^*$,

$$(X,a) \in \mathfrak{z}(e^{-\lambda/p}) \oplus \mathbf{R} \mapsto (X, \langle \frac{\lambda}{p}, X\rangle + a) \in \widetilde{\mathfrak{z}}(pd/dt + \lambda) \tag{4.37}$$

is an isomorphism of Lie algebras.

PROOF. This is clear by (4.34). □

REMARK 4.7. If $\lambda \in pP \cap \overline{CR}^*$, so that $e^{-\lambda/p} \in T_{\text{reg}}$, then $\mathfrak{z}(e^{-\lambda/p}) = \mathfrak{t}$. In this case, in (4.34), $\langle \lambda, [X,Y]\rangle = 0$. Therefore $(\lambda, a) \in \mathfrak{z}(e^{-\lambda/p}) \oplus \mathbf{R} \mapsto (X,a) \in \widetilde{\mathfrak{z}}(pd/dt + \lambda)$ is also an isomorphism of Lie algebras.

Observe that T is a maximal torus in $Z(e^{-\lambda/p})$. By [**14**, Corollaire 5.3.1], $Z(e^{-\lambda/p})/T$ is simply connected.

DEFINITION 4.8. Put

$$R_{e^{-\lambda/p}} = \{\alpha \in \mathbf{R}\ ;\ \langle \alpha, \lambda/p\rangle \in \mathbf{Z}\}. \tag{4.38}$$

We define $CR_{e^{-\lambda/p}}$ as in (1.136). Let $\overline{CR}_{e^{-\lambda/p}} \subset \mathfrak{t}$ be the lattice spanned by $CR_{e^{-\lambda/p}}$.

By [**15**, Theorem V.7.1],

$$\pi_1(Z(e^{-\lambda/p})) = \frac{\overline{CR}}{\overline{CR}_{e^{-\lambda/p}}}. \tag{4.39}$$

Put

$$\overline{R}^*_{e^{-\lambda/p}} = \{t \in \mathfrak{t}\ ;\ \text{if}\ \ \alpha \in R_{e^{-\lambda/p}}\ , \langle \alpha, t\rangle \in \mathbf{Z}\}. \tag{4.40}$$

By [15, Proposition V.7.16],

$$Z(Z(e^{-\lambda/p})) = \frac{\overline{R}^*_{e^{-\lambda/p}}}{\overline{CR}}. \tag{4.41}$$

In particular

$$-\lambda/p \in \frac{\overline{R}^*_{e^{-\lambda/p}}}{\overline{CR}}. \tag{4.42}$$

Now $\frac{\overline{R}^*_{e^{-\lambda/p}}}{\overline{CR}}$ maps into the group $\pi_1(Z(e^{-\lambda/p}))^*$. In particular the map

$$\psi(pd/dt+\lambda) : h \in \pi_1(Z(e^{-\lambda/p})) \mapsto e^{2i\pi\langle\lambda/p,h\rangle} \in \mathbf{C} \tag{4.43}$$

is well-defined. Since $\lambda \in \overline{CR}^*$, $e^{2i\pi\langle\lambda/p,h\rangle}$ is a pth root of unity.

Recall that pd/dt is identified to the 1-form $a \in \mathbf{R} \mapsto -pa \in \mathbf{R}$. Also $\lambda \in \overline{CR}^*$ is a left invariant 1-form on $\widetilde{L}G$.

By Proposition 1.56, $pd/dt+\lambda$ is a closed 1-form on $\widetilde{Z}(pd/dt+\lambda)$.

Let $\widetilde{Z}(e^{-\lambda/p})$ be the universal cover of $Z(e^{-\lambda/p})$. Then pd/dt is a closed form on $\widetilde{Z}(e^{-\lambda/p}) \times_{\psi(pd/dt+\lambda)} S^1$.

THEOREM 4.9. *We have the identity of Lie groups*

$$\widetilde{Z}(pd/dt+\lambda) \simeq \widetilde{Z}(e^{-\lambda/p}) \times_{\psi(pd/dt+\lambda)} S^1. \tag{4.44}$$

In particular $\widetilde{Z}(pd/dt+\lambda)$ contains a pth cover of $Z(e^{-\lambda/p})$. Under the identification (4.44),

$$pd/dt+\lambda \simeq pd/dt, \tag{4.45}$$

and the forms in (4.45) are closed and integral.

PROOF. Clearly by (4.37), $\widetilde{Z}(pd/dt+\lambda)$ is a locally trivial central extension of $Z(e^{-\lambda/p})$. Let $s \in S^1 \mapsto t_s \in T$ be a smooth loop. The corresponding loop with values in $Z(pd/dt+\lambda)$ is given by $s \in S^1 \mapsto (e^{-t\lambda/p}t_s e^{t\lambda/p}) \in LG$, i.e. by

$$s \in S^1 \mapsto t_s \in G \subset LG. \tag{4.46}$$

Now recall that $G \in \widetilde{L}G$, so that we have a loop

$$s \in S^1 \mapsto t_s \in G \cap \widetilde{Z}(pd/dt+\lambda). \tag{4.47}$$

By (4.37), the horizontal lift of $s \in S^1 \mapsto t_s \in G \subset LG$ in $\widetilde{Z}(pd/dt+\lambda)$ with respect to the flat connection on $\widetilde{Z}(pd/dt+\lambda) \to Z(e^{-\lambda/p})$ is given by

$$s \in S^1 \mapsto t_s \exp\left(2i\pi \int_{S^1} \langle \lambda/p, t_s^{-1}dt_s\rangle\right). \tag{4.48}$$

From (4.48), we get (4.44).

If S^1_p denotes the group of pth roots of unity in $\mathbf{C}$, $\widetilde{Z}(e^{-\lambda/p}) \times_{\psi(pd/dt+\lambda)} S^1_p$ is a subgroup of $\widetilde{Z}(pd/dt+\lambda)$, which is a p-covering of $Z(e^{-\lambda/p})$.

Observe that

$$\langle pd/dt+\lambda, (X, \langle\frac{\lambda}{p}, X\rangle + a)\rangle = -pa = \langle pd/dt, a\rangle. \tag{4.49}$$

From (4.49) we get (4.45).

By Proposition 1.56, the 1-form $pd/dt+\lambda$ is closed on $\widetilde{Z}(pd/dt+\lambda)$. It is even easier to see that pd/dt is closed on $\widetilde{Z}(e^{-\lambda/p}) \times_{\psi(pd/dt+\lambda)} S^1$. Finally it is trivial to verify that pd/dt is integral on $\widetilde{Z}(e^{-\lambda/p}) \times_{\psi(pd/dt+\lambda)} S^1$. The proof of our Theorem is completed. □

By Theorem 4.9, as in Proposition 1.58, the integral 1-form $pd/dt + \lambda$ defines a representation $\rho_1 : \widetilde{Z}(p\frac{d}{dt} + \lambda) \to S^1$. Also $\rho_2 : (g, x) \in \widetilde{Z}(e^{-\lambda/p}) \times_{\psi(pd/dt+\lambda)} S^1 \mapsto x^{-p} \in S^1$ is also a representation.

PROPOSITION 4.10. *Under the isomorphism (4.44),*

$$\rho_1 \simeq \rho_2 . \tag{4.50}$$

PROOF. By (4.45), we get (4.50). □

REMARK 4.11. Assume that $\lambda \in pP \cap \overline{CR}^*$, so that $Z(e^{-\lambda/p}) \simeq T$. Then

$$\widetilde{Z}(e^{-\lambda/p}) \simeq \mathfrak{t}. \tag{4.51}$$

Also the map

$$(f, a) \in \mathfrak{t} \times S^1 \mapsto (f, e^{2i\pi\langle\lambda/p, f\rangle} a) \in \mathfrak{t} \times S^1 \tag{4.52}$$

descends to an isomorphism

$$T \times S^1 \simeq \mathfrak{t} \times_{\psi(pd/dt+\lambda)} S^1 . \tag{4.53}$$

Then the 1-form $pd/dt + \lambda$ on $T \times S^1$ corresponds to pd/dt on $\mathfrak{t} \times_{\psi(pd/dt+\lambda)} S^1$.
In fact in this case,

$$\widetilde{Z}(pd/dt + \lambda) = T \times S^1, \tag{4.54}$$

and (4.53) is a special case of (4.44).

Clearly

$$\frac{\widetilde{LG}}{\widetilde{Z}(pd/dt + \lambda)} = \frac{LG}{Z(pd/dt + \lambda)} \simeq \mathcal{O}_{pd/dt+\lambda} . \tag{4.55}$$

DEFINITION 4.12. Let $H_{(\lambda,p)}$ be the line bundle on $\mathcal{O}_{pd/dt+\lambda}$,

$$H_{(\lambda,p)} = \widetilde{LG} \times_{\rho_1} S^1 . \tag{4.56}$$

Then $H_{(\lambda,p)}$ is a Hermitian line bundle. As in (1.198), we can equip $H_{(\lambda,p)}$ with a unitary connection $\nabla^{H_{(\lambda,p)}}$. By proceeding as in (1.199), we get

$$c_1(H_{(\lambda,p)}, \nabla^{H_{(\lambda,p)}}) = \sigma_{\mathcal{O}_{pd/dt+\lambda}} . \tag{4.57}$$

4.3. A central extension of $(LG)^s$.

DEFINITION 4.13. Put

$$P_s = \{(a_1, \dots, a_s) \in (S^1)^s \, , \, \prod_{j=1}^{s} a_j = 1\} . \tag{4.58}$$

Then P_s is a Lie subgroup of $(S^1)^s$, and its Lie algebra $\mathfrak{p}_s$ is given by

$$\mathfrak{p}_s = \{(b_1, \dots, b_s) \in \mathbf{R}^s \, , \, \sum_{j=1}^{s} b_j = 0\} . \tag{4.59}$$

Clearly $P_s \subset (S^1)^s$ is a Lie subgroup of $(\widetilde{LG})^s$.

DEFINITION 4.14. Put

$$(4.60)\qquad \begin{aligned}(\widetilde{LG})^s &= (\tilde{L}G)^s/P_s,\\ (\widetilde{L\mathfrak{g}})^s &= (\tilde{L}\mathfrak{g})^s/\mathfrak{p}_s.\end{aligned}$$

Then $(\widetilde{LG})^s$ is a central extension of $(LG)^s$ and $(\widetilde{L\mathfrak{g}})^s$ is its Lie algebra. Clearly

$$(4.61)\qquad (\widetilde{L\mathfrak{g}})^s = (L\mathfrak{g})^s \oplus \mathbf{R}.$$

Also if $(\alpha_1,\dots,\alpha_s,a),(\alpha'_1,\dots,\alpha'_s,a')\in(\widetilde{L\mathfrak{g}})^s$, then

(4.62)

$$[(\alpha_1,\dots,\alpha_s,a),(\alpha'_1,\dots,\alpha'_s,a')] = \left([\alpha_1,\alpha'_1],\dots,[\alpha_s,\alpha'_s]\ ,\ \sum_{j=1}^{s}\int_{S^1}\langle\alpha_j,d\alpha'_j\rangle\right).$$

Also $G^s\times S^1$ embeds as a subgroup of $(\widetilde{LG})^s$, and $\mathfrak{g}^s\oplus\mathbf{R}$ embeds as a Lie subalgebra of $(\widetilde{L\mathfrak{g}})^s$. Finally recall that in Section 4.1, we defined an action of S^1 on LG. Therefore S^1 acts on $(LG)^s$.

DEFINITION 4.15. Put

$$(4.63)\qquad \begin{aligned}(\widehat{LG})^s &= S^1\ltimes(LG)^s,\\ (\widehat{L\mathfrak{g}})^s &= R\oplus(L\mathfrak{g})^s.\end{aligned}$$

Then $(\widehat{L\mathfrak{g}})^s$ is the Lie algebra of $(\widehat{LG})^s$. If we embed $S^1,\mathbf{R}$ into $(S^1)^s,\mathbf{R}^s$ by the diagonal embeddings, then $(\widehat{LG})^s\subset(\hat{L}G)^s$, and $(\widehat{L\mathfrak{g}})^s$ is a Lie subalgebra of $(\hat{L}\mathfrak{g})^s$. Also by (4.10),

$$(4.64)\qquad (\hat{L}\mathfrak{g})^s\subset(\widetilde{L\mathfrak{g}})^{s*}.$$

4.4. The holonomy of $\tilde{L}G$. Let κ be the closed left invariant 3-form on G,

$$(4.65)\qquad \kappa(X,Y,Z)=\frac{1}{2}\langle X,[Y,Z]\rangle.$$

Then by [**47**, Proposition 4.4.5],

$$(4.66)\qquad \kappa\in H^3(G,\mathbf{Z}).$$

Recall that η is the left-invariant 2-form on LG,

$$(4.67)\qquad \eta=\frac{1}{2}\int_{S^1}\langle g^{-1}dg,\frac{\partial}{\partial t}g^{-1}dg\rangle dt.$$

Put

$$(4.68)\qquad \Delta=\{z\in\mathbf{C}\ ,\ |z|\le 1\}.$$

Let $g(s,t):S^1\times S^1\to G$ be a smooth map. We identify $g(s,t)$ with the smooth loop in LG $s\in S^1\mapsto g_s=g(s,\cdot)\in LG$.

Clearly

$$(4.69)\qquad \partial\Delta=S^1.$$

Since $\pi_i(G)=0$, $0\le i\le 2$, $g(s,t)$ extends to a smooth map $g(z,t):\Delta\times S^1\to G$. Therefore we get a map $\overline{g}:z\in\Delta\mapsto g(z,.)\in LG$.

THEOREM 4.16. *The following identity holds*

$$\frac{1}{2}\int_{S^1\times S^1}\langle g^{-1}\frac{\partial g}{\partial t}, g^{-1}\frac{\partial g}{\partial s}\rangle\, dtds - \int_{\Delta\times S^1} g^*\kappa + \int_\Delta \overline{g}^*\eta = 0. \tag{4.70}$$

PROOF. Let h be the left-invariant 1-form on LG,

$$h = \frac{1}{2}\int_{S^1}\langle g^{-1}\frac{\partial g}{\partial t}, g^{-1}dg\rangle dt. \tag{4.71}$$

A straightforward computation shows that

$$dh = -\eta + \frac{1}{2}\int_{S^1}\langle g^{-1}\frac{\partial g}{\partial t}, \frac{1}{2}[g^{-1}dg, g^{-1}dg]\rangle dt. \tag{4.72}$$

Using (4.72), we get (4.70). □

Clearly

$$\widetilde{L\mathfrak{g}} = L\mathfrak{g}\oplus \mathbf{R}. \tag{4.73}$$

The splitting (4.73) defines a connection on the S^1 bundle $\widetilde{LG}\xrightarrow{S^1} LG$. Let $s\in S^1\mapsto \widetilde{g}_s\in\widetilde{LG}$ be a horizontal lift of $s\in S^1\mapsto g_s\in LG$. Since $g_0=g_1$, then

$$\widetilde{g}_1\widetilde{g}_0^{-1}\in S^1. \tag{4.74}$$

THEOREM 4.17. *The following identity holds*

$$\begin{aligned} \widetilde{g}_1\widetilde{g}_0^{-1} &= \exp\left(2i\pi\int_\Delta \overline{g}^*\eta\right) \\ &= \exp\left(2i\pi\left(-\frac{1}{2}\int_{S^1}\langle g^{-1}\frac{\partial g}{\partial t}, g^{-1}\frac{\partial g}{\partial s}\rangle dsdt + \int_{\Delta\times S^1} g^*\kappa\right)\right). \end{aligned} \tag{4.75}$$

PROOF. The first identity is a straightforward consequence of (4.2), (4.67). The second identity follows from Theorem 4.16. □

4.5. The symplectic space of connections on Σ. Let X be a compact Riemann surface of genus g. Let $x_1,\dots,x_s$ be s distinct elements of X. Let $\Delta_1,\dots,\Delta_s$ be small non intersecting small open disks of center $x_1,\dots,x_s$.

Put

$$\Sigma = X\setminus\bigcup_{j=1}^{s}\Delta_j. \tag{4.76}$$

Then Σ is a compact Riemann surface with boundary $\partial\Sigma$. Also Σ being oriented, $\partial\Sigma$ is also naturally oriented.

Let G be a compact connected and simply connected compact simple Lie group. We use otherwise the same notation as in Section 1. In particular $\mathfrak{g}$ denotes the Lie algebra of G, and $\langle,\rangle$ denotes the basic scalar product on $\mathfrak{g}$ defined in Section 1.2.

Let $P=\Sigma\times G$ be the trivial G-bundle on Σ. Observe that since G is simply connected, a G-bundle on Σ is necessarily trivial.

DEFINITION 4.18. Let $\mathcal{A}$ be the affine space of G-connections on P.

Any element in $\mathcal{A}$ can be written in the form $d+A$, with $A \in \Omega^1(\Sigma, \mathfrak{g})$. Therefore $\mathcal{A}$ is an affine space with underlying vector space $\Omega^1(\Sigma, \mathfrak{g})$. Also

$$T\mathcal{A} = \mathcal{A} \times \Omega^1(\Sigma, \mathfrak{g}). \tag{4.77}$$

Let ΣG, $\Sigma \mathfrak{g}$ be the sets of smooth maps $\Sigma \to G$, $\Sigma \to \mathfrak{g}$. The $\Sigma\mathfrak{g}$ is a Lie algebra, which will be considered as the Lie algebra of ΣG.

DEFINITION 4.19. If $U, V \in \Omega^1(\Sigma, \mathfrak{g})$, put

$$\sigma(U, V) = \int_\Sigma -\langle U \wedge V\rangle. \tag{4.78}$$

Then σ is a symplectic form on $\mathcal{A}$.

Observe that ΣG acts on the right on $\mathcal{A}$, so that if $g \in \Sigma G$,

$$d + A \cdot g = g^{-1}(d + A)g, \tag{4.79}$$

i.e.

$$A \cdot g = g^{-1}Ag + g^{-1}dg. \tag{4.80}$$

Clearly ΣG preserves the symplectic form σ.

DEFINITION 4.20. If $\alpha \in \Sigma\mathfrak{g}$, let $\alpha^{\mathcal{A}}$ be the corresponding vector field on $\mathcal{A}$.

Let ∇^A be the covariant derivative associated to the connection $A \in \mathcal{A}$. Then

$$\alpha^{\mathcal{A}}_A = \nabla^A\alpha. \tag{4.81}$$

For simplicity, we now fix an oriented parametrization of the s circles in $\partial\Sigma$, so that

$$\partial\Sigma = (S^1)^s. \tag{4.82}$$

DEFINITION 4.21. Let r be the restriction map

$$\begin{aligned} g \in \Sigma G &\mapsto g_{|\partial\Sigma} \in (LG)^s, \\ \alpha \in \Sigma\mathfrak{g} &\mapsto \alpha_{|\partial\Sigma} \in (L\mathfrak{g})^s, \\ d + A \in \mathcal{A} &\mapsto (d + A)_{|\partial\Sigma} \in (\mathcal{A}^{S^1})^s. \end{aligned} \tag{4.83}$$

All these maps are equivariant. By (4.83), we get a map

$$d + A \mapsto \left(\frac{d}{dt} + A\left(\frac{d}{dt}\right)\right)_{|\partial\Sigma} \in (\widehat{L\mathfrak{g}})^s. \tag{4.84}$$

We still denote by ρ the projection $(\widetilde{LG})^s \to (LG)^s$.

DEFINITION 4.22. Put

$$\begin{aligned} \widetilde{\Sigma G} &= \{(g, g') \in \Sigma G \times (\widetilde{LG})^s\ ;\ g_{|\partial\Sigma} = \rho g' \text{ in } (LG)^s\}, \\ \widetilde{\Sigma\mathfrak{g}} &= \{(\alpha, \alpha') \in \Sigma\mathfrak{g} \times (\widetilde{L\mathfrak{g}})^s\ ;\ \alpha_{|\partial\Sigma} = \rho\alpha' \text{ in } (L\mathfrak{g})^s\}. \end{aligned} \tag{4.85}$$

Then $\widetilde{\Sigma\mathfrak{g}}$ is the Lie algebra of the Lie group $\widetilde{\Sigma G}$. More precisely

$$\widetilde{\Sigma\mathfrak{g}} = \Sigma\mathfrak{g} \oplus \mathbf{R}. \tag{4.86}$$

Let τ be the projection $\widetilde{\Sigma\mathfrak{g}} = \Sigma\mathfrak{g} \oplus \mathbf{R} \to \mathbf{R}$. If $(\alpha, a), (\alpha', a') \in \widetilde{\Sigma\mathfrak{g}}$, then

$$[(\alpha, a), (\alpha', a')] = \left([\alpha, \alpha'], \int_{\partial\Sigma} \langle\alpha, d\alpha'\rangle\right). \tag{4.87}$$

Also $G \times S^1$ embeds as a subgroup of $\widetilde{\Sigma G}$ by the map $(g,t) \mapsto (g,(g,\dots,g),t) \subset \Sigma G \times (\widetilde{LG})^s$.

Clearly the restriction map r extend to a map

$$\begin{aligned} \widetilde{\Sigma G} &\to (\widetilde{LG})^s\,, \\ \widetilde{\Sigma \mathfrak{g}} &\to (\widetilde{L\mathfrak{g}})^s\,. \end{aligned} \tag{4.88}$$

In particular, there is a dual map

$$(\widetilde{L\mathfrak{g}})^{s*} \to (\widetilde{\Sigma\mathfrak{g}})^*\,. \tag{4.89}$$

If $A \in \mathcal{A}$, then by (4.64),

$$\frac{d}{dt} + A(\partial/\partial t) \in (\widetilde{L\mathfrak{g}})^{s*}\,. \tag{4.90}$$

So by (4.89), (4.90), we may view $(d+A)_{|\partial\Sigma}$ as an element of $(\widetilde{\Sigma\mathfrak{g}})^*$.

As we just saw, ΣG acts on $\mathcal{A}$ and preserves the symplectic form σ. Therefore $\widetilde{\Sigma G}$ also acts symplectically on $\mathcal{A}$. Of course $S_1 \subset \widetilde{\Sigma G}$ acts trivially on $\mathcal{A}$. If $\tilde{\alpha} \in \widetilde{\Sigma\mathfrak{g}}$, let $\tilde{\alpha}^A$ be the associated vector field on $\mathcal{A}$. If $\alpha = \rho\tilde{\alpha}$,

$$\tilde{\alpha}^A = \alpha^A = \nabla^A_{\cdot}\alpha\,. \tag{4.91}$$

Clearly

$$\Sigma\mathfrak{g} = \Omega^0(\Sigma,\mathfrak{g})\,. \tag{4.92}$$

Also

$$\Omega^2(\Sigma,\mathfrak{g}) \subset \Omega^0(\Sigma,\mathfrak{g})^* = (\Sigma\mathfrak{g})^*\,. \tag{4.93}$$

DEFINITION 4.23. If $A \in \mathcal{A}$, let $F^A = dA + \frac{1}{2}[A,A] \in \Omega^2(\Sigma,\mathfrak{g})$ be the curvature of A.

By (4.93), if $A \in \mathcal{A}$, $F^A \subset (\Sigma\mathfrak{g})^* \subset (\widetilde{\Sigma\mathfrak{g}})^*$. Also by (4.89), (4.90) if $A \in \mathcal{A}$, $(d+A)_{|\partial\Sigma} \in (\widetilde{\Sigma\mathfrak{g}})^*$.

Recall that in Section 3.2, moment maps were defined with respect to symplectic actions of compact Lie groups on symplectic manifolds. Here we will use the same definition with respect to an action of $\widetilde{\Sigma G}$ on $\mathcal{A}$. We now state an extension of a result of Atiyah-Bott [**2**, Section 9].

THEOREM 4.24. *The map*

$$A \in \mathcal{A} \mapsto \mu(A) = F^A - (d+A)_{|\partial\Sigma} \in (\widetilde{\Sigma\mathfrak{g}})^* \tag{4.94}$$

is a moment map for the action of $\widetilde{\Sigma G}$ *on* $\mathcal{A}$.

PROOF. First we will prove that if $\tilde{\alpha} \in \widetilde{\Sigma\mathfrak{g}}$,

$$d\langle F^A - (d+A)_{|\partial\Sigma}, \tilde{\alpha}\rangle = i_{\tilde{\alpha}^A}\sigma\,. \tag{4.95}$$

Clearly if $\tilde{\alpha} = (\alpha, a_1, \dots, a_s) \in \widetilde{\Sigma\mathfrak{g}} \oplus \mathbf{R}^s$ is identified with the corresponding element in $\widetilde{\Sigma\mathfrak{g}}$,

$$\langle F^A - (d+A), \tilde{\alpha}\rangle = \int_\Sigma \langle F^A, \alpha\rangle - \int_{\partial\Sigma}\langle A, \alpha\rangle + \sum_{j=1}^s a_j\,. \tag{4.96}$$

If $B \in T\mathcal{A} \simeq \Omega^1(\Sigma, \mathfrak{g})$, then

$$\begin{aligned}(4.97)\qquad \langle B, d\langle F^A - (d+A)_{|\partial\Sigma}, \widetilde{\alpha}\rangle\rangle &= \int_\Sigma \langle D^A B, \alpha\rangle - \int_{\partial\Sigma}\langle B, \alpha\rangle \\ &= -\int_\Sigma \langle \nabla^A \alpha, B\rangle \\ &= i_{\widetilde{\alpha}^A}\sigma(B)\,.\end{aligned}$$

By (4.97), $\langle F^A - (d+A)_{|\partial\Sigma}, \widetilde{\alpha}\rangle$ is a Hamiltonian for the vector field $\widetilde{\alpha}^A$. Also by definition, if $g \in \widetilde{\Sigma G}$,

$$(4.98)\qquad F^{A\cdot g} - (d + A\cdot g)_{|\partial\Sigma} = (F^A - (d+A)_{|\partial\Sigma})\cdot g\,.$$

We have completed the proof of our Theorem. □

4.6. The canonical line bundle on $\mathcal{A}$. Now we describe the construction of the canonical line bundle L on $\mathcal{A}$. In the case where there are no marked points, our construction follows closely earlier work by Ramadas, Singer and Weitsman [**48**].

DEFINITION 4.25. Let $(L, \| \ \|_L)$ be the trivial Hermitian line bundle on $\mathcal{A}$. Let ∇^L be the unitary connection $(L, \| \ \|_L)$:

$$(4.99)\qquad \nabla^L = d - i\pi\int_\Sigma \langle .\,, A\rangle\,.$$

PROPOSITION 4.26. *The following identity holds*

$$(4.100)\qquad c_1(L, \nabla^L) = \sigma\,.$$

PROOF. This follows from (4.99). □

DEFINITION 4.27. If $\widetilde{\alpha} \in \widetilde{\Sigma\mathfrak{g}}$, put

$$(4.101)\qquad L_{\widetilde{\alpha}} = \nabla^L_{\alpha^A} - 2i\pi\langle \mu(A), \widetilde{\alpha}\rangle\,.$$

Clearly for $\widetilde{\alpha} \in \widetilde{\Sigma\mathfrak{g}}$, $L_{\widetilde{\alpha}}$ acts on $C^\infty(\mathcal{A}, L)$.

PROPOSITION 4.28. *If $\widetilde{\alpha} \in \widetilde{\Sigma\mathfrak{g}}$, then*

$$(4.102)\qquad [L_{\widetilde{\alpha}}, \nabla^L] = 0\,.$$

Also if $\widetilde{\alpha}, \widetilde{\beta} \in \widetilde{\Sigma\mathfrak{g}}$,

$$(4.103)\qquad [L_{\widetilde{\alpha}}, L_{\widetilde{\beta}}] = L_{[\widetilde{\alpha},\widetilde{\beta}]}\,.$$

PROOF. The identities (4.102) and (4.103) are trivial consequences of (3.120) and of Theorem 4.24. □

PROPOSITION 4.29. *If $\widetilde{\alpha} = (\alpha, a) \in \widetilde{\Sigma\mathfrak{g}} = \Sigma\mathfrak{g} \oplus \mathbf{R}$, then*

$$(4.104)\qquad L_{\widetilde{\alpha}} = \nabla_{\alpha^A} - i\pi\int_\Sigma \langle A, d\alpha\rangle - 2i\pi a\,.$$

PROOF. By (4.99), (4.101), we get

$$\begin{aligned}(4.105)\qquad L_{\widetilde{\alpha}} = \nabla_{\alpha^A} \quad - \quad & i\pi\int_\Sigma \langle \nabla^A\alpha, A\rangle - 2i\pi\int_\Sigma \langle F^A, \alpha\rangle \\ & +2i\pi\int_{\partial\Sigma}\langle A, \alpha\rangle - 2i\pi a\,.\end{aligned}$$

Also

$$(4.106)\quad \begin{aligned} \int_\Sigma \langle \nabla^A \alpha, A\rangle &= -\int_\Sigma \langle A, d\alpha\rangle - \int_\Sigma \langle [A,A],\alpha\rangle\,, \\ \int_\Sigma \langle F^A,\alpha\rangle &= \int_\Sigma \langle dA + \frac{1}{2}[A,A],\alpha\rangle \\ &= \int_{\partial\Sigma} \langle A,\alpha\rangle + \int_\Sigma (\langle A, d\alpha\rangle + \frac{1}{2}\langle [A,A],\alpha\rangle)\,. \end{aligned}$$

From (4.100), (4.106), we get (4.104). The proof of our Theorem is completed. □

REMARK 4.30. By (4.104), if $\alpha, \beta \in \Sigma\mathfrak{g}$,

$$(4.107)\quad [L_\alpha, L_\beta] = L_{[\alpha,\beta]} - 2i\pi \int_{\partial\Sigma} \langle \alpha, d\beta\rangle\,.$$

In view of (4.87), (4.104), (4.107) fits with (4.103).

Now we will show that the action of $\widetilde{\Sigma\mathfrak{g}}$ on L lifts to an action of $\widetilde{\Sigma G}$. This result has been proved by Ramadas, Singer and Weitsman [48] in the case where there are no marked points, so that $\widetilde{\Sigma G}$ is just ΣG.

THEOREM 4.31. *The action of $\widetilde{\Sigma\mathfrak{g}}$ on L defined in (4.101) lifts uniquely to a unitary action of $\widetilde{\Sigma G}$ on L which preserves ∇^L. In particular if $(g,t) \in G \times S^1 \subset \widetilde{\Sigma G}$, (g,t) acts on L by*

$$(4.108)\quad f \in L_A \mapsto f \cdot (g,t) = e^{2i\pi t} f \in L_{A\cdot g}\,.$$

PROOF. Let $g \in \Sigma G$. Since $\pi_i(G) = 0$, $0 \le i \le 2$, there is a smooth path $s \in [0,1] \mapsto g_s \in \Sigma G$ such that

$$(4.109)\quad \begin{aligned} g_0 &= 1\,, \\ g_1 &= g\,. \end{aligned}$$

Recall that

$$(4.110)\quad \widetilde{\Sigma\mathfrak{g}} = \Sigma\mathfrak{g} \oplus \mathbf{R}\,.$$

The splitting (4.110) defines a left-invariant connection on the S^1-bundle $\widetilde{\Sigma G} \xrightarrow{S^1} \Sigma G$. Let $s \in [0,1] \mapsto \widetilde{g}_s$ be the horizontal lift of $s \in [0,1] \mapsto g_s \in \Sigma G$ with respect to this connection, with $\widetilde{g}_0 = 1$. Put

$$(4.111)\quad \begin{aligned} \alpha_s &= g_s^{-1} \frac{dg_s}{ds} \in \Sigma\mathfrak{g}\,, \\ \widetilde{g} &= \widetilde{g}_1\,. \end{aligned}$$

Then

$$(4.112)\quad \widetilde{g}_s^{-1} \frac{d\widetilde{g}}{ds} = \alpha_s \in \widetilde{\Sigma\mathfrak{g}}\,.$$

Now we will show that if $A \in \mathcal{A}$, $f \in L_A$,

$$(4.113)\quad f \cdot \widetilde{g} = \exp\left(i\pi \int_0^1 ds \left[\int_\Sigma \langle A \cdot g_s, d\alpha_s\rangle\right]\right) f \in L_{A\cdot g}$$

is an unambigous formula for the action of $\widetilde{\Sigma G}$ on L. Clearly we may assume that $g_1 = g_0 = 1$, so that $s \in S_1 = \mathbf{R}/\mathbf{Z} \mapsto g_s \in \Sigma G$ is a loop in ΣG.

Recall that by(4.61),

$$(\widetilde{L\mathfrak{g}})^s = (L\mathfrak{g})^s \oplus \mathbf{R}. \tag{4.114}$$

The splitting (4.114) defines a connection on the S^1 bundle $(\widetilde{LG})^s \to (LG)^s$. Then by definition $s \in [0,1] \mapsto \widetilde{g}_{s|\partial\Sigma} \in (\widetilde{LG})^s$ is the horizontal lift of $s \in S^1 \mapsto g_s \in (LG)^s$. In particular $\widetilde{g}_{1|\partial\Sigma} \in S^1$.

By (4.104), we only need to verify that in this case,

$$\exp\left(i\pi \int_0^1 ds \left[\int_\Sigma \langle A \cdot g_s, d\alpha_s\rangle\right]\right) = \widetilde{g}_1 . \tag{4.115}$$

Observe that

$$A \cdot g_s = g_s^{-1} A g_s + g_s^{-1} dg_s . \tag{4.116}$$

Let θ^g, θ^d be the $\mathfrak{g}$-valued left and right invariant forms on G, i.e.

$$\theta^g = g^{-1}dg, \theta^d = dgg^{-1} . \tag{4.117}$$

Then

$$d\theta^g = -\frac{1}{2}[\theta^g,\theta^g] \quad d\theta^d = \frac{1}{2}[\theta^d,\theta^d] . \tag{4.118}$$

From (4.111), (4.118), we get

$$\begin{aligned} d\alpha_s &= \frac{\partial}{\partial s} g_s^*\theta^g + [\alpha_s, g_s^*\theta^g] \\ &= g_s^{-1}\frac{\partial}{\partial s}(g_s^*\theta^d) g_s . \end{aligned} \tag{4.119}$$

Therefore

$$\langle g_s^{-1} A g_s, d\alpha_s\rangle = \frac{\partial}{\partial s}\langle A, g_s^*\theta^d\rangle . \tag{4.120}$$

Since $g_0 = g_1 = 1$, by (4.120), we obtain

$$\int_0^1 ds \int_\Sigma \langle g_s^{-1} A g_s, d\alpha_s\rangle = 0 . \tag{4.121}$$

By (4.121),

$$\int_0^1 ds \int_\Sigma \langle A \cdot g_s, d\alpha_s\rangle = \int_{S^1} ds \int_\Sigma \langle g_s^{-1} dg_s, d\alpha_s\rangle \tag{4.122}$$

and so (4.122) does not depend on A. Observe that this last fact follows from general principles, and that the right-hand side of (4.122) is just the left-hand side evaluated at $A = 0$.

Also

$$\int_\Sigma \langle g_s^{-1}dg_s, d\alpha_s\rangle = -\int_{\partial\Sigma} \langle g_s^*\theta^g, \alpha_s\rangle - \frac{1}{2}\int_\Sigma \langle g_s^*[\theta^g,\theta^g], \alpha_s\rangle . \tag{4.123}$$

In view of (4.65), (4.123), we get

$$\frac{1}{2}\int_0^1 ds \int_\Sigma \langle g_s^{-1}dg_s, d\alpha_s\rangle = -\frac{1}{2}\int_{S^1} ds \int_{\partial\Sigma} \langle g_s^*\theta^g, \alpha_s\rangle - \int_{\Sigma\times S^1} g^*\kappa . \tag{4.124}$$

Recall that Σ is obtained from the Riemann surface X by deleting s disks $\Delta_1, \dots, \Delta_s$ centered at $x_1, \dots, x_s$. Since $\pi_i(G) = 0$, $0 \le i \le 2$, the smooth map $g : S^1 \times \Sigma \mapsto G$ extends to a smooth map $g : S^1 \times X \to G$ such that

$$g = 1 \text{ near } S^1 \times \{x_j\}\ ,\ 1 \le j \le s\,. \tag{4.125}$$

From (4.124), we get

(4.126)
$$\frac{1}{2}\int_0^1 ds \int_\Sigma \langle g_s^{-1} dg_s, d\alpha_s\rangle = -\frac{1}{2}\int_{S^1} ds \int_{\partial\Sigma} \langle g_s^* \theta^g, \alpha_s\rangle + \sum_{j=1}^s \int_{S^1\times\Delta_j} g^*\kappa - \int_{S^1\times\Delta_j} g^*\kappa.$$

Since $\kappa \in H^3(G, \mathbf{Z})$,

$$\int_{S^1\times X} g^*\kappa \in \mathbf{Z}. \tag{4.127}$$

For $0 \le r \le 1, t \in S^1$, then $rt \in \Delta$. If $(z,t) \in \Delta \times S_j^1$, put

$$h(z,t) = g(\frac{z}{|z|}, |z|t). \tag{4.128}$$

By (4.125), h is a smooth map from $\Delta \times S_j^1$ into G. Let $\overline{h} : \Delta \to (LG)^s$ be the smooth map defined the way we did after (4.69).

If we orient S_j^1 as a component of $\partial\Sigma$, we get

$$-\frac{1}{2}\int_{S^1} ds \int_{\partial\Sigma} \langle g_s^*\theta^g, \alpha_s\rangle = -\frac{1}{2}\sum_{j=1}^s \int_{S^1\times S_j^1} \langle h^{-1}\frac{\partial h}{\partial t}, h^{-1}\frac{\partial h}{\partial s}\rangle dtds. \tag{4.129}$$

Also for $1 \le j \le s$, by orienting S_j^1 as before,

$$\int_{S^1\times\Delta_j} g^*\kappa = \int_{\Delta\times S_j^1} h^*\kappa. \tag{4.130}$$

Using Theorem 4.16 and (4.129)-(4.130), we get

$$\begin{aligned} -\frac{1}{2}\int_{S^1} ds \int_{\partial\Sigma} \langle g_s^*\theta^g, \alpha_s\rangle + \textstyle\sum_{j=1}^s \int_{S^1\times\Delta_j} g^*\kappa \\ = \textstyle\int_\Delta \overline{h}^*\eta\,. \end{aligned} \tag{4.131}$$

Using now Theorem 4.17 and (4.122), (4.126), (4.127), (4.131), we get (4.115). So formula (4.113) for $f \cdot \widetilde{g}$ is unambiguous. Also by (4.102), $\widetilde{\Sigma G}$ preserves ∇^L.

Assume now that $g \in G$. Let $\alpha \in \mathfrak{g}$ be such that $g = \exp(\alpha)$. Put

$$g_s = \exp(s\alpha)\,. \tag{4.132}$$

Recall that $G \subset \widetilde{\Sigma G}$. Then for $s \in [0,1]$, $g_s \in \widetilde{\Sigma G}$. By (4.113), if $f \in L_A$,

$$f \cdot g = f \in L_{A\cdot g}\,. \tag{4.133}$$

Finally, by (4.104), if $t \in S^1$,

$$f \cdot t = e^{2i\pi t} f\,. \tag{4.134}$$

The proof of our Theorem is completed. □

REMARK 4.32. By (4.120), in (4.124), we get

$$
\begin{aligned}
(4.135) \qquad \frac{1}{2}\int_0^1 ds \int_\Sigma \langle A\cdot g_s, d\alpha_s\rangle = \frac{1}{2}\int_\Sigma \langle A, dg_1 g_1^{-1}\rangle \\
-\frac{1}{2}\int_0^1 ds \int_{\partial\Sigma} \langle g_s^{-1}dg, g_s^{-1}\frac{\partial g}{\partial s}\rangle \\
-\int_{\Sigma\times[0,1]} g^*\kappa .
\end{aligned}
$$

So (4.135) makes more explicit the dependence of (4.113) on A. Also if $f \in C^\infty(\mathcal{A}, L), \widetilde{g} \in \widetilde{\Sigma G}$, let $\widetilde{g}f \in C^\infty(\mathcal{A}, L)$ be given by

$$\widetilde{g}f(A) = g^{-1}f(A\cdot\widetilde{g}). \tag{4.136}$$

Then (4.136) defines a representation of $\widetilde{\Sigma G}$ on $C^\infty(\mathcal{A}, L)$. In this representation $t \in S^1$ acts like $e^{-2i\pi t}$. Also if $\widetilde{\alpha} \in \widetilde{\Sigma G}$,

$$\frac{d}{dt}e^{t\widetilde{\alpha}}\cdot f = L_{\widetilde{\alpha}^{\mathcal{A}}}f. \tag{4.137}$$

4.7. The non simply connected case. Assume now that G is a connected semisimple compact Lie group, which is not necessarily simply connected. Let $\widetilde{G}$ be the universal cover of G. Then $\pi_1(G) \subset Z(\widetilde{G})$ and $G = \widetilde{G}/\pi_1(G)$.

Let $P_{g,s}$ be a $4g+3s$ polygon covering Σ. The edges of $P_{g,s}$ are denoted $a_1, b_1, a_1^{-1}, b_1^{-1}, \ldots, a_g^{-1}, b_g^{-1}, c_1, d_1, c_1^{-1}, \ldots, c_s, d_s, d_s^{-1}$. For $1 \le j \le s$, set $w_j = c_j d_j c_j^{-1}$. Let q be an element of Σ which is a common point to the a_i, b_i, w_j. Then $\pi_1(q, \Sigma)$ is generated by the circles $a_1, b_1, \ldots, a_g, b_g, w_1, \ldots, w_s$. The case $g = 1, s = 1$ is represented in Figure 5.1.

Let $P \xrightarrow{G} \Sigma$ be a G-bundle on Σ. Take $x \in \Sigma \setminus (\partial\Sigma \bigcup \cup_{i=1}^g (a_i \cup b_i))$. Then $\Sigma\setminus\{x\}$ retracts on $(\cup_{i=1}^g a_i \cup b_i) \bigcup (\cup_{j=1}^s w_j)$. Since G is connected, the restriction of P to the 1-skeleton $(\cup_{i=1}^g a_i \cup b_i) \bigcup (\cup_{j=1}^s w_j)$ is trivial. Therefore $P_{\Sigma\setminus\{x\}}$ is trivial, i.e.

$$P_{|\Sigma\setminus\{x\}} \simeq \Sigma\setminus\{x\}\times G. \tag{4.138}$$

Let Δ be a small disk in Σ centered at x. We orient $\partial\Delta$ as part of $\partial(\Sigma\setminus\Delta)$. Then

$$P_{|\Delta} \simeq \Delta\times G. \tag{4.139}$$

Let $\sigma : \Delta\setminus\{x\} \to G$ be the transition map describing the G-bundle P on Σ, i.e.

$$(y, g) \in P_{|\Delta} \simeq (y, \sigma g) \in P_{|\Sigma\setminus\{x\}}, y \in \Delta\setminus\{x\}. \tag{4.140}$$

Then the homotopy classes of G-bundles are classified by the homotopy classes of maps : $\Delta\setminus\{x\} \to G$. Since $\Delta\setminus\{x\}$ retracts on $\partial\Delta \simeq S^1$, the homotopy classes of G bundles are specified by the element $[P]$ of $\pi_1(G)$ associated to $\sigma_{|\partial\Delta} : S^1 \to G$.

Clearly $[\sigma]$ does not depend on the trivialization of π on Δ.

The map $\pi \to [P] \in \pi_1(G)$ depends explicitly on the trivialization of P on $\Sigma\setminus\{x\}$. More precisely the homotopy classes of trivializations of P on the 1-skeleton are given by $\pi_1(G)^{2g+s}$. This just says that if a trivialization of P on $\Sigma\setminus\{x\}$ is given, all the other trivializations are given by the action of $\pi_1(G)^{2g+s}$ on this trivialization.

Let p be the map :

$$p:(a_1,b_1,\dots,a_g,b_g,c_1,\dots,c_s)\in\pi_1(G)^{2g+s}\mapsto\sum_{j=1}^{s}c_j\in\pi_1(G)\,. \tag{4.141}$$

Then if given $c\in\pi_1(G)^{2g+s}$, we replace the given trivialization of P on $\Sigma\setminus\{x\}$ by the action of c, $[P]$ is changed into $[P]-p(c)$.

Finally, with the above provisos, the map $P\to[P]\in\pi_1(G)$ does not depend on the choice of x.

Clearly, if $P\xrightarrow{G}\Sigma$ is trivial, P lifts to a $\widetilde{G}$-bundle. Conversely, if $P\xrightarrow{G}\Sigma$ lifts to a $\widetilde{G}$-bundle $Q\xrightarrow{\widetilde{G}}\Sigma$, since Q is trivial, and $P=Q/\pi_1(G)$, P is also trivial. Therefore $[P]\in\pi_1(G)$ describes the obstruction to a lifting of the G-bundle P to a $\widetilde{G}$-bundle Q.

As before we fix a trivialization of P on $\Sigma\setminus\{x\}$.

Let ∇^P be a connection on the G-bundle $P\xrightarrow{G}\Sigma$. Since

$$P_{|\Sigma\setminus\{x\}}\simeq\Sigma\setminus\{x\}\times G\,, \tag{4.142}$$

$P_{|\Sigma\setminus\{x\}}$ lifts to the trivial $\widetilde{G}$-bundle.

$$Q=\Sigma\setminus\{x\}\times\widetilde{G}\,. \tag{4.143}$$

The connection ∇^P lifts to a connection ∇^Q on Q. Clearly, this construction of Q depends on the choice of the trivialization of P on $\Sigma\setminus\{x\}$.

We will write that a family of circles Δ of center x tends to x if their radius tends to 0. Clearly as $\Delta\to x$, the holonomy of ∇^P around x tends to 1.

PROPOSITION 4.33. *As $\Delta\to x$, the holonomy of ∇^Q around $\partial\Delta\subset\partial(\Sigma\setminus\Delta)$ tends to $[P]\in\pi_1(G)\subset Z(\widetilde{G})$. In particular if ∇^P if flat, for Δ small enough, the holonomy of ∇^Q around $\partial\Sigma\setminus\Delta$ is equal to $[P]$.*

PROOF. We fix an origin in $S^1\simeq\partial\Delta$. Let $t\in[0,1]\mapsto\tau_t\in G$, $\tau_0=1$ be a horizontal lift of $S^1\subset\partial(\Sigma\setminus\Delta)$ in $P_{|\partial\Delta}$ with respect to ∇^P, in the trivialization $P_{|\Delta}\simeq\Delta\times G$. Then $t\in[0,1]\mapsto\sigma_t\tau_t\sigma_0^{-1}\in G$ denotes the corresponding horizontal lift of S^1 in the trivialization $P_{|\Sigma\setminus\{x\}}\simeq\Sigma\setminus\{x\}\times G$.

Let $t\in[0,1[\mapsto\widetilde{\sigma}_t\in\widetilde{G}$, $t\in[0,1]\mapsto\widetilde{\tau}_t\in\widetilde{G}$ be lifts of $t\in[0,1[\mapsto\sigma_t\in G$, $t\in[0,1[\mapsto\tau_t\in G$, with $\widetilde{\tau}_0=1$. Then $t\in[0,1]\mapsto\widetilde{\sigma}_t\widetilde{\tau}_t\widetilde{\sigma}_0^{-1}\in\widetilde{G}$ is a horizontal lift of $S^1\simeq\partial\Delta$ in $Q_{|\partial\Delta}$ in the trivialization $Q_{|\Sigma\setminus\{x\}}\simeq\Sigma\setminus\{x\}\times\widetilde{G}$, with respect to ∇^Q. In particular, parallel transport along $S^1\simeq\partial\Delta$ is given by $\widetilde{\sigma}_1\widetilde{\tau}_1\widetilde{\sigma}_0^{-1}$. Now recall that by definition

$$\widetilde{\sigma}_1=[P]\widetilde{\sigma}_0\;,\;[P]\in\pi_1(G)\subset Z(\widetilde{G})\,. \tag{4.144}$$

Therefore the above parallel transport is just $[P]\widetilde{\sigma}_0\widetilde{\tau}_1\widetilde{\sigma}_0^{-1}$.

Now as $\Delta\to x$, $\tau_1\to 1$. Therefore $\widetilde{\tau}_1\to 1$, and the parallel transport tends to $[P]$.

The proof of our Proposition is completed. □

4.8. The case of a connected subgroup of a simply connected group. Let now G be a connected and simply connected compact simple Lie group. Let $H \subset G$ be a connected compact semisimple Lie subgroup of G.

Let $P \xrightarrow{G} \Sigma$ be a G-bundle on Σ. Then

$$P \simeq \Sigma \times G\,. \tag{4.145}$$

Reducing the G-bundle P to a H-bundle Q is equivalent to finding a section of $P \times_G G/H \simeq \Sigma \times G/H$. Let $\dot{1} \in G/H$ be the image of $1 \in G$. Then if $\eta \in \Sigma(G/H)$, the corresponding H-bundle Q is given by

$$Q = \{(y,g) \in \Sigma \times G\ ,\ g^{-1}\eta = \dot{1}\}\,. \tag{4.146}$$

Therefore homotopy classes of H-reductions of P are just homotopy classes in $\Sigma(G/H)$.

Since $\pi_i(G) = 0$, $0 \leq i \leq 2$, we get

$$\begin{aligned} \pi_2(G/H) &\simeq \pi_1(H)\,, \\ \pi_i(G/H) &= 0\ ,\ 0 \leq i \leq 1\,. \end{aligned} \tag{4.147}$$

Take $x \in \Sigma \setminus (\partial\Sigma \bigcup \cup_{i=1}^{g}(a_i \cup b_i))$ as in Section 4.7. Recall that $\Sigma \setminus \{x\}$ is homotopy equivalent to $\cup_{i=1}^{g}(a_i \cup b_i) \bigcup \cup_{j=1}^{g} w_j$. Since $\pi_i(G/H) = 0$, $0 \leq i \leq 1$, there is only one homotopy class of sections $\Sigma \setminus \{x\} \to G/H$.

If η is a section of $\Sigma \times G/H$, we may and we will assume that $\eta = \dot{1}$ on $\Sigma \setminus \Delta$. In this case

$$Q_{|\Sigma\setminus\Delta} = \Sigma \times H, \tag{4.148}$$

and Q has a canonical section over $\Sigma \setminus \Delta$.

Therefore homotopy classes of H-reductions of P are classified by homotopy classes of maps $\eta : \Delta \to G/H$ such that $\eta_{|\partial\Delta} = \dot{1}$, i.e. by $\pi_2(G/H) \simeq \pi_1(H)$.

Take $z \in \pi_2(G/H) \simeq \pi_1(H)$. Let Q_z be the H-reduction of P associated to z.

By the above, $Q_{|\Sigma\setminus\{x\}}$ has been canonically trivialized on $\Sigma \setminus \{x\}$. Since Q is a H-bundle, the class $[Q_z] \in \pi_1(H)$ is well-defined.

PROPOSITION 4.34. *The following identity holds*

$$[Q_z] = z \text{ in } \pi_1(H)\,. \tag{4.149}$$

PROOF. Let $y \in \Delta \mapsto h(y) \in G/H$ be a smooth map such that $h_{|\partial\Delta} = \dot{1}$, representing z in $\pi_1(H) \simeq \pi_2(G/H)$. Then

$$Q_{z|\Delta} = \{(y,g) \in \Sigma \times G\ ;\ g^{-1}h(y) = \dot{1}\}\,. \tag{4.150}$$

To trivialize Q_z on Δ, we fix a connection ∇^{Q_z} on Q_z. If $(x,g) \in Q_{z,x}$, we parallel transport g along radial lines in Δ. This way, we obtain $g_1 : \partial\Delta \to H$. By definition

$$[g_1] = z \text{ in } \pi_1(H)\,. \tag{4.151}$$

Also $\sigma = g_1$ exactly defines the H-bundle Q_z on Σ in the sense of (4.140). From (4.151), we get (4.149). The proof of our Proposition is completed. □

4.9. The action of stabilizers on the line bundle L. Now we use the notation of Section 1, and more especially of Section 1.9.

Let $u \in C/\overline{CR}$. We will specialize the results of Section 4.8 to the case $H = Z(u)$. Clearly the map

$$g \in G/Z(u) \mapsto gug^{-1} \in \mathcal{O}_u \subset G \tag{4.152}$$

is one to one. It maps $\dot{1}$ into u.

If η is a section of $\Sigma \times \mathcal{O}_u$ over Σ, the corresponding $Z(u)$-bundle Q is given by

$$Q = \{(x, g) \in \Sigma \times G \;,\; g^{-1}\eta g = u\}\,. \tag{4.153}$$

Since $Q_{|\Sigma\setminus\{x\}} \simeq \Sigma\setminus\{x\} \times Z(u)$, Q has a section over $\Sigma\setminus\{x\}$. Equivalently, there is $g \in \Sigma\setminus\{x\} \times G$ such that

$$\eta = gug^{-1} \text{ on } \Sigma\setminus\{x\}\,. \tag{4.154}$$

Homotopy classes of trivializations of Q on $\Sigma\setminus\{x\}$ are just homotopy classes of $g \in \Sigma\setminus\{x\}$ such that (4.154) holds. We have already classified these homotopy classes by $\pi_1(Z(u))^{2g+s}$. By (4.147), we know that $\eta_{|\Sigma\setminus\{x\}}$ is homotopy equivalent to the constant $\eta = \dot{1} \simeq u$.

In the sequel, we will assume that $\eta = u$ on $\Sigma\setminus\Delta$ which corresponds to $\eta = \dot{1}$ on $\Sigma\setminus\Delta$.

Similarly

$$Q_{|\Delta} \simeq \Delta \times Z(u) \tag{4.155}$$

i.e. $Q_{|\Delta}$ has a section. Therefore there exists $g \in \Delta G$ such that

$$\eta = gug^{-1} \text{ on } \Delta\,. \tag{4.156}$$

In particular by (4.156), $g_{|\partial\Delta}$ takes its values in $Z(u)$, so that $g_{|\partial\Delta} \in LZ(u)$. Also there is only one homotopy class of $g \in \Delta G$ such that (4.156) holds.

Then $\sigma = g_{|\partial\Delta} \in LZ(u)$ is exactly the σ in (4.140) defining the $Z(u)$-bundle Q on Σ.

Recall that $[\sigma] \in \pi_1(Z(u)) \simeq \overline{CR}/\overline{CR}_u$. We identify $[\sigma]$ to a given element $[\sigma] \in \overline{CR}$. Let $s \in S^1 \mapsto b_s \in T$ be a loop which represents $[\sigma]$ in $\pi_1(Z(u))$. Put

$$g^0 = (bg^{-1})_{|\partial\Delta} \in LZ(u)\,. \tag{4.157}$$

Then $g^0_{|\partial\Delta}$ is homotopic to 0. Equivalently $g^0_{|\partial\Delta}$ defines an element of $L\widetilde{Z}(u)$. Since $\pi_i(\widetilde{Z}(u)) = 0$, $0 \le i \le 2$, g^0 extends to $g^0 \in \Sigma Z(u)$.

By conjugation of η by $g^0 \in \Sigma Z(u)$, we may and we will assume in the sequel that

$$\begin{aligned} &\eta = u \text{ on } \Sigma\setminus\Delta\,,\\ &\eta = gug^{-1} \text{ on } \Delta\,,\\ &g \in \Delta G,\ g_{|\partial\Delta} \in LT\,. \end{aligned} \tag{4.158}$$

Clearly $\int_{\partial\Sigma\setminus\Delta} g^{-1}dg \in \overline{CR}$ is the homotopy class of $g \in LT$.

Let now $\eta \in \Sigma G$. Put

$$\mathcal{A}^\eta = \{A \in \mathcal{A}\;;\; A\cdot\eta = A\}\,. \tag{4.159}$$

Assume that

$$\mathcal{A}^{\eta} \neq \emptyset . \tag{4.160}$$

Put

$$u = \eta(x) . \tag{4.161}$$

In the sequel we assume that

$$u \in C/\overline{CR} . \tag{4.162}$$

Then by (4.159), (4.160), it is clear that η is a section of the orbit $\mathcal{O} = \mathcal{O}_u$. In the sequel we assume that the conditions in (4.158) hold.

Let $B \in \mathfrak{t}$ be such that

$$u = \exp(B) . \tag{4.163}$$

For $s \in [0,1]$, $x \in \Sigma \setminus \Delta$, put

$$\overline{\eta}(x,s) = \exp(sB) . \tag{4.164}$$

Then since $\pi_i(G) = 0$, $0 \leq i \leq 2$, the map $\overline{\eta} : \Sigma \setminus \Delta \times [0,1] \to G$ extends to a map $\overline{\eta} : \Sigma \times [0,1] \to G$ such that

$$\begin{aligned} &\bullet\ \overline{\eta}_{|\Sigma\setminus\Delta\times[0,1]} = \exp(sB) , \\ &\bullet\ \overline{\eta}_{|\Sigma\times\{1\}} = \eta . \end{aligned} \tag{4.165}$$

For $s \in [0,1]$, put

$$\eta_s(x) = \overline{\eta}(x,s) . \tag{4.166}$$

Then $s \in [0,1] \mapsto \eta_s \in \Sigma G$ is a smooth path. Let $s \in [0,1] \in \widetilde{\eta}_s \in \widetilde{\Sigma G}$ be the corresponding horizontal lift in $\widetilde{\Sigma G}$.

THEOREM 4.35. *The following identity holds*

$$\widetilde{\eta}_1|_{L|\mathcal{A}^{\eta}} = \exp\left(-2i\pi\langle \int_{\partial\Sigma\setminus\Delta} g^{-1}dg, B\rangle\right) . \tag{4.167}$$

PROOF. We consider the splitting

$$\Sigma = \Sigma \setminus \Delta \cup \Delta . \tag{4.168}$$

Now both $\Sigma \setminus \Delta$ and Δ are objects like Σ. They carry the trivial G-bundles $P_{|\Sigma\setminus\Delta}$ and P_{Δ}. Therefore to $\Sigma \setminus \Delta$ and Δ, we associate the spaces of G-connections $\mathcal{A}^{\Sigma\setminus\Delta}$ and $\mathcal{A}^{\Delta}$, the line bundles $L^{\Sigma/\Delta}$ and L^{Δ}, equipped with the actions of $\widetilde{(\Sigma/\Delta)G}$ and $\widetilde{\Delta G}$.

Clearly $\mathcal{A}$ embeds into $\mathcal{A}^{\Sigma\setminus\Delta} \times \mathcal{A}^{\Delta}$, and ΣG embeds into $(\Sigma \setminus \Delta)G \times \Delta G$. First we claim that

$$L = (L^{\Sigma\setminus\Delta} \otimes L^{\Delta})_{|\mathcal{A}} \tag{4.169}$$

and the isomorphism (4.169) identifies the metrics and the connections. In fact this is clear by (4.99).

Put

$$\varepsilon = \{(x,y) \in S^1 \times S^1 \ , \ xy = 1\} . \tag{4.170}$$

Then $(\widetilde{(\Sigma \setminus \Delta)G} \times \widetilde{\Delta G})/\varepsilon$ acts naturally on $L^{\Sigma\setminus\Delta} \otimes L^{\Delta}$.

We claim that $\widetilde{\Sigma G}$ embeds as a subgroup of $((\widetilde{(\Sigma \setminus \Delta)G} \times \widetilde{\Delta G})/\varepsilon$. In fact recall that we orient $\partial\Sigma \setminus \Delta$ as the boundary of $\Sigma \setminus \Delta$. Also in Section 4.1, we defined the morphism $f \in \widetilde{LG} \mapsto \psi f \in \widetilde{LG}$. So if $f \in \widetilde{\Sigma G}$, let $\widetilde{g} \in \widetilde{LG}$ be such that $\rho\widetilde{g} = f_{|\partial\Sigma\setminus\Delta}$. Then to $f \in \widetilde{\Sigma G}$, we associate $((f_{|\Sigma\setminus\Delta G}, \widetilde{g}), (f_{\Delta G}, \psi\widetilde{g})) \in ((\widetilde{(\Sigma \setminus \Delta)G} \times \widetilde{\Delta G})/\varepsilon$. Since as we saw in (4.7), if $s \in S^1$, $\psi s = s^{-1}$, we find that this element of $((\widetilde{(\Sigma \setminus \Delta)G} \times \widetilde{\Delta G})/\varepsilon$ does not depend on the choice of $\widetilde{g}$.

Now $\widetilde{\Sigma G}$ acts on L. Similarly, by the above embedding, $\widetilde{\Sigma G}$ also acts on $(L^{\Sigma\setminus\Delta} \otimes L^{\Delta})_{|\mathcal{A}}$. We claim that both actions coincide. In fact this is obvious by the explicit formula in (4.113).

Let $\eta_s^{\Sigma\setminus\Delta}$ and η_s^{Δ} be the restrictions of η_s to $\Sigma \setminus \Delta$ and Δ. Let $s \in [0,1] \mapsto \widetilde{\eta}_s^{\Sigma\setminus\Delta} \in \widetilde{\Sigma \setminus \Delta G}$, $s \in [0,1] \mapsto \widetilde{\eta}_s^{\Delta} \in \widetilde{\Delta G}$ be the corresponding horizontal lifts. By the above, we get the equality in S^1,

$$\widetilde{\eta}_{1|L} = \widetilde{\eta}_{1|L^{\Sigma\setminus\Delta}}^{\Sigma\setminus\Delta} \widetilde{\eta}_{1|L^{\Delta}}^{\Delta} . \tag{4.171}$$

Now we will compute both terms in (4.171). By (4.165),

$$\eta_s^{\Sigma/\Delta} = \exp(sB) . \tag{4.172}$$

Using (4.113), we obtain

$$\widetilde{\eta}_{1|L^{\Sigma\setminus\Delta}}^{\Sigma\setminus\Delta} = 1 . \tag{4.173}$$

Consider the path $s \in [0,1] \mapsto \theta_s \in \Delta G$, with

$$\theta_s = g \exp(sB) g^{-1} . \tag{4.174}$$

Clearly by (4.158), (4.165), (4.174),

$$\eta_1^{\Delta} = \theta_1 . \tag{4.175}$$

Moreover since $g_{|\partial\Delta}$ takes its values in T,

$$\eta_{s|\partial\Delta}^{\Delta} = \theta_{s|\partial\Delta} . \tag{4.176}$$

Let $s \in [0,1] \mapsto \widetilde{\theta}_s \in \widetilde{\Delta G}$ be the horizontal lift of $s \in [0,1] \mapsto \theta_s \in \Delta G$. By (4.175), (4.176), we get

$$\widetilde{\eta}_1^{\Delta} = \widetilde{\theta}_1 . \tag{4.177}$$

Set

$$\kappa_s = \exp(sB) \in \Delta G . \tag{4.178}$$

Let $s \in [0,1] \mapsto \widetilde{\kappa}_s \in \widetilde{\Delta G}$ be the horizontal lift of $s \in [0,1] \mapsto \kappa_s \in \Delta G$.

Recall that

$$\widetilde{L}\mathfrak{g} = L\mathfrak{g} \oplus \mathbf{R} . \tag{4.179}$$

By [**47**, Proposition 4.3.2], we get

$$g(B,0)g^{-1} = \left(gBg^{-1}, -\int_{\partial\Delta} \langle g^{-1}dg, B\rangle\right) . \tag{4.180}$$

Using (4.180), we obtain

$$g\widetilde{\kappa}_1 g^{-1} = \widetilde{\theta}_1 \exp\left(-2i\pi \int_{\partial\Delta} \langle g^{-1}dg, B\rangle\right) . \tag{4.181}$$

Clearly

$$\kappa_1 = u\,, \tag{4.182}$$

and so

$$\theta_1 = g\kappa_1 g^{-1}\,. \tag{4.183}$$

If $A \in \mathcal{A}^{\Delta,\theta_1}$, then $A \cdot g \in \mathcal{A}^{\Delta,\kappa_1}$. By (4.181), we get

$$\widetilde{\kappa}_{1|L^\Delta} = \widetilde{\theta}_{1|L^\Delta} \exp\left(-2i\pi \int_{\partial\Delta} \langle g^{-1}dg, B\rangle\right)\,. \tag{4.184}$$

Finally using (4.113) again, we find

$$\widetilde{\kappa}_{1|L^\Delta} = 1\,. \tag{4.185}$$

From (4.177), (4.184), (4.185), we obtain

$$\begin{aligned} \widetilde{\eta}^{\Delta}_{1|L^\Delta} &= \exp\left(2i\pi \int_{\partial\Delta} \langle g^{-1}dg, B\rangle\right) \\ &= \exp\left(-2i\pi \int_{\partial\Sigma\setminus\Delta} \langle g^{-1}dg, B\rangle\right)\,. \end{aligned} \tag{4.186}$$

By (4.171), (4.173), (4.186), we get (4.167). The proof of our Theorem is completed. □

4.10. The action of stabilizers on the line bundle λ_p. Take now $\theta_1, \dots, \theta_s \in \overline{CR}^*$. Take $p \in \mathbf{Z}^*$. Let $L_1, \dots, L_s$ be the canonical line bundles constructed in Sections 1.14 and 4.2, which are associated to the $\widetilde{L}G$ orbits of $\left(-\frac{pd}{dt} + \theta_1, \dots, -\frac{pd}{dt} + \theta_s\right)$.

DEFINITION 4.36. Put

$$\begin{aligned} \mathcal{A}_p(\theta_1, \dots, \theta_s) &= \{A \in \mathcal{A}, -p\Big(\frac{d}{dt} + A\Big)_{|S^1_1} \in \mathcal{O}_{-p\frac{d}{dt}+\theta_1}, \dots, \\ &\qquad -p\Big(\frac{d}{dt} + A\Big)_{|S^1_s} \in \mathcal{O}_{-p\frac{d}{dt}+\theta_s}\,. \end{aligned} \tag{4.187}$$

Let $\nu_p : \mathcal{A}_p(\theta_1, \dots, \theta_s) \mapsto \prod_{j=1}^{s} \mathcal{O}_{-p\frac{d}{dt}+\theta_j}$ be given by

$$A \mapsto \nu_p(A) = \Big(-p(\frac{d}{dt} + A_{|S^1_1}), \dots, -p(\frac{d}{dt} + A_{|S^1_s})\Big)\,. \tag{4.188}$$

Equivalently, $\mathcal{A}_p(\theta_1, \dots, \theta_s)$ is the set of $A \in \mathcal{A}$ such that the holonomy of A on S^1_1 lies in the G-orbit of $e^{\theta_1/p}, \dots$ the holonomy of A on S^1_s lies in the G-orbit of $e^{\theta_s/p}$.

DEFINITION 4.37. Let λ_p be the line bundle on $\mathcal{A}_p(\theta_1, \dots, \theta_s)$,

$$\lambda_p = L^p \otimes \nu_p^*\Big(\otimes_{j=1}^{s} L_j\Big)\,. \tag{4.189}$$

Clearly ΣG acts on the right on $\mathcal{A}_p(\theta_1, \dots, \theta_s)$. By Theorem 4.31, this action lifts to an action on the right of $\widetilde{\Sigma G}$ on λ_p.

PROPOSITION 4.38. *The action of $\widetilde{\Sigma G}$ on λ_p descends to an action of ΣG on λ_p.*

PROOF. We only must show that $S^1 \subset \widetilde{\Sigma G}$ acts trivially on λ_p. However by (4.108), if $t \in S^1$, t acts on the right on L^p like $e^{2i\pi pt}$. Also t acts on the right on $\otimes_{j=1}^s L_j$ like $e^{-2i\pi pt}$. Our Proposition follows. □

Let $\eta \in \Sigma G$. Assume that

$$A^\eta \cap A_p(\theta_1, \dots, \theta_s) \neq \emptyset . \tag{4.190}$$

Put

$$u = \eta(x) . \tag{4.191}$$

By conjugating η by $g \in G$, we may and we will assume that $u \in T$. We identify u to a corresponding element in $\mathfrak{t}$, which we still denote by u. In the sequel, we assume that

$$u \in C/\overline{CR}, \tag{4.192}$$

so that $Z(u)$ is semisimple.

Now if $A \in A^\eta$, the G-bundle P on Σ reduces to a $Z(u)$-bundle Q on Σ.

Recall that by (4.138),

$$Q_{|\Sigma\setminus\{x\}} \simeq \Sigma \setminus \{x\} \times Z(u) . \tag{4.193}$$

Put

$$\widetilde{Q}_{|\Sigma\setminus\{x\}} = \Sigma \setminus \{x\} \times \widetilde{Z}(u) . \tag{4.194}$$

Then the connection A lifts to a $\widetilde{Z}(u)$ connection on $\widetilde{Q}_{|\Sigma\setminus\{x\}}$. Still two different trivializations of $Q_{|\Sigma\setminus\{x\}}$ produce two distinct $\widetilde{Z}(u)$ bundles $\widetilde{Q}_{|\Sigma\setminus\{x\}}$.

Recall that by (1.173),

$$\mathcal{O}_{\theta_j/p} \cap Z(u) = \bigcup_{w \in W_u\setminus W} \mathcal{O}_{Z(u)}(w\theta_j/p) . \tag{4.195}$$

To normalize $Q_{|\Sigma\setminus\{x\}}$, we impose that the holonomy of the connection A along S_j^1 be conjugate in $\widetilde{Z}(u)$ to $e^{w^j\theta_j/p} \in \widetilde{Z}(u)$, with $w^j \in W_u\setminus W$.

Let then $[Q] \in \pi_1(Z(u)) \simeq \frac{\overline{CR}}{\overline{CR}_u}$ be defined in Sections 4.7-4.9.

THEOREM 4.39. *The following identity holds*

$$\eta_{|\lambda_p|A^\eta\cap A_p}(\theta_1, \dots, \theta_s) = \exp\left(-2i\pi\langle\sum_{j=1}^s w^j\theta_j + p[Q], u\rangle\right) . \tag{4.196}$$

PROOF. As we saw in (4.158), by conjugating η by an element of ΣG, we may and we will assume that

$$\begin{aligned} \eta &= u \text{ on } \Sigma \setminus \Delta , \\ &= gug^{-1} \text{ on } \Delta \text{ with } g \in \Delta G \ , \ g_{|\partial\Delta} \in T . \end{aligned} \tag{4.197}$$

Then

$$Q_{|\Sigma\setminus\Delta} = \Sigma \setminus \Delta \times Z(u) . \tag{4.198}$$

Let ∇^P be a connection on P which preserves η. Then ∇^P induces a connection ∇^Q, which, in the trivialization associated to (4.198), is given by

$$\nabla^Q = d + A \ , \ A \in \Omega^1(\Sigma \setminus \Delta, \mathfrak{z}(u)) . \tag{4.199}$$

The connection ∇^Q lifts to a connection $\nabla^{\widetilde{Q}}$ given by

$$\nabla^{\widetilde{Q}} = d + A. \tag{4.200}$$

By our choice of $\widetilde{Q}$, we may and we will assume that the holonomy of $\nabla^{\widetilde{Q}}$ on S^1_j lies in the $\widetilde{Z}(u)$ conjugacy class of $e^{w^j\theta_j/p}$.

By [21, Section 3.2], [47, Proposition 4.3.6] for $1 \le j \le s$, there is $\widetilde{h}_j \in L\widetilde{Z}(u)$ such that

$$\widetilde{h}_j^{-1}\left(\frac{d}{dt} + A_{|S^1_j}\right)\widetilde{h}_j = \frac{d}{dt} - w_j\frac{\theta_j}{p}. \tag{4.201}$$

Since $\widetilde{Z}(u)$ is simply connected, there is $h \in \Sigma\widetilde{Z}(u)$ such that

$$\widetilde{h}_{|S^1_j} = \widetilde{h}_j\ ,\ 1 \le j \le s. \tag{4.202}$$

Let $h \in \Sigma Z(u)$ be the image of $\widetilde{h}$. Then $h_{|\partial\Delta} \in LZ(u)$ is homotopic to the constant loop 1. By proceeding as in (4.157), we may and we will assume that $h_{|\partial\Delta} \in LT$ is homotopic to the constant loop 1 in $Z(u)$.

Now we may as well replace η by $h^{-1}\eta h$ and $A \in \mathcal{A}^\eta$ by $A \cdot h \in \mathcal{A}^{h^{-1}\eta h}$. By (4.197),

$$\begin{aligned} h^{-1}\eta h_{|\Delta} &= h^{-1}gug^{-1}h_{|\Delta}, \\ h^{-1}g_{|\partial\Delta} &\in T. \end{aligned} \tag{4.203}$$

Finally the homotopy class of $h^{-1}g_{|S^1_j} \in LZ(u)$ is the same as the homotopy class of $g_{|S^1_j} \in LZ(u)$, $1 \le j \le s$.

So basically, we may and we will assume that η verifies the assumptions in (4.197), but we also have the extra assumptions

$$\left(\frac{d}{dt} + A\right)_{|S^1_j} = \frac{d}{dt} - w^j\frac{\theta_j}{p}\ ,\ 1 \le j \le s. \tag{4.204}$$

Now by definition, the right action of $u \in \widetilde{L}G$ on $L_{j,-pd/dt+w^j\theta_j}$ is given by $\exp(-2i\pi\langle u, w^j\theta_j\rangle) \in S^1$. Using (4.167), we see that

$$\eta_{|\lambda_p} = \exp\left(2i\pi\langle -\sum_{j=1}^s w^j\theta_j - p\int_{\partial\Sigma\setminus\Delta} g^{-1}dg, u\rangle\right). \tag{4.205}$$

By Proposition 4.34,

$$\int_{\partial\Sigma\setminus\Delta} g^{-1}dg = [Q] \in \pi_1(Z(u)). \tag{4.206}$$

From (4.205), (4.206), we get (4.196).

The proof of our Theorem is completed. □

REMARK 4.40. It should be pointed out that the expression (4.196) is natural. In fact it only depends on the $w^j \in W_u\backslash W$. Also as we saw after (4.141), we know how $[Q]$ changes if we change the trivialization of $Q_{|\Sigma\setminus\Delta}$. One verifies easily that (4.196) is compatible with this formula.

Also by Proposition 1.44, one verifies easily that (4.196) is compatible with the fact that ΣG acts as a group on λ_p.

5. The moduli space of flat bundles on a Riemann surface

The purpose of this Section is to construct the moduli space M/G of flat G-bundles over a Riemann surface Σ with marked points. We establish the Witten formula **[63, 64]** for the symplectic volume of M/G. Also we show that the formula of Witten **[64]** and Jeffrey-Kirwan **[28]** can be applied to M/G. In particular, we express the integrals of certain characteristic classes over M/G in terms of the action of certain differential operators on local polynomials over the maximal torus T. Also we give a formula for $c_1(TM/G)$. All these results will be needed in Section 6, when we apply the Theorem of Riemann-Roch-Kawasaki **[32, 33]** to M/G .

As explained in the introduction , our derivation of the Witten formula is closely related to earlier work by Liu **[39, 40]**.

This Section is organized as follows. In Section 5.1, we give the standard combinatorial description of Σ. In Section 5.2, we introduce the corresponding combinatorial complexes, which compute the absolute and relative cohomology of flat vector bundles over Σ. In Section 5.3, we introduce the G-equivariant map $\phi : X = G^{2g} \times \prod_1^s \mathcal{O}_j \to G$, such that $M = \{x \in X, \phi(x) = 1\}$. Also we relate the differential of ϕ to the combinatorial complexes on Σ which compute the absolute and relative cohomology of the flat adjoint vector bundle E. In Section 5.4, we give natural conditions on the orbits $\mathcal{O}_j$ under which 1 is a regular value of ϕ, so that M/G is an orbifold. In Section 5.5, we give conditions under which the set of elements in the fibres of ϕ or in X with non trivial stabilisers are of codimension ≥ 2. In Section 5.6, we describe TM/G and the symplectic form ω. In Section 5.7, we show that the symplectic volume form on M/G can be evaluated in terms of the corresponding combinatorial complexes. In Section 5.8, using the results of the previous subsections, we prove Witten's formula **[63, 64]** for the symplectic volume of the reductions of M/G.

In Section 5.9, we define logarithms from certain subsets of G into $\mathfrak{g}$. In Section 5.10, we show that the invariant open sets of X where G acts locally freely are naturally equipped with a symplectic form, and that the action of G on these sets has a moment map. In Section 5.11, we compute the integrals of certain characteristic classes on the moduli spaces associated to the centralizers $Z(u), u \in C/\overline{R}^*$. As we will see in Section 6.4, these moduli spaces correspond to the strata of M/G. In Section 5.12, we compute certain Euler characteristics. Finally, in Section 5.13, we give a formula for $c_1(TM/G)$.

5.1. Combinatorial description of the Riemann surface Σ. If a, b lie in a group Γ, put

$$[a, b] = aba^{-1}b^{-1} . \tag{5.1}$$

Let $g \in \mathbf{N}$, $s \in \mathbf{N}$, $g + s > 0$. Let Γ be the discrete group generated by 1, $u_1, v_1, \dots, u_g, v_g, w_1, \dots, w_s$, and the relation

$$\prod_{i=1}^{g} [u_i, v_i] \prod_{j=1}^{s} w_j = 1 . \tag{5.2}$$

Let X be an oriented connected compact surface of genus g. Let $x_1, \dots, x_s$ be s distinct elements of X. Put

$$X' = X \setminus \{x_1, \dots, x_s\} . \tag{5.3}$$

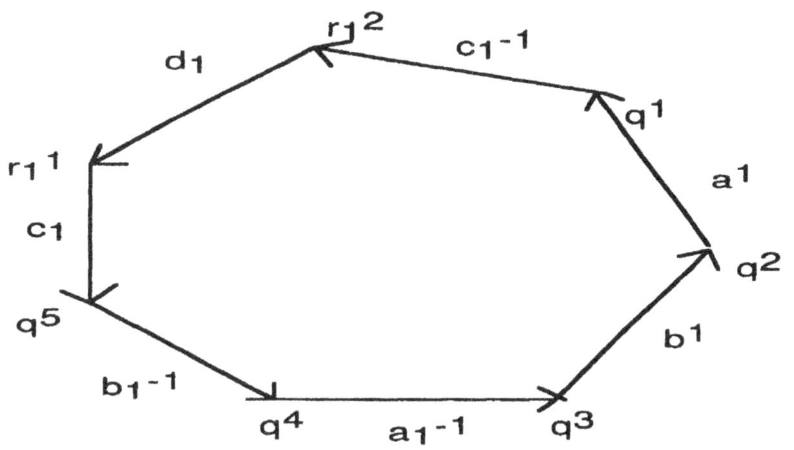

FIGURE 5.1

Let $\Delta_1, \ldots, \Delta_s$ be small non intersecting open disks in X, centered at $x_1, \ldots, x_s$. Let Σ be the surface with boundary

$$\Sigma = X \setminus \bigcup_{j=1}^{s} \Delta_j \,. \tag{5.4}$$

Let $q \in \Sigma$. Then Γ can be canonically identified with $\pi_1(q, \Sigma) = \pi_1(q, X')$. Let $P_{g,s}$ be a $4g+3s$ polygon covering Σ. The edges of $P_{g,s}$ are denoted $a_1, b_1, a_1^{-1}, b_1^{-1}, \ldots,$ $a_g^{-1}, b_g^{-1}, c_1, d_1, c_1^{-1}, \ldots, c_s, d_s, c_s^{-1}$. Then $P_{g,s}$ induces a cell decomposition of Σ with one two-cell, $2g+2s$ one-cells, $1+s$ zero-cells. The two-cell is the interior $\overset{\circ}{P}_{g,s}$ of $P_{g,s}$. The $2g+2s$ one-cells are the circles $a_1, b_1, \ldots, a_g, b_g, d_1, \ldots, d_s$, and the segments $c_1, \ldots, c_s$. The $1+s$ zero-cells consist of q which lies in the a_i, b_i, and is a boundary point for the c_i, and also $r_1, \ldots, r_s$ which lie respectively in $c_1, \ldots, c_s$ and in $d_1, \ldots, d_s$. The case $g=1, s=1$ is represented in Figure 5.1, where $q^1, \ldots, q^5$ represent q, and r_1^1, r_2^2 represent r_1. In the description given above, the group $\pi_1(q, \Sigma)$ is generated by the circles $u_1 = a_1, v_1 = b_1, \ldots, u_g = a_g, v_g = b_g$, $w_1 = c_1 d_1 c_1^{-1}, \ldots, w_s = c_s d_s c_s^{-1}$.

Also the above decomposition induces a cell decomposition of $\partial\Sigma$, with s 1-cells $d_1, \ldots, d_s$, and s 0-cells $r_1, \ldots, r_s$.

To the above cell-decompositions of Σ, we associate the corresponding complexes over $\mathbf{Z}$, $(C^{\Sigma}_{\cdot}, \partial)$, $(C^{\partial\Sigma}_{\cdot}, \partial)$ which calculate the homology groups $H_{\cdot}(\Sigma, \mathbf{Z}), H_{\cdot}(\partial\Sigma, \mathbf{Z})$. Let $(C^{\Sigma,r}_{\cdot}, \partial)$ be the quotient complex defined by the exact sequence

$$0 \to (C^{\partial\Sigma}_{\cdot}, \partial) \to (C^{\Sigma}_{\cdot}, \partial) \to (C^{\Sigma,r}_{\cdot}, \partial) \to 0 \,. \tag{5.5}$$

Then the homology of $(C^{\Sigma,r}_{\cdot}, \partial)$ is the relative homology $H((\Sigma, \partial\Sigma), \mathbf{Z})$.

Note in particular that $C_2^{\Sigma,r}$ is generated by $\overset{\circ}{P}$, $C_1^{\Sigma,r}$ by the $2g+s$ one-cells $a_1, b_1, \ldots, a_g, b_g, c_1, \ldots, c_s$, $C_0^{\Sigma,r}$ by the zero-cell q.

5.2. The combinatorial complexes on Σ. Let V be a finite dimensional complex vector space. Let $\tau : \Gamma \to \mathrm{Aut}(V)$ be a represention of Γ. Equivalently let

u_i, v_i $(1 \leq i \leq g)$, w_j $(1 \leq j \leq s)$ in $\mathrm{Aut}(V)$ such that

$$\prod_{i=1}^{i}[u_i, v_i]\prod_{j=1}^{s} w_j = 1 \tag{5.6}$$

Let $\pi : \widetilde{\Sigma} \to \Sigma$ be the universal cover of Σ. Put

$$F = \widehat{\Sigma} \times_\Gamma V\,. \tag{5.7}$$

Then F is a flat complex vector bundle on Σ. Let ∇^F be the flat connection on F.

To the above cell decompositions of $\Sigma, \partial\Sigma$, we associate the corresponding combinatorial complexes $(C^\Sigma_.(F), \partial)$, $(C^{\partial\Sigma}_.(F), \partial)$, $(C^{\Sigma,r}_.(F), \partial)$, whose homologies are respectively $H_.(\Sigma, F), H_.(\partial\Sigma, F), H_.((\Sigma, \partial\Sigma), F)$, and which fit in the exact sequence of complexes

$$0 \to (C^{\partial\Sigma}_.(F), \partial) \to (C^\Sigma_.(F), \partial) \to (C^{\Sigma,r}_.(F), \partial) \to 0\,. \tag{5.8}$$

For $s > 0$, we have the associated long exact sequence

$$\begin{aligned} 0 &\to H_2((\Sigma, \partial\Sigma), F) \to H_1(\partial\Sigma, F) \to H_1(\Sigma, F) \to H_1((\Sigma, \partial\Sigma), F) \\ &\to H_0(\partial\Sigma, F) \to H_0(\Sigma, F) \to 0 \end{aligned} \tag{5.9}$$

(in (5.9), we used the fact that if $s > 0$, $H_2(\Sigma, F) = 0$, $H_0((\Sigma, \partial\Sigma), F) = 0$).

Let p be the barycenter of the 2-cell $\overset{\circ}{P}$, let $q^1, \ldots, q^{4g+1}, r^1_1, r^2_1, \ldots, r^1_s, r^2_s$ be the vertices of P (see Figure 5.1 for the case $g = 1, s = 1$).

Then $C^\Sigma_2(F)$ is the space of flat sections of π^*F on $\overset{\circ}{P}$, and can be identified to F_p. Similarly $C^\Sigma_1(F)$ is the space of flat sections of F in the "interior" of the corresponding one-cells. We identify $(\pi^*F)_{q_1}$ to π^*F_p by parallel transport with respect to $\pi^*\nabla^F$ along a radial line connecting p to q^1. Along ∂P , we identify π^*F to $(\pi^*F)_{q_1}$ by clockwise parallel transport with respect to $\pi^*\nabla^F$.

Then we identify $C^\Sigma_1(F)$

- over a_1, with $(\pi^*F)_{q^1}$.
- over b_1, with $(\pi^*F)_{q^2}$.
- over a_2, with $(\pi^*F)_{q^5}$.
- over b_2, with $(\pi^*F)_{q^6}$.

$\vdots$

- over c_1, with $(\pi^*F)_{q^{4g+1}}$.
- over d_1, with $(\pi^*F)_{r^1_1}$

$\vdots$

Of course ultimately, $(\pi^*F)_{q^1}$, $(\pi^*F)_{q^2}$, $(\pi^*F)_{q^5}, \ldots$ are identified to π^*F_p as indicated before.

Finally $C^\Sigma_0(F)$ is identified to $(\pi^*F)_{q^1} \oplus (\bigoplus_{j=1}^s F_{r^1_j})$, which itself is identified to a sum of $1 + s$ copies of $(\pi^*F)_p$.

So $C^\Sigma_2(F)$ is just $(\pi^*F)_p$, $C^\Sigma_1(F)$ is a sum of $2g + 3s$ copies of $(\pi^*F)_p$, and $C^\Sigma_0(F)$ a sum of $1 + s$ copies of $(\pi^*F)_p$.

Then one verifies easily that if $f \in C_2^\Sigma(F) \simeq (\pi^* F)_p$,

$$\begin{aligned} \partial f &= (1 - u_1 v_1^{-1} u_1^{-1}) f_{|a_1} + (1 - v_1 u_1 v_1^{-1}) u_1^{-1} f_{|b_1} + \\ &(1 - u_2 v_2^{-1} u_2^{-1})[v_1, u_1] f_{|a_2} + (1 - v_2 u_2 v_2^{-1}) u_2^{-1} [v_1, u_1] f_{|b_2} + \ldots + \\ &(1 - w_1^{-1})[v_g, u_g] \ldots [v_1, u_1] f_{|c_1} + [v_g, u_g] \ldots [v_1, u_1] f_{|d_1} \\ &+ (1 - w_2^{-1}) w_1^{-1} [v_g, u_g] \ldots [v_1, u_1] f_{|c_2} + w_1^{-1} [v_g, u_g] \ldots [v_1, u_1] f_{|d_2} + \ldots \end{aligned} \tag{5.10}$$

Similarly $\partial : C_1^\Sigma(F) \to C_0^\Sigma(F)$ is such that if $f \in (\pi^* F)_p$,

$$\begin{aligned} \partial(f_{|a_i}) &= (1 - u_i^{-1}) f_{|q^1}, \\ \partial(f_{|b_i}) &= (1 - v_i^{-1}) f_{|q^1}, \\ \partial(f_{|c_j}) &= f_{|q^1} - f_{|r_j^1}, \\ \partial(f_{|d_j}) &= (1 - w_j^{-1}) f_{|r_j^1}. \end{aligned} \tag{5.11}$$

In view of (5.6), (5.10), (5.11) one verifies that, as should be the case, $\partial^2 = 0$.

By restriction to $\partial\Sigma$, we obtain the chain map $\partial : C_1^{\partial\Sigma}(F) \to C_0^{\partial\Sigma}(F)$ given by

$$\partial f_{|d_j} = (1 - w_j^{-1}) f_{|r_j^1}. \tag{5.12}$$

The complex $(C^{\Sigma,r}(F), \partial)$ is obtained by making formally $f_{|r_j^1} = 0$, $f_{|d_j} = 0$ in (5.10), (5.11).

Note that for $s > 0$, we recover the fact that

$$H_2(\Sigma, F) = 0\,, \quad H_0((\Sigma, \partial\Sigma), F) = 0\,. \tag{5.13}$$

PROPOSITION 5.1. *The following identity holds*

$$\begin{aligned} H_2((\Sigma, \partial\Sigma), F) \simeq \{ f \in (\pi^* F)_p, &(u_1 - 1) f = 0, (v_1 - 1) f = 0, \ldots, \\ &(u_g - 1) f = 0, (v_g - 1) f = 0, \\ &(w_1 - 1) f = 0, \ldots, (w_s - 1) f = 0 \}. \end{aligned} \tag{5.14}$$

PROOF. By (5.10), if $f \in C_2^{\Sigma,r}(F)$, the condition $\partial f = 0$, can be written as

$$\begin{aligned} &u_1 v_1^{-1} u_1^{-1} f = f\,, \\ &[v_1, u_1] f = u_1^{-1} f\,, \\ &u_2 v_2^{-1} u_2^{-1} [v_1, u_1] f = [v_1, u_1] f\,, \\ &[v_2, u_2][v_1, u_1] f = u_2^{-1} [v_1, u_1] f, \\ &\vdots \\ &(1 - w_1^{-1})[v_g, u_g] \ldots [v_1, u_1] f = 0\,, \\ &(1 - w_2^{-1}) w_1^{-1} [v_g, u_g] \ldots [v_1, u_1] f = 0\,. \\ &\vdots \end{aligned} \tag{5.15}$$

From the first two equalities in (5.15),

$$[v_1, u_1] f = v_1 f = u_1^{-1} f\,, \tag{5.16}$$

so that

$$v_1^{-1} u_1^{-1} f = f\,. \tag{5.17}$$

From (5.16), (5.17), we find that the first two equalities in (5.15) are equivalent to

$$u_1 f = v_1 f = f . \tag{5.18}$$

By proceeding as before, we get from (5.15),

$$u_i f = f \ , \ v_i f = f \ , \ 1 \leq i \leq g , \tag{5.19}$$

so that

$$w_j f = f \ , \ 1 \in j \leq s . \tag{5.20}$$

The proof of our Proposition is completed. □

Let F^* be the flat vector bundle on Σ, which is dual to F. Then if V^* is dual to V, F^* is the flat bundle on Σ associated to the dual representation $\tau^* \ : \Gamma \to \mathrm{Aut}(V^*)$.

Let $(C^{\Sigma \cdot}(F), \partial)$, $(C^{\partial\Sigma \cdot}(F), \partial)$, $(C^{\Sigma, r}(F), \partial)$ be the complexes dual to $(C^{\Sigma}_{\cdot}(F^*), \partial)$, $(C^{\partial\Sigma}_{\cdot}(F^*), \partial)$, $(C^{\Sigma, r}_{\cdot}(F^*), \partial)$. Then we have the exact sequence of complexes

$$0 \to (C^{\Sigma, r \cdot}(F), \partial) \to (C^{\Sigma \cdot}(F), \partial) \to (C^{\partial\Sigma \cdot}(F), \partial) \to 0 . \tag{5.21}$$

The associated cohomology groups are respectively $H^{\cdot}((\Sigma, \partial\Sigma), F)$, $(H^{\cdot}(\Sigma, F)$, $H(\partial\Sigma, F)$. For $s > 0$, the corresponding long exact sequence

$$\begin{aligned} 0 \to H^0(\Sigma, F) \to H^0(\partial\Sigma, F) \to H^1((\Sigma, \partial\Sigma), F) \to H^1(\Sigma, F) \to \\ H^1(\partial\Sigma, F) \to H^2((\Sigma, \partial\Sigma), F) \to 0 \end{aligned} \tag{5.22}$$

is dual to the exact sequence (5.9) for F^*.

Now we describe the chain map ∂ in the complexes (5.21). We trivialize the flat vector bundle F as indicated above. In particular the complexes in (5.21) are now direct sums of copies of $(\pi^* F)_p$. By (5.10), (5.11), we find that if $f \in \pi^* F_p$, $\partial : C^{\Sigma, 0}(F) \to C^{\Sigma, 1}(F)$ is such that

$$\begin{aligned} \partial(f | q^1) &= \sum_{i=1}^{g} ((1 - u_i) f | a_i + (1 - v_i) f | b_i) + \sum_{j=1}^{s} f | c_j , \\ \partial(f | r_j^1) &= -f | c_j + (1 - w_j) f | d_j , \end{aligned} \tag{5.23}$$

and $\partial : C^{\Sigma, 1}(F) \to C^{\Sigma, 2}(F)$ by

$$\begin{aligned} \partial(f | a_i) &= [u_1, v_1] \dots [u_{i-1}, v_{i-1}](1 - u_i v_i u_i^{-1}) f , \\ \partial(f | b_i) &= [u_1, v_1] \dots [u_{i-1}, v_{i-1}] u_i (1 - v_i u_i^{-1} v_i^{-1}) f , \\ \partial(f | c_j) &= [u_1, v_1] \dots [u_g, v_g] w_1 \dots w_{j-1} (1 - w_j) f , \\ \partial(f | d_j) &= [u_1, v_1] \dots [u_g, v_g] w_1 \dots w_{j-1} f . \end{aligned} \tag{5.24}$$

The complex $(C^{\Sigma, r, \cdot}(F), \partial)$ is obtained formally from (5.23), (5.24) by making the components indexed by d_j or r_j^1 equal to 0. Finally $\partial : C^{0, \partial\Sigma}(F) \to C^{1, \partial\Sigma}(F)$ is given by

$$\partial(f | r_j^1) = (1 - w_j) f | d_j . \tag{5.25}$$

Observe that by (5.23), (5.24), it is clear that for $s > 0$,

$$H^2(\Sigma, F) = 0 \ , \ H^0((\Sigma, \partial\Sigma), F) = 0 . \tag{5.26}$$

PROPOSITION 5.2. *The following identity holds,*

$$H^0(\Sigma, F) = \{f|q^1 + \sum_{j=1}^{s} f|r_j^1 \ , \ \textit{with } f \in \pi^* F_p \textit{ such that} \tag{5.27}$$
$$(1-u_i)f = 0, (1-v_i)f = 0 \ (1 \le i \le g) \ , \ (1-w_j)f = 0 \ (1 \le j \le s)\}.$$

PROOF. Our identity follows from (5.23). □

REMARK 5.3. By Poincaré duality

$$H_2((\Sigma, \partial\Sigma), F) \simeq H^0(\Sigma, F). \tag{5.28}$$

Using Propositions 5.1 and 5.2, the identification (5.28) has been made explicit. Another interpretation is to view the complex $(C^{\Sigma\cdot}(F), \partial)$ as the relative homology complex associated to the dual cell decomposition of Σ.

Suppose that $s \ge 1$. Let $(K^\cdot, \partial)$ be the trivial complex concentrated in degree 1,

$$K^1 = \bigoplus_{j=1}^{s} \ker(1-w_j)|c_j \simeq H^0(\partial\Sigma, F). \tag{5.29}$$

Then by (5.24), $(K^\cdot, \partial)$ is a subcomplex of $(C^{\Sigma,r\cdot}(F), \partial)$.

We have the obvious exact sequences

$$0 \longrightarrow \ker(1-w_j) \longrightarrow \pi^* F_p \xrightarrow[T_j = 1 - w_j]{} \operatorname{Im}(1-w_j) \longrightarrow 0. \tag{5.30}$$

The exact sequences (5.30) induce an exact sequence of complexes

$$0 \longrightarrow (K^\cdot, \partial) \longrightarrow (C^{\Sigma,r\cdot}(F), \partial) \xrightarrow[T]{} (\widehat{C}^{\Sigma\cdot}(F), \partial) \longrightarrow 0. \tag{5.31}$$

By (5.23), (5.24),

$$\begin{aligned} \widehat{C}^{\Sigma,k}(F) &= C^{\Sigma,r,k}(F), k = 0 \text{ or } 2, \\ &= \bigoplus_{i=1}^{g} \pi^* F_{p|a_i} \oplus \pi^* F_{p|b_i} \oplus \bigoplus_{j=1}^{s} \operatorname{Im}(1-w_j)_{|c_j}, k = 1. \end{aligned} \tag{5.32}$$

By (5.23), (5.24), if $f \in \pi^* F_p$, $\partial : \widehat{C}^{\Sigma,0}(F) \to \widehat{C}^{\Sigma,1}(F)$ is given by

$$\partial f = \sum_{1}^{g} ((1-u_i) f_{|a_i} + (1-v_i) f_{|b_i}) + \sum_{j=1}^{s} (1-w_j) f_{|c_j}, \tag{5.33}$$

and $\partial : \widehat{C}^{\Sigma,1}(F) \to \widehat{C}^{\Sigma,2}(F)$ by

$$\begin{aligned} \partial(f|a_i) &= [u_1, v_1] \dots [u_{i-1}, v_{i-1}](1 - u_i v_i u_i^{-1}) f, \\ \partial(f|b_i) &= [u_1, v_1] \dots [u_{i-1}, v_{i-1}] u_i (1 - v_i u_i^{-1} v_i^{-1}) f, \\ \partial(f|c_j) &= [u_1, v_1] \dots [u_g, v_g] w_1 \dots w_{j-1} f. \end{aligned} \tag{5.34}$$

Let $\widehat{H}^{\cdot}(\Sigma, F)$ be the cohomology of $(\widehat{C}^{\Sigma \cdot}(F), \partial)$. By (5.31), for $s > 0$, we have the long exact sequence

(5.35)

$$0 \longrightarrow \widehat{H}^0(\Sigma, F) \longrightarrow K^1 = H^0(\partial\Sigma, F) \longrightarrow H^1((\Sigma, \partial\Sigma), F)$$

$$\xrightarrow[T]{} \widehat{H}^1(\Sigma, F) \longrightarrow 0 \longrightarrow H^2((\Sigma, \partial\Sigma), F) \longrightarrow \widehat{H}^2(\Sigma, F) \longrightarrow 0.$$

THEOREM 5.4. *For $s > 0$, the following identities hold,*

$$\widehat{H}^0(\Sigma, F) = H^0(\Sigma, F), \qquad \widehat{H}^2(\Sigma, F) = H^2((\Sigma, \partial\Sigma), F). \tag{5.36}$$

Also $\widehat{H}^1(\Sigma, F)$ is the image of $H^1((\Sigma, \partial\Sigma), F)$ in $H^1(\Sigma, F)$. In (5.35), the map $\widehat{H}^0(\Sigma, F) \to K^1$ is just the map $H^0(\Sigma, F) \to H^0(\partial\Sigma, F)$, the map $K^1 \to H^1((\Sigma, \partial\Sigma), F)$ is minus the map $H^0(\partial\Sigma, F) \to H^1((\Sigma, \partial\Sigma), F)$, and $T : H^1((\Sigma, \partial\Sigma), F) \to \widehat{H}^1(\Sigma, F)$ is induced by the map $H^1((\Sigma, \partial\Sigma), F) \to H^1(\Sigma, F)$.

PROOF. By (5.33),

$$\widehat{H}^0(\Sigma, F) = \{f \in \pi^* F_p, (1 - u_i)f = 0, (1 \le i \le g) \quad (1 - v_i)f = 0, (1 - w_j)f = 0, 1 \le j \le s\}. \tag{5.37}$$

Using Proposition 5.2, and (5.37), we get the first equality in (5.36). The second equality in (5.36) follows from (5.24), (5.34).

Inspection of (5.23), (5.24), (5.37) shows that the map $\widehat{H}^0(\Sigma, F) \to K^1$ is the canonical $H^0(\Sigma, F) \to H^0(\partial\Sigma, F)$. Take $f \in \ker(1 - w_j)$. By (5.23), if $f|r_j^1$ is viewed as lying in $C^0(\Sigma, F)$, then

$$\partial f|r_j^1 = -f|c_j. \tag{5.38}$$

Therefore the map $K^1 \to H^1((\Sigma, \partial\Sigma), F)$ is minus the map $H^0(\partial\Sigma, F) \to H^1((\Sigma, \partial\Sigma), F)$.

By (5.35), we have the canonical isomorphism

$$\widehat{H}^1(\Sigma, F) \simeq \frac{H^1((\Sigma, \partial\Sigma), F)}{\operatorname{Im} H^0(\partial\Sigma, F)}. \tag{5.39}$$

By (5.22), (5.39), we find that $\widehat{H}^1(\Sigma, F)$ is exactly the image of $H^1((\Sigma, \partial\Sigma), F)$ in $H^1(\Sigma, F)$.

The proof of our Theorem is completed. □

REMARK 5.5. Assume that F and F^* are naturally isomorphic as flat bundles. This is the case when F is real and carries a flat scalar product. Then $H^0(\Sigma, F)$ and $H^2((\Sigma, \partial\Sigma), F)$ are Poincaré dual, and the image of $H^1((\Sigma, \partial\Sigma), F)$ in $H^1(\Sigma, F)$ is equipped with a non degenerate 2-form, the intersection product. The conclusion is that in this case, the cohomology of $(\widehat{C}^{\Sigma \cdot}(F), \partial)$ exhibits Poincaré duality.

5.3. The map ϕ. Let G be a compact connected semisimple Lie group. Let $Z(G)$ be the center of G. Then $Z(G)$ is a finite subgroup of G. Let $G' = G/Z(G)$ be the adjoint group. Then G' acts naturally on G by conjugation. If $u \in G'$, $v \in G$, we will write

$$u \cdot v = uvu^{-1}. \tag{5.40}$$

Similarly G' acts on G^{2g+s} by conjugation. We will use the same notation for the action of G on G or on G^{2g+s} by conjugation.

Take $g, s \in \mathbf{N}$, with $g + s > 0$.

DEFINITION 5.6. Let $\phi : G^{2g+s} \to G$ be the map

$$\phi(u_1, v_1, \dots, u_g, v_g, w_1, \dots, w_s) = \prod_{i=1}^{g} [u_i, v_i] \prod_{j=1}^{s} w_j \,. \tag{5.41}$$

Then ϕ is G'-equivariant, i.e. if $g \in G'$, $x \in G^{2g+s}$,

$$\phi(g \cdot x) = g \cdot \phi(x) \,. \tag{5.42}$$

Let $\mathfrak{g}$ be the Lie algebra of G. If $g \in G$, we identify T_gG to $\mathfrak{g} = T_eG$ via the right multiplication operator R_{g*}. Then if $z \in G^{2g+s}$,

$$T_zG^{2g+s} = \mathfrak{g}^{2g+s} \,. \tag{5.43}$$

DEFINITION 5.7. Take $x \in G^{2g+s}$. Let $\delta : \mathfrak{g} \to \mathfrak{g}^{2g+s}$ be the derivative at $g = 1$ of the map $g \in G \mapsto g \cdot x \in G^{2g+s}$. Similarly let $\delta : \mathfrak{g}^{2g+s} \to \mathfrak{g}$ be the derivative at $x \in G^{2g+s}$ of $x' = G^{2g+s} \mapsto \phi(x') \in G$.

Clearly G acts on $\mathfrak{g}$ by the adjoint action.

PROPOSITION 5.8. *Take $x = (u_1, v_1, \dots, u_g, v_g, w_1, \dots, w_s) \in G^{2g+s}$. Then if $f \in \mathfrak{g}$,*

$$\begin{aligned}\delta f = ((1-u_1)f, (1-v_1)f, \dots, (1-u_g)f, (1-v_g)f,\\ (1-w_1)f, \dots, (1-w_s)f) \,.\end{aligned} \tag{5.44}$$

Similarly, if $(f_1, \dots, f_{2g+s}) \in \mathfrak{g}^{2g+s}$,

$$\begin{aligned}\delta(f_1, \dots, f_{2g+s}) = \sum_{i=1}^{g} [u_1, v_1] \dots [u_{i-1}, v_{i-1}]\\ ((1 - u_i v_i u_i^{-1}) f_{2i-1} + u_i(1 - v_i u_i^{-1} v_i^{-1}) f_{2i})\\ + \sum_{j=1}^{s} [u_1, v_1] \dots [u_g, v_g] w_1 \dots w_{j-1} f_{2g+j} \,.\end{aligned} \tag{5.45}$$

PROOF. This is a trivial computation, which is left to the reader. □

Let $\mathcal{O} \subset G$ be an orbit in G. More precisely, take $g \in G$, and put

$$\mathcal{O} = \{g' \cdot g, g' \in G\} \,. \tag{5.46}$$

Of course $g \in \mathcal{O}$. If $Z(g) \subset G$ is the centralizer of g,

$$\mathcal{O} \simeq G/Z(g) \,. \tag{5.47}$$

Clearly the tangent space $T_g\mathcal{O} \subset T_gG \simeq \mathfrak{g}$ is given by

$$T_g\mathcal{O}_g = \mathrm{Im}(1 - g) \,. \tag{5.48}$$

Also $T_eZ(g) \subset \mathfrak{g}$ is given by

$$T_eZ(g) = \ker(1 - g) \,. \tag{5.49}$$

Then the exact sequence

$$0 \to T_eZ(g) \to T_eG \to T_g\mathcal{O}_g \to 0 \tag{5.50}$$

corresponds to the obvious

$$0 \longrightarrow \ker(1-g) \longrightarrow \mathfrak{g} \xrightarrow[1-g]{} \operatorname{Im}(1-g) \longrightarrow 0\,. \tag{5.51}$$

Let $\mathcal{O}_1, \dots, \mathcal{O}_s$ be s adjoint orbits of G in G. Put

$$X = G^{2g} \times \prod_{j=1}^{s} \mathcal{O}_j\,. \tag{5.52}$$

If $x = (u_1, v_1, \dots, u_g, v_g, w_1, \dots, w_s) \in X$, $\delta : \mathfrak{g} \to \mathfrak{g}^{2g+s}$ is in effect a linear map : $\mathfrak{g} \to \mathfrak{g}^{2g} \oplus \bigoplus_{j=1}^{s} \operatorname{Im}(1-w_j)$. By restriction we obtain the linear maps

$$(D,\delta) : 0 \longrightarrow \mathfrak{g} \xrightarrow[\delta]{} \mathfrak{g}^{2g} \oplus \bigoplus\nolimits_{j=1}^{s} \operatorname{Im}(1-w_j) \xrightarrow[\delta]{} \mathfrak{g} \longrightarrow 0\,. \tag{5.53}$$

Note that in general

$$\delta^2 \neq 0\,. \tag{5.54}$$

Also observe that if $g \in G'$, g maps $(D, \partial)_x$ into $(D, \partial)_{gx}$.

DEFINITION 5.9. If $x \in X$, let $Z(x) \subset G$, $Z'(x) \subset G'$ be the stabilizers of x, and let $\mathfrak{z}(x)$ be the Lie algebra of $Z(x)$ and $Z'(x)$.

Of course $Z'(x) = Z(x)/Z(G)$.

PROPOSITION 5.10. *For $x \in X$, then*

$$\{f \in \mathfrak{g}, \delta f = 0\} = \mathfrak{z}(x)\,. \tag{5.55}$$

PROOF. This is obvious by (5.44). □

Let $\langle\ ,\ \rangle$ be a G-invariant scalar product on $\mathfrak{g}$. If $\mathfrak{h}$ is a vector subspace of $\mathfrak{g}$, let $\mathfrak{h}^{\perp}$ be the orthogonal space to $\mathfrak{h}$ in $\mathfrak{g}$.

PROPOSITION 5.11. *For $x \in X$, then*

$$\left[\delta\Big(\mathfrak{g}^{2g} \oplus \bigoplus_{j=1}^{s} \operatorname{Im}(1-w_j)\Big)\right]^{\perp} = \mathfrak{z}(x)\,. \tag{5.56}$$

PROOF. Let $\delta^* : \mathfrak{g} \to \mathfrak{g}^{2g} \oplus \bigoplus_{j=1}^{s} \operatorname{Im}(1-w_j)$ be the adjoint of $\delta : \mathfrak{g}^{2g} \oplus \bigoplus_{j=1}^{s} \operatorname{Im}(1-w_j) \to \mathfrak{g}$. Clearly

$$\delta(\mathfrak{g}^{2g} \oplus \bigoplus_{j=1}^{s} \operatorname{Im}(1-w_j))^{\perp} = \ker \delta^*\,. \tag{5.57}$$

Let $P^{\operatorname{Im}(1-w_j)}$ be the orthogonal projection operator $\mathfrak{g} \to \operatorname{Im}(1-w_j)$. Then

$$(1-w_j^{-1})P^{\operatorname{Im}(1-w_j)} = (1-w_j^{-1}), \tag{5.58}$$

and $1-w_j^{-1}$ is one to one from $\operatorname{Im}(1-w_j)$ into itself. Using 5.45, we obtain a formula for δ^* involving the projection $P^{\operatorname{Im}(1-w_j)}$. By (5.58), to calculate $\ker\delta^*$, we may replace $P^{\operatorname{Im}(1-w_j)}$ by $(1-w_j)$. By proceeding as in Proposition 5.1, we get (5.56). □

Recall that $\phi : X \to G$ is said to be regular at $x \in X$ if $d\phi(x) : T_x X \to T_{\phi(x)} G$ is surjective.

THEOREM 5.12. *The map* $\phi : X \to G$ *is regular at* $x \in X$ *if and only if* $\mathfrak{z}(x) = 0$.

PROOF. Our Theorem is a trivial consequence of Proposition 5.11. □

DEFINITION 5.13. Put

$$M = \{x \in X,\ \phi(x) = 1\}\,. \tag{5.59}$$

If $x = (u_1, \dots, w_s) \in M$, then x defines a morphism $\Gamma \to G$. Also G acts on $\mathfrak{g}$ via the adjoint representation. Let E be the flat real vector bundle on Σ constructed as in (5.7), i.e.

$$E = \widehat{\Sigma} \times_\Gamma \mathfrak{g}\,. \tag{5.60}$$

The following result was first obtained by Liu [**39, 40**].

THEOREM 5.14. *If* $x \in M$, *then in (5.53),* $\delta^2 = 0$, *i.e.* $(D, \partial)_x$ *is a complex. The complex* $(D, \delta)_x$ *can then be canonically identified to the complex* $(\widehat{C}^{\Sigma\cdot}(E), \partial)_x$.

PROOF. By (5.42), if $x \in M$,

$$\phi(gx) = 1\,. \tag{5.61}$$

By (5.61), we find that $\delta^2 = 0$. Comparing (5.33), (5.34) and (5.44), (5.45), our Theorem follows. □

REMARK 5.15. By Theorem 5.4, if $x \in M$,

$$\begin{aligned} \widehat{H}^0(\Sigma, E) &= H^0(\Sigma, E) \\ \widehat{H}^2(\Sigma, E) &= H^2((\Sigma, \partial\Sigma), E)\,. \end{aligned} \tag{5.62}$$

In view of Propositions 5.1 and 5.10, we find that if $x \in M$,

$$\begin{aligned} \widehat{H}^0(\Sigma, E) &= \mathfrak{z}(x)\,, \\ \widehat{H}^2(\Sigma, E) &= \mathfrak{z}(x)^*\,. \end{aligned} \tag{5.63}$$

So when $x \in M$, Proposition 5.11 follows from (5.63).

5.4. The set of regular values of the map ϕ**.** Recall that G and G' act by the adjoint action on $\mathfrak{g}$. Then one verifies easily that if $x \in M, \mathfrak{g} \in G$,

$$(\widehat{C}_x^{\Sigma\cdot}(E), \partial) \xrightarrow[g]{} (\widehat{C}_{g\cdot x}^{\Sigma\cdot}(E), \partial) \tag{5.64}$$

is an isomorphism of complexes. In particular the induced map

$$\widehat{H}_x^{\Sigma\cdot}(E) \xrightarrow[g]{} \widehat{H}_{g\cdot x}^{\Sigma\cdot}(E) \tag{5.65}$$

is an isomorphism.

Let T be a maximal torus in G, let $\mathfrak{t}$ be the Lie algebra of T. Let W be the Weyl group of G with respect to T. Let $R = \{\alpha\} \subset \mathfrak{t}^*$ be the root system of G, let $CR = \{h_\alpha\} \subset \mathfrak{t}$ be the corresponding coroots.

Now we recall the definition of $S \subset T$ given in Definition 2.23.

DEFINITION 5.16. Let $S \subset T$ be given by

$$S = \{t \in T; t = \sum_{\alpha \in R} t^\alpha h_\alpha\,; \text{the } h_\alpha \text{ for which } t^\alpha \neq 0 \text{ do not span } \mathfrak{t}\}\,. \tag{5.66}$$

Let $t_1, \dots, t_s \in T$. Let $\mathcal{O}_1, \dots, \mathcal{O}_s$ be the orbits of $t_1, \dots, t_s$ in G. By [**15**, Lemma IV.2.5],

$$\mathcal{O}_j \cap T = W t_j \ , \ 1 \leq j \leq s \,. \tag{5.67}$$

DEFINITION 5.17. We will say that $(t_1, \dots, t_s)$ verify (A) if for any $(w^1, \dots, w^s) \in W^s$,

$$\sum_1^s w^j t_j \notin S \,. \tag{5.68}$$

THEOREM 5.18. *For $g \geq 1$, the map $\phi : X \to G$ is surjective. For $g \geq 1$, $t \in T$ is a regular value of ϕ if and only if $(t_1, t_2, \dots, t_s, -t) \in T^{s+1}$ verify (A).*

For $g = 0$, $t \in T$ is a regular value of ϕ in the following two cases :

(5.69)

- $t \notin \phi(X)$.
- $t \in \phi(X)$, *and* $(t_1, \dots, t_s, -t)$ *verify* (A).

PROOF. By [**14**, Corollaire 4.5], since G is semisimple, the map $(u, v) \in G^2 \mapsto [u, v] \in G$ is surjective. Therefore, for $g \geq 1$, the map ϕ is surjective.

First we will prove the remainder of our Theorem for $t = 1$. By Theorem 5.12, $1 \in G$ is a non regular value of ϕ if and only if there exists $x \in M$ such that $\mathfrak{z}(x) \neq 0$. Equivalently there exists $p \in \mathfrak{g}$, $p \neq 0$ and $x = (u_1, \dots, w_s) \in M$ such that $x \in Z(p)^{2g+s}$. By [**15**, Theorem IV.2.3], $Z(p)$ is a connected Lie subgroup of G, which is not semisimple, since $\mathbf{R}p$ is contained in the Lie algebra of its center. Let $T_0 = Z_0(Z(p))$ be the connected component of the identity in $Z(Z(p))$. Then we have the exact sequence

$$1 \to T_0 \to Z(p) \to Z(p)/T_0 \to 1, \tag{5.70}$$

and $Z(p)/T_0$ is semisimple. Put

$$\gamma = \pi_1(Z(p)/T_0) \,, \tag{5.71}$$

and let $\widetilde{Z(p)/T_0}$ be the universal cover of $Z(p)/T_0$. Then (5.70) fits in the diagram

$$\begin{array}{ccccccccc}
 & & 1 & & 1 & & 1 & & \\
 & & \downarrow & & \downarrow & & \downarrow & & \\
1 & \longrightarrow & 1 & \longrightarrow & \gamma & \longrightarrow & \gamma & \longrightarrow & 1 \\
 & & \downarrow & & \downarrow & & \downarrow & & \\
1 & \longrightarrow & T_0 & \longrightarrow & \widehat{Z}(p) & \longrightarrow & \widetilde{Z(p)/T_0} & \longrightarrow & 1 \\
 & & \downarrow & & \downarrow & & \downarrow & & \\
1 & \longrightarrow & T_0 & \longrightarrow & Z(p) & \longrightarrow & Z(p)/T_0 & \longrightarrow & 1 \\
 & & \downarrow & & \downarrow & & \downarrow & & \\
 & & 1 & & 1 & & 1 & &
\end{array} \tag{5.72}$$

In (5.72), $\widehat{Z}(p) \to Z(p)$ is the obvious γ covering of $Z(p)$.

Since $\widetilde{Z(p)/T_0}$ is semisimple and simply connected, any central extension of $\widetilde{Z(p)/T_0}$ by a torus is trivial, so that

$$\widehat{Z}(p) = T_0 \times \widetilde{Z(p)/T_0}\,. \tag{5.73}$$

Therefore $\gamma \subset \widetilde{Z(p)/T_0}$ embeds as a finite subgroup of $T_0 \times \widetilde{Z(p)/T_0}$.

Let $\widetilde{T}(p)$ be a maximal torus in $\widetilde{Z(p)/T_0}$. Then $T_0 \times \widetilde{T}(p)$ is a maximal torus in $\widehat{Z}(p)$. From (5.72), (5.73), we get the complex

$$\begin{array}{ccccccccc}
 & & 1 & & 1 & & 1 & & \\
 & & \downarrow & & \downarrow & & \downarrow & & \\
1 & \longrightarrow & 1 & \longrightarrow & \gamma & \longrightarrow & \gamma & \longrightarrow & 1 \\
 & & \downarrow & & \downarrow & & \downarrow & & \\
1 & \longrightarrow & T_0 & \longrightarrow & T_0 \times \widetilde{T}(p) & \longrightarrow & \widetilde{T}(p) & \longrightarrow & 1 \\
 & & \downarrow & & \downarrow & & \downarrow & & \\
1 & \longrightarrow & T_0 & \longrightarrow & T & \longrightarrow & T/T_0 & \longrightarrow & 1 \\
 & & \downarrow & & \downarrow & & \downarrow & & \\
 & & 1 & & 1 & & 1 & &
\end{array} \tag{5.74}$$

Now we use the notation of Theorem 1.52. Let $\mathfrak{t}_0 = \mathfrak{z}(Z(p))$ be the Lie algebra of T_0. Let $\mathfrak{t}_1 \subset \mathfrak{t}$ be the vector subspace of $\mathfrak{t}$ spanned by the $\{h_\alpha\}_{\alpha \in R_p}$. Clearly

$$\mathfrak{t} = \mathfrak{t}_0 \oplus \mathfrak{t}_1\,. \tag{5.75}$$

Also $\mathfrak{t}_1$ is the Lie algebra of T/T_0 or of $\widetilde{T}(p)$.

Let $[\widehat{Z}(p), \widehat{Z}(p)]$ be the commutator subgroup $\widehat{Z}(p)$, i.e. the group spanned by commutators in $\widehat{Z}(p)$. By (5.73), since $\widetilde{Z(p)/T_0}$ is semisimple,

$$[\widehat{Z}(p), \widehat{Z}(p)] = \widetilde{Z(p)/T_0}\,. \tag{5.76}$$

Clearly $[\widehat{Z}(p), \widehat{Z}(p)]$ maps onto $[Z(p), Z(p)]$. Therefore $[Z(p), Z(p)]$ is a compact connected subgroup of $Z(p)$, and so it is a connected Lie subgroup of $Z(p)$.

Let now $t \in T \cap [Z(p), Z(p)]$. By the above, there is $a \in \widetilde{Z(p)/T_0}$, $b = (b_0, b_1) \in T_0 \times \widetilde{T}(p)$ which map to t. Therefore, if $c = ab^{-1}$, then $c \in \gamma$. Clearly

$$c = (b_0^{-1}, ab_1^{-1})\,. \tag{5.77}$$

By (5.74), (5.77),

$$ab_1^{-1} \in \gamma \subset \widetilde{T}(p)\,. \tag{5.78}$$

From (5.77), we get

$$a \in \widetilde{T}(p)\,. \tag{5.79}$$

We thus find that if $t \in T \cap [Z(p), Z(p)]$, t is the projection of an element of $\widetilde{T}(p)$. Equivalently t can be represented by $\bar{t} \in \mathfrak{t}_1$.

Recall that

$$\prod_{i=1}^{g}[u_i, v_i]\prod_{j=1}^{s} w_j = 1\,, \tag{5.80}$$

and $u_i, v_i \in Z(p)$, $1 \leq i \leq g$, $w_j \in Z(p)$, $1 \leq j \leq s$. Clearly T is a maximal torus in $Z(p)$. So there is $g \in Z(p)$ such that $gw_jg^{-1} \in T$. Since $gw_jg^{-1} \in T \cap \mathcal{O}_{t_j}$, by (5.67), there is $w^j \in W$ such that

$$gw_jg^{-1} = w^j t_j\,. \tag{5.81}$$

Therefore

$$w_j = w^j t_j \text{ in } Z(p)/[Z(p), Z(p)]\,. \tag{5.82}$$

So by (5.80), (5.82), we get

$$\sum_{j=1}^{s} w^j t_j \in T \cap [Z(p), Z(p)]\,. \tag{5.83}$$

By the above, we find that

$$\sum_{j=1}^{s} w^j t_j = \sum_{\alpha\in R_p} s^\alpha h_\alpha \text{ in } T\,. \tag{5.84}$$

Now since $p \neq 0$, $\{h_\alpha, \alpha \in R_p\}$ does not span $\mathfrak{t}$, i.e. $(t_1, \ldots, t_s)$ does not verify (A).

Conversely, suppose that $g \geq 1$, and $(t_1, \ldots, t_s)$ does not verify (A). Then there is $(w^1, \ldots, w^s) \in W^s$, such that $\sum_{j=1}^{s} w^j t_j$ can be expressed as a linear combination of $\{h_\alpha\}$ which do not span $\mathfrak{t}$. Let $p \in \mathfrak{t}$, $p \neq 0$ be orthogonal to the corresponding $\{\alpha\}$. By the above,

$$\sum_{j=1}^{s} w^j t_j \in [Z(p), Z(p)]\,. \tag{5.85}$$

Using the fact that in a compact semisimple Lie group, any element is a commutator [**14**, Corollaire 4.5], and also the considerations after (5.76), by (5.85), there is $u_g, v_g \in Z(p)$ such that

$$[u_g, v_g]\prod_{j=1}^{s} w^j t_j = 1\,. \tag{5.86}$$

If $x = (1, \ldots, 1, u_g, v_g, w^1t_1, \ldots, w^st_s)$, then $x \in M$, and $p \in \mathfrak{z}(x)$, so that $\mathfrak{z}(x) \neq \{0\}$. By Theorem 5.12, we find that 1 is not a regular value of ϕ.

Let $t = t_{s+1}^{-1} \in T$. Let $\mathcal{O}_{s+1} \subset G$ be the G-orbit of t_{s+1}. Put $X_{+1} = G^{2g} \times \prod_{j=1}^{s+1} \mathcal{O}_j$. More generally we denote with the index $+1$ the objects we constructed above, which are associated to X_{+1}. Clearly

$$X_{+1} = X \times \mathcal{O}_{s+1}\,. \tag{5.87}$$

Also if $(x, w_{s+1}) \in X_{+1}$,

$$\phi_{+1}(x, w_{s+1}) = \phi(x)w_{s+1}\,. \tag{5.88}$$

Clearly $\phi(x) = t$ if and only if $(x, t_{s+1}) \in M_{+1}$. By (5.88), if ϕ is regular at x, ϕ_{+1} is regular at (x, t_{s+1}). Conversely, suppose that ϕ_{+1} is regular at (x, t_{s+1}). By (5.88),

$$\operatorname{Im}(d\phi_{+1})(x, t_{s+1}) = \operatorname{Im}(d\phi\phi^{-1})_x + (t_{s+1} - 1)(\mathfrak{g}) . \tag{5.89}$$

By (5.42), we find that

$$(t_{s+1} - 1)(\mathfrak{g}) \subset \operatorname{Im}(d\phi\phi^{-1})_x . \tag{5.90}$$

From (5.89), (5.90), we find that ϕ is regular at t if and only if ϕ_{+1} is regular at (x, t_{s+1}). Now we use our Theorem for $t = 1$ and we obtain the stated result in full generality. □

REMARK 5.19. Instead of studying the case $t = 1$ first, we could as well have used Theorem 5.12 and proceeded directly.

5.5. The stabilizers $Z'(x)$.

THEOREM 5.20. *If G is simply connected, under one of the following three conditions, if $t \in T$ is a regular value of ϕ, $\{x \in \phi^{-1}\{t\}, Z'(x) \neq 1\}$ is included in a union of submanifolds of $\{x \in \phi^{-1}(t)\}$ of codimension ≥ 2 :*

- *For $g \geq 2$.*
- *For $g = 1$, if one of the t_j's or t is regular.*
- *For $g = 0$, $s = 1$, in which case $\phi^{-1}\{t\}$ is empty, or if at least 3 of the elements $\{t_1, \dots, t_s, t\}$ are regular.*

If G is only connected, if $t_1, \dots, t_s$ are very regular, if $t \in T$ is a regular value of ϕ, $\{x \in \phi^{-1}(t), Z'(x) \neq 1\}$ is included in a union of submanifolds of $\{x \in \phi^{-1}(t)\}$ of codimension ≥ 2 :

- *For $g \geq 2$.*
- *For $g = 1$, if $s \geq 1$ or if t is very regular.*
- *For $g = 0$, $s = 1$, in which case $\phi^{-1}(t)$ is empty, if $s = 2$ and t is very regular, or if $s \geq 3$.*

PROOF. We will prove our Theorem in various stages.

- *The case where G is simply connected and $t = 1$.*

If 1 is a regular value of ϕ, either $M = \phi^{-1}\{1\}$ is empty, or it is a smooth submanifold of X.

By Theorem 5.12, when M is non empty, the group G acts locally freely on M. Therefore M/G is an orbifold. If $\dim(M/G)$ is the dimension of the maximal statum of M/G, then

$$\dim M/G = (2g - 2)\dim(\mathfrak{g}) + \sum_{j=1}^{s} \dim(\mathcal{O}_j) . \tag{5.91}$$

Of course the same observation applies to the action of G' on M. Clearly $M/G \simeq M/G'$. In the sequel we will use the notation M/G or M/G' indifferently. However in Section 6.3, we will construct an orbifold G-line bundle on M/G, which may well not be an orbifold G'-line bundle.

Take now $x \in M$ such that $Z'(x) \neq 1$. Let $u \in Z'(x)$, $u \neq 1$. If $x = (u_1, \dots, w_s)$, then $u_1, \dots, w_s \in Z(u)$. By conjugation, we may as well assume that $u \in T' = T/Z(G)$. Since G acts locally freely on M, the group $Z(u)$ is semisimple. By Theorem 1.38, $u \in C/\overline{R}^*$. Of course here $u \neq 0$.

Since G is simply connected, $Z(u)$ is connected. By Theorem 1.50, for any $t \in T$,

$$\mathcal{O}(t) \cap Z(u) = \bigcup_{w \in W_u \backslash W} \mathcal{O}_{Z(u)}(wt)\,. \tag{5.92}$$

For $1 \leq j \leq s$, let $w^j \in W$ such that

$$w_j \in \mathcal{O}_{Z(u)}(w^j t_j),\ 1 \leq j \leq s\,. \tag{5.93}$$

Put

$$X^u = Z(u)^{2g-2} \times \prod_{j=1}^{s} \mathcal{O}_{Z(u)}(w^j t_j)\,. \tag{5.94}$$

We define $\phi_u : X^u \to Z(u)$ as in (5.41). Let $M^u = \phi_u^{-1}\{1\}$. Then $x \in M^u$. Moreover $Z(u)$ acts locally freely on M^u. Therefore, by Theorem 5.12, 1 is a regular value of ϕ_u, i.e. M^u is a smooth submanifold of M. Note that this also follows from the fact that M is a smooth manifold, G acts on M, and M^u is a component of the fixed point set of u in M. Then

$$\dim M^u/Z(u) = (2g-2)\dim \mathfrak{z}(u) + \sum_{j=1}^{s} \dim \mathcal{O}_{Z(u)}(w^j t_j)\,. \tag{5.95}$$

Now by Theorem 1.52,

$$\dim \mathfrak{z}(u) \leq \dim(\mathfrak{g}) - 2\,. \tag{5.96}$$

Also

$$\dim \mathcal{O}_{Z(u)}(w^j t_j) \leq \dim \mathcal{O}_j\,. \tag{5.97}$$

Now we consider 3 cases :

- If $g \geq 2$, by (5.91), (5.95)-(5.97),

$$\dim M^u/Z(u) \leq \dim M/G - 2\,. \tag{5.98}$$

By (5.98), $M^u/Z(u)$ maps to a submanifold of codimension ≥ 2 in M/G. Therefore the G-orbit of M^u in M is a submanifold of codimension ≥ 2 in M.

- If $g = 1$, assume that $t_1 \in T$ is regular. Then

$$\begin{aligned} \dim \mathcal{O}_{t_1} &= \dim(\mathfrak{g}) - \dim(\mathfrak{t})\,, \\ \dim \mathcal{O}_{Z(u)}(w^1 t_1) &= \dim \mathfrak{z}(u) - \dim(\mathfrak{t})\,. \end{aligned} \tag{5.99}$$

By (5.91), (5.95), (5.96), (5.97), (5.99), we find again that

$$\dim M^u/Z(u) \leq \dim M/G - 2\,. \tag{5.100}$$

- If $g = 0$, if $s = 2$, then

$$\dim \mathcal{O}_{t_j} \leq \dim(\mathfrak{g}) - \dim(\mathfrak{t})\,;\ s = 1, 2\,, \tag{5.101}$$

so that using (5.91),

$$\dim M/G < 0, \tag{5.102}$$

i.e. $M = \emptyset$.

If $s \geq 3$, assume that t_1, t_2, t_3 are regular. Then

$$\dim M/G = \dim(\mathfrak{g}) - 3\dim(\mathfrak{t}) + \sum_{j=4}^{s} \dim(\mathcal{O}_j)\,, \tag{5.103}$$

$$\dim M^u/Z(u) = \dim \mathfrak{z}(u) - 3\dim(\mathfrak{t}) + \sum_{j=4}^{s} \dim \mathcal{O}_{Z(u)}(w^j t_j)\,.$$

By (5.96), (5.97), (5.103)

$$\dim M^u/Z(u) \leq \dim M/G - 2\,. \tag{5.104}$$

So we have proved our Theorem in this case.

- *The case where G is non simply connected and $t = 1$.*

First we proceed as above. In this case for $u \in C/\overline{R}^*$, $u \neq 1$, $Z(u)$ is in general non connected. However by (1.174), since $t_1, \dots, t_s$ are very regular,

$$\mathcal{O}_j \cap Z(u) = \bigcup_{w^j \in W_{Z(u)_o} \backslash W} \mathcal{O}_{Z(u)_o}(w^j t_j)\,. \tag{5.105}$$

For $1 \leq j \leq s$, let $w^j \in W$ be such that

$$w_j \in \mathcal{O}_{Z(u)_o}(w^j t_j)\,. \tag{5.106}$$

Again we define X^u by (5.94), and we define ϕ_u as before. Put $M^u = \phi_u^{-1}\{1\}$. By the same argument as before, 1 is a regular value of ϕ_u. Then (5.95) still holds. If $g \geq 2$, (5.104) is still true. If $g = 1$, since t_1 is very regular in G, then t_1 lies in $Z(u)_0$ and is very regular in $Z(u)$, so that (5.100) still holds.

If $g = 0$, under the stated assumptions, the argument in the proof above can still be reproduced.

The proof of our Theorem is completed in this case.

- *The case where $t \in T$ is an arbitrary regular value of ϕ.*

Let $t = t_{s+1}^{-1} \in T$. We use the notation in the proof of Theorem 5.18. Then t is a regular value of ϕ if and only if 1 is a regular value of ϕ_{+1}. By the above, we obtain our Theorem in full generality.

The proof of our Theorem is completed. □

THEOREM 5.21. *If G is simply connected, under one of the following conditions, $\{x \in X; Z'(x) \neq 1\}$ is included in a union of submanifolds of codimension ≥ 2:*

- *$g \geq 1$.*
- *$g = 0$, $s \geq 2$, and at least 2 of the t_j's are regular.*

If G is non necessarily simply connected, if all the t_j's are very regular, then $\{x \in X, Z'(x) \neq 1\}$ is included in a union of submanifolds of codimension ≥ 2 if

- *$g \geq 1$.*
- *$g = 0$, $s \geq 2$, and at least 2 of the t_j's are very regular.*

PROOF. By [**24**, Proposition 27.4], the set of conjugacy classes of stabilizers $Z'(x)$ is finite. Take $x \in X$ and assume that $Z'(x) \neq e$. Let $G'(x)$ be the orbit of G' through x. Then

$$G'(x) \simeq G'/Z'(x)\,. \tag{5.107}$$

Let N_x be the normal bundle to $G'(x)$ at x. Then $Z'(x)$ acts on N_x. By [24, Proposition 27.2], there is a G'-invariant open neighborhood of $G'(x)$ in X which can be identified as a G-space to a neighborhood of the zero section in $N = G' \times_{Z'(x)} N_x$.

Let N_x^f be the invariant part of N_x under $Z'(x)$. Then N_x^f extends to a vector subbundle of the vector bundle N on $G'/Z'(x)$, of the form $G'/Z'(x) \times N_x^f$.

If $y \in N_x$, then $Z'(y) \subset Z'(x)$ and $Z'(y) = Z'(x)$ if and only if $y \in N_x^f$. In particular $Z'(y)$ is conjugate to $Z'(x)$ if and only if $y \in N_x^f$, in which case $Z'(y) = Z'(x)$. It follows that near $G'(x)$, the elements of X whose stabilizer is conjugate to $Z'(x)$ form a neighborhood of the zero section of N^f.

Let $u \in Z'(x)$, $u \neq 1$. We may and we will assume that $u \in T$. Let $(G'/Z'(x))^u$ be the fixed point set of u in $G'/Z'(x)$. Let $\mathrm{codim}((G'/Z'(x))^u, G'/Z'(x))$ be the codimension of $(G'/Z'(x))^u$ in $G'/Z'(x)$. Clearly

$$\mathrm{codim}((G'/Z'(x))^u, G'/Z'(x)) \leq \dim(\mathfrak{g}) - \dim(\mathfrak{z}(u)) . \tag{5.108}$$

Let N_x^u be the vector subspace of N_x fixed under u. Then $N_x^f \subset N_x^u$. If N^u is the fixed point set of N under u, then N^u is a vector bundle over $(G'/Z'(x))^u$, with fibre modelled on N_x^u.

Let $\dim(N^f) = \dim(G'/Z'(x)) + \dim N_x^f$ be the dimension of the total space of N^f. Similarly let $\dim(N^u)$ be the dimension of the total space of N_u. By (5.108), we get

$$\dim(N^f) \leq \dim(N^u) + \dim(\mathfrak{g}) - \dim \mathfrak{z}(u) . \tag{5.109}$$

Clearly

$$\dim X = 2g \dim(\mathfrak{g}) + \sum_{j=1}^{s} \dim(\mathcal{O}_j) . \tag{5.110}$$

Let X^u be the fixed point set of X under u. Then

$$X^u = Z(u)^{2g} \times \prod_{j=1}^{s} (\mathcal{O}_j \cap Z(u)), \tag{5.111}$$

so that

$$\dim(X^u) = 2g \dim \mathfrak{z}(u) + \sum_{j=1}^{s} \dim(\mathcal{O}_j \cap Z(u)) . \tag{5.112}$$

By (5.109), (5.112),

$$\dim(N^f) \leq (2g-1) \dim \mathfrak{z}(u) + \dim \mathfrak{g} + \sum_{j=1}^{s} \dim(\mathcal{O}_j \cap Z(u)) . \tag{5.113}$$

If $g \geq 1$, using (5.96), (5.110), (5.113),

$$\dim(N^f) \leq \dim(X) - 2 . \tag{5.114}$$

Assume now that $g = 0$, that $s \geq 2$, that G is simply connected and t_1, t_2 are regular, or more generally that t_1, t_2 are very regular. By Theorem 1.50,

$$\begin{aligned} \dim \mathcal{O}_j &= \dim(\mathfrak{g}) - \dim(\mathfrak{t}) , \\ \dim \mathcal{O}_j \cap Z(u) &= \dim \mathfrak{z}(u) - \dim \mathfrak{t} . \end{aligned} \tag{5.115}$$

So if $g = 0$, using (5.96), (5.110), (5.113), (5.115), we get

$$(5.116)\quad \dim(N^f) \leq \dim \mathfrak{z}(u) + \dim \mathfrak{g} - 2\dim(\mathfrak{t}) + \sum_{j=3}^{s} \dim(\mathcal{O}_j \cap Z(u)) \leq \dim(X) - 2.$$

From (5.114), (5.116), our Theorem follows. □

5.6. The tangent bundle to the moduli space and its symplectic form. We make the same assumptions as in Sections 5.1-5.4.

If $x \in X$, let $G'(x)$ be the orbit of G' at x, and let $T_xG'(x)$ be the tangent space to $G'(x)$ at x. Recall that $T_xX \subset \mathfrak{g}^{2g+s}$, so that $T_xG'(x) \subset \mathfrak{g}^{2g+s}$.

PROPOSITION 5.22. *If $x \in M$,*

$$(5.117)\qquad T_xG'(x) = \partial(\widehat{C}_x^{\Sigma,0}(E)) \subset \widehat{C}_x^{\Sigma,1}(E).$$

If ϕ is regular at $x \in M$, then M is a submanifold of X near x, and

$$(5.118)\qquad T_xM = \ker \partial_{|\widehat{C}^{\Sigma,1}(E) \to \widehat{C}^{\Sigma,2}(E)}.$$

PROOF. By Theorem 5.14, the first part of our Proposition is clear. If ϕ is regular at $x \in M$,

$$(5.119)\qquad T_xM = \ker d\phi(x).$$

Using Theorem 5.14 and (5.119), we get (5.118). □

REMARK 5.23. Of course, if ϕ is regular at x, then $T_xG'(x) \subset T_xM$. This fits with (5.117), (5.118), because $\partial^2 = 0$. By Proposition 5.10 and Theorem 5.12, $T_xG'(x)$ has dimension $\dim \mathfrak{g}$.

Let $\pi : M \to M/G$ be the obvious projection. If ϕ is regular at $x \in M$, $Z'(x)$ is a finite group.

THEOREM 5.24. *If ϕ is regular at $x \in M$, M is smooth near x, and M/G is an orbifold near $\pi(x)$. More precisely, near $\pi(x) \in M/G$,*

$$(5.120)\qquad TM/G = M \times_G \widehat{H}^{\Sigma,1}(E).$$

Also near $\pi(x)$,

$$(5.121)\qquad M/G \simeq \widehat{H}_x^1(\Sigma, E)/Z'(x).$$

PROOF. These are classical facts from the theory of orbifolds (see [?, Proposition 27.7]. □

REMARK 5.25. By Theorems 5.4 and 5.24, TM/G is the image of $H^1((\Sigma, \partial\Sigma), E)$ into $H^1(\Sigma, E)$. This fact was also observed in Guruprasad, Huebschamnn,Jeffrey and Weinstein [**25**].

Now we assume temporarily that G is simply connected. We use the notation of Section 4.

DEFINITION 5.26. Put

$$(5.122)\quad \mathcal{A}^{\text{flat}}(t_1, \dots, t_s) = \{A \in \mathcal{A}, F^A = 0, -(\frac{d}{dt} + A)_{S_j^1} \in \mathcal{O}_{-\frac{d}{dt}+t_j}\}, 1 \leq j \leq s.$$

Equivalently, $\mathcal{A}^{\text{flat}}(t_1, \ldots, t_s)$ is the set of flat connections A such that for $1 \le j \le s$, the holonomy w_j of A along S^1_j lies in the orbit $\mathcal{O}_j$.

Clearly ΣG acts on the right on $\mathcal{A}^{\text{flat}}(t_1, \ldots, t_s)$.

DEFINITION 5.27. Let $\Sigma_q G$ be the group of smooth maps $g : \Sigma \to G$ such that $g_q = 1$.

It is easy to see that $\Sigma_q G$ acts freely on $\mathcal{A}^{\text{flat}}(t_1, \ldots, t_s)$. Also there is an obvious continuous map $f : \mathcal{A}^{\text{flat}}(t_1, \ldots, t_s) \mapsto M$ which, to A, associates $(u_1, v_1, \ldots w_1, \ldots,$
$w_s)$, the holonomies of A along the circles $a_1, b_1, \ldots, c_1 d_1 c_1^{-1}, \ldots, c_s d_s c_s^{-1}$.

We equip $\mathcal{A}^{\text{flat}}(t_1, \ldots, t_s)/\Sigma_q G$ with the quotient topology.

PROPOSITION 5.28. *The map f induces an identification of compact spaces*

$$\mathcal{A}^{\text{flat}}(t_1, \ldots, t_s)/\Sigma_q G \simeq M\,. \tag{5.123}$$

PROOF. Clearly f descends to a one to one map $\mathcal{A}^{\text{flat}}(t_1, \ldots, t_s)/\Sigma_q G \simeq M$. This map is continuous. By [**18**, Proposition 2.2.3], f is a pointwise identification. The fact that the topologies coincide follow for instance from the techniques of [**18**, Section 4.2], in a much simpler context. □

Assume now that $(t_1, \ldots, t_s)$ verify (A). Then by Theorem 5.18, M is a smooth manifold, possibly empty if $g = 0$. Using the techniques of [**18**, Section 4.2], we find that $\mathcal{A}^{\text{flat}}(t_1, \ldots, t_s)$ is a smooth manifold. Note that here, we use explicitly the fact that for any $x \in M$, $\widehat{H}^2(\Sigma, E) = 0$. By [**18**, Proposition 4.2.23], $\mathcal{A}^{\text{flat}}(t_1, \ldots t_s)/\Sigma G$ is a smooth manifold. Also $M \to M/G$ is a G-orbifold. By the above, ΣG acts locally freely on $\mathcal{A}^{\text{flat}}(t_1, \ldots t_s)$.

PROPOSITION 5.29. *We have the identification of orbifolds,*

$$\mathcal{A}^{\text{flat}}(t_1, \ldots t_s)/\Sigma G \simeq M/G. \tag{5.124}$$

PROOF. By proceeding as in [**18**, Proposition 4.2.23], we find easily that the identification f in Proposition 5.28 is an identification of smooth manifolds. Our Proposition follows. □

We do no longer assume G to be simply connected.

Let $\langle , \rangle$ be a G-invariant bilinear symmetric form on $\mathfrak{g}$. Then E is equipped with the corresponding flat bilinear symmetric form. Recall that by Theorem 5.4, $\widehat{H}^1(\Sigma, E)$ is the image of $H^1((\Sigma, \partial\Sigma), E)$ into $H^1(\Sigma, E)$.

DEFINITION 5.30. If $x \in M, \alpha, \alpha' \in \widehat{H}^1_x(\Sigma, E)$, put

$$\omega_x(\alpha, \alpha') = \int_\Sigma -\langle \alpha, \alpha' \rangle. \tag{5.125}$$

Then ω_x is an intersection form, so that it is non degenerate.

THEOREM 5.31. *The 2-form ω is G-invariant. It descends to a symplectic 2-form on the orbifold M/G.*

PROOF. Suppose first that G is simply connected. By Proposition 5.28, the space $\mathcal{A}^{\text{flat}}(t_1, \ldots t_s)/\Sigma G$ is an orbifold. Then the results of Section 4.1 and Theorem 4.24 show that $\mathcal{A}^{\text{flat}}(t_1, \ldots t_s)/\Sigma G$ is a symplectic reduction of the symplectic manifold $\mathcal{A}$. Since it is a symplectic reduction of the symplectic affine space $\mathcal{A}$, it carries a symplectic form. One verifies easily that ω is just this form. Therefore ω is closed.

If G is not simply connected, the homotopy types of G-bundles have been classified in Section 4.7. Let $\widetilde{G}$ be the universal cover of G. By introducing an extra holonomy $h \in Z(\widetilde{G})$, we can then replace G by $\widetilde{G}$. The fact that ω is closed is now a consequence of the corresponding result for $\widetilde{G}$.

The proof of our Theorem is completed. □

Recall that G acts on the right and on the left on X. If $x \in M, a \in \Gamma$, then $x(a) \in G$. We define a right action of Γ on X by the formula

$$x.a = x(a).x,\ x \in M,\ a \in \Gamma \tag{5.126}$$

Then if $a \in \Gamma, g \in G, x \in M$,

$$(x.g).a = (x.g)(a).(x.g) = (x.a).g, \tag{5.127}$$

i.e. the actions of Γ and G on M commute.

Also $\Gamma \times G$ acts on the right on $\widehat{\Sigma}$ (of course, the factor G acts trivially on $\widehat{\Sigma}$). Therefore $\Gamma \times G$ acts on the right on $M \times \widehat{\Sigma}$. Also $\Gamma \times G$ acts locally freely on M, and Γ acts freely on $\widehat{\Sigma}$.

Let V be a complex vector space. Let $\rho : G \to \mathrm{Aut}(V)$ be a representation of G. Then ρ induces a representation $\Gamma \times G \to \mathrm{Aut}(V)$. Put

$$F = (M \times \widehat{\Sigma}) \times_{\Gamma \times G} V. \tag{5.128}$$

Then F is an orbifold vector bundle on $M/G \times \Sigma$. It is obtained via the identification

$$(x, \sigma, f) \simeq (x.g, \sigma.a, g^{-1}x(a)^{-1}f), (x, \sigma, f) \in M \times \widehat{\Sigma} \times V, (a, g) \in \Gamma \times G. \tag{5.129}$$

For a given $x \in M$, the restriction of F to the fibre Σ is exactly the flat vector bundle considered in Section 5.2.

If $\sigma \in \Sigma$, then $\sigma^* F$ is an orbifold vector bundle on M/G. If $\sigma, \sigma' \in \Sigma$, if $t \in [0,1] \mapsto \sigma_t \in \Sigma$ is a smooth path, with $\sigma_0 = \sigma, \sigma_1 = \sigma'$, parallel transport with respect to the flat connection identifies $\sigma^* F$ and $\sigma'^* F$.

Over Σ, there are two distinguished points p and q. Recall that G acts on $\mathfrak{g}$ by conjugation. When $V = \mathfrak{g}$, let E be the corresponding real vector bundle on $M/G \times \Sigma$.

DEFINITION 5.32. Let $\mathcal{E}$ be the vector bundle on M/G,

$$\mathcal{E} = p^* E. \tag{5.130}$$

Also G acts by conjugation on G and on the $\mathcal{O}_j$'s. Put

$$\widehat{G} = M \times_G G,\ \widehat{\mathcal{O}}_j = M \times_G \mathcal{O}_j. \tag{5.131}$$

Then $\widehat{G} \to M/G$ is a G-bundle, and $\widehat{\mathcal{O}}_j \to M/G$ is a $\mathcal{O}_j$-bundle. Also $u_1, \dots, v_g$ are sections of $\widehat{G}$ and $w_1, \dots, w_s$ are sections of $\widehat{\mathcal{O}}_1, \dots, \widehat{\mathcal{O}}_s$.

Elements of $\widehat{G}$ act naturally on $\mathcal{E}$. As explained in Section 5.2, u_1 can be considered as the parallel transport operator along the closed curve $a_1 \dots$

For $1 \le j \le s$, let $T_{w_j}\widehat{\mathcal{O}}_j/(M/G)$ be the relative tangent bundle to the fibre $\widehat{\mathcal{O}}_j$. As we saw in (5.48),

$$T_{w_j}\widehat{\mathcal{O}}_j/M = \mathrm{Im}(1 - w_j) \subset \mathcal{E}. \tag{5.132}$$

As explained in Theorem 5.14, on M/G, we have the bundle of complexes

$$0 \to \widehat{C}^{\Sigma,0}(E) \to \widehat{C}^{\Sigma,1}(E) \to \widehat{C}^{\Sigma,2}(E) \to 0, \tag{5.133}$$

which, by (5.53), (5.132) can also be written as

$$0 \to \mathcal{E} \to \mathcal{E}^{2g} \oplus (\bigoplus_{j=1}^{s} T_{w_j}(\widehat{O}_j/(M/G))) \to \mathcal{E} \to 0. \tag{5.134}$$

THEOREM 5.33. *On M/G, we have the identity*

$$TM/G = \mathcal{E}^{2g} \oplus (\bigoplus_{j=1}^{s} T_{w_j}\widehat{O}_j/(M/G)) \ominus \mathcal{E}^2 \text{ in } K(M/G). \tag{5.135}$$

PROOF. By (5.120),

$$TM/G = \widehat{H}^1(\Sigma, E). \tag{5.136}$$

Also over M, by (5.63),

$$\widehat{H}^j(\Sigma, E) = 0,\ j = 0, 2. \tag{5.137}$$

Hence, over M/G, by (5.134), (5.136), (5.137),

$$\widehat{H}^1(\Sigma, E) = \mathcal{E}^{2g} \oplus (\bigoplus_{j=1}^{s} T_{w_j}\widehat{O}_j/(M/G)) \ominus \mathcal{E}^2 \text{ in } K(M/G). \tag{5.138}$$

From (5.137), (5.138), we get (5.135). □

THEOREM 5.34. *If the orbits $O_1, \dots, O_s$ are very regular (resp. regular), then*

$$TM/G = \mathcal{E}^{2(g-2)+s} \ominus \mathbf{R}^{s \dim \mathfrak{t}} \text{ in } K(M/G) \text{ (resp. in } K(M/G) \otimes_{\mathbf{Q}} \mathbf{Z}.) \tag{5.139}$$

PROOF. Let $t \in T$ be very regular, let O be the orbit of t in G. Then $Z(t) = T$. If $g \in G$,

$$Z(g.t) = g.T,\ \mathfrak{z}(g.t) = g.\mathfrak{t}. \tag{5.140}$$

If $w \in O$, we have the splitting

$$\mathfrak{g} = \operatorname{Im}(1 - w) \oplus \ker(1 - w), \tag{5.141}$$

which corresponds to the splitting

$$\mathcal{E} = T_w O \oplus N_{O/\mathcal{E}}. \tag{5.142}$$

By (5.140), the vector bundle $\mathfrak{z}(w) = \ker(1 - w)$ is trivial on O. The action of G on the orbit O lifts to an action on the vector bundle $\mathfrak{z}(w)$. From (5.141), (5.142), we find that the normal bundle $N_{O/\mathcal{E}}$ is equivariantly trivial. From (5.135), (5.142), we get (5.139).

If $t \in T$ is only regular, then $\mathfrak{z}(t) = \mathfrak{t}$. Also $O = G/Z(t)$. Then if $w = g.t$, $g \in G \mapsto g.\mathfrak{z}(t) = \mathfrak{z}(w)$ induces a G-invariant flat connection on $\mathfrak{z}(w) \simeq N_{O/\mathfrak{g}}$. Consider the splitting of vector bundles

$$\mathcal{E} = T_{w_j}(\widehat{O}_j/(M/G)) \oplus \ker(1 - w_j). \tag{5.143}$$

If the orbit O_j is very regular, $\ker(1 - w_j)$ is a trivial vector bundle on M/G. Therefore

$$T_{w_j}\widehat{O}_j/(M/G) = \mathcal{E} \ominus \mathbf{R}^{\dim \mathfrak{t}} \text{ in } K(M/G). \tag{5.144}$$

If the orbit $\mathcal{O}_j$ is only regular, then one finds easily by the above that $\ker(1-w_j)$ is equipped over M/G with a flat connection. We thus get (5.139) in this case. The proof of our Theorem is completed. □

5.7. A metric on the determinant of TM/G. If λ is a line, let λ^{-1} be the dual line. If E is a vector space, set

$$\det(E) = \Lambda^{\max}(E)\,. \tag{5.145}$$

If $E = \bigoplus_{i=0}^{m} E_i$ is a $\mathbf{Z}$ graded vector space, set

$$\det(E) = \bigotimes_{i=0}^{m} (\det E_i)^{(-1)^i}\,. \tag{5.146}$$

Let G be a compact connected semisimple Lie group.

DEFINITION 5.35. If $x \in M$, let λ_x be the real line

$$\lambda_x = (\det \widehat{H}_x(\Sigma, E))^{-1}. \tag{5.147}$$

Let $\langle,\rangle$ be a G-invariant scalar product on $\mathfrak{g}$. Then E is equipped with a fibrewise flat scalar product. By Theorem 5.4 and Remark 5.5,

$$\widehat{H}^2(\Sigma, E) \simeq (\widehat{H}^0(\Sigma, E))^*,\ \widehat{H}^1(\Sigma, E) \simeq (\widehat{H}^1(\Sigma, E))^*. \tag{5.148}$$

Observe that in (5.148), the identifications depend explicitly on $\langle,\rangle$. By (5.148), for $x \in M$,

$$\lambda_x^2 \simeq \mathbf{R}. \tag{5.149}$$

Now $\mathbf{R}$ carries a canonical metric $\|\ \|_{\mathbf{R}}$ such that $\|1\|_{\mathbf{R}} = 1$.

DEFINITION 5.36. Let $\|\ \|_{\lambda_x}$ be the metric on λ_x such that the identification (5.149) is an isometry.

In the sequel, we assume that $(t_1, \dots, t_s)$ verify (A).

PROPOSITION 5.37. *If $x \in M$, the metric $\|\ \|_{\lambda_x}$ defines a volume element on $\Lambda^{\max} T_x(M/G)$, which is exactly the volume associated to the symplectic form ω_x.*

PROOF. If $x \in M$, then $\widehat{H}^0_x(\Sigma, E) = 0$, $\widehat{H}^1_x(\Sigma, E) = 0$, so that

$$\lambda_x = \det \widehat{H}^1_x(\Sigma, E)\,. \tag{5.150}$$

By (5.120),

$$T_x M/G = \widehat{H}^1_x(\Sigma, E)\,. \tag{5.151}$$

Finally in (5.148), the identification $\widehat{H}^1_x(\Sigma, E) \simeq (\widehat{H}^1_x(\Sigma, E))^*$ is done via the symplectic form ω_x. Our Proposition follows. □

Now recall that $\widehat{H}_x(\Sigma, E)$ is the cohomology of the complex $(\widehat{C}^{\Sigma}_x(E), \partial)$. Put

$$\widehat{\lambda}_x = (\det(\widehat{C}^{\Sigma}_x(E)))^{-1}. \tag{5.152}$$

Then by **[35]**, there is a canonical isomorphism

$$\lambda_x \simeq \widehat{\lambda}_x. \tag{5.153}$$

Also, since E is equipped with a flat scalar product, $\widehat{\lambda}_x$ is also naturally equipped with a metric $\|\ \|_{\widehat{\lambda}_x}$, which also depends on $\langle,\rangle$. Let $\|\ \|_{\widehat{\lambda_x}}$ be the corresponding metric on λ_x via the canonical isomorphism (5.153). .

DEFINITION 5.38. For $1 \leq j \leq s$, put

$$|\sigma_{\mathcal{O}_j}| = |\det(1-w_j)_{|\mathrm{Im}(1-w_j)}|^{1/2},\ w_j \in \mathcal{O}_j. \tag{5.154}$$

As the notation indicates, $|\sigma_{\mathcal{O}_j}|$ is a constant on $\mathcal{O}_j$. We make the convention that if $\mathrm{Im}(1-w_j) \subset \mathfrak{g}$ is reduced to 0,

$$|\sigma_{\mathcal{O}_j}| = 1 \tag{5.155}$$

REMARK 5.39. Recall that the function $\sigma : G \to \mathbf{C}$ was defined in (1.42). If $\mathcal{O}_j$ is regular, if $w_j \in \mathcal{O}_j$,

$$|\sigma_{\mathcal{O}_j}| = |\sigma(w_j)|. \tag{5.156}$$

The following result has been proved by Witten [**63**, Section 4], who exhibited the role of the Reidemeister torsion [**49**] in this context.

THEOREM 5.40. *For any $x \in M$,*

$$\|\cdot\|_{\widehat{\lambda_x}} = \prod_{j=1}^{s} |\sigma_{\mathcal{O}_j}| \|\cdot\|_{\lambda_x}. \tag{5.157}$$

PROOF. By (5.31),

$$\det(\widehat{C}^{\Sigma}(E)) \simeq \det(C^{\Sigma,r}(E)) \otimes (\det(K))^{-1}, \tag{5.158}$$

which can also be written in the form,

$$\det(\widehat{C}^{\Sigma}(E)) \simeq \det(C^{\Sigma,r}(E)) \otimes \det(H^0(\partial\Sigma, E)). \tag{5.159}$$

By (5.29),

$$H^0(\partial\Sigma, E) = \bigoplus_{j=1}^{s} \ker(1-w_j). \tag{5.160}$$

Let $\|.\|_{\det H^0(\partial\Sigma,E)}$ be the metric on $\det H^0(\partial\Sigma, E)$ induced by the scalar product of E. By (5.31) , we find that under the isomorphism (5.159),

$$\| \|_{\det \widehat{C}^{\Sigma}(E)} = \| \|_{\det C^{\Sigma,r}(E)} \| \|_{\det H^0(\partial\Sigma,E)} \left[\prod_{j=1}^{s} \det(1-w_j)_{|\mathrm{Im}(1-w_j)}\right]^{-1}. \tag{5.161}$$

Also by (5.21), there is a canonical isomorphism

$$\det C^{\Sigma}(E) \simeq \det C^{\Sigma,r}(E) \otimes \det C^{\partial\Sigma}(E). \tag{5.162}$$

If $\| \|_{\det C^{\Sigma}(E)}$ denote the obvious metric on $\det(C^{\Sigma}(E))$, by (5.21), (5.162), we get

$$\| \|_{\det C^{\Sigma}(E)} = \| \|_{\det C^{\Sigma,r}(E)} \| \|_{\det C^{\partial\Sigma}(E)}. \tag{5.163}$$

Also, we have a canonical isomorphism

$$\det(C^{\partial\Sigma}(E)) \simeq \det(H(\partial\Sigma, E)). \tag{5.164}$$

By Poincaré duality,

$$H^1(\partial\Sigma, E) = (H^0(\partial\Sigma, E))^*, \tag{5.165}$$

so that by (5.164), (5.165),

$$\det(C^{\partial\Sigma}(E)) \simeq \det(H^0(\partial\Sigma, E))^2. \tag{5.166}$$

By (5.25), (5.166),

$$\| \, \|_{\det(C^{\theta\Sigma}(E))} = \| \, \|^2_{\det H^0(\partial\Sigma,E)} \left[\prod_{j=1}^{s} \det(1-w_j)_{|\mathrm{Im}(1-w_j)} \right]^{-1}. \tag{5.167}$$

Using (5.159)-(5.166),we get

$$(\det \widehat{C}^{\Sigma}(E))^2 \simeq \det C^{\Sigma}(E) \otimes \det C^{\Sigma,r}(E). \tag{5.168}$$

Also under the isomorphism (5.168),

$$\| \, \|^2_{\det \widehat{C}^{\Sigma}(E)} = \| \, \|_{\det C^{\Sigma}(E)} \| \, \|_{\det C^{\Sigma,r}(E)} \left[\prod_{j=1}^{s} \det(1-w_j)_{|\mathrm{Im}(1-w_j)} \right]^{-1}. \tag{5.169}$$

Now recall that by [**35**], there are canonical isomorphisms

$$\begin{aligned} \det(C^{\Sigma}(E)) &\simeq \det(H(\Sigma,E)), \\ \det(C^{\Sigma,r}(E)) &\simeq \det(H((\Sigma,\partial\Sigma),E). \end{aligned} \tag{5.170}$$

By Poincaré duality,

$$\det H(\Sigma,E) \otimes \det H((\Sigma,\partial\Sigma),E) \simeq \mathbf{R}. \tag{5.171}$$

So by (5.170), (5.171),

$$\det C^{\Sigma}(E) \otimes \det C^{\Sigma,r}(E) \simeq \mathbf{R}. \tag{5.172}$$

Let $\| \, \|_{\mathbf{R}}$ be the trivial metric on $\mathbf{R}$, such that $\|1\|_{\mathbf{R}} = 1$. We claim that under the Poincaré duality isomorphism (5.172),

$$\| \, \|_{\det C^{\Sigma}(E)} \| \, \|_{\det C^{\Sigma,r}(E)} = \| \, \|_{\mathbf{R}}. \tag{5.173}$$

In fact (5.173) is a simple consequence of the existence of the Reidemeister torsion [**49**], or more precisely of the Reidemeister metric [**13**, Section 1] on $\det H(\Sigma,E))$ and $\det H((\Sigma,\partial\Sigma),E)$. In fact given any cell decomposition of the manifold with boundary Σ, one can construct Reidemeister metrics on $\det H(\Sigma,E)$ and $\det H((\Sigma,\partial\Sigma),E)$ by the procedure indicated in [**13**]. The basic fact is that these metrics do not depend on the choice of the cell decomposition. By applying this argument to a cell decomposition and the corresponding dual decomposition, one deduces immediately that the Reidemeister metrics on $\det H(\Sigma,E)$ and $\det H((\Sigma,\partial\Sigma),E)$ correspond by Poincaré duality. As a consequence, (5.173) follows.

From (5.169), (5.173), we get (5.157). The proof of our Theorem is completed. □

From (5.156), (5.157), it follows that if $\mathcal{O}_1,\dots,\mathcal{O}_s$ are regular, if $x \in B$,

$$\| \, \|^{\widehat{}}_{\lambda_x} = \prod_{j=1}^{s} |\sigma(w_j)| \| \, \|_{\lambda_x}. \tag{5.174}$$

5.8. The Witten formula for the symplectic volume distribution. Let G be a compact connected semisimple Lie group. Let $\langle\ ,\ \rangle$ be a G-invariant scalar product on $\mathfrak{g}$. Let dg be the Haar measure on G associated to $\langle\ ,\ \rangle$. We us otherwise the same notation as in Section 5.7. In particular $T \subset G$ is a maximal torus in G.

Let $\mathcal{O}_1, \dots, \mathcal{O}_s$ be s adjoint orbits of G in G. Take $w_j \in \mathcal{O}_j$. Let $Z(w_j)$ be the centralizer of w_j. Then the map

$$a \in G/Z(w_j) \mapsto a \cdot w_j \in \mathcal{O}_j \tag{5.175}$$

is an identification of smooth manifolds.

The scalar product $\langle\ ,\ \rangle$ induces a volume form $dv_{\mathcal{O}_j}$ on $\mathcal{O}_j \subset G$.

DEFINITION 5.41. Let dv_X be the volume form on $X = G^{2g} \times \prod_{j=1}^{s} \mathcal{O}_j$,

$$dv_X(u_1, v_1, \dots, u_g, v_g, w_1, \dots, w_s) = \prod_{i=1}^{2g} du_i dv_i \prod_{j=1}^{s} dv_{\mathcal{O}_j}(w_j). \tag{5.176}$$

Let dt be the Lebesgue measure on $\mathfrak{t}$ induced by $\langle,\rangle$.

DEFINITION 5.42. If $g \in G$, let $X_{\mathcal{O}_g} \subset G^{2g} \times \prod_{j=1}^{s} \mathcal{O}_j \times \mathcal{O}_g$ be the set M in (5.59) associated to the orbits $\mathcal{O}_1, \dots, \mathcal{O}_s, \mathcal{O}_g$.

More precisely

$$X_{\mathcal{O}_g} = \Big\{(u_1, v_1, \dots, w_1, \dots, w_s, w) \in G^{2g} \times \prod_{j=1}^{s} \mathcal{O}_j \times \mathcal{O}_g, \tag{5.177}$$

$$\prod_{i=1}^{g}[u_i, v_i] \prod_{j=1}^{s} w_j w = 1\Big\}. \tag{5.178}$$

Clearly, $X_{\mathcal{O}_g}$ can be identified with $\{x \in X,\ \phi(x) \in \mathcal{O}_{g^{-1}}\}$. Set

$$X_g = \{x \in X; \phi(x) = g^{-1}\}. \tag{5.179}$$

Now G acts naturally on $X_{\mathcal{O}_g}$, and $Z(g)$ acts on X_g. Then one has the obvious

$$X_{\mathcal{O}_g}/G \simeq X_g/Z(g). \tag{5.180}$$

Clearly, if $x \in X$, $\int_G f(x \cdot g) dg$ depends only on $\pi x \in X/G$.

Let G_{vreg} (resp. G_{reg}) be the set of very regular (resp. regular) elements in G. If $g \in G_{\text{vreg}}$, let dt_g be the Haar measure on the maximal torus $Z(g)$, which is associated to $\langle\ ,\ \rangle$.

By Sard's theorem, a.e. every $g \in G$ is a regular value of ϕ. For such a $g \in G$, $X_{\mathcal{O}_g}$ is smooth, and G acts locally freely on $X_{\mathcal{O}_g}$. Also by Theorem 5.31, the orbifold $X_{\mathcal{O}_g}/G$ is equipped with a symplectic 2 form ω_g. Let $dv_{X_{\mathcal{O}_g}/G}$ be the corresponding volume element on $X_{\mathcal{O}_g}/G$. Since $X_{\mathcal{O}_g}/G = X_g/Z(g)$, let $dv_{X_g/Z(g)}$ be the associated volume on $X_g/Z(g)$. Observe here that $dv_{X_g/G}$ is unsensitive to orientation. In particular the integral of a nonnegative function with respect to $dv_{X_{\mathcal{O}_g}/G}$ is nonnegative.

In the sequel, we make the following assumptions:

- G is simply connected, and one of the following assumptions is verified :
  - $g \geq 2$.
  - $g \geq 1$, $s \geq 1$ and at least one of the t_j's is regular.
  - $g = 0$, $s \geq 3$, and at least 3 of the t_j's are regular.

or

- G is connected, the t_j's are very regular , and either
  - $g \geq 2$.
  - $g \geq 1,\ s \geq 1$.
  - $g = 0,\ s \geq 3$.

Now we will get an analogue of Theorem 3.10.

THEOREM 5.43. *Let $f : X \to \mathbf{R}$ be a bounded measurable function. Then*

$$
\begin{aligned}
(5.181)\qquad \int_X f(x)dv_X(x) &= \frac{\prod_{j=1}^s |\sigma_{\mathcal{O}_j}|}{|Z(G)|}\int_{T/W}|\sigma(t)|dt\int_{X_t/T}dv_{X_t/T}(x)\\
&\qquad \int_G f(x\cdot g)dg\,,\\
\int_M f(x)dv_X(x) &= \frac{\prod_{j=1}^s |\sigma_{\mathcal{O}_j}|}{|Z(G)|}\int_G \frac{dg}{|\sigma(g)|}\int_{X_g/Z(g)}dv_{X_g/Z(g)}(x)\\
&\qquad \int_{Z(g)} f(x\cdot t)dt_g\,.
\end{aligned}
$$

PROOF. By Theorem 5.21, we know that a.e., $Z'(x) = 1$. By Theorem 5.12, for a.e. $x \in X, d\phi(x)$ is surjective. Therefore the image $\phi_* dv_X$ of dv_X by ϕ is absolutely continuous with respect to dg. Also on the set $\{x \in X,\ \phi$ is regular at $x,\ Z'(x) = 1\}$, which has full measure and is stable by G, we can use the implicit function theorem, and also integrate along the fibres of the action of G , which are diffeomorphic to G.

Let $p : G_{\mathrm{vreg}} \to T_{\mathrm{vreg}}/W$ be the obvious projection. Since $\phi_* dv_X$ is absolutely continuous with respect to dg, $\phi^{-1}(G_{\mathrm{vreg}})$ is an open set in X, whose complement is dv_X negligible. Then $x \in \phi^{-1}(G_{\mathrm{vreg}}) \mapsto p\phi(x) \in T_{\mathrm{vreg}}/W$ is a smooth map. Also if $x \in \phi^{-1}(G_{\mathrm{vreg}}),\ g \in G$,

$$(5.182)\qquad p\phi(x\cdot g) = p\phi(x)\,.$$

By (5.182), if $x \in \phi^{-1}(G_{\mathrm{vreg}})$, the obvious analogue of (D,δ) in (5.53) is the complex

$$(5.183)\qquad (C'_x,\partial) : 0 \longrightarrow \mathfrak{g} \xrightarrow[\delta]{} \mathfrak{g}^{2g} \oplus \bigoplus_{j=1}^{s} \mathrm{Im}(1-w_j) \xrightarrow[d(p\phi)_*]{} \mathfrak{t} \longrightarrow 0\,.$$

By Proposition 5.10 and Theorem 5.12, if $x \in G$ is such that $Z'(x) = 1$, the cohomology of the complex (5.183) is concentrated in degree 1, and the first cohomology group of (5.183) is isomorphic to $T_{\pi(x)}X_{p\phi(x)}/T$. In particular,

$$(5.184)\qquad (\det C'_x)^{-1} \simeq \det(T_{\pi(x)}X_{p\phi(x)}/T)\,.$$

Now the complex C'_x is equipped with a scalar product. Let $\|\ \|_{\det(C'_x)}$ be the obvious metric on $\det(C'_x)$, and let $\|\ \|_{\det(T_{\pi(x)}X_{p\phi(x)}/T)}$ be the corresponding metric on $\det(T_{\pi(x)}X_{p\phi(x)}/T)$. Then $\|\ \|_{\det(T_{\pi(x)}X_{p\phi(x)}/T)}$ defines a smooth volume form on $(X_{p\phi(x)}/T)_{\mathrm{reg}}$, which will be denoted $\widetilde{dv}_{X_{p\phi(x)}/T}$. Then the formula of change of variable asserts that

$$(5.185)\qquad \int_X f(x)dv_X(x) = \frac{1}{|Z(G)|}\int_{T/W}dt\int_{X_t/T}\widetilde{dv}_{X_t/T}(x)\int_G f(x\cdot g)dg\,.$$

If $x \in \phi^{-1}(T_{\mathrm{vreg}})$, put $t = [\phi(x)]^{-1} \in T_{\mathrm{vreg}}$. Then we have the double complex

(5.186)
$$\begin{array}{ccccccccc}
 & & 0 & & 0 & & 0 & & \\
 & & \uparrow & & \uparrow & & \uparrow & & \\
0 & \longrightarrow & \mathfrak{g} & \longrightarrow & \mathfrak{g}^{2g} \oplus \bigoplus_{j=1}^{s} \mathrm{Im}(1-w_j) & \xrightarrow{d(p\phi)} & \mathfrak{t} & \longrightarrow & 0 \\
 & & \uparrow & & \uparrow & & \uparrow & & \\
0 & \longrightarrow & \mathfrak{g} & \longrightarrow & \mathfrak{g}^{2g} \oplus \bigoplus_{j=1}^{s} \mathrm{Im}(1-w_j) \oplus \mathrm{Im}(1-t) & \longrightarrow & \mathfrak{g} & \longrightarrow & 0 \\
 & & \uparrow & & \uparrow & & \uparrow & & \\
0 & \longrightarrow & 0 & \longrightarrow & \mathrm{Im}(1-t) & \xrightarrow{t^{-1}} & \mathrm{Im}(1-t) & \longrightarrow & 0 \\
 & & \uparrow & & \uparrow & & \uparrow & & \\
 & & 0 & & 0 & & 0 & &
\end{array}$$

Let $x \in \Sigma \backslash \partial\Sigma$, and let Δ be a small disk of center x. Put $\Sigma_{+1} = \Sigma \backslash \Delta$. In (5.186), starting from below, the first row is trivial. The second row is just the complex $(\widehat{C}^{\Sigma_{+1}}_{(x,t)}(E), \partial)$ we constructed in (5.31) (with s replaced by $s+1$). The third row is the complex (C'_x, ∂). Also the columns are acyclic.

Let us now explain why the diagram commutes. By (5.34), if $x = (u_1, v_1, \dots, w_s)$,

$$\begin{aligned}
\partial(f_1, \dots, f_{2g+s+1}) = {} & \sum_{i=1}^{g} ([u_1, v_1] \dots [u_{i-1}, v_{i-1}](1 - u_i v_i u_i^{-1}) f_{2i-1} \\
& + u_i(1 - v_i u_i v_i^{-1}) f_{2i}) + \sum_{j=1}^{s} [u_1, v_1] \dots [u_g, v_g] w_1 \dots w_{j-1} f_{2g+j} + \\
& \prod_{i=1}^{g} [u_i, v_i] \prod_{j=1}^{s} w_j f_{2g+s+1} \,.
\end{aligned} \tag{5.187}$$

By construction,

$$\prod_{i=1}^{g} [u_i, v_i] \prod_{j=1}^{s} w_j = t^{-1} \,. \tag{5.188}$$

So in (5.187),

$$\partial(f_{2g+s+1}) = t^{-1} f_{2g+s+1} \,. \tag{5.189}$$

So from (5.189), we find the diagram (5.186) commutes.

We equip the vector spaces which appear in (5.186) with the scalar products induced by the scalar product of $\mathfrak{g}$. The determinants of the columns of (5.186) are canonically trivial, and the norms of the associated canonical sections are equal to 1. Also since the first row is acyclic, the cohomology groups of the second and third rows are canonically identified, using the obvious long exact sequence. Finally the first row is acyclic, and the norm of the canonical section of the determinant of the first row is equal to 1.

By an obvious extension of [**11**, Theorem 1.10], it follows that if $x \in \phi^{-1}(T_{\mathrm{vreg}})$ is regular, then the metric on $\det(\widehat{H}^1(\Sigma_{+1}, E))$ (where $\widehat{H}^1(\Sigma_{+1}, E)$ is the first cohomology group of the complexes (C'_x, ∂) or $(\widehat{C}^{\Sigma_{+1}}_{(x,t)}(E), \partial))$ induced by the complexes (C'_x, ∂) or $(\widehat{C}^{\Sigma_{+1}}_{(x,t)}(E), \partial)$ are identical. Using now (5.156) and Theorem 5.40, we get

$$\widetilde{dv}_{X_t/T} = \prod_{j=1}^{s} |\sigma_{O_j}||\sigma(t)| dv_{X_t/T}\,. \tag{5.190}$$

By (5.185), (5.190), we obtain the first equality in (5.181).

The scalar product $\langle\ ,\ \rangle$ induces a scalar product on $\mathfrak{g}/\mathfrak{t} = \mathfrak{t}^{\perp}$. Let $d\dot{g}$ be the corresponding volume element on T/G. Tautologically if $k : G \to \mathbf{R}$ is bounded and measurable

$$\int_G k(g)dg = \int_{T\backslash G} d\dot{g} \int_T k(tg)dt\,. \tag{5.191}$$

Also Weyl's integration formula [**15**, Theorem IV.1.11] asserts that

$$\int_G k(g)dg = \frac{1}{\mathrm{Vol}(T)} \int_{T/W} |\sigma|^2(t)dt \int_G f(t\cdot g)dg\,. \tag{5.192}$$

Using the first identity in (5.181) and (5.191), (5.192), we obtain

$$\begin{aligned}\int_X f(x)dv_X(x) = \frac{\prod_{j=1}^{s}|\sigma_{O_j}|}{|Z(G)|}\int_{T/W}|\sigma(t)|dt \\ \int_{X_t/T} dv_{X_t/T}(\dot{x}) \int_{T\backslash G} d\dot{g} \int_T f(x\cdot t' g)dt' \\ = \frac{\prod_{j=1}^{s}|\sigma_{O_j}|}{|Z(G)|}\int_G \frac{dg}{|\sigma(g)|}\int_{X_g/Z(g)} dv_{X_g/Z(g)}(x)\int_{Z(g)} f(xt)dt_g\,.\end{aligned} \tag{5.193}$$

The proof of our Theorem is completed. □

Definition 5.44. If $t \in T$, let $|V(t)|$ be the absolute value of volume of $X_t/Z(t)$ with respect to ω_t.

Then $V(t)$ is a W-invariant function on T, which extends to a central function on G.

Now we prove a result which was already essentially proved by Liu [**39, 40**].

Theorem 5.45. *Let $f : G \to \mathbf{R}$ be a bounded measurable function. Then*

$$\int_X f(\phi^{-1}(x))dv_X(x) = \frac{\prod_{j=1}^{s}|\sigma_{O_j}|}{|Z(G)|}\int_{T/W}\Big[\int_G f(g\cdot t)dg\Big]|\sigma(t)||V(t)|dt\,, \tag{5.194}$$

$$\int_X f(\phi^{-1}(x))dv_X(x) = \frac{\mathrm{Vol}(T)\prod_{j=1}^{s}|\sigma_{O_j}|}{|Z(G)|}\int_G \frac{f(g)|V(g)|}{|\sigma(g)|}dg\,.$$

Proof. Our formula follows from Theorem 5.43. □

Remark 5.46. In [**39, 40**], Liu uses an essentially similar argument in his proof of the Witten formula [**63, 64**]. In fact let $p_t(g)$ be the convolution heat kernel on G. Liu considers the quantity $\int_X p_t(\phi^{-1}(x))dv_X$, and following Witten [**63**], he studies its limit as $t \to 0$. The arguments he uses to evaluate the limit as being (up to a normalization constant) the absolute value of the symplectic volume of M/G are essentially the ones which are used in the proofs of Theorems 5.43 and 5.45.

Let A be the set of regular values of ϕ^{-1} in G. Then A is an open dense set, such that ${}^cA$ is negligible. Also on A, $\phi_* dv_X$ has a smooth density with respect to dg. By Theorem 5.45, it follows that $\frac{|V(g)|}{|\sigma(g)|}$ is smooth on $A \cap G_{\text{vreg}}$.

Assume temporarily that $(t_1, \ldots, t_s)$ verify (A). By Theorem 5.18, 1 is a regular value of ϕ. By Theorem 5.20, $Z'(x) = 1$ a.e. on $M = X_1$. Using Theorem 5.45, and proceeding as in the proof of Theorem 3.13, we get

$$\lim_{\substack{g \in G_{\text{vreg}} \\ g \to 1}} \frac{|V(g)|}{|\sigma(g)|} = \frac{\text{Vol}(G)}{\text{Vol}(T)} V(1). \tag{5.195}$$

Using (5.195) we find that

$$\lim_{t \to 0} \int_X p_t(\phi^{-1}(x)) dv_X = \frac{\text{Vol}(G) \prod_{j=1}^s |\sigma_{\mathcal{O}_j}|}{|Z(G)|} V(1), \tag{5.196}$$

which is a formula obtained by Liu [**39, 40**] by arguments essentially similar to the ones we gave in our proof of Theorems 5.43 and 5.45.

Let K be a positive Weyl chamber in $\mathfrak{t}$. Let $\Lambda_+ = \Lambda \cap \overline{K}$ be the set of non-negative weights. Then the irreducible representations of G are indexed by Λ_+. If $\lambda \in \Lambda_+$, let χ_λ be the character of the corresponding representation of G.

By Theorem 5.45, $\frac{|V(g)|}{|\sigma(g)|}$ defines a L_1 G-invariant function on G. Therefore, it defines an invariant distribution on G, which can be expanded as a linear combination of the characters χ_λ of G.

Let $w_j \in \mathcal{O}_j$. Then $\mathcal{O}_j \simeq G/Z(w_j)$. Also $\text{Vol}(Z(w_j))$ does not depend on the choice of w_j. Finally χ_λ takes the complex value $\chi_\lambda(t_j)$ on the orbit $\mathcal{O}_j$.

Now we prove a result of Witten [**63, 64**].

THEOREM 5.47. *The following identity of G-invariant distributions holds on G,*

$$\frac{|V(g)|}{|\sigma(g)| \prod_{j=1}^s |\sigma_{\mathcal{O}_j}|} = |Z(G)| \frac{\text{Vol}(G)^{2g+s-1}}{\text{Vol}(T) \prod_{j=1}^s \text{Vol}(Z(t_j))} \sum_{\lambda \in \Lambda_+} \frac{\prod_{j=1}^s \chi_\lambda(t_j) \chi_\lambda(g)}{\chi_\lambda(1)^{2g+s-1}}. \tag{5.197}$$

PROOF. By Theorem5.45, we get

$$\frac{1}{\text{Vol}(G)} \int_G \overline{\chi}_\lambda(g) \frac{|V(g)|}{|\sigma(g)|} dg = \frac{|Z(G)|}{\text{Vol}(G)\text{Vol}(T)} \frac{1}{\prod_{j=1}^s |\sigma_{\mathcal{O}_j}|} \int_X \chi_\lambda(\phi(x)) dv_X(x). \tag{5.198}$$

To evaluate the integral in the right-hand side of (5.198), we follow Witten [**63**]. By [**15**, Theorem II.4.5], if $a, b \in G$,

$$\int_G \chi_\lambda(agbg^{-1}) dg = \frac{\text{Vol}(G)}{\chi_\lambda(1)} \chi_\lambda(a) \chi_\lambda(b), \tag{5.199}$$

$$\int_G \chi_\lambda(ag) \chi_{\lambda'}(g^{-1}b) dg = \text{Vol}(G) \delta_{\lambda, \lambda'} \frac{\chi_\lambda(ab)}{\chi_\lambda(1)}.$$

Finally, if $h : \mathcal{O}_j \to \mathbf{R}$ is a bounded measurable function, we have the easy formula

$$\int_{\mathcal{O}_j} h(g)dv_{\mathcal{O}_j}(g) = \frac{|\sigma_{\mathcal{O}_i}|^2}{\mathrm{Vol}(Z(t_j))}\int_G h(gt_jg^{-1})dg\,. \tag{5.200}$$

By (5.199), (5.200), we obtain

$$\int_X \chi_\lambda(\phi(x))dv_X(x) = \frac{\mathrm{Vol}(G)^{2g+s}\prod_{j=1}^s|\sigma_{\mathcal{O}_j}|^2}{\prod_{j=1}^s\mathrm{Vol}(Z(t_j))}\frac{\chi_\lambda(t_1)\dots\chi_\lambda(t_s)}{\chi_\lambda^{2g+s-1}(1)}\,. \tag{5.201}$$

So by (5.198), (5.201), we get

$$\begin{aligned}&\frac{1}{\mathrm{Vol}(G)}\int_G \overline{\chi}_\lambda(g)\frac{|V(g)|}{|\sigma(g)|}dg \\ &= |Z(G)|\frac{\mathrm{Vol}(G)^{2g+s-1}}{\mathrm{Vol}(T)}\prod_{j=1}^s\frac{|\sigma_{\mathcal{O}_j}|}{\mathrm{Vol}(Z(t_j))}\chi_\lambda(t_j)\frac{1}{\chi_\lambda(1)^{2g+s-1}}\,.\end{aligned} \tag{5.202}$$

Now (5.202) is exactly the λ-Fourier coefficient of the invariant distribution $\frac{|V(g)|}{|\sigma(g)|}$. The proof of our Theorem is completed. □

Clearly the distribution $\frac{|V(g)|}{|\sigma(g)|}$ in C^∞ at the regular values of the function ϕ. Now we will make a crude analysis of the Sobolev regularity of $|V(g)|$.

PROPOSITION 5.48. *If $p < 2g - 1 - \dim(\mathfrak{t})/2$, the invariant distribution in the right-hand side of (5.197) lies in $H^p(G)$. If $2g - 2 - \dim(\mathfrak{t}) > 0$, this distribution is continuous.*

PROOF. Let ρ be the half-sum of the positive roots. By Weyl's dimension formula [**15**, Theorem VI.1.7], if R_+ is the set of positive roots,

$$\chi_\lambda(1) = \prod_{\alpha\in R_+}\frac{\langle\lambda+\rho,\alpha\rangle}{\langle\rho,\alpha\rangle}\,. \tag{5.203}$$

By [**15**, Proposition V.4.12], if $\alpha \in R_+$, $\langle\rho,\alpha\rangle > 0$. Therefore there is $c > 0$ such that if $\alpha \in R_+$, $\lambda \in \Lambda_+$,

$$\langle\lambda+\rho,\alpha\rangle \geq \sup(c, \langle\lambda,\alpha\rangle)\,. \tag{5.204}$$

By (5.203), (5.204), we find that since the $\alpha \in R_+$ form a basis of $\mathfrak{t}^*$, there is $C > 0$ such that

$$\chi_\lambda(1) \geq C\|\lambda\|\,. \tag{5.205}$$

Also

$$|\chi_\lambda| \leq \chi_\lambda(1)\,. \tag{5.206}$$

Let Δ_G be the Casimir operator on G [**47**, Section 9.4]. By [**47**, Proposition 9.4.2],

$$\Delta_G\chi_\lambda = \frac{1}{2}(\|\lambda+\rho\|^2 - \|\rho\|^2)\chi_L\,. \tag{5.207}$$

Using (5.207), we get

$$(\Delta^G)^{p/2}\sum_{\lambda\in\Lambda_+}\frac{\prod_{j=1}^{s}\chi_\lambda(t_j)\chi_\lambda(g)}{(\chi_\lambda(1))^{2g+s-1}} = \frac{1}{2^{p/2}}\sum_{\lambda\in\Lambda_+}\frac{(\|\lambda+\rho\|^2-\|\rho\|^2)^{p/2}}{\chi_\lambda(1)^{2g+s-1}}\prod_{j=1}^{s}\chi_\lambda(t_j)\chi_\lambda(g). \tag{5.208}$$

By (5.205), (5.206),

$$\frac{|(\|\lambda+\rho\|^2-\|\rho\|^2)^{p/2}\prod_1^s\chi_\lambda(t_j)|}{\chi_\lambda(1)^{2g+s-1}} \leq C(1+|\lambda|)^{-(2g-1-p)}. \tag{5.209}$$

Also

$$\int_{\mathfrak{t}}\frac{d\lambda}{(1+|\lambda|)^{2(2g-1-p)}} < +\infty \tag{5.210}$$

if and only if $p < 2g-1-\frac{\dim(\mathfrak{t})}{2}$. From (5.209), (5.210), we get the first part of our Proposition. Also,

$$\int_{\mathfrak{t}}\frac{d\lambda}{(1+|\lambda|)^{2g-2}} < +\infty \tag{5.211}$$

if and only if $2g-2-\dim(\mathfrak{t}) > 0$. By (5.210), we obtain the second part of our Proposition. □

5.9. Logarithms. Let U be a nonempty open set in G, stable by the adjoint action of G. We assume that there is a well-defined logarithm $\log : U \to \mathfrak{g}$, i.e. a smooth function $U \to \mathfrak{g}$ such that if $g \in U$, $g' \in G$,

$$\begin{aligned} g &= \exp(\log(g)), \\ \log(g'gg'^{-1}) &= g'\cdot\log(g). \end{aligned} \tag{5.212}$$

In particular by (5.212), $\log(g)$ is $Z(g)$-invariant.

EXAMPLE 5.49. If U is a small ad-invariant open neighborhood of a central element in G, a logarithm is well-defined on U.

EXAMPLE 5.50. Let $\mathcal{O} \subset G$ be a very regular orbit. Then $\mathcal{O}\cap T$ consists of $|W|$ distinct elements. Let $t \in \mathcal{O}\cap T$, and let $h \in \mathfrak{t}$ such that $\exp(h) = t$. Then $t \mapsto h$ extends into a well defined logarithm $\mathcal{O} \to \mathcal{G}$.

EXAMPLE 5.51. Suppose that G is simply connected. Let K be a Weyl Chamber in $\mathfrak{t}$, let P be the alcove in K whose closure contains 0. Then by [**15**, Proposition V.7.10]

$$W\backslash T_{\mathrm{reg}} = P. \tag{5.213}$$

Also by [**15**, Proposition V.7.11], the map

$$(g,t) \in G/T\times P \mapsto g\exp(t)g^{-1} \in G_{\mathrm{reg}} \tag{5.214}$$

is one to one. Let $\log : G_{\mathrm{reg}} \to \mathfrak{g}$ be the G-invariant function on G_{reg} such that if $t \in P$,

$$\log(\exp(t)) = t \in P. \tag{5.215}$$

Then $\log : G_{\mathrm{reg}} \to \mathfrak{g}$ is a logarithm.

We still assume that G is simply connected and simple. Let $u \in C/\overline{R}^*$. Let $Z(u) \subset G$ be the centralizer of u. Let $Z(u)_{\mathrm{vreg}}$ be the set of very regular elements in $Z(u)$. Clearly

$$Z(u) \cap G_{\mathrm{reg}} \subset Z(u)_{\mathrm{vreg}} . \tag{5.216}$$

Also the function log maps $Z(u) \cap G_{\mathrm{reg}}$ into $\mathfrak{z}(u)$, and gives a logarithm on $Z(u)_{\mathrm{vreg}}$.

Let $\mathcal{B}_U$ be the set of connections on the trivial G bundle over S_1, whose holonomy w lies in U. Needless to say, in a given trivialization, any element of $\mathcal{B}_u$ can be written in the form

$$\frac{D}{Dt} = \frac{d}{dt} + a_t \, , \quad a_t \in L\mathfrak{g}_t . \tag{5.217}$$

Let τ_t^0 be the parallel transport operator along $s \in [0,t]$, so that $w = \tau_1^0$. Set

$$w_t = \tau_t^0 \tau_1^0 (\tau_t^0)^{-1} . \tag{5.218}$$

Then w_t is just w under translation of the origin in S_1 by t.

Clearly $\log(w_t)$ is well-defined, and

$$\log(w_t) = \tau_t^0 \log(w) (\tau_t^0)^{-1} . \tag{5.219}$$

Also

$$\frac{D}{Dt} \log(w_t) = 0 . \tag{5.220}$$

By (5.219),

$$e^{t \log(w_t)} = \tau_t^0 e^{t \log(w)} (\tau_t^0)^{-1} . \tag{5.221}$$

Put

$$\frac{{}^0 D}{Dt} = e^{-t \log(w_t)} \frac{D}{Dt} e^{t \log(w_t)} . \tag{5.222}$$

By (5.220), (5.222),

$$\frac{{}^0 D}{Dt} = \frac{D}{Dt} + \log(w_t) . \tag{5.223}$$

From (5.220), (5.223), we get

$$\frac{{}^0 D}{Dt} \log(w_t) = 0 . \tag{5.224}$$

The parallel transport operator ${}^0\tau_t^0$ for $\frac{{}^0 D}{Dt}$ is given by

$${}^0\tau_t^0 = e^{-t \log(w_t)} \tau_t^0 , \tag{5.225}$$

so that

$${}^0\tau_1^0 = 1 . \tag{5.226}$$

By (5.226), the parallel transport trivialization with respect to $\frac{{}^0 D}{Dt}$ is globally defined on S^1, and in this trivialization,

$$\frac{{}^0 D}{Dt} = \frac{d}{dt} . \tag{5.227}$$

5.10. A symplectic structure on an open set in X. In the sequel, we assume that the assumptions before Theorem 5.43 are in force, and also that $(t_1,\dots,t_s)$ verify (A).

Let $U_1,\dots,U_s$ be ad-invariant open subsets of G on which a logarithm is well defined. Then G acts on the right on $G^{2g}\times\prod_{j=1}^s U_j$.

Recall that $x\in G^{2g+s}\mapsto\phi(x)\in G$ was defined in (5.41). Let $x\in G^{2g}\times\prod_{j=1}^s U_j$ be such that $\phi(x)=1$, and that G acts locally freely at x. Let U be an open neighbhorhood of 1 in G, such that a logarithm is still well-defined on U.

Let $\mathcal{V}$ be an open neighbhorhood of x in $G^{2g}\times\prod_{j=1}^s U_j$ such that $h(\mathcal{V})\subset U$. Then we can find an open neighborhood $A\subset G^{2g}\times\prod_{j=1}^s U_j$ of x, which is G-invariant, such that $h(A)\subset U$, and on which G acts locally freely.

Let $\phi_{+1}:G^{2g+s+1}\to G$ be given as in (5.41), with s replaced by $s+1$. Set

$$A'=\{x'\in G^{2g+s+1};\phi_{+1}(x')=1\}. \tag{5.228}$$

Clearly $x\in G^{2g+s}\mapsto(x,\phi^{-1}(x))\in A'$ is a one to one G-equivariant map. Let $\widetilde{A}\subset A'$ be the image of A by this map.

Let Σ_{+1} be the Riemann surface Σ with $s+1$ small disks deleted. Equivalently Σ_{+1} is obtained from Σ by deleting an extra small disk. Let $\widehat{\Sigma}_{+1}$ be the universal cover of Σ_{+1}. Put $\Gamma_{+1}=\pi_1(\Sigma_{+1})$. Then Γ_{+1} is generated by $u_1,v_1,\dots,u_g,v_g,w_1,\dots,w_{s+1}$, and the relation

$$\prod_{i=1}^{g}[u_i,v_i]\prod_{j=1}^{s+1}w_j=1. \tag{5.229}$$

As in (5.126), we find that Γ_{+1} and G both act on the right on $\widetilde{A}$, and these actions commute. Also $\Gamma_{+1}\times G$ acts on the right on $\widehat{\Sigma}_{+1}$ (and the action of G on $\widehat{\Sigma}_{+1}$ is trivial). Then $\Gamma_{+1}\times G$ acts on the right on $\widetilde{A}\times\widehat{\Sigma}_{+1}$. Since Γ_{+1} acts freely on $\widehat{\Sigma}_{+1}$, $\Gamma_{+1}\times G$ acts locally freely on $\widetilde{A}\times\widehat{\Sigma}_{+1}$.

Set

$$C=(\widetilde{A}\times\widehat{\Sigma}_{+1})/\Gamma_{+1}. \tag{5.230}$$

Then C is a fibre bundle over Σ_{+1} with fibre $\widetilde{A}$. Also G acts locally freely on $\widetilde{A}$ and this action descends to a locally free action on C. We can then form the orbifolds $\widetilde{A}/G$ and $C/G=\widetilde{A}/G\times\Sigma_{+1}$.

Let V be a complex vector space. Let $\rho:G\to\operatorname{Aut}(V)$ be a representation of G. We still denote by ρ the corresponding representation of $\Gamma_{+1}\times G$ in $\operatorname{Aut}(V)$.

Put

$$F=\widetilde{A}\times\widehat{\Sigma}_{+1}\times_{\Gamma_{+1}\times G}V. \tag{5.231}$$

Then F is a vector bundle on $\widetilde{A}/G\times\Sigma_{+1}$. Also,verifies easily that

$$F=C\times_G V. \tag{5.232}$$

Moreover F is obained by the identification in (5.129), i.e.

$$(x,\sigma,f)\simeq(xg,\sigma a,g^{-1}x(a^{-1})f)\,,\ x\in\widetilde{A},\quad \sigma\in\widetilde{\Sigma}_{+1}\,,\ f\in V\,,\ g\in G\,,\ a\in\Gamma_{+1}. \tag{5.233}$$

Recall that $r_1, \dots, r_{s+1}$ are the origin in $S_1^1, \dots, S_{s+1}^1$. For $1 \le j \le s+1$, put

$$\frac{{}^0D_j}{Dt} = \frac{D}{Dt} + \log(w_{j,t})\,. \tag{5.234}$$

Then $\frac{{}^0D_j}{Dt}$ is a connection on the G-bundle P on S_j, with holonomy 1.

For $1 \le j \le s+1$, let ∇_j be a connection on the G-bundle $\widetilde{A} \to \widetilde{A}/G$. Consider the G-bundle $\widetilde{A} \times \widehat{\Sigma}_{+1} \xrightarrow{G} \widetilde{A}/G \times \widehat{\Sigma}_{+1}$. Along the fibres $\widehat{\Sigma}_{+1}$, we can equip this G-bundle with the trivial connection. This connection is Γ_{+1} invariant. Therefore it descends to a G-connection on $(\widetilde{A} \times \widehat{\Sigma}_{+1})/\Gamma_{+1} \xrightarrow{G} \widetilde{A}/G \times \Sigma_{+1}$ along the fibres Σ_{+1}. This connection along Σ_{+1} is exactly the flat G-connection associated to the given element of $\widetilde{A}$.

Along S_j^1, $1 \le j \le s+1$, we trivialize the G-bundle $(\widetilde{A} \times \widehat{\Sigma}_{+1})/\Gamma_{+1} \xrightarrow{G} \widetilde{A}/G \times \Sigma_{+1}$ with respect to $\frac{{}^0D_j}{Dt}$. Then over $\widetilde{A}/G \times S_j^1$, $1 \le j \le s+1$, the connection ∇_j induces a G-connection on $(\widetilde{A} \times \widehat{\Sigma}_{+1})/\Gamma_{+1} \xrightarrow{G} \widetilde{A}/G \times \Sigma_{+1}$ along $\widehat{A}/G$. Since the $\widetilde{A} \times S_j^1$ are disjoint, we can extend this connection to a G-connection on $(\widetilde{A} \times \widehat{\Sigma}_{+1})/\Gamma_{+1} \xrightarrow{G} \widetilde{A}/G \times \widehat{\Sigma}_{+1}$ along $\widetilde{A}/G$.

Ultimately the G-bundle $(\widetilde{A} \times \widehat{\Sigma}_{+1})/\Gamma_{+1} \xrightarrow{G} \widetilde{A}/G \times \Sigma_{+1}$ is equipped with a G-connection ∇. Let F be the curvature of this connection. Then F is a G-equivariant basic 2-form on $(\widetilde{A} \times \widehat{\Sigma}_{+1})/\Gamma_{+1}$ with values in $\mathfrak{g}$.

Let E be the orbifold vector bundle on $\widetilde{A}/G \times \Sigma_{+1}$ associated to the adjoint representation $G \to \mathrm{Aut}(\mathfrak{g})$. Let ∇^E be the induced connection on E. Then F descends to a 2-form on $\widetilde{A}/G \times \Sigma_{+1}$ with values in E. Set

$$E_j = r_j^* E \; ; \; 1 \le j \le s+1. \tag{5.235}$$

Then E_j is a vector bundle on $\widetilde{A}/G$. Let ∇^{E_j} be the connection induced by ∇^E (or ∇_j) on E_j. Observe that for $1 \le j \le s+1$, $\log(w_j)$ is a section of E_j.

Recall that $C/G = \widetilde{A}/G \times \Sigma_{+1}$. Then

$$\Lambda(T^*C/G) = \Lambda(T^*\widetilde{A}/G)\widehat{\otimes}\Lambda(T^*\Sigma_{+1})\,. \tag{5.236}$$

If $\omega \in \Lambda(T^*C)$, we can write ω in the form

$$\omega = \Sigma\omega^{(p,q)} \;,\; \omega^{(p,q)} \in \Lambda^p(T^*\widetilde{A}/G)\widehat{\otimes}\Lambda^q(T^*\Sigma_{+1})\,. \tag{5.237}$$

Let k be the embedding $\widetilde{A}/G \times \partial\Sigma_{+1} \to C/G$. Let Θ_j be the curvature of ∇_j.

THEOREM 5.52. *The following identities hold*

$$\begin{aligned} F^{(0,2)} &= 0\,, \\ k^*F &= \Theta_j - \nabla^{E_j}\log(w_j)dt \text{ on } \widetilde{A}/G \times S_j^1 \;,\; 1 \le j \le s+1\,. \end{aligned} \tag{5.238}$$

PROOF. By construction the first identity holds in (5.238). Also on $\widetilde{A}/G \times S_j^1$,

$$k^*F = k^*\left(\nabla^{E_j} + dt\Big(\frac{{}^0D}{Dt} - \log(w_j)\Big)\right)^{2,(1,1)} = \Theta_j - \nabla^{E_j}\log(w_j)dt\,. \tag{5.239}$$

The proof of our Theorem is completed. □

Let $\langle\, ,\, \rangle$ be a G-invariant bilinear symmetric form on $\mathfrak{g}$. Clearly $F^2 \in \Lambda^4(T^*C)\otimes \mathfrak{g}\otimes\mathfrak{g}$. Then $\langle F^2\rangle \in \Lambda^4(T^*C)$.

Let π, π' be the projections $\widetilde{A}/G \times \Sigma_{+1} \to \widetilde{A}/G$, $\widetilde{A}/G \times \partial\Sigma_{+1} \to \widetilde{A}/G$. Let ρ be the projection $\widetilde{A} \to \widetilde{A}/G$.

For $1 \leq j \leq s+1$, let θ_j be the connection 1-form on the G-bundle $\widetilde{A} \to \widetilde{A}/G$ associated to the connection ∇_j. Then θ_j is a 1 form with values in $\mathfrak{g}^*$.

DEFINITION 5.53. Let α be the 2-form on $\widetilde{A}$,

$$\alpha = \rho^*\left[\pi_*\left[\frac{\langle F^2\rangle}{2}\right] + \sum_{j=1}^{s+1}\langle \log(w_j), \Theta_j\rangle\right] - \sum_{j=1}^{s+1} d\langle \log(w_j), \theta_j\rangle\,. \tag{5.240}$$

THEOREM 5.54. *The 2-form α is G-invariant and closed. It does not depend on the choices made in its construction.*

PROOF. By Chern-Weil theory, the form $\frac{\langle F^2\rangle}{2}$ is closed on $\widetilde{A}/G \times \Sigma_{+1}$. Using Stokes formula, we get

$$d\pi_*\left[\frac{\langle F^2\rangle}{2}\right] = \pi'_*\left[\frac{k^*\langle F^2\rangle}{2}\right]\,. \tag{5.241}$$

By (5.238), (5.241), we get

$$d\pi_*\left[\frac{\langle F^2\rangle}{2}\right] = -\sum_{j=1}^{s+1}\langle \nabla^{E_j}\log(w_j), \Theta_j\rangle\,. \tag{5.242}$$

Also by Bianchi's identity,

$$d\langle \log(w_j), \Theta_j\rangle) = \langle \nabla^{E_j}\log(w_j), \Theta_j\rangle\,. \tag{5.243}$$

By (5.242), (5.243), we find that the form α is closed.

Now we replace $\widehat{\Sigma}_{+1}$ by $\widehat{\Sigma}_{+1} \times \mathbf{R}$. We still denote by π the projection $\widetilde{A} \times \widehat{\Sigma}_{+1} \times \mathbf{R} \to \widetilde{A}/G \times \mathbf{R}$ with fibre $\widehat{\Sigma}_{+1}$. The G- bundle $\widetilde{A} \to \widetilde{A}/G$ is replaced by $\widetilde{A} \times \mathbf{R} \to \widetilde{A}/G \times \mathbf{R}$. We consider a smooth family of data, which are used to construct the form α_ℓ. In particular the connections ∇_j depends on ℓ. We extend these fibrewise connections: ∇_j $\widetilde{A} \to \widetilde{A}/G$ to a connection $d\ell\frac{\partial}{\partial\ell} + \nabla_j$ on $\widetilde{A} \times \mathbf{R} \to \widetilde{A}/G \times \mathbf{R}$. We will denote with a $\widetilde{\ }$ the analogue of the above objects over $\widetilde{A} \times \widehat{\Sigma}_{+1} \times \mathbf{R}$. This way, we obtain a form $\widetilde{\alpha}$ on $\widetilde{A} \times \mathbf{R}$ such that

$$\widetilde{\alpha} = \alpha_\ell + d\ell \wedge \beta_\ell\,, \tag{5.244}$$

where β_ℓ is a 1-form on $\widetilde{A}/G$. By the above arguments, $\widetilde{\alpha}$ is closed.

Now we will show that

$$\beta_\ell = 0\,. \tag{5.245}$$

This will imply that

$$\frac{\partial \alpha_\ell}{\partial \ell} = 0\,. \tag{5.246}$$

So we will have established our claim that α_ℓ does not depend on ℓ.

Recall that along the fibres $\widehat{\Sigma}_{+1}$, the connection on the considered G-bundle does not depend on $\ell \in \mathbf{R}$, and is flat. It then follows easily that $\pi_*\left[\frac{\langle \widetilde{F}^2\rangle}{2}\right]$ does not contain $d\ell$.

Also the connection form $\widetilde{\theta}_j$ does not contain $d\ell$, i.e. it is of the form

$$\widetilde{\theta}_j = \theta_{j,\ell}\,. \tag{5.247}$$

Then

$$\widetilde{\Theta}_j = \Theta_{j,\ell} + d\ell\frac{\partial}{\partial\ell}\theta_{j,\ell}\,. \tag{5.248}$$

Therefore

$$\begin{aligned}\langle \log(w_j), \widetilde{\Theta}_j\rangle &= \langle \log(w_j), \widetilde{\Theta}_{j,\ell} + d\ell\frac{\partial}{\partial\ell}\theta_{j,\ell}\rangle,\\ \widetilde{d}\langle \log(w_j), \widetilde{\theta}_j\rangle &= \langle d\log(w_j), \theta_{j,\ell}\rangle + \langle \log(w_j), d\theta_{j,\ell}\rangle,\\ &\quad + \langle \log(w_j), d\ell\frac{\partial}{\partial\ell}\theta_{j,\ell}\rangle\,.\end{aligned} \tag{5.249}$$

From (5.142), we find the sum of the last terms in (5.240) does not contain $d\ell$ either.

The proof of our Theorem is completed. □

Now we fix $t_1 \in T \cap U_1, \ldots, t_s \in T \cap U_s$. For $1 \le j \le s$, let $\mathcal{O}_j \subset G$ be the orbit of t_j.

Put

$$\widehat{X} = \{x \in \widetilde{A} w_j \in \mathcal{O}_j \ , \ 1 \le j \le s\}\,. \tag{5.250}$$

Then $\widehat{X} \subset X$ is stable under G. Let m be one of the embeddings $\widehat{X} \to \widetilde{A}$, $\widehat{X}/G \to \widetilde{A}/G$.

For $1 \le j \le s$, let $\widetilde{\mathcal{O}}_j \subset \mathfrak{g}$ be the G-orbit of $\log(t_j)$. Since there is a well-defined logarithm on U_j, $\mathcal{O}_j$ and $\widetilde{\mathcal{O}}_j$ are in one to one correspondance.

For $1 \le j \le s$, let σ_j be the canonical symplectic form on the orbit $\widetilde{\mathcal{O}}_j$. If $Y \in \mathfrak{g}$, let $Y^{\widetilde{\mathcal{O}}_j}$ be the corresponding vector field associated to the right action of G on $\widetilde{\mathcal{O}}_j$. Then if $Y, Y' \in \mathfrak{g}$, $p \in \widehat{\mathcal{O}}_j$, as in (1.193),

$$\sigma_{j,p}(Y,Y') = \langle p, [Y,Y']\rangle\,. \tag{5.251}$$

DEFINITION 5.55. Put

$$\sigma = m^*\alpha + \sum_{j=1}^{s} \log(w_j)^*\sigma_j\,. \tag{5.252}$$

Clearly σ is a G-invariant closed 2-form on $\widehat{X}$. We will calculate σ.

For $1 \le j \le s$, the G-bundle $\widehat{X} \to \widehat{X}/G$ can be reduced to the $Z(t_j)$-bundle $\{x \in \widehat{X}, w_j = t_j\} \mapsto \{x \in \widehat{X}, w_j = t_j\}/Z(t_j)$. Let ∇_j be a $Z(t_j)$-connection on this last bundle. This connection lifts to a G-connection on $\widehat{X} \to \widehat{X}/G$. We will make this choice of ∇_j in our construction of σ. Then for $1 \le j \le s$,

$$\begin{aligned}&\nabla_j w_j = 0\,,\\ &\nabla_j \log(w_j) = 0\,.\end{aligned} \tag{5.253}$$

THEOREM 5.56. *The following identity holds*

$$\begin{aligned}\sigma = m^*\rho^*\Big(\pi_*\frac{\langle F^2\rangle}{2}\Big) + \langle \log(w_{s+1}), \Theta_{s+1}\rangle\Big)\\ -d\langle \log(w_{s+1}), \theta_{s+1}\rangle\,.\end{aligned} \tag{5.254}$$

PROOF. The second identity in (5.253) can be written in the form

$$d\log(w_j) + [\theta_j, \log(w_j)] = 0. \tag{5.255}$$

By (5.251), we get

$$\log(w_j)^* \sigma_j = \langle \log(w_j), \frac{1}{2}[\theta_j, \theta_j] \rangle. \tag{5.256}$$

Also

$$d\theta_j = -\frac{1}{2}[\theta_j, \theta_j] + \Theta_j. \tag{5.257}$$

From (5.255)-(5.257), we obtain for $1 \leq j \leq s$,

$$\begin{aligned} &\rho^* \langle \log(w_j), \Theta_j \rangle - d\langle \log(w_j), \theta_j \rangle + \log(w_j)^* \sigma_j \\ &= -\langle d\log(w_j), \theta_j \rangle + \langle \log w_j, [\theta_j, \theta_j] \rangle = 0. \end{aligned} \tag{5.258}$$

The proof of our Theorem is completed. □

Set

$$\widehat{M} = \{x \in \widehat{X},\ w_{s+1} = 1\}. \tag{5.259}$$

Our notation in (5.259) is compatible with (5.59). By Propositions 5.10 and 5.11, we know that $\widehat{M}$ is a submanifold of $\widehat{X}$, and that G acts locally freely on $\widehat{M}$. Let $i : \widehat{M} \to \widehat{X}$, $\widehat{M}/G \to \widehat{X}/G$ be the obvious embeddings.

By Theorem 5.31, $\widehat{M}/G$ carries a symplectic form ω. Recall that $\mathfrak{g}$ is equipped with the scalar product $\langle , \rangle$. We identify $\mathfrak{g}$ and $\mathfrak{g}^*$ by this scalar product.

THEOREM 5.57. *If U is small enough, σ is a symplectic form on $\widehat{X}$. Also $x = (u_1, v_1, \ldots, u_g, v_g, w_1, \ldots, w_{s+1}) \in \widehat{X} \mapsto \log(w_{s+1}) \in \mathfrak{g}$ is a moment map for the action of G on X with respect to the 2-form σ. The associated symplectic form on the symplectic reduction $\widehat{M}/G$ coincides with $i^* \pi_* (\frac{\langle F^2 \rangle}{2})$ and with ω.*

PROOF. By (5.254),

$$i^* \sigma = i^* \rho^* \pi_* \left(\frac{\langle F^2 \rangle}{2} \right). \tag{5.260}$$

Let ∇^H, ∇^V be the components of ∇ along $\widetilde{A}/G, \Sigma_{+1}$. Since ∇ is flat along Σ_{+1},

$$F = \nabla^{H,2} + [\nabla^H, \nabla^V]. \tag{5.261}$$

From (5.261) we get

$$\pi_* (\frac{\langle F^2 \rangle}{2}) = \pi_* (\frac{[\nabla^H, \nabla^V]}{2}). \tag{5.262}$$

Moreover by (5.238), (5.253),

$$k^* F^{(1,1)} = 0. \tag{5.263}$$

From (5.261)-(5.263), we get easily

$$i^* \pi_* (\frac{\langle F^2 \rangle}{2}) = \rho_* \omega. \tag{5.264}$$

From (5.260), (5.264), we obtain

$$i^* \sigma = \rho^* \omega. \tag{5.265}$$

Now by Theorem 5.31, ω is a symplectic form. Therefore by (5.254), (5.265), if U is small enough, σ is also a symplectic form.

If $Y \in \mathfrak{g}$, let $Y^{\widehat{X}}$ be the corresponding vector field on $\widehat{X}$. Then by (5.254),

$$\begin{aligned} i_{Y^X}\sigma &= -i_{Y^{\widehat{X}}} d\langle \log(w_{s+1}), \theta_{s+1}\rangle \\ &= d i_{Y^{\widehat{X}}} \langle \log(w_{s+1}), \theta_{s+1}\rangle \\ &= d\langle \log(w_{s+1}), Y\rangle . \end{aligned} \tag{5.266}$$

From (5.266), we find that $x \in X \mapsto \log(w_{s+1}) \in \mathfrak{g} \simeq \mathfrak{g}^*$ is a moment map for the action of G on $\widehat{X}$.

The proof of our Theorem is completed. □

By Theorem 5.12, since G acts locally freely on $\widehat{X}$, the derivative of $x \in \widehat{X} \mapsto w_{s+1}^{-1} \in G$ is surjective.

Take $t \in U$ close enough to 0. Put

$$X_t = \{x \in \widehat{X}, w_{s+1} = t\} . \tag{5.267}$$

Then $Z(t)$ acts locally freely on X_t . So $X_t/Z(t)$ can be equipped with the symplectic form σ_t, the reduction of the symplectic form σ. Also recall that by Theorem 5.31, $M_t/Z(t)$ is equipped with a symplectic form ω_t.

THEOREM 5.58. *For $t \in U$,*

$$\sigma_t = \omega_t . \tag{5.268}$$

PROOF. Clearly the G-bundle $\widetilde{A} \to \widetilde{A}/G$ reduces to $X_t \to X_t/Z(t)$. Let ∇_{s+1} be a $Z(t)$ connection on this last bundle. Then ∇_{s+1} lifts to a G-connection on $\widetilde{A} \to \widetilde{A}/G$. We will use ∇_{s+1} to calculate the restriction of σ to X_t. On X_t,

$$\begin{aligned} &\nabla^{E_{s+1}} \log(w_{s+1}) = 0 , \\ &d\log(w_{s+1}) + [\theta_{s+1}, \log(w_{s+1})] = 0 . \end{aligned} \tag{5.269}$$

Over $\log(w_{s+1}) = -\log(t)$, we get

$$[\theta_{s+1}, \log(w_{s+1})] = 0 . \tag{5.270}$$

Then over $\log(w_{s+1}) = -\log(t)$, using (5.257), (5.269), (5.270),

$$\begin{aligned} d\langle \log(w_{s+1}), \theta_{s+1}\rangle &= \langle \log(w_{s+1}), \Theta_{s+1}\rangle - \langle \log(w_{s+1}), \frac{1}{2}[\theta_{s+1}, \theta_{s+1}]\rangle \\ &= \langle \log(w_{s+1}), \Theta_{s+1}\rangle . \end{aligned} \tag{5.271}$$

From (5.254), (5.271), we find that over X_t,

$$\sigma = \rho^* \pi_* \left(\frac{\langle F^2 \rangle}{2} \right) . \tag{5.272}$$

Using (5.272) and proceeding as in the proof of Theorem 5.57, our Theorem follows. □

Take j, $1 \leq j \leq s$. Put

$$\widehat{X}_j = \{x \in \widetilde{A}; w_{j'} \in \mathcal{O}_{j'}, j' \neq j, w_j \in U_j, w_{s+1} = 1\} . \tag{5.273}$$

Let n_j be the embedding $\widehat{X}_j \to \widetilde{A}$.

DEFINITION 5.59. Put

$$\kappa_j = n_j^* \alpha + \sum_{j'=1\ ,\ j' \neq j}^{s} \log(w_j)^* \sigma_j \,. \tag{5.274}$$

We will choose the connection $\nabla_{j'}$, $1 \leq j \leq s, j' \neq j$ as in (5.253). Then by proceeding as in the proof of Theorem 5.56, we get

$$\begin{aligned} \kappa_j &= n_j^* \rho^* \left(\pi_* \left(\frac{\langle F^2 \rangle}{2} \right) + \langle \log(w_j), \Theta_j \rangle \right) \\ &\quad - d\langle \log(w_j), \theta_j \rangle \,. \end{aligned} \tag{5.275}$$

THEOREM 5.60. *If U_j is small enough, κ_j is a symplectic form on $\widehat{X}_j$. Also $x = (u_1, v_1, \dots, u_g, v_g, w_1, \dots, w_s) \in \widehat{X}_j \mapsto \log(w_j) \in \mathfrak{g}$ is a moment map for the action of G on $\widehat{X}_j$ with respect to κ_j. Finally for $t_j \in U_j$, the symplectic form on the symplectic reduction $\{x \in \widehat{X}_j, w_j = t_j\}/Z(t_j)$ is the symplectic form defined in Theorem 5.31.*

PROOF. The proof of our Theorem is the same as the proof of Theorems 5.57 and 5.58. □

REMARK 5.61. If $t_1, \dots, t_s$ are restricted to be very regular, we may and will assume that $\theta_1, \dots, \theta_s$ are T-connections. By Theorems 5.58 and 5.60, we find that $\pi_*(\frac{\langle F^2 \rangle}{2})$ restricts to the symplectic form of Theorem 5.31. Also by (5.275), the cohomology class of

$$\omega + \sum_{j=1}^{s} \langle \log(w_j), \Theta_j \rangle \tag{5.276}$$

is locally constant, which is a consequence of the theory of the moment map for torus actions obtained by Duistermaat-HeckmanDH.

5.11. The integral of certain characteristic classes on the strata of M/G. Let G be a compact connected and simply connected simple Lie group. We use the notation of Section 1. In particular, $\langle , \rangle$ denotes the basic scalar product on $\mathfrak{g}$.

Let $t_1, \dots, t_s$ be regular elements in T. Then $t_1, \dots, t_s$ are very regular.

We assume that $s \geq 1$, and that

- $(t_1, \dots, t_s)$ verify (A).
- If $g = 0$, then $s \geq 3$.

In particular

$$2g - 2 + s \geq 1 \,. \tag{5.277}$$

Let $u \in C/\overline{R}^*$. Recall that $\pi_u : \widetilde{Z}(u) \to Z(u)$ is the universal cover of $Z(u)$, with fibre $\pi_1(Z(u)) \simeq \overline{CR}/\overline{CR}_u \subset Z(\widetilde{Z}(u))$. Also $\widetilde{T}_u = \mathfrak{t}/\overline{CR}_u$ is a maximal torus in $\widetilde{Z}(u)$.

Remember that T_{reg} is the set of regular elements in T with respect to G. Let $t \in T_{\text{reg}}$. Then $Z(t) = T$, t is very regular in $Z(u)$, and $\mathcal{O}_{Z(u)}(t) \simeq Z(u)/T$.

Let $\tilde{t} \in \widetilde{T}_u$ be a lift of t in $\widetilde{Z}(u)$. Then $\tilde{t}$ is still regular in $\widetilde{Z}(u)$. Since $Z(t) = T$, the centralizer $Z(\tilde{t})$ of $\tilde{t}$ in $\widetilde{Z}(u)$ is just $Z(\tilde{t}) = \mathfrak{t}/\overline{CR}_u$. Then $\mathcal{O}_{\widetilde{Z}(u)}(\tilde{t}) \simeq$

$\widetilde{Z}(u)/(\mathfrak{t}/\overline{CR}_u) \simeq Z(u)/T$. Equivalently the projection $\mathcal{O}_{\widetilde{Z}(u)}(\tilde{t}) \to \mathcal{O}_{Z(u)}(t)$ is one to one.

As before, we identify $t_1, \dots, t_s$ with corresponding elements of G-alcoves in $\mathfrak{t}$ whose closure contains 0. This way, we get elements of $\widetilde{T}_u = \mathfrak{t}/\overline{CR}_u$, which lift $t_1, \dots, t_s$ unambiguously in $\widetilde{Z}(u)$. We still denote these elements by $t_1, \dots, t_s$.

Clearly

$$\widetilde{Z}'(u) = Z'(u). \tag{5.278}$$

Set

$$\begin{aligned} X_u &= Z(u)^{2g} \times \prod_{j=1}^{s} \mathcal{O}_{Z(u)}(t_j), \\ \widetilde{X}_u &= \widetilde{Z}(u)^{2g} \times \prod_{j=1}^{s} \mathcal{O}_{\widetilde{Z}(u)}(t_j). \end{aligned} \tag{5.279}$$

Then $Z(u)$ acts on X_u and on $\widetilde{X}_u$. Moreover $(\overline{CR}/\overline{CR}_u)^{2g}$ acts freely on $\widetilde{X}_u$. Namely if $f = (\alpha_1, \beta_1, \dots \alpha_g, \beta_g) \in (\overline{CR}/\overline{CR}_u)^{2g}$, if $x = (\tilde{u}_1, \tilde{v}_1, \dots \tilde{u}_g, \tilde{v}_g, \tilde{w}_1, \dots \tilde{w}_s) \in \widetilde{X}_u$, put

$$f.x = (\tilde{u}_1\alpha_1, \tilde{v}_1\beta_1, \dots, \tilde{u}_g\alpha_g, \tilde{v}_g\beta_g, \tilde{w}_1, \dots \tilde{w}_s). \tag{5.280}$$

The actions of $Z(u)$ and of $(\overline{CR}/\overline{CR}_u)^{2g}$ commute. Also the map $\pi_u : \widetilde{Z}(u) \to Z(u)$ extends to a map $\widetilde{X}_u \to X_u$. Clearly π_u is $Z(u)$-equivariant. Also if $f \in (\overline{CR}/\overline{CR}_u)^{2g}$,

$$\pi_u f = \pi_u. \tag{5.281}$$

More precisely π_u is a $(\overline{CR}/\overline{CR}_u)^{2g}$ cover.

DEFINITION 5.62. Let $\tilde{\phi}_u : \widetilde{X}_u \to \widetilde{Z}(u)$ and $\phi_u : X_u \to Z(u)$ be the analogues of ϕ defined in (5.41).

Clearly

$$\phi_u\pi_u = \pi_u\tilde{\phi}_u. \tag{5.282}$$

Also $\tilde{\phi}_u$ and ϕ_u are $Z(u)$- equivariant .

Clearly $(t_1, \dots, t_s)$ verify (A) with respect to $Z(u)$ or $\widetilde{Z}(u)$. Therefore by Theorem 5.18, 1 is a regular value of ϕ_u. Equivalently, by Theorem 5.18, 1 is a regular value of ϕ. By Theorem 5.12, G acts locally freely on $M = \phi^{-1}(1)$, and so by Theorem 5.12, 1 is a regular value of ϕ_u.

PROPOSITION 5.63. *Any $h \in \overline{CR}/\overline{CR}_u$ is a regular value of $\tilde{\phi}_u$.*

PROOF. Since 1 is a regular value of ϕ_u, any $h \in \pi_u^{-1}(1)$ is a regular value of $\tilde{\phi}_u$. □

Recall that the Lie algebra $\mathfrak{z}(u)$ is equipped with the scalar product induced by the scalar product $\langle , \rangle$ on $\mathfrak{g}$.

Let U be a G- invariant open neighborhood of 1 in G, such that a logarithm $\log : U \to \mathfrak{g}$ is defined, with

$$\log(1) = 0. \tag{5.283}$$

Put

$$U_u = U \cap Z(u),\ \widetilde{U}_u = \pi_u^{-1}(U_u). \tag{5.284}$$

Clearly log maps U_u into $\mathfrak{z}(u)$. Then $\widetilde{U}_u$ is an open neighborhood of $\overline{CR}/\overline{CR}_u$ in $\widetilde{Z}(u)$, which only consists of regular values of $\tilde{\phi}_u$.

In the sequel, we will view $\exp(\log(\phi_u(x))$ as an element of $\widetilde{Z}(u)$. Observe that if $\tilde{x} \in \widetilde{X}_u$ is such that $\tilde{x} \in \tilde{\phi}_u^{-1}(\widetilde{U}_u)$,

$$\pi_u \left[\tilde{\phi}_u(\tilde{x}) \exp\left(-\log(\phi_u \pi_u(\tilde{x}))\right)\right] = 1. \tag{5.285}$$

Therefore

$$\tilde{\phi}_u(\tilde{x}) \exp\left(-\log(\phi_u \pi_u)(\tilde{x})\right)) \in \overline{CR}/\overline{CR}_u. \tag{5.286}$$

Finally note that if $x \in \phi_u^{-1}(U)$, if $\tilde{x}$ is such that $\pi_u(\tilde{x}) = x$, then $\tilde{\phi}_u(\tilde{x})$ does not depend on $\tilde{x}$. Therefore $\tilde{\phi}_u(\tilde{x}) \exp\left(-\log(\phi_u \pi_u(\tilde{x}))\right) \in \overline{CR}/\overline{CR}_u$ does not depend on $\tilde{x}$.

DEFINITION 5.64. If $h \in \overline{CR}/\overline{CR}_u$, let $\phi_u^{-1}(U_u)_h \subset \phi_u^{-1}(U)$ be given by

$$\phi_u^{-1}(U_u)_h = \{x \in \phi_u^{-1}(U_u),\ \text{if } \pi_u(\tilde{x}) = x, \tilde{\phi}_u(\tilde{x}) \exp\left(-\log(\phi_u \pi_u(\tilde{x}))\right) h = 1\}. \tag{5.287}$$

Similarly, let $\tilde{\phi}_u^{-1}(\widetilde{U}_u)_h \subset \tilde{\phi}_u^{-1}(\widetilde{U}_u)$ be given by

$$\tilde{\phi}_u^{-1}(\widetilde{U}_u)_h = \{\tilde{x} \in \tilde{\phi}_u^{-1}(\widetilde{U}_u), \tilde{\phi}_u(\tilde{x}) \exp\left(-\log(\phi_u \pi_u(\tilde{x}))\right) h = 1\}. \tag{5.288}$$

Then we have the disjoint union

$$\phi_u^{-1}(U_u) = \bigcup_{h \in \overline{CR}/\overline{CR}_u} \phi_u^{-1}(U_u)_h, \tag{5.289}$$

$$\tilde{\phi}_u^{-1}(\widetilde{U}_u) = \bigcup_{h \in}$$

$\overline{CR}/\overline{CR}_u \tilde{\phi}_u^{-1}(\widetilde{U}_u)_h$,and for any $h \in \overline{CR}/\overline{CR}_u, \pi_u : \tilde{\phi}_u^{-1}(\widetilde{U}_u)_h \to \phi_u^{-1}(U_u)_h$ is a $\left(\overline{CR}/\overline{CR}_u\right)^{2g}$-cover. Moreover $Z(u)$ preserves $\tilde{\phi}_u^{-1}(\widetilde{U}_u)_h$ and $\phi_u^{-1}(U_u)_h$, and $\left(\overline{CR}/\overline{CR}_u\right)^{2g}$ acts freely on $\tilde{\phi}_u^{-1}(\widetilde{U}_u)_h$.

If U is small enough, there is an open neignborhood $\widetilde{V}_u$ of 1 in $\widetilde{Z}(u)$ such that $\pi_u : (\widetilde{V}_u) \to U_u$ is one to one. Also if $t \in U$, and $\tilde{t} \in \widetilde{V}_u$ is the lift of t, then π_u is one to one from $\mathcal{O}_{\widetilde{Z}(u)}(\tilde{t})$ into $\mathcal{O}_{Z(u)}(t)$.

DEFINITION 5.65. If $h \in \overline{CR}/\overline{CR}_u, t \in U_u, \tilde{t} \in \widetilde{V}_u$, with $\pi_u(\tilde{t}) = t$, put

$$M_t(Z(u), \mathcal{O}_{Z(u)}(t_1), \dots, \mathcal{O}_{Z(u)}(t_s), h) = \{x \in \phi_u^{-1}(U_u)_h, \phi_u(x) = t^{-1}\}, \tag{5.290}$$
$$M_t(\widetilde{Z}(u), \mathcal{O}_{\widetilde{Z}(u)}(t_1), \dots, \mathcal{O}_{\widetilde{Z}(u)}(t_s), h) = \{\tilde{x} \in \tilde{\phi}_u^{-1}(\widetilde{U}_u)_h, \tilde{\phi}_u(\tilde{x}) = \tilde{t}^{-1} h^{-1}\}.$$

Then $\pi_u : M_t(\widetilde{Z}(u), \mathcal{O}_{\widetilde{Z}(u)}(t_1), \dots, \mathcal{O}_{\widetilde{Z}(u)}(t_s), h) \to M_t(Z(u), \mathcal{O}_{Z(u)}(t_1), \dots, \mathcal{O}_{Z(u)}(t_s), h)$ is a $|\overline{CR}/\overline{CR}_u|^{2g}$ cover. If $Z_u(t)$ is the centralizer of t in $Z(u)$, π_u is a $Z_u(t)$ equivariant map.

By Theorem 5.12 and Proposition 5.63, $Z_u(t)$ acts locally freely on $M_t(\widetilde{Z}(u), \mathcal{O}_{\widetilde{Z}(u)}(t_1), \dots, \mathcal{O}_{\widetilde{Z}(u)}(t_s), h)$ and on $M_t(Z(u), \mathcal{O}_{Z(u)}(t_1), \dots, \mathcal{O}_{Z(u)}(t_s), h)$.

PROPOSITION 5.66. *The map* $\pi_u : M_t(\widetilde{Z}(u), \mathcal{O}_{\widetilde{Z}(u)}(t_1), \dots, \mathcal{O}_{\widetilde{Z}(u)}(t_s), h)/Z_u(t) \to M_t(Z(u), \mathcal{O}_{Z(u)}(t_1), \dots, \mathcal{O}_{Z(u)}(t_s), h)/Z_u(t)$ *is a* $|\overline{CR}/\overline{CR}_u|^{2g}$ *cover on the regular part of the orbifold* $M_t(\widetilde{Z}(u), \mathcal{O}_{\widetilde{Z}(u)}(t_1), \dots, \mathcal{O}_{\widetilde{Z}(u)}(t_s), h)/Z_u(t)$.

PROOF. Let $\tilde{x} \in M_t(\widetilde{Z}(u), \mathcal{O}_{\widetilde{Z}(u)}(t_1), \dots, \mathcal{O}_{\widetilde{Z}(u)}(t_s), h)$, and let $f \in (\overline{CR}/\overline{CR}_u)^{2g}$, $g \in Z_u(t)$ be such that

$$f.\tilde{x} = \tilde{x}.g. \tag{5.291}$$

Then by (5.281), (5.291),

$$\pi_u(\tilde{x}) = \pi_u(\tilde{x}).g. \tag{5.292}$$

If $x = \pi_u(x)$ is in the regular part of $M_t(Z(u), \mathcal{O}_{Z(u)}(t_1), \dots, \mathcal{O}_{Z(u)}(t_s), h)$, from (5.292), we get

$$g = 1. \tag{5.293}$$

So by (5.291), (5.293),

$$f = 1. \tag{5.294}$$

The proof of our Proposition is completed. □

For $t \in U_u$, let σ_t^u be the symplectic form on $M_t(Z(u), \mathcal{O}_{Z(u)}(t_1), \dots, \mathcal{O}_{Z(u)}(t_s), h)/Z_u(t)$, and let $\tilde{\sigma}_t^u$ be the symplectic form on $M_t(\widetilde{Z}(u), \mathcal{O}_{\widetilde{Z}(u)}(t_1), \dots, \mathcal{O}_{\widetilde{Z}(u)}(t_s), h)/Z_u(t)$.

DEFINITION 5.67. For $u \in C/\overline{R}^*$, $t \in U_u$, put

$$V_u(t_1, \dots t_s, t, h) = \left| \int_{M_t(Z(u), \mathcal{O}_{Z(u)}(t_1), \dots, \mathcal{O}_{Z(u)}(t_s), h)/Z'_u(t)} e^{\tilde{\sigma}_t^u} \right|, \tag{5.295}$$

$$\widetilde{V}_u(t_1, \dots t_s, t, h) = \left| \int_{M_t(\widetilde{Z}(u), \mathcal{O}_{\widetilde{Z}(u)}(t_1), \dots, \mathcal{O}_{\widetilde{Z}(u)}(t_s), h)/Z'_u(t)} e^{\tilde{\sigma}_t^u} \right|.$$

PROPOSITION 5.68. *The following identity holds*

$$\tilde{\sigma}_t^u = \pi_u^* \sigma_t^u. \tag{5.296}$$

Moreover

$$|V_u|(t_1, \dots t_s, t, h) = \frac{1}{|\overline{CR}/\overline{CR}_u|^{2g}} |\widetilde{V}_u|(t_1 \dots t_s, t, h). \tag{5.297}$$

PROOF. Equation (5.296) is trivial. Using (5.296), we get (5.297). □

Let $K \subset \mathfrak{t}$ be a Weyl chamber for G which is fixed once and for all. Then if $u \in C/\overline{R}^*$, K is included in a unique Weyl chamber K_u for $Z(u)$ or $\widetilde{Z}(u)$. Put

$$\overline{CR}^*_{u,+} = \overline{CR}^*_u \cap \overline{K}_u. \tag{5.298}$$

Then the irreducible representations of $\widetilde{Z}(u)$ are parametrized by $\overline{CR}^*_{u,+}$. If $\lambda \in \overline{CR}^*_{u,+}$, let $\chi_\lambda^{\widetilde{Z}(u)}$ be the representation of $\widetilde{Z}(u)$ with highest weight λ. For $t \in \widetilde{T}_u = \mathfrak{t}/\overline{CR}_u$, put

$$\sigma_{Z(u)}(\widetilde{T}) = \prod_{\alpha \in R_{u,+}} \left(e^{i\pi\langle \alpha, t \rangle} - e^{-i\pi\langle \alpha, t \rangle} \right). \tag{5.299}$$

Then $|\sigma_{Z(u)}(t)|$ is well defined on $\mathfrak{t}/\overline{R}^*$, and so is well defined on $T = \mathfrak{t}/\overline{CR}$.

THEOREM 5.69. *For any $u \in C/\overline{R}^*, h \in \overline{CR}/\overline{CR}_u$, the following identity of $Z(u)$-invariant distributions in the variable $t \in U$ holds,*

$$\frac{|\widetilde{V}_u|(t_1, \dots t_s, h, t)}{|\sigma_{Z(u)}(t+h) \prod_{j=1}^s |\sigma_{Z(u)}(t_j)|} = |Z(\widetilde{Z}(u))| \mathrm{Vol}(t/\overline{CR}_u)|^{2g-2} \left|\frac{\mathrm{Vol}(\widetilde{Z}(u)}{\mathrm{Vol}(\widetilde{T}_u)}\right|^{2g+s-1} \sum_{\lambda \in \overline{CR}^*_{u,+}} \frac{\prod_{j=1}^s \chi_\lambda^{\widetilde{Z}(u)}(t_j)\chi_\lambda^{\widetilde{Z}(u)}(t+h)}{\chi_\lambda^{\widetilde{Z}(u)}(1)^{2g+s-1}}. \tag{5.300}$$

PROOF. Recall that $t \in U$ is identified to the corresponding element in $\widetilde{V}_u \subset \widetilde{Z}(u)$. Since $t_j \in \widetilde{Z}(u)$ is very regular, the centralizer of t_j in $\widetilde{Z}(u)$ is equal to $\widetilde{T}_u$. Then by (5.156) and by Theorem 5.47, we get (5.300) . The proof of our Theorem is completed. □

REMARK 5.70. Note that $\overline{CR}^* \cap \overline{K}_u$ is exactly the set of nonnegative weights for $Z(u)$. Also since $h \in \overline{CR}/\overline{CR}_u$ lies in $Z(\widetilde{Z}(u))$,

$$\chi_\lambda^{\widetilde{Z}(u)}(t+h) = e^{2i\pi\langle\lambda,h\rangle}\chi_\lambda^{\widetilde{Z}(u)}(t). \tag{5.301}$$

From (5.300), (5.301), we get easily,

$$\frac{1}{|\overline{CR}/\overline{CR}_u|^{2g}} \frac{\sum_{h\in\overline{CR}/\overline{CR}_u} |\widetilde{V}_u|(t_1, \dots, t_s, h, t)}{|\sigma_{Z(u)}(t)| | \prod_{j=1}^s |\sigma_{Z(u)}(t_j)|} = |Z(\widetilde{Z}(u))| |\mathrm{Vol}(t/\overline{CR}_u)|^{2g-2} \left|\overline{CR}/\overline{CR}_u\right|^{1-2g} \left|\frac{\mathrm{Vol}(\widetilde{Z}_u)}{\mathrm{Vol}(\widetilde{T}_u)}\right|^{2g+s-1} \sum_{\lambda\in\overline{CR}^*\cap\overline{K}_u} \frac{\prod_{j=1}^s \chi_\lambda^{Z(u)}(t_j)\chi_\lambda^{Z(u)}(t)}{\chi_\lambda^{Z(u)}(1)^{2g+s-1}}. \tag{5.302}$$

Clearly

$$\begin{aligned} Z(Z(u)) &= Z(\widetilde{Z}(u))/(\overline{CR}/\overline{CR}_u), \\ \mathrm{Vol}(t/\overline{CR}_u) &= \mathrm{Vol}(T)|\overline{CR}/\overline{CR}_u|, \\ \frac{\mathrm{Vol}(\widetilde{Z}(u))}{\mathrm{Vol}(\widetilde{T}_u)} &= \frac{\mathrm{Vol}(Z(u))}{\mathrm{Vol}(T)}. \end{aligned} \tag{5.303}$$

From (5.302), (5.303), we get

$$\frac{1}{|\overline{CR}/\overline{CR}_u|^{2g}} \frac{\sum_{h\in\overline{CR}/\overline{CR}_u} \widetilde{V}_u(t_1, \dots, t_s, h, t)}{|\sigma_{Z(u)}(t)| | \prod_{j=1}^s |\sigma_{Z(u)}(t_j)|} = |Z(Z(u))| \mathrm{Vol}(t/\overline{CR})^{2g-2} \left|\frac{\mathrm{Vol}(Z(u)}{\mathrm{Vol}(T)}\right|^{2g+s-1} \sum_{\lambda\in\overline{CR}^*\cap\overline{K}_u} \frac{\prod_{j=1}^s \chi_\lambda^{Z(u)}(t_j)\chi_\lambda^{Z(u)}(t)}{\chi_\lambda^{Z(u)}(1)^{2g+s-1}}. \tag{5.304}$$

Identity (5.304) fits with (5.197) and with (5.297).

Since $t_1, \dots, t_s$ are very regular in $\widetilde{Z}(u)$, there are $Z(u)$-invariant open neighborhoods $U_1^u, \dots, U_s^u$ of $t_1, \dots, t_s$ in $\widetilde{Z}(u)$ on which a logarithm is well defined.

Since $Z(u)$ acts locally freely on $M_0(\widetilde{Z}(u), \mathcal{O}_{\widetilde{Z}(u)}(t_1), \dots, \mathcal{O}_{\widetilde{Z}(u)}(t_s), h)$, all the results of Section 5.10 can be used in this situation.

In particular, by Theorem 5.57, there is a $Z(u)$-invariant open neighborhood $\widehat{X}_{u,h}$ of $M_0(\widetilde{Z}(u), \mathcal{O}_{\widetilde{Z}(u)}(t_1), \dots, \mathcal{O}_{\widetilde{Z}(u)}(t_s), h)$ in $\widetilde{X}_u$, equipped with a symplectic form $\bar{\sigma}^u$, such that the $\bar{\sigma}^u_t$ are the symplectic reductions of $\bar{\sigma}^u$. So we may use the results of Section 3 in this situation.

Let $T'_u = \mathfrak{t}/\overrightarrow{R_u^*}$ be the obvious maximal torus in $Z'(u) = Z(u)/Z(Z(u))$. Now we will use the notation in Section 3.6. As in Section 3.6, the choice of a Weyl chamber$K$ and of the corresponding Weyl chamber K_u for $Z(u)$ defines an orientation on the $M_t(\widetilde{Z}(u), \mathcal{O}_{\widetilde{Z}(u)}(t_1), \dots, \mathcal{O}_{\widetilde{Z}(u)}(t_s), h)/T'_u$, for $t \in T \cap U_u$.

DEFINITION 5.71. For $t \in T \cap U_u$, put

$$H_{(u,g,s)}(t_1, \dots t_s, t) = \int_{M_t(\widetilde{Z}(u), \mathcal{O}_{\widetilde{Z}(u)}(t_1), \dots, \mathcal{O}_{\widetilde{Z}(u)}(t_s), h)/T'_u} e^{\bar{\sigma}^u_t}. \tag{5.305}$$

Using Theorems 3.15, 5.57 and 5.60 , we know that $H_{(u,g,s)}(t_1, \dots t_s, t)$ is locally a polynomial in $t_1, \dots, t_s, t$. Also recall that $\widetilde{P}_{u,2g+s-1}(t)$, $t \in \widetilde{T}_u$ was defined in Definition 2.39. By Theorem 2.37, $\widetilde{P}_{u,2g+s-1}(t)$ is a polynomial on $\widetilde{T}_u \backslash \widetilde{S}_u$. Finally remember that in (2.33) , we set $\ell_u = |R_{u,+}|$.

THEOREM 5.72. *The following identity of local polynomials in* $(t_1, \dots, t_s, t)$ *holds*

$$\begin{aligned} H_{(u,g,s)}(t_1, \dots, t_s, h, t) = (-1)^{\ell_u(g-1)+1} |Z(\widetilde{Z}(u))| \\ |\mathrm{Vol}(\mathfrak{t}/\overline{CR}_u)|^{2g-2} \prod_{j=1}^{s} \mathrm{sgn}\big((-i)^{\ell_u} \sigma_{Z(u)}(t_j)\big) e^{-2i\pi\langle \rho_u, h\rangle} \\ \sum_{(w'^1, \dots, w'^s) \in W_u^s} \prod_{j=1}^{s} \epsilon_{w'^j} \widetilde{P}_{u,2g+s-1}(t + h + \sum_{j=1}^{s} w'^j t_j). \end{aligned} \tag{5.306}$$

PROOF. Clearly, if $t \in U_u \cap T_{\mathrm{reg}}$, then $Z'(t) = T'$, so that $Z_u(t) = T'_u$. By (5.295), (5.305),

$$|H_{(u,g,s)}(t_1, \dots, t_s, h, t)| = \widetilde{V}_u(t_1, \dots t_s, h, t) \text{ on } U_u \cap T_{\mathrm{reg}}. \tag{5.307}$$

Also

$$|\sigma_{Z(u)}(t)| = \mathrm{sgn}\big((-i)^{\ell_u} \sigma_{Z(u)}(t)\big)(-i)^{\ell_u} \sigma_{Z(u)}(t). \tag{5.308}$$

Moreover, by (1.43), if $h \in \overline{CR}/\overline{CR}_u$,

$$\sigma_{Z(u)}(t + h) = e^{2i\pi\langle \rho_u, h\rangle} \sigma_{Z(u)}(t), \tag{5.309}$$

By Theorem 1.41,

$$e^{2i\pi\langle \rho_u, h\rangle} = \pm 1. \tag{5.310}$$

Moreover for t close enough to 0, $(-i)^{\ell_u} \sigma_{Z(u)}(t)$ and $\pi_u(t/i)$ have the same sign. Also by (3.129), (3.133),

$$\frac{\chi_\lambda^{\widetilde{Z}(u)}(1)}{\mathrm{Vol}(\widetilde{Z}(u))/\mathrm{Vol}(\widetilde{T}_u)} = \pi_u(\frac{\rho_u + \lambda}{i}). \tag{5.311}$$

Finally by Theorem 3.15, for t close enough to 0, $H_{(u,g,s)}(t)$ and $\pi_u(t/i)$ either vanish together, or they are nonzero, and then they have the same sign. Using (5.300), (5.310), (5.311) , we get the identity of distributions

$$H_{(u,g,s)}(t_1,\dots,t_s,h,t) = (-1)^{\ell_u(g-1)}|Z(\widetilde{Z}(u))||\mathrm{Vol}(\mathfrak{t}/\overline{CR}_u)|^{2g-2} \prod_{j=1}^{s} \operatorname{sgn}\big((-i)^{\ell_u}\sigma_{Z(u)}(t_j)\big)e^{-2i\pi\langle\rho_u,h\rangle} \sum_{\lambda\in\overline{CR}^*_{u,+}} \frac{\prod_{j=1}^{s}\big(\sigma_{Z(u)}(t_j)\chi_\lambda^{\widetilde{Z}(u)}(t_j)\big)\sigma_{Z(u)}(t+h)\chi_\lambda^{\widetilde{Z}(u)}(t+h)}{\big(\pi_u(\rho_u+\lambda)\big)^{2g+s-1}}. \tag{5.312}$$

By (1.94), Theorem 1.38 and by (5.312), we obtain

$$H_{(u,g,s)}(t_1,\dots,t_s,h,t) = (-1)^{\ell_u(g-1)}|Z(\widetilde{Z}(u))|\mathrm{Vol}(\mathfrak{t}/\overline{CR}_u)|^{2g-2} \prod_{j=1}^{s}\operatorname{sgn}\big((-i)^{\ell_u}\sigma_{Z(u)}(t_j)\big)e^{-2i\pi\langle\rho_u,h\rangle} \sum_{\lambda\in\overline{CR}^*_{u,+}} \sum_{(w'^1,\dots,w'^s,w)\in W_u^{s+1}} \frac{\prod_{j=1}^{s}\epsilon_{w'^j}\epsilon_w^{s-1}xp(2i\pi\langle w(\lambda+\rho),t+h+\sum_{j=1}^{s}w'^j t_j\rangle)}{\big(\pi_u(\rho_u+\lambda)\big)^{2g+s-1}} \tag{5.313}$$

Now by (1.186), if $w\in W_u$,

$$\pi_u(w(\rho_u+\lambda)) = \epsilon_w\pi_u(\rho_u+\lambda). \tag{5.314}$$

Also by using in particular [**15**, Note V.4.14],

$$\{w(\rho_u+\lambda)\}_{\substack{w\in W_u,\\ \lambda\in\overline{CR}^*_{u,+}}} = \{\lambda\in\overline{CR}^*_u, \pi_u(\lambda)\neq 0\}. \tag{5.315}$$

Using (2.158), ((5.313)-(5.315), we get the identity of distributions on U_u,

$$H_{(u,g,s)}(t_1,\dots,t_s,h,t) = (-1)^{\ell_u(g-1)+1}|Z(\widetilde{Z}(u))||\mathrm{Vol}(\mathfrak{t}/\overline{CR}_u)|^{2g-2} \prod_{j=1}^{s}\operatorname{sgn}\big((-i)^{\ell_u}\sigma_{Z(u)}(t_j)\big)e^{-2i\pi\langle\rho_u,h\rangle} \sum_{(w'^1,\dots,w'^s)\in W_u^s} \prod_{j=1}^{s}\epsilon_{w'^j}\tilde{P}_{u,2g+s-1}\Big(t+h+\sum_{j=1}^{s}w'^j t_j\Big). \tag{5.316}$$

Now by Theorems 2.37, 3.15, 5.57, 5.60, we know that both sides of (5.316) are local polynomials of $(t_1,\dots,t_s,t)$. Therefore (5.316) extends to an identity of local polynomials. The proof of our Theorem is completed. □

Put

$$M(Z(u),\mathcal{O}_{Z(u)}(t_1),\dots,\mathcal{O}_{Z(u)}(t_s),h) = M_0(Z(u),\mathcal{O}_{Z(u)}(t_1),\dots,\mathcal{O}_{Z(u)}(t_s),h). \tag{5.317}$$

Then $M(Z(u),\mathcal{O}_{Z(u)}(t_1),\dots,\mathcal{O}_{Z(u)}(t_s),h)\subset M$, and $Z(u)$ acts locally freely on $M(Z(u),\mathcal{O}_{Z(u)}(t_1),\dots,\mathcal{O}_{Z(u)}(t_s),h)$.

Let θ be a $Z(u)$ connection form on the $Z(u)$-bundle $M(Z(u),\mathcal{O}_{Z(u)}(t_1),\dots,\mathcal{O}_{Z(u)}(t_s),h)\to M(Z(u),\mathcal{O}_{Z(u)}(t_1),\dots,\mathcal{O}_{Z(u)}(t_s),h)/Z(u)$, and let Θ be its curvature. Let $\theta_1,\dots,\theta_s$ be connection forms taken as in Section 5.10, and such that (5.253) holds, and let $\Theta_1,\dots,\Theta_s$ be their curvatures. Then $\Theta_1,\dots,\Theta_s$ take their values in $\mathfrak{t}$.

Let Q be a $Z(u)$-invariant C^∞ function on $\mathfrak{z}(u)$, let $Q_1, \ldots, Q_s$ be C^∞ functions on $\mathfrak{t}$. Recall that ω is the canonical symplectic form on M/G which is associated to the basic scalar product $\langle , \rangle$ on $\mathfrak{g}$.

PROPOSITION 5.73. *If $p \in \mathbf{R}^*$, the following identity holds*

$$\int_{M(Z(u),\mathcal{O}_{Z(u)}(t_1),\ldots,\mathcal{O}_{Z(u)}(t_s),h)/Z(u)} Q(-\Theta) \textstyle\prod_{j=1}^{s} Q_j(-\Theta_j) e^{p\omega} = \tag{5.318}$$
$$\frac{1}{|\overline{CR}/\overline{CR}_u|^{2g}} \frac{p^{(g-1)\dim \mathfrak{z}(u) + \frac{s}{2}\dim(\mathfrak{z}(u)/\mathfrak{t})}}{|W_u|}$$
$$Q(\frac{\partial/\partial t}{p}) \prod_{j=1}^{s} Q_j(\frac{\partial/\partial t_j}{p}) \pi_u(\frac{\partial/\partial t}{2i\pi}) H_{(u,g,s)}(t_1, \ldots, t_s, h, t).$$

PROOF. Clearly

$$\int_{M(Z(u),\mathcal{O}_{Z(u)}(t_1),\ldots,\mathcal{O}_{Z(u)}(t_s),h)/Z(u)} Q(-\Theta) e^{p\omega} \prod_1^s Q_j(-\Theta_j) = \tag{5.319}$$
$$p^{(\dim M(Z(u),\mathcal{O}_{Z(u)}(t_1),\ldots,\mathcal{O}_{Z(u)}(t_s),h)/Z(u))/2}$$
$$\int_{M(Z(u),\mathcal{O}_{Z(u)}(t_1),\ldots,\mathcal{O}_{Z(u)}(t_s),h)/Z(u)} Q(-\Theta/p) \prod_1^s Q_j(-\Theta_j/p) e^{\omega}.$$

Also by Theorem 5.57 and Proposition 5.66,

$$\int_{M(Z(u),\mathcal{O}_{Z(u)}(t_1),\ldots,\mathcal{O}_{Z(u)}(t_s),h)/Z(u)} Q(-\Theta/p) e^{\omega} = \tag{5.320}$$
$$\frac{1}{|\overline{CR}/\overline{CR}_u|^{2g}} \int_{M(\widetilde{Z}(u),\mathcal{O}_{\widetilde{Z}(u)}(t_1),\ldots,\mathcal{O}_{\widetilde{Z}(u)}(t_s),h)/Z(u)Z(u)} Q(-\Theta/p) e^{\bar{\sigma}_0^u}.$$

Finally our assumptions on the t_j's, Theorems 5.20 and 5.21 guarantee that $Z'_u(x) = 1$ a.e. on $M(\widetilde{Z}(u), \mathcal{O}_{\widetilde{Z}(u)}(t_1), \ldots, \mathcal{O}_{\widetilde{Z}(u)}(t_s), h)$. We can then apply Theorem 3.21 to (5.320) and get (5.318). The proof of our Theorem is completed. □

REMARK 5.74. In [**39, 40**], Liu derived the above formulas for the intersection numbers of the corresponding moduli spaces.

5.12. An evaluation of certain Euler characteristics. Recall that $x \in \mathbf{R} \mapsto [x] \in [0,1[$ is the periodic function of period 1 such that for $x \in [0,1[$, $[x] = x$.

PROPOSITION 5.75. *Let $m \in \mathbf{N}$. Put $z = e^{\frac{2i\pi}{m}}$. Then for $\ell \in \mathbf{Z}$,*

$$\frac{1}{m} \sum_{k=1}^{m-1} \frac{z^{k\ell}}{1 - z^{-k}} = \frac{1}{2} - \left[\frac{\ell}{m}\right] - \frac{1}{2m}. \tag{5.321}$$

PROOF. Take $\rho \in]-1, +1[$, and ℓ with $0 \leq \ell < m$. Then

$$\sum_{k=1}^{m-1} \frac{z^{-k\ell}}{1 - \rho z^k} = \sum_{n=0}^{+\infty} \sum_{k=1}^{m-1} \rho^n z^{k(n-\ell)} = \sum_{n=0}^{+\infty} \rho^n \left(\sum_{k=0}^{m-1} z^{k(n-\ell)} - 1 \right) \tag{5.322}$$
$$= \sum_{n=0}^{+\infty} \rho^n m 1_{m|(n-\ell)} - \frac{1}{1-\rho}$$
$$= \frac{m\rho^\ell}{1 - \rho^m} - \frac{1}{1-\rho}.$$

By making $\rho \to 1$ in (5.322), we get

$$\sum_{k=1}^{m-1} \frac{z^{-k\ell}}{1-z^k} = \frac{m-1}{2} - \ell, \tag{5.323}$$

which is equivalent to

$$\frac{1}{m}\sum_{k=1}^{m-1} \frac{z^{k\ell}}{1-z^{-k}} = \frac{1}{2} - \left[\frac{\ell}{m}\right] - \frac{1}{2m}. \tag{5.324}$$

So we have established (5.321) for $0 \leq \ell < m$. Similarly if $-m \leq \ell < 0$, using (5.324), we obtain

$$\begin{aligned}\frac{1}{m}\sum_{k=1}^{m-1} \frac{z^{k\ell}}{1-z^{-k}} &= \frac{1}{m}\sum_{k=1}^{m-1} \frac{z^{k(m+\ell)}}{1-z^{-k}} = \frac{1}{2} - \left[1+\frac{\ell}{m}\right] - \frac{1}{2m} \\ &\frac{1}{2} - \left[\frac{\ell}{m}\right] - \frac{1}{2m}.\end{aligned} \tag{5.325}$$

The proof of our Proposition is completed. □

Let Σ_g be a Riemann surface of genus g. Here, we have fixed a complex structure on Σ_g. Let $x_1, \dots, x_s$ be s distinct points in Σ_g. Let D be the divisor

$$D = \sum_{j=1}^{s} x_j. \tag{5.326}$$

Let $[D]$ be the corresponding holomorphic line bundle on Σ_g. Let σ_D be the canonical section of $[D]$. Clearly

$$\int_{\Sigma_g} c_1([D]) = s. \tag{5.327}$$

Let $m \in \mathbf{N}$ such that m divides s. Let λ be a holomorphic line bundle on Σ_g such that

$$\lambda^m = [D]. \tag{5.328}$$

Put

$$\Sigma_g^b = \{t \in \lambda\, ,\ \sigma_D = t^m\}. \tag{5.329}$$

Then $p = \Sigma_g^b \to \Sigma_g$ is a branched covering of order m, with branching points $x_1, \dots, x_s$. Also by Hurwitz's formula, the genus g' of Σ_g^b is given by

$$g' = mg + \frac{1}{2}(m-1)(s-2). \tag{5.330}$$

If $\ell \in \mathbf{Z}/m\mathbf{Z}, t \in \Sigma_g^b$, put

$$\ell(t) = e^{\frac{2i\pi\ell}{m}} t. \tag{5.331}$$

Then (5.331) defines an action of $\mathbf{Z}/m\mathbf{Z}$ over Σ_g^b such that $p\ell = p$. Also if $\ell \in \mathbf{Z}/m\mathbf{Z}$, $l \neq 0$, then $x_1, \dots, x_s \in \Sigma_g^b$ are the only fixed points of ℓ.

Let Σ be the Riemann surface with boundary, which is obtained from Σ_g by deleting s small disks centered at $x_1, \dots, x_s$. Set

$$\Sigma^b = p^{-1}\Sigma. \tag{5.332}$$

Then Σ^b is also a Riemann surface with boundary, obtained from Σ_g^b by deleting s small disks $\Delta_1, \ldots, \Delta_s$ centered at $x_1, \ldots, x_s$.

Let G be a compact connected and simply connected simple Lie group.

DEFINITION 5.76. We will say that $g \in G$ is of order m if $g^m = 1$. More generally if $\mathcal{O} \subset G$ is an adjoint orbit, $\mathcal{O}$ will be said to be of order m if for one (or any) element g$\in \mathcal{O}$, $g^m = 1$.

In the sequel, we assume that $\mathcal{O}_1, \ldots \mathcal{O}_s$ are of order m.

Let $x \in M$. Then the trivial G-bundle P over Σ is equipped with the corresponding flat G-connection. Therefore the G-bundle $p^*P \to \Sigma^b$ is equipped with the corresponding flat connection. Moreover $\mathbf{Z}/m\mathbf{Z}$ acts naturally on this G-bundle and preserves the flat connection.

Observe that for $1 \leq j \leq s$, the holonomy of the flat connection over the circle $p^{-1}(S_j^1)$ —which is a m cover of S_j— is $w_j^m = 1$, i.e. π^*P has trivial holonomy around $\pi^{-1}(S_j)$. Therefore the flat connection on Σ^b extends to a flat connection on the trivial G-bundle $p^*P \to \Sigma_g^b$.

We claim the action of $\mathbf{Z}/m\mathbf{Z}$ extends to the bundle $p^*P \to \Sigma_g^b$. To define the flat bundle p^*P near x_j, we use the identification by parallel transport along the circle S_j^1. In fact recall that for $1 \leq j \leq s$, w_j is the holonomy of the flat connection along S_j^1 considered as lying in $\partial\Sigma$. The holonomy along S_j considered as the boundary of the disk Δ_j is w_j^{-1}. If $\ell \in \mathbf{Z}/m\mathbf{Z}$, $t \in \Sigma$, $f \in P$, then

$$\ell(t,f) = \left(e^{\frac{2i\pi\ell}{m}}t, f\right) . \tag{5.333}$$

However using the trivialization of parallel transport along $p^*S_j^1$, in this trivialization

$$\ell(t,f) = \left(e^{\frac{2i\pi\ell}{m}}t, w_j^\ell f\right) . \tag{5.334}$$

In particular the action of ℓ on $p^*P_{x_j}$ is given by $f \in P \mapsto w_j^\ell f \in P$.

Let V be a complex vector space. Let $\rho : G \to \operatorname{Aut}(V)$ be a representation of G. Let F be the flat vector bundle on Σ

$$F = P \times_G V . \tag{5.335}$$

Then by the above construction, p^*F extends to a flat vector bundle on Σ_g^b, on which $\mathbf{Z}/m\mathbf{Z}$ acts. In particular $\mathbf{Z}/m\mathbf{Z}$ acts on $H^\cdot(\Sigma_g^b, p^*F)$. Let $[H(\Sigma_g^b, p^*F)]^{\mathbf{Z}/m\mathbf{Z}}$ be the invariant part of $H(\Sigma_g^b, p^*F)$ under the action of $\mathbf{Z}/m\mathbf{Z}$.

Let $\langle\ ,\ \rangle$ be a G-invariant bilinear symmetric form on V. Let $\langle\ ,\ \rangle_{\widehat{H}^1(\Sigma,F)}$ and $\langle\ ,\ \rangle_{H^1(\Sigma_g^b,p^*F)}$ be the coresponding intersection forms on $\widehat{H}^1(\Sigma, F)$ and $H^1(\Sigma_g^b, p^*F)$.

PROPOSITION 5.77. *The following identity holds*

$$\widehat{H}^\cdot(\Sigma, F) = [H^\cdot(\Sigma_g^b, p^*F)]^{\mathbf{Z}/m\mathbf{Z}}. \tag{5.336}$$

Under the identification (5.336), if $\alpha, \alpha' \in \widehat{H}^1(\Sigma, F)$,

$$\langle\alpha,\alpha'\rangle_{\widehat{H}^1(\Sigma,F)} = \frac{1}{m}\langle\alpha,\alpha'\rangle_{H^1(\Sigma_g^b,p^*F)} . \tag{5.337}$$

PROOF. Clearly, $[H^0(\Sigma_g^b, p^*F)]^{\mathbf{Z}/m\mathbf{Z}}$ consists of flat $\mathbf{Z}/m\mathbf{Z}$-invariant sections of p^*F on Σ_g^b. In particular by for $1 \leq j \leq s$, $w_j f_{|x_j} = f_{|x_j}$. Therefore these sections descend to flat sections of F on Σ. Using (5.27) and (5.36), (5.336) holds in degree 0. By Theorem 5.4, using Poincaré duality, (5.336) holds in degree 2.

Let α be a $\mathbf{Z}/m\mathbf{Z}$-invariant closed form in $\Omega^1(\Sigma_g^b, p^*F)$ representing $[\alpha] \in [H^1(\Sigma_g^b,$
$p^*F)]^{\mathbf{Z}/m\mathbf{Z}}$. We may and will assume that α vanish near $x_1, \dots, x_s$. Then α descends to a smooth closed 1 form on Σ, which vanishes on $\partial\Sigma$. Also α is defined up to the coboundary of a $\mathbf{Z}/m\mathbf{Z}$-invariant form in $\Omega^0(\Sigma_g^b, p^*F)$. Using Theorem 5.4, we find there is a well-defined map $[H^1(\Sigma_g^b, p^*F)]^{\mathbf{Z}/m\mathbf{Z}} \to \widehat{H}^1(\Sigma, F)$. Conservely, if β is closed in $\Omega^1(\Sigma, F)$ and vanishes near $\partial\Sigma$, then $p^*\beta$ is a smooth $\mathbf{Z}/m\mathbf{Z}$-invariant closed 1 form in $\Omega^1(\Sigma_g^b, p^*F)$. So we have defined a map $\widehat{H}^1(\Sigma, F) \to [H^1(\Sigma_g^b, p^*F)]^{\mathbf{Z}/m\mathbf{Z}}$. It is now easy to verify these two maps are inverse to each other. The identity (5.337) follows trivially.

The proof of our Theorem is completed. □

Let $\widehat{\chi}(F)$ be the Euler characteristic of the complex $(\widehat{C}^\Sigma(F), \partial)$. Then

$$\begin{aligned}\widehat{\chi}(F) &= \sum_{i=0}^{2}(-1)^i \dim(\widehat{H}^i(\Sigma, F)) \\ &= \sum_{i=0}^{2}(-1)^i \dim(\widehat{C}^{\Sigma,i}(F)) .\end{aligned} \tag{5.338}$$

So by (5.32), (5.338),

$$\widehat{\chi}(F) = (2-2g)\dim F - \sum_{j=1}^{s}\dim(1-w_j)(F)) . \tag{5.339}$$

Let $\chi^{\mathbf{Z}/m\mathbf{Z}}(p^*F)$ be the invariant Euler characteric of p^*F on Σ_g^b, i.e.

$$\chi^{\mathbf{Z}/m\mathbf{Z}}(p^*F) = \sum_{i=0}^{2}(-1)^i \dim[H^i(\Sigma_g^b, p^*F)]^{\mathbf{Z}/m\mathbf{Z}} . \tag{5.340}$$

By Proposition 5.77,

$$\widehat{\chi}(F) = \chi^{\mathbf{Z}/m\mathbf{Z}}(p^*F) . \tag{5.341}$$

We will prove (5.341) again using the Theorem of Riemann-Roch-Kawasaki [**32, 33**], stated in Theorem 6.8. We get

$$\chi^{\mathbf{Z}/m\mathbf{Z}}(p^*F) = \frac{1}{m}\left[\dim(F)\int_{\Sigma_g^b} e(T\Sigma_g^b) + \sum_{j=1}^{s}\mathrm{Tr}^F\left[\sum_{k=1}^{m-1} w_j^k\right]\right] . \tag{5.342}$$

Now using (5.330), we get

$$\int_{\Sigma_g^b} e(T\Sigma_g^b) = 2-2g' = m(2-2g) - (m-1)s . \tag{5.343}$$

Moreover

$$\frac{1}{m}\sum_{k=1}^{m-1} w_j^k = \frac{1}{m}\sum_{k=0}^{m-1} w_j^k - \frac{1}{m} . \tag{5.344}$$

Now $\frac{1}{m}\sum_1^{m-1} w_j^k$ is a projection on $\{f \in F, w_j f = f\}$. So by (5.344),

$$\frac{1}{m}\operatorname{Tr}^F\left[\sum_1^{m-1} w_j^k\right] = \dim\ker((w_j - 1)_{|F}) - \frac{\dim F}{m}. \tag{5.345}$$

By (5.342)-(5.345),

$$\chi^{\mathbf{Z}/m\mathbf{Z}}(p^*F) = (2-2g)\dim F - \sum_{j=1}^{s}\dim((1-w_j)(F)), \tag{5.346}$$

which fits with (5.339), (5.341).

Now we assume that $\rho : G \to \operatorname{Aut}(V)$ is a real representation of G. We will now give another proof of (5.346). Let $\chi^{h,\mathbf{Z}/m\mathbf{Z}}(p^*F)$ be the holomorphic invariant Euler characteristic

$$\chi^{h,\mathbf{Z}/m\mathbf{Z}}(p^*F) = \sum_{i=0}^{1}\dim\left[H^{0,i}(\Sigma_g^b, p^*F)\right]^{\mathbf{Z}/m\mathbf{Z}}. \tag{5.347}$$

By Hodge theory,

$$\chi^{\mathbf{Z}/m\mathbf{Z}}(p^*F) = 2\chi^{h,\mathbf{Z}/m\mathbf{Z}}(p^*F). \tag{5.348}$$

Also by the Riemann-Roch-Kawasaki theorem [**32**, **33**],

$$\chi^{h,\mathbf{Z}/m\mathbf{Z}}(p^*F) = \frac{1}{m}\Big[\dim(F)\int_{\Sigma_g^b}\frac{1}{2}c_1(T\Sigma_g^b) + \tag{5.349}$$

$$\sum_{j=1}^{s}\sum_{k=1}^{m-1}\frac{1}{1-e^{-2i\pi k/m}}\operatorname{Tr}^F[w_j^k].$$

By (5.343),

$$\frac{1}{m}\int_{\Sigma_g^b}\frac{1}{2}c_1(T\Sigma_g^b) = 1 - g - \frac{1}{2}(1-\frac{1}{m})s. \tag{5.350}$$

The eigenvalues of $w_{j|V}$ have absolute value 1. Since V is real, they are either ± 1, or they come as complex conjugate pairs. By Proposition 5.75, we get

$$\frac{1}{m}\sum_{k=1}^{m-1}\frac{1}{1-e^{-2i\pi k/m}}\operatorname{Tr}^V[w_j^k] = -\frac{1}{2m}\dim F + \frac{1}{2}\dim\ker(w_j - 1)_{|F}. \tag{5.351}$$

By (5.349)-(5.351), we obtain

$$\chi^{h,\mathbf{Z}/m\mathbf{Z}}(p^*F) = (1-g)\dim F - \frac{1}{2}\sum_1^{s}\dim((1-w_j)(F)), \tag{5.352}$$

which fits with (5.346).

5.13. Evaluation of $c_1(TM/G)$. In the sequel, we assume that G is a connected simply connected compact simple Lie group. Otherwise we use the notation in Section 1. In particular $\langle\ ,\ \rangle$ denotes the basic scalar product on $\mathfrak{g}$ defined in Section 1.2.

As in Section 5.10, we construct a connection on the G-bundle $\frac{M\times\widehat{\Sigma}}{\Gamma} \xrightarrow{G} M/G \times \Sigma$, such that the assumptions after (5.252) hold. In particular, for $1 \leq j \leq s$,

$$\begin{aligned} \nabla_{j.} w_j &= 0, \\ \nabla_{j.} \log(w_j) &= 0. \end{aligned} \tag{5.353}$$

Also since t_j lies in an alcove P, it determines $\rho_j \in \overline{CR}^* \cap P$ by formula (1.35). Over M/G, t_j, ρ_j descend to sections of E_j. By (5.353), for $1 \leq j \leq s$,

$$\nabla_j \rho_j = 0. \tag{5.354}$$

Let θ_j be the connection form on associated to ∇_j, let Θ_j be its curvature. Then θ_j can be considered as a t-connection, and Θ_j is a t-valued 2 form on M/G.

Recall that TM/G is a symplectic vector bundle. Let $J^{TM/G}$ be any almost complex structure polarizing σ, i.e. σ is $J^{TM/G}$ invariant, and $U, V \in TM/G \mapsto \sigma(J^{TM/G}U, V)$ is a scalar product. Such $J^{TM/G}$ exist and are homotopic. Therefore $c_1(TM/G)$ is a well-defined element of $H^2(M/G, \mathbf{Q})$.

Recall that c is the dual Coxeter number defined in Definition 1.7.

THEOREM 5.78. *The following identity hold,*

$$c_1(TM/G) = 2(c\omega + \sum_{j=1}^{s} \langle ct_j - \rho_j, \Theta_j \rangle). \tag{5.355}$$

PROOF. First we assume that m$\in \mathbf{N}$, that $t_1, \ldots, t_s$ are of order m, and $s|m$. Then the G-bundle $(M \times \widehat{\Sigma})/\Gamma \xrightarrow{G} M/G \times \Sigma$ lifts to a G-bundle $Q \xrightarrow{G} M/G \times \Sigma_g^b$ on which $\mathbf{Z}/m\mathbf{Z}$ acts naturally. In fact, we only need to make the lifting constructions of Section 5.12 fibrewise.

Recall that a complex structure has been fixed on Σ_g and Σ_g^b, and that $\mathbf{Z}/m\mathbf{Z}$ acts holomorphically on Σ_g^b.

By (5.120), we have the identity

$$T_{\mathbf{R}}M/G = \widehat{H}^1(\Sigma, E). \tag{5.356}$$

Using Proposition 5.77 and (5.356), we get

$$T_{\mathbf{R}}M/G = \left[H^1(\Sigma_g^b, p^*E)\right]^{\mathbf{Z}/m\mathbf{Z}}. \tag{5.357}$$

Now

$$H^1(\Sigma_g^b, p^*E) \otimes_{\mathbf{R}} \mathbf{C} = H^{(1,0)}(\Sigma_g^b, p^*E) \oplus H^{(0,1)}(\Sigma_g^b, p^*E), \tag{5.358}$$

and the splitting (5.358) is $\mathbf{Z}/m\mathbf{Z}$ invariant. By (5.356)-(5.359), we get

$$\widehat{H}^1(\Sigma, E) \otimes_{\mathbf{R}} \mathbf{C} = \left[H^{(1,0)}(\Sigma_g^b, p^*E)\right]^{\mathbf{Z}/m\mathbf{Z}} \oplus \left[H^{(0,1)}(\Sigma_g^b, p^*E)\right]^{\mathbf{Z}/m\mathbf{Z}}. \tag{5.359}$$

Let J be the complex structure on $H^1(\Sigma_g^b, p^*E)$ which is i on $H^{(1,0)}(\Sigma_g^b, p^*E)$, $-i$ on $H^{(0,1)}(\Sigma_g^b, p^*E)$. We claim that J polarizes the symplectic form ω on $\widehat{H}^1(\Sigma, E) = T_{\mathbf{R}}M/G$. In fact by if $\alpha, \alpha' \in \widehat{H}^1(\Sigma, E)$ are represented by the forms $\eta, \eta' \in \Omega^1(\Sigma_j^b, p^*E)$ which are closed and $\mathbf{Z}/m\mathbf{Z}$ invariant, then by Proposition 5.77,

$$\omega(\alpha, \alpha') = -\frac{1}{m}\int_{\Sigma_g^b} \langle \eta, \eta' \rangle. \tag{5.360}$$

It is now trivial to verify that J polarizes ω.

So by (5.356), (5.359),

$$c_1(TM/G) = c_1\left(\left[H^{(1,0)}(\Sigma_g^b, p^*E)\right]^{\mathbf{Z}/m\mathbf{Z}}\right), \tag{5.361}$$

which is equivalent to

$$c_1(TM/G) = -c_1\left(\left[H^{(0,1)}(\Sigma_g^b, p^*E)\right]^{\mathbf{Z}/m\mathbf{Z}}\right). \tag{5.362}$$

Let θ be a $\mathbf{Z}/m\mathbf{Z}$ invariant connection on the G-bundle $Q \xrightarrow{G} \Sigma_g^b \times M/G$. Observe that $(\{x_j\} \times M/G)_{1\le j\le s}$ are exactly the fixed point of the action of $\mathbf{Z}/m\mathbf{Z}$ over $\Sigma_g^b \times M/G$. Since the connection θ is $\mathbf{Z}/m\mathbf{Z}$-invariant, and since, for $1 \le j \le s$, $1 \in \mathbf{Z}/m\mathbf{Z}$ acts on E_{x_j} like w_j, we find that over M/G

$$\nabla w_{j_{|x_j}} = 0. \tag{5.363}$$

By (5.63),

$$H^0(\Sigma, E) = 0. \tag{5.364}$$

By (5.336), (5.364),

$$[H^0(\Sigma_g^b, p^*E)]^{\mathbf{Z}/m\mathbf{Z}} = 0. \tag{5.365}$$

Now we will use an equivariant version of the curvature theorem of Bismut and Freed [**10**]. Namely we equip $T\Sigma_g^b$ with a $\mathbf{Z}/m\mathbf{Z}$-invariant metric. Recall that p^*E is also equipped with a $\mathbf{Z}/m\mathbf{Z}$-invariant metric. Since $\mathbf{Z}/m\mathbf{Z}$ is a finite group, the construction sof [**9**, Definition 2.2] provide us with a like metric on the line bundle $\det([H^{(0,\cdot)}(\Sigma_g^b, p^*E)]^{\mathbf{Z}/m\mathbf{Z}})$. By proceeding as in [**10**], we also obtain a unitary connection on this line bundle. Incidently, observe that since all our data are holomorphic, we could instead use the holomorphic constructions of [**11**] in an equivariant context. The curvature of the connection on $\det([H^{(0,\cdot)}(\Sigma_g^b, p^*E)]^{\mathbf{Z}/m\mathbf{Z}})$ is obtained by applying the techniques of [**10**] or [**11**]. An important technical point is to prove an equivariant version of the local families index theorem of [**8**]. A large part of the steps which are needed in extending the results of [**8**] is already done in [**9**]. By mixing the techniques used in Lefschetz fixed point theory and in the prooof of the local families index theorem as in [**8**], [**6**], [**9**], one finds easily that the curvature is given by a differential form version of the theorem of Riemann-Roch-Kawasaki [**32, 33**].

Using (5.365) and the above considerations, we obtain

$$c_1\left(\left[H^{(0,1)}(\Sigma_g^b, E)\right]^{\mathbf{Z}/m\mathbf{Z}}\right) = -\frac{1}{m}\left(\int_{\Sigma_g^b} \mathrm{Td}(T\Sigma_g^b)\mathrm{ch}(p^*E, \nabla^{p^*E}) + \sum_{j=1}^{s}\sum_{k=1}^{m-1} \frac{1}{1-e^{-2i\pi k/m}} \mathrm{Tr}^E\left[w_j^k \exp\left(-\frac{\Theta_j}{2i\pi}\right)\right]\right)^{(2)}. \tag{5.366}$$

By (1.37) and by Theorems 5.55 and 5.56,

$$\left(\frac{1}{m}\int_{\Sigma_g^b} \mathrm{Td}(T\Sigma_g^b)\mathrm{ch}(p^*E, \nabla^{p^*E})\right)^{(2)} = 2c\omega. \tag{5.367}$$

Recall that

$$\mathfrak{g} \otimes_{\mathbf{R}} \mathbf{C} = \mathfrak{t} \oplus \left(\bigoplus_{\alpha \in \mathbf{R}} \mathfrak{g}_\alpha \right) . \tag{5.368}$$

Let S_j be the operator acting on $\mathfrak{g} \otimes_{\mathbf{R}} \mathbf{C}$,

$$S_j = \frac{1}{m} \sum_{k=1}^{m-1} \frac{1}{1 - e^{-2i\pi k/m}} \tau(t_j^k) . \tag{5.369}$$

By Proposition 5.75, we find that

$$\begin{aligned} S_{j|\mathfrak{t}} &= \frac{1}{2} - \frac{1}{2m} , \\ S_{j|\mathfrak{g}_\alpha} &= \frac{1}{2} - [\langle \alpha, t_j \rangle] - \frac{1}{2m}, \ \alpha \in R . \end{aligned} \tag{5.370}$$

By conjugation by an element G, we may and we will assume that $w_j = t_j$. Then

$$\left(\sum_{k=1}^{m-1} \frac{1}{1 - e^{-2i\pi k/m}} \mathrm{Tr}^E \left[w_j^k \exp \left(- \frac{\Theta_j}{2i\pi} \right) \right] \right)^{(2)} = -\mathrm{Tr}^{\mathfrak{g}} \left[S_j \frac{\Theta_j}{2i\pi} \right] . \tag{5.371}$$

Also because $w_j = t_j$, $\nabla_{x_j}^{E,2}$ is a 2 form on M/G with values in $\mathfrak{t}$. From (5.370), we get easily

$$\begin{aligned} -\mathrm{Tr}^{\mathfrak{g}} \left[S_j \frac{\Theta_j}{2i\pi} \right] &= -\sum_{\alpha \in \mathbf{R}} \left(\frac{1}{2} - [\langle \alpha, t_j \rangle] \right) \langle \alpha, \Theta_j \rangle \\ &= \sum_{\alpha \in \mathbf{R}} [\langle \alpha, t_j \rangle] \langle \alpha, \Theta_j \rangle . \end{aligned} \tag{5.372}$$

Now by Proposition 1.20, and by (5.372), we get

$$-\mathrm{Tr}^{\mathfrak{g}} \left[S_j \frac{\nabla_{x_j}^{E,2}}{2i\pi} \right] = 2 \langle c t_j - \rho_j, \Theta_j \rangle . \tag{5.373}$$

From (5.362), (5.366), (5.367), (5.371)-(5.373), we get

$$c_1(TM/G) = 2 (c\omega + \sum_{j=1}^{s} \langle c t_j - \rho_j, \Theta_j \rangle) . \tag{5.374}$$

We claim now that in the special case when the orbits $\mathcal{O}_1, \dots, \mathcal{O}_m$ are of order m, (5.374) is exactly (5.355). In spite of the formal simultanities, the objects introduced in both equation are not exactly of the same kind. However we leave to the reader the verification that they indeed coincide .

Now we establish (5.355) in full generality. Clearly if $(t_1, \dots, t_s)$ are regular and verify (A), they can be approximated by a sequence of regular elements $(t_1^m, \dots, t_s^m)$ which verify (A) and are of order m. Let $s_m \geq s$ be such that $m | s_m$. Then we consider the above situation, with s replaced by s_m. At $x_1, \dots, x_s$, we assume that the holonomies are $w_1, \dots, w_s$ and at $x_{s+1}, \dots, x_{s_m}$, they are 1. Needless to say, since 1 is far from being regular, we do not impose any restriction on the connection $\nabla_j, j \geq s+1$. Since for $j \geq s+1$, $\mathrm{Tr}^{\mathfrak{g}}[\Theta_j] = 0$, it is clear that in the final formula, the $j \geq s+1$ do not contribute, so that (5.374) still hold. Then by the above (5.355) holds. A trivial limit procedure shows that (5.355) still holds in full generality. □

REMARK 5.79. By Remark 5.66, the cohomology class of $\omega+\sum_{j=1}^{s}\langle\log(w_j),\Theta_j\rangle$ is locally constant. Observe that in (5.355), $\langle\rho_j,\Theta_j\rangle$ is a closed form on M/G whose cohomology class does not depend locally on $t_1,\dots,t_s$. So Theorem 5.78 fits with the above considerations.

Assume temporarily that some t_j lie in $Z(G)=\overline{R}^*/\overline{CR}$. By (5.372), we find that such t_j do not contribute to formula (5.355) for $c_1(TM/G)$. In other words if the t_j are either regular or central, formula (5.355) still holds, where in the right-hand side, the summation is limited to a sum over the regular orbits.

6. The Riemann-Roch-Kawasaki formula on the moduli space of flat bundles

The purpose of this Section is to give formula for the index of a Dirac operator on the moduli space M/G of flat vector G bundles, by using the theorem of Riemann-Roch-Kawasaki [**32, 33**]. To do this, we describe the strata of M/G and we express the contribution of each stratum as a residue in several variables, using the results of Sections 2 and 5. The results of this Section were already obtained by Szenes [**53**] for $G = \mathrm{SU}(3)$ and Jeffrey-Kirwan [**30**] in the case $G = \mathrm{SU}(n), s = 1$, with a central holonomy at the marked point for which M/G is smooth.

This Section is organized as follows. In Section 6.1, we describe the strata of a general orbifold, and we introduce various associated characteristic classes. In Section 6.2, we state the theorem of Riemann-Roch-Kawasaki for almost complex orbifolds. In Section 6.3, we construct the orbifold line bundle λ^p on the orbifold M/G. In Section 6.4, we describe the strata of the moduli space M/G as moduli spaces associated to semisimple centralizers in G. In Section 6.5, we compute the Atiyah-Bott-Lefschetz-Todd class of a given stratum. In Section 6.6, we compute the dimension of certain vector spaces which appear naturally in the evaluation of the Atiyah-Bott-Lefschetz class.

Then we make a number of genericity assumptions on the t_j's. In Section 6.7, we compute the contribution of a stratum to the Riemann-Roch-Kawasaki formula in terms of differential operators acting on symplectic volumes. In Section 6.8, we briefly show that under an obvious condition on the holonomies $t_j, 1 \leq j \leq s$, the index of the considered Dirac operator vanishes identically. In Section 6.9, we give a residue formula for the index. In Section 6.10, we give another related asymptotic formula for $|p|$ large.

Then we drop the genericity assumptions. In Section 6.11, we compute the index of a Dirac operator on a perturbed moduli space, for which the genericity assumptions hold. The point is that, as we shall see in Section 7, the index of the Dirac operator for the perturbed moduli space is exactly given by the Verlinde formula. Under genericity assumptions, this is only true asymptotically for the given moduli space M/G.

6.1. Almost complex orbifolds. Let M be a smooth compact manifold. Let G be a compact connected Lie group, and let $\mathfrak{g}$ be its Lie algebra. We assume that G acts on M on the left. If $X \in \mathfrak{g}$, let $X^M \in \mathrm{Vect}(M)$ be the corresponding vector field.

We assume that G acts locally freely on M, i.e. for any non zero $X \in \mathfrak{g}$, X^M is a non vanishing vector field on M.

We will use here the notation of Section 3.1, with X replaced by M. Then M/G is an orbifold.

DEFINITION 6.1. If $g \in G$, put

$$M^g = \{x \in M\,, gx = x\}\,. \tag{6.1}$$

Set

$$H = \{g \in G\,; M^g \neq \phi\}\,. \tag{6.2}$$

Then H is a finite union of conjugacy classes in G. Let (H) be the corresponding finite set of conjugacy classes.

If $g \in G$, then $Z(g)$ acts locally freely on M^g. We can then apply the above constructions to M^g. Let $H_g \subset Z(g)$ be the generic stabilizer of M^g.

Observe that if $g' \in G$ is conjugate to g, the above constructions correspond by conjugation.

Take $g \in H$. Let $N_{M^g/M} \simeq TM/TM^g$ be the normal bundle to M^g in M. Then we have the complex of $Z(g)$ vector bundles over M^g,

$$
\begin{array}{ccccccccc}
 & & 0 & & 0 & & 0 & & \\
 & & \uparrow & & \uparrow & & \uparrow & & \\
0 & \longrightarrow & \mathfrak{g}^M/\mathfrak{z}^M(g) & \longrightarrow & N_{M^g/M} & \longrightarrow & N_{M^g/M}/(\mathfrak{g}^M/\mathfrak{z}^M(g)) & \longrightarrow & 0 \\
 & & \uparrow & & \uparrow & & \uparrow & & \\
0 & \longrightarrow & \mathfrak{g}^M & \longrightarrow & TM & \longrightarrow & TM/\mathfrak{g}^M & \longrightarrow & 0 \\
 & & \uparrow & & \uparrow & & \uparrow & & \\
0 & \longrightarrow & \mathfrak{z}^M(g) & \longrightarrow & TM^g & \longrightarrow & TM^g/\mathfrak{z}^M(g) & \longrightarrow & 0 \\
 & & \uparrow & & \uparrow & & \uparrow & & \\
 & & 0 & & 0 & & 0 & &
\end{array}
\tag{6.3}
$$

and the rows and columns in (6.3) are acyclic.

Clearly if $g' \in G$, g' maps M^g into $M^{g'gg'^{-1}}$. Also g' acts on the complex (6.3).

Put

$$N^g = N_{M^g/M}/(\mathfrak{g}^M/\mathfrak{z}^M(g)). \tag{6.4}$$

Then the third column in (6.3) is the exact sequence of $Z(g)$ vector bundles on M^g,

$$0 \to TM^g/Z(g) \to TM/G \to N^g \to 0. \tag{6.5}$$

Equivalently N^g is the "normal bundle" to $TM^g/Z(g)$ into M/G.

Clearly g acts on each of the vector bundles in (6.3). Then g acts like 1 on $\mathfrak{z}^M(g), TM^g$ and $TM^g/Z(g)$. In particular , there are locally constants $\theta, 0 < \theta < \pi$ on M^g such that

$$N_{M^g/M} \otimes_{\mathbf{R}} \mathbf{C} = \bigoplus_{0<\theta<\pi} (N^{\theta}_{M^g/M} \oplus N^{-\theta}_{M^g/M}) \oplus N^{\pi}. \tag{6.6}$$

In (6.6), the θ's are distinct, g acts on the left on $N^{\theta}_{M^g/M}, N^{-\theta}_{M^g/M}$ by multiplication by $e^{i\theta}, e^{-i\theta}$, and on N^{π} by multiplication by -1. Therefore $N^g \otimes_{\mathbf{R}} \mathbf{C}$ splits as

$$N^g \otimes_{\mathbf{R}} \mathbf{C} = \bigoplus_{0<\theta<\pi} (N^{g,\theta} \oplus N^{g,-\theta}) \oplus N^{g,\pi}. \tag{6.7}$$

DEFINITION 6.2. The orbifold M/G will be said to be almost complex if $TM/G = TM/\mathfrak{g}^M$ is equipped with a G-invariant almost complex structure $\mathcal{J}^{TM/G}$.

Since in (6.5), $TM^g/Z(g)$ is the $+1$ eigenspace of g on TM/G, we find that for any $g \in G$, $M^g/Z(g)$ is an almost complex orbifold. Therefore N^g is also equipped with an almost complex structure.

Now we denote with the superscript $(1,0)$ the $+i$ eigenspace of the given complex structure. In particular $T^{(1,0)}M^g/Z(g)$ is well defined, and $N^{g,(1,0)}$ splits as

$$N^{g,(1,0)} = \bigoplus_{\theta \in]-\pi,\pi]\setminus\{0\}} N^{g,(1,0),\theta}. \tag{6.8}$$

In the sequel, we will assume that the orbifold M/G is almost complex.

6.2. The Theorem of Riemann-Roch-Kawasaki. Let $\nabla^{T^{(1,0)}M_g/Z(g)}, \ldots, \nabla^{N^{g,(1,0),\theta}}$ be $Z(g)$ invariant horizontal connections on $T^{(1,0)}M_g/Z(g), N^{g,(1,0),\theta}$.

Let E be a complex G-vector bundle on M, equipped with a G-invariant horizontal connection ∇^E. Let F^E be the curvature of ∇^E. Then $(E,\nabla^E)_{|M^g}$ is a $Z(g)$-vector bundle equipped with a $Z(g)$ invariant connection.

DEFINITION 6.3. Let $\mathrm{ch}_g(E,\nabla^E)$ be the closed form on $M^g/Z(g)$

$$\mathrm{ch}_g(E,\nabla^E) = \mathrm{Tr}\left[g \exp\left(\frac{-F^E}{2i\pi}\right)\right]. \tag{6.9}$$

In (6.9), g denotes the left action of the given element of G on on E. Let $\mathrm{ch}_g(E)$ be the cohomology class associated to the form $\mathrm{ch}_g(E,\nabla^E)$.

DEFINITION 6.4. If B is a square matrix, put

$$\begin{aligned} \mathrm{Td}(B) &= \det\left(\frac{B}{1-e^{-B}}\right), \\ \widehat{\mathrm{Td}}(B) &= \det\left(\frac{1}{1-e^{-B}}\right), \\ \widehat{A}(B) &= \det\left(\frac{B}{2\sinh(B/2)}\right). \end{aligned} \tag{6.10}$$

Observe that

$$\mathrm{Td}(B) = \widehat{A}(B)e^{\frac{1}{2}\mathrm{Tr}[B]} \tag{6.11}$$

DEFINITION 6.5. Put

$$\begin{aligned} \mathrm{Td}(T^{(1,0)}M_g/Z_g, \nabla^{T^{(1,0)}M_g/Z_g}) &= \mathrm{Td}\left(-\frac{F^{T^{(1,0)}M_g/Z_g}}{2i\pi}\right), \\ \widehat{\mathrm{Td}}(N^{g(1,0)}, \nabla^{N^{g(1,0)}}) &= \prod_{\theta\in]-\pi,\pi]\setminus\{0\}} \widehat{\mathrm{Td}}\left(i\theta - \frac{F^{N^{g,(1,0),\theta}}}{2i\pi}\right). \end{aligned} \tag{6.12}$$

Then the forms in (6.12) are closed on $M_g/Z(g)$. Let $\mathrm{Td}(T^{(1,0)}M_g/Z_g), \widehat{\mathrm{Td}}(N^{g,(1,0)})$ be the corresponding cohomology classes.

DEFINITION 6.6. Put

$$L(g,E) = \mathrm{Td}(T^{(1,0)}M_g/Z_g)\widehat{\mathrm{Td}}(N^{g,(1,0)})\mathrm{ch}_g(E). \tag{6.13}$$

Then $\int_{M^g/Z(g)} L(g,E)$ depends only on the conjugacy class of g in G.

Let $h^{T^{(1,0)}M}, h^E$ be G-invariant metrics on $T^{(1,0)}M, E$. Let $\nabla^{T^{(1,0)}M/G}, \nabla^E$ be G-invariant unitary horizontal connections on $T^{(1,0)}M, E$. Then $\Lambda T^{*(1,0)}M/G \otimes E$ is naturally equipped with a G-invariant unitary horizontal connection $\nabla^{\Lambda T^{*(1,0)}M/G\otimes E}$.

Recall that by [**6**, p 135] , $\Lambda T^{*(1,0)}M/G \otimes E$ is a TM/G Clifford module. If $X \in TM(G)$, let $c(X)$ be the corresponding Clifford multiplication operator.

Let dv be the volume element on M_{reg}/G associated to $h^{T^{(1,0)}M/G}$.

Let $K = K_+ \oplus K_-$ be the vector space of G-invariant C^∞ sections of $\Lambda(T^{*(0,1)}M/G) \otimes E = (\Lambda^{\mathrm{even}}(T^{*(0,1)}M/G) \otimes E) \oplus (\Lambda^{\mathrm{odd}}(T^{*(0,1)}M/G) \otimes E)$. We equip H with the Hermitian product

$$s, s' \in H \mapsto \langle s, s' \rangle = \int_{M_{\mathrm{reg}}/G} \langle s, s' \rangle_{\Lambda(T^{*(0,1)}M/G)\otimes E} dv \,. \tag{6.14}$$

Let $e_1, \dots, e_n$ be an orthonormal basis of TM/G.

DEFINITION 6.7. Let D^M be the Dirac operator acting on K

$$D^M = \sum_1^n c(e_i) \nabla_{e_i}^{\Lambda(T^{*(0,1)}M/G)\otimes E} . \tag{6.15}$$

Then D^M is a formally self-adjoint operator which exchanges K_+ and K_-. Let D_+^M be the restriction of D^M to K_+. Then we write D^M in matrix form as

$$D^M = \begin{bmatrix} 0 & D_-^M \\ D_+^M & 0 \end{bmatrix} \tag{6.16}$$

Then by [**32, 33**], D_+^M is a Fredholm operator. Its index $\mathrm{Ind}(D_+^M)$ is given by

$$\mathrm{Ind}(D_+^M) = \dim\ker D_+^M - \dim\ker D_-^M . \tag{6.17}$$

If $\gamma \in (H)$, let $g_\gamma \in \gamma$ be any representative in G of the conjugacy class γ. Now we state the theorem of Riemann-Roch-Kawasaki [**32, 33**].

THEOREM 6.8. *The following identity holds*

$$\mathrm{Ind}(D_+^M) = \sum_{\gamma \in (H)} \int_{M^{g_\gamma}/Z(g_\gamma)} \frac{1}{|H_{g_\gamma}|} L(g_\gamma, E) . \tag{6.18}$$

6.3. The line bundle λ^p. From now on, we suppose that all the assumptions of Section 5.11 are in force. Also we fix once and for all a positive Weyl chamber K and the corresponding alcove $P \subset K$ whose closure contains 0. Finally we may and we will assume that

$$t_j \in P,\ 1 \le j \le s. \tag{6.19}$$

DEFINITION 6.9. Let $\mathbf{M} \in \mathbf{Z}$ be given by

$$\mathbf{M} = \{p \in \mathbf{Z}, pt_1, \dots, pt_s \in \overline{CR}^*\}. \tag{6.20}$$

In the sequel we assume that $\mathbf{M}$ is not reduced to 0. Then there is $p_0 \in \mathbf{N}^*$ such that

$$\mathbf{M} = p_0 \mathbf{Z}. \tag{6.21}$$

Let $p \in \mathbf{M}$. Put

$$\theta_j = pt_j,\ 1 \le j \le s. \tag{6.22}$$

Recall that the set of connections $\mathcal{A}_p(\theta_1, \dots, \theta_s)$ was defined in Definition 4.36. Also the Hermitian line bundle with unitary connection $(\lambda_p, \nabla^{\lambda_p})$ on $\mathcal{A}_p(\theta_1, \dots, \theta_s)$ was defined in Definition 4.37. By Proposition 4.38, ΣG acts on the right on λ_p and preserves ∇^{λ_p}. Finally $\mathcal{A}^{\mathrm{flat}}(t_1, \dots, t_s)$ was defined in Definition 5.26.

Clearly if $p \in \mathbf{M}$, if $\theta_1, \dots \theta_s$ are given by (6.22), then

$$\mathcal{A}^{\mathrm{flat}}(t_1, \dots, t_s) \subset \mathcal{A}_p(\theta_1, \dots, \theta_s). \tag{6.23}$$

Therefore the line bundle λ_p restricts to $\mathcal{A}^{\mathrm{flat}}(t_1, \dots, t_s)$. Also ΣG acts on $\mathcal{A}^{\mathrm{flat}}(t_1, \dots, t_s)$. Finally by Proposition 5.28,

$$\mathcal{A}^{\mathrm{flat}}(t_1, \dots, t_s)/\Sigma_q G \simeq M. \tag{6.24}$$

DEFINITION 6.10. Let $\lambda^p, \nabla^{\lambda^p}$ be the Hermitian line bundle with unitary connection over M of the $\Sigma_q G$-invariant sections of λ_p on $\mathcal{A}^{\mathrm{flat}}(t_1, \dots, t_s)$.

The notation for λ^p is justified by the fact that if $p \in \mathbf{M}, p' \in \mathbf{Z}$, then

$$\lambda^{pp'} = (\lambda^p)^{\otimes p'}. \tag{6.25}$$

It is then clear that the action of G on M lifts to λ^p. Recall that ω is the canonical symplectic form on M/G which was defined in Definition 5.30, which is associated to the basic scalar product $\langle , \rangle$ on $\mathfrak{g}$.

PROPOSITION 6.11. *The following identity of closed 2-forms holds on M,*

$$c_1(\lambda^p, \nabla^{\lambda^p}) = p\omega. \tag{6.26}$$

PROOF. Let $A \in \mathcal{A}^{\mathrm{flat}}(t_1, \dots, t_s)$. Let α, α' be 2 closed forms in $\Omega^1(\Sigma, E)$, which are exact on $\partial\Sigma$, i.e.. there are $\beta, \beta' \in \Omega^0(\partial\Sigma, E)$ such that $\alpha_{|\partial\Sigma} = \nabla^A \beta, \alpha'_{|\partial\Sigma} = \nabla^A \beta'$. By (4.24),(4.78), (4.100), (4.189),

$$c_1(\lambda^p, \nabla^{\lambda^p})(\alpha, \alpha') = p \int_\Sigma -\langle \alpha, \alpha' \rangle + p \int_{\partial\Sigma} \langle \beta, \nabla^A \beta' \rangle. \tag{6.27}$$

If $[\alpha], [\alpha']$ are the classes of α, α' in $\widehat{H}^1(\Sigma, E)$, from (6.27), we get

$$c_1(\lambda^p, \nabla^{\lambda^p})(\alpha, \alpha') = p\omega([\alpha], [\alpha']). \tag{6.28}$$

The proof of our Theorem is completed. □

Since ω is a symplectic form, there is an almost complex structure J on TM/G which polarizes ω, i.e. $\omega(JX, Y)$ is a Riemannian metric on TM/G. Also J is unique up to homotopy. In the sequel, we will always equip TM/G with such a complex structure.

Then we will apply the theorem of Riemann-Roch-Kawasaki [**32, 33**] the orbifold M/G and the orbifold line bundle λ^p. Up to now, we have made G act on M or on λ^p on the right. However to fit with the formalism of Sections 6.1 and 6.2 , we will now make G act on the left by setting $gx = xg^{-1}$.

6.4. The Theorem of Riemann-Roch-Kawasaki on the moduli space of flat bundles. Recall that

$$M = \{x \in G^{2g} \times \prod_{j=1}^{s} \mathcal{O}_j \,, h(x) = 1\}. \tag{6.29}$$

We will use the notation

$$M = M(G, \mathcal{O}_1, \dots, \mathcal{O}_s). \tag{6.30}$$

Let τ be projection $T = \mathfrak{t}/\overline{CR} \to T' = \mathfrak{t}/\overline{R}^*$.

THEOREM 6.12. *The following identity holds*

$$(H) = W\backslash C/\overline{CR}. \tag{6.31}$$

Also if $v \in C/\overline{CR}$, *if* $u = \tau v \in C/\overline{R}^*$, $M^v = M^u$. *Moreover*

$$M^u = \bigcup_{(w^1,\dots,w^s)\in(W_u\backslash W)^s} M(Z(u), \mathcal{O}_{Z(u)}(w^1t_1), \dots, \mathcal{O}_{Z(u)}(w^st_s)), \tag{6.32}$$

and the union in (6.32) is disjoint.Finally, if $v \in C/\overline{CR}$,

$$H_v = Z(Z(v)). \tag{6.33}$$

PROOF. Clearly, if $u \in G'$, then $ugu^{-1} = g$ if and only if $g \in Z(u)$. It is then clear that if $u \in G'$,

$$M^u = \{x \in Z(u)^{2g} \times \prod_{j=1}^{s}(\mathcal{O}_j \cap Z(u)),\ h(x) = 1\}. \tag{6.34}$$

Since G acts locally freely on M, if $M^u \neq$, $Z(u)$ is semisimple. By Theorem 1.38, we get (6.31). If $u \in C/\overline{CR}$, using Theorem 1.50 and (6.34), we get (6.32). Finally by Theorem 5.20, we obtain (6.33). The proof of our Theorem is completed. □

REMARK 6.13. By Proposition 1.40, if $G = \mathrm{SU}(n)$, $n \geq 2$, if $u \in C$, then $u \in \overline{R}^*$. From Theorem 6.12, it follows that G acts freely on M, so that M/G is a smooth manifold.

Clearly $Z(G) = \overline{R}^*/\overline{CR} \subset T$ is fixed by W. Therefore if $v \in T$, W_v depends only on $u = \tau v \in \mathfrak{t}/\overline{R}^*$. We will then write W_u instead of W_v.

We use the notation of Section 1. In particular $\pi_u : \widetilde{Z}(u) \to Z(u)$ is the universal cover of $Z(u)$.

Let $u \in C/\overline{R}^*$, let $t \in T = \mathfrak{t}/\overline{CR}$ be regular. Then $Z(t) = T$, and $\mathcal{O}_{Z(u)}(t) \simeq Z(u)/T$.

Let $\widetilde{t} \in \mathfrak{t}/\overline{CR}_u$ be a lift of t in $\widetilde{Z}(u)$. Then $\widetilde{t}$ is still regular in $\widetilde{Z}(u)$. Since $Z(t) = T$, the centralizer $Z(\widetilde{t})$ in $\widetilde{Z}(u)$ is just $Z(\widetilde{t}) = \mathfrak{t}/\overline{CR}_u$. Then $\mathcal{O}_{\widetilde{Z}(u)}(\widetilde{t}) \simeq \widetilde{Z}(u)\big/(\mathfrak{t}/\overline{CR}_u) \simeq Z(u)/T$. Equivalently the projection $\mathcal{O}_{\widetilde{Z}(u)}(\widetilde{t}) \to \mathcal{O}_{Z(u)}(t)$ is one to one.

Take $u \in C/\overline{R}^*$, $(w^1, \dots, w^s) \in W^s$. Recall that here $t_1, \dots, t_s$ are also considered as elements of $\mathfrak{t}$, so that $w^1t_1, \dots, w^st_s \in \mathfrak{t}$. Ultimately, we may consider $w^1t_1, \dots, w^st_s$ as element of $\widetilde{Z}(u)$. Then by the above, $\mathcal{O}_{\widetilde{Z}(u)}(w^1t_1), \dots, \mathcal{O}_{\widetilde{Z}(u)}(w^st_s)$ lift $\mathcal{O}_{Z(u)}(w^1t_1), \dots, \mathcal{O}_{Z(u)}(w^st_s)$, and the projection π_u identifies the corresponding orbits.

Let $x = (u_1, v_1, \dots, u_g, v_g, w_1, \dots, w_s) \in M(Z(u), \mathcal{O}_{Z(u)}(w^1t_1), \dots, \mathcal{O}_{Z(u)}(w^st_s))$. Let $\widetilde{u}_1, \widetilde{v}_1, \dots, \widetilde{u}_g, \widetilde{v}_g \in \widetilde{Z}(u)$ be lifts of $u_1, v_1, \dots, u_g, v_g \in Z(u)$. Also $w_1, \dots, w_s \in \mathcal{O}_{Z(u)}(w^1t_1), \dots, \mathcal{O}_{Z(u)}(w^st_s)$ lift uniquely to $\widetilde{w}_1, \dots, \widetilde{w}_s \in \mathcal{O}_{\widetilde{Z}(u)}(w^1t_1), \dots, \mathcal{O}_{\widetilde{Z}(u)}(w^st_s)$.

PROPOSITION 6.14. *The element* $\prod_{i=1}^{g}[\widetilde{u}_i, \widetilde{v}_i] \prod_{j=1}^{s} \widetilde{w}_j \in \widetilde{Z}(u)$ *does not depend on* $\widetilde{u}_1, \dots, \widetilde{v}_g$. *It lies in* $\pi_1(Z(u)) = \overline{CR}/\overline{CR}_u$.

PROOF. Since $\pi_1(Z(u)) \subset Z(\widetilde{Z}(u))$, the first part of the Proposition is trivial. Also,

$$\pi_u \left[\prod_{i=1}^{g} [\widetilde{u}_i, \widetilde{v}_i] \prod_{j=1}^{s} \widetilde{w}_j \right] = \prod_{i=1}^{g} [u_i, v_i] \prod_{j=1}^{s} w_j = 1 . \tag{6.35}$$

From (6.35), we get the second part of the Proposition. □

Using Proposition 6.14 , we can now define:

DEFINITION 6.15. If $u \in C/\overline{CR}$, $(w^1, \dots, w^s) \in W^s$, $h \in \pi_1(Z(u)) = \overline{CR}/\overline{CR}_u$ put

$$\begin{gathered} M(Z(u), \mathcal{O}_{Z(u)}(w^1 t_1), \dots, \mathcal{O}_{Z(u)}(w^s t_s), h) = \\ \Big\{ x = (u_1, \dots, u_g, v_g, w_1, \dots, w_s) \in M(Z(u), \mathcal{O}_{Z(u)}(w^1 t_1), \dots, \mathcal{O}_{Z(u)}(w^s t_s)), \\ \textstyle\prod_{i=1}^{g} [\widetilde{u}_i, \widetilde{v}_i] \prod_{j=1}^{s} \widetilde{w}_j h = 1 \Big\} . \end{gathered} \tag{6.36}$$

Clearly, we have the disjoint union

$$\begin{gathered} M(Z(u), \mathcal{O}_{Z(u)}(w^1 t_1), \dots, \mathcal{O}_{Z(u)}(w^s t_s)) = \\ \textstyle\bigcup_{h \in \overline{CR}/\overline{CR}_u} M(Z(u), \mathcal{O}_{Z(u)}(w^1 t_1), \dots, \mathcal{O}_{Z(u)}(w^s t_s), h) . \end{gathered} \tag{6.37}$$

Also since $\overline{CR}/\overline{CR}_u \subset Z(\widetilde{Z}(u))$, $Z'(u)$ preserves each $M(Z(u), \mathcal{O}_{Z(u)}(w^1 t_1), \dots, \mathcal{O}_{Z(u)}(w^s t_s), h)$.

By Definition 6.10, if $p \in \mathbf{M}$, there is a well-defined G-orbifold line bundle λ^p on M/G. Let D_p be the corresponding Dirac operator acting on smooth sections of $\Lambda(T^{*(0,1)} M/G) \otimes \lambda^p$ over M/G.

THEOREM 6.16. *For $p \in \mathbf{M}$, the following identity holds*

$$\operatorname{Ind}(D_{p,+}) = \sum_{u \in C/\overline{CR}} \frac{|W_u|}{|W|} \frac{1}{|Z(Z(u))|} \int_{M^u/Z(u)} L(u, \lambda^p) . \tag{6.38}$$

PROOF. By Theorems 6.8 and 6.12,

$$\operatorname{Ind}(D_{p,+}) = \sum_{u \in W \backslash C/\overline{CR}} \frac{1}{|Z(Z(u))|} \int_{M^u/Z(u)} L(u, \lambda^p) \tag{6.39}$$

from which (6.38) follows. The proof of our Theorem is completed. □

6.5. Evaluation of the Atiyah-Bott-Lefschetz Todd class on a stratum of the moduli space. Now we take $u \in C/\overline{R}^*$, $x \in M^u$. Then u descends to a flat section of the G bundle $\widehat{G}$, which acts naturally on the left on the vector bundle E as a flat section of $\operatorname{Aut}(E)$. In particular, on M^u, the vector bundle $\mathcal{E} \otimes_{\mathbf{R}} \mathbf{C}$ splits as a direct sum of vector bundles

$$\mathcal{E} \otimes_{\mathbf{R}} \mathbf{C} = \bigoplus_{0<\theta<\pi} (\mathcal{E}^\theta \oplus \mathcal{E}^{-\theta}) \oplus \mathcal{E}^\pi . \tag{6.40}$$

In (6.40), for $-\pi < \theta \le \pi$, u acts on $\mathcal{E}^\theta$ like $e^{i\theta}$.

For $0 < \theta < \pi$, $\mathcal{E}^\theta$ is a complex vector bundle on $M^u/Z'(u)$. Let $\prod\Big(e^{\frac{x+i\theta}{2}} - e^{-\frac{x+i\theta}{2}}\Big)(\mathcal{E}^\theta$ be the corresponding characteristic class. Also $\mathcal{E}^\pi$ is a real vector bundle on $M^u/Z'(u)$. Moreover $2i\cosh(\frac{x}{2})$ is an even function of x. Let $\prod\Big(e^{\frac{x+i\pi}{2}} -$

$e^{-\frac{x+i\pi}{2}}\big)(\mathcal{E}^\pi) = \prod 2i\cosh(\frac{x}{2})(\mathcal{E}^\pi)$ be the corresponding Pontryagin class of $\mathcal{E}^\pi$. Needless to say, if $\mathcal{E}^\pi$ has a complex structure , then this class is exactly $\prod 2i\cosh(\frac{x}{2})(\mathcal{E}^{\pi,(1,0)})$.

THEOREM 6.17. *For $u \in C/\overline{R}^*$, the following identity of characteristic classes holds on $M^u/Z'(u)$,*

$$\mathrm{Td}(T^{(1,0)}M^u/Z'(u)) = \widehat{A}^{2g-2+s}(\mathcal{E}^0)e^{\frac{1}{2}c_1(T^{(1,0)}M^u/Z'(u))}, \tag{6.41}$$

$$\widehat{\mathrm{Td}}(N^{u(1,0)}) = \left[\frac{1}{\prod_{0<\theta\leq\pi}\Pi\left(e^{\frac{x+i\theta}{2}} - e^{\frac{-(x+i\theta)}{2}}\right)(\mathcal{E}^\theta)}\right]^{2g-2+s}$$

$$e^{\frac{1}{2}c_1(N^{u(1,0)})}e^{\frac{i}{2}\sum_{\theta\in]-\pi,\pi]\setminus\{0\}}\theta\dim(N^{u(1,0),\theta})}(-1)^{\sum_{-\pi<\theta<0}\dim(N^{u(1,0),\theta})}.$$

PROOF. Using Theorem 5.34 and (6.11), we get the first identity in (6.41). Clearly

$$\widehat{\mathrm{Td}}(N^{u(1,0)}) = \left[\prod_{\theta\in]-\pi,\pi]\setminus\{0\}}\left(e^{\frac{1}{2}(x+i\theta)} - e^{-\frac{1}{2}(x+i\theta)}\right)(N^{u(1,0),\theta})\right]^{-1} \tag{6.42}$$
$$c^{\frac{1}{2}c_1(N^{u(1,0)})+\frac{i}{2}\sum_{\theta\in]-\pi,\pi]\setminus\{0\}}\theta\dim(N^{u(1,0),\theta})}.$$

Also, for $-\pi < \theta < 0$,

$$N^{u(1,0),\theta} = \overline{N^{u(0,1),-\theta}} \tag{6.43}$$

From (6.43), we get

$$\begin{aligned}\prod_{\theta\in]-\pi,\pi]\setminus\{0\}}&\left(e^{\frac{1}{2}(x+i\theta)} - e^{-\frac{1}{2}(x+i\theta)}\right)(N^{u(1,0),\theta}) \\ &= \prod_{0<\theta\leq\pi}\left(e^{\frac{1}{2}(x+i\theta)} - e^{-\frac{1}{2}(x+i\theta)}\right)(N^{u(1,0),\theta}) \\ &\quad\prod_{0<\theta<\pi}\left(e^{-\frac{1}{2}(x+i\theta)} - e^{\frac{1}{2}(x+i\theta)}\right)(N^{u(0,1),\theta}) \\ &= \prod_{0<\theta<\pi}\left(e^{\frac{1}{2}(x+i\theta)} - e^{-\frac{1}{2}(x+i\theta)}\right)(N^{u,\theta})\prod(ie^{x/2} + ie^{-x/2})(N^{u(1,0),\pi}) \\ &\quad(-1)^{\sum_{-\pi<\theta<0}\dim(N^{u(1,0),\theta})}.\end{aligned} \tag{6.44}$$

Also one finds easily that over $M^u/Z(u)$, equality (5.139) of Theorem 5.34 can be split according to the values of θ. So we find that for $0 < \theta < \pi$,

$$\begin{aligned}&\Pi\left(e^{\frac{1}{2}(x+i\theta)} - e^{-\frac{1}{2}(x+i\theta)}\right)(N^{u,\theta}) \\ &= \left[\Pi\left(e^{\frac{1}{2}(x+i\theta)} - e^{-\frac{1}{2}(x+i\theta)}\right)(\mathcal{E}^\theta)\right]^{2g-2+s}.\end{aligned} \tag{6.45}$$

Also $2i\cosh(x/2)$ is an even function of x. Then

$$\begin{aligned}&\prod(ie^{x/2} + ie^{-x/2})(N^{u(1,0),\pi}) = \prod(2i\cosh(x/2))(N^{u,\pi}) \\ &= \left[\prod(2i\cosh(\tfrac{x}{2})\right]^{2g-2+s}(\mathcal{E}^\pi) = \prod\left(e^{\frac{x+i\pi}{2}} - e^{-\frac{(x+i\pi)}{2}}\right)^{2g-2+s}(\mathcal{E}^\pi).\end{aligned} \tag{6.46}$$

By (6.42)-(6.46), we get the second identity in (6.41). The proof of our Theorem is completed. □

6.6. The dimensions of the splitting of the normal bundle to the strata. Let J be a G-invariant almost complex structure on TM/G which polarizes the symplectic form ω. We know that J is unique up to homotopy.

Take $u \in C/\overline{R}^*$. Let $x \in M^u$. Then by (5.120),

$$TM/G = \widehat{H}^1(\Sigma, E). \tag{6.47}$$

Recall that u defines a flat section of the bundle $\widehat{G} \xrightarrow{G} \Sigma$. Clearly

$$\mathfrak{g} \otimes_{\mathbf{R}} \mathbf{C} = (\mathfrak{z}(u) \otimes_{\mathbf{R}} \mathbf{C}) \oplus \Big(\bigoplus_{\alpha \in R \setminus R_u} \mathfrak{g}_\alpha \Big). \tag{6.48}$$

Let $x \in \mathbf{R} \mapsto [x]' \in]-1/2, +1/2]$ be the function periodic of period 1, such that

$$[x]' = x \text{ for } x \in]-1/2, +1/2]. \tag{6.49}$$

Clearly u acts on the left on $\mathfrak{z}(u)$ like the identity and on $\mathfrak{g}_\alpha$ like $e^{2i\pi\langle\alpha,u\rangle}$. For $-\pi < \theta \leq \pi$, put

$$\mathfrak{g}^\theta = \bigoplus_{\substack{\alpha \in R \setminus R_u \\ [\langle \alpha, u\rangle]' = \frac{\theta}{2\pi}}} \mathfrak{g}^\alpha, \tag{6.50}$$

with the convention that if $\theta = 0$, $\mathfrak{g} = \mathfrak{z}(u)$.

Then (6.48) can be written as

$$\mathfrak{g} \otimes_{\mathbf{R}} \mathbf{C} = \bigoplus_{-\pi < \theta \leq \pi} \mathfrak{g}^\theta. \tag{6.51}$$

Also (6.51) is a $Z(u)$-invariant splitting. It induces a corresponding flat splitting of the flat bundle E of the form

$$E = \bigoplus_{-\pi < \theta \leq \pi} E^\theta. \tag{6.52}$$

In (6.52), E^0 is just the analogue of E when replacing G by $Z(u)$.

By (6.52), we get

$$\widehat{H}^1(\Sigma, E) \otimes_{\mathbf{R}} \mathbf{C} = \bigoplus_{-\pi < \theta \leq \pi} \widehat{H}^1(\Sigma, E^\theta). \tag{6.53}$$

Also $TM/G = \widehat{H}^1(\Sigma, E)$, and J acts on $\widehat{H}^1(\Sigma, E)$. Since J is G-invariant, J commutes with u. In particular J acts on each $\widehat{H}^1(\Sigma, E^\theta)$. Let $\widehat{H}^{(1,0)}(\Sigma, E^\theta)$, $\widehat{H}^{(0,1)}(\Sigma, E^\theta)$ be the $+i, -i$ eigenspaces of J, so that

$$\widehat{H}^1(\Sigma, E^\theta) = \widehat{H}^{(1,0)}(\Sigma, E^\theta) \oplus \widehat{H}^{(0,1)}(\Sigma, E^\theta). \tag{6.54}$$

Observe that since $\widehat{H}^1(\Sigma, E^\theta)$ is equipped with the u-invariant symplectic form ω, then

$$\begin{aligned} [\widehat{H}^{(1,0)}(\Sigma, E^\theta)]^* &= \widehat{H}^{(0,1)}(\Sigma, E^\theta), \text{ if } \theta = 0, \pi, \\ [\widehat{H}^{(1,0)}(\Sigma, E^\theta)]^* &= \widehat{H}^{(0,1)}(\Sigma, E^{-\theta}), \ \theta \in]-\pi, \pi[. \end{aligned} \tag{6.55}$$

Since J is unique up to homology, the dimensions of the vector spaces which appear in (6.55) do not depend on J.

Since $Z(u)$ is semisimple, for $-\pi < \theta \leq \pi$,

$$\sum_{\substack{\alpha \in R \\ [\langle \alpha, u \rangle]' = \frac{\theta}{2\pi}}} \alpha = 0 . \tag{6.56}$$

Therefore for any $t \in \mathfrak{t}$,

$$\sum_{[\langle \alpha, u \rangle]' = \frac{\theta}{2\pi}} [\langle \alpha, t \rangle] \in \mathbf{Z} . \tag{6.57}$$

THEOREM 6.18. *For $-\pi < \theta \leq \pi$, over M^u,*

$$\dim \widehat{H}^{(0,1)}(\Sigma, E^\theta) = (g-1) \dim \mathfrak{g}^\theta + \sum_{j=1}^{s} \sum_{\substack{\alpha \in R \\ [\langle \alpha, u \rangle]' = \frac{\theta}{2\pi}}} [\langle \alpha, t_j \rangle] . \tag{6.58}$$

PROOF. First assume that $m \in \mathbf{N}$, that $t_1, \ldots, t_s$ are of order m, and $m|s$. Then we make the construction in Section 5.12. In particular u lifts to a $\mathbf{Z}/m\mathbf{Z}$-invariant parallel section over Σ_g^b. Then we have the $\mathbf{Z}/m\mathbf{Z}$ invariant splitting

$$H^1(\Sigma_g^b, E) \otimes_{\mathbf{R}} \mathbf{C} = H^{(1,0)}(\Sigma_g^b, E) \oplus H^{(0,1)}(\Sigma_g^b, E) . \tag{6.59}$$

By arguing as in the proof of Theorem 5.78, when taking the $\mathbf{Z}/m\mathbf{Z}$ invariant part of (6.59), we get a complex structure on $\widehat{H}^1(\Sigma, E)$ which polarizes σ. It is then feasible to take

$$\widehat{H}^{(0,1)}(\Sigma, E^\theta) = [H^{(0,1)}(\Sigma, E^\theta)]^{\mathbf{Z}/m\mathbf{Z}} . \tag{6.60}$$

By construction,

$$[\widehat{H}^0(\Sigma, E^\theta)] = 0 . \tag{6.61}$$

Therefore

$$\dim [H^{0,1}(\Sigma, E^\theta)]^{\mathbf{Z}/m\mathbf{Z}} = -\chi^{h, \mathbf{Z}/m\mathbf{Z}}(\Sigma_g^b, p^* E^\theta) . \tag{6.62}$$

Using (5.321) and the theorem of Riemann-Roch-Kawasaki [**32, 33**] as in (5.349), (5.350), and also (6.62), we get (6.58).

Let us now consider the general case. As in the proof of Theorem 5.78, we approximate $(t_1, \ldots, t_s)$ by regular $(t_1^m, \ldots, t_s^m)$ which are of order m and such that (A) holds. Recall that over Σ_g^b, we still have a $Z(u)$ bundle, so that we may and we will assume that at x_j $(j \geq s+1)$, we have in fact a $Z(u)$ connection. Since $Z(u)$ is semisimple, for $j \geq s+1$,

$$\operatorname{Tr}^{E_\theta}[\Theta_j] = 0 . \tag{6.63}$$

Again the extra points $x_{s+1}, \ldots, x_{s_m}$ do not contribute to the computation, so that (6.58) still holds. Since for $1 \leq j \leq s$, $[\langle \alpha, t_j^m \rangle] \to [\langle \alpha, t_j \rangle]$, we get (6.58) in full generality.

The proof of our Theorem is completed. □

THEOREM 6.19. *For $\theta \in]-\pi, \pi]$,*

$$\dim \widehat{H}^{(1,0)}(\Sigma, E^\theta) = (g-1) \dim \mathfrak{g}^\theta + \sum_{j=1}^{s} \sum_{\substack{\alpha \in R \\ [\langle \alpha, u \rangle]' = \frac{\theta}{2\pi}}} (1 - [\langle \alpha, t_j \rangle]) . \tag{6.64}$$

PROOF. By (6.55), for $\theta \in]-\pi,\pi[$,

$$\dim H^{1,0}(\Sigma, E^{\theta}) = \dim H^{0,1}(\Sigma, E^{-\theta})\,. \tag{6.65}$$

So by Theorem 6.18, we get

$$\dim \widehat{H}^{1,0}(\Sigma, E^{\theta}) = (g-1)\dim \mathfrak{g}^{-\theta} + \sum_{j=1}^{s} \sum_{\substack{\alpha \in R \\ [\langle \alpha, u\rangle]' = \frac{-\theta}{2\pi}}} [\langle \alpha, t_j\rangle]\,. \tag{6.66}$$

Now for $\theta \in]-\pi,\pi[$,

$$\dim \mathfrak{g}^{\theta} = \dim \mathfrak{g}^{-\theta}\,. \tag{6.67}$$

Also since for $1 \leq j \leq s$, $\alpha \in R$, then $\langle \alpha, t_j\rangle \notin \mathbf{Z}$,

$$\sum_{\substack{\alpha \in R \\ [\langle \alpha, u\rangle]' = \frac{-\theta}{2\pi}}} [\langle \alpha, t_j\rangle] = \sum_{\substack{\alpha \in R \\ [\langle \alpha, u\rangle]' = \frac{\theta}{2\pi}}} [\langle -\alpha, t_j\rangle] = \sum_{\substack{\alpha \in R \\ [\langle \alpha, u\rangle]' = \frac{\theta}{2\pi}}} (1 - [\langle \alpha, t_j\rangle])\,. \tag{6.68}$$

From (6.66), (6.68), we get (6.64) when $\theta \in]-\pi,\pi[$.

Also E^{π} is a real vector bundle. Then

$$\dim \widehat{H}^{(1,0)}(\Sigma, E^{\pi}) = \dim \widehat{H}^{(0,1)}(\Sigma, E^{\pi})\,. \tag{6.69}$$

By Theorem 6.18, we get

$$\dim H^{(1,0)}(\Sigma, E^{\pi}) = (g-1)\dim \mathfrak{g}^{\pi} + \sum_{j=1}^{s} \sum_{\substack{\alpha \in R \\ [\langle \alpha, u\rangle]' = \frac{1}{2}}} [\langle \alpha, t_j\rangle]\,. \tag{6.70}$$

Observe that if $\alpha \in R$, then $[\langle \alpha, u\rangle]' = \frac{1}{2}$ if and only if $[-\langle \alpha, u\rangle]' = \frac{1}{2}$. Also when changing α into $-\alpha$, $[\langle \alpha, t_j\rangle]$ is changed into $1 - [\langle \alpha, t_j\rangle]$. So for $\theta = \pi$, we find that (6.70) is still equivalent to (6.64).

The proof of our Theorem is completed. □

THEOREM 6.20. *For $u \in C/\overline{R}^{*}$, θ, $\theta \in]-\pi,\pi]\backslash\{0\}$,*

$$\dim N^{u(1,0),\theta} = (g-1)\dim \mathfrak{g}^{\theta} + \sum_{j=1}^{s} \sum_{\substack{\alpha \in R \\ [\langle \alpha, u\rangle]' = \frac{\theta}{2\pi}}} (1 - [\langle \alpha, t_j\rangle])\,. \tag{6.71}$$

PROOF. Recall that

$$TM/G = \widehat{H}^{1}(\Sigma, E)\,. \tag{6.72}$$

Also (6.72) is an identification of u-spaces. Then our Theorem is just another formulation of Theorem 6.19. □

6.7. The contribution of a stratum of the moduli space. Recall that by Theorem 1.41,

$$h \in \overline{CR}/\overline{CR}_u \mapsto \exp(2i\pi\langle \rho_u, h\rangle) \in S_1 \tag{6.73}$$

is a character with values in ± 1.

By Theorem 5.18 and by Proposition 5.63, for $g \geq 1$, $M(Z(u), \mathcal{O}_{Z(u)}(w^1 t_1), \dots, \mathcal{O}_{Z(u)}(w^s t_s), h)$ is a non empty smooth manifold, and for $g = 0$, $M(Z(u), \mathcal{O}_{Z(u)}(w^1 t_1), \dots, \mathcal{O}_{Z(u)}(w^s t_s), h)$ is either non empty and smooth, or it is empty.

In the sequel, it will be understood that any geometric statement about the empty set is empty.

Recall that by [**34**], [**6**, Lemma 7.22] and (1.193), the orbits $\mathcal{O}(w^j t_j)$ carry a natural complex structure, which polarizes the canonical symplectic form $\sigma_{\mathcal{O}(w^j t_j)}$ on $\mathcal{O}(w^j t_j)$. Let $N^{(1,0)}_{\mathcal{O}_{Z(u)}(w^j t_j)/\mathcal{O}(w^j t_j)}$ be the holomorphic normal bundle to $\mathcal{O}_{Z(u)}(w^j t_j)$ in $\mathcal{O}(w^j t_j)$.

As we saw after (5.305), $H_{u,g,s}(t_1, \dots, t_s, t)$ is locally a polynomial of $(t_1, \dots, t_s, t)$.

THEOREM 6.21. *For $u \in C/\overline{CR}$, $p \in \mathbf{M}, p \neq -c$, $(w^1, \dots, w^s) \in W^s$, $h \in \frac{\overline{CR}}{\overline{CR}_u}$,*

$$\begin{aligned}
\int_{M(Z(u), \mathcal{O}_{Z(u)}(w^1 t_1), \dots, \mathcal{O}_{Z(u)}(w^s t^s), h)/Z(u)} & L(u, \lambda^p) = (p+c)^{(g-1)\dim \mathfrak{z}(u) + \frac{s}{2}\dim \mathfrak{z}(u)/t} \\
& \frac{(-1)^{\frac{(g-1)}{2}(\dim(\mathfrak{g}) - \dim(\mathfrak{z}(u)))}}{|W_u|} \prod_{j=1}^{s} e^{i\pi \sum_{\alpha \in R_+}[\langle w^j \alpha, u\rangle] + 2i\pi\langle w^j p t_j + ph, u\rangle} \\
& \Bigg[\prod_{\alpha \in R_{u,+}} \widehat{A}\left(\langle \alpha, \frac{\partial/\partial t}{p+c}\rangle\right) \\
& \prod_{\alpha \in R_+ \setminus R_{u,+}} \frac{1}{2\sinh\left(\frac{1}{2}(\langle \alpha, \frac{\partial/\partial t}{p+c}\rangle + 2i\pi[\langle \alpha, u\rangle])\right)} \Bigg]^{2g-2+s} \\
& \prod_{j=1}^{s} e^{\frac{1}{p+c}\langle(\rho - c t_j), \partial/\partial t_j\rangle} \pi_u\left(\frac{\partial/\partial t}{2\pi i}\right) \frac{1}{|\frac{\overline{CR}}{\overline{CR}_u}|^{2g}} H_{u,g,s}(w^1 t_1, \dots, w^s t_s, h, t).
\end{aligned} \tag{6.74}$$

Also

$$\begin{aligned}
& \left\{\widehat{\mathrm{Td}}(N^{u(1,0)})_{|M(Z(u), \mathcal{O}_{Z(u)}(w^1 t_1), \dots, \mathcal{O}_{Z(u)}(w^s t_s))/Z'(u)}\right\}^{(0)} = \\
& \prod_{\alpha \in R_+ \setminus R_{u,+}} \left(\frac{1}{4\sin^2(\pi\langle \alpha, u\rangle)}\right)^{g-1} \prod_{j=1}^{s} \frac{1}{\det(1 - u^{-1})_{|N^{(1,0)}_{\mathcal{O}_{Z(u)}(w^j t_j)/\mathcal{O}(w^j t_j)}|w^j t_j}}.
\end{aligned} \tag{6.75}$$

PROOF. First we consider the case where $g = 0$, and $M(Z(u), \mathcal{O}_{Z(u)}(w^1 t_1), \dots, \mathcal{O}_{Z(u)}(w^s t_s), h) = \emptyset$. By definition, the left-hand side of (6.74) is 0. Also for $t \in T$ close enough to 1, $M_t(Z(u), \mathcal{O}_{Z(u)}(w^1 t_1, \dots, \mathcal{O}_{Z(u)}(w^s t_s), h)$ is also empty. By (5.305), it follows that $H_{u,g,s}(t_1, \dots, t_s, t)$ vanishes identically near $t = 1$. Then (6.74) is just the identity $0 = 0$.

Now we assume that $M(Z(u), \mathcal{O}_{Z(u)}(w^1 t_1), \dots, \mathcal{O}_{Z(u)}(w^s t_s), h)$ is non empty. If $w = (w^1, \dots, w^s) \in W^s$, put

$$M^u_{w,h} = M(Z(u), \mathcal{O}_{Z(u)}(w^1 t_1), \dots, \mathcal{O}_{Z(u)}(w^s t_s), h). \tag{6.76}$$

We use the notation of Section 4.7. If $x \in M^u_{w,h}$, the corresponding flat connection reduces the G-bundle P to a $Z(u)$-bundle we still denote by P. This $Z(u)$-bundle lifts to a $\widetilde{Z}(u)$-bundle Q. By Proposition 4.33, and by (6.36),

$$[P] = e^h. \tag{6.77}$$

By Theorem 4.39, the right action of u on λ^p over $M^u_{w,h}$ is given by $e^{-2i\pi p\langle\sum_{j=1}^s w^j t_j + h, u\rangle}$. The corresponding left action is then given by $e^{2i\pi p\langle\sum_{j=1}^s w^j t_j + h, u\rangle}$. By (6.41) , and by the above result,we get

$$\begin{gathered}\int_{M^u_{w,h}/Z(u)} L(u,\lambda^p) = \int_{M^u_{w,h}/Z(u)} \left[\widehat{A}(\mathcal{E}^0)\frac{1}{\prod\limits_{0<\theta\leq\pi} \Pi\left(e^{\frac{x+i\theta}{2}} - e^{-\frac{(x+i\theta)}{2}}\right)(\mathcal{E}^\theta)}\right]^{2g-2+s} \\ e^{\frac{1}{2}c_1(T^{1,0}M/G)+p\omega} e^{\frac{i}{2}\sum_{\theta\in]-\pi,\pi]\setminus\{0\}} \theta \dim(N^{u(1,0),\theta})} \\ (-1)^{\sum_{-\pi<\theta<0}\dim(N^{u(1,0),\theta})} \prod_{j=1}^s e^{2i\pi\langle w^j p t_j, u\rangle} e^{2i\pi\langle pu, h\rangle}.\end{gathered} \tag{6.78}$$

Now we use Theorem 5.78, which gives us a formula for $c_1(TM/G)$. By Proposition 5.73 and by (6.78),

$$\begin{gathered}\int_{M^u_{w,h}/Z(u)} L(u,\lambda^p) = \frac{1}{|W_u|} e^{\frac{i}{2}\sum_{\theta\in]-\pi,\pi]\setminus\{0\}} \theta \dim(N^{u(1,0),\theta})} \\ (-1)^{\sum_{-\pi<\theta<0}\dim(N^{u(1,0),\theta})} (p+c)^{(g-1)\dim\mathfrak{z}(u)+\frac{s}{2}\dim(\mathfrak{z}(u)/\mathfrak{t})} \left[\prod_{\alpha\in R_{u,+}} \widehat{A}\left(\langle\alpha, \frac{\partial/\partial t}{p+c}\rangle\right)\right. \\ \left.\prod_{\substack{\alpha\in R \\ 0<[\langle\alpha,u\rangle]<\frac{1}{2}}} \frac{1}{2\sinh\left(\frac{1}{2}\left(\langle\alpha, \frac{\partial/\partial t}{p+c}\rangle + 2i\pi[\langle\alpha,u\rangle]\right)\right)} \prod_{\substack{\alpha\in R_+ \\ [\langle\alpha,u\rangle]=\frac{1}{2}}} \frac{1}{2i\cosh\left(\frac{1}{2}\langle\alpha, \frac{\partial/\partial t}{p+c}\rangle\right)}\right]^{2g-2+s} \\ \prod_{j=1}^s e^{\frac{1}{p+c}\langle\rho - ct_j, \partial/\partial t_j\rangle} \pi_u\left(\frac{\partial/\partial t}{2\pi i}\right) \frac{1}{\left|\frac{\overline{CR}}{\overline{CR}_u}\right|^{2g}} H_{u,g,s}(w^1 t_1, \ldots, w^s t_s, h, t) \prod_{j=1}^s e^{2i\pi\langle w^j p t_j, u\rangle} e^{2i\pi\langle ph, u\rangle}.\end{gathered} \tag{6.79}$$

Observe that if $\alpha \in R\setminus R_u$, when changing α into $-\alpha$, $2\sinh\left(\frac{1}{2}\left(\langle\alpha, \frac{\partial/\partial t}{p+c}\rangle + 2i\pi[\langle\alpha,u\rangle]\right)\right)$ is unchanged. Therefore

$$\begin{gathered}\prod_{\substack{\alpha\in R \\ 0<[\langle\alpha,u\rangle]<\frac{1}{2}}} \frac{1}{2\sinh\left(\frac{1}{2}\left(\langle\alpha, \frac{\partial/\partial t}{p+c}\rangle + 2i\pi[\langle\alpha,u\rangle]\right)\right)} \prod_{\substack{\alpha\in R_+ \\ [\langle\alpha,u\rangle]=\frac{1}{2}}} \\ \frac{1}{2i\cosh\frac{1}{2}\left(\langle\alpha, \frac{\partial/\partial t}{p+c}\rangle\right)} = \prod_{\alpha\in R_+\setminus R_{u,+}} \frac{1}{2\sinh\left(\frac{1}{2}\left(\langle\alpha, \frac{\partial/\partial t}{p+c}\rangle + 2i\pi[\langle\alpha,u\rangle]\right)\right)}.\end{gathered} \tag{6.80}$$

By Theorem 6.20,

$$\begin{gathered}\exp(i/2 \sum_{\theta\in]-\pi,\pi]\setminus\{0\}} \theta\dim(N^{u(1,0),\theta})) = \\ \exp\left(i/2(\sum_{\theta\in]-\pi,\pi]} \theta(\dim\mathfrak{g}^\theta(g-1) + \sum_{j=1}^s \sum_{\substack{\alpha\in R \\ [\langle\alpha,u\rangle]'=\theta/2\pi}} (1 - [\langle\alpha, w^j t_j\rangle])))\right).\end{gathered} \tag{6.81}$$

Now the $\theta \in]-\pi,\pi[\backslash\{0\}$ come by opposite pairs. Therefore

$$\sum_{\theta\in]-\pi,\pi]\backslash\{0\}} \theta \dim \mathfrak{g}^{\theta} = \pi \dim \mathfrak{g}^{\pi}, \tag{6.82}$$

and so

$$e^{\frac{i}{2}\Sigma\theta \dim \mathfrak{g}^{\theta}(g-1)} = (-1)^{\frac{\dim \mathfrak{g}^{\pi}}{2}(g-1)}. \tag{6.83}$$

For $1 \leq j \leq s$, put

$$\begin{aligned} t'_j &= w^j t_j \\ R_{j,+} &= w^j R_+ \end{aligned} \tag{6.84}$$

Then $R_{j,+}$ is the positive root system associated to t'_j. For $1 \leq j \leq s$,

$$\begin{aligned} &\textstyle\sum_{s\in]-1/2,1/2]} s \sum_{\substack{\alpha\in R\\ [\langle\alpha,u\rangle]'=s}} (1-[\langle\alpha,t'_j\rangle]) \\ &= \textstyle\sum_{\alpha\in R}[\langle\alpha,u\rangle]'(1-[\langle\alpha,t'_j\rangle]) \\ &= \textstyle\sum_{\alpha\in R_{j,+}} \left([\langle\alpha,u\rangle]'(1-\langle\alpha,t'_j\rangle) + [-\langle\alpha,u\rangle]'\langle\alpha,t'_j\rangle\right) \\ &= -\textstyle\sum_{\alpha\in R}[\langle\alpha,u\rangle]'\langle\alpha,t'_j\rangle + \sum_{\alpha\in R_{j,+}}[\langle\alpha,u\rangle]'. \end{aligned} \tag{6.85}$$

Clearly,

$$\sum_{\alpha\in R}[\langle\alpha,u\rangle]'\langle\alpha,t'_j\rangle = \sum_{s} s\langle \sum_{\substack{\alpha\in R\\ [\langle\alpha,u\rangle]'=s}} \alpha, t'_j\rangle. \tag{6.86}$$

Now for a given s, the $\{\alpha \in R, [\langle\alpha,u\rangle]' = s\}$ are exactly the weights of the representation of $Z(u)$ on $\{f \in \mathfrak{g}\otimes_{\mathbf{R}} \mathbf{C}, uf = e^{2i\pi s}f\}$. Since $Z(u)$ is semisimple,

$$\sum_{\substack{\alpha\in R\\ [\langle\alpha,u\rangle]'=s}} \alpha = 0. \tag{6.87}$$

From (6.81)-(6.87) , we get

(6.88)

$$\exp(\frac{i}{2} \sum_{\theta\in]-\pi,\pi]\backslash\{0\}} \theta\dim(N^{u(1,0),\theta})) = (-1)^{\frac{\dim \mathfrak{g}^{\pi}}{2}(g-1)} \prod_{j=1}^{s} \exp(i\pi \sum_{\alpha\in R_+} [\langle w^j\alpha,u\rangle]').$$

Also by Theorem 6.20, using (6.87) , we get

$$\begin{aligned} &\textstyle\sum_{-\pi<\theta<0} \dim N^{u(1,0),\theta} \\ &= (g-1)\textstyle\sum_{-\pi<\theta<0} \dim\mathfrak{g}^{\theta} + \sum_{j=1}^{s} \sum_{\substack{\alpha\in R\\ -1/2<[\langle\alpha,u\rangle]'<0}} (1-[\langle\alpha,t'_j\rangle]) \\ &= (g-1)|\{\alpha\in R, -\tfrac{1}{2} < [\langle\alpha,u\rangle]' < 0\}| \\ &+ \textstyle\sum_{j=1}^{s}\Big(-\sum_{-1/2<s<0}\langle\sum_{\substack{\alpha\in R\\ [\langle\alpha,u\rangle]'=s}} \alpha, t'_j\rangle + |\{\alpha\in R, -\tfrac{1}{2} < [\langle\alpha,u\rangle]' < 0, \langle\alpha,t'_j\rangle > 0\}|\Big) \\ &= (g-1)|\{\alpha\in R, -\tfrac{1}{2} < [\langle\alpha,u\rangle]' < 0\}| \\ &+ \textstyle\sum_{j=1}^{s} |\{\alpha\in R, -\tfrac{1}{2} < [\langle\alpha,u\rangle]' < 0, \langle\alpha,t'_j\rangle > 0\}| \\ &= (g-1)|\{\alpha\in R, -\tfrac{1}{2} < [\langle\alpha,u\rangle]' < 0\}| + \textstyle\sum_{j=1}^{s} |\{\alpha\in R_+, -\tfrac{1}{2} < [\langle w^j\alpha,u\rangle]' < 0\}. \end{aligned} \tag{6.89}$$

So by (6.88),(6.89),

$$
\begin{aligned}
(6.90)\quad & \exp\Big(\tfrac{i}{2}\sum_{\theta\in]-\pi,\pi]\setminus\{0\}} \theta \dim(N^{u(1,0),\theta})\Big)(-1)^{\sum_{-\pi<\theta<0}\dim(N^{u,(1,0),\theta})} = \\
& (-1)^{(g-1)(|\{\alpha\in R, 0<[\langle\alpha,u\rangle]'<\frac{1}{2}\}|+\frac{1}{2}|\{\alpha\in R,0<[\langle\alpha,u\rangle]'=\frac{1}{2}\}|)} \\
& \prod_{j=1}^{s}\exp\Big(i\pi\sum_{\alpha\in R_+}[\langle w^j\alpha,u\rangle]\Big)\,.
\end{aligned}
$$

Moreover

(6.91)

$$
|\{\alpha\in R, 0<[\langle\alpha,u\rangle]'<\frac{1}{2}\}|+\frac{1}{2}|\{\alpha\in R,[\langle\alpha,u\rangle]'=\frac{1}{2}\}| = \frac{1}{2}(\dim(\mathfrak{g})-\dim\mathfrak{z}(u))\,.
$$

From (6.79), (6.80), (6.90), (6.91) , we get (6.74).

Using (6.74) , or by proceeding directly, we get

$$
\begin{aligned}
\Big\{\widehat{\mathrm{Td}}(N^{u(1,0)})_{|M^u_{w,h}/Z'(u)}\Big\}^{(0)} &= (-1)^{(g-1)(\dim\mathfrak{g}-\dim\mathfrak{z}(u))/2}\prod_{j=1}^{s}\exp\Big(i\pi\sum_{\alpha\in R_+}[\langle w^j\alpha,u\rangle]\Big) \\
(6.92)\qquad & \Big(\prod_{\alpha\in R_+\setminus R_{u,+}}\frac{1}{2\sinh(i\pi[\langle\alpha,u\rangle])}\Big)^{2g-2+s}\,.
\end{aligned}
$$

Using the invariance of $\sinh(i\pi[\langle\alpha,u\rangle])$ when $\alpha\in R\backslash R_u$ is changed into $-\alpha$, from (6.92), we get

$$
\begin{aligned}
(6.93)\quad \Big\{\widehat{\mathrm{Td}}(N^{u(1,0)})_{|M^u_{w,h}/Z'(u)}\Big\}^{0} &= \Big(\prod_{\alpha\in R_+\setminus R_{u,+}}\frac{1}{4\sin^2(\pi\langle\alpha,u\rangle)}\Big)^{g-1} \\
& \prod_{j=1}^{s}\prod_{\substack{\alpha\in R_+\\ \alpha\notin w_j^{-1}R_u}}\frac{e^{i\pi[\langle\alpha,(w^j)^{-1}u\rangle]}}{2\sinh(i\pi[\langle\alpha,(w^j)^{-1}u\rangle])} = \\
&= \prod_{\alpha\in R_+\setminus R_{u,+}}\Big[\frac{1}{4\sin^2(\pi\langle\alpha,u\rangle)}\Big]^{g-1}\prod_{j=1}^{s}\prod_{\substack{\alpha\in w_jR_+,\\ \alpha\notin R_u}}\frac{1}{1-e^{-2i\pi\langle\alpha,u\rangle}} = \\
&= \prod_{\alpha\in R_+\setminus R_{u,+}}\Big[\frac{1}{4\sin^2(\pi\langle\alpha,u\rangle)}\Big]^{g-1}\prod_{j=1}^{s}\frac{1}{\det(1-u^{-1})_{|N^{(1,0)}_{\mathcal{O}_{Z(u)}(w^jt_j)/\mathcal{O}(w^jt_j)|w^jt_j}}}\,,
\end{aligned}
$$

which is just (6.75). The proof of our Theorem is completed. □

Recall that

$$
\begin{aligned}
(6.94)\qquad \ell &= |R_+| \\
\ell_u &= |R_{u,+}|\,, \\
r &= \dim\mathfrak{t}\,.
\end{aligned}
$$

Our assumptions on g,s guarantee that $2g-2+s\geq 1$. Also by (2.100) and by Theorem 2.45, the function $P_{u,p+c,2g-2+s}(t)$ is a polynomial on $T\setminus S$.

THEOREM 6.22. *For* $u \in C/\overline{CR}$, $p \in \mathbf{M}, p \neq -c$, *then*

$$\int_{M^u/Z(u)} L(u,\lambda^p) = \mathrm{Vol}(T)^{2g-2}\frac{|Z(Z(u))|}{|W_u|}(p+c)^{(g-1)\dim(\mathfrak{z}(u))+\frac{s}{2}\dim(\mathfrak{z}(u)/\mathfrak{t})}$$

$$(6.95)\qquad (-1)^{\ell(g-1)+1}\prod_{\alpha\in R_{u,+}}$$

$$\left[\widehat{A}\left(\langle\alpha,\frac{\partial/\partial t\rangle}{p+c}\right)\prod_{\alpha\in R_+\setminus R_{u,+}}\frac{1}{2\sinh\left(\frac{1}{2}\left(\langle\alpha,\frac{\partial/\partial t}{p+c}\rangle+2i\pi[\langle\alpha,u\rangle]\right)\right)}\right]^{2g-2+s}$$

$$\sum_{(w^1,\dots,w^s)\in W^s}\prod_{j=1}^{s}\left[\mathrm{sgn}((-i)^{\ell_u}\sigma_{Z(u)}(w^jt_j))e^{\langle w^j(\rho-ct_j),\frac{\partial/\partial t}{p+c}\rangle+}\right.$$

$$\left.+i\pi\sum_{\alpha\in R_+}[\langle w^j\alpha,u\rangle]+2i\pi\langle w^jpt_j,u\rangle\right]P_{u,2g-2+s,p+c}(t)_{|t=\sum_1^s w^jt_j}$$

PROOF. By (6.32), (6.37),

$$(6.96)\qquad \int_{M^u/Z(u)} L(u,\lambda^p) =$$

$$\sum_{\substack{(w^1,\dots,w^s)\in(W_u\setminus W)^s\\ h\in\overline{CR}/\overline{CR_u}}}\int_{M(Z(u),\mathcal{O}_{Z(u)}(w^1t_1),\dots,\mathcal{O}_{Z(u)}(w^st_s),h)/Z(u)} L(u,\lambda^p)\,.$$

We use Theorem 6.21 to calculate the term in the right-hand side of (6.96) . Also by (2.131) and by Theorem 5.72,

$$(6.97)\qquad \pi_u\left(\frac{\partial/\partial t}{2i\pi}\right)H_{u,g,s}(w^1t_1,\dots,w^st_s,h,t) = (-1)^{\ell_u(g-1)+1}$$

$$|Z(\widetilde{Z}(u))|\mathrm{Vol}(\mathfrak{t}/\overline{CR_u})|^{2g-2}\sum_{(w'^1,\dots,w'^s)\in W_u^s}$$

$$\prod_{j=1}^{s}\varepsilon_{w'^j}\mathrm{sgn}((-i)^{\ell_u}\sigma_{Z(u)}(w^jt_j))e^{-2i\pi\langle\rho_u,h\rangle}\widetilde{P}_{u,2g-2+s}(t+h+\sum_{j=1}^{s}w'^jw^jt_j)\,.$$

Clearly

$$(6.98)\qquad Z(Z(u)) = \frac{Z(\widetilde{Z}(u))}{\overline{CR}/\overline{CR_u}},$$

and so

$$(6.99)\qquad |Z(\widetilde{Z}(u))| = |Z(Z(u)|\left|\frac{\overline{CR}}{\overline{CR_u}}\right|.$$

Moreover

$$(6.100)\qquad \mathrm{Vol}(\mathfrak{t}/\overline{CR_u}) = \mathrm{Vol}(\mathfrak{t}/\overline{CR})\left|\frac{\overline{CR}}{\overline{CR_u}}\right| = \mathrm{Vol}(T)\left|\frac{\overline{CR}}{\overline{CR_u}}\right|.$$

By (6.97)-(6.100) , we get

$$(6.101)\qquad \pi_u\left(\frac{\partial/\partial t}{2i\pi}\right)\frac{H_{u,g,s}}{\left|\frac{\overline{CR}}{\overline{CR_u}}\right|^{2g}}(w^1t_1,\dots,w^st_s,h,t) = (-1)^{\ell_u(g-1)+1}$$

$$|Z(Z(u))|\mathrm{Vol}(T)^{2g-2}\sum_{(w'^1,\dots,w'^s)\in W^s(u)}\prod_{j=1}^{s}\varepsilon_{w'^j}((-i)^{\ell u}\sigma_{Z(u)})(w^jt_j)$$

$$\exp(-2i\pi\langle\rho_u,h\rangle)\frac{\widetilde{P}_{u,2g-2+s}}{\left|\frac{\overline{CR}}{\overline{CR_u}}\right|}(t+h+\sum_{j=1}^{s}w'^jw^jt_j)\,.$$

Also by Theorem 1.41 , for $h\in\overline{CR}/\overline{CR_u}$,

$$(6.102)\qquad \exp(-2i\pi\langle\rho_u,h\rangle) = \exp(2i\pi\langle cu,h\rangle)\,.$$

Now observe that if $w'^j \in W_u$, by (1.45),

$$\varepsilon_{w'^j} = \frac{\sigma_{Z(u)}(w'^j w^j t_j)}{\sigma_{Z(u)}(w^j t_j)}, 1 \leq j \leq s. \tag{6.103}$$

Therefore

$$\operatorname{sgn}\left((-i)^{\ell_u}\sigma_{Z(u)}(w^j t_j)\right)\varepsilon_{w'^j} = \operatorname{sgn}\left((-i)^{\ell_u}\sigma_{Z(u)}(w'^j w^j t_j)\right). \tag{6.104}$$

Also

$$\ell_u + (\dim(\mathfrak{g}) - \dim(\mathfrak{z}(u)))/2 = \ell. \tag{6.105}$$

From (2.166), (6.74), (6.96), (6.101)-(6.105), we get (6.95). The proof of our Theorem is completed. □

REMARK 6.23. As shown in Section 2.11, the function $\exp(2i\pi\langle qu, t\rangle)P_{u,2g-2+s,p+c}(t)$ descends to a function which is well-defined on $T = \mathfrak{t}/\overline{CR}$. Ultimately, this explains why equation (6.95) is unambiguous.

Recall that the function $\sigma(t) = \prod_{i=1}^{\ell}\left(e^{i\pi\langle\alpha,t\rangle} - e^{-i\pi\langle\alpha,t\rangle}\right)$ is well-defined on $T = \mathfrak{t}/\overline{CR}$, and the function $\dfrac{e^{2i\pi\langle\rho,t\rangle}}{\sigma(t)}$ is well-defined on $T' = \mathfrak{t}/\overline{R}^*$.

PROPOSITION 6.24. *For any $u \in C/\overline{R}^*$, for any $s \in \mathfrak{t}_{\text{reg}}$, $x \in P$, $w \in W$,*

$$\operatorname{sgn}((-i)\sigma_{Z(u)}(wx)) \prod_{\alpha\in R_+} \frac{1}{2\sinh(\frac{1}{2}(\langle\alpha,s\rangle + 2i\pi[\langle\alpha,u\rangle]))} \tag{6.106}$$
$$\exp(i\pi\sum_{\alpha\in R_+}[\langle w\alpha,u\rangle]) = \varepsilon_w \prod_{\alpha\in R_+}\frac{1}{2\sinh(\frac{1}{2}\langle\alpha, s+2i\pi u\rangle)} e^{2i\pi\langle w\rho,u\rangle}.$$

PROOF. As we just saw, both sides are unchanged when replacing u by $u+\gamma$, $\gamma \in \overline{CR}$. By [**15**, Proposition V.7.10], we may as well assume that for any $\alpha \in R$,

$$|\langle\alpha,u\rangle| \leq 1. \tag{6.107}$$

If $\alpha \in R\backslash R_u$, $\sinh(\frac{1}{2}(\langle\alpha,s\rangle + 2i\pi[\langle\alpha,u\rangle]))$ is unchanged when α is replaced by $-\alpha$. Therefore using (1.45), we get

$$\begin{aligned}\prod\nolimits_{\alpha\in R_+}\tfrac{1}{2\sinh(\frac{1}{2}(\langle\alpha,s\rangle+2i\pi[\langle\alpha,u\rangle]))} &= (-1)^{|R_+\cap w^{-1}(-R_{u,+})|}\\ \prod\nolimits_{\alpha\in wR_+}\tfrac{1}{2\sinh(\frac{1}{2}(\langle\alpha,s\rangle+2i\pi[\langle\alpha,u\rangle]))} &=\\ (-1)^{|R_+\cap w^{-1}(-R_{u,+})|+|\{\alpha\in R_+,-1<\langle w\alpha,u\rangle<0\}|+|\{\alpha\in R_+,\langle w\alpha,u\rangle=\pm1\}|}&\\ \prod\nolimits_{\alpha\in R_+}\tfrac{1}{2\sinh(\frac{1}{2}\langle\alpha,w^{-1}(s+2i\pi u)\rangle)} &=\\ \varepsilon_w(-1)^{|R_+\cap w^{-1}(-R_{u,+})|+|\{\alpha\in R_+,-1<\langle w\alpha,u\rangle<0\}|+|\{\alpha\in R_+,\langle w\alpha,u\rangle=\pm1\}|}&\\ \prod\nolimits_{\alpha\in R_+}\tfrac{1}{2\sinh(\frac{1}{2}\langle\alpha,s+2i\pi u\rangle)}&.\end{aligned} \tag{6.108}$$

Moreover, for $x \in P$,

$$\operatorname{sgn}((-i)^{\ell_u}\sigma_{Z(u)}(wx)) = (-1)^{|R_+\cap w^{-1}(-R_{u,+})|}. \tag{6.109}$$

Also by (6.107),

$$\begin{aligned}(6.110)\quad &\exp\Big(i\pi\sum_{\alpha\in R_+}[\langle w\alpha,u\rangle]\Big)= \\ &\exp\Big(i\pi\langle w\sum_{\alpha\in R_+}\alpha,u\rangle\Big)(-1)^{|\{\alpha\in R_+,-1<\langle w\alpha,u\rangle<0\}|+|\{\alpha\in R_+,\langle w\alpha,u\rangle=\pm1\}|}= \\ &e^{2i\pi\langle w\rho,u\rangle}(-1)^{|\{\alpha\in R_+,-1<\langle w\alpha,u\rangle<0\}|+|\{\alpha\in R_+,\langle w\alpha,u\rangle=\pm1\}|}.\end{aligned}$$

From (6.108)-(6.110) , we get (6.106). The proof of our Proposition is completed. □

REMARK 6.25. It is clear that the right-hand side of (6.106) only depends on the class of u in $C/\overline{R}^*$.

Observe that the function

$$(6.111)\qquad s\in\mathfrak{t}\mapsto\frac{\prod_{\alpha\in R_{u,+}}\langle\alpha,s\rangle e^{2i\pi\langle w\rho,u\rangle}}{\prod_{\alpha\in R_+}2\sinh(\frac{1}{2}\langle\alpha,s+2i\pi u\rangle)}\in\mathbf{C}$$

is a well-defined holomorphic function near $s=0$.

THEOREM 6.26. *For any $u\in C/\overline{CR}, p\in\mathbf{M}, p\neq -c$, the following identity holds,*

$$\begin{aligned}&\int_{M^u/Z(u)}L(u,\lambda^p)=\mathrm{Vol}(T)^{2g-2}\tfrac{|Z(Z(u))|}{|W_u|}(p+c)^{(g-1)\dim(\mathfrak{z}(u))+\frac{s}{2}\dim(\mathfrak{z}(u)/\mathfrak{t})} \\ (6.112)\qquad &(-1)^{\ell(g-1)+1}\left[\frac{\prod_{\alpha\in R_{u,+}}\langle\alpha,\frac{\partial/\partial t}{p+c}\rangle}{\prod_{\alpha\in R_+}2\sinh\left(\frac{1}{2}\langle\alpha,\frac{\partial/\partial t}{p+c}+2i\pi u\rangle\right)}\right]^{2g-2+s} \\ &\textstyle\sum_{(w^1,\dots,w^s)\in W^s}\prod_{j=1}^s\varepsilon_{w^j}\exp(\langle w^j(\rho-ct_j),\frac{\partial/\partial t}{p+c}\rangle \\ &+2i\pi\langle w^j(\rho+pt_j),u\rangle)P_{u,2g-2+s,p+c}(t)_{|t=\sum_{j=1}^s w^jt_j}.\end{aligned}$$

PROOF. Clearly, since $2g-2$ is even,

$$\begin{aligned}(6.113)\qquad &\left[\prod_{\alpha\in R_{u,+}}\widehat{A}\Big(\langle\alpha,\frac{\partial/\partial t}{p+c}\rangle\Big)\right. \\ &\left.\prod_{\alpha\in R_+\setminus R_{u,+}}\frac{1}{2\sinh(\frac{1}{2}(\langle\alpha,\frac{\partial/\partial t}{p+c}\rangle+2i\pi[\langle\alpha,u\rangle]))}\right]^{2g-2} \\ &=\left[\frac{\prod_{\alpha\in R_{u,+}}\langle\alpha,\frac{\partial/\partial t}{p+c}\rangle}{\prod_{\alpha\in R_+}2\sinh(\frac{1}{2}\langle\alpha,\frac{\partial/\partial t}{p+c}+2i\pi u\rangle)}\right]^{2g-2}\end{aligned}$$

Also, since $t_j \in P$, by Proposition 6.24 , we get

$$\prod_{\alpha\in R_{u,+}} \widehat{A}\Big(\langle\alpha, \tfrac{\partial/\partial t}{p+c}\rangle\Big) \prod_{\alpha\in R_+\setminus R_{u,+}} \frac{1}{2\sinh(\frac{1}{2}(\langle\alpha, \frac{\partial/\partial t}{p+c}\rangle + 2i\pi[\langle\alpha,u\rangle]))}$$

$$\operatorname{sgn}((-i)^{\ell_u}\sigma_{Z(u)}(w^j t_j))e^{i\pi\sum_{\alpha\in R_+}[\langle w^j\alpha,u\rangle]} = \tag{6.114}$$

$$\varepsilon_w \frac{\prod_{\alpha\in R_{u,+}} \langle\alpha, \frac{\partial/\partial t}{p+c}\rangle}{\prod_{\alpha\in R_+} 2\sinh(\frac{1}{2}\langle\alpha, \frac{\partial/\partial t}{p+c} + 2i\pi u\rangle)} e^{2i\pi\langle w^j\rho,u\rangle}.$$

By Theorem 6.22 and by (6.113),(6.114) , we get (6.112). The proof of our Theorem is completed. □

REMARK 6.27. Suppose temporarily that $t_1, \dots, t_{s-1}$ verify assumption (A), and that $\epsilon > 0$ is small enough so that if $|t_s| < \epsilon$, $(t_1, \dots, t_s)$ still verify (A). Then (6.112) can be written in the form

$$\int_{M^u/Z(u)} L(u,\lambda^p) = \tag{6.115}$$

$$\operatorname{Vol}(T)^{2g-2}\frac{|Z(Z(u))|}{|W_u|}(p+c)^{(g-1)\dim(\mathfrak{z}(u))+\frac{s}{2}\dim(\mathfrak{z}(u)/\mathfrak{t})}$$

$$(-1)^{\ell(g-1)+1}\left[\frac{\prod_{\alpha\in R_{u,+}}\langle\alpha, \frac{\partial/\partial t}{p+c}\rangle}{\prod_{\alpha\in R_+} 2\sinh\left(\frac{1}{2}\langle\alpha, \frac{\partial/\partial t}{p+c} + 2i\pi u\rangle\right)}\right]^{2g-2+s}.$$

$$\sum_{w\in W}\varepsilon_w e^{\langle w(\rho+pt_s), \frac{\partial/\partial t}{p+c}+2i\pi u\rangle} \sum_{(w^1,\dots,w^{s-1})\in W^{s-1}}$$

$$\prod_{j=1}^{s-1}\varepsilon_{w^j} e^{\langle w^j(\rho-ct_j), \frac{\partial/\partial t}{p+c}\rangle + 2i\pi\langle w^j(\rho+pt_j),u\rangle} P_{u,2g-2+s,p+c}(t)_{|t=\sum_{j=1}^{s-1} w^j t_j}.$$

Using (1.94), (2.131), since $pt_s = \theta_s \in \overline{CR}_+^*$, we may rewrite (6.115) in the form

$$
(6.116)\quad \int_{M^u/Z(u)} L(u,\lambda^p) =
$$

$$
\mathrm{Vol}(T)^{2g-2}\frac{|Z(Z(u))|}{|W_u|}(p+c)^{(g-1)\dim\mathfrak{z}(u)+\frac{(s-1)}{2}\dim(\mathfrak{z}(u)/\mathfrak{t})}
$$

$$
(-1)^{\ell(g-1)+1}\left[\frac{\prod_{\alpha\in R_{u,+}}\langle\alpha,\frac{\partial/\partial t}{p+c}\rangle}{\prod_{\alpha\in R_+}2\sinh\left(\frac{1}{2}\langle\alpha,\frac{\partial/\partial t}{p+c}+2i\pi u\rangle\right)}\right]^{2g-2+s-1} \chi_{pt_s}\left(e^{\frac{\partial/\partial t}{2i\pi(p+c)}+u}\right)
$$

$$
\sum_{(w^1,\dots,w^{s-1})\in W^{s-1}}\prod_{j=1}^{s-1}\varepsilon_{w^j}e^{\langle w^j(\rho-ct_j),\frac{\partial/\partial t}{p+c}\rangle+2i\pi\langle w^j(\rho+pt_j),u\rangle}
$$

$$
P_{u,2g-2+s-1,p+c}(t)_{|t=\sum_{j=1}^{s-1}w^jt_j}\,.
$$

With respect to (6.112) , s has been replaced by $s-1$, and we have the extra differential operator $\chi_{pt_s}\left(e^{\frac{\partial/\partial t}{2i\pi(p+c)}+u}\right)$.

Let M' be the manifold attached to G, $\mathcal{O}_1,\dots,\mathcal{O}_{s-1}$. Let F be the vector bundle over M'/G associated to the representation of highest weight $pt_s \subset \overline{CR}^*$ of G. Then if we still denote by $L(u,\lambda^p)$ the corresponding class (6.13) over $M'^u/Z(u)$, in view of (5.318), we can rewrite (6.115),(6.116) in the form

$$
(6.117)\qquad \int_{M^u/Z(u)} L(u,\lambda^p) = \int_{M'^u/Z(u)} L(u,\lambda^p)\mathrm{ch}_u(F)\,.
$$

Also by Theorem 5.57, for $|t_s|$ small enough, the orbifold M/G fibres over M'/G with fibre G/T. Let ρ be the projection $M/G \to M'/G$. Then one verifies easily that

$$
(6.118)\qquad \rho_* L(u,\lambda^p) = L(u,\lambda^p)\mathrm{ch}_u(E),
$$

which makes (6.117) tautological.

REMARK 6.28. If $u \in C/\overline{CR}$, $w \in W$, one should have the equality

$$
(6.119)\qquad \int_{M^u/Z(u)} L(u,\lambda^p) = \int_{M^{wu}/Z(wu)} L(wu,\lambda^p)\,.
$$

We will briefly explain why the right-hand side of (6.112) is unchanged when replacing u by wu. Put

$$
(6.120)\quad C = \mathrm{Vol}(T)^{2g-2}\frac{|Z(Z(u))|}{|W_u|}(p+c)^{(g-1)\dim\mathfrak{z}(u))+s/2\dim(\mathfrak{z}(u)/\mathfrak{t})}(-1)^{\ell(g-1)+1}\,.
$$

Then using (1.45), (2.167), (6.112), we get

$$\int_{M^{wu}/Z(wu)} L(wu,\lambda^p) = C\left[\frac{\prod_{\alpha\in R_{u,+}} \frac{\langle w\alpha, \partial/\partial t\rangle}{p+c}}{\prod_{\alpha\in R_+} 2\sinh\left(\frac{1}{2}\langle w\alpha, \frac{\partial/\partial t}{p+c} + 2i\pi wu\rangle\right)}\right]^{2g-2+s}$$

$$\left(\varepsilon_w(-1)^{|R_+\cap w(-R_{u,+})|}\right)^s \sum_{(w^1,\dots,w^s)\in W^s} \prod_{j=1}^s \varepsilon_{w^j} e^{\langle w^j(\rho-ct_j), \frac{\partial/\partial t}{p+c}\rangle_+ + 2i\pi\langle w^j(\rho+pt_j), wu\rangle}$$

$$P_{wu,2g-2+s,p+c}(t)|_{t=\sum_1^s w^j t_j} = C\left[\frac{\prod_{\alpha\in R_{u,+}} \frac{\langle \alpha, \partial/\partial t\rangle}{p+c}}{\prod_{\alpha\in R_+} 2\sinh\left(\frac{1}{2}\langle \alpha, \frac{\partial/\partial t}{p+c} + 2i\pi u\rangle\right)}\right]^{2g-2+s}$$

$$\sum_{(w^1,\dots,w^s)\in W^s} \prod_{j=1}^s \varepsilon_{w^j} e^{\langle w^j(\rho-ct_j), \frac{\partial/\partial t}{p+c}\rangle_+ + 2i\pi\langle w^j(\rho+pt_j), u\rangle}$$

$$P_{u,2g-2+s,p+c}(t)|_{t=\sum_1^s w^j t_j} = \int_{M^u/Z(u)} L(u,\lambda^p)\,. \tag{6.121}$$

So we find that (6.112) is compatible with (6.119).

6.8. The case where $\sum_{j=1}^s pt_j \notin \overline{R}$. Recall that in Section 1.8, we saw that if $\sum_{j=1}^s pt_j \in \overline{R}$, for any $(w^1,\dots,w^s)\in W^s$, then $\sum_{j=1}^s w^j pt_j \in \overline{R}$.

THEOREM 6.29. *If $\sum_1^s pt_j \notin \overline{R}$, if $v \in C/\overline{R}^*$, then*

$$\sum_{\substack{u\in C/\overline{CR}\\ \tau u = v}} \int_{M^u/Z(u)} L(u,\lambda^p) = 0. \tag{6.122}$$

In particular, if $p\in \mathbf{M}, p\neq -c$,

$$\operatorname{Ind}(D_{p,+}) = 0 \tag{6.123}$$

PROOF. To establish (6.122), we use Theorem 6.26. In fact, in the right hand-side of formula (6.112), we observe that the various terms depend on the image $v=\tau u$, with the exception of $\prod_{j=1}^s \exp(2i\pi\langle w^j pt_j, u\rangle)$. Also

$$\sum_{v\in \overline{R}^*/\overline{CR}} \exp\left(2i\pi\langle \sum_{j=1}^s w^j pt_j, v\rangle\right) = 0 \qquad \text{if} \sum_{j=1}^s w^j pt_j \notin \overline{R}, \tag{6.124}$$

$$= \left|\frac{\overline{R}^*}{\overline{CR}}\right| \qquad \text{if} \sum_{j=1}^s w^j pt_j \in \overline{R}. \tag{6.125}$$

It is now clear that (6.122) holds. Equation (6.123) follows from Theorem 6.16 and from (6.122). □

6.9. A residue formula for the index. In view of Theorem 6.29, we may and we will assume that

$$\sum_{j=1}^s pt_j \in \overline{R}. \tag{6.126}$$

We now use the notation of Section 2. Recall that $\overline{S}\subset \mathfrak{t}/\overline{CR}$ has been defined in (2.111).

DEFINITION 6.30. We will say that $(t_1, \dots, t_s)$ verify assumption $(\overline{A})$ if for any $(w^1, \dots, w^s) \in W^s$,

$$\sum_{j=1}^{s} w^j t_j \notin \overline{S}. \tag{6.127}$$

Clearly assumption $(\overline{A})$ is stronger than assumption (A) of Definition 5.17. In the sequel, we assume that $(t_1, \dots, t_s)$ verify $(\overline{A})$.

THEOREM 6.31. *For any $u \in C/\overline{R}^*$, $p \in \mathrm{M}, p \neq -c$, the following identity holds*

$$\int_{M^u/Z(u)} L(u, \lambda^p) = \left|\frac{\overline{R}}{\overline{CR}}\right| \mathrm{Vol}(T)^{2g-2} \frac{|Z(Z(u))|}{|W_u|}(p+c)^{(g-1)r}(-1)^{\ell(g-1)+r}$$

$$\sum_{\substack{I=(i_1,\dots,i_r)\in\mathcal{I}_u \\ f\in\overline{CR}/d\overline{R}}} \frac{1}{\langle\alpha_{i_1}, \dots, \alpha_{i_r}\rangle} \mathrm{Res}^I_{x=0}\left[\prod_{\alpha\in R_+} \frac{1}{2\sinh\left(\frac{1}{2}\langle\alpha, \frac{x^I}{p+c} + 2i\pi u\rangle\right)}\right]^{2g-2+s}$$

$$\sum_{(w^1,\dots,w^s)\in W^s} \varepsilon_{w^1}\dots\varepsilon_{w^s} \exp\Big(\langle\textstyle\sum_{j=1}^s w^j(\rho - ct_j), \frac{x^I}{p+c}\rangle \tag{6.128}$$
$$+2i\pi\langle\textstyle\sum_{j=1}^s w^j(\rho+pt_j) + (p+c)f, u\rangle\Big) \exp\Big(d\sum_{j=1}^r \frac{\langle p^I_{j-1}\alpha_{i_j}, x^I\rangle}{\langle p^I_{j-1}\alpha_{i_j}, e^j\rangle}$$
$$\Big[\tfrac{1}{d}\langle p^I_{j-1}(\textstyle\sum_{k=1}^s w^k t_k + f), e^j\rangle\Big]\Big)\frac{1}{\prod_{j=1}^r \exp\left(d\frac{\langle p^I_{j-1}\alpha_{i_j}, x^I\rangle}{\langle p^I_{j-1}\alpha_{i_j}, e^j\rangle}\right)-1}.$$

PROOF. If $(w^1, \dots, w^s) \in W^s, \sum_{j=1}^s w^j t_j \notin \overline{S}$. Then we use Theorems 2.50 and 6.26 and we get (6.128). □

THEOREM 6.32. *If $p \in \mathrm{M}, p \neq -c$, then*

$$\mathrm{Ind}(D_{p,+}) = \left|\frac{\overline{R}}{\overline{CR}}\right| \mathrm{Vol}(T)^{2g-2}\frac{|Z(G)|}{|W|}(p+c)^{(g-1)r}(-1)^{\ell(g-1)+r} \sum_{u\in C/\overline{R}^*} \tag{6.129}$$

$$\sum_{\substack{I=(i_1,\dots,i_r)\in\mathcal{I}_u \\ f\in\overline{CR}/d\overline{R}}} \frac{1}{\langle\alpha_{i_1}, \dots, \alpha_{i_r}\rangle} \mathrm{Res}^I_{x=0}\left[\prod_{\alpha\in R_+} \frac{1}{2\sinh\left(\frac{1}{2}\langle\alpha, \frac{x^I}{p+c} + 2i\pi u\rangle\right)}\right]^{2g-2+s}$$

$$\sum_{(w^1,\dots,w^s)\in W^s} \varepsilon_{w^1}\dots\varepsilon_{w^s} \exp\Big(\langle\sum_{j=1}^s w^j(\rho - ct_j), \frac{x^I}{p+c}\rangle$$

$$+2i\pi\sum_{j=1}^s w^j(\rho+pt_j) + (p+c)f, u\rangle\Big) \exp\Big(d\sum_{j=1}^r \frac{\langle p^I_{j-1}\alpha_{i_j}, x^I\rangle}{\langle p^I_{j-1}\alpha_{i_j}, e^j\rangle}$$

$$\Big[\frac{1}{d}\langle p^I_{j-1}(\sum_{k=1}^s w^k t_k + f), e^j\rangle\Big]\Big)\frac{1}{\prod_{j=1}^r\left(\exp\left(d\frac{\langle p^I_{j-1}\alpha_{i_j}, x^I\rangle}{\langle p^I_{j-1}\alpha_{i_j}, e^j\rangle}\right)-1\right)}.$$

PROOF. Recall that $Z(G) = \overline{R}^*/\overline{CR}$. Using Theorems 6.16, (6.124) and 6.31, we get (6.129). The proof of our Theorem is completed. □

REMARK 6.33. As we saw in Proposition 1.40 and Remark 2.7, for $n \geq 2$ and $G = \mathrm{SU}(n)$, then $C/\overline{R}^* = 0$, and we can choose $d = 1$. Then formula (6.129) has an especially simple form.

6.10. A formula for the index for large p.

THEOREM 6.34. *For $p \in \mathbf{M}$, and $|p|$ large enough, the following identity holds*

$$\mathrm{Ind}(D_{p,+}) = \left|\frac{\overline{R}}{\overline{CR}}\right| \mathrm{Vol}(T)^{2g-2} \frac{|Z(G)|}{|W|}(p+c)^{(g-1)r}(-1)^{\ell(g-1)+r} \sum_{u \in C/\overline{R}^*}$$

$$\sum_{\substack{I=(i_1,\dots,i_r)\in\mathcal{I}_u \\ f\in\overline{CR}/d\overline{R}}} \frac{1}{\langle \alpha_{i_1},\dots,\alpha_{i_r}\rangle} \mathrm{Res}^I_{x=0}\left[\prod_{\alpha\in R_+}\frac{1}{2\sinh\left(\frac{1}{2}\langle\alpha,\frac{x^I}{p+c}+2i\pi u\rangle\right)}\right]^{2g-2+s}$$

$$(6.130)\quad \sum_{(w^1,\dots,w^s)\in W} \varepsilon_{w^1}\dots\varepsilon_{w^s}\exp\Big(\langle\sum_{j=1}^s w^j\frac{\rho}{p+c},x^I\rangle$$

$$+2i\pi\langle\sum_{j=1}^s w^j(\rho+pt_j+(p+c)f,u\rangle\Big)$$

$$\exp\left(d\sum_{j=1}^r\frac{\langle p^I_{j-1}\alpha_{i_j},x^I\rangle}{\langle p^I_{j-1}\alpha_{i_j},e^j\rangle}\left[\frac{1}{d}\langle p^I_{j-1}(\sum_{k=1}^s\frac{w^kpt_k}{p+c}+f),e^j\rangle\right]\right)$$

$$\frac{1}{\prod_{j=1}^r\left(\exp\left(d\frac{\langle p^I_{j-1}\alpha_{i_j},x^I\rangle}{\langle p^I_{j-1}\alpha_{i_j},e^j\rangle}\right)-1\right)}.$$

PROOF. Clearly, for $p\to+\infty$, $\sum_1^s\frac{w^jct_j}{p+c}\to 0$. Also by Proposition 2.14, since $\sum_{j=1}^s w^jt_j\notin\overline{S}$, if $f\in\overline{R}$,

$$(6.131)\qquad \langle p^I_{j-1}\Big(\sum_{k=1}^s w^kt_k+f\Big),e^j\rangle\notin\mathbf{Z}.$$

Therefore by (2.13) , for $|p|$ large enough,

$$(6.132)\quad d\sum_{j=1}^r\frac{\langle p^I_{j-1}\alpha_{i_j},x^I\rangle}{\langle p^I_{j-1}\alpha_{i_j},e^j\rangle}\left[\frac{1}{d}\langle p^I_{j-1}\Big(\sum_{k=1}^s\frac{w^kpt_k}{p+c}+f\Big),e^j\rangle\right]=$$

$$d\sum_{j=1}^r\frac{\langle p^I_{j-1}\alpha_{i_j},x^I\rangle}{\langle p^I_{j-1}\alpha_{i_j},e^j\rangle}\left[\frac{1}{d}\langle p^I_{j-1}\Big(\sum_{k=1}^s w^kt_k+f),e^j\rangle\right]-\langle\sum_{j=1}^s\frac{cw^jt_j}{p+c},x^I\rangle.$$

By (2.115), (6.132) , we find that for $|p|$ large enough,

$$\exp\left(\sum_{j=1}^r\frac{\langle p^I_{j-1}\alpha_{i_j},x^I\rangle}{\langle p^I_{j-1}\alpha_i,e^j\rangle}\left[\frac{1}{d}\langle p^I_{j-1}\Big(\sum_{k=1}^s\frac{w^kpt_k}{p+c}+f\Big),e^j\rangle\right]\right)$$

$$(6.133)\qquad =\exp\left(\langle-\sum_{j=1}^s\frac{w^jct_j}{p+c},x^I\rangle\right)$$

$$\exp\left(d\sum_{j=1}^r\frac{\langle p^I_{j-1}\alpha_{i_j},x^I\rangle}{\langle p^I_{j-1}\alpha_{i_j},e^j\rangle}\left[\frac{1}{d}\langle p^I_{j-1}\Big(\sum_{k=1}^s w^kt_k+f),e^j\rangle\right]\right).$$

From (6.129), (6.133) , we get (6.130). The proof of our Theorem is completed. □

REMARK 6.35. The formulas in Theorems 6.32 and 6.34 are essentially identical.The point is that, as we will see in Theorems 7.23 and 8.1, the right-hand side of (6.130) is just Verlinde's formula.

6.11. A perturbation of the index problem. We still assume that $s \geq 1$ and that if $g = 0$, then $s \geq 3$, so that $2g - 2 + s \geq 1$. Note that these assumptions are almost irrelevant, since by adding as many marked points as one wishes with holonomy equal to 1, one can make s arbitrarily large.

Recall that, by Proposition 1.23, $\overline{P}$ embeds into T. Also the orbit under the Weyl group W of any element in T always intersects $\overline{P}$.

Let $t_1, \ldots, t_s$ be s elements in T. We may and we will assume that they all lie in $\overline{P}$. In the sequel we will consider $t_1, \ldots, t_s$ as elements of $\mathfrak{t}$. We still define $\mathbf{M} \subset \mathbf{Z}$ as in (6.20), and we assume that $\mathbf{M}$ is not reduced to 0. For $p \in \mathbf{M}$, set

$$\theta_j = p\theta_j,\ 1 \leq j \leq s. \tag{6.134}$$

Then $\theta_j \in \overline{CR}^*$, $1 \leq j \leq s$.

Let $\delta = (\delta_1, \ldots, \delta_s) \in \mathfrak{t}^s$ and assume that $t_1 + \delta_1, \ldots, t_s + \delta_s$ lie in P and verify $(\overline{A})$. Put

$$X^\delta = G^{2g} \times \prod_{j=1}^{s} \mathcal{O}_{t_j + \delta_j}, \tag{6.135}$$

and let M^δ correspond to X^δ as in (5.59). Then M^δ is a smooth submanifold of X^δ.

We will briefly show how to equip M^δ with an orbifold line bundle λ^p. We use the notation of Section 4.10. For $1 \leq j \leq s$, consider the orbit $\mathcal{O}_{-pd/dt + p(t_j + \delta_j)} \subset \widehat{L\mathfrak{g}}$. By (4.30),

$$\mathcal{O}_{-pd/dt+p(t_j+\delta_j)} \simeq LG/T. \tag{6.136}$$

Also as in Definition 4.3, $\widetilde{LG}/(T \times S^1) \simeq LG/T$ can be equipped with the line bundle $H_{(\theta_j, -p)}$ of weight (θ_j, p). Therefore $\mathcal{O}_{-pd/dt+p(t_j+\delta_j)}$ can be equipped with the corresponding line bundle L_j. Recall that $pt_j = \theta_j$. Then we define the line bundle λ_p as in (4.189), and the line bundle λ^p on M^δ as in Definition 6.10.

We can then define a Dirac operator $D_p^{(t_j+\delta_j)}$ in the same way as in Sections 6.3 and 6.4.

THEOREM 6.36. *For $p \in \mathrm{M}, p \neq -c$,*

$$\operatorname{Ind}(D_{p,+}^{(t_j+\delta_j)}) = \left|\frac{\overline{R}}{\overline{CR}}\right| \operatorname{Vol}(T)^{2g-2} \frac{1}{|W|}(p+c)^{(g-1)r} \tag{6.137}$$

$$(-1)^{\ell(g-1)+r} \sum_{u \in C/\overline{CR}} \sum_{\substack{I=(i_1,\dots,i_r)\in\mathcal{I}_u \\ f\in\overline{R}/d\overline{R}}} \frac{1}{\langle \alpha_{i_1}, \dots, \alpha_{i_r}\rangle}$$

$$\operatorname{Res}_{x=0}^{I}\left[\prod_{\alpha\in R_+} \frac{1}{2\sinh\left(\frac{1}{2}\langle\alpha, \frac{x^I}{p+c} + 2i\pi u\rangle\right)}\right]^{2g-2+s}$$

$$\sum_{(w^1,\dots,w^s)\in W^s} \varepsilon_{w^1}\dots\varepsilon_{w^s} \exp\Big(\langle\sum_{j=1}^{s} w^j\Big(\frac{\rho - ct_j}{p+c} - \delta_j\Big), x^I\rangle$$

$$+2i\pi\langle\sum_{j=1}^{s} w^k(\rho + pt_j) + (p+c)f, u\rangle\Big)$$

$$\exp\left(d\sum_{j=1}^{r}\frac{\langle p_{j-1}^I\alpha_{i_j}, x^I\rangle}{\langle p_{j-1}^I\alpha_{i_j}, e^j\rangle}\left[\frac{1}{d}\langle p_{j-1}^I\left(\sum_{k=1}^{s} w^k(t_k+\delta_k) + f\right), e^j\rangle\right]\right)$$

$$\frac{1}{\prod_{j=1}^{r}\left(\exp\left(\frac{d\langle p_{j-1}^I\alpha_{i_j}, x^I\rangle}{\langle p_{j-1}^I\alpha_{i_j}, e^j\rangle}\right) - 1\right)}.$$

PROOF. The proof of our Theorem follows the same steps as the proof of Theorem 6.32, where the case $\delta = 0$ was considered. We briefly describe the main steps where the proof of our more general result differs.

- Instead of (6.26) in Proposition 5.11, we have

$$c_1(\lambda^p, \nabla^{\lambda^p}) = p(\omega + \sum_{j=1}^{s}\langle\delta_j, \Theta_j\rangle). \tag{6.138}$$

  This follows from an easy computation which is left to the reader.
- In formula (6.74) in Theorem 6.21, the main point is in fact that $\prod_{j=1}^{s} e^{2i\pi\langle w^j pt_j + ph, u\rangle}$ is unchanged. The argument is in fact the same as in the proof of Theorem 6.21. Also using (6.138) instead of Theorem 5.78, one finds that in formula (6.74) in Theorem 6.21, $\prod_{j=1}^{s} e^{\frac{1}{p+c}\langle(\rho - ct_j), \partial/\partial t_j\rangle}$ should be replaced by $\prod_{j=1}^{s} e^{\langle\frac{\rho - ct_j}{p+c} - \delta_j, \partial/\partial t_j\rangle}$.

By proceeding otherwise as before, we get (6.137). The proof of our Theorem is completed. □

REMARK 6.37. Using (6.137) and proceeding as in the proof of Theorem 6.29, we find that (6.123) is still true. So in the sequel, we may and we will assume that (6.126) holds.

Now we will slightly modify the statement of Theorem 6.36. Recall that $t_1, \dots, t_s$ all lie in $\overline{P}$. Therefore for $p \in \mathrm{M}, p \geq 0$, for $1 \leq j \leq s$, $pt_j/(p+c)$ still lies in $\overline{P}$.

THEOREM 6.38. *For $p \in \mathbf{M}, p \geq 0$, if $\varepsilon_1, \ldots, \varepsilon_s \in \mathfrak{t}$ are such that $\frac{pt_1}{p+c} + \varepsilon_1, \ldots \frac{pt_s}{p+c} + \varepsilon_s \in P$, and that, for any $(w^1, \ldots, w^s) \in W^s$, $\sum_{j=1}^s w^j \left(\frac{pt_j}{p+c} + \varepsilon_j\right) \notin \overline{S}$, then*

$$\operatorname{Ind}(D_{p,+}^{(\frac{pt_j}{p+c}+\varepsilon_j)}) = \left|\frac{\overline{R}}{\overline{CR}}\right| \operatorname{Vol}(T)^{2g-2} \frac{|Z(G)|}{|W|}(p+c)^{(g-1)r}$$
$$(-1)^{\ell(g-1)+r} \sum_{u \in C/\overline{R}^*} \sum_{\substack{I=(i_1,\ldots,i_r)\in \mathcal{I}_u \\ f \in \overline{R}/d\overline{R}}} \frac{1}{\langle \alpha_{i_1}, \ldots, \alpha_{i_r}\rangle}$$
$$\operatorname{Res}^I_{x=0}\left[\prod_{\alpha\in R_+} \frac{1}{2\sinh\left(\frac{1}{2}\langle \alpha, \frac{x^I}{p+c} + 2i\pi u\rangle\right)}\right]^{2g-2+s} \tag{6.139}$$
$$\sum_{(w^1,\ldots,w^s)\in W^s} \varepsilon_{w^1}\ldots\varepsilon_{w^s} \exp\Big(\langle \sum_{j=1}^s w^j\Big(\frac{\rho}{p+c} - \varepsilon_j\Big), x^I\rangle$$
$$+2i\pi\langle \sum_{j=1}^s w^j(\rho + pt_j) + (p+c)f, u\rangle\Big)$$
$$\exp\left(d\sum_{j=1}^r \frac{\langle p^I_{j-1}\alpha_{i_j}, x^I\rangle}{\langle p^I_{j-1}\alpha_{i_j}, e^j\rangle}\left[\frac{1}{d}\langle p^I_{j-1}\left(\sum_{k=1}^s w^k\Big(\frac{pt_k}{p+c} + \varepsilon_k\Big) + f\right), e^j\rangle\right]\right)$$
$$\frac{1}{\prod_{j=1}^r \left(\exp\left(\frac{d\langle p^I_{j-1}\alpha_{i_j}, x^I\rangle}{\langle p^I_{j-1}\alpha_{i_j}, e^j\rangle}\right) - 1\right)}.$$

PROOF. In Theorem 6.36, we take

$$\delta_j = \frac{-ct_j}{p+c} + \varepsilon_j \;,\; 1 \leq j \leq s. \tag{6.140}$$

Then we get (6.139). □

PROPOSITION 6.39. *For $p \in \mathbf{M}, p \geq 0$, there is $\varepsilon_1, \ldots, \varepsilon_s \in \mathfrak{t}$ such that for $\ell \in]0,1]$, $1 \leq j \leq s$, $\frac{pt_j}{p+c} + \ell\varepsilon_j \in T$ is regular, and for any $(w^1, \ldots, w^s) \in W^s, \ell \in]0,1]$, $\sum_{j=1}^s w^j \left(\frac{pt_j}{p+c} + \ell\varepsilon_j\right) \notin \overline{S}$.*

PROOF. Recall that $\overline{S}$ is a union of hypertori in T. Therefore $\overline{S} \cap \overline{P}$ is the intersection of $\overline{P} \subset \mathfrak{t}$ with a union of hyperplanes in $\mathfrak{t}$. Our Proposition now follows easily. □

Let $[x]_+, [x]_-$ be the functions defined on $\mathbf{R}$ with values in $[0,1]$, which are periodic of period 1 and such that

$$\begin{aligned} [x]_+ &= x \text{ for } x \in [0,1[, \\ [x]_- &= x \text{ for } x \in]0,1]. \end{aligned} \tag{6.141}$$

Observe that by 2.14 , for $\ell \in]0,1]$, $f \in \overline{R}/d\overline{R}$, $1 \leq j \leq r$,

$$\langle p^I_{j-1}\left(\sum_{k=1}^s w^k\left(\frac{pt_k}{p+c} + \ell\varepsilon_k\right) + f\right), e_j\rangle \notin \mathbf{Z}. \tag{6.142}$$

From (6.142) , it follows that either

$$\langle p^I_{j-1}\left(\sum_{j=1}^{s} w^j\left(\frac{pt_j}{p+c}\right)+f\right), e^j\rangle \notin \mathbf{Z}, \tag{6.143}$$

or

$$\begin{aligned}&\langle p^I_{j-1}\left(\sum_{j=1}^{s} w^j\left(\frac{pt_j}{p+c}\right)+f\right), e^j\rangle \in \mathbf{Z}\\ &\langle p^I_{j-1}\left(\sum_{j=1}^{s} w^j\varepsilon^j\right), e^j\rangle \neq 0.\end{aligned} \tag{6.144}$$

DEFINITION 6.40. Given $(w^1,\dots,w^s)\in W^s$, for $f\in \overline{R}/d\overline{R}$, $I=(i_1,\dots,i_r)$, put

$$\begin{aligned}\eta_j^{(w^1,\dots,w^s)}(f,I) &= +\text{ if } \langle p^I_{j-1}\left(\sum_{k=1}^{s} w^k\left(\frac{pt_j}{p+c}\right)+f\right), e^j\rangle \notin \mathbf{Z},\\ &\quad\text{or if } \langle p^I_{j-1}\left(\sum_{k=1}^{s} w^k\left(\frac{pt_k}{p+c}\right)+f\right), e^j\rangle \in \mathbf{Z},\\ &\quad\langle p^I_{j-1}\left(\sum_{k=1}^{s} w^k\varepsilon_k\right), e^j\rangle > 0\\ &= -\text{ if } \langle p^I_{j-1}\left(\sum_{k=1}^{s} w^k\left(\frac{pt_k}{p+c}\right)+f\right), e^j\rangle \in \mathbf{Z},\\ &\quad\langle p^I_{j-1}\left(\sum_{k=1}^{s} w^k\varepsilon_k\right), e^j\rangle < 0.\end{aligned} \tag{6.145}$$

Observe that if $f=\sum_{k=1}^{r} f^j e_j$

$$\langle p^I_{j-1}\left(\sum_{k=1}^{s} w^k\left(\frac{pt_k}{p+c}+f\right), e^j\rangle = \langle p^I_{j-1}\left(\sum_{k=1}^{s} w^k\frac{pt_k}{p+c}+\sum_{k=1}^{j} f^k e_k\right), e^j\rangle. \tag{6.146}$$

By (6.146), it is clear that $\eta_j^{(w^1,\dots,w^s)}(f,I)$ depend only on $f^1,\dots,f^j,i_1,\dots,i_{j-1}$.

THEOREM 6.41. *Given $p \in \mathbf{M}, p \geq 0$, the integer* Ind $D_{p,+}^{(\frac{pt_j}{p+c}+\ell\varepsilon_j)}$ *does not depend on $\ell \in]0,1]$. More precisely*

$$\operatorname{Ind} D_{p,+}^{(\frac{pt_i}{p+c}+\ell\varepsilon_i)} = \left|\frac{\overline{R}}{\overline{\overline{CR}}}\right| \operatorname{Vol}(T)^{2g-2}\frac{|Z(G)|}{|W|}(p+c)^{(g-1)r} \tag{6.147}$$

$$(-1)^{\ell(g-1)+r} \sum_{u\in C/\overline{R}^*} \sum_{\substack{I=(i_1,\dots,i_r)\in\mathcal{I}_u \\ f\in\overline{CR}/d\overline{R}}} \frac{1}{\langle\alpha_{i_1},\dots,\alpha_{i_r}\rangle}$$

$$\operatorname{Res}^I_{x=0}\left[\prod_{\alpha\in R_+}\frac{1}{2\sinh\left(\frac{1}{2}\langle\alpha,\frac{x^I}{p+c}+2i\pi u\rangle\right)}\right]^{2g-2+s}$$

$$\sum_{(w^1,\dots,w^s)\in W^s}\varepsilon_{w^1}\dots\varepsilon_{w^s}\exp\Big(\langle\sum_{j=1}^s w^j\Big(\frac{\rho}{p+c}\Big),x^I\rangle$$

$$+2i\pi\langle\sum_1^s w^j(\rho+pt_j)+(p+c)f,u\rangle\Big)$$

$$\exp\left(d\sum_{j=1}^s\frac{\langle p^I_{j-1}\alpha_{i_j},x^I\rangle}{\langle p^I_{j-1}\alpha_{i_j},e^j\rangle}\right.$$

$$\left.\left[\frac{1}{d}\langle p^I_{j-1}\left(\sum_{k=1}^s w^k\Big(\frac{pt_k}{p+c}\Big)+f\right),e_j\rangle\right]_{\eta_j^{(w^1,\dots,w^s)}(f,I)}\right)$$

$$\frac{1}{\prod_{j=1}^s\left(\exp\Big(\frac{d\langle p^I_{j-1}\alpha_{i_j},x^I\rangle}{\langle p^I_{j-1}\alpha_{i_j},e^j\rangle}\Big)-1\right)}.$$

PROOF. Clearly, for $\ell\in]0,1]$,

$$\left[\frac{1}{d}\langle p^I_{j-1}\sum_{k=1}^s w^k\left(\frac{pt_k}{p+c}+\ell\varepsilon_k\right)+f,e_j\rangle\right] \tag{6.148}$$

depends continuously on $\ell\in]0,1]$, and moreover as $\ell\to 0$, it converges to

$$\left[\frac{1}{d}\langle p^I_{j-1}\left(\sum_1^s\frac{w^jt_j}{p+c}+f\right),e_j\rangle\right]_{\eta_j^{(w^1,\dots,w^s)}(f,I)}. \tag{6.149}$$

By (6.139) and the above, Ind $D_{p,+}^{(\frac{pt_j}{p+c}+\ell\varepsilon_j)}$ depends continuously on $\ell\in]0,1]$, and so remains constant. This last fact should also be clear by the construction of $D_{p,+}^{(\frac{pt_j}{p+c}+\ell\varepsilon_j)}$.

Using (6.139), (6.149) , we get (6.147). The proof of our Theorem is completed. □

REMARK 6.42. Observe that in our direct index theoretic computations, we have avoided introducing central holonomies, because they are non generic. We dealt with this case by a perturbation argument. However we may assume that one of the t_j, say t_s, is equal to $h_o\in Z(G)$, while the other holonomies are generic.

By Proposition 1.24, we can identify h_o to a unique element in $\overline{P}$, which we still denote h_o, so that $h_o \in \overline{P} \cap \overline{R}^*$. When the moduli space M/G is an orbifold (which is the case when condition $(\overline{A})$ holds), all the computations we did in Sections 6.1-6.9 can be done directly. By proceeding as in Remark 6.27, it is obvious that in both cases, we compute the same index. This is less obvious at the level of explicit computations. Using in particular equations (1.78) and (1.140), we find that equation (6.112) is now replaced by

$$\begin{aligned}
&\int_{M^u/Z(u)} L(u,\lambda^p) = \epsilon_{w_{h_o}} \operatorname{Vol}(T)^{2g-2} \frac{|Z(Z(u))|}{|W_u|} \\
&\quad (p+c)^{(g-1)\dim(\mathfrak{z}(u)) + \frac{(s-1)}{2}\dim(\mathfrak{z}(u)/\mathfrak{t})} \\
&\quad (-1)^{\ell(g-1)+1} \left[\frac{\prod_{\alpha \in R_{u,+}} \langle \alpha, \frac{\partial/\partial t}{p+c} \rangle}{\prod_{\alpha \in R_+} 2\sinh\left(\frac{1}{2}\langle \alpha, \frac{\partial/\partial t}{p+c} + 2i\pi u \rangle\right)} \right]^{2g-2+s-1} \\
&\quad \sum_{(w^1,\dots,w^{s-1}) \in W^{s-1}} \prod_{j=1}^{s-1} \varepsilon_{w^j} \exp(\langle w^j(\rho - ct_j), \tfrac{\partial/\partial t}{p+c} \rangle \\
&\quad + 2i\pi \langle w^j(\rho + pt_j), u \rangle) \exp(2i\pi \langle (p+c)h_o, u \rangle) \\
&\quad P_{u,2g-2+s-1,p+c}(t)_{|t = \sum_{j=1}^{s-1} w^j t_j + h_o}.
\end{aligned} \tag{6.150}$$

The fact that ultimately, the two explicit index formulas coincide will be verified via the Verlinde formulas in Remark 8.4, by using in particular Theorem 1.33.

7. Residues and the Verlinde formula

In this Section, we apply residues techniques to the finite Verlinde sum, we express it as a residue in several complex variables, and we prove that for p large enough, the Riemann-Roch number of M/G is given by the Verlinde formula. Without any condition on p, the Verlinde formula is the Riemann-Roch number of some perturbation of the moduli space of M/G.

As indicated in the introduction, higher cohomology groups on M/G may well not vanish. This would account for the discrepancy between the index and the Verlinde formula for small p.

This Section is organized as follows. In Section 7.1, we give a residue formula for a Fourier series over $\overline{R}^*/q\overline{R}^*$, $L_q(t,x)$. In Section 7.2, we consider a related series $M_q(t,x)$, which we also evaluate as a sum of residues, the sum being indexed by the semisimple centralizers $Z(u)$. In Section 7.3, we introduce the Verlinde sums $V_{g,q}(\theta_1,\dots,\theta_s)$. In Section 7.4, we express the Verlinde sums as residues.

Finally in Sections 7.5-7.7, we "localize" the Verlinde formulas to put them a form which is very similar to the form we obtained in Section 6 for $\operatorname{Ind}(D_{p,+})$. This is done using the techniques we already developed in Section 2.

7.1. A residue formula for $L_q(t,x)$.

DEFINITION 7.1. For $q\in\mathbf{N}^*$, $t\in\overline{R}/q\overline{R}$, $x\in(\mathbf{C}\backslash 2i\pi\mathbf{Z})^\ell$, put

$$L_q(t,x)=\sum_{\lambda\in\overline{R}^*/q\overline{R}^*}\frac{\exp(2i\pi\langle\lambda,t/q\rangle)}{\prod_{i=1}^{\ell}2q\tanh\left(\frac{2i\pi\langle\alpha_i,\lambda\rangle-x_i}{2q}\right)}. \tag{7.1}$$

If $t\in\overline{R}/q\overline{R}$, then $t/q\in(\overline{R}/q)\big/\overline{R}\subset\mathfrak{t}/\overline{R}$. Recall that the subset H of $\mathfrak{t}/\overline{R}$ was defined in (2.44), (2.45). Then by (2.45),

$$\{t\in\overline{R}/q\overline{R},t/q\in H\}=\Big\{t\in\overline{R}/q\overline{R},t=\sum_{j\in\mathcal{J}\subset\{1,\dots,r\}}t^j\alpha_j\text{ in }\mathfrak{t}/q\overline{R}, \tag{7.2}$$
$$\text{and the }\{\alpha_j,j\in\mathcal{J}\}\text{ do not span }\mathfrak{t}^*\simeq\mathfrak{t}\Big\}.$$

As in Section 2.4, we identify $\overline{R}/d\overline{R}$ to $\{0,1,\dots,d-1\}^r$, i.e. $f\in\overline{R}/d\overline{R}$ is identified to $\sum_1^r f^ie_i$, $f^i\in\{0,\dots,d-1\}$.

By (2.14), if $I=(i_1,\dots,i_r)$ is generic, if $x\in\mathbf{Z}^\ell$, then $d\langle e_j,x^I\rangle\in\mathbf{Z}$, $1\le j\le r$, and so for $1\le i\le\ell$, $d\langle\alpha_i,x^I\rangle\in\mathbf{Z}$.Therefore the function $x\in\mathbf{C}^\ell\mapsto\tanh\left(\frac{\langle\alpha_i,x^I\rangle-x_i}{2q}\right)$ is periodic of period $(2i\pi qd,\dots,2i\pi qd)$.

Similarly if $t\in\overline{R}$, we claim that the function

$$x\in\mathbf{C}^\ell\mapsto\exp\left(d\sum_{j=1}^r\frac{\langle p^I_{j-1}\alpha_{i_j},x^I\rangle}{\langle p^I_{j-1}\alpha_{i_j},e^j\rangle}\left[\frac1d\langle p^I_{j-1}t,\frac{e^j}{q}\rangle\right]\right) \tag{7.3}$$

is periodic of period $(2i\pi qd,\dots,2i\pi qd)$. In fact by (2.13), (2.28), (2.29), if $x\in(2i\pi\mathbf{Z})^\ell$,

$$\exp\left(d\sum_{j=1}^r\frac{\langle p^I_{j-1}\alpha_{i_j},x^I\rangle}{\langle p^I_{j-1}\alpha_{i_j},e^j\rangle}\left[\frac1d\langle p^I_{j-1}t,\frac{e^j}{q}\rangle\right]\right)=\exp\left(\langle\frac tq,x^I\rangle\right). \tag{7.4}$$

Also, as we just saw, if $t \in \overline{R}$, $z \in \mathbf{Z}^\ell$, $d\langle t, x^I\rangle \in \mathbf{Z}$. From (7.4), it follows that if $x \in (2i\pi qd\mathbf{Z})^\ell, t \in \overline{R}$, (4.4) is equal to 1, which proves the periodicity of (7.3).

If $g = (g_1, \dots, g_r) \in \mathbf{Z}^r$, if $I = (i_1, \dots, i_r) \subset \{1, \dots, \ell\}$, let $g_I \subset \mathbf{C}^\ell$ be such that

$$(g_I)_{i_j} = g_j, \quad 1 \le j \le r, \qquad (g_I)_i = 0, \quad i \notin I. \tag{7.5}$$

THEOREM 7.2. *For generic values of $x \in (\mathbf{C}\backslash 2i\pi\mathbf{Z})^\ell$, for $t \in \overline{R}/q\overline{R}$, $t/q \notin H$, if we still denote by t a representative of t in $\overline{R}$, then*

$$L_q(t,x) = \frac{(-1)^r}{d^r} \sum_{\substack{I=(i_1,\dots,i_r)\subset\{1,\dots,\ell\}, I \text{ generic} \\ f\in\overline{R}/d\overline{R}, g\in(\mathbf{Z}/d\mathbf{Z})^r}} \operatorname{sgn}\left(\langle \alpha_{i_1}, \dots, \alpha_{i_r}\rangle\right) \left[\exp\left(d\sum_{j=1}^{r} \frac{\langle p^I_{j-1}\alpha_{i_j}, x^I\rangle}{\langle p^I_{j-1}\alpha_{i_j}, e^j\rangle}\left[\frac{1}{d}\langle p^I_{j-1}(t+qf), \frac{e^j}{q}\rangle\right]\right) \frac{1}{\prod_{i\in{}^cI} 2q\tanh\left(\frac{\langle\alpha_i, x^I\rangle - x_i}{2q}\right)}\right](x + 2i\pi q g_I) \prod_{j=1}^{r} \frac{1}{\exp\left(\frac{d\langle p^I_{j-1}\alpha_{i_j}, x^I\rangle}{\langle p^I_{j-1}\alpha_{i_j}, e^j\rangle}\right) - 1}. \tag{7.6}$$

PROOF. Take $1 \le j \le r$, $I = (i_1, \dots, i_{j-1}) \subset \{1, \dots, \ell\}$ such that $\langle\alpha_{i_1}\rangle \ne 0$, $\langle\alpha_{i_1}, \dots, \alpha_{i_{j-1}}\rangle \ne 0$. We use otherwise the notation and the conventions in the proof of Theorem 2.19, in particular in equation (2.63).

For $\widehat{k} = (k_{j+1}, \dots, k_r) \subset \mathbf{Z}^{r-j}$, we identify $\widehat{k}$ to $\sum_{i=j+1}^{r} k^i e_i \in \overline{R}^*$. For $a \in \mathbf{C}$, $x \in \mathbf{C}^\ell$, $t \in \overline{R}$, put

$$S_I(a,x) = \sum_{\substack{0\le g_p<d \\ 1\le p\le j-1}} \left\{\exp\left(a\langle p^I_{j-1}t, \frac{e^j}{q}\rangle + d\sum_{n=1}^{j-1}\frac{\langle p^I_{n-1}\alpha_{i_n}, x^I\rangle}{\langle p^I_{n-1}\alpha_{i_n}, e^n\rangle}\left[\frac{1}{d}\langle p^I_{n-1}t, \frac{e^n}{q}\rangle\right]\right) \frac{1}{\prod_{i\in{}^cI} 2q\tanh\left(\frac{\langle p^I_{j-1}\alpha_i, ae^j + 2i\pi\widehat{k} - x^I\rangle}{2q}\right)}\right\}\left(x + 2i\pi q(g_1, \dots, g_{j-1})_I\right). \tag{7.7}$$

We claim that as a function of $x \in \mathbf{C}^\ell$, each of the expressions $\{\quad\}$, with $(g_1, \dots, g_{j-1}) = 0$, is periodic of period $2i\pi qd, \dots, 2i\pi qd$. In fact by (2.63), if $x \in \mathbf{Z}^\ell$,

$$d\langle p^I_{j-1}, \alpha_i, x^I\rangle \in \mathbf{Z}. \tag{7.8}$$

Also by (2.28), (2.63), for $1 \le n \le j-1$, $x \in \mathbf{Z}^\ell$

$$\frac{d\langle p^I_{n-1}\alpha_{i_n}, x^I\rangle}{\langle p^I_{n-1}\alpha_{i_n}, e^n\rangle} \in \mathbf{Z}, \tag{7.9}$$

so that if $x \in \mathbf{Z}^\ell$, using (2.12),

$$\begin{aligned}(7.10)\qquad &\exp\left(2i\pi d \sum_{n=1}^{j-1} \frac{\langle p_{n-1}^I \alpha_{i_n}, x^I\rangle}{\langle p_{n-1}^I \alpha_{i_n}, e^n\rangle}\left[\frac{1}{d}\langle p_{n-1}^I t, \frac{e^n}{q}\rangle\right]\right)\\ &= \exp\left(\frac{2i\pi}{q}\sum_{n=1}^{j-1} \frac{\langle p_{n-1}^I \alpha_{i_n}, x^I\rangle}{\langle p_{n-1}^I \alpha_{i_n}, e^n\rangle}\langle p_{n-1}^I t, e^n\rangle\right)\\ &\exp\left(\frac{2i\pi}{q}\left(\langle t, x^I\rangle - \langle p_{j-1}^I t, x^I\rangle\right)\right).\end{aligned}$$

Also by (2.14), (2.63), if $t \in \overline{R}$, $x \in \mathbf{Z}^\ell$,

$$\begin{aligned}(7.11)\qquad & d\langle t, x^I\rangle \in \mathbf{Z},\\ & d\langle p_{j-1}^I t, x^I\rangle \in \mathbf{Z}.\end{aligned}$$

By (7.11), it follows that if $x \in qd\mathbf{Z}^\ell$, (7.10) is equal to 1.

Ultimately, we find that indeed, each of the expression $\{\ \}$ in (7.7) is periodic in $x \in \mathbf{C}^\ell$ of period $2i\pi qd, \dots, 2i\pi qd$. Therefore, we will write (7.7) in the form

$$\begin{aligned}(7.12)\qquad S_I(a,x) = \sum_{g\in(\mathbf{Z}/d\mathbf{Z})^{j-1}} &\left\{\exp\left(a\langle p_{j-1}^I t, \frac{e^j}{q}\rangle + d\right.\right.\\ &\left.\sum_{n=1}^{j-1} \frac{\langle p_{n-1}^I \alpha_{i_n}, x^I\rangle}{\langle p_{n-1}^I \alpha_{i_n}, e^n\rangle}\left[\frac{1}{d}\langle p_{n-1}^I t, \frac{e^n}{q}\rangle\right]\right)\\ &\left.\frac{1}{\prod_{i\in e_I} 2q \tanh\left(\frac{\langle p_{j-1}^I \alpha_i, ae^j + 2i\pi\widehat{k} - x^I\rangle}{2q}\right)}\right\}(x + 2i\pi q g_I).\end{aligned}$$

In particular $S_I(a,x)$ is periodic in $x \in \mathbf{C}^\ell$ of period $2i\pi q, \dots, 2i\pi q$.

Now we claim that $S_I(a,x)$ is periodic in $a \in \mathbf{C}$, with period $2i\pi q$. In fact, since $e^j = (\widetilde{a}e^j)^I$,

$$(7.13)\qquad \langle p_{j-1}^I \alpha_i, e^j\rangle = \langle p_{j-1}^I \alpha_i, (\widetilde{a}e^j)^I\rangle.$$

Also by (2.12),

$$(7.14)\qquad \langle p_{j-1}^I t, e^j\rangle = \langle t, e^j\rangle - \sum_{n=1}^{j-1} \frac{\langle p_{n-1}^I \alpha_{i_n}, (\widetilde{a}e^j)^I\rangle}{\langle p_{n-1}^I \alpha_{i_n}, e^n\rangle}\langle p_{n-1}^I t, e^n\rangle.$$

Since $t \in \overline{R}$, then $\langle t, e^j\rangle \in \mathbf{Z}$ and so, by (7.14), we get

$$(7.15)\qquad \langle p_{j-1}^I t, e^j\rangle = -\sum_{n-1}^{j-1} \frac{\langle p_{n-1}^I \alpha_{i_n}, (\widetilde{a}e^j)^I\rangle}{\langle p_{n-1}^I \alpha_{i_n}, e^n\rangle}\langle p_{n-1}^I t, e^n\rangle \mod (\mathbf{Z}).$$

From (7.12), (7.13), (7.15), we find that

$$\begin{aligned}(7.16)\qquad S_I(a + 2i\pi q, x) &= S_I(a, x - 2i\pi q\widetilde{a}e^j)\\ &= S_I(a,x),\end{aligned}$$

i.e. $S_I(a,x)$ is periodic in a with period $2i\pi q$.

Put

$$(7.17)\qquad L_I = \sum_{k\in\mathbf{Z}/q\mathbf{Z}} S_I(2i\pi k, x).$$

Clearly

$$L_I = \frac{1}{d} \sum_{k \in \mathbf{Z}/dq\mathbf{Z}} S_I(2i\pi k, x). \tag{7.18}$$

From (7.18), we obtain

$$L_I = \frac{1}{d^2} \sum_{\substack{0 \le f < d \\ k \in \mathbf{Z}/d^2 q\mathbf{Z}}} \exp\left(\frac{2i\pi fk}{d}\right) S_I\left(\frac{2i\pi k}{d}, x\right). \tag{7.19}$$

By (7.12), for generic x, we rewrite (7.19) in the form

$$\begin{aligned} L_I = \frac{1}{d^2} \sum_{k \in \mathbf{Z}/d^2 q\mathbf{Z}} \Bigg\{ \sum_{\substack{0 \le f < d \\ g \in (\mathbf{Z}/d\mathbf{Z})^{j-1}}} \operatorname{Res}_{a=2i\pi k} \Bigg[\exp\Bigg(a \left[\frac{\langle p_{j-1}^I (t + qfe_j), e^j/q \rangle}{d} \right] \\ + d \sum_{n=1}^{j-1} \frac{\langle p_{n-1}^I \alpha_{i_n}, x^I \rangle}{\langle p_{n-1}^I \alpha_{i_n}, e^n \rangle} \left[\frac{1}{d} \langle p_{n-1}^I t, e^n/q \rangle \right] \Bigg) \\ \frac{1}{\prod_{i \in {}^c I} 2q \tanh\left(\frac{\langle p_{j-1}^I \alpha_i, ae^j/d + 2i\pi \widehat{k} - x^I \rangle}{2q} \right)} \frac{1}{e^a - 1} \Bigg] \Bigg\} (x + 2i\pi q g_I). \end{aligned} \tag{7.20}$$

Now we just saw that the function of $a, \sum_{\substack{0 \le f < d \\ g \in (\mathbf{Z}/d\mathbf{Z})^{j-1}}} \dots$, which appears in the right-hand side of(7.20), is periodic of $2i\pi qd$.

We claim that for f, $0 \le f < d$, $g \in (\mathbf{Z}/d\mathbf{Z})^{j-1}$, the function of a,

$$h_{f,g}(a) = \frac{\exp\left(a \left[\frac{p_{j-1}^I (t+qfe_j), e^j/q \rangle}{d} \right] \right)}{\prod_{i \in {}^c I} 2q \tanh\left(\frac{\langle p_{j-1}^I \alpha_i, ae^j/d + 2i\pi \widehat{k} - x^I \rangle}{2q} \right)} \frac{1}{e^a - 1} \tag{7.21}$$

is periodic of period $2i\pi d^2 q$. This follows from the fact that since $t \in \overline{R}$,

$$\begin{aligned} &d \langle p_{j-1}^I (t + qfe_j, e^j \rangle \in \mathbf{Z}, \\ &d \langle p_{j-1}^I \alpha_i, e^j \rangle \in \mathbf{Z}. \end{aligned} \tag{7.22}$$

By Proposition 2.14, if $t/q \notin H$,

$$0 < \left[\frac{\langle p_{j-1}^I (t + qfe^j), e^j/q \rangle}{d} \right] < 1. \tag{7.23}$$

Then for given $f, 0 \le f < d$, $g \in (\mathbf{Z}/d\mathbf{Z})^{j-1}$, we will apply the theorem of residues to the function $h_{f,g}(a)$ on the domain given in Figure 7.1. By (2.65), for at least one $i_j \notin I$, $\langle p_{j-1}^I \alpha_{i_j}, e^j \rangle \neq 0$. For generic x, $\mod (2i\pi d^2 q)$, the poles of the function $h_{f,g}(a)$ other than the $2i\pi k$ are simple and given by

$$\begin{aligned} &a = \frac{d}{\langle p_{j-1}^I \alpha_{i_j}, e^j \rangle} \left(\langle p_{j-1}^I \alpha_{i_j}, x^I - 2i\pi \widehat{k} \rangle + 2i\pi g^j q \right), \\ &0 \le g^j < d |\langle p_{j-1}^I \alpha_{i_j}, e^j \rangle| \,,\ i_j \notin I, \langle p_{j-1}^I \alpha_{i_j}, e^j \rangle \neq 0. \end{aligned} \tag{7.24}$$

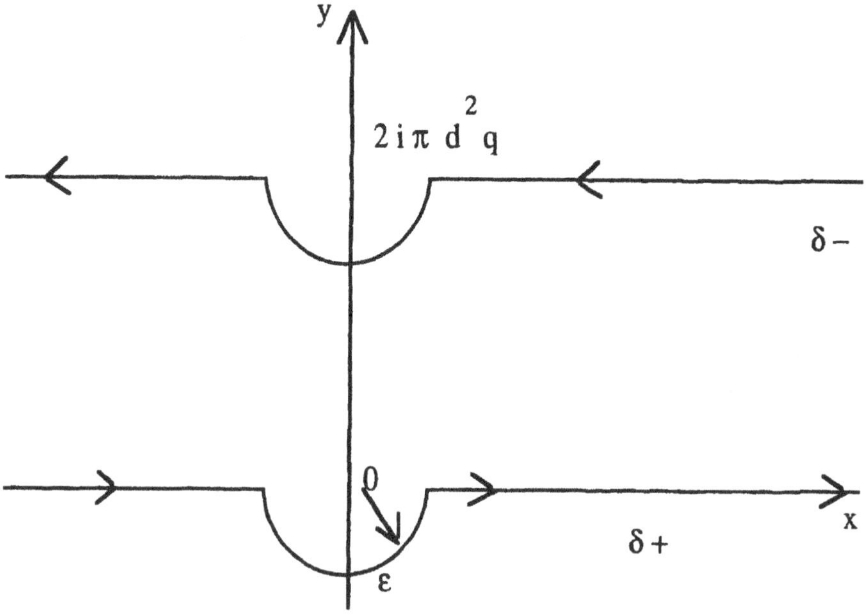

FIGURE 7.1

Needless to say, in (7.24), $d\langle p^I_{j-1}\alpha_{i_j}, e^j\rangle \in \mathbf{Z}$. In view of (2.28), we rewrite (7.24) in the form

$$(7.25)\qquad a = \frac{d}{\langle p^I_{j-1}\alpha_{i_j}, e^j\rangle}\langle p^I_{j-1}\alpha_{i_j}, \left(x + 2i\pi q(g^j)_{I,i_j}\right)^I - 2i\pi\widehat{k}\rangle ,$$
$$0 \le g^j < d|\langle p^I_{j-1}\alpha_{i_j}, e^j\rangle| .$$

By periodicity, the integrals of $h_{g,j}(a)$ over δ_+ and δ_- cancel each other.

Using the theorem of residues and (2.70), (2.76), (2.78), (7.20), (7.23), (7.25), we get

$$(7.26)\qquad L_I = -\frac{1}{d} \sum_{\substack{0\le f<d\ ,g\in(\mathbf{Z}/d\mathbf{Z})^{j-1} \\ i_j:(\alpha_{i_1},\dots,\alpha_{i_j})\neq 0\ ,0\le g^j<d|\langle p^I_{j-1}\alpha_{i_j}, e^j\rangle|}}$$
$$\left\{\frac{1}{\langle p^I_{j-1}\alpha_{i_j}, e^j\rangle} \exp\left(2i\pi\langle p_j^{(I,i_j)}e_j, \widehat{k}\rangle\langle p^I_{j-1}(t+qfe_j), e^j/q\rangle\right.\right.$$
$$\left. + d\sum_{n=1}^{j} \frac{\langle p^I_{n-1}\alpha_{i_n}, x^I\rangle}{\langle p^I_{n-1}\alpha_{i_n}, e^n\rangle}\left[\frac{1}{d}\langle p^I_{n-1}(t+qfe_j), e^n/q\rangle\right]\right)$$
$$\left.\frac{1}{\prod_{i\notin(I,i_j)} 2q\tanh\left(\frac{\langle p_j^{(I,i_j)}\alpha_i, 2i\pi\widehat{k} - x^I\rangle}{2q}\right)}\right\}(x + 2i\pi q(g,g_j)_{I,i_j})$$
$$\frac{1}{\exp\left(\frac{d\langle p^I_{j-1}\alpha_{i_j}, x^I\rangle}{\langle p^I_{j-1}\alpha_{i_j}, e^j\rangle}\right) - 1} .$$

We claim that as functions of x_{i_j}, the terms in the right-hand side of (7.26) are periodic of period $2i\pi qd$. In fact this follows from what we saw after (7.7), by replacing $j-1$ by j. We also claim these functions, are periodic as functions of x_{i_j}, with period $2i\pi qd\langle p^I_{j-1}\alpha_{i_j}, e^j\rangle$. In fact, using (2.29),

$$\frac{d\langle p^I_{j-1}\alpha_{i_j}, e^j\rangle}{\langle \alpha_{i_1}, \dots, \alpha_{i_j}\rangle} = \frac{d}{\langle \alpha_{i_1}, \dots, \alpha_{j-1}\rangle} \in \mathbf{Z}. \tag{7.27}$$

In view of (2.63), (7.27) takes care of the terms $\tanh\left(\frac{\langle p^{(I,i_j)}_{j-1}\alpha_i, 2i\pi\widehat{k} - x^I\rangle}{2q}\right)$. Also among the term $\langle p^I_{n-1}\alpha_{i_n}, x^I\rangle$, $n \le j$, only $\langle p^I_{j-1}\alpha_{i_j}, x^I\rangle$ depends on x_{i_j}. More precisely by (2.63),

$$\frac{\partial}{\partial x_{i_j}}\langle p^I_{j-1}\alpha_{i_j}, x^I\rangle = 1. \tag{7.28}$$

Clearly,

$$\begin{aligned} qd\frac{\langle p^I_{j-1}\alpha_{i_j}, e^j\rangle}{\langle p^I_{j-1}\alpha_{i_j}, e^j\rangle} d\left[\frac{1}{d}\langle p^I_{j-1}(t+qfe_j), e^j/q\rangle\right] = \\ d\langle p^I_{j-1}(t+qfe_j), e^j\rangle \mod (\mathbf{Z}). \end{aligned} \tag{7.29}$$

Using (2.29), we find that (7.29) is $0 \mod (\mathbf{Z})$, i.e. the left-hand side of (7.29) lies in $\mathbf{Z}$. Ultimately, this guarantees the periodicity of the functions appearing in (7.26) in x_{i_j}, with period $2i\pi qd\langle p^I_{j-1}\alpha_{i_j}, e^j\rangle$.

Now by (2.29),

$$d\langle p^I_{j-1}\alpha_{i_j}, e^j\rangle = \frac{d\langle \alpha_{i_1}, \dots, \alpha_{i_j}\rangle}{\langle \alpha_{i_1}, \dots, \alpha_{i_{j-1}}\rangle}. \tag{7.30}$$

From (7.30), we find that

$$d\langle p^I_{j-1}\alpha_{i_j}, e^j\rangle \mid d^2. \tag{7.31}$$

Since the functions in (7.26) are periodic in x_{i_j} of period $2i\pi qd\langle p^I_{j-1}\alpha_{i_j}, e^j\rangle$, using (7.31), in (7.26), we may and we will replace the condition $0 \le g^j < d|\langle p^I_{j-1}\alpha_{i_j}, e^j\rangle|$ by the condition $0 \le g^j < d^2$, and this introduces a correcting factor $\frac{|\langle p^I_{j-1}\alpha_{i_j}, e^j\rangle|}{d}$ in the right-hand side of (7.26). Using the periodicity in x_{i_j} of period $2i\pi qd$, we may finally replace the condition $0 \le g^j < d^2$ by $0 \le g^j < d$, with a new correcting factor d. Ultimately in (7.26), we replace the condition $0 \le g^j < d|\langle p^I_{j-1}\alpha_{i_j}, e^j\rangle|$ by $0 \le g^j < d$, with a correcting factor $|\langle p^I_{j-1}\alpha_{i_j}, e^j\rangle|$. Also by (7.30),

$$\operatorname{sgn}\langle p^I_{j-1}\alpha_{i_j}, e^j\rangle = \operatorname{sgn}\langle \alpha_{i_1}, \dots, \alpha_{i_j}\rangle \operatorname{sgn}\langle \alpha_{i_1}, \dots, \alpha_{j-1}\rangle. \tag{7.32}$$

So, from (7.26), (7.32), we get

$$L_I = \frac{-1}{d} \sum_{\substack{0\le f<d, g\in(\mathbf{Z}/d\mathbf{Z})^j \\ i_j:\langle\alpha_{i_1},\dots,\alpha_{i_j}\rangle\neq 0}} \operatorname{sgn}(\langle\alpha_{i_1},\dots,\alpha_{i_{j-1}}\rangle\langle\alpha_{i_1},\dots,\alpha_{i_j}\rangle) \tag{7.33}$$

$$\Bigg\{\exp\left(2i\pi\langle p_j^{(I,i_j)}e_j,\widehat{k}\rangle\langle p_{j-1}^I(t+qfe_j),e^j\rangle + d\sum_{n=1}^{j}\frac{\langle p_{n-1}^I\alpha_{i_n},x^I\rangle}{\langle p_{n-1}^I\alpha_{i_n},e^n\rangle}\right.$$

$$\left.\left[\frac{1}{d}\langle p_{n-1}^I(t+qfe_j),e^n/q\rangle\right]\right)$$

$$\frac{1}{\prod_{i\notin(I,i_j)} 2q\tanh\left(\frac{\langle p_j^{(I,i_j)}\alpha_i,2i\pi\widehat{k}-x^I\rangle}{2q}\right)}\Bigg\}(x+2i\pi q g_{(I,i_j)})$$

$$\frac{1}{\exp\left(\frac{d\langle p_{j-1}^I\alpha_{i_j},x^I\rangle}{\langle p_{j-1}^I\alpha_{i_j},e^j\rangle}\right)-1}.$$

Clearly

$$L_q(t,x) = \sum_{k=(k_1,\dots,k_r)\in(\mathbf{Z}/q\mathbf{Z})^r} \frac{\exp(2i\pi\sum_{j=1}^r k_j\langle t,e^j/q\rangle)}{\prod_{i=1}^{\ell} 2q\tanh\left(\frac{2i\pi\langle\alpha_i,\sum_{j=1}^r k_je^j\rangle-x_i}{2q}\right)}. \tag{7.34}$$

Also with the notation in (2.63),

$$\langle p_0^{\phi}\alpha_i,x^I\rangle = x_i\,. \tag{7.35}$$

Then we use (7.33), so that if $t\in\overline{R}/q\overline{R}$, $t/q\notin H$,

$$L_q(t,x) = \frac{-1}{d} \sum_{\substack{0\le f^1<d, g\in\mathbf{Z}/d\mathbf{Z} \\ \langle\alpha_{i_1}\rangle\neq 0, k=(k^2,\dots,k^r)\in(\mathbf{Z}/q\mathbf{Z})^{r-1}}} \tag{7.36}$$

$$\operatorname{sgn}\langle\alpha_{i_1}\rangle\exp\left(2i\pi\langle p_1^{(i_1)}(t+qf^1e_1),k\rangle + d\frac{x_{i_1}}{\langle\alpha_{i_1}\rangle}\left[\frac{\langle t+qf'e_1,e^1/q\rangle}{d}\right]\right)$$

$$\frac{1}{\prod_{i\neq i_1} 2q\tanh\left(\frac{\langle p_1^{(i_1)}\alpha_i,2i\pi k-x^I\rangle}{2q}\right)}(x+2i\pi q g_{i_1})\frac{1}{\exp\left(\frac{dx_{i_1}}{\langle\alpha_{i_1}\rangle}\right)-1}.$$

Using (7.33), (7.36) and an obvious iteration, in view of (2.83), we obtain (7.6). The proof of our Theorem is completed. □

7.2. A residue formula for $M_q(t,x)$.

DEFINITION 7.3. For $q\in\mathbf{N}^*$, $t\in\overline{R}/q\overline{CR}$, $x\in(\mathbf{C}\backslash 2i\pi\mathbf{Z}/m)^{\ell}$, put

$$M_q(t,x) = \sum_{\lambda\in\overline{CR}^*/q\overline{R}^*} \frac{\exp(2i\pi\langle\lambda,t/q\rangle)}{\prod_{i=1}^{\ell} 2q\tanh\left(\frac{2i\pi\langle\alpha_i,\lambda\rangle-x_i}{2q}\right)}. \tag{7.37}$$

Now we will adapt the formalism of Sections 2.5 and 2.6, with $t/\overline{CR}$ replaced by $\overline{R}/q\overline{CR}$. If $f(t)$ is a function on $\overline{R}/q\overline{CR}$, if $\mu\in\overline{CR}^*/\overline{R}^*$, put

$$\widehat{f}_\mu(t) = \frac{1}{|\overline{R}/\overline{CR}|}\sum_{v\in\overline{R}/\overline{CR}} e^{-2i\pi\langle\mu,v\rangle}f(t+qv)\,. \tag{7.38}$$

Then by (2.92),

$$f(t) = \sum_{\mu \in \overline{CR}^*/\overline{R}^*} \widehat{f}_\mu(t)\,. \tag{7.39}$$

PROPOSITION 7.4. *If $\mu \in \overline{CR}^*/\overline{R}^*$, if $\lambda_\mu \in \overline{CR}^*$ represents μ, for $t \in \overline{R}/q\overline{CR}$,*

$$(\widehat{M_q})_\mu(t,x) = \exp(2i\pi\langle\lambda_\mu, t/q\rangle)L_q(t, x - 2i\pi\widetilde{a}\lambda_\mu)\,. \tag{7.40}$$

PROOF. If $\lambda \in \overline{CR}^*/q\overline{R}^*$, then

$$\begin{aligned} \exp(2i\pi\widehat{\langle\lambda, t/q\rangle})_\mu &= \exp(2i\pi\langle\lambda, t/q\rangle) \text{ if } \lambda \text{ maps to } \mu \\ &\qquad \text{in } \overline{CR}^*/\overline{R}^* \\ &\quad 0 \qquad \text{otherwise}\,. \end{aligned} \tag{7.41}$$

Then

$$(\widehat{M_q})_\mu(t,x) = \sum_{\lambda \in \overline{R}^*/q\overline{R}^*} \frac{\exp(2i\pi\langle\lambda + \lambda_\mu, t/q\rangle}{\prod_{i=1}^{\ell} 2q\tanh\left(\frac{2i\pi\langle\alpha_i,\lambda\rangle - (x_i - 2i\pi\langle\alpha_i,\lambda_\mu\rangle)}{2q}\right)}\,, \tag{7.42}$$

which is equivalent to (7.40). □

PROPOSITION 7.5. *For $q \in \mathbf{N}^*$, $t \in \overline{R}/q\overline{CR}$, $x \in (\mathbf{C}\backslash 2i\pi\mathbf{Z}/m)^\ell$,*

$$M_q(t,x) = \sum_{\mu \in \overline{CR}^*/\overline{R}^*} \exp(2i\pi\langle\lambda_\mu, t/q\rangle)L_q(t, x - 2i\pi\widetilde{a}\lambda_\mu)\,. \tag{7.43}$$

PROOF. This follows from (7.39), (7.40). □

Recall that τ is the projection $\mathfrak{t}/\overline{CR} \to \mathfrak{t}/\overline{R}$. Then τ induces the projection $\overline{R}/q\overline{CR} \to \overline{R}/q\overline{R}$. Also recall that the set $\overline{S} \subset \mathfrak{t}/\overline{CR}$ was defined in (2.111). By (2.45), (2.111),

$$\begin{aligned} \{t \in \overline{R}/q\overline{CR}, t/q \in \overline{S}\} &= \{t \in \overline{R}/q\overline{CR}, \tau t = \sum_{j \in \mathcal{J}} t^j \alpha_j\,, \\ &\qquad \{\alpha_j, j \in \mathcal{J}\} \text{ does not span } \mathfrak{t}^*\}\,. \end{aligned} \tag{7.44}$$

THEOREM 7.6. *For generic values of $x \in (\mathbf{C}\backslash 2i\pi\mathbf{Z}/m)^\ell$, $t \in \overline{R}/q\overline{CR}$, $t/q \notin \overline{S}$, if we still denote by t a representative of t in $\overline{R}$, then*

$$\begin{aligned} M_q(t,x) = \\ \frac{(-1)^r}{d^r} \sum_{\substack{I=(i_1,\dots,i_r)\subset\{1,\dots,\ell\}, I \text{ generic} \\ f \in \overline{R}/d\overline{R}, g \in (\mathbf{Z}/d\mathbf{Z})^r, \mu \in \overline{CR}^*/\overline{R}^*}} & \operatorname{sgn}(\langle\alpha_{i_1},\dots,\alpha_{i_r}\rangle)\exp(2i\pi\langle\lambda_\mu, t/q\rangle) \\ \left[\exp\left(d\sum_{j=1}^{r}\frac{\langle p_{j-1}^I\alpha_{i_j}, x^I\rangle}{\langle p_{j-1}^I\alpha_{i_j}, e^j\rangle}\left[\frac{1}{d}\langle p_{j-1}^I(t+qf), e^j/q\rangle\right]\right)\right. \\ \left.\frac{1}{\prod_{i \in {}^cI} 2q\tanh\left(\frac{\langle\alpha_i, x^I\rangle - x_i}{2q}\right)}\right] & (x + 2i\pi q g_I - 2i\pi\widetilde{a}\lambda_\mu) \\ \frac{1}{\prod_{j=1}^{r}\left(\exp\left(\frac{d\langle p_{j-1}^I\alpha_{i_j}, x^I\rangle}{\langle p_{j-1}^I\alpha_{i_j}, e^j\rangle}\right) - 1\right)} & (x - 2i\pi\widetilde{a}\lambda_\mu)\,. \end{aligned} \tag{7.45}$$

PROOF. This follows from Theorem 7.2 and Proposition 7.5. □

DEFINITION 7.7. If $I = (i_1, \dots, i_r)$ is generic, let H^I be the lattice in $\mathfrak{t}$ generated by $\alpha^{i_1}, \dots, \alpha^{i_r}$.

Recall that $C \subset \overline{R}^*$ was defined in Definition 1.35.

PROPOSITION 7.8. *The following identity holds*

$$C = \bigcup_{I \text{ generic}} H^I. \tag{7.46}$$

In particular for I generic,

$$dH^I \subset \overline{R}^*. \tag{7.47}$$

PROOF. The identity (7.46) follows from the definition of C. Then (7.47) follows from (2.21), (7.46). The proof of our Proposition is completed. □

By Proposition 7.8, we have a natural projection

$$h^I : H^I/dH^I \to H^I/\overline{R}^*, \tag{7.48}$$

whose kernel is just

$$\ker h^I = \overline{R}^*/dH^I. \tag{7.49}$$

PROPOSITION 7.9. *For any generic I,*

$$|\overline{R}^*/dH^I| = \frac{d^r}{|\langle \alpha_{i_1}, \dots, \alpha_{i_r} \rangle|}. \tag{7.50}$$

PROOF. Recall that $e_1, \dots, e_r$ span $\overline{R}$, and $e^1, \dots, e^r$ span $\overline{R}^*$. Therefore

$$\begin{aligned} |\overline{R}^*/dH^I| &= d^r |\langle \alpha^{i_1} \wedge \dots \wedge \alpha^{i_r}, e_1 \wedge \dots \wedge e_r \rangle| \\ &= \frac{d^r}{|\langle \alpha_{i_1} \wedge \dots \wedge \alpha_{i_r} \rangle|}. \end{aligned} \tag{7.51}$$

The proof of our Proposition is completed. □

THEOREM 7.10. *For generic values of $x \in (\mathbf{C}\backslash 2i\pi\mathbf{Z}/m)^\ell$, for $t \in \overline{R}/q\overline{CR}$, $t/q \notin \overline{S}$, if we still denote by t a representative of t in $\overline{R}$, then*

$$\begin{aligned} M_q(t,x) = &\left|\frac{\overline{R}}{\overline{CR}}\right| (-1)^r \sum_{u \in C/\overline{R}^*} \sum_{\substack{I=(i_1,\dots,i_r) \subset I_u, I \text{ generic} \\ f \in \overline{CR}/d\overline{R}}} \\ &\frac{1}{\langle \alpha_{i_1}, \dots, \alpha_{i_r} \rangle} \frac{\exp\left(2i\pi\left(\langle u, t+qf \rangle\right)\right)}{\prod_{i \in {}^cI} 2q \tanh\left(\frac{\langle \alpha_i, x^I + 2i\pi q u \rangle - x_i}{2q}\right)} \\ &\exp\left(d \sum_{j=1}^r \frac{\langle p^I_{j-1}\alpha_{i_j}, x^I \rangle}{\langle p^I_{j-1}\alpha_{i_j}, e^j \rangle} \left[\frac{1}{d} \langle p^I_{j-1}(t+qf), e^j/q \rangle\right]\right) \\ &\frac{1}{\prod_{j=1}^r \left(\exp\left(\frac{d\langle p^I_{j-1}\alpha_{i_j}, x^I \rangle}{\langle p^I_{j-1}\alpha_{i_j}, e^j \rangle}\right) - 1\right)}. \end{aligned} \tag{7.52}$$

PROOF. Clearly the map $g \in (\mathbf{Z}/d\mathbf{Z})^\ell \mapsto g_I^I \in H^I/dH^I$ is one to one. Also if $t \in \overline{R}$, $f \in \overline{R}$, by (2.13), (2.28), (2.29),

$$\exp\left(d\sum_{j=1}^{r}\frac{\langle p_{j-1}^I\alpha_{i_j}, 2i\pi q g_I^I\rangle}{\langle p_{j-1}^I\alpha_{i_j}, e^j\rangle}\left[\frac{1}{d}\langle p_{j-1}^I(t+qf), e^j/q\rangle\right]\right) \tag{7.53}$$
$$= \exp(2i\pi\langle t+qf, g_I^I\rangle) = \exp(2i\pi\langle t+qf, h^I g_I^I\rangle)\,.$$

Moreover by (2.115),

$$\langle\alpha_i, (\tilde{a}\lambda_\mu)^I\rangle - (\tilde{a}\lambda_\mu)_i = 0\,. \tag{7.54}$$

Also if $i \notin I$, using (7.47),

$$\langle\alpha_i, g_I^I\rangle - (g_I)_i = \langle\alpha_i, g_I^I\rangle\,, \tag{7.55}$$
$$q\langle\alpha_i, g_I^I\rangle = q\langle\alpha_i, h^I g_I^I\rangle \text{ in } \mathbf{Z}/q\mathbf{Z}\,.$$

Finally by Proposition 7.8, the map $h^I : H^I/dH^I \to C/\overline{R}^*$ surjects on $\{u \in C/\overline{R}^*, I \subset I_u\}$, and by (7.49), (7.50), the fibre has $\frac{d^r}{|\langle\alpha_{i_1},\dots,\alpha_{i_r}\rangle|}$ elements.

From (2.13), (2.119A), (7.45), (7.53)-(7.55), and proceeding as in the proof of Theorem 2.28, we get (7.52). The proof of our Theorem is completed. □

7.3. The Verlinde sums. Recall that by Proposition 1.27, if $\lambda \in \overline{CR}^*$, the character χ_λ of G does not depend on the choice of a Weyl chamber K such that $\lambda \in \overline{K}$. Also by Proposition 1.28, if $w \in W$

$$\chi_{w\lambda} = \chi_\lambda\,. \tag{7.56}$$

Moreover the function $(i^\ell\sigma)^2(t)$ is well defined on $T' = \mathfrak{t}/\overline{R}^*$.

Now we introduce the Verlinde sums [62], [3].

DEFINITION 7.11. For $g \in \mathbf{N}$, $q \in \mathbf{N}^*$, $\theta_1, \dots, \theta_s \in \overline{CR}^*$, put

$$V_{g,q}(\theta_1,\dots,\theta_s) = \left|\frac{\overline{CR}^*}{q\overline{CR}}\right|^{g-1}\frac{1}{|W|}\sum_{\substack{\lambda\in\overline{CR}^*/q\overline{CR}\\ \sigma^2(\lambda/q)\neq 0}} \tag{7.57}$$
$$\frac{1}{(i^\ell\sigma(\lambda/q))^{2g-2}}\prod_{j=1}^{s}\chi_{\theta_j}(e^{\lambda/q})\,.$$

By Proposition 1.12, we get

$$V_{g,q}(\theta_1,\dots,\theta_s) = \left|\frac{\overline{CR}^*}{q\overline{CR}}\right|^{g-1}\sum_{\lambda\in qP\cap\overline{CR}^*}\frac{1}{(i^\ell\sigma(\lambda/q))^{2g-2}}\prod_{j=1}^{s}\chi_{\theta_j}(e^{\lambda/q})\,. \tag{7.58}$$

Recall that in Section 1.8, it was shown that $\sum_{j=1}^s\theta_j \in \overline{R}$, if and only if for $(w^1,\dots,w^s) \in W^s$, $\sum_{j=1}^s w^j\theta_j \in \overline{R}$.

By Proposition 1.29, observe that if $\sum_{j=1}^s\theta_j \in \overline{R}$, then $\prod_{j=1}^s\chi_{\theta_j}(e^{\lambda/q})$ is a well-defined function of $\lambda \in \overline{CR}^*/q\overline{R}^*$.

THEOREM 7.12. *If* $\sum_{j=1}^s\theta_j \notin \overline{R}$, *then*

$$V_{g,q}(\theta_1,\dots,\theta_s) = 0\,. \tag{7.59}$$

If $\sum_{j=1}^{s} \theta_j \in \overline{R}$, *then*

$$V_{g,q}(\theta_1, \ldots, \theta_s) = \left|\frac{\overline{CR}^*}{q\overline{CR}}\right|^{g-1} \frac{|Z(G)|}{|W|} \sum_{\substack{\lambda \in \overline{CR}^*/q\overline{R}^* \\ \sigma^2(\lambda/q) \neq 0}} \frac{1}{(i^\ell \sigma(\lambda/q))^{2g-2}} \prod_{j=1}^{s} \chi_{\theta_j}(e^{\lambda/q}). \tag{7.60}$$

PROOF. Clearly

$$V_{g,q}(\theta_1, \ldots, \theta_s) = \left|\frac{\overline{CR}^*}{q\overline{CR}}\right|^{g-1} \frac{1}{|W|} \sum_{\substack{\lambda \in \overline{CR}/q\overline{R}^* \\ \sigma^2(\lambda/q) \neq 0}} \frac{1}{(i^\ell \sigma(\lambda/q))^{2g-2}} \sum_{\mu \in \overline{R}^*/\overline{CR}} \prod_{j=1}^{s} \chi_{\theta_j}(e^{\lambda/q+\mu}). \tag{7.61}$$

Using Propositions 1.29 and 1.30, (7.61), and the fact that $Z(G) = \overline{R}^*/\overline{CR}$, we get our Theorem. □

THEOREM 7.13. *The following identity holds,*

$$V_{g,c}(\theta_1, \ldots, \theta_s) = \prod_{j=1}^{s} \chi_{\theta_j}(e^{\rho/c}). \tag{7.62}$$

If one of the θ_j*'s does not lie in* $\overline{R}$,

$$V_{g,c}(\theta_1, \ldots, \theta_s) = 0. \tag{7.63}$$

More generally

$$V_{g,c}(\theta_1, \ldots, \theta_s) = 0\,,\ +1 \text{ or } -1. \tag{7.64}$$

PROOF. By Proposition 1.10 and 1.11, we get (7.62). By Proposition 1.31 or by Theorem 7.12 when $s = 1$ and by (7.62), we get (7.63). By (7.62) and by Theorem 1.32, (7.64) follows. The proof of our Theorem is completed. □

Let K be a Weyl chamber.

DEFINITION 7.14. If $\theta \in \overline{CR}^*$, put

$$\overline{\chi}_\theta^K(t) = \frac{1}{\sigma(t)} \sum_{w \in W} \varepsilon_w e^{2i\pi\langle w(\rho+\theta), t\rangle}. \tag{7.65}$$

By (3.115) , if $\rho + \theta$ does not lie in a Weyl chamber,

$$\overline{\chi}_\theta^K(t) = 0. \tag{7.66}$$

If $\rho + \theta$ lies in the Weyl chamber $w_0 K$, then

$$\overline{\chi}_\theta^K(t) = \chi_{\rho+\theta-w_0\rho}(t). \tag{7.67}$$

Also if $w \in W$,

$$\overline{\chi}_{w\theta}^{wK}(t) = \overline{\chi}_\theta^K(t). \tag{7.68}$$

DEFINITION 7.15. For $g \in \mathbf{N}$, $q \in \mathbf{Z}^*$, $\theta_1, \dots, \theta_s, \theta_{s+1} \in \overline{CR}^*$, put

$$\overline{V}^K_{g,q}(\theta_1, \dots, \theta_s, \theta_{s+1}) = \left|\frac{\overline{CR}^*}{q\overline{CR}}\right|^{g-1} \frac{1}{|W|} \sum_{\substack{\lambda \in \overline{CR}^*/q\overline{CR} \\ \sigma^2(\lambda/q) \neq 0}} \frac{1}{(i^\ell \sigma(\lambda/q))^{g-1}} \prod_{j=1}^{s} \chi_{\theta_j}(e^{\lambda/q}) \overline{\chi}^K_{\theta_{s+1}}(e^{\lambda/q}) . \tag{7.69}$$

THEOREM 7.16. *For $g \in \mathbf{N}$, $q \in \mathbf{N}^*$, the following identity holds,*

$$V_{g,q}(\theta_1, \dots, \theta_s) = \frac{(-1)^\ell}{\left|\frac{\overline{CR}^*}{q\overline{CR}}\right|} \sum_{(v^1,v^2) \in W^2} \varepsilon_{v^1} \varepsilon_{v^2} \overline{V}^K_{g+1,q}(\theta_1, \dots, \theta_s, v^1\rho + v^2\rho) . \tag{7.70}$$

PROOF. Recall that

$$\chi_0 = 1 . \tag{7.71}$$

By (7.57), (7.71),

$$V_{g,q}(\theta_1, \dots, \theta_s) = \left|\frac{\overline{CR}^*}{q\overline{CR}}\right|^{g-1} \frac{1}{|W|} \sum_{\substack{\lambda \in \overline{CR}^*/q\overline{CR} \\ \sigma^2(\lambda/q) \neq 0}} \frac{1}{(i^\ell \sigma(\lambda/q))^{2g-2}} \prod_{j=1}^{s} \chi_{\theta_j}(e^{\lambda/q}) \chi_0^3(e^{\lambda/q}) . \tag{7.72}$$

By (7.56), we may and we will assume that $\theta_1, \dots, \theta_s \in \overline{CR}^*_+ = \overline{CR}^* \cap \overline{K}$.
Using (1.94), (7.72) we obtain

$$\begin{aligned} V_{g,q}(\theta_1, \dots, \theta_s) = {} & \frac{\left|\frac{\overline{CR}^*}{q\overline{CR}}\right|^{g-1}}{|W|} \sum_{\substack{\lambda \in \overline{CR}^*/q\overline{CR} \\ \sigma^2(\lambda/q) \neq 0}} \frac{1}{(i^\ell \sigma(\lambda/q))^{2g-2}} \frac{1}{(\sigma(\lambda/q))^{s+3}} \sum_{(w^1, \dots, w^s) \in W^{s+3}} \prod_{j=1}^{s+3} \varepsilon_{w^j} \\ & \exp\left(2i\pi \langle \sum_{j=1}^{s} w^j(\rho + \theta_j) + \sum_{j=s+1}^{s+3} w^j \rho, \lambda/q \rangle\right) \\ = {} & (-1)^\ell \frac{|\overline{CR}^*/q\overline{CR}|^{g-1}}{|W|} \sum_{\substack{\lambda \in \overline{CR}^*/q\overline{CR} \\ \sigma^2(\lambda/q) \neq 0}} \frac{1}{(i^\ell \sigma(\lambda/q))^{2g}} \frac{1}{(\sigma(\lambda/q))^{s+1}} \\ & \sum_{(v^1,v^2) \in W^2} \varepsilon_{v^1} \varepsilon_{v^2} \sum_{(w^1, \dots, w^{s+1}) \in W^{s+1}} \prod_{j=1}^{s+1} \varepsilon_{w^j} \\ & \exp\left(2i\pi \langle \sum_{j=1}^{s} w^j(\rho + \theta_j) + w^{s+1}(\rho + v^1\rho + v^2\rho), \lambda/q \rangle\right) , \end{aligned} \tag{7.73}$$

which is just (7.70).

The proof of our Theorem is completed. □

REMARK 7.17. If $w \in W$, $\rho + w\rho \in \overline{R}$. Therefore (7.59) and (7.70) are compatible. Also observe that the fusion rules [**3**], [**62**] express $V_{g,q}(.)$ in terms of $V_{g-1,q}(.)$. Here equation (7.70) goes the opposite way.

7.4. A residue formula for the Verlinde sums. In the sequel, we fix a Weyl chamber $K \subset \mathfrak{t}$, and we use the notation in Section 2.

Also by (7.56) and (7.59), we may and we will assume that

$$\theta_j \in \overline{CR}^*_+ = \overline{CR}^* \cap \overline{K} \ , \ 1 \le j \le s, \qquad \sum_{j=1}^{s} \theta_j \in \overline{R}. \tag{7.74}$$

DEFINITION 7.18. If $w \in W$, let $\sigma_w \in \mathcal{S}_\ell$, let $\varepsilon_w(1), \dots, \varepsilon_w(\ell) \in \{-1, +1\}$ be such that for $1 \le i \le \ell$,

$$w^{-1}\alpha_i = \varepsilon_w(i)\alpha_{\sigma_w(i)}\,. \tag{7.75}$$

By [**15**, Corollary V.4.6 and Lemma V.4.10] , we know that

$$\varepsilon_w = \prod_{i=1}^{\ell} \varepsilon_w(i)\,. \tag{7.76}$$

DEFINITION 7.19. For $x \in \mathbf{C}^\ell$, $w \in W$, put

$$\varphi_w(x) = \prod_{i=1}^{\ell} \frac{1}{1 - e^{-\varepsilon_w(i)x_i}},. \tag{7.77}$$

Equivalently, by (7.76),

$$\varphi_w(x) = \varepsilon_w \prod_{i=1}^{\ell} \frac{1}{2\sinh(\frac{x_i}{2})} e^{\frac{1}{2}\sum_1^\ell \varepsilon_w(i)x_i}\,. \tag{7.78}$$

THEOREM 7.20. *For $g \ge 2$, $q \in \mathbf{N}^*$, the following identity holds,*

$$\begin{aligned} V_{g,q}(\theta_1, \dots, \theta_s) = {} & \mathrm{Vol}(T)^{2g-2} q^{(g-1)r} \frac{|Z(G)|}{|W|} \\ & (-1)^{\ell(g-1)} \mathrm{Res}^{(1,\dots,\ell)}_{x=0} \frac{1}{\left[\prod_{i=1}^{\ell} 2\sinh(\frac{x_i}{2q})\right]^{2g-2}} \sum_{(w^1,\dots,w^s)\in W^s} \prod_{j=1}^{s} \varphi_{w^j}(x/q) \\ & M_q\left(\sum_{j=1}^{s} w^j\theta_j, x\right)\,. \end{aligned} \tag{7.79}$$

For $g \ge 0$, for s even, $2g - 2 + s \ge 1$ and $q \in \mathbf{N}^*$, the following identity holds,

$$\begin{aligned} V_{g,q}(\theta_1, \dots, \theta_s) = {} & \mathrm{Vol}(T)^{2g-2} q^{(g-1)r} \frac{|Z(G)|}{|W|} \\ & (-1)^{\ell(g-1)} \mathrm{Res}^{(1,\dots,\ell)}_{x=0} \frac{1}{\left[\prod_{i=1}^{\ell} 2\sinh(\frac{x_i}{2q})\right]^{2g-2+s}} \sum_{(w^1,\dots,w^s)\in W^s} \prod_{j=1}^{s} \epsilon_{w^j} \\ & M_q(\sum_{j=1}^{s} w^j(\rho + \theta_j), x). \end{aligned} \tag{7.80}$$

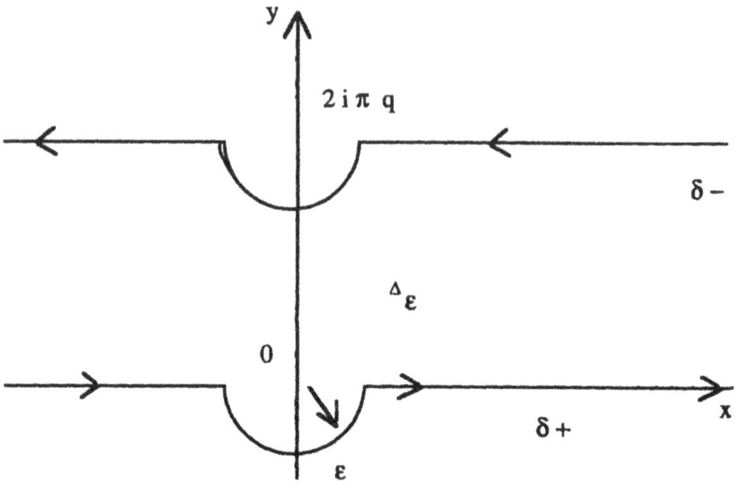

FIGURE 7.2

PROOF. First we assume that $g \geq 2$. For $\lambda \in \overline{CR}^* / q\overline{R}^*$, $w = (w^1, \dots, w^s) \in W^s$, $x \in (\mathbf{C} \backslash 2i\pi \mathbf{Z}/m)^\ell$, put

$$U_{\lambda,w}(x) = \frac{1}{\left[\prod_{i=1}^{\ell} 2\sinh(\frac{x_i}{2q})\right]^{2g-2}} \prod_{j=1}^{s} \varphi_{w^j}(x/q) \tag{7.81}$$

$$\prod_{i=1}^{\ell} \frac{1}{2q \tanh\left(\frac{2i\pi\langle\lambda,\alpha_i\rangle - x_i}{2q}\right)} .$$

Then $U_{\lambda,w}(x)$ is a meromorphic function of $x_1, \dots, x_\ell$, which is periodic of period $2i\pi q, \dots, 2i\pi q$.

For $\varepsilon > 0$, let Δ_ε be the domain in $\mathbf{C}$ given in Figure 7.2. For $\varepsilon > 0$ small enough, as a function of $x_i \in \Delta_\varepsilon$, the function $U_{\lambda,w}(x)$ has poles at 0, and also at $2i\pi q \left[\frac{\langle\lambda,\alpha_i\rangle}{q}\right]$. Here $q\left[\frac{\langle\lambda,\alpha_i\rangle}{q}\right]$ is the real number in $[0, q[$ which represents $\langle\lambda, \alpha_i\rangle$ mod q. For $\varepsilon > 0$ small enough, the poles do not meet the boundary $\partial\Delta_\varepsilon = \delta_- \cup \delta_+$.

Assume first that $\left[\frac{\langle\lambda,\alpha_i\rangle}{q}\right] \neq 0$. Then $2i\pi q \left[\frac{\langle\lambda,\alpha_i\rangle}{q}\right]$ is a simple pole of $U_{\lambda,w}(x)$. We now use the theorem of residues on Δ_ε. Since $g \geq 2$, as $x_i \to \pm\infty$ inside Δ_ε, the function $U_{\lambda,w}(x)$ tends to 0. Also since $U_{\lambda,w}(x)$ is periodic of period $2i\pi q$ in the variable x_i, the boundary integrals cancel each other. Therefore

$$\operatorname{Res}_{x_i=0} U_{\lambda,w}(x) + \operatorname{Res}_{x_i = 2i\pi q\left[\frac{\langle\lambda,\alpha_i\rangle}{q}\right]} U_{\lambda,w}(x) = 0 . \tag{7.82}$$

Clearly

$$\text{(7.83)}\quad \operatorname{Res}_{x_i=2i\pi q\left[\frac{\langle\lambda,\alpha_i\rangle}{q}\right]} U_{\lambda,w}(x) =$$

$$-\left[\frac{1}{2i\sin\left(\frac{\langle\lambda,\alpha_i\rangle\pi}{q}\right)}\prod_{j\neq i}\frac{1}{2\sinh\left(\frac{x_j}{2q}\right)}\right]^{2g-2}$$

$$\prod_{j=1}^{s}\varphi_{w^j}\left(\frac{x_1}{q}\dots\frac{x_{i-1}}{q},\frac{2i\pi\langle\lambda,\alpha_i\rangle}{q},\dots,\frac{x_\ell}{q}\right)\prod_{j\neq i}\frac{1}{2q\tanh\left(\frac{2i\pi\langle\lambda,\alpha_i\rangle-x_i}{2q}\right)}.$$

Assume now that $\left[\frac{\langle\lambda,\alpha_i\rangle}{q}\right]=0$. Then as a function of x_i, the function $U_{\lambda,w}(x)$ has a single pole at $x_i=0$. By the theorem of residues used as before, we get

$$\text{(7.84)}\qquad \operatorname{Res}_{x_i=0}U_{\lambda,w}(x)=0\,.$$

By (7.83), (7.84), we obtain

$$\text{(7.85)}\qquad (-1)^{\ell(g-1)}\operatorname{Res}^{1,\dots,\ell}_{x=0}\frac{1}{\left[\prod_1^\ell 2\sinh\left(\frac{x_i}{2q}\right)\right]^{2g-2}}\sum_{(w^1,\dots,w^s)\in W^s}\varphi_{w^j}\left(\frac{x}{q}\right)$$

$$M_q\left(\sum_{j=1}^{s}w^j\theta_j,x\right)=\sum_{\substack{\lambda\in\overline{CR}^*/q\overline{R}^*,\sigma^2(\lambda/q)\neq0\\(w^1,\dots,w^s)\in W^s}}\frac{1}{(i^\ell\sigma(\lambda/q))^{2g-2}}$$

$$\prod_{j=1}^{s}\varphi_{w^j}\left(2i\pi\frac{\langle\lambda,\alpha_1\rangle}{q},\dots,2i\pi\frac{\langle\lambda,\alpha_\ell\rangle}{q}\right)\exp\left(2i\pi\langle w^j\theta_j,\lambda/q\rangle\right).$$

Now by (1.93), (7.75), (7.78), for $1\leq j\leq s$,

$$\text{(7.86)}\qquad \sum_{w^j\in W}\varphi_{w^j}\left(2i\pi\frac{\langle\lambda,\alpha_i\rangle}{q},\dots,2i\pi\frac{\langle\lambda,\alpha_1\rangle}{q}\right)\exp\left(2i\pi\langle w^j\theta_j,\lambda/q\rangle\right)$$
$$=\chi_{\theta_j}(e^{\lambda/q})\,.$$

From (1.17), (7.85), (7.86), we get (7.79).

Now we consider the case where $g\geq 0$, s is even and $2g-2+s\geq 1$. Put

$$\text{(7.87)}\quad H_{\lambda,w}(x)=\frac{1}{\left[\prod_{i=1}^{\ell}2\sinh(\frac{x_i}{2q})\right]^{2g-2+s}}\prod_{i=1}^{\ell}\frac{1}{2q\tanh\left(\frac{2i\pi\langle\lambda,\alpha_i\rangle-x_i}{2q}\right)}.$$

Then $H_{\lambda,w}(x)$ is a meromorphic function of $x_1,\dots,x_\ell$. Since s is even, it is periodic of period $2i\pi q,\dots,2i\pi q$ in $x_1,\dots,x_\ell$.

Assume that $\left[\frac{\langle\lambda,\alpha_i\rangle}{q}\right]\neq 0$. Since $2g-2+s\geq 1$, as x_i tends to $\pm\infty$ inside Δ_ϵ, $H_{\lambda,w}(x)$ tends to 0. We thus find that the analogue of (7.82) holds, with $U_{\lambda,w}$ replaced by $H_{\lambda,w}$. The analogue of (7.83) is now

$$\operatorname{Res}_{x_i=2i\pi q\left[\frac{\langle\lambda,\alpha_i\rangle}{q}\right]}H_{\lambda,w}(x)=-\left[\frac{1}{2i\sin\left(\frac{\langle\lambda,\alpha_i\rangle\pi}{q}\right)}\prod_{j\neq i}\frac{1}{2\sinh\left(\frac{x_j}{2q}\right)}\right]^{2g-2+s}$$

$$\text{(7.88)}\qquad \prod_{j\neq i}\frac{1}{2q\tanh\left(\frac{2i\pi\langle\lambda,\alpha_i\rangle-x_i}{2q}\right)}.$$

Also the analogue of (7.84) still holds. So we get the analogue of (7.85),

$$(7.89)\quad (-1)^{\ell(g-1)}\operatorname{Res}^{1,\dots,\ell}_{x=0}\frac{1}{\left[\prod_1^\ell 2\sinh\left(\frac{x_i}{2q}\right)\right]^{2g-2+s}}\sum_{(w^1,\dots,w^s)\in W^s}\epsilon_{w^1}\dots\epsilon_{w^s}$$
$$M_q\left(\sum_{j=1}^s w^j(\rho+\theta_j),x\right)=\sum_{\substack{\lambda\in\overline{CR}^*/q\overline{R}^*,\sigma^2(\lambda/q)\neq 0\\(w^1,\dots,w^s)\in W^s}}\frac{1}{(i^\ell\sigma(\lambda/q))^{2g-2}}\frac{1}{(\sigma(\lambda/q))^s}$$
$$\prod_{j=1}^s\epsilon_{w^j}\exp\left(2i\pi\langle w^j(\rho+\theta_j),\lambda/q\rangle\right).$$

Now by (1.94), for $1\leq j\leq s$,

$$(7.90)\qquad \frac{1}{\sigma(\lambda/q)}\sum_{w^j\in W}\epsilon_{w^j}\exp(2i\pi\langle w^j(\rho+\theta_j),\lambda/q\rangle)=\chi_{\theta_j}(e^{\lambda/q}).$$

From (1.17), (7.89), (7.90), we get (7.80).

The proof of our Theorem is completed. □

REMARK 7.21. It is crucial that in (7.80), s is even. In fact under the given conditions on the θ_j's, $\sum_{j=1}^s w^j\theta_j\in\overline{R}$, but we know that $\sum_{j=1}^s w^j\rho\in\overline{R}$ only if s is even.

7.5. The generic case with $g\geq 2$.

DEFINITION 7.22. We will say that $(\theta_1,\dots,\theta_s)$ verify assumption $(\overline{A}_q)$ if for any $(w^1,\dots,w^s)\in W^s$, $\frac{1}{q}\sum_{j=1}^s w^j\theta_j\notin\overline{S}$.

In the sequel, if $u\in C/\overline{R}^*$, we choose once and for all a representative of u in C, which we still denote by u.

THEOREM 7.23. *If $q \in \mathbf{N}^*$, $g \geq 2$, if $(\theta_1, \dots, \theta_s)$ verify $(\overline{A}_q)$, then*

$$V_{g,q}(\theta_1, \dots, \theta_s) = \left|\frac{\overline{R}}{\overline{CR}}\right| \mathrm{Vol}(T)^{2g-2} \frac{|Z(G)|}{|W|} q^{(g-1)r} (-1)^{\ell(g-1)+r}$$

$$\sum_{u \in C/\overline{R}^*} \sum_{\substack{I=(i_1,\dots,i_r)\in\mathcal{I}_u \\ f\in\overline{CR}/d\overline{R}}} \frac{1}{\langle \alpha_{i_1}, \dots, \alpha_{i_r}\rangle}$$

$$\mathrm{Res}^I_{x=0} \left[\prod_{\alpha\in R_+} \frac{1}{2\sinh\left(\frac{1}{2}\langle\alpha, x^I/q + 2i\pi u\rangle\right)}\right]^{2g-2+s}$$

$$\sum_{(w^1,\dots,w^s)\in W^s} \varepsilon_{w^1}\dots\varepsilon_{w^s} \exp\left(\langle\sum_{j=1}^s w^j\rho, x^I/q\rangle\right. \tag{7.91}$$

$$\left. +2i\pi\langle\sum_{j=1}^s w^j(\rho+\theta^j) + qf, u\rangle\right)$$

$$\exp\left(d\sum_{j=1}^r \frac{\langle p^I_{j-1}\alpha_{i_j}, x^I\rangle}{\langle p^I_{j-1}\alpha_{i_j}, e^j\rangle}\left[\frac{1}{d}\langle p^I_{j-1}\Big(\sum_{k=1}^s w^k\theta_k/q + f\Big), e^j\rangle\right]\right)$$

$$\frac{1}{\prod_{j=1}^r\left(\exp\left(\frac{d\langle p^I_{j-1}\alpha_{i_j}, x^I\rangle}{\langle p^I_{j-1}\alpha_{i_j}, e^j\rangle}\right)-1\right)}.$$

PROOF. Take $t \in \overline{R}$. By Theorem 7.10, if $t \in \overline{R}$, $t/q \notin \overline{S}$, then

$$\mathrm{Res}^{(1,\dots,\ell)}_{x=0} \frac{1}{\left[\prod_{i=1}^{\ell} 2\sinh\left(\frac{x_i}{2q}\right)\right]^{2g-2}} \prod_{j=1}^s \varphi_{w^j}\left(\frac{x}{q}\right) M_q(t,x)$$

$$= \left|\frac{\overline{R}}{\overline{CR}}\right| (-1)^r \sum_{u\in C/\overline{R}^*} \sum_{\substack{I=(i_1,\dots,i_r)\subset\mathcal{I}_u, I \text{ generic} \\ f\in\overline{CR}/d\overline{R}}}$$

$$\mathrm{Res}^{(1,\dots,\ell)}_{x=0}\left\{\frac{1}{\left[\prod_{i=1}^{\ell} 2\sinh\left(\frac{x_i}{2q}\right)\right]^{2g-2}} \prod_{j=1}^s \varphi_{w^j}(x/q)\right. \tag{7.92}$$

$$\frac{1}{\prod_{i\in{}^cI} 2q\tanh\left(\frac{\langle\alpha_i, x^I + 2i\pi q u\rangle - x_i}{2q}\right)}$$

$$\frac{1}{\langle\alpha_{i_1}, \dots, \alpha_{i_r}\rangle}\exp(2i\pi\langle u, t+qf\rangle)$$

$$\exp\left(d\sum_{j=1}^r \frac{\langle p^I_{j-1}\alpha_{i_j}, x^I\rangle}{\langle p^I_{j-1}\alpha_{i_j}, e^j\rangle}\left[\frac{1}{d}\langle p^I_{j-1}(t+qf), \frac{e^j}{q}\rangle\right]\right)$$

$$\frac{1}{\prod_{j=1}^r\left(\exp\left(\frac{d\langle p^I_{j-1}\alpha_{i_j}, x^I\rangle}{\langle p^I_{j-1}\alpha_{i_j}, e^j\rangle}\right)-1\right)}.$$

Observe that in the right-hand side of (7.92), the expressions starting after $\exp(2i\pi\langle u,t+qf\rangle)$ does not depend on the x_i, $i\notin I$.

Now we explain how to evaluate the residue in (7.92). We use the notation in Definition 2.47.

1. *The case* $i\notin I$, $i\notin \mathcal{I}_u$ *or* $i\notin I$, $i\in\mathcal{I}_u$, $\alpha_i\notin\{\alpha_{\sigma^I(1)},\dots,\alpha_{\sigma^I(p_i)}\}$.

Note that the above condition just says that $i\notin I$, and $\langle\alpha_i,u\rangle\in\mathbf{Z}$, $\alpha_i\notin\{\alpha_{\sigma^I(1)},\dots,\alpha_{\sigma^I(p_i)}\}$, or $\langle\alpha_i,u\rangle\notin\mathbf{Z}$. If $\langle\alpha_i,u\rangle\in\mathbf{Z}$, $\alpha_i\notin\{\alpha_{\sigma^I(1)},\dots,\alpha_{\sigma^I(p_i)}\}$, for at least one $k\in I$, $k>i$, then $\langle\alpha_i,\alpha^k\rangle\neq 0$. So since the $\{x_k\}_{k>i}$ have not yet been made "small", for generic x, $\langle\alpha_i,x^I\rangle$ is "large". So we find that under the assumptions we made in 1., for generic x, mod $(2i\pi q\mathbf{Z})$, $\langle\alpha_i,x^I+2i\pi qu\rangle$ should be considered as different of 0. Then by proceeding as in (7.82) and using the fact that $g\geq 2$, we get

$$\operatorname{Res}_{x_i=0}\left[\frac{1}{\left[2\sinh\left(\frac{x_i}{2q}\right)\right]^{2g-2+s}}e^{\frac{1}{2q}\sum_{j=1}^{s}\varepsilon_{w^j}(i)x_i}\right.\left.\frac{1}{2q\tanh\left(\frac{\langle\alpha_i,x^I+2i\pi qu\rangle-x_i}{2q}\right)}\right]=\left[\frac{1}{2\sinh\left(\frac{\langle\alpha_i,x^I+2i\pi qu\rangle}{2q}\right)}\right]^{2g-2+s}e^{\frac{1}{2q}\langle\sum_{j=1}^{s}\varepsilon_{w^j}(i)\alpha_i,x^I+2i\pi qu\rangle}. \tag{7.93}$$

2. *The case where* $i\notin I$, $i\in\mathcal{I}_u$ *and* $\alpha_i\in\{\alpha_{\sigma^I(1)},\dots,\alpha_{\sigma^I(p_i)}\}$.

In this case $\langle\alpha_i,u\rangle\in\mathbf{Z}$, and so

$$\frac{1}{2q\tanh\left(\frac{\langle\alpha_i,x^I+2i\pi qu\rangle-x_i}{2q}\right)}=\frac{1}{2q\tanh\left(\frac{\langle\alpha_i,x^I\rangle-x_i}{2q}\right)}=-\frac{1}{2q}\frac{1-\tanh\left(\frac{x_i}{2q}\right)\tanh\left(\frac{\langle\alpha_i,x^I\rangle}{2q}\right)}{\tanh\left(\frac{x_i}{2q}\right)\left(1-\frac{\tanh\left(\frac{\langle\alpha_i,x^I\rangle}{2q}\right)}{\tanh\left(\frac{x_i}{2q}\right)}\right)}. \tag{7.94}$$

Now since $\alpha_i\in\{\alpha_{\sigma^I(1)},\dots,\alpha_{\sigma^I(p_i)}\}$, $\langle\alpha_i,x^I\rangle$ is a linear combination of the $x_{\sigma^I(k)}$, $k\leq p_i$, which have been made "small". Therefore we get the expression

$$\frac{1}{1-\frac{\tanh\left(\frac{\langle\alpha_i,x^I\rangle}{2q}\right)}{\tanh(x_i/2q)}}=1+\frac{\tanh\left(\frac{\langle\alpha_i,x^I\rangle}{2q}\right)}{\tanh(x_i/2q)}+\frac{\tanh^2\left(\frac{\langle\alpha_i,x^I\rangle}{2q}\right)}{\tanh^2(x_i/2q)}+\dots. \tag{7.95}$$

In view of (7.92), (7.94), (7.95), for $k\geq 1$, we should evaluate

$$\operatorname{Res}_{x_i=0}\left[\frac{1}{\left[2\sinh(x_i/2q)\right]^{2g-2+s}}e^{\frac{1}{2q}\sum_{j=1}^{s}\varepsilon_{w^j}(i)x_i}\frac{1}{\tanh^k(x_i/2q)}\right]. \tag{7.96}$$

The function of x_i which appears in (7.96) is periodic of period $2i\pi q$. Recall that the contour Δ_ε was defined in the proof of Theorem 7.20. For $\varepsilon>0$, 0 is the only pole inside Δ_ε. By using the theorem of residues as in (7.84), and the fact that $g\geq 2$, we find that (7.96) vanishes.

From (7.94)-(7.96), we see that if $i \notin I$, $i \in \mathcal{I}_u$, $\alpha_i \in \{\alpha_{\sigma^I(1)}, \dots, \alpha_{\sigma^I(p_i)}\}$, then

$$\operatorname{Res}_{x_i=0}\left[\frac{1}{\left[2\sinh(x_i/2q)\right]^{2g-2+s}} e^{\frac{1}{2q}\sum_{j=1}^{s}\varepsilon_{w^j}(i)x_i} \frac{1}{2q\tanh\left(\frac{\langle\alpha_i, x^I+2i\pi qu\rangle - x_i}{2q}\right)}\right] = 0. \tag{7.97}$$

Using (7.79) in Theorem 7.20 and (7.92), (7.93), (7.97), we get

$$\begin{aligned} V_{g,q}(\theta_1,\dots,\theta_s) = {} & \left|\frac{\overline{R}}{\overline{CR}}\right| \operatorname{Vol}(T)^{2g-2}\frac{|Z(G)|}{|W|}q^{(g-1)r}(-1)^{\ell(g-1)+r} \\ & \sum_{u\in C/\overline{R}^*}\sum_{\substack{I=(i_1,\dots,i_r)\in\mathcal{I}_u \\ f\in\overline{CR}/d\overline{R}}}\frac{1}{\langle\alpha_{i_1},\dots,\alpha_{i_r}\rangle} \\ & \operatorname{Res}^I_{x=0}\left\{\left[\prod_{i\in I}\frac{1}{2\sinh(\frac{x_i}{2q})}\prod_{i\notin I}\frac{1}{2\sinh\left(\frac{\langle\alpha_i,x^I+2i\pi qu\rangle}{2q}\right)}\right]^{2g-2+s}\right. \\ & \sum_{(w^1,\dots,w^s)\in W^s}\varepsilon_{w^1}\dots\varepsilon_{w^s}\exp\left(\frac{1}{2}\sum_{j=1}^{s}\left(\langle\sum_{i\notin I}\varepsilon_{w^j}(i)\alpha_i,\frac{x^I}{q}\rangle + \frac{\sum_{i\in I}\varepsilon_{w^j}(i)x_i}{q}\right)\right) \\ & \exp\left(2i\pi\langle u,\sum_{j=1}^{s}(\frac{\sum_{i\notin I}\varepsilon_{w^j}(i)\alpha_i}{2} + w^j\theta_j) + qf\rangle\right) \\ & \exp\left(d\sum_{j=1}^{r}\frac{\langle p^I_{j-1}\alpha_{i_j},x^I\rangle}{\langle p^I_{j-1}\alpha_{i_j},e^j\rangle}\left[\frac{1}{d}\langle p^I_{j-1}\left(\sum_{k=1}^{s}w^k\theta_k+qf\right),e^j/q\rangle\right]\right) \\ & \left.\frac{1}{\prod_{j=1}^{r}\left(\exp\left(\frac{d\langle p^I_{j-1}\alpha_{i_j},x^I\rangle}{\langle p^I_{j-1}\alpha_{i_j},e^j\rangle}\right)-1\right)}\right\}. \end{aligned} \tag{7.98}$$

Now observe that if $i \in I$,

$$\langle\alpha_i, x^I\rangle = x_i\,. \tag{7.99}$$

Therefore

$$\begin{aligned} \sum_{i\notin I}\langle\varepsilon_{w^j}(i)\alpha_i, x^I\rangle + \sum_{i\in I}\varepsilon_{w^j}(i)x_i = {} & \\ \langle\sum_{i=1}^{\ell}\varepsilon_{w^j}(i)\alpha_i, x^I\rangle = \langle w^j\sum_{i=1}^{\ell}\alpha_i, x^I\rangle = 2\langle w^j\rho, x^I\rangle\,. & \end{aligned} \tag{7.100}$$

Also if $i \in I \subset \mathcal{I}_u$, then $\langle\alpha_i, u\rangle \in \mathbf{Z}$, and so

$$\begin{aligned} & \left[\frac{1}{2\sinh\left(\frac{\langle\alpha_i,x^I+2i\pi qu\rangle}{2q}\right)}\right]^{2g-2} = \left[\frac{1}{2\sin(\frac{x_i}{2q})}\right]^{2g-2}, \\ & \frac{1}{2\sinh\left(\frac{\langle\alpha_i,x^I+2i\pi qu\rangle}{2q}+i\pi\right)}\exp\left(\frac{2i\pi\langle u,\varepsilon_{w^j}(i)\alpha_i\rangle}{2}\right) = \frac{1}{2\sinh\left(\frac{x_i}{2q}\right)}\,. \end{aligned} \tag{7.101}$$

From (7.98)-(7.101), we get (7.91). The proof of our Theorem is completed. □

REMARK 7.24. Clearly,

$$\left[\frac{1}{2\sin\left(\frac{\langle\alpha_i, x^I/q+2i\pi u\rangle}{2}\right)}\right]^{2g-2} \tag{7.102}$$

does not depend on the representative u in C. Also

$$\prod_{\alpha\in R_+}\frac{1}{2\sinh\left(\langle\alpha_i, x^I/q+2i\pi u\rangle\right)}\varepsilon_{w^j}\exp\left(\langle w^j\rho, x^I/q+2i\pi u\rangle\right) = \prod_{\alpha\in R_+}\frac{1}{1-\exp\left(-\langle w^j\alpha, x^I/q+2i\pi u\rangle\right)} \tag{7.103}$$

does not depend either on the representative $u\in C$. This explains why the terms in the right-hand side of (7.91) do not depend on the choice of the representatives of u.

7.6. The non generic case for $g \geq 2$. Recall that the functions $[x]_+, [x]_-$ were defined in (6.141).

Let $\eta_1, \dots, \eta_r$ be r functions from $\overline{R}/d\overline{R} \times \{I, I\text{generic}\}$ into $\{+,-\}$ such that if $f = \sum_1^r f^j e_j \in \overline{R}/d\overline{R}$, $I = (i_1, \dots, i_r)$, then $\eta_j(f, I)$ depends only on $f^1, \dots, f^j, i_1, \dots, i_{j-1}$.

First, we will extend Theorem 7.2 to the case where $t \in \overline{R}/q\overline{R}$ is arbitrary.

THEOREM 7.25. *For any $t\in\overline{R}/q\overline{R}$, there are meromorphic functions $\varphi_1(t, x_2, \dots, x_\ell), \dots, \varphi_i(t, x_1, \dots, x_{i-1}, x_{i+1}, \dots, x_\ell)\dots$ which vanish identically when $t/q \notin H$, such that*

$$\begin{aligned} L_q(t,x) = \frac{(-1)^r}{d^r} &\sum_{\substack{I=(i_1,\dots,i_r)\subset\{I,\dots,\ell\}, I \text{ generic}\\ f\in\overline{R}/d\overline{R}, g\in(\mathbf{Z}/d\mathbf{Z})^r}} \operatorname{sgn}\left(\langle\alpha_{i_1},\dots,\alpha_{i_r}\rangle\right) \\ &\left[\exp\left(d\sum_{j=1}^r \frac{\langle p^I_{j-1}\alpha_{i_j}, x^I\rangle}{\langle p^I_{j-1}\alpha_{i_j}, e^j\rangle}\left[\frac{1}{d}\langle p^I_{j-1}(t+qf), \frac{e^j}{q}\rangle\right]_{\eta_j(f,I)}\right)\right. \\ &\left.\frac{1}{\prod_{i\in{}^cI} 2q\tanh\left(\frac{\langle\alpha_i,x^I\rangle - x_i}{2q}\right)}\right](x+2i\pi q g_I)\prod_{j=1}^r\frac{1}{\exp\left(\frac{d\langle p^I_{j-1}\alpha_{i_j}, x^I\rangle}{\langle p^I_{j-1}\alpha_{i_j}, e^j\rangle}\right)-1} \\ &+\sum_{i=1}^{\ell}\varphi_i(t, x_1,\dots,x_{i-1},x_{i+1},\dots,x_\ell)\,. \end{aligned} \tag{7.104}$$

PROOF. We use the notation in the proof of Theorem 7.2. Let $\eta_1, \dots, \eta_{j-1} \in \{+,-\}$. We define $S_I(a,x)$ as in (7.7), except that $\left[\frac{1}{d}\langle p^I_{i_{n-1}} t, \frac{e^n}{q}\rangle\right]$ is replaced by $\left[\frac{1}{d}\langle p^I_{i_{n-1}} t, \frac{e^n}{q}\rangle\right]_{\eta_n}$, $1 \leq n \leq j-1$.

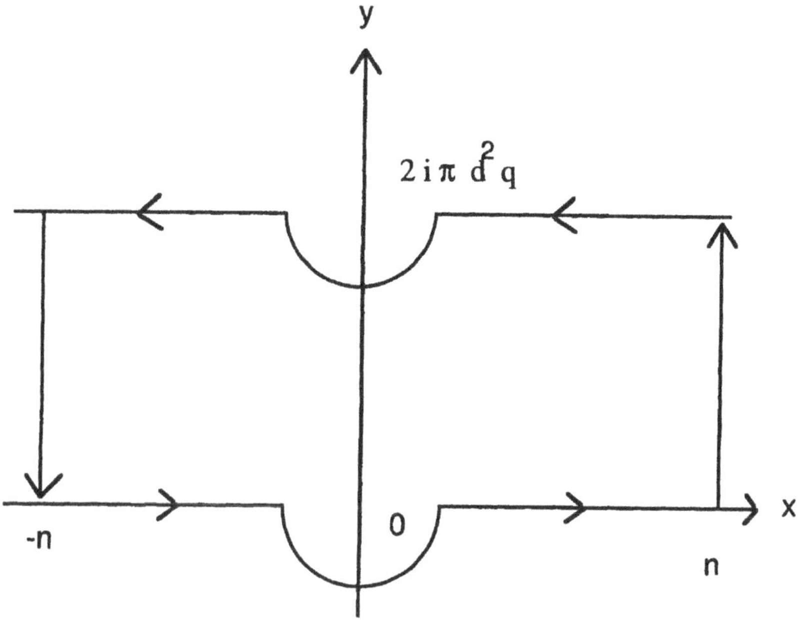

FIGURE 7.3

Take $\eta_j(f) : \{0, \ldots, d-1\} \mapsto \{+,-\}$. Then we rewrite (7.20) in the form

$$L_I = \frac{1}{d^2} \sum_{k \in \mathbf{Z}/d^2q\mathbf{Z}} \Bigg\{ \sum_{\substack{0 \le f < d \\ g \in (\mathbf{Z}/d\mathbf{Z})^{j-1}}} \tag{7.105}$$
$$\mathrm{Res}_{a=2i\pi k} \Bigg[\exp\Bigg(a \left[\frac{\langle p^I_{j-1}(t+qfe_j), e^j/q\rangle}{d} \right]_{\eta_j(f)}$$
$$+ d \sum_{n=1}^{j-1} \frac{\langle p^I_{n-1}\alpha_{i_n}, x^I\rangle}{\langle p^I_{n-1}\alpha_{i_n}, e^n\rangle} \left[\frac{1}{d}\langle p^I_{n-1}t, e^n/q\rangle \right]_{\eta_n} \Bigg)$$
$$\frac{1}{\prod_{i \in {}^c I} 2q \tanh\left(\frac{\langle p^I_{j-1}\alpha_i, ae^j/d + 2i\pi\widehat{k} - x^I\rangle}{2q} \right)} \frac{1}{e^a - 1} \Bigg] \Bigg\} (x + 2i\pi q g_I).$$

Instead of (7.21), we now set

$$h_{f,g}(a) = \frac{\exp\left(a \left[\frac{\langle p^I_{j-1}(t+qfe_j), \frac{e^j}{q}\rangle}{d} \right]_{\eta_j(f)} \right)}{\prod_{i \in {}^c I} 2q \tanh\left(\frac{\langle p^I_{j-1}\alpha_i, \frac{ae^j}{d} + 2i\pi\widehat{k} - x^I\rangle}{2q} \right)} \frac{1}{e^a - 1}. \tag{7.106}$$

The key point is that we do no longer assume that (7.23) holds. Equivalently $\frac{\langle p^I_{j-1}(t+qfe_j), e^j/q\rangle}{d}$ may now well be an integer.

For $n \in \mathbf{N}$, we now replace the contour in Figure 7.1 by the contour of Figure 7.3. We apply the theorem of residues to $h_{f,g}(a)$ on this contour, and let $n \to +\infty$.

Clearly, if $\frac{\langle p_{j-1}^I(t+q/e_j),e^j/q\rangle}{d}$ is an integer, if $\eta_j(f)=\pm$, then

$$\lim_{a\to\pm\infty} h_{f,g}(a)=0\,, \tag{7.107}$$

$$\lim_{a\to\mp\infty} h_{f,g}(a)=\mp\frac{1}{\prod\limits_{\substack{i\in{}^cI\\ \langle p_{j-1}^I\alpha_i,e^j\rangle=0}}\tanh\left(\frac{\langle p_{j-1}^I\alpha_i,2i\pi\widehat{k}-x^I\rangle}{2q}\right)}$$

$$\frac{1}{(2q)^{|{}^cI|}}(-1)^{|\{i\in{}^cI,\langle p_{j-1}^I\alpha_i,e^j\rangle\neq0,\ \mathrm{sgn}\langle p_{j-1}^I\alpha_i,e^j\rangle=\pm1\}|}\,.$$

Observe here that by (2.65), $\{i\in{}^cI;\langle p_{j-1}^I\alpha_i,e^j\rangle\neq0\}$ is non empty. Then by (2.63), the right-hand side of the second equation in (7.107) does not depend on $\{x_i\}_{i\in{}^cI\,,\langle p_{j-1}^I\alpha_i,e^j\rangle\neq0}$.

So by proceeding as in (7.21)-(7.25), we find that

- if $\frac{\langle p_{j-1}^I(t+qfe_j),e^j/q\rangle}{d}$ is not an integer, (7.26) still holds, with $[\ ]$ replaced by $[\ ]_{\eta_1},\cdots,[\ ]_{\eta_{j-1}},[\ ]_{\eta_j(f)}$.
- if $\frac{\langle p_{j-1}^I(t+qfe_j),e^j/q\rangle}{d}$ is an integer, writing η_j instead of $\eta_j(f)$, then (7.26) is replaced by

$$L_I=-\frac{1}{d}\sum_{\substack{0\le f<d\ ,g\in(\mathbf{Z}/d\mathbf{Z})^{j-1}\\ i_j:(\alpha_{i_1},\dots,\alpha_{i_j})\neq0\ ,0\le g^j<d|\langle p_{j-1}^I\alpha_{i_j},e^j\rangle|}} \tag{7.108}$$

$$\Bigg\{\frac{1}{\langle p_{j-1}^I\alpha_{i_j},e^j\rangle}\exp\Bigg(2i\pi\langle p_j^{(I,i_j)}e_j,\widehat{k}\rangle\langle p_{j-1}^I(t+qfe_j),e^j/q\rangle$$

$$+d\sum_{n=1}^{j}\frac{\langle p_{n-1}^I\alpha_{i_n},x^I\rangle}{\langle p_{n-1}^I\alpha_{i_n},e^n\rangle}\left[\frac{1}{d}\langle p_{n-1}^I(t+qfe_j),e^n/q\rangle\right]_{\eta_n}\Bigg)$$

$$\frac{1}{\prod_{i\notin(I,i_j)}2q\tanh\left(\frac{\langle p_j^{(I,i_j)}\alpha_i,2i\pi\widehat{k}-x^I\rangle}{2q}\right)}\Bigg\}(x+2i\pi q(g,g_j)_{I,i_j})$$

$$\frac{1}{\exp\left(\frac{d\langle p_{j-1}^I\alpha_{i_j},x^I\rangle}{\langle p_{j-1}^I\alpha_{i_j},e^j\rangle}\right)-1}+q\sum_{\substack{0\le f<d\\ g\in(\mathbf{Z}/d\mathbf{Z})^{g-1}}}$$

$$\Bigg[\exp\Bigg(d\sum_{n=1}^{j-1}\frac{\langle p_{n-1}^I\alpha_{i_n},x^I\rangle}{\langle p_{n-1}^I\alpha_i,e^n\rangle}\left[\frac{1}{d}\langle p_{n-1}^It,e^n/q\rangle\right]_{\eta_n}\Bigg)$$

$$\frac{1}{\prod\limits_{\substack{i\in{}^cI\\ \langle p_{j-1}^I\alpha_i,e^j\rangle=0}}\tanh\left(\frac{\langle p_{j-1}^I\alpha_i,2i\pi\widehat{k}-x^I\rangle}{2q}\right)}\Bigg](x+2i\pi qg_I)\frac{1}{(2q)^{|{}^cI|}}$$

$$(-1)^{|\{i\in{}^cI,\langle p_{j-1}^I\alpha_i,e^j\rangle\neq0,\ \mathrm{sgn}\langle p_{j-1}^I\alpha_i,e^j\rangle=\eta_j(f)\}|}\,,$$

the key point being that the last sum in (7.108) is a sum of functions which do not depend on the x_i, $i\in{}^cI$, $\langle p_{j-1}^I\alpha_i,e_j\rangle\neq0)$.

Using (7.108) and proceeding by recursion as in the proof of Theorem 7.2, we get (7.104). The proof of our Theorem is completed. □

REMARK 7.26. The last term is the right-hand side of (7.108) depends in a non trivial way on the choice of η_j. More precisely, it is a sum of terms depending on η_j up to sign. As a corollary, we find that the function $\sum_1^\ell \varphi_i(t, x_1, \dots, x_{i-1}, x_{i+1}, \dots, x_\ell)$ also depends on the choice of $\eta_1, \dots, \eta_r$ in a non trivial way.

Now we extend Theorem 7.6 to the general case. We still take $\eta_1, \dots, \eta_r$ as in Theorem 7.25.

THEOREM 7.27. *For any $t \in \overline{R}/q\overline{R}$, there are meromorphic functions $\psi_1(t, x_2, \dots, x_\ell), \dots, \psi_i(t, x_1, \dots, x_{i-1}, x_{i+1}, \dots, x_\ell)$ which vanish identically when $t/q \notin \overline{S}$, such that*

(7.109)
$$\begin{aligned}
&M_q(t,x) = \\
&\frac{(-1)^r}{d^r} \sum_{\substack{I=(i_1,\dots,i_r)\subset\{1,\dots,\ell\}, I \text{ generic} \\ f\in\overline{R}/d\overline{R}, g\in(\mathbb{Z}/d\mathbb{Z})^r, \mu\in\overline{CR^*}/\overline{R}^*}} \operatorname{sgn}(\langle\alpha_{i_1},\dots,\alpha_{i_r}\rangle)\exp(2i\pi\langle\lambda_\mu, t/q\rangle) \\
&\left[\exp\left(d\sum_{j=1}^r \frac{\langle p^I_{j-1}\alpha_{i_j}, x^I\rangle}{\langle p^I_{j-1}\alpha_{i_j}, e^j\rangle}\left[\frac{1}{d}\langle p^I_{j-1}(t+qf), e^j/q\rangle\right]_{\eta_j(f,I)}\right]\right. \\
&\frac{1}{\prod_{i\in {}^cI} 2q\tanh\left(\frac{\langle\alpha_i, x^I\rangle - x_i}{2q}\right)}(x + 2i\pi q g_I - 2i\pi\widetilde{a}\lambda_\mu) \\
&\prod_{j=1}^r \frac{1}{\exp\left(\frac{d\langle p^I_{j-1}\alpha_{i_j}, x^I\rangle}{\langle p^I_{j-1}\alpha_{i_j}, e^j\rangle}\right) - 1}(x - 2i\pi q\widetilde{a}\lambda_\mu) + \\
&\sum_{i=1}^\ell \psi_i(t, x_1, \dots, x_{i-1}, x_{i+1}, \dots, x_\ell).
\end{aligned}$$

PROOF. Using Proposition 7.5 and Theorem 7.25, we get (7.109). □

Given $(w^1, \dots, w^s) \in W^s$, let $\eta_1^{(w^1,\dots,w^s)}, \dots, \eta_s^{w^1,\dots,w^s)}$ be s functions from $\overline{R}/d\overline{R} \times \{I, I \,\text{generic}\}$ into $\{+,-\}$ having the properties listed before Theorem 7.25.

THEOREM 7.28. *If $q \in \mathbf{N}^*$, $g \geq 2$, then*

$$V_{g,q}(\theta_1, \ldots, \theta_s) = \left|\frac{\overline{R}}{\overline{CR}}\right| \mathrm{Vol}(T)^{2g-2} \frac{|Z(G)|}{|W|} q^{(g-1)r} (-1)^{\ell(g+1)+r} \tag{7.110}$$

$$\sum_{u \in C/\overline{R}^*} \sum_{\substack{I=(i_1,\ldots,i_r)\in \mathcal{I}_u \\ f \in \overline{CR}/d\overline{R}}} \frac{1}{\langle \alpha_{i_1}, \ldots, \alpha_{i_r}\rangle}$$

$$\mathrm{Res}^I_{x=0} \left[\prod_{\alpha \in R_+} \frac{1}{2\sinh\left(\frac{1}{2}\langle \alpha_i, x^I/q + 2i\pi u\rangle\right)} \right]^{2g-2+s}$$

$$\sum_{(w^1,\ldots,w^s)\in W^s} \varepsilon_{w^1} \ldots \varepsilon_{w^s} \exp\left(\langle \sum_{j=1}^s w^j \rho, x^I/q\rangle \right.$$

$$\left. + 2i\pi \langle \sum_{j=1}^s w^j(\rho + \theta^j) + qf, u\rangle \right)$$

$$\exp\left(d \sum_{j=1}^r \frac{\langle p^I_{j-1}\alpha_{i_j}, x^I\rangle}{\langle p^I_{j-1}\alpha_{i_j}, e^j\rangle} \left[\frac{1}{d} \langle p^I_{j-1} \left(\sum_{k=1}^s \frac{w^k \theta_k}{q} + f \right), e^j\rangle \right]_{\eta_j^{(w^1,\ldots,w^s)}(f,I)} \right)$$

$$\frac{1}{\prod_{j=1}^r \left(\exp\left(\frac{d\langle p^I_{j-1}\alpha_{i_j}, x^I\rangle}{\langle p^I_{j-1}\alpha_{i_j}, e^j\rangle} \right) - 1 \right)}.$$

PROOF. We use (7.79) in Theorem 7.20 and 7.27, and also the arguments in the proof of Theorem 7.23. We will show that for $1 \leq i \leq \ell$,

(7.111)

$$\mathrm{Res}^{1,\ldots,\ell}_{x=0} \frac{1}{\left[\prod_1^\ell 2\sinh\left(\frac{x_i}{2q}\right)\right]^{2g-2}} \prod_{j=1}^s \varphi_{w^j}(x/q) \psi_i(x_1, \ldots, x_{i-1}, x_{i+1}, \ldots, x_\ell) = 0$$

from which (7.110) follows. To establish (7.111), we only need to show that

$$\mathrm{Res}_{x_i=0} \frac{1}{\left[2\sinh\left(\frac{x_i}{2q}\right)\right]^{2g-2}} \prod_{j=1}^s \varphi_{w^j}(x/q) = 0\,. \tag{7.112}$$

The proof of (7.112) is the same as the proof of (7.84).

We have thus established our Theorem. □

7.7. The general case. We will now establish another form of Theorem 7.28.

THEOREM 7.29. *If $g \geq 0$, if s is even and if $2g-2+s \geq 1$, the following identity holds*

$$(7.113)\quad V_{g,q}(\theta_1,\dots,\theta_s) = \left|\frac{\overline{R}}{\overline{CR}}\right| \mathrm{Vol}(T)^{2g-2}\frac{|Z(G)|}{|W|}q^{(g-1)r}(-1)^{\ell(g+1)+r}$$

$$\sum_{u\in C/\overline{R}^*}\ \sum_{\substack{I=(i_1,\dots,i_r)\in\mathcal{I}_u\\ f\in\overline{CR}/d\overline{R}}} \frac{1}{\langle\alpha_{i_1},\dots,\alpha_{i_r}\rangle}$$

$$\mathrm{Res}^I_{x=0}\left[\prod_{\alpha\in R_+}\frac{1}{2\sinh\left(\frac{1}{2}\langle\alpha_i, x^I/q+2i\pi u\rangle\right)}\right]^{2g-2+s}$$

$$\sum_{(w^1,\dots,w^s)\in W^s} \varepsilon_{w^1}\dots\varepsilon_{w^s}\ \exp(2i\pi\langle\sum_{j=1}^{s} w^j(\rho+\theta^j)+qf,u\rangle)$$

$$\exp\left(d\frac{\sum_{j=1}^r\langle p^I_{j-1}\alpha_{i_j},x^I\rangle}{\langle p^I_{j-1}\alpha_{i_j},e^j\rangle}\left[\frac{1}{d}\langle p^I_{j-1}\Big(\sum_{k=1}^{s}\frac{w^k(\rho+\theta_k)}{q}+f\Big),e^j\rangle\right]_{\eta_j^{(w^1,\dots,w^s)}(f,I)}\right)$$

$$\frac{1}{\prod_{j=1}^r\left(\exp\left(\frac{d\langle p^I_{j-1}\alpha_{i_j},x^I\rangle}{\langle p^I_{j-1}\alpha_{i_j},e^j\rangle}\right)-1\right)}.$$

PROOF. We proceed as in the proof of Theorems 7.23 and 7.28. Instead of (7.79) in Theorem 7.20, we use (7.80) and we still use Theorem 7.27. In particular the obvious analogues of (7.93) and (7.97) still hold because s is even and $2g-2+s \geq 1$. For the same reason, the analogue of (7.111), (7.112), with the φ_{w^j} replaced by 1 still holds.

The proof of our Theorem is completed. □

8. The Verlinde formulas

In this Section, we prove the Verlinde formulas, in the restricted sense which was described in the Introduction. Namely, we show essentially that for p large enough, the Riemann-Roch number of M/G is given by the Verlinde formulas. Also we show that the Riemann-Roch numbers of suitable perturbations of M/G are given by the Verlinde formulas.

This Section is organized as follows. In Section 8.1, we establish our results under a suitable genericity assumption on the holonomies. In Sections 8.2 and 8.3, we consider suitable perturbations of the moduli space.

8.1. The generic case. Here we will assume that $s \geq 1$, $2g-2+s \geq 1$, that $t_1, \dots, t_s$ are regular and lie in the alcove P, and that $(t_1, \dots, t_s)$ verify assumption $(\overline{A})$ of Section 6.9. Namely we assume that for any $(w^1, \dots, w^s) \in W^s$, $\sum_{j=1}^{s} w^j t_j \notin \overline{S}$.

Recall that $\mathbf{M} \subset \mathbf{Z}$ was defined in (6.20). Here we assume that $\mathbf{M}$ is not reduced to 0.

THEOREM 8.1. *For $p \in \mathbf{M}$, $p \geq 0$ large enough,*

$$\operatorname{Ind}(D_{p,+}) = V_{g,p+c}(pt_1, \dots, pt_s). \tag{8.1}$$

PROOF. By Theorem 6.29 and by Theorem 7.12, we know that if $\sum_{j=1}^{s} pt_j \notin \overline{R}$, then both sides of (8.1) are equal to 0. So we may as well assume that $\sum_{j=1}^{s} pt_j \in \overline{R}$.

First we consider the case where $g \geq 2$. Observe that for $p \in \mathbf{M}$ large enough, $(pt_1, \dots, pt_s)$ verify $(\overline{A}_{p+c})$. By comparing formula (6.130) in Theorem 6.34 and formula (7.91) in Theorem 7.23, we get (8.1).

Now we consider the case $g = 1$. By Theorem 7.16, we get

$$V_{1,p+c}(pt_1, \dots, pt_s) = \frac{(-1)^{\ell}}{\left|\frac{CR^{*}}{(p+c)CR}\right|} \sum_{(v^1,v^2) \in W^2} \epsilon_{v^1} \epsilon_{v^2} \overline{V}^{K}_{2,p+c}(pt_1, \dots, pt_s, v^1\rho + v^2\rho). \tag{8.2}$$

Observe that for p large enough, $(pt_1, \dots, pt_s, v^1\rho + v^2\rho)$ verify $(\overline{A}_{p+c})$. Using (1.17), Theorem 7.23 with $g = 2$ and (8.2), we obtain

$$(8.3)\qquad V_{1,p+c}(pt_1, \dots, pt_s) = \frac{|Z(G)|}{|W|}(-1)^r$$
$$\sum_{u\in C/\overline{R}^*} \sum_{\substack{I=(i_1,\dots,i_r)\in\mathcal{I}_u \\ f\in\overline{R}/d\overline{R},\mu\in\overline{CR}^*/\overline{R}^*}} \frac{1}{\langle\alpha_{i_1}, \dots, \alpha_{i_r}\rangle}$$
$$\operatorname{Res}^I_{x=0}\left[\prod_{\alpha\in R_+}\frac{1}{2\sinh\left(\frac{1}{2}\langle\alpha, x^I/q + 2i\pi u\rangle\right)}\right]^{s+3} \sum_{(w^1,\dots,w^{s+3})\in W^{s+3}} \varepsilon_{w^1}\dots\varepsilon_{w^{s+3}}$$
$$\exp\left(\langle\sum_{j=1}^{s+1} w^j\rho, x^I/(p+c)\rangle + 2i\pi\langle\sum_{j=1}^{s} w^j pt_j/(p+c) + \sum_{j=s+2}^{s+3} w^j\rho/(p+c), \lambda_\mu\rangle\right.$$
$$\left.+2i\pi\langle\sum_{j=1}^{s} w^j(\rho + pt^j) + \sum_{j=s+1}^{s+3} w^j\rho + (p+c)f, u\rangle\right)$$
$$\left\{\exp\left(d\sum_{j=1}^{r}\frac{\langle p^I_{j-1}\alpha_{i_j}, x^I\rangle}{\langle p^I_{j-1}\alpha_{i_j}, e^j\rangle}\left[\frac{1}{d}\langle p^I_{j-1}\Big(\sum_{k=1}^{s} w^k pt_k/(p+c) +\right.\right.\right.$$
$$\left.\left.\sum_{k=s+2}^{s+3} w^k\rho/(p+c) + f\Big), e^j\rangle\right]\right)$$
$$\left.\frac{1}{\prod_{j=1}^r\left(\exp\left(\frac{d\langle p^I_{j-1}\alpha_{i_j}, x^I\rangle}{\langle p^I_{j-1}\alpha_{i_j}, e^j\rangle}\right) - 1\right)}\right\}(x - 2i\pi\widetilde{a}\lambda_\mu).$$

Since $(t_1, \dots, t_s)$ verify $(\overline{A})$, by Proposition 2.14,

$$(8.4)\qquad \langle p^I_{j-1}\sum_{k=1}^{s} w^k t_k + f, e^j\rangle \notin \mathbf{Z}.$$

By proceeding as in (6.132) and using (8.4), we get for $p \in \mathbf{M}$ large enough,

$$(8.5)\qquad [\frac{1}{d}\langle p^I_{j-1}(\sum_{k=1}^{s} w^k pt_k/(p+c) + \sum_{k=s+2}^{s+3} w^k\rho/(p+c) + f), e^j\rangle] =$$
$$[\frac{1}{d}\langle p^I_{j-1}(\sum_{k=1}^{s} w^k pt_k/(p+c) + f), e^j\rangle] + \frac{1}{d}\langle p^I_{j-1}\sum_{k=s+2}^{s+3} w^k\rho/(p+c), e^j\rangle.$$

Also by (2.13),

$$(8.6)\qquad \sum_{j=1}^{s}\frac{\langle p^I_{j-1}\alpha_{i_j}, x^I\rangle}{\langle p^I_{j-1}\alpha_{i_j}, ej\rangle}\langle p^I_{j-1}\sum_{k=s+2}^{s+3} w^k\rho/(p+c), e^j\rangle = \langle\sum_{k=s+2}^{s+3} w^k\rho/(p+c), x^I\rangle.$$

So by (8.3), (8.5), (8.6), we get

$$
\begin{aligned}
(8.7)\ & V_{1,p+c}(pt_1, \ldots, pt_s) = \frac{|Z(G)|}{|W|}(-1)^r \\
& \sum_{u \in C/\overline{R}^*} \sum_{\substack{I=(i_1,\ldots,i_r)\in \mathcal{I}_u \\ f\in \overline{R}/d\overline{R}, \mu \in \overline{CR}^*/\overline{R}^*}} \frac{1}{\langle \alpha_{i_1}, \ldots, \alpha_{i_r}\rangle} \\
& \mathrm{Res}^I_{x=0} \left[\prod_{\alpha\in R_+} \frac{1}{2\sinh\left(\frac{1}{2}\langle \alpha, x^I/(p+c) + 2i\pi u\rangle\right)} \right]^{s+3} \sum_{(w^1,\ldots,w^{s+3})\in W^{s+3}} \varepsilon_{w^1}\ldots\varepsilon_{w^{s+3}} \\
& \exp\Bigg(\langle \sum_{j=1}^{s} w^j\rho, x^I/(p+c)\rangle + \langle \sum_{j=s+1}^{s+3} w^j\rho, \frac{x^I}{p+c} + 2i\pi u\rangle + \\
& 2i\pi\langle \sum_{j=1}^{s} w^j pt_j/(p+c), \lambda_\mu\rangle + 2i\pi\langle \sum_{j=1}^{s} w^j(\rho + pt^j) + (p+c)f, u\rangle \Bigg) \\
& \Bigg\{ \exp\left(d\sum_{j=1}^{r} \frac{\langle p^I_{j-1}\alpha_{i_j}, x^I\rangle}{\langle p^I_{j-1}\alpha_{i_j}, e^j\rangle} \left[\frac{1}{d}\langle p^I_{j-1}\Big(\sum_{k=1}^{s} w^k pt_k/(p+c) + f\Big), e^j\rangle \right] \right) \\
& \frac{1}{\prod_{j=1}^{r}\left(\exp\left(\frac{d\langle p^I_{j-1}\alpha_{i_j}, x^I\rangle}{\langle p^I_{j-1}\alpha_{i_j}, e^j\rangle}\right) - 1\right)} \Bigg\} (x - 2i\pi\widetilde{a}\lambda_\mu).
\end{aligned}
$$

Now by (1.94), (7.71),

$$
(8.8) \qquad \frac{1}{\prod_{\alpha\in R_+} \sinh\left(\frac{1}{2}\langle \alpha, \frac{x^I}{p+c} + 2i\pi u\rangle\right)} \sum_{w\in W} \epsilon_w \exp(\langle w\rho, \frac{x^I}{p+c} + 2i\pi u\rangle) = 1.
$$

By (8.7), (8.8), we obtain

$$(8.9)\qquad V_{1,p+c}(pt_1,\dots,pt_s)=\frac{|Z(G)|}{|W|}(-1)^r$$

$$\sum_{u\in C/\overline{R}^*}\ \sum_{\substack{I=(i_1,\dots,i_r)\in\mathcal{I}_u\\ f\in\overline{R}/d\overline{R},\mu\in\overline{CR}^*/\overline{R}^*}}\frac{1}{\langle\alpha_{i_1},\dots,\alpha_{i_r}\rangle}$$

$$\mathrm{Res}^I_{x=0}\left[\prod_{\alpha\in R_+}\frac{1}{2\sinh\left(\frac{1}{2}\langle\alpha,x^I/(p+c)+2i\pi u\rangle\right)}\right]^s\sum_{(w^1,\dots,w^s)\in W^s}\varepsilon_{w^1}\dots\varepsilon_{w^s}$$

$$\exp\left(\langle\sum_{j=1}^s w^j\rho,x^I/(p+c)\rangle+2i\pi\langle\sum_{j=1}^s w^j pt_j/(p+c),\lambda_\mu\rangle\right.$$

$$\left.+2i\pi\langle\sum_{j=1}^s w^j(\rho+pt^j)+(p+c)f,u\rangle\right)$$

$$\left\{\exp\left(d\sum_{j=1}^r\frac{\langle p^I_{j-1}\alpha_{i_j},x^I\rangle}{\langle p^I_{j-1}\alpha_{i_j},e^j\rangle}\left[\frac{1}{d}\langle p^I_{j-1}\left(\sum_{k=1}^s w^k\frac{pt_k}{p+c}+f\right),e^j\rangle\right]\right)\right.$$

$$\left.\frac{1}{\prod_{j=1}^r\left(\exp\left(\frac{d\langle p^I_{j-1}\alpha_{i_j},x^I\rangle}{\langle p^I_{j-1}\alpha_{i_j},e^j\rangle}\right)-1\right)}\right\}(x-2i\pi\tilde{a}\lambda_\mu)\,.$$

By comparing (6.130) and (8.9), we get (8.1) for g = 1.

A similar proof can be given for the case $g=0$. The proof of our Theorem is completed. □

8.2. The perturbed case at $\frac{pt_j}{p+c}$. Now we make the same assumptions as in Section 6.11. Namely take $p\in\mathbf{M},p\geq 0$. We assume that $\epsilon_1,\dots,\epsilon_s\in\mathfrak{t}$ are such that form $\ell\in]0,1]$, $1\leq j\leq s$, $\frac{pt_j}{p+c}+\ell\epsilon_j\in T$ is regular, and for any $(w^1,\dots,w^s)\in W^s,\ell\in]0,1]$, $\sum_{j=1}^s w^j\frac{pt_j}{p+c}+\ell\epsilon_j\notin\overline{S}$.

THEOREM 8.2. *For $g\geq 2$,*

$$(8.10)\qquad \mathrm{Ind}(D_{p,+}^{(\frac{pt_j}{p+c}+\ell\epsilon_j)})=V_{g,p+c}(pt_1,\dots,pt_s).$$

For $g=0$ or $g=1$, if $(t_1,\dots,t_s)$ verify $(\overline{A})$, for $p\in\mathbf{M}$ and $p\geq 0$ large enough, equation (8.10) still holds.

PROOF. By Remark 6.37, we find that if $\sum_{j=1}^s pt_j\notin\overline{R}$, then the left-hand side of (8.10) vanishes. By (7.59) in Theorem 7.12, the right hand-side also vanishes. As in the proof of Theorem 8.1, we may and we will assume that $\sum_{j=1}^s pt_j\in\overline{R}$.

First we consider the case $g\geq 2$. Then (8.10) follows by comparing (6.147) in Theorem 6.41, and (7.110) in Theorem 7.28.

In the case $0\leq g\leq 1$, we proceed as in the proof of Theorem 8.1, and use (8.10) in the case $g\geq 2$.

The proof of our Theorem is completed. □

8.3. The perturbed case at $\frac{\rho+pt_j}{p+c}$. Observe that if $t \in \overline{P}$, for $p \in \mathbf{N}$, $\frac{\rho+pt}{p+c} \in P$, i.e. $\frac{\rho+pt}{p+c}$ is regular. This follows follows from (1.33), (1.34), (1.36).

The same argument as in the proof of Proposition 6.39 shows that if $s \geq 1$, given $p \geq 0, p \in \mathbf{M}$, there exist $\epsilon_1, \ldots, \epsilon_s \in \mathfrak{t}$ such that for $\ell \in]0,1], 1 \leq j \leq s$, $\frac{pt_j+\rho}{p+c} + \ell\epsilon_j \in T$ is regular, and for any $(w^1, \ldots, w^s) \in W^s, \ell \in]0,1]$, $\sum_{j=1}^s w^j(\frac{pt_j}{p+c} + \ell\epsilon_j) \notin \overline{S}$.

In Theorem 6.36 we now take

$$\delta_j = \frac{\rho - ct_j}{p+c} + \ell\epsilon_j. \tag{8.11}$$

By the construction of Section 6.11, we get a Dirac operator $D_{p,+}^{\frac{\rho+pt_j}{p+c}}$.

THEOREM 8.3. *For $g \geq 0$, and $2g-2+s \geq 1$, for $\ell \in]0,1]$,*

$$\operatorname{Ind}(D_{p,+}^{(\frac{\rho+pt_j}{p+c}+\ell\epsilon_j)}) = V_{g,p+c}(pt_1, \ldots, pt_s). \tag{8.12}$$

PROOF. As in the proof of Theorems 8.1 and 8.2, we may restrict ourselves to the case where $\sum_{j=1}^s pt_j \in \overline{R}$. First assume that s is even. Then (8.12) follows from (6.137) in Theorem 6.36, with δ_j given by (8.11), and from (7.113) in Theorem 7.29. When s is odd, first we perturb the s holonomies as indicated before. Then we add an extra marked point x_{s+1}, with holonomy equal to 1. We perturb the holonomy 1 to a generic holonomy, and we use the result we just proved with $s+1$ marked points. We can then gently make our holonomy t_{s+1} tend to 1. This is in fact possible because the perturbation of the first s holonomies are generic and verify condition $(\overline{A})$. The proof of our Theorem is completed. □

REMARK 8.4. Suppose temporarily that the assumptions of Remark 6.42 are satisfied. In particular we assume the holonomy t_s to be central and represented by $h_o \in \overline{P}$. A direct treatment would show that an analogue of (8.12) would still hold, where, in the Verlinde sum (7.57), the term $\chi_{ph_o}(e^{\lambda/(p+c)}$ should apparently be replaced by $\epsilon_{w_{h_o}} \exp(2i\pi\langle h_o, \lambda\rangle)$. However Theorem 1.33 tells us that indeed, this is just $\chi_{ph_o}(e^{\lambda/(p+c)}$, so that we recover the standard Verlinde formula.

REMARK 8.5. By Teleman's vanishing results **[55, 56]**, if $(t_1, \ldots, t_s)$ verify (A), for any $p \in \mathbf{M}$, one should have

$$\operatorname{Ind}(D_{p,+}) = \dim H^0(\mathcal{M}, \lambda^p). \tag{8.13}$$

Since $\dim H^0(\mathcal{M}, \lambda^p)$ is given by Verlinde's formula, then one should have for any $p \in \mathbf{M}$,

$$\operatorname{Ind}(D_{p,+}) = V_{g,p+c}(pt_1, \ldots, pt_s). \tag{8.14}$$

The arguments of Teleman also show that more generally, if $\delta_1, \ldots, \delta_s$ are taken as in Theorem 6.36, for any $p \in \mathbf{M}$,

$$\operatorname{Ind}(D_{p,+}^{t_j+\delta_j}) = V_{g,p+c}(pt_1, \ldots, pt_s). \tag{8.15}$$

Here, Theorem 8.1 only asserts that if $(t_1, \ldots, t_s)$ verify $(\overline{A})$, for p large enough, (8.13) holds. For a given $p \in \mathbf{M}$, Theorems 8.2 and 8.3 give a corresponding equality for a suitable perturbation of the moduli space, depending on p, the perturbation being smaller as $p \to +\infty$. A proof of (8.14), (8.15) for any $p \in \mathbf{M}$ would be possible if one could modify Theorem 7.20, so that in (7.79) or (7.80), the term where the function M_q appears would be instead $M_q(\frac{q}{p}\sum_{j=1}^s w^j\theta_j, x)$. To establish such a result, one would need to prove the vanishing of a certain residue. This vanishing result is true for g large enough, but not obvious for small values of g.

References

[1] Alekseev A., Malkin A., Meinrenken E. : Lie group valued moment maps. *dg-gal/9707021* (1997).

[2] Atiyah M.F., Bott R. : The Yang-Mills equations over Riemann surfaces. *Phil. Trans. Roy. Soc. London*, **A 308**, 523-615 (1983).

[3] Beauville A. : Conformal blocks, fusion rules and the Verlinde formula. In *Proceedings of the Hirzebruch 65 Conference in Algebraic Geometry*, Ramat-Gan 1993, 75-96. Israël Math. Conf. Proc. **9**. Ramat-Gan: Bar-Ilan University 1996.

[4] Beauville A., Laszlo Y. : Conformal blocks and generalized theta functions. *Comm. Math. Phys.*, **164**, 385-419 (1994).

[5] Berline N., Vergne M. : Zéros d'un champ de vecteurs et classes caractéristiques équivariantes. *Duke Math. J.*, **50**, 539-549 (1983).

[6] Berline N., Getzler E., Vergne M. : *Heat kernels and Dirac operators*. Grundl. Math. Wiss. **298**. Berlin-Heldelberg-New-York: Springer 1992.

[7] Bertram A., Szenes A. : Hilbert polynomials of moduli spaces of rank 2 vector bundles. II. *Topology*, **32**, 599-609 (1993).

[8] Bismut, J.-M. : The index Theorem for families of Dirac operators : two heat equation proofs. Invent. Math. ,**83**, 91-151 (1986).

[9] Bismut J.-M. : Equivariant immersions and Quillen metrics. *J. Diff. Geom.*, **41**,53-157 (1995).

[10] Bismut, J.-M., Freed D. : The analysis of elliptic families. I. Metrics and connections on determinant bundles. *Comm. Math. Phys.*, **106**, 159-176 (1986). II. Eta invariants and the holonomy theorem. *Comm. Math. Phys.* , **107**, 103-163 (1986).

[11] Bismut J.-M., Gillet H., Soulé C. : Analytic torsion and holomorphic determinant bundles I. *Comm. Math. Phys.*, **115**, 49-78 (1988). II. *Comm. Math. Phys.*, **115**, 79-126 (1988). III. *Comm. Math. Phys.*, **115**, 301-351 (1988).

[12] Bismut J.-M., Labourie F. : Formules de Verlinde pour les groupes simplement connexes et géométrie symplectique. *C.R. Acad. Sci. Paris*, **325**, Série I, 1009-1014 (1997).

[13] Bismut J.-M., Zhang W.P. : *An extension of a theorem by Cheeger and Müller*. Astérisque **205**. Paris SMF 1992.

[14] Bourbaki N. : *Eléments de Mathématique*. Chapitre 9. Paris: Masson 1982.

[15] Bröcker T., Dieck T. : *Representations of compact Lie groups*. Graduate texts in Mathematics **98**. Berlin-Heidelberg-New-York: Springer 1995.

[16] Chang S.X.: Fixed point formulas and loop group actions. *alg-geom/9812148* (1996).

[17] Donaldson S.K. : Gluing techniques in the cohomology of moduli spaces. In *Topological methods in modern mathematics*, L. Goldberg and A. Phillips eds., 137-170. Houston: Publish or Perish 1993.

[18] Donaldson S.K., Kronheimer P. : *The geometry of Four-Manifolds*. Oxford: Clarendon Press 1990.

[19] Duistermaat J.J., Heckman G. : On the variation of the cohomology of the reduced phase space. *Invent. Math.*, **69**, 256-268 (1982). Addendum , **72**, 153-158 (1983).

[20] Faltings G. : A proof of the Verlinde formula. *J. of Alg. Geom.* **3**, 347-374 (1994).

[21] Frenkel I.B. : Orbital theory for affine Lie algebras. *Invent. Math.*, **77**, 301-352 (1984).

[22] Fuchs J. : *Affine Lie Algebras and Quantum Groups*. Cambridge Monographs in Math. Phys.. Cambridge: Cambridge Univ. Press 1992.

[23] Guillemin V., Sternberg S. : Geometric quantization and multiplicities of group representations. *Invent. Math.*, **67**, 515-538 (1982).

[24] Guillemin V., Sternberg S. : *Symplectic techniques in Physics*. Cambridge: Cambridge University Press 1990.

[25] Guruprasad K., Huebschmann J., Jeffrey L., Weinstein A.: Group systems, groupoids, and moduli spaces of parabolic bundles. *Duke Math. J.*, **89**, 377-412 (1997).

[26] Hörmander L. : *The analysis of linear partial differential operators, Vol. 1*, Grundl. der Math. Wiss. Band **256**. Berlin-Heidelberg-New-York: Springer 1983.

[27] Jeffrey L. : Extended moduli spaces of flat connections on Riemann surfaces. *Math. Annalen* **298**, 667-692 (1994).

[28] Jeffrey L., Kirwan F. : Localization for non abelian group actions. *Topology*, **34**, 291-327 (1995).

[29] Jeffrey-Kirwan F. : Localization and the quantization conjecture. *Topology*, **36**, 647-693 (1977).

[30] Jeffrey L., Kirwan F. : Intersection theory on moduli spaces of holomorphic bundles of arbitrary rank on a Riemann surface. *alg-geom*/9608029 (1996).

[31] Jeffrey L., Weitsman J. : Bohr-Sommerfeld orbits in the moduli space of flat connections and the Verlinde dimension formula. *Comm. Math. Phys.* **150**, 593-630 (1992).

[32] Kawasaki T. : The Riemann-Roch theorem for complex V-manifolds. *Osaka J. Math.*, **16**, 151-159 (1979).

[33] Kawasaki T. : The index of elliptic operators on V-manifolds. *Nagoya Math. J.*, **9**, 135-157 (1981).

[34] Kirillov A.A. : *Elements of the theory of representations*. Grundl. Math. Wiss. **220**. Berlin-Heidelberg-New-York: Springer 1976.

[35] Knudsen, F.F., Mumford, D. : The projectivity of the moduli space of stable curves, I: Preliminaries on "det" and "div". *Math Scand.* ,**39**, 19-55 (1976).

[36] Kostant B. : On Macdonald's η-function formula, the Laplacian and generalized exponents. *Adv. in Math.*, **20**, 179-212 (1976).

[37] Kumar S., Narasimhan M.S. : Picard group of the moduli space of G-bundles. *Math. Annal.*, **308**, 155-173 (1997).

[38] Kumar S., Narasimhan M.S., Ramanathan A. : Infinite Grassmannian and moduli spaces of G-bundles. *Math. Annal.*, **300**, 41-75 (1994).

[39] Liu K. : Heat kernel and moduli spaces . *Math. Res. Letters*, **3**, 743-762 (1996).

[40] Liu K. : Heat kernel and moduli spaces II. *Math. Res. Letters*, **4**, 569-588 (1997).

[41] Meinrenken E. : On Riemann-Roch formulas for multiplicities. *J. Am. Math. Soc.* , **9**, 373-389 (1996).

[42] Meinrenken E. : Symplectic surgery and the Spin^c Dirac operator. *Adv. in Math.*, to appear.

[43] Meinrenken E., Woodward C. : A symplectic proof of Verlinde factorization. *dg-ga*/9612018 (1996).

[44] Meinrenken E., Woodward C. : Fusion of Hamiltonian loop group manifolds and cobordism. Preprint, July 1997.

[45] Milnor J.M., Stasheff J.D. : *Characteristic classes*. Annals of Mathematics Studies **76**. Princeton: Princeton University Press 1974.

[46] Narasimhan M.S., Seshadri C.S. : Stable and unitary vector bundles on a compact Riemann surface. *Ann. Math.*, **82**, 540-567 (1965).

[47] Pressley A., Segal G. : *Loop groups*. Oxford Mathematical Monographs. Oxford: Clarendon Press 1986.

[48] Ramadas T.R., Singer I.M., Weitsman J. : Some comments on Chern-Simons theory. *Comm. Math. Phys.* **126**, 409-420 (1989).

[49] Reidemeister K. : Homotopieringe und Linsenraüme. *Hamburger Abhandl.*, **11**, 102-109 (1935).

[50] Rossmann W. : Kirillov's character formula for reductive Lie groups. *Invent. Math.*, **48**, 207-220 (1978).

[51] Serre J.P. : *Cours d'arithmétique*. Paris: Presses Universitaires de France 1970.

[52] Szenes A. : Hilbert polynomials of moduli spaces of rank 2 vector bundles. I. *Topology*, **32**, 587-597 (1993).

[53] Szenes A. : The combinatorics of the Verlinde formula. In *Vector bundles in Algebraic geometry*, 241-253. LMS Lecture Notes Series **208**. Cambridge: Cambridge University Press 1995.

[54] Szenes A. : Iterated residues and multiple Bernouilli polynomials. *Preprint* MIT 1998.

[55] Teleman C. : Borel-Weil-Bott theory on the moduli stack of G-bundles over a curve. *Invent. Math.*, **134**, 1-57 (1998).

[56] Teleman C. : The quantization conjecture revisited. *alg-geom*/9808029 v2 (1998).

[57] Thaddeus M. : Stable pairs, linear systems and the Verlinde formula. *Inv. Math.* **117**, 317-353 (1994).

[58] Tian Y., Zhang W. : An analytic proof of the geometric quantization conjecture of Guillemin-Sternberg. *Invent. Math.*, **132**, 229-259 (1998).

[59] Tsuchiya, A., Ueno K., Yamada Y. : Conformal field theory and universal family of stable curves with gauge symmetry. *Adv. Studies in Pure Math.*, **19**, 459-566 (1989).

[60] Vergne M. : On Rossmann's character formula for discrete series. *Invent. Math.*, **54**, 11-14 (1979).

[61] Vergne M. : A note on the Jeffrey-Kirwan-Witten localization formula. *Topology*, **35**, 243-266 (1996).

[62] Verlinde E. : Fusion rules and modular transformations in $2D$-conformal field theory. *Nuclear Physics B*, **300**, 360-376 (1988).

[63] Witten E. : On quantum gauge theories in two dimensions. *Comm. Math. Phys.* **141**, 153-209 (1991).

[64] Witten E. : Two dimensional gauge theories revisited. *J. Geom. Phys.* **9**, 303-368 (1992).

Département de Mathématique, Université Paris-Sud, Bâtiment 425, 91405 Orsay, France
E-mail address: bismut@topo.math.u-psud.fr

Département de Mathématique, Université Paris-Sud, Bâtiment 425, 91405 Orsay, France
E-mail address: labourie@topo.math.u-psud.fr

Surveys in Differential Geometry, vol. 5

Counting curves on irrational surfaces

Jim Bryan
Naichung Conan Leung

ABSTRACT. In this paper we survey recent results and conjectures concerning enumeration problems on irrational surfaces.

CONTENTS

1.	Introduction	313
2.	Counting curves via the method of Yau and Zaslow	314
3.	Problems and conjectures	316
4.	Gromov-Witten invariants	319
5.	Modified invariants and the case of $K3$ and Abelian surfaces	325
6.	Acknowledgments	335
References		335

1. Introduction

Problems in enumerative algebraic geometry associated to counting curves on projective varieties have been heavily influenced in recent years by the introduction of stable maps, Gromov-Witten invariants, and quantum cohomology. Using these new ideas as well as classical methods, much progress has been made for many enumeration problems on rational surfaces.

In this survey paper, we focus on the situation for irrational surfaces. The first breakthrough in this direction was the work of Yau and Zaslow [**91**] in 1995 which gave a formula for the number of rational curves on a $K3$ surface. Their formula expresses a generating function for the number of curves as a modular form. Their method is motivated from mirror symmetry considerations and it is strikingly different both the "classical" algebro-geometric approach and the methods using quantum cohomology or Gromov-Witten invariants.

Methods from quantum cohomology and Gromov-Witten theory are not very useful in solving enumeration problems on most irrational surfaces. For instance, a generic $K3$ surface has no curves at all and so its ordinary Gromov-Witten invariants are zero. Nevertheless, some modifications of Gromov-Witten theory can be made and have been recently applied with success in the case of $K3$ and Abelian surfaces.

In [**34**], Göttsche gave an intriguing generalization of the Yau-Zaslow formula which conjecturally applies to any surface and any genus. The Göttsche-Yau-Zaslow formula has been verified to order eight by the work of Vainsencher [**82**] and Kleiman-Piene [**51**] and it has also been proved in the case of $K3$ and Abelian surfaces [**15**][**17**]. The conjecture is also consistant with various recursive computations on rational surfaces [**18**][**83**]. The Göttsche-Yau-Zaslow formula conjecturally provides a very nice answer to a fairly general set of enumeration problems.

Our survey is organized as follows.

We begin in Section 2 by describing the method of counting rational curves due to Yau and Zaslow which led to their discovery of the presence of modular forms in enumeration problems on surfaces.

In Section 3 we formulate the various kinds of enumeration problems on surfaces and then we focus on the primary problem of interest: counting curves in a linear system passing through a fixed number of points. We describe in detail Göttsche's generalization of the Yau-Zaslow formula.

In Section 4 we start with a short general exposition of the Gromov-Witten invariants including a general enough version to include families of symplectic structures. We then discuss the problems with the Gromov-Witten invariants on surfaces whose geometric genus and/or irregularity is non-zero. We discuss Taubes' "Seiberg-Witten equals Gromov-Witten" theorem and describe its relation to enumerative geometry. Finally, we discuss when the Gromov-Witten invariants are "enumerative".

In Section 5 we give an expository account of our computation of modified Gromov-Witten invariants to prove the Göttsche-Yau-Zaslow formula for $K3$ and Abelian surfaces. We include a description of how to use a "matching technique" to compute the contribution of multiple covers of nodal rational curves to the invariants. We end the section with a brief description of recent work of Behrend and Fantechi who give a purely algebraic modification of the Gromov-Witten invariants that generalize the (non-algebraic) modifications used in the $K3$ and Abelian surface case.

2. Counting curves via the method of Yau and Zaslow

The first big breakthrough for counting curves on irrational surfaces came in the 1995 paper of Yau and Zaslow [**91**] who discovered the unexpected link between modular forms and enumerating curves. In [**91**], Yau and Zaslow describe a method to count rational curves on $K3$ surfaces (for an exposition see Beauville [**8**]). They found that the numbers are given by the coefficients of the series

$$\frac{q}{\Delta(q)} = \prod_{m=1}^{\infty}(1-q^m)^{-24}. \tag{1}$$

They prove this formula under the assumption that all the rational curves are reduced, irreducible, and nodal. More generally, if the curves are all irreducible and reduced (but possibly having complicated singularities), their argument counts curves with multiplicities that are shown by Fantechi-Göttsche-van Straten [**26**] to be related to Gromov-Witten multiplicities (see also Chen [**19**]).

The formula was later generalized by Göttsche [**34**] to a conjectural formula that applies to all surfaces and any genus (see Section 3).

Although it is not clear if Yau and Zaslow's argument can be generalized to other situations, it is so beautiful and strikingly different that we feel it is worthwhile to describe it here.

Let $\mathbf{P}^r$ be an r-dimensional linear system on a surface X with a finite number of rational curves (for example $\mathbf{P}^r$ could be the sublinear system of a complete linear system $|C|$ obtained by imposing the appropriate number of point conditions). Consider the compactified family of Jacobians:

$$\pi : \mathcal{J} \to \mathbf{P}^r$$

so that $\pi^{-1}(p)$ is $\mathrm{Jac}(C)$ if p is a point representing a smooth divisor C and if C is singular then $\pi^{-1}(p)$ is $\overline{\mathrm{Jac}}(C)$, a compactification of the Jacobian of C (such a family exists by [**4**][**3**], c.f. [**1**][**2**]). Yau and Zaslow show that if we assume all the rational curves in the linear system are nodal, then

$$\#\text{ of rational curves in the linear system } \mathbf{P}^r = \text{Euler characteristic}(\mathcal{J}).$$

The crucial observation is that the fibers $\overline{\mathrm{Jac}}(C)$ have Euler characteristic zero unless C is rational[1]. We denote the Euler characteristic of a space M by $e(M)$. Recall that the Euler characteristic of a fiber bundle is the product of the Euler characteristics of the fiber and the base and if $X = U \cup U^c$ is a disjoint union of an algebraic space X into a Zariski open set U and its compliment, then $e(X) = e(U) + e(U^c)$. The linear system $\mathbf{P}^r$ has a natural stratification given by the geometric genus of the corresponding divisor. The map $\pi : \mathcal{J} \to \mathbf{P}^r$ is a fiber bundle restricted to each strata so that $e(\mathcal{J})$ is given as the sum over all the strata of the product of the Euler characteristic of the fiber times the Euler characteristic of the corresponding strata.

We see that only the strata corresponding to rational curves contribute to the Euler characteristic. Consequently, we get

$$e(\mathcal{J}) = \sum_{\text{rational curves } C \text{ in } \mathbf{P}^r} e(\overline{\mathrm{Jac}}(C)).$$

If C is a nodal rational curve, then $e(\overline{\mathrm{Jac}}(C)) = 1$ and so $e(\mathcal{J})$ is exactly the number of rational curves. In general, $e(\mathcal{J})$ counts the rational curves with multiplicities given by $e(\overline{\mathrm{Jac}}(C))$. Fantechi-Göttsche-van Straten [**26**] show that if C is irreducible, then $e(\overline{\mathrm{Jac}}(C))$ coincides with the length of the (zero dimensional) moduli space of genus 0 stable maps with image C and so the Yau-Zaslow multiplicities agree with the multiplicities arising in Gromov-Witten theory (in the language of Subsection 4.4 we can thus say $e(\mathcal{J})$ is *weakly enumerative*).

In general, there is no easy way to compute $e(\mathcal{J})$, but for $K3$ surfaces we can utilize some very special properties of $\mathcal{J}$. A r-dimensional complete linear system $|C|$ on a $K3$ surface X has a finite number of rational curves. If we assume that the linear system consists of only reduced and irreducible curves (for example, if $\mathcal{O}(C)$ generates $\mathrm{Pic}(X)$), then the associated family of compactified Jacobians

---

[1] This is because $\overline{\mathrm{Jac}}(C)$ fibers over $\mathrm{Jac}(\tilde{C})$, the Jacobian of the normalization and $e(\mathrm{Jac}(\tilde{C})) = 0$ if $g(\tilde{C}) \neq 0$, *i.e.* C is not rational (see section 2 in Beauville [**8**]).

$\mathcal{J} \to |C|$ is a *smooth hyperkähler* manifold (see Mukai [**68**] ex. 0.5). Furthermore, it is birational[2] to $X^{[r]}$, The Hilbert scheme of r points on X. $X^{[r]}$ is also a *smooth hyperkähler* manifold and by a result of Batyrev [**7**], the Betti numbers (and hence the Euler characteristics) of two birationally equivalent, smooth hyperkähler manifolds agree.

The Euler characteristic of the Hilbert scheme $X^{[r]}$ was determined in [**35**] by Göttsche using Deligne's proof of the Weil conjectures. Equation 1 then follows from his calculation.

3. Problems and conjectures

In this section we describe the general set up for enumeration problems on surfaces and we explain the Göttsche-Yau-Zaslow formula which conjecturally gives the answer to a very general set of enumerative problems. The discussion of this section for the most part applies equally well to rational surfaces as well as irrational.

3.1. Formulation of the problem. Since the only interesting subvarieties of a fixed algebraic surface X are curves, the general enumeration problem for X is to count the number of curves on X satisfying some set of prescribed properties. It is natural to begin by fixing the geometric genus g of the curves to be counted and to fix the homology class $[C] \in H_2(X,\mathbf{Z})$ of their image[3]. The set of curves with fixed geometric genus and homology class will in general form a positive dimensional family and so to get a well defined counting problem one imposes additional conditions. Some typical conditions on a curve C are given below:

Point: Require C to pass through a prescribed set of fixed points. The condition that a curve pass through a single fixed point is a codimension one condition on the family of all curves.

FLS: Require C to lie in a fixed linear system (FLS), by considering C as a divisor. Equivalently, fix the holomorphic structure on the line bundle $\mathcal{O}(C)$. For irregular surfaces ($H^1(X,\mathcal{O}) \neq 0$), this imposes a non-trivial constraint of codimension equal to $\dim H^1(X,\mathcal{O})$.

Loop: Require C to pass through fixed loops in X representing non-trivial elements in $H^1(X;\mathbf{Z})$ (this also is a non-trivial constraint only on irregular surfaces). Each loop imposes a *real* codimension one condition on the family of curves. In Section 4 we will show that these loop constraints are directly related to the FLS constraint (see Theorem 4.1).

Multi-point: Require C to pass through a fixed set of points with a prescribed set of multiplicities. This can be reformulated as a homological condition on the blown-up manifold; *i.e.* curves on X in the class $[C]$ passing through fixed

[2]The birational morphism between $\mathcal{J}$ and $X^{[r]}$ can be seen as follows. The generic point of $\mathcal{J}$ corresponds to a smooth genus r curve $C \subset X$ and an element of Jac(C). Since Jac(C) is generically isomorphic to Sym$^r(C)$ the generic point in $\mathcal{J}$ gives us r (unordered) points on C and hence r points in X. Conversely, r generic points in X are contained in a unique smooth genus r curve C in $|C|$; the points on C then further determine an element of Jac(C) via the map Sym$^r(C) \to$ Jac(C).

[3]Notation: if C is a curve in X, then we denote its homology class in $H_2(X,\mathbf{Z})$ by $[C]$, its Poincaré dual in $H^2(X,\mathbf{Z})$ by $[C]^\vee$, the corresponding line bundle $\mathcal{O}(C)$, and the linear system $|C| := \mathbf{P}(H^0(X,\mathcal{O}(C)))$. We will also use the shorthand $C^2 := [C]\cdot[C]$ and $KC := c_1(TX)([C])$. In subscripts for moduli spaces and invariants we will just write C for the homology class, *e.g.* $\mathcal{M}_{g,C}(X) := \mathcal{M}_{g,[C]}(X)$.

points $x_1, \ldots, x_l$ with prescribed multiplicities α_i are in one-to-one correspondence with curves on $Bl_{\{x_1,\ldots,x_l\}}X$ in the class $[C] - \sum_{i=1}^{l} \alpha_i[E_i]$ where $Bl_{\{x_1,\ldots,x_l\}}(X)$ is the blow-up of X at the points $x_1, \ldots, x_l$ and $E_1, \ldots, E_l$ are the the exceptional divisors.

Tangency: Fix an auxiliary smooth curve D and require that C meet D with a prescribed degree of tangency. This can be reformulated in terms of curves on a certain family of blow-ups (see [**16**]). See also [**24**][**25**] where the problem was considered on $\mathbf{P}^2$ and attacked using certain stable maps to the incidence variety.

CX structure: One can also impose conditions on the complex structure of C itself. For example, one can count only curves with a fixed complex structure; for $g \geq 2$, this imposes $3g-3$ constraints (for example see [**42**],[**69**]). For another example, one can count only those curves that are hyper-elliptic. The expected number of constraints this imposes is $g-2$ since the codimension of the hyper-elliptic locus in the moduli space of curves is $g-2$ (for example see [**37**]).

For the most part, we will focus on the most straight forward problem: counting genus g curves in a fixed linear system passing through a fixed set of points (with multiplicity one), although we will comment on the other kinds of constraints as they come up. We formalize the problem below:

Let C be a curve on a surface X and let $|C|$ denote the corresponding complete linear system and let K denote the canonical class. We assume that $\mathcal{O}(C-K)$ is ample so that the dimension of $|C|$ is determined by the Riemann-Roch formula.

Define the Severi variety $V_g(C)$ to be the closure of the set of curves in $|C|$ with geometric genus g. The condition that a divisor passes through a fixed point imposes a linear condition on $|C|$. We can thus interpret the degree of $V_g(C) \subset |C|$ as the number of genus g curves in $|C|$ passing through r points where r is the expected dimension of $V_g(C)$. The expected dimension of $V_g(C)$ is given by

$$\begin{aligned} \dim V_g(C) &= r \\ &= -KC + g - 1 + p_g - q \\ &= -KC + g - 2 + \chi(\mathcal{O}_X) \end{aligned}$$

where $p_g = \dim H^2(X, \mathcal{O})$, $q = \dim H^1(X, \mathcal{O})$, and $\chi(\mathcal{O}_X) = 1 - q + p_g$ is the holomorphic Euler characteristic of $\mathcal{O}_X$.

Main Enumeration Problem: The main problem we consider is computing $N_g(X, C)$ which is defined to be the number (when finite) of genus g curves in the linear system $|C|$ passing through $r = -KC + g - 1 + p_g - q$ generic points. Equivalently, $N_g(X, C)$ is the degree of $V_g(C)$ (when $V_g(C)$ is of the expected dimension).

3.2. The Göttsche-Yau-Zaslow formula. *A priori*, $N_g(X, C)$ depends heavily on the complex structure of X and the choice of the linear system $|C|$. Under sufficient ampleness conditions on C it is conjectured (essentially in Vainsencher [**82**]) that $N_g(X, C)$ only depends on the numbers g, C^2, CK, K^2, and $c_2(X)$, where $c_2(X)$ is the Euler characteristic of X.

Göttsche has found a remarkable generalization of the Yau-Zaslow formula. The formula is a generating function for the numbers $N_g(X, C)$ in terms of five universal power series and the numbers g, C^2, CK, K^2, and $c_2(X)$. Three of the five universal power series are describes explicitly as (quasi-)modular forms while the remaining two series have coefficients that can be determined recursively.

CONJECTURE 1 ([**34**]). *Let C be a sufficiently ample*[4] *divisor on X. Then $N_g(X,C)$, the number of genus g curves in $|C|$ passing through*

$$r = -KC + g - 2 + \chi(\mathcal{O}_X)$$

points, is given as the coefficient of $q^{\frac{1}{2}C(C-K)}$ in the following power series in q:

$$B_1^{K^2} B_2^{CK} (DG_2)^r \frac{D^2G_2}{(\Delta(D^2G_2))^{\chi(\mathcal{O}_X)/2}} \tag{2}$$

where $D = q\frac{d}{dq}$, G_2 is the Eisenstein series:

$$G_2(q) = -\frac{1}{24} + \sum_{k>0}(\sum_{d|k} d)q^k,$$

Δ is the discriminant:

$$\Delta(q) = q\prod_{k>0}(1-q^k)^{24},$$

and $B_i(q)$ are universal power series whose first terms are

$$\begin{aligned} B_1(q) &= 1 - q - 5q^2 + 39q^3 - 345q^4 + \cdots \\ B_2(q) &= 1 + 5q + 2q^2 + 35q^3 - 140q^4 + \cdots \end{aligned}$$

*(see [**34**] for the coefficients of B_i to order 20).*

Note that when K is numerically trivial (X is a $K3$, Abelian, Enriques, or hyperelliptic surface), then the power series (2) does not depend on B_1 or B_2 and so is given by an explicit (quasi-)modular form. In the case when X is a $K3$ or Abelian surface, the conjecture was proved in [**15**] and [**17**] to hold for *all* C representing a primitive homology class (using a slightly modified definition of $N_g(X,C)$, see section 5).

The impetus for Göttsche's generalization stemmed largely from two sources: the work of Vainsencher [**82**] (and its subsequent generalization by Kleiman and Piene [**51**]), and the formula of Yau and Zaslow [**91**].

In [**82**], Vainsencher gives universal polynomials that count the number of curves with 6 or fewer nodes, passing through the appropriate number of points, in a sufficiently ample linear system on *any* surface. In other words, he computes $N_g(X,C)$ for any g, X, and C provided that C is sufficiently ample and

$$\frac{1}{2}C(C+K) - g \leq 5.$$

Kleiman and Piene have refine the methods of Vainsencher to extend his results up to eight nodes. They also provide explicit bounds for the power of the ample class required to guarantee that the formulas hold. The methods used to obtain these results are classical in the sense that they do not use physics, quantum cohomology, or even stable maps. They also provide precise enumerative information as oppose to

[4]For the precise formulation of the ampleness condition see [**34**]. Depending on X, the conjecture is expected to hold under weaker assumptions on C.

the "virtual" or "weakly enumerative" count sometimes determined by the Gromov-Witten invariants (c.f. subsection 4.4), in particular, they show that when the ampleness conditions are satisfied, all the curves passing through a generic choice of points are nodal. We refer the reader to **[82]** and **[51]** for more details.

Göttsche's crucial observation is that if there exist universal polynomial formulas for $N_g(X, C)$ that apply to any surface, then they must satisfy very strong multiplicative properties. The reason is that they should also apply to disconnected surfaces, and Göttsche shows that the obvious relationship

$$N_g(X_1 \coprod X_2, C_1 \coprod C_2) = \sum_{g_1+g_2=g} N_{g_1}(X_1, C_1) N_{g_2}(X_2, C_2)$$

forces the formulas to take on a very special form. In particular, the polynomials for $N_g(X, C)$ must be determined by five universal power series.

On the other hand, it was the work of Yau and Zaslow that suggested that the universal power series may be related to modular forms. Using ideas from physics, Yau and Zaslow gave predictions for the number of rational curves with n nodes on a $K3$ surface (see Section 2). The numbers appear as the coefficients of the Fourier expansion of a well known modular form, the discriminant. This work is in a sense complimentary to the Vainsencher work; while Vainsencher's formulas apply to any surface but for only a small number of nodes, the Yau-Zaslow formula applies to only genus 0 and a $K3$ surface, but to any number of nodes. Thus, while Vainsencher's formulas determine the first few coefficients of each of the five power series, the Yau-Zaslow formula provides a closed form for a certain product of three of the power series.

By building on this knowledge along with other known results (particularly the recursive formulas of Caporaso-Harris **[18]** and Vakil **[83]**) and using some remarkable pattern recognition, Göttsche arrived at Conjecture 1 and verified it to a fairly high degree of redundancy. Closed formulas for the series B_1 and B_2 are unknown, but a recursive scheme for the coefficients can be derived from the Caporaso-Harris or Vakil formulas.

One intriguing aspect of the conjecture is the appearance of modular forms. The underlying "reason" for the modularity is currently a mystery.

4. Gromov-Witten invariants

In this section we begin with a short general exposition of the Gromov-Witten invariants including a general enough version to include families of symplectic structures. We then discuss the problems with the Gromov-Witten invariants on surfaces whose geometric genus and/or irregularity is non-zero. We discuss Taubes' "Seiberg-Witten equals Gromov-Witten" theorem and describe its relation to enumerative geometry. Finally, we discuss when the Gromov-Witten invariants are "enumerative".

Many papers have been written on Gromov-Witten theory, for the reader's convenience we give an extensive list in the bibliography: **[63]** **[61]** **[62]** **[43]** **[12]** **[6]** **[73]** **[85]** **[69]** **[50]** **[52]** **[87]** **[67]** **[40]** **[30]** **[11]** **[65]** **[27]** **[89]** **[49]** **[21]** **[9]** **[33]** **[66]** **[39]** **[86]** **[14]** **[57]** **[48]** **[46]** **[70]** **[54]** **[32]** **[29]** **[74]** **[23]** **[76]** **[58]** **[56]** **[38]** **[31]** **[75]** **[20]** **[13]** **[47]** **[45]** **[41]** **[53]** **[64]** **[72]** **[71]** **[55]**.

4.1. Gromov-Witten invariants. Gromov-Witten invariants have their origins in symplectic geometry and conformal field theory but have been recently defined purely algebro-geometrically **[11][62]**. The basic object of study is the moduli space $\mathcal{M}_{g,n,C}(X)$ of ***stable maps*** of n-marked, genus g curves to X in the class C. In general, Gromov-Witten invariants are certain intersection numbers of cycles on $\mathcal{M}_{g,n,C}(X)$ which are shown to be invariant under deformations of the (almost) Kähler structure of X.

We briefly outline the framework of the Gromov-Witten invariants in order to fix notations and we refer the reader to (for example) **[5]** or **[28]** for complete accounts. We include here a straight forward generalization of the usual framework to include families of (almost) Kähler structures (see **[15][17]**).

Let (X,ω) be any compact (almost) Kähler manifold. Recall that an n-marked, genus g ***stable map*** of degree $[C] \in H_2(X,\mathbf{Z})$ is a (pseudo)-holomorphic map $f : \Sigma \to X$ from an n-marked nodal curve $(\Sigma, x_1, \ldots, x_n)$ of geometric genus g to X with $f_*([\Sigma]) = [C]$ that has no infinitesimal automorphisms. Two stable maps $f : \Sigma \to X$ and $f' : \Sigma' \to X$ are equivalent if there is a biholomorphism $h : \Sigma \to \Sigma'$ such that $f = f' \circ h$. We write $\mathcal{M}_{g,n,C}(X,\omega)$ for the moduli space of equivalence classes of genus g, n-marked, stable maps of degree $[C]$ to X. We will often drop the ω or X from the notation if they are understood and we sometimes will drop the n from the notation when it is 0. If B is a family of almost Kähler structures, we denote parameterized version of the moduli space:

$$\mathcal{M}_{g,n,C}(X,B) = \coprod_{t\in B} \mathcal{M}_{g,n,C}(X,\omega_t).$$

If B is a compact, connected, oriented manifold then $\mathcal{M}_{g,n,C}(X,B)$ has a fiduciary cycle $[\mathcal{M}_{g,n,C}(X,B)]^{vir}$ called the virtual fundamental cycle (see **[15]** and the fundamental papers of Li and Tian **[62][59][60]**, or alternatively Behrend-Fantechi and Siebert **[11][77]**). The dimension of the cycle is

$$\dim_{\mathbf{R}}[\mathcal{M}_{g,n,C}(X,B)]^{vir} = -2KC + (6 - \dim_{\mathbf{R}} X)(g-1) + 2n + \dim_{\mathbf{R}} B.$$

The invariants are defined by evaluating cohomology classes of $\mathcal{M}_{g,n,C}$ on the virtual fundamental cycle. The cohomology classes are defined via incidence relations of the maps with cycles in X. The framework is as follows. There are maps

$$\begin{array}{ccc} \mathcal{M}_{g,1,C} & \xrightarrow{ev} & X \\ \big\downarrow{\scriptstyle ft} & & \\ \mathcal{M}_{g,C} & & \end{array}$$

called the ***evaluation*** and ***forgetful*** maps defined by $ev(\{f : (\Sigma, x_1) \to X\}) = f(x_1)$ and $ft(\{f : (\Sigma, x_1) \to X\}) = \{f : \Sigma \to X\}$.[5] The diagram should be regarded as the universal map over $\mathcal{M}_{g,C}$.

Given geometric cycles $\alpha_1, \ldots, \alpha_l$ in X representing classes $[\alpha_1], \ldots, [\alpha_l] \in H_*(X,\mathbf{Z})$ with Poincaré duals $[\alpha_1]^\vee, \ldots, [\alpha_l]^\vee$, we can define the Gromov-Witten

[5]There is some subtlety to making this definition rigorous since forgetting the point may make a stable map unstable, but it can be done.

invariant

$$(3) \qquad \Phi_{g,C}^{(X,B)}(\alpha_1,\dots,\alpha_l) = \int_{[\mathcal{M}_{g,C}(X,B)]^{vir}} ft_*ev^*([\alpha_1]^\vee) \cup \dots \cup ft_*ev^*([\alpha_l]^\vee).$$

$\Phi_{g,C}^{(X,B)}(\alpha_1,\dots,\alpha_l)$ counts the number of genus g, degree $[C]$ maps which are pseudo-holomorphic with respect to some almost Kähler structure in B and such that the image of the map intersects each of the cycles $\alpha_1,\dots,\alpha_l$.[6] The Gromov-Witten invariants are multi-linear in the α's and they are symmetric for α's of even degree and skew symmetric for α's of odd degree. If $p_1,\dots,p_k$ are points in a path-connected X, we will use the shorthand

$$\Phi_{g,C}^{(X,B)}(\text{pt.}^k,\alpha_{k+1},\dots,\alpha_l) := \Phi_{g,C}^{(X,B)}(p_1,\dots,p_k,\alpha_{k+1},\dots,\alpha_l).$$

Now suppose that X is a Kähler surface. The only non-trivial constraints arising from intersecting with cycles are those coming from zero and one dimensional cycles, *i.e.* points and loops. The constraints imposed by intersecting two dimensional cycles (divisors) are determined purely homologically (the so-called "divisor equation") and cycles of dimension three and four impose no constraints.

The **loop** constraints (C passes through a fixed set of loops) are related to the **FLS** constraint (fixing the linear system of C) by the following:

THEOREM 4.1 (Thm. 2.1 of [**17**]). *Let $\gamma_1,\dots,\gamma_{b_1}$ be loops representing an oriented basis of $H_1(X,\mathbf{Z})$. Then the invariant:*

$$\Phi_{g,C}(\gamma_1,\dots,\gamma_{b_1},pt.^l)$$

counts the number of genus g maps whose image lies in the fixed linear system $|C|$ and passes through l points.

We can formulate a more precise version of this. In order to count curves in a fixed linear system $|C|$ one would like to restrict the integral of Equation 3 to the cycle defined by $\Psi_{\Sigma_0}^{-1}(0)$ where Ψ_{Σ_0} is the map

$$\Psi_{\Sigma_0} : \mathcal{M}_{g,C}(X,\omega) \to \mathrm{Pic}^0(X)$$

given by

$$f \mapsto \mathcal{O}(\mathrm{Im}(f) - \Sigma_0)$$

where $\Sigma_0 \in |C|$ is a fixed divisor. Dually, one can add the pullback by Ψ_{Σ_0} of the volume form on $\mathrm{Pic}^0(X)$ to the integrand defining the invariant:

$$\int_{[\mathcal{M}_{g,C}(X)]^{vir}} \Psi_{\Sigma_0}^*([\text{pt.}]^\vee) \cup ft_*ev^*([\alpha_1]^\vee) \cup \dots \cup ft_*ev^*([\alpha_l]^\vee).$$

The class $\Psi_{\Sigma_0}^*([\text{pt.}]^\vee)$ can be expressed in terms of classes arising from the constraints imposed by loops:

THEOREM 4.2. *Let X be a Kähler surface and let $[\gamma] \in H_1(X,\mathbf{Z})$ and let $\tilde{\gamma}$ be the corresponding class in $H^1(\mathrm{Pic}^0(X),\mathbf{Z})$ induced by the identification $\mathrm{Pic}^0(X) \cong H^1(X,\mathbf{R})/H^1(X,\mathbf{Z})$. Then*

$$\Psi_{\Sigma_0}^*(\tilde{\gamma}) = ft_*ev^*([\gamma]^\vee).$$

COROLLARY 4.3. $\Psi_{\Sigma_0}^*([pt.]^\vee) = ft_*ev^*([\gamma_1]^\vee) \cup \dots \cup ft_*ev^*([\gamma_{b_1}]^\vee)$.

[6]The integral is defined to be 0 if the integrand is not a form of the correct degree.

PROOF: Theorem 4.2 is proved in the appendix of [**17**]. Roughly the idea is this: We view classes in $H^1(\mathcal{M}_{g,C}(X);\mathbf{Z})$ as homotopy classes of circle valued functions on $\mathcal{M}_{g,C}(X)$. The values of a certain circle valued function on $\mathcal{M}_{g,C}(X)$ representing the class $ft_*ev^*([\gamma]^\vee)$ are given as integrals of a form representing $[\gamma]^\vee$ over 3-cycles that are obtained by sweeping out a path of curves in X. On the other hand, it is shown that the values of a circle valued function representing $\Phi^*_{\Sigma_0}(\tilde{\gamma})$ are given by an integral over X of a certain 1-form wedged with $[\gamma]^\vee$. The equality of the two circle valued functions is then established with essentially a residue calculation.

The corollary follows immediately and is essentially a restatement of Theorem 4.1.

4.2. Difficulties with ordinary Gromov-Witten invariants. Gromov-Witten invariants have been remarkably effective in answering many questions in enumerative geometry for rational surfaces (see for example [**84**][**83**][**36**][**56**][**24**][**37**] and many others). Rational surfaces all have $p_g = q = 0$; however, the ordinary Gromov-Witten invariants are not very effective for counting curves when p_g or q are not zero[7]. One basic reason is that the moduli space of stable maps fails to be a good model for a linear system (and the corresponding Severi varieties) for dimensional reasons: For an effective divisor C such that $C-K$ is ample, the dimension of the Severi variety $V_g(C)$ (the closure of the set of geometric genus g curves in the complete linear system $|C|$) is

$$\dim_{\mathbf{C}} V_g(C) = -KC + g - 1 + p_g - q.$$

On the other hand, the virtual dimension of the moduli space $\mathcal{M}_{g,C}(X)$ of stable maps of genus g in the class $[C]$ is

$$\operatorname{virdim}_{\mathbf{C}} \mathcal{M}_{g,C}(X) = -KC + g - 1.$$

If the virtual dimension of the stable maps doesn't match the number of constraints required for $N_g(X,C)$, the corresponding Gromov-Witten invariant *must be zero*.

The discrepancy $p_g - q$ arises from two sources. Since the image of maps in $\mathcal{M}_{g,C}(X)$ are divisors not only in $|C|$ but also potentially in every linear system in $\mathrm{Pic}^{[C]}(X)$, one would expect $\dim \mathcal{M}_{g,C}(X)$ to exceed $\dim V_g(C)$ by $q = \dim \mathrm{Pic}^{[C]}(X)$.[8] As we discussed in the previous section, this discrepancy can be accounted for within the framework of the usual Gromov-Witten invariants using loop constraints (see Theorem 4.1).

However, even if we consider $\mathcal{M}_{g,C}$ as a model for the parameterized Severi varieties

$$V_g([C]) \equiv \coprod_{\mathcal{O}(C')\in \mathrm{Pic}^{[C]}(X)} V_g(C'),$$

there is still a p_g dimensional discrepancy (Donaldson also discusses this in detail [**22**]).

[7]Not much is known about the enumerative geometry of irrational surfaces with $p_g = q = 0$ even with quantum cohomology and the usual Gromov-Witten invariants at our disposal. However, it may be possible to adapt the methods of Subsections 5.1 and 5.2 to the case of the Enriques surface (which is irrational and has $p_g = q = 0$).

[8]we use $\mathrm{Pic}^{[C]}(X)$ to denote the component of $\mathrm{Pic}(X)$ corresponding to line bundles with first Chern class $[C]^\vee$.

The reason is the following. The virtual dimension of $\mathcal{M}_{g,C}(X)$ is the dimension of the space of curves that persist as ***pseudo-holomorphic*** curves when we perturb the Kähler structure to a generic almost Kähler structure. The difference of p_g in the dimensions of $\mathcal{M}_{g,C}$ and $V_g([C])$ means that only a codimension p_g subspace of $V_g([C])$ persists as pseudo-holomorphic curves when we perturb the Kähler structure. Algebraically, this arises as the obstruction to those infinitesimal deformations of the map that deform the image of the map in the $H^{0,2}(X)$ direction (see Subsection 5.3). One way to rectify this situation is to find a compact p_g-dimensional[9] *family* of almost Kähler structures that has the property that the only almost Kähler structure in the family that supports pseudo-holomorphic curves in the class $[C]$ is the original Kähler structure. If T is such a family, then the moduli space $\mathcal{M}_{g,C}(X,T)$ of stable maps for the family T is a better model for the space $V_g([C])$ in the sense that its dimension is stable under generic perturbations of the family $T \mapsto T'$.

Given the existence of a p_g-dimensional family as described above, these invariants can be used to answer enumerative geometry questions for the corresponding surface and linear system. In general, it is not clear when such a family will exist; however, if X has a hyperkähler metric g (*i.e.* X is an Abelian or $K3$ surface), then there is a natural candidate for T, namely the hyperkähler family of Kähler structures. We call this family the ***twistor family*** associated to the metric g and we denote it T_g. It is parameterized by a 2-sphere and so $\dim_{\mathbf{R}} T_g = 2 = 2p_g$ as it should. Furthermore, the property that all the curves in $\mathcal{M}_{g,C}(X,T_g)$ are holomorphic for the original complex structure can be proved with Hodge theory (of course this need no longer be the case for a perturbation of T_g to a generic family of almost Kähler structure.

We will discuss this case in further detail in Section 5 and in Subsection 5.3 we discuss recent work of Behrend and Fantechi that in many cases fixes the "p_g discrepancy" purely algebraically.

4.3. Relationship between the Seiberg-Witten and Gromov-Witten invariants.

In a series of papers [**78**][**79**][**80**][**81**] in 1994 and 1995, Taubes proved that the Seiberg-Witten invariants of a symplectic 4-manifold are given by a certain set of Gromov-Witten invariants. This work has had a profound impact on symplectic and smooth topology and has found numerous applications; however, from the point of view of enumerative algebraic geometry, it gives us no information. The Seiberg-Witten invariants are topological invariants only when $b_2^+(X) > 1$; for a Kähler surface this is equivalent to $p_g(X) > 0$. As we explained in the last section, the Gromov-Witten invariants are *zero* in any class $[C]$ with $\mathcal{O}(C-K)$ ample, so there is only a small range of possible classes with non-zero invariants. In fact, in the Kähler case Taubes' theorem reduced to a fact already explained in Witten's original paper [**88**]: the Seiberg-Witten invariants only "count" connected components of the canonical divisor. That is, the only spin^c structures with non-zero solutions to the Seiberg-Witten equations correspond to the classes in $H_2(X,\mathbf{Z})$ that are the various sums of the components of the canonical

[9] By this we really mean a real $2p_g$ dimensional family; the parameter space for the family need not have a complex structure or even an almost complex structure.

divisor[10]. In particular, on a minimal surface of general type, the Seiberg-Witten and Gromov-Witten invariants are non-trivial only for the canonical class (and the zero class).

Taubes shows in general that the Seiberg-Witten invariants count pseudo-holomorphic curves that are always *smoothly embedded* (though possibly disconnected). From the point of view of enumerative geometry, smoothly embedded curves are uninteresting (the corresponding Severi variety is a linear system and so $N_g(X, C) = 1$, *i.e.* there is exactly one smooth curve in $|C|$ passing through the appropriate number of points).

It is worth noting here that although the enumerative information in the Seiberg-Witten/Gromov-Witten invariants is trivial, the actual Seiberg-Witten/Gromov-Witten multiplicities can be somewhat subtle. Non-trivial multiplicities for curves with a smooth image occur when the curve is a multiple of a square-zero curve of genus one. Multiplicities for the ordinary Gromov-Witten invariants arise because of the possibility of many different maps multiply covering the same image. Taubes' defines his own version of Gromov-Witten invariants which count embedded curves (rather than maps) and allow for the possibility of many components. His multiplicities are defined using the spectral flow of a certain operator that arises naturally in the Seiberg-Witten context. The exact relationship between the two definitions of Gromov-Witten invariants was clarified by Ionel and Parker **[43]** who showed that the invariants contain equivalent information.

4.4. When are Gromov-Witten invariants "enumerative"? By counting maps to X rather than subvarieties of X, the Gromov-Witten invariants acquire many advantageous properties such as deformation invariance and numerous relations. The disadvantage to enumerative applications is that the maps may contract or multiply-cover some of their components. Thus Gromov-Witten invariants may count genus g maps whose image does not have geometric genus g. Furthermore, a given curve in X may be the image of many different maps, possibly even a family of maps. Thus the Gromov-Witten invariants may count a given isolated curve with a non-trivial multiplicity that may be negative and/or non-integral.

For these reasons, the Gromov-Witten invariants are said to give a *virtual* count of curves. When the count defined by a Gromov-Witten invariant coincides with the actual number of curves, the invariant is said to be *enumerative*. Here we introduce an intermediate notion:

DEFINITION 4.4. *A Gromov-Witten invariant is said to be* weakly enumerative *if it counts only curves with geometric genus g each with positive, integral multiplicity that is one for curves with (at worst) nodal singularities.*

This notion is particularly applicable to surfaces where one can often rule out maps that collapse or multiply-cover components by dimensional arguments. The

[10]On a symplectic manifold there are two different ways to identify spin^c structures with elements of $H_2(X, \mathbf{Z})$. One can always take the dual to the first Chern class of the bundle of positive spinors, *i.e.* $c_1(W^+)^\vee \in H_2(X, \mathbf{Z})$; alternatively, on a symplectic 4-manifold every spin^c structure can be obtained by twisting the canonical spin^c structure W_0^+ by a line bundle L and so we can consider the class $c_1(L)^\vee \in H_2(X, \mathbf{Z})$. The latter correspondence is more natural from the point of view of Gromov-Witten theory since on a Kähler surface, a smooth curve C corresponds to a solution of the Seiberg-Witten equations for the spin^c structure $\mathcal{O}(C) \otimes W_0^+$.

basic reason for this is the dependence of the dimension of the space of stable maps on $\dim X$ and g. When X is complex dimension 2 or less, the dimension of the space of stable maps grows linearly with g.

For surfaces, a Gromov-Witten invariant will be weakly invariant if all the maps counted by the invariant are birational isomorphisms onto their image. For example, Göttsche and Pandharipande **[36]** show that all the genus 0 Gromov-Witten invariants of $\mathbf{P}^2$ blown-up at n generic points are enumerative for $n \leq 10$ but their arguments also imply that the invariants are weakly enumerative for *all* n. If the location of the blow-up points are not generic, then the invariants may fail to even be weakly enumerative (c.f. Subsection 5.2). Other examples include the modified Gromov-Witten invariants for $K3$ and Abelian surfaces discussed in Section 5 which are weakly enumerative for generic choices of the $K3$ or Abelian surface.

If a genus 0 Gromov-Witten invariant on a surface is weakly enumerative, then the work of Fantechi-Göttsche-Van Staten **[26]** shows that the multiplicities of the irreducible curves are determined solely by the type and number of singularities (see also Section 2).

5. Modified invariants and the case of $K3$ and Abelian surfaces

In this section we give an expository account of our use of modified Gromov-Witten invariants to prove the Göttsche-Yau-Zaslow formula for $K3$ and Abelian surfaces. We include a description of how to use a "matching technique" to compute the contribution of multiple covers of nodal rational curves to the invariants. We end the section with a brief description of recent work of Behrend and Fantechi who give a purely algebraic modification of the Gromov-Witten invariants that generalizes the (non-algebraic) modifications used in the case of $K3$ and Abelian surfaces.

5.1. The $K3$ and Abelian surface case. In this section we explain the proof of the Göttsche-Yau-Zaslow formula for primitive classes in $K3$ and Abelian surfaces **[15][17]**. Let X be a $K3$ or Abelian surface and let $C \subset X$ be a curve representing a primitive homology class. To verify the conjecture we need to show first that the numbers $N_g(X, C)$ only depend on g and $[C]^2$ (and whether X is a $K3$ or Abelian surface) and then show the numbers are given as the coefficients of the predicted modular forms.

Since $p_g(X) = 1$, there are difficulties with the ordinary Gromov-Witten invariants (see Subsection 4.2). In fact, it is easy to see the following:

LEMMA 5.1. *All the (ordinary) Gromov-Witten invariants of X are zero.*

PROOF: Gromov-Witten invariants are invariant under deformations of the (almost) Kähler structure. Any $K3$ or Abelian surface can be deformed to a (non-algebraic) Kähler surface that has *no* holomorphic curves at all. *q.e.d.*

However, this problem can be rectified and a proof of the Göttsche-Yau-Zaslow formula can be obtained. The following two theorems are the main theorems:

THEOREM 5.2. *There exists a family Gromov-Witten invariant that computes $N_g(X,C)$. Furthermore, the invariant only depends on C^2 and g (and possibly the divisibility of $[C]$). The invariant is weakly enumerative for generic X.*[11]

The following theorem computes the invariant of Theorem 5.2 thus verifying the Göttsche-Yau-Zaslow formula:

THEOREM 5.3. *The Göttsche-Yau-Zaslow formula holds for all $C \subset X$ representing a primitive homology class. That is, if X is a $K3$ surface then $N_g(X,C)$ is the coefficient of $q^{\frac{1}{2}C^2}$ in the series*

$$\begin{aligned} & q^{-1}\prod_{m=1}^{\infty}(1-q^m)^{-24}\left(\sum_{k=1}^{\infty}k\sigma(k)q^k\right)^g \\ = \; & \Delta^{-1}(DG_2)^g \end{aligned}$$

and if X is an Abelian surface then $N_g(X,C)$ is the coefficient of $q^{\frac{1}{2}C^2}$ in the series

$$\begin{aligned} & \left(\sum_{k=1}^{\infty}k^2\sigma(k)q^k\right)\left(\sum_{k=1}^{\infty}k\sigma(k)q^k\right)^{g-2} \\ = \; & D^2G_2(DG_2)^{g-2} \end{aligned}$$

where $\sigma(k)=\sum_{d|k}d$ is the sum of the divisors of k.

REMARK 5.4. There is also a formula for the number of genus g curves on an Abelian surface passing through g points (without imposing the **FLS** condition). The numbers are the coefficients of the series $g(DG_2)^{g-1}$ (see [**17**]).

We first outline the proof of Theorem 5.2. The family of Kähler structures used in Theorem 5.2 is provided by the existence of hyperkähler metrics on X. Since $c_1(TX)=0$, Yau's proof of the Calabi conjecture [**90**] provides a Ricci flat Kähler Einstein metric, which for surfaces is a hyperkähler metric.

A hyperkähler metric is characterized by a 2-sphere's worth of Kähler structures $a\omega_I + b\omega_J + c\omega_K$ where

$$S^2 = \{(a,b,c)\in \mathbf{R}^3 : a^2+b^2+c^2=1\}$$

and the associated complex structures I, J, and K satisfy the algebra of the imaginary quaternions. We call this family the twistor family associated to a hyperkähler metric. The following theorem lists the key properties of the twistor family which allow us to use it with the family version of Gromov-Witten invariants to compute $N_g(X,C)$.

THEOREM 5.5. *Denote the twistor family associated to a hyperkähler metric h by T_h. Then,*

1. *For any two hyperkähler metrics h and h', T_h is deformation equivalent to $T_{h'}$. We denote the deformation equivalence class by simply T.*
2. *For any orientation preserving diffeomorphism f the family $f^*(T)$ is deformation equivalent to T.*

[11] By generic in this setting we mean that X is generic among those $K3$ surfaces that admit a curve in the class $[C]$. See subsection 4.4 for the definition of weakly enumerative.

3. *For any class $[C] \in H_2(X;\mathbf{Z})$ with $[C]^2 \geq -2$, there is exactly one member of the family T_h which admits holomorphic curves in the class $[C]$.*

If $C \subset X$ is a curve on a $K3$ surface, choose a hyperkähler metric h on X and consider the invariant $\Phi_{g,C}^{(X,T_h)}(\text{pt.}^g)$. Property 1 of the previous theorem shows that this invariant is independent of the chosen hyperkähler structure on X. The orientation preserving diffeomorphisms act transitively on elements of $H_2(X;\mathbf{Z})$ with the same square and divisibility and so by property 2, the invariant $\Phi_{g,C}^{(X,T)}(\text{pt.}^g)$ is a universal number that *only depends on g, C^2, and the divisibility of $[C]$*. Finally, property 3 shows that $\Phi_{g,C}^{(X,T)}(\text{pt.}^g)$ only counts curves that are holomorphic with respect to a single complex structure on X.

With a little further work, one can show that $\Phi_{g,C}^{(X,T)}(\text{pt.}^g)$ is weakly enumerative for generic X. That is, for a generic $K3$ surface X,

$$N_g(X,C) = \Phi_{g,C}^{(X,T)}(\text{pt.}^g)$$

as long as we understand that curves should be counted with certain positive integral multiplicities if the singularities are other than nodes.

A similar discussion applies to an Abelian surface X with the addition complication due to non-trivial $q = \dim H^1(X,\mathcal{O}_X)$. Since we wish to count curves in a fixed linear system passing through $g-2$ points, we combine Theorem 4.1 and Theorem 5.2 to show that for a generic X and loops $\gamma_1,\dots,\gamma_4$ giving an oriented basis of $H_1(X,\mathbf{Z})$,

$$N_g(X,C) = \Phi_{g,C}^{(X,C)}(\gamma_1,\dots,\gamma_4,\text{pt.}^{g-2}) \tag{4}$$

(again, in the weakly enumerative sense).

This explains the proof of Theorem 5.2. Then with the invariants in hand, it suffices to compute the invariant for some particular choice of X and C to prove Theorem 5.3. We choose X to be elliptically fibered with a section and we take the class $[C]$ to be the section class plus a multiple of the fiber class. This is a primitive homology class and so our computation will give us a proof of the Göttsche-Yau-Zaslow formula for all primitive classes (Theorem 5.3). We do not know how to make the analogous computation for multiples of this class (or any other non-primitive class), but we make a brief remark about this case below:

REMARK 5.6. To even formulate the enumeration problem for non-primitive classes on a $K3$ or Abelian surface, one must always deal with non-reduced curves. On these surfaces, there is a curve in a class $d[C]$ if and only if there is a curve in the class $[C]$. So for example, when counting rational curves in the class $d[C]$ one must decide how to count those rational curves in the class $[C]$ with their non-reduced structure. Our definition of $N_g(X,dC)$ using the family Gromov-Witten invariant applies in the non-primitive case as well, but its enumerative significance is less clear. When we show that our Gromov-Witten invariant is weakly enumerative for generic X and primitive $[C]$ we use the fact that all the curves will be irreducible and reduced. *A priori*, the family Gromov-Witten invariant may assign a negative and/or non-integral multiplicity to non-reduced curves.

We illustrate the computation of Theorem 5.3 with the $K3$ case and afterwards we discuss the new issues involved for the Abelian surface calculation. A elliptically fibered $K3$ surface with a section generically has 24 singular fibers consisting of nodal rational curves $N_1, \ldots, N_{24}$ and a section S which is a rational curve of square -2. We choose our g points $x_1, \ldots, x_g$ away from the section and lying on g distinct, generic, smooth fibers $F_1, \ldots, F_g$:

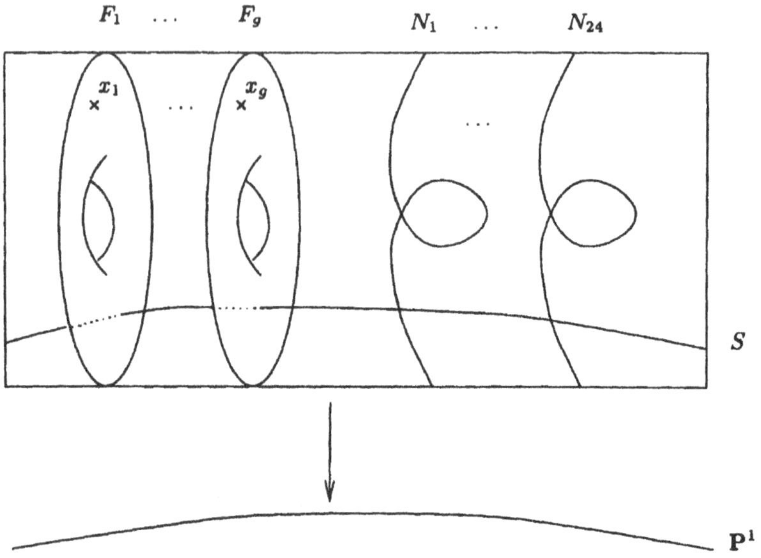

Fix $n \geq 0$ and let $[C_n] = [S] + (n+g)[F]$. Note that $[C_n]^2 = 2g - 2 + 2n$ and so to verify the Göttsche-Yau-Zaslow formula we need to show:

$$\sum_{n=0}^{\infty} \Phi_{g,C_n}^{(X,T)}(\mathrm{pt.}^g) q^{g-1+n} = q^{-1} \prod_{m=1}^{\infty} (1-q^m)^{-24} \left(\sum_{k=1}^{\infty} k\sigma(k) q^k \right)^g .$$

The advantage of our choice of X and C_n is that we can really *see* all the curves in the linear system $|C_n|$. They are all reducible and their components consists of the section S along with $n+g$ fiber curves (possibly non-reduced). In order for a curve in $|C_n|$ to pass through the points $x_1, \ldots, x_g$ *and* have geometric genus no more than g (as it must in order to be the image of a genus g map) it must consist of the section, the g fibers $F_1, \ldots, F_g$ (possibly with multiplicities), and some number of the rational fibers $N_1, \ldots, N_{24}$ (again possibly with multiplicities). In other words, the curve must be

$$S + \sum_{i=1}^{g} b_i F_i + \sum_{j=1}^{24} a_j N_j$$

where the g-tuple $\mathbf{b} = (b_1, \dots, b_g)$ and the 24-tuple $\mathbf{a} = (a_1, \dots, a_{24})$ satisfy $b_i \geq 1$, $a_j \geq 0$, and $\sum_{i=1}^{g} b_i + \sum_{j=1}^{24} a_j = n + g$.

To compute the invariant $\Phi^{(X,T)}_{g,C_n}(\mathrm{pt.}^g)$ we must compute the number of *maps* with the various images determined by $\mathbf{a}$ and $\mathbf{b}$. What we mean by this, strictly speaking, is that we must compute the virtual fundamental class of the moduli space of stable maps with image given by $\mathbf{a}$ and $\mathbf{b}$. The virtual dimension of the moduli space is zero and so the virtual fundamental class is a number, but the moduli space may actually be higher dimensional. In the case at hand, we show that the moduli space splits as a product of other moduli spaces and the virtual fundamental class splits into a product of virtual classes coming from each of the factors. The factors in this product can be identified with the moduli spaces of maps whose images multiple cover a single fiber F_i or N_j and the corresponding factor of the virtual class is (essentially[12]) the usual virtual class of each factor.

In this way we show that the contribution to the invariant from maps with image corresponding to $\mathbf{a}$ and $\mathbf{b}$ is a product of "local" contributions from multiple covers of the fibers $F_1, \dots, F_g$ and $N_1, \dots, N_{24}$. The invariant is thus of the form

$$\Phi^{(X,T)}_{g,C_n}(\mathrm{pt.}^g) = \sum_{\substack{(\mathbf{a},\mathbf{b}):\\ \sum_j a_j + \sum_i b_i = n+g}} \prod_{i=1}^{g} r(b_i) \prod_{j=1}^{24} p(a_j)$$

where $r(b)$ and $p(a)$ are the "local contributions" of b-fold covers of a smooth fiber and a-fold covers of a nodal rational fiber respectively. Multiplying both sides of the equation by q^{g-1+n} we get

$$\Phi^{(X,T)}_{g,C_n}(\mathrm{pt.}^g) q^{g-1+n} = q^{-1} \sum_{\substack{(\mathbf{a},\mathbf{b}):\\ \sum_j a_j + \sum_i b_i = n+g}} \prod_{i=1}^{g} r(b_i) q^{b_i} \prod_{j=1}^{24} p(a_j) q^{a_j}$$

and then summing over n:

$$\begin{aligned} \sum_{n=0}^{\infty} \Phi^{(X,T)}_{g,C_n}(\mathrm{pt.}^g) q^{g-1+n} &= q^{-1} \sum_{(\mathbf{a},\mathbf{b})} \prod_{i=1}^{g} r(b_i) q^{b_i} \prod_{j=1}^{24} p(a_j) q^{a_j} \\ &= q^{-1} \left(\sum_{b=1}^{\infty} r(b) q^b \right)^g \left(\sum_{a=0}^{\infty} p(a) q^a \right)^{24}. \end{aligned}$$

To prove the theorem then, it remains to be shown (which we will do in the next subsection) that

1. the local contribution $r(b)$ of b-fold covers of a smooth fiber is given by $b\sigma(b)$, and
2. the local contribution $p(a)$ of a-fold covers of a nodal fiber is given by the number of partitions of a since the generating function for the number of

[12]The only difference is the role that the section plays. It turns out that the obstruction to deforming the section is one dimensional and exactly cancels the obstruction to deforming the complex structure on X in the direction of the twistor family. The end result is the same as if we pretend the section is a square -1 curve and there is no family.

partitions is given by

$$\prod_{m=1}^{\infty}(1-q^m)^{-1}.$$

The computation for an Abelian surface follows by similar methods by again assuming that the Abelian surface is elliptically fibered, *i.e.* it is a product of two elliptic curves $E_1 \times E_2$. The only new complication is the use of loops as geometric constraints (see Equation 4). Using a careful choice of loops, all the possible images of the genus g maps satisfying the geometric constraints are easily identified. The invariant is then again computed by calculating the contribution of the various multiple covers of a given image.

5.2. Local contributions. In this subsection we compute the local contributions $r(b)$ and $p(a)$ of multiple covers of smooth and nodal fibers.

The contribution $r(b)$ is easy to understand. It is the virtual fundamental class of the moduli space of genus 1, 2-marked, degree b maps to a fixed smooth genus 1, 2-marked curve that send the marked points to the marked points. The two marked points correspond to the intersection with the section and the point x_i. By declaring one of the marked points to be the origin, we give the domain and range the structure of elliptic curves and then the map must be a homomorphism. Thus $r(b)$ is the number of elliptic curves with a non-zero marked point that admit a degree b homomorphism onto a fixed elliptic curve with a non-zero marked point mapping the marked point to the marked point. The number of elliptic curves admitting a degree b homomorphism to a fixed elliptic curve is the number of index b sublattices of $\mathbf{Z}\oplus\mathbf{Z}$ which is classically known to be $\sigma(b)$. The number of choices for the location of the marked point in the domain is then just b and so the moduli space of b-fold covers of a smooth fiber is a discrete space consisting of $b\sigma(b)$ points.

It is more difficult to directly see the contribution $p(a)$ of a-fold covers of a nodal fiber. This moduli space does not have the expected dimension zero so we need to compute its virtual fundamental class.

Define $\mathcal{M}(S+aN)$ to be the moduli space of genus 0 maps to X in the class $[S]+a[N]$ with image $S+aN_j$ for some fixed N_j (it doesn't matter which one—their neighborhoods are all biholomorphic). Denote its virtual fundamental class by $[\mathcal{M}(S+aN)]^{vir}$. The moduli space $\mathcal{M}(S+aN)$ has a number of different path components arising from the possible "jumping" behavior of the map at the node. The basic phenomenon is illustrated below for degree two maps. The moduli space $\mathcal{M}(S+2N)$ has different components depending on whether the map factors through the normalization or not. In the figure below, points labeled by A and B are mapped to corresponding points labeled A and B and the normalization map identifies A and B to the nodal point. Notice that the bottom map cannot be factored through the normalization (the dashed map doesn't exist) even though the map to the nodal curve is well defined.

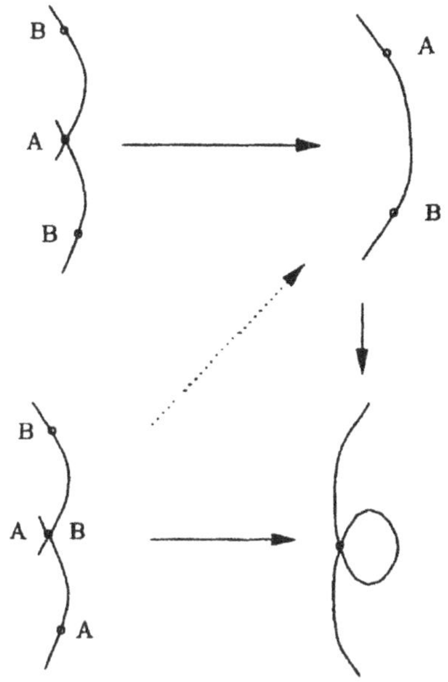

In fact, the space $\mathcal{M}(S+2N)$ has three connected components: one for maps that factor through the normalization and the two maps that consist of the bottom map in the figure along with the section attached (it can attach to either of the components).

For higher degree, maps that factor through the bottom map of the figure will lie in a different component from maps that factor through the normalization. In general, the components of $\mathcal{M}(S+aN)$ are labeled by sequences $\ldots, s_{-2}, s_{-1}, s_0, s_1, \ldots$ of non-negative integers with $\sum s_i = a$. This can be seen as follows. Consider the nodal rational curve Σ whose dual graph is:

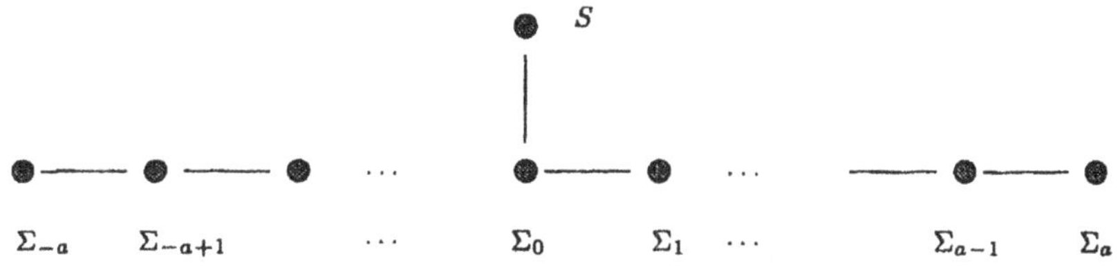

The curve Σ maps to $S+N$ by sending the component S isomorphically to S and each component Σ_i by a degree one map to N in such a way that the nodes connecting Σ_i to Σ_{i+1} are mapped by a local isomorphism onto the node of N. It can be shown that every map in $\mathcal{M}(S+aN)$ factors uniquely through a map to Σ. The components of $\mathcal{M}(S+aN)$ are determined by the various ways the degree of the map can distribute among the Σ_i components of Σ. We denote by $\mathcal{M}(S+\sum_i s_i\Sigma_i)$ the moduli space of genus 0 maps to Σ in the class $[S]+\sum_i s_i[\Sigma_i]$.

The preceding discussion leads to

$$\begin{aligned} p(a) &= [\mathcal{M}(S+aN)]^{vir} \\ &= \sum_{\{s_i\}:\sum_i s_i = a} [\mathcal{M}(S+\sum_i s_i\Sigma_i)]^{vir} \end{aligned}$$

where by $[\mathcal{M}(S+\sum_i s_i\Sigma_i)]^{vir}$ we mean the virtual fundamental class pulled back from $\mathcal{M}(S+aN)$. This fundamental class is induced by assigning normal bundles $\mathcal{O}(-2)$ to the Σ_i components and $\mathcal{O}(-1)$ to the S component.

We can now compute $[\mathcal{M}(S+\sum_i s_i\Sigma_i)]^{vir}$ indirectly by identifying it with another moduli-obstruction problem arising from certain blow-ups of $\mathbf{P}^2$. It will turn out that $[\mathcal{M}(S+\sum_i s_i\Sigma_i)]^{vir}$ is always zero or one depending on whether the sequence $\{s_i\}$ satisfies a certain property or not.

To identify the moduli-obstruction problem with one coming from a blow-up of $\mathbf{P}^2$, we must blow-up $\mathbf{P}^2$ in such a way that it contains a configuration of rational curves isomorphic to Σ with normal bundles $\mathcal{O}(-2)$ on the Σ_i curves and $\mathcal{O}(-1)$ on the S curve. To see that this can be done, begin by blowing up a line at three points. Its proper transform is a -2 curve which we identify with Σ_0; it meets three -1 curves, one of which we identify with S. The remaining two -1 curves can be made into -2 curves by blowing up a point on each of them. We identify their proper transforms with Σ_{-1} and Σ_1 and we repeat this process with the new -1 curves, continuing until we have the configuration Σ.

We then consider the Gromov-Witten invariants of this blow-up of $\mathbf{P}^2$ in the class corresponding to $S+\sum_i s_i\Sigma_i$. Assuming that all the rational curves in the blown up $\mathbf{P}^2$ in the class $[S]+\sum_i s_i[\Sigma_i]$ actually lie in the configuration Σ,[13] the number $[\mathcal{M}(S+\sum_i s_i\Sigma_i)]^{vir}$ is given by the corresponding invariant on the blown up $\mathbf{P}^2$.

Genus 0 Gromov-Witten invariants of blow ups of $\mathbf{P}^2$ have been thoroughly studied in general by Göttsche and Pandharipande **[36]** but the particular invariants arising in this moduli-obstruction problem can be computed from elementary properties. The key property we use is the invariance of the Gromov-Witten invariants under Cremona transformations. By successive applications of the Cremona transformation, one shows that the invariant corresponding to $[\mathcal{M}(S+\sum_i s_i\Sigma_i)]^{vir}$ is either zero or equivalent to the number of lines in the plane through two points, *i.e.* one. The latter case occurs if and only if the sequence $\{s_i\}$ is such that for each $i \geq 0$, s_{i+1} is either s_i or $s_i - 1$ and for each $i \leq 0$, s_{i-1} is either s_i or $s_i - 1$. Following **[15]** we call a sequence with this property 1-admissible.

By the preceding arguments we conclude that

$$p(a) = \#\text{ of 1-admissable sequences } \{s_i\} \text{ with } \sum_i s_i = a$$

and we simply need to see that 1-admissible sequences of total sum a are in one-to-one correspondence with partitions of a. This is achieved by exhibiting a bijection between 1-admissible sequences of total sum a and Young diagrams of size a (which

[13] Some extra work is required to justify this assumption. One can perform the blow-ups in such a way that the resulting surface has an action of $\mathbf{C}^*$ preserving the configuration Σ. This action provides the tool needed to show that there are no curves outside Σ in the relevant homology class. One assumes that such a curve exists and studies its limits under the $\mathbf{C}^*$ action; the desired contradiction is then arrived at by a homological argument.

are well known to correspond bijectively to partitions). Given a Young diagram define a 1-admissible sequence $\{s_i\}$ by setting s_0 equal to the number of blocks on the diagonal, s_1 equal to the number of blocks on the first lower diagonal, s_2 equal to the number of blocks on the second lower diagonal, and so on, doing the same for $s_{-1}, s_{-2}, \ldots$ with the upper diagonals. It is easily seen that this defines a bijection and thus concludes the proof of Theorem 5.3.

5.3. Other modifications via algebraic means. Although the results of the previous section are purely algebraic, the use of the twistor family to modify the usual Gromov-Witten invariants is a non-algebraic tool. Behrend and Fantechi have recently announced [**10**] a purely algebraic modification of the Gromov-Witten invariants that apply to any smooth algebraic variety with $H^{0,2}(X) > 0$, and in particular to surfaces with $p_g > 0$. For $K3$ and Abelian surfaces, their modification is equivalent to the twistor family invariant. In this section we outline their modification for surfaces in general.

In the usual algebraic definition of Gromov-Witten invariants, one defines a virtual fundamental class on the space of stable maps that is invariant under deformations. The ingredients in Behrend and Fantechi's approach to this [**11**] are the intrinsic normal cone and the obstruction complex of the moduli space of stable maps $\mathcal{M}_{g,n,C}(X)$. Behrend and Fantechi modify the usual obstruction complex so that the resulting virtual fundamental class has its dimension larger than the usual dimension by p_g and is invariant under deformations of X preserving the $(1,1)$ type of $[C]$.

The tangent-obstruction complex for $\mathcal{M}_{g,n,C}(X)$ is built from the tangent-obstruction complex of the Deligne-Mumford moduli space of stable curves $\overline{\mathcal{M}}_{g,n}$ and the relative tangent-obstruction complex of

$$\mathcal{M}_{g,n,C}(X) \to \overline{\mathcal{M}}_{g,n}.$$

The relative tangent and obstruction spaces at the map $\{f : C \to X\}$ can be identified with $H^0(C, f^*TX)$ and $H^1(C, f^*TX)$ respectively. An infinitesimal deformation of X together with f is automatically obstructed unless the class $[\mathrm{Im}\, f]$ remains type $(1,1)$. For such deformations, the obstructions always lie in the kernel of

$$H^1(C, f^*TX) \to H^2(X, \mathcal{O}).$$

We can define this map via its dual map

$$H^0(X, \Omega_X^2) \to H^0(C, f^*\Omega_X^1 \otimes \omega_C)$$

which is induced by the composition of $H^0(X, \Omega_X^2) \to H^0(C, f^*\Omega_X^2)$ and the map induced by

$$f^*\Omega_X^2 \to f^*\Omega_X^1 \otimes f^*\Omega_X^1 \to f^*\Omega_X^1 \otimes \Omega_C^1 \to f^*\Omega_X^1 \otimes \omega_C.$$

Behrend and Fantechi modify the usual tangent-obstruction complex by replacing the relative obstruction space by the kernel of $H^1(C, f^*TX) \to H^2(X, \mathcal{O})$. In order for their machinery to work, the obstruction complex must be "perfect", that is, equivalent in the derived category to a two term complex of vector bundles. Thus, if $H^1(C, f^*TX) \to H^2(X, \mathcal{O})$ is surjective (or of constant rank) for all f, then the modification leads to an invariant. The theorem is as follows:

THEOREM 5.7. *Let X be a surface and $[C]$ a class of type $(1,1)$. There is a (modified) virtual class $[\mathcal{M}_{g,C}]^{vir}_{mod}$ of dimension $-KC+g-1+p_g$. If the map*

$$H^1(C, f^*TX) \to H^2(X, \mathcal{O})$$

is surjective for every $\{f : C \to X\}$ in $\mathcal{M}_{g,C}$, then $[\mathcal{M}_{g,C}]^{vir}_{mod}$ defines modified Gromov-Witten invariants that are invariant under deformations of X preserving the $(1,1)$ type of $[C]$.

The hypotheses of the theorem can be shown to hold for $K3$ and Abelian surfaces and also for more general surfaces with appropriate ampleness conditions on C.

REMARK 5.8. Using these invariants and Theorem 4.2, one can define $N_g(X, C)$ using Gromov-Witten theory and perhaps use them to devise a proof of the Göttsche-Yau-Zaslow formula.

REMARK 5.9. As with the families version of Gromov-Witten invariants, it is not clear what the analogues of the composition law and quantum cohomology should be.

REMARK 5.10. Except for the cases where X admits a hyperkähler structure, it is not clear if there is a symplectic version of Behrend and Fantechi's modified invariants.

The computations for $K3$ and Abelian surfaces described in section 5 easily extend to compute the modified invariants of any elliptic surface with a section in the class $[C_n] = [S] + n[F]$ where S is the section and $[F]$ is the class of the fiber. For example, let $E(m)$ be a generic elliptic surface over $\mathbf{P}^1$ with a section and Euler characteristic $12m$.[14] With this convention, $E(1)$ is a rational elliptic surface, $E(2)$ is a $K3$ surface, and etcetera so that $p_g(E(m)) = m-1$. Let $\tilde{\Phi}$ be the modified invariants of Behrend-Fantechi. Then the methods of Subsections 5.1 and 5.2 can be used to compute $\tilde{\Phi}^{E(m)}_{g,C_n}(\mathrm{pt.}^g)$:

$$\sum_{n=0}^{\infty} \tilde{\Phi}^{E(m)}_{g,C_n}(\mathrm{pt.}^g) q^{g+n} = \left(\frac{q}{\Delta}\right)^{m/2} (DG_2)^g. \tag{5}$$

There are a few remarks worth making about this formula.

REMARK 5.11. The formula *is different* from the Göttsche-Yau-Zaslow formula except in the case $m = 2$ (the $K3$ case). In general, the classes $[S] + n[F]$ do not satisfy Göttsche's ampleness conditions so there is no contradiction with the conjecture. It does show that the ampleness conditions cannot be removed.

REMARK 5.12. The invariants $\tilde{\Phi}^{E(m)}_{g,C_n}$ only have enumerative significance for $E(1)$ and $E(2)$. For $m > 2$, one cannot deform away the elliptic fibration and so $\tilde{\Phi}^{E(m)}_{g,C_n}$ always gives the virtual count done in the computation of Subsections 5.1 and 5.2. For $E(1)$ and $E(2)$ (*i.e.* $\mathbf{P}^2$ blown up at nine points and $K3$) a generic deformation will no longer be elliptically fibered and $\tilde{\Phi}^{E(m)}_{g,C_n}$ will be weakly enumerative. In fact, for $E(1)$, $\tilde{\Phi}^{E(1)}_{g,C_n}$ is just the ordinary Gromov-Witten invariants for $\mathbf{P}^2$ blown-up at nine points. For genus zero they are determined recursively from

[14] The classes $[C_n]$ are exactly characterized as those classes $[C]$ such that $[C] \cdot K = m-2$.

the quantum cohomology and are enumerative in the strong sense (for a generic choice of blow-up points). Furthermore, in [**15**] it is shown that the Gromov-Witten invariant of $E(1)$ for *any* class C with $CK = -1$ is equivalent to the invariant for some C_n.

The fact that there is a closed formula (different from the Göttsche-Yau-Zaslow formula!) for these invariants of $E(1)$ in terms of modular forms is somewhat of a surprise. Although the genus zero numbers are determined recursively, it is not clear how to obtain the closed formula from the recursion or how the modularity is reflected in the structure of the quantum cohomology. Ionel and Parker [**44**] have recently outlined a new proof of the genus zero formula for $E(1)$ that is perhaps the most transparent. They use a formula for the invariants of a fiber sum along with a topological recursion relation to show that the left hand side of Equation 5 satisfies a differential equation that is solved by the modular form on the right hand side.

6. Acknowledgments

We would like to thank the many people with whom we have had valuable conversations about Gromov-Witten theory and enumerative geometry. In particular we thank K. Behrend, A. Bertram, H. Clemens, O. DeBarre, R. Donagi, S. Donaldson, B. Fantechi, E. Getzler, A. Givental, L. Göttsche, T. Graber, E. Ionel, A. K. Liu, S. Katz, S. Kleiman, J. Li, T. Parker, R. Pandharipande, Y.-B. Ruan, B. Siebert, C. Taubes, M. Thaddeus, G. Tian, A. Todorov, C. Vafa, I. Vainsencher, R. Vakil, S.-T. Yau, and E. Zaslow.

We would also like to thank the organizers of the conferences that made it possible to talk to many of the above people. In particular, Y. Eliashberg and L. Traynor for organizing symplectic geometry program in IAS/Park City summer institute (1997); H. Clemens, E. Arbarello, and J. Harer for organizing the workshop on "Hodge theory, mirror symmetry, and quantum cohomology" in Pisa/Cortona, and A. Bertram and Y.-B. Ruan for organizing the AMS summer workshop on Quantum Cohomology.

References

[1] Allen B. Altman, Anthony Iarrobino, and Steven L. Kleiman. *Irreducibility of the compactified Jacobian*, pages 1–12. Sijthoff and Noordhoff, Alphen aan den Rijn, 1977.
[2] Allen B. Altman and Steven L. Kleiman. Compactifying the Jacobian. *Bull. Amer. Math. Soc.*, 82(6):947–949, 1976.
[3] Allen B. Altman and Steven L. Kleiman. Compactifying the Picard scheme. II. *Amer. J. Math.*, 101(1):10–41, 1979.
[4] Allen B. Altman and Steven L. Kleiman. Compactifying the Picard scheme. *Adv. in Math.*, 35(1):50–112, 1980.
[5] Paolo Aluffi. *Quantum cohomology at the Mittag-Leffler Institute.* Scuola Normale Superiore, Pisa, 1997.
[6] Michèle Audin. Cohomologie quantique. *Astérisque*, (241):Exp. No. 806, 3, 29–58, 1997. Séminaire Bourbaki, Vol. 1995/96.
[7] Victor Batyrev. On the Betti numbers of birationally isomorphic projective varieties with trivial canonical bundles. Preprint, alg-geom//9710020.
[8] A. Beauville. Counting rational curves on $K3$ surfaces. Preprint alg-geom/9701019, 1997.
[9] K. Behrend. Gromov-Witten invariants in algebraic geometry. *Invent. Math.*, 127(3):601–617, 1997.
[10] K. Behrend and B. Fantechi. In preparation.
[11] K. Behrend and B. Fantechi. The intrinsic normal cone. *Invent. Math.*, 128(1):45–88, 1997.

[12] K. Behrend and Yu. Manin. Stacks of stable maps and Gromov-Witten invariants. *Duke Math. J.*, 85(1):1–60, 1996.

[13] Aaron Bertram, Georgios Daskalopoulos, and Richard Wentworth. Gromov invariants for holomorphic maps from Riemann surfaces to Grassmannians. *J. Amer. Math. Soc.*, 9(2):529–571, 1996.

[14] Paul Biran. Connectedness of spaces of symplectic embeddings. *Internat. Math. Res. Notices*, 1996(10):487–491.

[15] Jim Bryan and Naichung Conan Leung. The enumerative geometry of $K3$ surfaces and modular forms. alg-geom/9711031, 1997.

[16] Jim Bryan and Naichung Conan Leung. Counting curves with tangency conditions on the nine-fold blow-up of $\mathbf{P}^2$. In preparation, 1998.

[17] Jim Bryan and Naichung Conan Leung. Generating functions for the number of curves on abelian surfaces. math.AG/9802125, 1998.

[18] Lucia Caporaso and Joe Harris. Counting plane curves of any genus. *Inventiones Mathematicae*, 131(2):345–392, 1998.

[19] Xi Chen. Rational curves on $K3$ surfaces. Preprint math.AG/9804075, 1998.

[20] Ionuţ Ciocan-Fontanine. Quantum cohomology of flag varieties. *Internat. Math. Res. Notices*, 1995(6):263–277.

[21] P. Deligne. Local behavior of Hodge structures at infinity. In *Mirror symmetry, II*, volume 1 of *AMS/IP Stud. Adv. Math.*, pages 683–699. Amer. Math. Soc., Providence, RI, 1997.

[22] S. K. Donaldson. Yang-Mills invariants of 4-manifolds. In S. K. Donaldson and C. B. Thomas, editors, *Geometry of Low-Dimensional Manifolds: Gauge Theory and Algebraic Surfaces*, number 150 in London Mathematical Society Lecture Note Series. Cambridge University Press, 1989.

[23] Boris Dubrovin. Geometry of 2D topological field theories. In *Integrable systems and quantum groups (Montecatini Terme, 1993)*, volume 1620 of *Lecture Notes in Math.*, pages 120–348. Springer, Berlin, 1996.

[24] Lars Ernström and Gary Kennedy. Recursive formulas for the characteristic numbers of rational plane curves. preprint alg-geom/9604019, 1996.

[25] Lars Ernström and Gary Kennedy. Contact cohomology of the projective plane. preprint alg-geom/9703013, 1998.

[26] Barbara Fantechi, Lothar Göttsche, and Duco van Straten. Euler number of the compactified Jacobian and multiplicity of rational curves. Preprint, alg-geom/9708012.

[27] Sergey Fomin, Sergei Gelfand, and Alexander Postnikov. Quantum Schubert polynomials. *J. Amer. Math. Soc.*, 10(3):565–596, 1997.

[28] William Fulton and Rahul Pandharipande. Notes on stable maps and quantum cohomology. In *Algebraic geometry—Santa Cruz 1995*, volume 62 of *Proc. Sympos. Pure Math.*, pages 45–96, Providence, RI, 1997. Amererican Mathematical Society.

[29] E. Gava, K. S. Narain, and C. Vafa, editors. *S-duality and mirror symmetry*, Amsterdam, 1996. North-Holland Publishing Co. Nuclear Phys. B Proc. Suppl. **46** (1996).

[30] E. Getzler. Intersection theory on $\overline{m}_{1,4}$ and elliptic Gromov-Witten invariants. *J. Amer. Math. Soc.*, 10(4):973–998, 1997.

[31] A. B. Givental Homological geometry. I. Projective hypersurfaces. *Selecta Math. (N.S.)*, 1(2):325–345, 1995.

[32] Alexander B. Givental. Equivariant Gromov-Witten invariants. *Internat. Math. Res. Notices*, 1996(13):613–663.

[33] Alexander B. Givental. Homological geometry and mirror symmetry. In *Proceedings of the International Congress of Mathematicians, Vol. 1, 2 (Zürich, 1994)*, pages 472–480, Basel, 1995. Birkhäuser.

[34] L. Göttsche. A conjectural generating function for numbers of curves on surfaces. alg-geom/9711012.

[35] L. Göttsche. The Betti numbers of the Hilbert scheme of points on a smooth projective surface. *Math. Ann.*, pages 193–207, 1990.

[36] L. Göttsche and R. Pandharipande. The quantum cohomology of blow-ups of $\mathbf{P}^2$. Preprint, alg-geom/9611012, 1996.

[37] Tom Graber. *Enumerative Geometry of Hyperelliptic Plane Curves*. PhD thesis, University of Chicago, 1998.

[38] B. Greene and S.-T. Yau, editors. *Mirror symmetry. II*, volume 1 of *AMS/IP Studies in Advanced Mathematics*. American Mathematical Society, Providence, RI, 1997.

[39] Walter Greiner and Joachim Reinhardt. *Field quantization*. Springer-Verlag, Berlin, 1996. Translated from the German, With a foreword by D. A. Bromley.

[40] Nigel Hitchin. Frobenius manifolds. In *Gauge theory and symplectic geometry (Montreal, PQ, 1995)*, volume 488 of *NATO Adv. Sci. Inst. Ser. C Math. Phys. Sci.*, pages 69–112. Kluwer Acad. Publ., Dordrecht, 1997. With notes by David Calderbank.

[41] Petr Hořava. Equivariant topological sigma models. *Nuclear Phys. B*, 418(3):571–602, 1994.

[42] E. Ionel. *Genus one enumerative invariants in* $\mathbf{P}^n$ *with fixed* $\jmath$ *invariant*. PhD thesis, Michigan State, 1996.

[43] Eleny-Nicoleta Ionel and Thomas H. Parker. The Gromov invariants of Ruan-Tian and Taubes. *Math. Res. Lett.*, 4(4):521–532, 1997.

[44] Eleny-Nicoleta Ionel and Thomas H. Parker. Gromov-Witten invariants of symplectic sums. Preprint math.SG/9806013, 1998.

[45] C. Itzykson. Counting rational curves on rational surfaces. *Internat. J. Modern Phys. B*, 8(25-26):3703–3724, 1994. Perspectives on solvable models.

[46] Masao Jinzenji and Yi Sun. Calculation of Gromov-Witten invariants for cp^3, cp^4 and $\mathrm{gr}(2,4)$. *Internat. J. Modern Phys. A*, 11(1):171–202, 1996.

[47] Sheldon Katz. Rational curves on Calabi-Yau manifolds: verifying predictions of mirror symmetry. In *Projective geometry with applications*, volume 166 of *Lecture Notes in Pure and Appl. Math.*, pages 231–239. Dekker, New York, 1994.

[48] Sheldon Katz. Gromov-Witten invariants via algebraic geometry. *Nuclear Phys. B Proc. Suppl.*, 46:108–115, 1996. S-duality and mirror symmetry (Trieste, 1995).

[49] Ralph Kaufmann. The intersection form in $h^*(\overline{m}_{0n})$ and the explicit Künneth formula in quantum cohomology. *Internat. Math. Res. Notices*, 1996(19):929–952.

[50] Bumsig Kim. On equivariant quantum cohomology. *Internat. Math. Res. Notices*, 1996(17):841–851.

[51] S. Kleiman and R. Piene. Polynomials enumerating curves with eight nodes or less. In preparation.

[52] János Kollár, Robert Lazarsfeld, and David R. Morrison, editors. *Algebraic geometry—Santa Cruz 1995*, volume 62 of *Proceedings of Symposia in Pure Mathematics*, Providence, RI, 1997. American Mathematical Society.

[53] M. Kontsevich and Yu. Manin. Gromov-Witten classes, quantum cohomology, and enumerative geometry. *Comm. Math. Phys.*, 164(3):525–562, 1994.

[54] M. Kontsevich and Yu. Manin. Quantum cohomology of a product. *Invent. Math.*, 124(1-3):313–339, 1996. With an appendix by R. Kaufmann.

[55] M. Kontsevich and Yu. Manin. Gromov-Witten classes, quantum cohomology, and enumerative geometry [MR 95i:14049]. In *Mirror symmetry, II*, volume 1 of *AMS/IP Stud. Adv. Math.*, pages 607–653. Amer. Math. Soc., Providence, RI, 1997.

[56] Maxim Kontsevich. Enumeration of rational curves via torus actions. In *The moduli space of curves (Texel Island, 1994)*, volume 129 of *Progr. Math.*, pages 335–368. Birkhäuser Boston, Boston, MA, 1995.

[57] Maxim Kontsevich. Homological algebra of mirror symmetry. In *Proceedings of the International Congress of Mathematicians, Vol. 1, 2 (Zürich, 1994)*, pages 120–139, Basel, 1995. Birkhäuser.

[58] Maxim Kontsevich. Rigorous results in topological σ-model. In *XIth International Congress of Mathematical Physics (Paris, 1994)*, pages 47–59. Internat. Press, Cambridge, MA, 1995.

[59] J. Li and G. Tian. Virtual moduli cycles and Gromov-Witten invariants of general symplectic manifolds. Preprint, alg-geom/9608032, 1996.

[60] J. Li and G. Tian. Comparison of the algebraic and the symplectic Gromov-Witten invariants. Preprint, alg-geom/9712035, 1997.

[61] Jun Li and Gang Tian. Algebraic and symplectic geometry of Gromov-Witten invariants. In *Algebraic geometry—Santa Cruz 1995*, volume 62 of *Proc. Sympos. Pure Math.*, pages 143–170. Amer. Math. Soc., Providence, RI, 1997.

[62] Jun Li and Gang Tian. Virtual moduli cycles and Gromov-Witten invariants of algebraic varieties. *J. Amer. Math. Soc.*, 11(1):119–174, 1998.

[63] Bong H. Lian, Kefeng Liu, and Shing-Tung Yau. Mirror principle. I. *Asian J. Math.*, 1(4):729–763, 1997.

[64] Dusa McDuff and Dietmar Salamon. *J-holomorphic curves and quantum cohomology*, volume 6 of *University Lecture Series*. American Mathematical Society, Providence, RI, 1994.

[65] Klaus Mohnke. On Seiberg-Witten equations on symplectic 4-manifolds. In *Symplectic singularities and geometry of gauge fields (Warsaw, 1995)*, volume 39 of *Banach Center Publ.*, pages 89–104. Polish Acad. Sci., Warsaw, 1997.

[66] David R. Morrison. Making enumerative predictions by means of mirror symmetry. In *Mirror symmetry, II*, volume 1 of *AMS/IP Stud. Adv. Math.*, pages 457–482. Amer. Math. Soc., Providence, RI, 1997.

[67] David R. Morrison. Mathematical aspects of mirror symmetry. In *Complex algebraic geometry (Park City, UT, 1993)*, volume 3 of *IAS/Park City Math. Ser.*, pages 265–327. Amer. Math. Soc., Providence, RI, 1997.

[68] Shigeru Mukai. Symplectic structure of the moduli space of sheaves on an Abelian or $K3$ surface. *Inventiones Mathematicae*, 77(1):101–116, 1984.

[69] Rahul Pandharipande. Counting elliptic plane curves with fixed j-invariant. *Proc. Amer. Math. Soc.*, 125(12):3471–3479, 1997.

[70] L. Polterovich. Gromov's K-area and symplectic rigidity. *Geom. Funct. Anal.*, 6(4):726–739, 1996.

[71] Yong Bin Ruan. Symplectic topology on algebraic 3-folds. *J. Differential Geom.*, 39(1):215–227, 1994.

[72] Yong Bin Ruan and Gang Tian. A mathematical theory of quantum cohomology. *Math. Res. Lett.*, 1(2):269–278, 1994.

[73] Yongbin Ruan. Symplectic topology and complex surfaces. In *Geometry and analysis on complex manifolds*, pages 171–197. World Sci. Publishing, River Edge, NJ, 1994.

[74] Yongbin Ruan. Topological sigma model and Donaldson-type invariants in Gromov theory. *Duke Math. J.*, 83(2):461–500, 1996.

[75] Yongbin Ruan and Gang Tian. A mathematical theory of quantum cohomology. *J. Differential Geom.*, 42(2):259–367, 1995.

[76] Yongbin Ruan and Gang Tian. Higher genus symplectic invariants and sigma model coupled with gravity. *Turkish J. Math.*, 20(1):75–83, 1996.

[77] Bernd Siebert. Algebraic and symplectic Gromov-Witten invariants coincide. Preprint: math.AG/9804108, 1998.

[78] C. H. Taubes. The Seiberg-Witten invariants and symplectic forms. *Mathematical Research Letters*, 1:809–822, 1994.

[79] C. H. Taubes. The Seiberg-Witten and Gromov invariants. *Mathematical Research Letters*, 2(2), 1995.

[80] C. H. Taubes. Counting pseudo-holomorphic submanifolds in dimension 4. *Journal of Differential Geometry*, 44(4):818–893, 1996.

[81] C. H. Taubes. SW⇒Gr: from the Seiberg-Witten equations to pseudo-holomorphic curves. *Journal of the American Mathematical Society*, 9(3), 1996.

[82] Isreal Vainsencher. Enumeration of n-fold tangent hyperplanes to a surface. *Journal of Algebraic Geometry*, 4(3):503–526, 1995.

[83] Ravi Vakil. Genus g Gromov-Witten invariants of Del Pezzo surfaces: Counting plane curves with fixed multiple points. preprint: alg-geom/9709004, 1997.

[84] Ravi Vakil. Enumerative geometry of plane curves of low genus. preprint: math.AG/9803007, 1998.

[85] Claire Voisin. A mathematical proof of a formula of Aspinwall and Morrison. *Compositio Math.*, 104(2):135–151, 1996.

[86] Claire Voisin. *Symétrie miroir*, volume 2 of *Panoramas et Synthèses [Panoramas and Syntheses]*. Société Mathématique de France, Paris, 1996.

[87] P. M. H. Wilson. The role of c_2 in Calabi-Yau classification—a preliminary survey. In *Mirror symmetry, II*, volume 1 of *AMS/IP Stud. Adv. Math.*, pages 381–392. Amer. Math. Soc., Providence, RI, 1997.

[88] E. Witten. Monopoles and four-manifolds. *Mathematical Research Letters*, 1:769–796, 1994.

[89] Edward Witten. The Verlinde algebra and the cohomology of the Grassmannian. In *Geometry, topology, & physics*, Conf. Proc. Lecture Notes Geom. Topology, IV, pages 357–422. Internat. Press, Cambridge, MA, 1995.
[90] S.-T. Yau. On the Ricci curvature of a compact Kähler manifold and the complex Monge Ampère equation I. *Com. Pure and Appl. Math*, 31:339–411, 1978.
[91] S.-T. Yau and E. Zaslow. BPS states, string duality, and nodal curves on $K3$. *Nuclear Physics B*, 471(3):503–512, 1996. Also: hep-th/9512121.

Addresses:

DEPARTMENT OF MATHEMATICS
TULANE UNIVERSITY
6823 ST. CHARLES AVENUE
NEW ORLEANS, LA 70118

SCHOOL OF MATHEMATICS
UNIVERSITY OF MINNESOTA
MINNEAPOLIS, MN 55455

Special Lagrangian Fibrations II: Geometry.

A Survey of Techniques in the Study of Special Lagrangian Fibrations

Mark Gross*

Mathematics Institute
University of Warwick
Coventry, CV4 7AL
mgross@maths.warwick.ac.uk

§0. Introduction.

This paper is a progress report on work surrounding the Strominger-Yau-Zaslow mirror symmetry conjecture [28]. Roughly put, this conjecture suggests the following program for attacking an appropriate form of the mirror conjecture. Let X be a Calabi-Yau n-fold, with a large complex structure limit point p in a compactification of the complex moduli space of X. One expects mirrors of X to be associated to such boundary points of the complex moduli space of X. For complex structures on X in an open neighborhood of the boundary point p and suitable choice of a Ricci flat metric on X, one attempts to construct the mirror of X via the following program:

(1) There is an n-torus representing a homology class in $H_n(X, \mathbf{Z})$ which is invariant under all monodromy transformations about the discriminant locus passing through the point p. (See [14], §3 for details of this representative). The first task is to find an homologous n-torus which is special Lagrangian.

(2) Having found one special Lagrangian torus, show that it deforms to yield a fibration $f : X \rightarrow B$ all of whose fibres are special Lagrangian and whose general fibre is an n-torus.

(3) Construct the dual n-torus fibration as follows. Let $B_0 \subseteq B$ be the complement of the discriminant locus of f, $f_0 : X_0 \rightarrow B_0$ the restriction of f to $X_0 = f^{-1}(B_0)$. The dual n-torus fibration over B_0 is $\check{X}_0 = R^1 f_{0*}\mathbf{R}/R^1 f_{0*}\mathbf{Z} \rightarrow B_0$. Find a suitable compactification of $\check{X}_0$ to a manifold $\check{X}$ along with a fibration $\check{f} : \check{X} \rightarrow B$.

(4) Show that X and $\check{X}$ satisfy a topological form of mirror symmetry. It is not clear what this means in arbitrary dimension, but for threefolds, this will be

* Supported in part by NSF grant DMS-9700761

an isomorphism $H^{even}(X,\mathbf{Q}) \cong H^{odd}(\check{X},\mathbf{Q})$ and $H^{odd}(X,\mathbf{Q}) \cong H^{even}(\check{X},\mathbf{Q})$. This in particular implies the usual interchange of Hodge numbers for threefolds. One might also dare to hope these isomorphisms also hold over $\mathbf{Z}$; we will see this will often be the case in Theorem 3.10.

(5) Put a complex and Kähler structure on $\check{X}$. The choice of such structures determines the mirror map. One expects that the complex structure on X should entirely determine the Kähler structure on $\check{X}$, while the Kähler structure on X along with a choice of the B-field, a cohomology class in $H^2(X,\mathbf{R}/\mathbf{Z})$, or a related cohomology group, will determine the complex structure on $\check{X}$. In [16], a somewhat more precise conjecture was given as to how this interchange of structures should look on the level of cohomology. Specifically, let ω, Ω be the Kähler form and holomorphic n-form on X with Ω normalised so that $\int_{X_b}\Omega = 1$. In addition, one is given a choice of B-field, which right now we'll take to be a cohomology class $\mathbf{B} \in H^1(B, R^1 f_*\mathbf{R})$. (The B-field will always be denoted by a bold-face $\mathbf{B}$ to differentiate it typographically from the base B of the fibration.) The choice of Kähler and complex structures on the mirror, determined by forms $\check{\omega}$ and $\check{\Omega}$, should satisfy the following relationship: using the identifications $H^1(B,R^1 f_*\mathbf{R}) \cong H^1(B,R^{n-1}\check{f}_*\mathbf{R})$ and $H^1(B,R^{n-1}f_*\mathbf{R}) \cong H^1(B,R^1\check{f}_*\mathbf{R})$ which conjecturally hold, the following identities should hold in these cohomology groups:

$$[\check{\omega}] = [\operatorname{Im}\Omega]$$
$$[\operatorname{Im}\check{\Omega}] = [\omega]$$
$$[\operatorname{Re}\check{\Omega}] - [\sigma_0] = \mathbf{B}$$

where σ_0 is a chosen zero-section of $\check{f} : \check{X} \to B$.

(6) Show that the above procedure yields the correct enumerative predictions for Gromov-Witten invariants of X and $\check{X}$.

This program is still a long way from completion, and this paper represents only one small step in this direction. Not much is known about items (1) and (2) yet; this may well prove to be the hardest part of the program. In [16], we gave examples of special Lagrangian T^3-fibrations with a degenerate metric. In [32], T^n-fibrations are constructed on Calabi-Yau hypersurfaces in smooth toric varieties. These fibrations are constructed as deformations of the natural T^n-fibration on the large complex structure limit given by the moment map. Unfortunately these tori are neither special nor Lagrangian, but this fibration may be sufficient for a purely topological version of mirror symmetry and provides evidence for the SYZ conjecture.

We will not address issues (1) and (2) further in this paper. Instead, throughout this paper, we assume the existence of a special Lagrangian fibration $f : X \to B$ on X, much as we did in [14]. However, we wish to delve more deeply into the properties of such fibrations, and to do so, we need to make reasonable guesses as to what kind of regularity properties such fibrations will possess. We discuss these issues and assumptions in §1. In [14], we did not make real use of the fact that $f : X \to B$ was special Lagrangian, but rather only used the topological fact that f was a torus fibration, along with a technical condition we called *simplicity* to control the cohomological contributions of the singular fibres. In this paper, we try to make more serious use of the special Lagrangian condition. We find that we make very heavy use of the Lagrangian condition, but as yet we do not use the additional special condition in a profound way. We do however use the fact that special Lagrangian submanifolds are in fact volume minimizing. This allows us to provide moral guidelines as to what we might expect of special Lagrangian fibrations, by drawing on the wealth of material known about volume minimizing rectifiable currents. Some of these ideas are discussed in §1.

Of course, the study of Lagrangian torus fibrations is a familiar one in the subject of completely integrable Hamiltonian systems. In §2 of this paper we review and generalise slightly for our purposes Duistermaat's work on global action-angle coordinates [12]. Simplifying a bit, if $f : X \to B$ is a Lagrangian T^n fibration with a Lagrangian section and if $X^\#$ is the complement of the critical locus of f in X, then one can canonically write $X^\#$ as a quotient of the cotangent bundle of B by a possibly degenerating family of lattices suitably embedded in $\mathcal{T}_B^*$. Furthermore, the canonical symplectic form on $\mathcal{T}_B^*$ descends to the symplectic form on $X^\#$. Thus if X is a Calabi-Yau manifold, the existence of a special Lagrangian fibration gives us coordinates on a large open subset of X on which it is easy to write the symplectic form. I think of these coordinates as *special Lagrangian coordinates*. These can be compared with traditional complex coordinates, where it is easy to write the holomorphic n-form $(dz_1 \wedge \cdots \wedge dz_n)$, but very difficult to write down the Kähler form of a Ricci flat metric. Thus we should expect that in the coordinates given by a special Lagrangian fibration, the difficulty will be to write down the holomorphic n-form, or equivalently, the complex structure.

Taking this analogy further, we note that complex coordinates give a filtration on de Rham cohomology, namely the Dolbeault cohomology groups. Similarly, special Lagrangian coordinates can be thought of as giving rise to a filtration on cohomology, namely that given by the Leray spectral sequence associated to the special Lagrangian fibration. Just as the Dolbeault cohomology groups give rise to the Hodge filtration, we saw in [14] that the Leray filtration should, modulo some conjectures about monodromy, give rise to the monodromy weight filtration associated to the large complex structure limit point.

In [14], we studied the Leray spectral sequence of f with coefficients in $\mathbf{Q}$. We now show in §3, modulo suitable regularity assumptions stated in §1, that three-dimensional special Lagrangian fibrations satisfy the condition of $\mathbf{Z}$-simplicity introduced in [14]. This is a stronger hypothesis then was used in [14], and as a result, we can analyse the Leray spectral sequence over $\mathbf{Z}$. This leads to some interesting results as to the role torsion in cohomology plays in mirror symmetry. In particular, it provides an explanation of the phenomenon proposed in [5] of the role of "discrete torsion" (torsion in $H^3(X, \mathbf{Z})$) in mirror symmetry. This is discussed in §3. In addition, the analysis of the Leray spectral sequence over $\mathbf{Z}$ sheds light on item (4) of the program proposed above.

Moving on, we next address the subject of putting symplectic and complex structures on the mirror $\check{X}$. This subject has already been discussed to some extent by Hitchin in [18]. Our approach is inspired by Hitchin's, and is a generalisation of that approach. In fact, one of the goals accomplished is to restate Hitchin's constructions in a more coordinate independent form, so as to allow us to understand the cohomological ramifications of these constructions.

One should in fact consider any special Lagrangian submanifold $M \subseteq X$, and consider the moduli space of deformations of this submanifold. Calling this moduli space B, the D-brane moduli space of M is then the set of pairs (M', α) where α is a flat $U(1)$ connection modulo gauge equivalence on M' a deformation of M. This is a T^s-bundle over B, where $s = b_1(M)$. Specifically, if $f : \mathcal{U} \to B$ is the universal family of special Lagrangian submanifolds parametrized by $B, \mathcal{U} \subseteq B \times X$, then the D-brane moduli space is $\mathcal{M} = R^1 f_* \mathbf{R}/R^1 f_* \mathbf{Z} \to B$. In addition, Mclean [22] gives us a canonical isomorphism between the tangent bundle of B, $\mathcal{T}_B$, and $R^1 f_* \mathbf{R} \otimes C^\infty(B)$. This gives a canonical embedding of $R^1 f_* \mathbf{Z}$ in $\mathcal{T}_B$. However, $\mathcal{T}_B$ of course does not carry a canonical symplectic form; rather, it is $\mathcal{T}_B^*$ which does. So to find a symplectic form on $\mathcal{M}$, we need to reembed $R^1 f_* \mathbf{Z}$ in $\mathcal{T}_B^*$. There are two ways of doing this. One is to use periods integrals related to $\operatorname{Im} \Omega$, the imaginary part of the holomorphic n-form on X; the other is to use a canonical metric introduced by McLean on $\mathcal{T}_B$ to identify $\mathcal{T}_B$ with $\mathcal{T}_B^*$. In fact these two methods give the same embedding of $R^1 f_* \mathbf{Z}$ in $\mathcal{T}_B^*$. This allows us to define a symplectic form on $\mathcal{M}$ by writing $\mathcal{M}$ as $\mathcal{T}_B^*/R^1 f_* \mathbf{Z}$ and taking the form on $\mathcal{M}$ induced by the canonical symplectic form on $\mathcal{T}_B^*$. This is the same method as proposed by Hitchin, recast in a slightly more invariant way, which makes it easy to see that the cohomology class of the symplectic form defined in this manner is as predicted by the conjecture of [16]. Thus, if we can understand how to extend this symplectic form to the compactification of the D-brane moduli space, we will solve the first half of item (5). This circle of ideas is discussed in §4.

Because it is easy to write the symplectic form in special Lagrangian coordinates, we expect it to be very difficult to write the complex structure. We take up this issue in §§5 and 6. First we explore what data is necessary to place a complex structure on a Lagrangian torus fibration in order to make the fibration special Lagrangian and the induced metric Ricci-flat. A moment's thought shows that given knowledge of ω, to specify an almost complex structure compatible with ω, it is enough to give for each point $b \in B$ a metric g on the fibre X_b, and for each point $x \in X_b$ the Lagrangian subspace $J(\mathcal{T}_{X_b,x}) \subseteq \mathcal{T}_{X,x}$. Then J is completely determined by the requirement that $g(v, w) = \omega(Jv, w)$. The collection of subspaces $J(\mathcal{T}_{X_b,x})$ can be thought of as the horizontal subspaces of an Ehresmann connection on $f : X^\# \to B$. We then need to ask when this data determines an integrable complex structure which induces a Ricci-flat metric in which the fibres of f are special Lagrangian.

Following [18], it is easier to determine this by describing the holomorphic n-form Ω. In local coordinates $y_1, \ldots, y_n$ on the base, and canonical coordinates $x_1, \ldots, x_n$ on the fibres of $\mathcal{T}_B^*$, we can write the general form of Ω as

$$\Omega = V \bigwedge_{i=1}^{n} (dx_i + \sum_{j=1}^{n} \beta_{ij} dy_j).$$

where β_{ij} is a complex-valued function on $\mathcal{T}_B^*$ and V is a real-valued function. The Ehresmann connection is in fact encoded in $\operatorname{Re} \beta$, while $\operatorname{Im} \beta$ is just the inverse of the metric on the fibres. The integrability and Ricci-flatness conditions are then easy to write down. In fact, we show that one needs the conditions:

(1) The matrix $\beta = (\beta_{ij})$ is symmetric, $\operatorname{Im}\beta$ is positive definite, and $V = 1/\sqrt{\det \operatorname{Im}\beta}$.

(2) $d\Omega = 0$.

The first condition is of course easily achieved, but the second condition is a quite subtle condition. The second condition is what requires real effort, and understanding it is really at the heart of the SYZ program. It turns out that (2) is equivalent to the following three conditions:

(1) The almost complex structure is integrable. (Note that $d\Omega = 0$ is a much stronger condition than integrability. Given any Lagrangian fibration on a Calabi-Yau manifold with a holomorphic n-form Ω, by replacing Ω by $e^{i\theta(x)}\Omega$ for a suitable function $\theta(x)$, one can ensure that $\operatorname{Im}\Omega$ restricted to each fibre is zero. However now Ω is no longer holomorphic.)

(2) $dx_1, \ldots, dx_n$ are harmonic 1-forms on X_b for each $b \in B$. (That this is a necessary condition follows from Mclean's results.)

(3) The volume form $V dx_1 \wedge \cdots \wedge dx_n$ on fibres is parallel under translation via the Ehresmann connection.

In [18], Hitchin constructed a complex structure on the D-brane moduli space by specifying the holomorphic n-form Ω. The form he constructed had very special properties: he used $\operatorname{Re}\beta = 0$ and $\operatorname{Im}\beta$ constant along fibres. As a result, the integrability condition $d\Omega = 0$ was much easier to check. In the general case, it requires a great deal more effort to analyse this equation, and the results of §5 and 6 represent only a beginning. The conditions (1)-(3) above should perhaps be thought of as mirror equations to the usual complex Monge-Ampère equations which arise in the study of Ricci-flat metrics. Understanding their solution should be one of the key steps in the Strominger-Yau-Zaslow program. We analyse the solution in several simple cases. For example, we show that if the Ehresmann connection is in fact flat, then the metric on each fibre must be flat. This explains why in Hitchin's situation, where the connection was trivial, it would be impossible to obtain any solutions which are not flat along the fibres. In particular, the connection clearly cannot be flat if the fibration possesses singular fibres. We also give a whole family of solutions related to Hitchin's, but which take the B-field into account. This gives some hint as to the role the B-field plays and leads us to a refined form of the mirror symmetry conjecture (Conjecture 6.6). We will argue that the B-field should not be thought of as an element of $H^2(X, \mathbf{R}/\mathbf{Z})$ but rather as an element of $H^1(B, R^1 f_* \mathbf{R}/\mathbf{Z})$. Thus the group the B-field takes values in should depend not just on X but on the fibration f. These two groups can in fact be different, so this suggestion is quite a serious modification of previous interpretations of the B-field (for example that of [5]).

In §7 we apply much of what we have done to the situation of K3 surfaces, by way of an extended example. Here we know that special Lagrangian fibrations do exist, and we know the precise construction of the mirror map. We are able to show that the various recipes and constructions given in earlier sections of the paper can be carried out completely for K3 surfaces. In particular, we obtain an almost purely differential geometric description of mirror symmetry for K3 surfaces.

Finally, in §8, we give a brief discussion of results about the Strominger-Yau-Zaslow conjecture obtained since the initial version of this paper was prepared.

I would like to thank P. Aspinwall, D. Calderbank, N. Hitchin, M. Micallef, A. Todorov, P. Wilson, S.-T. Yau, and E. Zaslow for useful discussions. In addition, I would like to thank S.-T. Yau for his hospitality at Harvard, where some of this work was carried out.

Convention. For a p-form α and tangent vectors $v_1, \ldots, v_q$, $\iota(v_1, \ldots, v_q)\alpha$ denotes the $p-q$-form $\alpha(v_1, \ldots, v_q, \cdot)$.

§1. Special Lagrangian Fibrations.

In what follows, X will denote a complex Calabi-Yau manifold with a Ricci-flat Kähler metric g, Kähler form ω, and holomorphic n-form Ω. The form Ω is always normalised to be of unit length, i.e.

$$\omega^n/n! = (-1)^{n(n-1)/2}(i/2)^n\Omega \wedge \bar{\Omega}.$$

This only fixes Ω up to a phase factor. As the metric is Ricci-flat, we can take the metric to be real analytic.

Recall

Definition-Proposition 1.1. *[17]* $\mathrm{Re}\,\Omega$ *is a calibration, called the special Lagrangian calibration. An n-dimensional real submanifold $M \subseteq X$ is special Lagrangian if* $\mathrm{Re}\,\Omega|_M = Vol(M)$. *Modulo orientation, a submanifold $M \subseteq X$ of real dimension n is special Lagrangian if and only if $\omega|_M = 0$ and* $\mathrm{Im}\,\Omega|_M = 0$.

It is natural to extend the notion of special Lagrangian submanifolds to special Lagrangian integral currents, as done in [17]. For an introduction to integral and rectifiable currents and geometric measure theory, see [23], and for more rigourous treatments see [13] and [27]. We will not make much use of the language of geometric measure theory here except to justify some of the assumptions made on special Lagrangian fibrations. The reader unfamiliar with this language should just keep in mind that passing from submanifolds to integral currents means extending the class of submanifolds to subsets with integer multiplicities attached, over which one can still integrate forms. There are natural compactness theorems for integral currents, enabling one to easily construct volume minimizing currents. One can then try to control the singularities of such currents. The biggest regularity result of this nature is Almgren's monumental

Theorem 1.2. *[1,2] Suppose N is an $m+l$ dimensional submanifold of $\mathbf{R}^{m+n}$ of class $k+2$ and that T is an m-dimensional rectifiable current in $\mathbf{R}^{m+n}$ which is absolutely area minimizing with respect to N. Then there is an open subset U of $\mathbf{R}^{m+n}$ such that $Supp(T) \cap U$ is an m-dimensional minimal submanifold of N of class k and the Hausdorff dimension of $Supp(T) - (U \cup Supp(\partial T))$ does not exceed $m-2$.*

The proof [2] is unpublished and is over 1500 pages long in preprint form. Thus an urgent question is

Question 1.3. *Are there nice regularity theorems for special Lagrangian currents?*

For example, [19] shows that holomorphic integral currents are obtained by integration over complex analytic subvarieties, and of course the singular locus of such varieties is well-behaved. This is much stronger than Almgren's result, which doesn't even guarantee finite $m-2$ dimensional Hausdorff measure of the singular set. We do not want the theory of special Lagrangian currents to have to depend on Almgren's result.

Next we should consider what is a reasonable definition of a special Lagrangian fibration. It might be reasonable to say that

Definition 1.4. *If B is a topological space and $f : X \rightarrow B$ is a continuous map, we say f is a special Lagrangian fibration if for all $b \in B$, $X_b := f^{-1}(b)$ is the support of a special Lagrangian integral current T with $\partial T = 0$.*

Even this might be too strong; one might insist that the fibres be special Lagrangian on a dense subset of B. This would allow some fibres to jump dimension. Nevertheless, we do not expect fibres to decrease in dimension as this would suggest the cohomology class of the general fibre X_b was trivial, which contradicts $\int_{X_b} \Omega \neq 0$. I do not want to give a rigorous argument of this sort here as it requires being clearer about concepts such as the dimension of the fibre. However it is clear that special Lagrangian fibrations cannot behave as many completely integrable Hamiltonian systems do, in which some fibres are tori of smaller dimension. On the other hand, it is not as clear that we want to rule out the possibility of fibres jumping up in dimension, something which often happens in algebro-geometric contexts. Nevertheless, we will stick to Definition 1.4.

We are interested in very specific sorts of special Lagrangian T^n fibrations. As argued in [14], §3, we are looking for special Lagrangian fibrations on Calabi-Yau manifolds near a specific large complex structure limit point in the boundary of complex moduli space, and the homology class of a fibre should be represented by a specific vanishing cycle associated to the boundary point. It was argued in [14], Observation 3.4 that, in the three dimensional case, if this homology class is primitive, then for *general* choice of complex moduli near the large complex structure limit point, all fibres of $f : X \rightarrow B$ must be irreducible. Let us be more precise. There is a notion of *indecomposable integral current* ([13], 4.2.25) which is analagous to the notion of irreducibility in algebraic geometry, and we will say a fibre $f^{-1}(b)$ is irreducible if the current T_b obtained by integrating forms over $f^{-1}(b)$ (with orientation induced by Ω) is indecomposable.

Definition 1.5. *A special Lagrangian fibration is integral if each fibre is irreducible and the currents $\{T_b | b \in B\}$ all represent the same integral homology class.*

This last condition rules out fibres which need to be thought of as multiple fibres. However, it is difficult to define multiplicity of a fibre in the context of a continuous map. Here the term "integral" is being used in its algebro-geometric sense: each fibre is irreducible and "reduced".

Now a special Lagrangian fibration need not be integral any more than an elliptic fibration need only have integral fibres. But this is the generic behaviour of special Lagrangian fibrations in dimensions ≤ 3, and the assumption of integrality vastly decreases the range of potential singular fibres. Without assumptions of integrality, it would be very difficult to relate the cohomology of a T^n-fibration and its dual. But one should keep in mind that even in dimension 3, integrality should fail for special values of the complex structure. Furthermore, if the homology class of a fibre is not primitive, then there is still the possibility that integrality will fail, and perhaps even multiple fibres appear. I have no argument to rule this out, but as we shall see in later sections, it is special Lagrangian fibrations with sections which are most important, and for these the homology class of a fibre is primitive.

The next natural question is whether we expect B itself to be a manifold. I would like to give a rough argument that this is a natural expectation if the fibration $f : X \rightarrow B$ is integral. Indeed, given a fibre X_b, X_b will be smooth at a general point $x \in X_b$, and in a neighborhood of x, using the exponential map, the deformations of X_b can be identified with deformations of X_b inside its normal bundle near x. Thus locally the fibre of the normal bundle of X_b at x yields a natural local section for $f : X \rightarrow B$, and hence gives a manifold structure on B. This construction hopefully in addtion yields a C^∞ structure on B. Thus we will feel justified in making the following assumption, to be in force throughout the remainder of the paper.

Assumption 1.6. *All special Lagrangian fibrations $f : X \rightarrow B$ will be C^∞ maps of C^∞ manifolds. Furthermore, f will be assumed to have a local section at each point $b \in B$. In addition, if f is assumed to be integral, we will assume that for any point $x \in X_b - Sing(X_b)$, there is a local C^∞ section passing through x.*

Again, if the fibration is not integral, I would not be surprised if singular B can arise naturally. Also, since the metric on X is real analytic, we can hope that with suitable coordinates on B, f will in fact be real analytic. We will not assume that here, however.

Next, we pass to the nature of the discriminant locus. Given f integral as in Assumption 1.6, we can now consider the rank of the differential $f_* : \mathcal{T}_{X,x} \rightarrow \mathcal{T}_{B,f(x)}$ at various points $x \in X$. At points with a local C^∞ section, $\operatorname{rank} f_* = n$. It will be shown in Proposition 2.2 that if $\operatorname{rank} f_* = \nu$, then the fibre in fact contains a submanifold of dimension ν on which $\operatorname{rank} f_* = \nu$. Thus the existence of points $x \in X$ for which $\operatorname{rank} f_* = n - 1$ contradicts Almgren's theorem. Even if one doesn't accept Almgren's theorem, it is not difficult to rule out the existence of such points in low dimension using regularity results for minimizing hypersurfaces. Thus it is quite safe to assume that $\operatorname{rank} f_* \neq n-1$ for any point $x \in X$. This gives a stratification of the discriminant locus

$$\Delta_\nu = f(\{x \in X | \operatorname{rank} f_* : \mathcal{T}_{X,x} \rightarrow \mathcal{T}_{B,f(x)} \leq \nu\},$$

with $\Delta_0 \subseteq \cdots \subseteq \Delta_{n-2} = \Delta$, the discriminant locus. Then Federer's generalization of Sard's Theorem, [13], 3.4.3, states that $\mathcal{H}^{\nu+(2n-\nu)/k}(\Delta_\nu) = 0$, where $\mathcal{H}^n$ denotes the n-dimensional Hausdorff measure. In particular, if $k = \infty$, $\mathcal{H}^{\nu+\epsilon}(\Delta_\nu) = 0$ for all $\epsilon > 0$, so Δ_ν is of Hausdorff dimension $\leq \nu$. We need however stronger information than Hausdorff dimension to reach any cohomological conclusions. Thus the following assumptions will be used at various points of this paper; unlike Assumption 1.6, we will assume these only when we need them, mentioning them specifically.

Assumption 1.7. *For an integral special Lagrangian fibration $f : X \to B$, $b \in \Delta_\nu - \Delta_{\nu-1}$, there is a dense subset $\mathcal{L}$ of the real Grassmannian of $n-\nu$ dimensional subspaces of $\mathcal{T}_{B,b}$ such that one can find locally, for each $L \in \mathcal{L}$, a submanifold B' of B passing through b such that $\mathcal{T}_{B',b} = L$ and $B' \cap \Delta_\nu = \{b\}$.*

This requires reasonable regularity results about the discriminant, which certainly hold if $f : X \to B$ is real analytic, so that Δ_ν is a sub-analytic set. A stronger form of this assumption which we will need in §3 and will comment on there is

Assumption 1.7′. *In addition in Assumption 1.7, B' can be found so that $f^{-1}(B')$ is a submanifold in a neighborhood of the fibre X_b.*

Finally, we will require some assumptions on $Sing(X_b)$. Almgren's theorem that the Hausdorff dimension of $Sing(X_b)$ is no more than $n-2$ is not sufficiently strong for most purposes, and at the very least, we will frequently need to use

Assumption 1.8. *If $f : X \to B$ is an integral special Lagrangian fibration then*

$$H^i(Sing(X_b), \mathbf{Z}) = 0$$

for $i > n-2$, for all $b \in B$.

This is a restriction on homological dimension. If f is real analytic, then this should hold given Almgren's result.

§2. Action-angle Coordinates.

There is a standard theory of global action-angle coordinates due to Duistermaat [12]. We will extend this slightly so as to include information about the smooth part of the singular fibres. In this section, $f : X \to B$ denotes any special Lagrangian fibration satisfying Assumption 1.6. However, many of the results in this section apply to C^k Lagrangian fibrations $f : X \to B$ with B a manifold and f having a local section in a neighborhood of each point $b \in B$.

The first observation is that there is an action of $\mathcal{T}^*_B$, the cotangent bundle of B, on X.

Proposition-Definition 2.1. *There is a C^∞ action of $\Gamma(U, \mathcal{T}^*_B)$ on $f^{-1}(U)$ for any $U \subseteq B$, which we write for any $\alpha \in \Gamma(U, \mathcal{T}^*_B)$ as a map $T_\alpha : f^{-1}(U) \to f^{-1}(U)$. This satisfies the following properties:*

(1) If $d\alpha = 0$, then T_α is a symplectomorphism.

(2) T_α acts fibrewise, and $T_\alpha|_{f^{-1}(b)} : f^{-1}(b) \to f^{-1}(b)$ only depends on the value of α at b.

(3) $T_\alpha \circ T_\beta = T_{\alpha+\beta}$.

Proof. This is standard: see for example [12], §1, or [3], Chap. 10. We review the definition of these maps however. If α is a compactly supported 1-form on Y, then $f^*\alpha$ is a compactly supported 1-form on $f^{-1}(U)$. There is then a vector field v_α on $f^{-1}(U)$ with $\iota(v_\alpha)\omega = f^*\alpha$. This generates a flow $\phi_t : f^{-1}(U) \to f^{-1}(U)$ for all t, and we take $T_\alpha = \phi_1$. It is then standard that if $d\alpha = 0$, v is locally Hamiltonian and ϕ_t then is a 1-parameter family of symplectomorphisms.

If locally $d\alpha = 0$, we can write $\alpha = dH$ for some function H on B, and if G is any other function on B, $\{G\circ f, H\circ f\} = 0$ since $f : X \to B$ is a Lagrangian fibration. But then ϕ_t is the Hamiltonian flow associated to H, and

$$0 = \{G\circ f, H\circ f\}(x) = \frac{d}{dt}\Big|_{t=0} (G\circ f)(\phi_t(x))$$

so $G\circ f$ is a constant on $\phi_t(x)$ for any x. Thus ϕ_t acts on fibres. Clearly $v_\alpha|_{f^{-1}(b)}$ depends only on the value of α at b, so the action of ϕ_t on $f^{-1}(b)$ depends only on the value of α at b. In particular ϕ_t acts on fibres for arbitrary α, not just compactly supported α. •

Next, a standard analysis of the orbits of this action.

Proposition 2.2. *If $b\in B$, $x\in f^{-1}(b)$, then the orbit $\{T_\alpha(x)|\alpha\in \mathcal{T}^*_{B,b}\}$ is diffeomorphic to $\mathbf{R}^l\times T^s$ and $l+s$ coincides with $rank(f^* : \mathcal{T}^*_{B,b}\to \mathcal{T}^*_{X,x}) = rank(f_* : \mathcal{T}_{X,x}\to \mathcal{T}_{B,b})$.*

Proof. If $\alpha\in\mathcal{T}^*_{B,b}$ is in the kernel of f^*, then v_α is zero at x, so $T_\alpha(x) = x$. Thus for any $\alpha\in\mathcal{T}^*_{B,b}$, $T_\alpha(x)$ depends only on α modulo $\ker f^*$, and the orbit of x is homeomorphic to a quotient of $V = \mathcal{T}^*_{B,b}/\ker f^*$ by a subgroup Γ, via the map $\alpha\in V\mapsto T_\alpha(x)$. The differential of this map at $0\in V$ is injective, and hence the map $V\to X$ is a diffeomorphism of an open neighborhood of $0\in V$ with its image in X. Thus Γ is a discrete subgroup of V, and the orbit of x is diffeomorphic to V/Γ. •

Suppose first that $f : X\to B$ is equipped with a C^∞ section $\sigma_0 : B\to X$. We define a C^∞ map $\pi : \mathcal{T}^*_B\to X$ by, for $\alpha\in\mathcal{T}^*_{B,b}$, $\pi(\alpha) = T_\alpha(\sigma_0(b))$. The image of the zero section of $\mathcal{T}^*_B$ is $\sigma_0(B)$. Let $\Lambda\subseteq\mathcal{T}^*_B$ be given by $\Lambda = \pi^{-1}(\sigma_0(B))$. Then $\Lambda_b\subseteq\mathcal{T}^*_{B,b}$ is the discrete subgroup of the vector space $\mathcal{T}^*_{B,b}$ given by $\Lambda_b = \pi^{-1}(\sigma_0(b))$. Let $X_0^\#$ be the image of the map π, $f^\# : X_0^\#\to B$ the projection. Then clearly Λ_b is canonically isomorphic to $H_1((f^\#)^{-1}(b),\mathbf{Z})$, which is isomorphic to $H^{n-1}_c((f^\#)^{-1}(b),\mathbf{Z})$ by Poincaré duality. Here we use the fact that f is special Lagrangian to give a canonical orientation on the fibres of $f^\#$. Also, Λ, as a subset of the total space $\mathcal{T}^*_B$, is closed as π is continuous. Since the map π is a local isomorphism, Λ is also étale over B. Thus we can think of Λ as the éspace étalé of $R^{n-1}_c f^\#_*\mathbf{Z}$, and in particular, we obtain an exact sequence of sheaves of abelian groups

$$0\to R^{n-1}_c f^\#_*\mathbf{Z}\to\mathcal{T}^*_B\to X_0^\#\to 0,$$

where this now defines the group structure on $X_0^\#$. Observe also that since Λ is étale over B, $R^{n-1}_c f^\#_*\mathbf{Z}$ has no sections with support in proper closed subsets of B. In particular, $\underline{H}^0_\Delta(B, R^{n-1}_c f^\#_*\mathbf{Z}) = 0$.

We recall the notion of canonical coordinates on the total space of the bundle $\mathcal{T}^*_B$. Given $U\subseteq B$ an open set with coordinates $y_1,\ldots,y_n$, *canonical coordinates* on $\mathcal{T}^*_U$ are $y_1,\ldots,y_n,x_1,\ldots,x_n$, where $(y_1,\ldots,y_n,x_1,\ldots,x_n)$ is the coordinate representation of the differential form $\sum x_i dy_i\in\mathcal{T}^*_{U,(y_1,\ldots,y_n)}$. We will use canonical coordinates consistently throughout this paper, whenever we use local coordinates to perform calculations. Note that $\mathcal{T}_B^*$ always carries a standard symplectic form, which in canonical coordinates can be written as $\sum_{i=1}^n dx_i \wedge dy_i$. (This is the opposite of some sign conventions).

Proposition 2.3. *With notation as above, let $y_1, \ldots, y_n$ be local coordinates on a neighborhood $U \subseteq B$. Then on $\mathcal{T}_U^*$,*

$$\pi^*\omega = \sum_i dx_i \wedge dy_i + \sum_{i,j} a_{ij} dy_i \wedge dy_j$$

where the a_{ij} are functions depending only on $y_1, \ldots, y_n$. Furthermore, if $\sigma_0 : B \to X$ is a Lagrangian section, then

$$\pi^*\omega = \sum_i dx_i \wedge dy_i$$

on U, and thus $\pi^*\omega$ is the standard symplectic form on $\mathcal{T}_B^*$. Finally, if $H^2(B, \mathbf{R}) = 0$, then every section of f is homotopic to a Lagrangian section.

Proof. Since the fibres of $\mathcal{T}_B^* \to B$ are Lagrangian with respect to $\pi^*\omega$, we can locally write

$$\pi^*\omega = \sum_i dx_i \wedge \theta_i + \sum_{i,j} a_{ij} dy_i \wedge dy_j$$

with the θ_i 1-forms not involving the dx_i's and a_{ij} functions on $\mathcal{T}_B^*$. Now the function y_i induces on X a Hamiltonian vector field which, by definition of the map π, must be $\pi_*\partial/\partial x_i$. Thus $\iota(\partial/\partial x_i)(\pi^*\omega) = dy_i$, from which we see that $\theta_i = dy_i$.

The condition $d(\pi^*\omega) = 0$ then implies that the functions a_{ij} are independent of $x_1, \ldots, x_n$.

If σ_0 is Lagrangian with respect to ω, then the zero section of $\mathcal{T}_B^*$ is Lagrangian, from which we see that $a_{ij} = 0$.

If σ_0 is not Lagrangian, let $\omega' = \sum_i dx_i \wedge dy_i$ (locally) be the standard symplectic form on $\mathcal{T}_B^*$. Then $\pi^*\omega - \omega'$ is a closed 2-form locally given by $\sum_{i,j} a_{ij} dy_i \wedge dy_j$, and hence is the pull-back of a closed 2-form on B. Thus if $H^2(B, \mathbf{R}) = 0$, there exists a one-form θ on B with $d\theta = \pi^*\omega - \omega'$. Then $-\theta$ defines a section of $\mathcal{T}_B^*$ which is Lagrangian with respect to $\pi^*\omega$, and this maps to a Lagrangian section of f homotopic to σ_0. •

This allows us to prove a result stated in [14]. Recall that $X^\#$ is the complement of the critical locus of f in X.

Theorem 2.4. *(Theorem 3.6 of [14]) Let X be a Calabi-Yau n-fold, B a smooth real n-dimensional manifold, with $f : X \to B$ a C^∞ special Lagrangian torus fibration such that $R^n f_* \mathbf{Q} = \mathbf{Q}_B$ and such that the singular locus of each singular fibre has cohomological dimension $\leq n-2$. Suppose furthermore that f has a C^∞ section σ_0. Then $X^\#$ has the structure of a fibre space of groups with σ_0 the zero section. In fact there is an exact sequence of sheaves of abelian groups*

$$0 \to R^{n-1} f_* \mathbf{Z} \to \mathcal{T}_B^* \to X^\# \to 0.$$

Given a section $\sigma \in \Gamma(U, X^{\#})$, one obtains a C^∞ diffeomorphism $T_\sigma : f^{-1}(U) \cap X^{\#} \to f^{-1}(U) \cap X^{\#}$ given by $x \mapsto x + \sigma(f(x))$, and this diffeomorphism extends to a diffeomorphism $T_\sigma : f^{-1}(U) \to f^{-1}(U)$.

Proof. This follows from the previous discussion if we can show that the hypotheses imply two things:

(1) $X^{\#} = X_0^{\#}$

(2) $R_c^{n-1} f_*^{\#}\mathbf{Z} \cong R^{n-1} f_* \mathbf{Z}$.

To show (1) and (2), let X_b be a fibre, and let $Z \subseteq X_b$ be the singular locus of $Supp(X_b)$. We then have an exact sequence

$$H^{n-2}(Z, \mathbf{Q}) \to H_c^{n-1}(X_b - Z, \mathbf{Q}) \to H^{n-1}(X_b, \mathbf{Q}) \\ \to H^{n-1}(Z, \mathbf{Q}) = 0 \to H_c^n(X_b - Z, \mathbf{Q}) \to H^n(X_b, \mathbf{Q}) \to 0.$$

Here $H^{n-1}(Z, \mathbf{Q}) = 0$ by the assumption on the cohomological dimension of Z. Thus, since $H^n(X_b, \mathbf{Q}) = \mathbf{Q}$ by assumption, $X_b - Z$ can have only one connected component. Since the $\mathcal{T}_{B,b}^*$-orbit of $\sigma_0(b)$ is already one connected component of $X_b - Z$, we see that $X_b - Z = (X_0^{\#})_b$. But as $X_b^{\#} \subseteq X_b - Z$, we must have $X_0^{\#} = X^{\#}$.

We also see from the above exact sequence that there is a surjection

$$R_c^{n-1} f_*^{\#}\mathbf{Z} \to R^{n-1} f_* \mathbf{Z} \to 0.$$

This is an isomorphism outside of Δ. But since $\underline{H}_\Delta^0(B, R_c^{n-1} f_*^{\#}\mathbf{Z}) = 0$, we conclude this surjection is in fact an isomorphism. •

Note that the hypotheses of Theorem 2.4 hold if $f : X \to B$ is integral and satisfies Assumption 1.8. Another useful observation about the topology of singular fibres:

Lemma 2.5. *If $f : X \to B$ is integral and satisfies Assumption 1.8, and $b \in B$ with $Z = Sing(X_b)$, then Z is connected.*

Proof. Let $U = X_b - Z$. As argued above, $U \cong \mathbf{R}^k \times T^{n-k}$ for some k. Assuming Z is non-empty, we have an exact sequence

$$0 = H_c^0(U, \mathbf{Z}) \to H^0(X_b, \mathbf{Z}) \to H^0(Z, \mathbf{Z}) \to H_c^1(U, \mathbf{Z}) \to H^1(X_b, \mathbf{Z})$$

and we wish to show $H^0(Z, \mathbf{Z}) = \mathbf{Z}$, which is equivalent to the injectivity of $H_c^1(U, \mathbf{Z}) \to H^1(X_b, \mathbf{Z})$. If $H_c^1(U, \mathbf{Z}) = 0$, then there is no problem, so the only possiblity is that $k = 1$ and $H_c^1(U, \mathbf{Z}) = \mathbf{Z}$. In this case, $H_c^{n-1}(U, \mathbf{Z}) = \mathbf{Z}^{n-1}$ and thus in a small neighborhood $V \subseteq B$ of b, the sheaf $R_c^{n-1} f_*^{\#}\mathbf{Z}$ contains $\mathbf{Z}^{n-1}$ as a subsheaf. Thus over V, there is a T^{n-1} bundle $T \to B$, $T \subseteq X^{\#}$. The fundamental class of each fibre yields a non-zero section σ of $R_c^1 f_*^{\#}\mathbf{Z}$ over B, and under the map $R_c^1 f_*^{\#}\mathbf{Z} \to R^1 f_* \mathbf{Z}$, σ maps to a section of $R^1 f_* \mathbf{Z}$ which is non-zero on $V - \Delta$. But the map $H_c^1(U, \mathbf{Z}) \to H^1(X_b, \mathbf{Z})$ is the induced map on stalks, and hence is non-zero, thus injective. •

We now address the situation when $f : X \to B$ does not have a Lagrangian section but we assume f is integral. We recall though that Assumption 1.6 is always in force, giving the existence of a local section. This theory was developed in Duistermaat's paper [12], and is completely analagous to Kodaira's theory of elliptic surfaces, or to Ogg-Shafarevich theory.

Let $X^\# = X - Crit(f)$ as usual. We now obtain a map

$$\psi : R_c^{n-1} f_*^\# \mathbf{Z} \to \mathcal{T}_B^*$$

as follows. Given a fibre $X_b^\#$, $\gamma \in H_1(X_b^\#, \mathbf{Z}) \cong H_c^{n-1}(X_b^\#, \mathbf{Z})$, map γ to the differential

$$v \mapsto - \int_\gamma \iota(v)\omega$$

where we choose any lifting of $v \in \mathcal{T}_{B,b}$ to $X^\#$.

Now in the case f did have a section, we previously constructed an embedding $R_c^{n-1} f_*^\# \mathbf{Z} \hookrightarrow \mathcal{T}_B^*$. We compare these two constructions. Let $U \subseteq B$ be an open set where $f^{-1}(U) \to U$ possesses a section, which we can take to be Lagrangian. Using this section as the zero section we obtain an exact sequence

$$0 \to \Lambda = R_c^{n-1} f_*^\# \mathbf{Z}|_U \xrightarrow{\psi'} \mathcal{T}_U^* \xrightarrow{\pi} f^{-1}(U)^\# \to 0$$

as before. Now if $\lambda \in \Lambda_b = H_c^{n-1}(X_b^\#, \mathbf{Z}) = H_1(X_b^\#, \mathbf{Z}) \subseteq \mathcal{T}_{B,b}^*$ via the map ψ', then in local coordinates $(y_1, \ldots, y_n)$ on B, $\lambda = \sum \lambda_i dy_i$. Now

$$\begin{aligned} -\int_\lambda \iota(\partial/\partial y_i)\omega &= -\int_{(0,\ldots,0)}^{(\lambda_1,\ldots,\lambda_n)} \iota(\partial/\partial y_i)\pi^*\omega \\ &= \int_{(0,\ldots,0)}^{(\lambda_1,\ldots,\lambda_n)} dx_i \\ &= \lambda_i \end{aligned}$$

so the two maps $\psi, \psi' : R_c^{n-1} f_*^\# \mathbf{Z}|_U \to \mathcal{T}_U^*$ coincide. Since $\Lambda \subseteq \mathcal{T}_U^*$ is Lagrangian, we conclude that the image of $R_c^{n-1} f_*^\# \mathbf{Z}$ in $\mathcal{T}_B^*$ under ψ is Lagrangian. Thus there is an exact sequence

$$0 \to R_c^{n-1} f_*^\# \mathbf{Z} \to \mathcal{T}_B^* \to J^\# \to 0$$

defining $J^\#$ in which $J^\#$ inherits the standard symplectic form from $\mathcal{T}_B^*$. Since $f : X^\# \to B$ is locally isomorphic to $J^\# \to B$, a standard argument [12] shows that one can obtain $X^\# \to B$ from $J^\# \to B$ by regluing using a Čech 1-cocycle $\{(U_i, \sigma_i)\}$ where σ_i is a Lagrangian section of $J^\# \to B$ over U_i. We call $j : J^\# \to B$ the *Jacobian fibration* of $f : X^\# \to B$, in analogy with the theory of elliptic curves. This gives a one-to-one correspondence between the group $H^1(B, \Lambda(J^\#))$, where $\Lambda(J^\#)$ is the sheaf of Lagrangian sections of $J^\#$, and the set

$$\{f : Y^\# \to B \text{ a Lagrangian fibration with local section and Jacobian } j : J^\# \to B\}/\cong .$$

In fact, this can be extended to the compactifications. We phrase this more generally for Lagrangian fibrations.

Theorem 2.6. *Let $f : X \to B$ be a proper Lagrangian fibration with connected fibres, with a local section everywhere. Then there is a symplectic manifold J, called the Jacobian of f, and a (proper) Lagrangian fibration $j : J \to B$ which is locally isomorphic to $f : X \to B$, and which has a Lagrangian section. Furthermore, there is a one-to-one correspondence between the sets*

$$\{f : Y \to B \text{ a Lagrangian fibration with local section and Jacobian } j : J \to B\}/\cong$$

and $H^1(B, \Lambda(J^\#))$.

Proof. To construct J, choose an open covering $\{U_i\}$ such that $f^{-1}(U_i) \to U_i$ has a Lagrangian section σ_i for each i. Then $f^{-1}(U_i \cap U_j)$ has a symplectomorphism we will write as $T_{i,j}$ obtained by treating σ_i as the zero-section and then translating by σ_j, so that $T_{i,j}$ takes σ_i into σ_j. We construct J by identifying $f^{-1}(U_i)$ and $f^{-1}(U_j)$ along $f^{-1}(U_i \cap U_j)$, using $T_{i,j}$ to identify $f^{-1}(U_i \cap U_j) \subseteq f^{-1}(U_i)$ and $f^{-1}(U_i \cap U_j) \subseteq f^{-1}(U_j)$. These identifications are compatible over $U_i \cap U_j \cap U_k$ because $T_{j,k} \circ T_{i,j} = T_{i,k}$. Thus we obtain a Lagrangian fibration $j : J \to B$ with a section, as desired. The usual regluing construction gives the 1-1 correspondence. •

For $j : J \to B$ special Lagrangian and integral, one computes $H^1(B, \Lambda(J^\#))$ by using the exact sequence

$$0 \to R_c^{n-1} j_*^\# \mathbf{Z} \to \Lambda(\mathcal{T}_B^*) \to \Lambda(J^\#) \to 0.$$

$\Lambda(\mathcal{T}_B^*)$ is just the kernel of exterior differentiation acting on $\mathcal{T}_B^*$, so $H^i(B, \Lambda(\mathcal{T}_B^*)) = H^{i+1}(B, \mathbf{R})$ for $i \geq 1$. From this we obtain the sequence

$$H^2(B, \mathbf{R}) \to H^1(B, \Lambda(J^\#)) \to H^2(B, R_c^{n-1} j_*^\# \mathbf{Z}) \to H^3(B, \mathbf{R}).$$

In any dimension, if $H^2(B, \mathbf{R}) = 0$, then $H^2(B, R_c^{n-1} j_*^\# \mathbf{Z})_{tors} \subseteq H^1(B, \Lambda(J^\#))$. Duistermaat [12] observed that if an element $\alpha \in H^1(B, \Lambda(J^\#))$ comes from an element $[\alpha'] \in H^2(B, \mathbf{R})$, then the corresponding $f : X \to B$ can be obtained by choosing a 2-form α' on B representing $[\alpha']$, and taking $X = J$ with symplectic form $\omega + f^*\alpha'$. Any two choices of α' can be related by translation by a section.

Example 2.7. If $n = 3$ and $H^2(B, \mathbf{R}) = 0$, then we have

$$H^2(B, R_c^2 f_*^\# \mathbf{Z})_{tors} \subseteq H^1(B, \Lambda(J^\#)) \subseteq H^2(B, R_c^2 f_*^\# \mathbf{Z}).$$

If $n = 2$ and J is a K3 surface, then $H^2(B, R_c^1 f_*^\# \mathbf{Z}) = 0$ and so there is a sequence

$$H^1(B, R_c^1 f_*^\# \mathbf{Z}) \to H^2(B, \mathbf{R}) \to H^1(B, \Lambda(J^\#)) \to 0.$$

Remark 2.8. If $U \subseteq B - \Delta$ is a simply connected set, then there is no monodromy in the local system $(R^{n-1} f_* \mathbf{Z})|_U \subseteq \mathcal{T}_U^*$. Thus, if $\lambda_1, \ldots, \lambda_n$ are sections of $\mathcal{T}_U^*$ generating $(R^{n-1} f_* \mathbf{Z})|_U$, the fact that $d\lambda_i = 0$ shows there are functions u_i such that $du_i = \lambda_i$ on U. The $u_1, \ldots, u_n$ form local coordinates on U, since $du_1, \ldots, du_n$ are independent. This is the standard construction of action coordinates on U. We will say that $u_1, \ldots, u_n$ are *action coordinates* for $f : X \to B$ on U. Canonical coordinates $u_1, \ldots, u_n, x_1, \ldots, x_n$ are called action-angle coordinates. These are also the coordinates that Hitchin introduces after [18], Prop. 1. The advantage of working in this coordinate system is that the periods are now just the constant periods $du_1, \ldots, du_n$.

§3. Simplicity and the Leray spectral sequence revisited.

Recall from [14] that a special Lagrangian T^n-fibration $f : X \to B$ was said to be *G-simple* if

$$i_* R^p f_{0*} G = R^p f_* G$$

for all p, where $i : B - \Delta \to B$ is the inclusion, $f_0 = f|_{f^{-1}(B-\Delta)}$, and G is an abelian group. This condition was crucial for getting a handle on the topology of X and the relationship between the topology of X and its dual. In [14], we only made use of $\mathbf{Q}$-simplicity, while here, we will go further. We are interested in a broader range of groups G. In particular, $G = \mathbf{R}, \mathbf{Q}, \mathbf{Z}, \mathbf{Z}/m\mathbf{Z}$, or $\mathbf{R}/\mathbf{Z}$ will be of relevance for us. Clearly $\mathbf{Z}$-simplicity implies $\mathbf{Q}$-simplicity or $\mathbf{R}$-simplicity, but $\mathbf{Z}/m\mathbf{Z}$ or $\mathbf{R}/\mathbf{Z}$-simplicity provides extra information about monodromy modulo m which may be valuable in trying to classify possible monodromy transformations about the discriminant locus.

If f is integral, then f has connected fibres, and $G = f_* G = i_* f_{0*} G$. We have seen in §2 that if Assumption 1.8 holds then $R^n f_* \mathbf{Z} = R^n_c f^{\#}_* \mathbf{Z} = \mathbf{Z}$, and thus by the universal coefficient theorem, $R^n f_* G = G$. Thus for integral fibrations satisfying Assumption 1.8, the simplicity condition holds for $p = 0$ and n and any abelian group G.

We now prove it holds for $p = 1$, under the additional Assumption 1.7′.

Definition 3.1. *We say a point $b \in B$ is a rank k point if*

$$k = \min_{x \in X_b} \operatorname{rank} f_* : \mathcal{T}_{X,x} \to \mathcal{T}_{B,b}.$$

In particular b is rank n if and only if f is smooth over b.

We first comment on when Assumption 1.7′ might hold.

Lemma 3.2. *If b is a rank k point, $k \neq 0$, then Assumption 1.7 implies Assumption 1.7′ at b if X_b contains only a finite number of orbits of the action of $\mathcal{T}^*_{B,b}$. Note Assumption 1.7 automatically implies Assumption 1.7′ at a rank 0 point.*

Proof. Consider the set $\mathcal{L}' = \{\operatorname{Im} f_* : \mathcal{T}_{X,x} \to \mathcal{T}_{B,b} | x \in X_b\}$ of subspaces of $\mathcal{T}_{B,b}$. Note that if $x, y \in X_b$ with $T_\sigma(x) = y$ for some σ, then $f_{*,x} = f_{*,y} \circ (T_\sigma)_*$ where $f_{*,x}$ denotes the pushforward of tangent vectors at x. Thus $\operatorname{Im} f_{*,x} = \operatorname{Im} f_{*,y}$. By the assumption that X_b is a union of a finite number of orbits, we see then that $\mathcal{L}'$ is finite, and $\min_{L \in \mathcal{L}'} \dim L = k$. Because $\mathcal{L}'$ is finite we can choose a subspace $T \subseteq \mathcal{T}_{B,b}$ of dimension $n - k$ in the set $\mathcal{L}$ given by Assumption 1.7 which intersects every element of $\mathcal{L}'$ in the expected dimension. It then follows that $\dim f_{*,x}^{-1}(T) = 2n - k$ for all $x \in X_b$. Now if B' is taken to be a submanifold of B with tangent space T at b, we see that the implicit function theorem implies that $f^{-1}(B')$ is a manifold in a neighborhood of X_b. •

This makes Assumption 1.7$'$ appear quite reasonable, at least in the three-dimensional case, as the finiteness of the number of orbits for rank 1 fibres would follow from finiteness of the $\mathcal{H}^1$-measure of the singular locus. (As mentioned in §1 however, such a result is not actually known.)

Theorem 3.3. *If $f : X \to B$ is an integral special Lagrangian fibration satisfying Assumption 1.7$'$, then $i_* R^1 f_{0*} G = R^1 f_* G$.*

Proof. Let

$$\Delta_0 \subseteq \Delta_1 \subseteq \cdots \subseteq \Delta_{n-2} = \Delta \subseteq B$$

be the stratification of Δ given in §1. Let $i_k : B - \Delta_k \to B - \Delta_{k-1}$ be the inclusion. We will show using descending induction that

$$i_{k*}(R^1 f_* G)|_{B-\Delta_k} = (R^1 f_* G)|_{B-\Delta_{k-1}}$$

for each k. One always has a functorial map

$$(R^1 f_* G)|_{B-\Delta_{k-1}} \to i_{k*} i_k^* (R^1 f_* G)|_{B-\Delta_{k-1}} = i_{k*}(R^1 f_* G)|_{B-\Delta_k},$$

and we just need to show this is an isomorphism on the level of stalks at each point $b \in \Delta_k$, $b \notin \Delta_{k-1}$.

We can choose a B' through the point b using Assumption 1.7$'$, so that $\Delta_k \cap B' = \{b\}$ and $X' = f^{-1}(B')$ is a manifold. This gives a diagram

$$\begin{array}{ccc} B' - \{b\} & \xrightarrow{i'} & B' \\ \downarrow{\scriptstyle j'} & & \downarrow{\scriptstyle j} \\ B - \Delta_k & \xrightarrow{i_k} & B - \Delta_{k-1} \end{array}$$

i', j', j the inclusions. Let $f' : X' \to B'$ be the restriction of f. Then $j^* R^1 f_* G = R^1 f'_* G$ and in particular the stalks of $R^1 f_* G$ and $R^1 f'_* G$ at b are the same. On the other hand, there is a natural map $(i_{k*} i_k^* R^1 f_* G)_b \to (i'_* i'^* R^1 f'_* G)_b$. Indeed, an element of $(i_{k*} i_k^* R^1 f_* G)_b$ represented by a germ (U, α), $\alpha \in \Gamma(U - \Delta_k, R^1 f_* G)$, is mapped to $(U \cap B', \alpha|_{(U\cap B')-\{b\}})$, $\alpha|_{(U\cap B')-\{b\}} \in \Gamma((U \cap B') - \{b\}, R^1 f'_* G)$. This map is in fact injective. Indeed, by descending induction, $(R^1 f_* G)|_{B-\Delta_k}$ has no sections over any open subset of $B - \Delta_k$ supported on a proper closed subset. Thus the restriction maps of the sheaf $(R^1 f_* G)|_{B-\Delta_k}$ are injective, and it follows that the map $(i_{k*} i_k^* R^1 f_* G)_b \to (i'_* i'^* R^1 f'_* G)_b$ is injective. We then have a diagram

$$\begin{array}{ccc} (R^1 f_* G)_b & \xrightarrow{\cong} & (R^1 f'_* G)_b \\ \downarrow & & \downarrow \\ (i_{k*} i_k^* R^1 f_* G)_b & \hookrightarrow & (i'_* i'^* R^1 f_* G)_b \end{array}$$

so $(R^1 f_* G)_b \to (i_{k*} i_k^* R^1 f_* G)_b$ is an isomorphism if $(R^1 f'_* G)_b \to (i'_* i'^* R^1 f'_* G)_b$ is. Thus we need to show that

$$\varinjlim_{b \in U \subseteq B'} H^1(f'^{-1}(U), G) \to \varinjlim_{b \in U \subseteq B'} \Gamma(U - \{b\}, R^1 f'_* G)$$

is an isomorphism. We can take the direct limit over contractible U.

We will need to know what $H^2_{X_b}(X',G)$ is. By the universal coefficient theorem, there is an exact sequence

$$0 \to H^i_{X_b}(X',\mathbf{Z}) \otimes_{\mathbf{Z}} G \to H^i_{X_b}(X',G) \to \mathrm{Tor}^1_{\mathbf{Z}}(H^{i+1}_{X_b}(X',\mathbf{Z}),G) \to 0.$$

Now by a suitable form of Poincaré duality ([10], V 9.3),

$$H^i_{X_b}(X',\mathbf{Z}) \cong H_{2n-k-i}(X_b,\mathbf{Z}),$$

where the latter group is Borel-Moore homology. This is computed via the exact sequence ([10] V, §3, (9))

$$0 \to \mathrm{Ext}^1(H^{2n-k-i+1}_c(X_b,\mathbf{Z}),\mathbf{Z}) \to H_{2n-k-i}(X_b,\mathbf{Z}) \to \mathrm{Hom}(H^{2n-k-i}_c(X_b,\mathbf{Z}),\mathbf{Z}) \to 0.$$

Given that $H^n_c(X_b,\mathbf{Z}) = \mathbf{Z}$ and $H^{n-1}_c(X_b,\mathbf{Z}) = H^{n-1}_c(X^{\#}_b,\mathbf{Z})$ is free, we see that

$$H^2_{X_b}(X',\mathbf{Z}) = \begin{cases} \mathbf{Z} & \text{if } k = n-2; \\ 0 & \text{if } k < n-2, \end{cases}$$

and $H^3_{X_b}(X',\mathbf{Z})$ is free, so that

$$H^2_{X_b}(X',G) = \begin{cases} G & \text{if } k = n-2; \\ 0 & \text{if } k < n-2. \end{cases}$$

Of course $H^i_{X_b}(X',G) = 0$ for $i < 2$.

There are two cases:

Case 1: $k < n-2$. For $b \in U \subseteq B'$, $H^1(f'^{-1}(U),G) \cong H^1(f'^{-1}(U-\{b\}),G)$, by the relative cohomology long exact sequence. On the other hand, the Leray spectral sequence for $f' : X^* = f'^{-1}(U-\{b\}) \to B^* = U - \{b\}$ yields the exact sequence

$$0 = H^1(B^*,G) \to H^1(X^*,G) \to H^0(B^*,R^1f'_*G) \to H^2(B^*,G) \to H^2(X^*,G).$$

This last map is injective as $X^* \to B^*$ can be assumed to have a section. Thus we obtain the isomorphism

$$H^1(f'^{-1}(U),G) \to H^0(B^*,R^1f'_*G),$$

and taking direct limts gives the desired isomorphism.

Case 2. $k = n-2$. In this case, the relative cohomology exact sequence gives for $b \in U \subseteq B'$, $X^* = f'^{-1}(U-\{b\}), B^* = U - \{b\}$,

$$0 \to H^1(f'^{-1}(U),G) \to H^1(X^*,G) \to H^2_{X_b}(f'^{-1}(U),G) \to H^2(f'^{-1}(U),G).$$

In this case $H^2_{X_b}(f'^{-1}(U),G) = G$. We have a commutative square

$$\begin{array}{ccc} H^1(B^*,G) & \xrightarrow{\cong} & H^2_{\{b\}}(U,G) \\ \downarrow & & \downarrow\cong \\ H^1(X^*,G) & \longrightarrow & H^2_{X_b}(f'^{-1}(U),G) \end{array}$$

showing the map $H^1(X^*, G) \to H^2_{X_b}(f'^{-1}(U), G)$ is surjective, yielding

$$0 \to H^1(f'^{-1}(U), G) \to H^1(X^*, G) \to G \to 0.$$

On the other hand, the Leray spectral sequence for $f' : X^* \to B^*$ gives

$$0 \to H^1(B^*, G) \cong G \to H^1(X^*, G) \to H^0(B^*, R^1 f'_* G) \to 0.$$

Putting these two sequences together one finds that the map

$$H^1(f'^{-1}(U), G) \to H^0(B^*, R^1 f'_* G)$$

is an isomorphism, and hence

$$\varinjlim H^1(f'^{-1}(U), G) \to \varinjlim \Gamma(U - \{b\}, R^1 f'_* G)$$

is an isomorphism. •

We next try to understand $R^{n-1} f_* \mathbf{Z}$. In any event, in §2 we have seen that if $f : X \to B$ is integral and satisfies Assumption 1.8 then $R_c^{n-1} f_*^{\#} \mathbf{Z} \cong R^{n-1} f_* \mathbf{Z}$ and $\underline{H}^0_\Delta(B, R_c^{n-1} f_*^{\#} \mathbf{Z}) = 0$. Since there is an exact sequence

$$0 \to \underline{H}^0_\Delta(B, R^{n-1} f_* \mathbf{Z}) \to R^{n-1} f_* \mathbf{Z} \to i_* R^{n-1} f_{0*} \mathbf{Z} \to \underline{H}^1_\Delta(B, R^{n-1} f_* \mathbf{Z}) \to 0,$$

we have already shown at least that the natural map $R^{n-1} f_* \mathbf{Z} \to i_* R^{n-1} f_{0*} \mathbf{Z}$ is injective. However, to show surjectivity, we need a more delicate understanding of the inductive structure of the singular locus. If $b \in B$ is a rank k point, it would, for example, be sufficient to show that there is a fixed-point-free Hamiltonian T^k action in a neighborhood of X_b in order to achieve a sufficiently strong inductive description of the singular fibres. However failing to prove such a result, we will make an ad hoc argument in the $n = 3$ case. Nevertheless, in any dimension we have

Lemma 3.4. *If f is integral and satisfies Assumptions 1.7 and 1.8, Δ_0 the set of rank 0 points of B, $i_0 : B - \Delta_0 \hookrightarrow B$ the inclusion, then*

$$i_{0*} i_0^* R^{n-1} f_* \mathbf{Z} = R^{n-1} f_* \mathbf{Z}.$$

Proof. If U is a contractible open neighborhood of $b \in \Delta_0$, we have an exact sequence

$$0 \to H^{n-1}(f^{-1}(U), \mathbf{Z}) \to H^{n-1}(f^{-1}(U - \{b\}), \mathbf{Z}) \to H^n_{X_b}(f^{-1}(U), \mathbf{Z}) \cong \mathbf{Z} \to 0.$$

In addition, the map

$$\varinjlim_{b \in U} H^{n-1}(f^{-1}(U), \mathbf{Z}) \to \varinjlim_{b \in U} \Gamma(U - \{b\}, R^{n-1} f_* \mathbf{Z})$$

is injective. We just need to show this map is surjective. ¿From the Leray spectral sequence for $f : X^* = f^{-1}(U - \{b\}) \to B^* = U - \{b\}$, we obtain a map φ :

$H^{n-1}(X^*,\mathbf{Z}) \to \Gamma(B^*, R^{n-1}f_*\mathbf{Z})$. We first show this map is surjective. Indeed, given a section $\sigma \in \Gamma(B^*, R^{n-1}f_*\mathbf{Z}) \subseteq \Gamma(B^*, \mathcal{T}_B^*)$, let $M \subseteq X^*$ be the circle bundle over B^* whose fibre at $b \in B^*$ is $\mathbf{R}\sigma(b)/\mathbf{Z}\sigma(b) \subseteq X_b^\# = \mathcal{T}_{B,b}^*/(R^{n-1}f_*\mathbf{Z})_b$. Let $[M] \in H^{n-1}(X^*,\mathbf{Z})$ be the image of $1 \in H_M^{n-1}(X^*,\mathbf{Z}) \cong H^0(M,\mathbf{Z})$ in $H^{n-1}(X^*,\mathbf{Z})$. It is then clear that $\varphi([M]) = \sigma$. Thus φ is surjective.

We now have a diagram

$$\begin{array}{ccccccccc}
 & & 0 & & & & & & \\
 & & \downarrow & & & & & & \\
0 & \longrightarrow & \lim_{\to} H^{n-1}(f^{-1}(U),\mathbf{Z}) & \xrightarrow{\beta} & \lim_{\to} H^{n-1}(X^*,\mathbf{Z}) & \longrightarrow & \mathbf{Z} & \longrightarrow & 0 \\
 & & \downarrow{\scriptstyle\varphi'} & & \downarrow{\scriptstyle\varphi} & & \downarrow{\scriptstyle\varphi''} & & \\
0 & \longrightarrow & \operatorname{Im}\varphi\circ\beta & \xrightarrow{\alpha} & \lim_{\to}\Gamma(B^*, R^{n-1}f_*\mathbf{Z}) & \longrightarrow & \operatorname{coker}\alpha & \longrightarrow & 0 \\
 & & \downarrow & & \downarrow & & & & \\
 & & 0 & & 0 & & & &
\end{array}$$

and we wish to show that $\operatorname{coker}\alpha = 0$, so that $\alpha\circ\varphi' = \varphi\circ\beta$ is surjective as desired. By the snake lemma, $\ker\varphi \cong \ker\varphi''$. By the Leray spectral sequence for $f : X^* \to B^*$, the image of $H^{n-1}(B^*,\mathbf{Z}) \to H^{n-1}(X^*,\mathbf{Z})$ is contained in $\ker\varphi$. On the other hand, by the commutativity of

$$\begin{array}{ccc}
H^{n-1}(B^*,\mathbf{Z}) & \xrightarrow{\cong} & H^n_{\{b\}}(U,\mathbf{Z}) \\
\downarrow & & \downarrow{\scriptstyle\cong} \\
H^{n-1}(X^*,\mathbf{Z}) & \longrightarrow & H^n_{X_b}(f^{-1}(U),\mathbf{Z})
\end{array}$$

it is then clear that $\ker\varphi$ surjects onto $H^n_{X_b}(f^{-1}(U),\mathbf{Z}) = \mathbf{Z}$, so $\ker\varphi'' = \mathbf{Z}$ and $\operatorname{coker}\alpha = 0$. •

Theorem 3.5. *If $\dim X = 3$, $f : X \to B$ an integral special Lagrangian fibration satisfying Assumptions 1.7 and 1.8, and if the fibres X_b for rank 1 points b are a union of a finite number of $\mathcal{T}_{B,b}^*$-orbits, then f is $\mathbf{Z}$-simple.*

Proof. The hypothesis of this theorem implies the hypothesis of Theorem 3.3, so the simplicity condition holds for $p = 0, 1$ and 3. In the notation of the proof of Theorem 3.3, we need to show $i_{1*}i_1^*(R^2f_*\mathbf{Z})|_{B-\Delta_0} = (R^2f_*\mathbf{Z})|_{B-\Delta_0}$, as Lemma 3.4 then allows us to complete the proof of simplicity. Choose $b \in \Delta_1$, $b \notin \Delta_0$, and choose a 2-dimensional disk B' passing through b as in the proof of Theorem 3.3, $f' : X' = f^{-1}(B') \to B'$. As in that proof, we just need to show that

$$H^2(X_b,\mathbf{Z}) \cong \varinjlim_{b\in U\subseteq B'} H^2(f'^{-1}(U),\mathbf{Z}) \to \varinjlim_{b\in U\subseteq B'} \Gamma(U-\{b\}, R^2f'_*\mathbf{Z})$$

is an isomorphism, and we already know this map is injective. Note also that the cokernel of this map is torsion-free: since $R^2f'_*\mathbf{Z} \subseteq \mathcal{T}_B^*|_{B'}$, if an integer multiple of a section of $R^2f'_*\mathbf{Z}$ over $U-\{b\}$ extends to a section over U, the section itself extends. Now $B^* = B' - \{b\}$ is a punctured disk, and hence $(R^2f'_*\mathbf{Z})|_{B^*}$ is a local system determined by a single monodromy transformation T. By Poincaré duality, $(R^1f'_*\mathbf{Z})|_{B^*}$ has monodromy ${}^tT$. Since $\Gamma(B^*, R^2f'_*\mathbf{Z}) = \ker(T-I)$, we see that

$\operatorname{rank}\Gamma(B^*, R^2 f'_* \mathbf{Z}) = \operatorname{rank}\Gamma(B^*, R^1 f'_* \mathbf{Z}) = \operatorname{rank} H^1(X_b, \mathbf{Z})$ by simplicity for $p = 1$. Thus it will be sufficient to show that $\operatorname{rank} H^1(X_b, \mathbf{Z}) = \operatorname{rank} H^2(X_b, \mathbf{Z})$ to show the above map is an isomorphism.

Since b is a rank 1 point, the singular locus of X_b is a union of circles, each being a closed orbit of the action of $\mathcal{T}^*_{B,b}$ on X_b. Since there are assumed to be only a finite number of such orbits, each S^1 is a connected component of $Sing(X_b)$. But since $Sing(X_b)$ is connected by Corollary 2.5, $Z := Sing(X_b) \cong S^1$. Now we also have exact sequences

$$0 \to H^i_c(X_b - \mathbf{Z}, \mathbf{Z}) \to H^i(X_b, \mathbf{Z}) \to H^i(Z, \mathbf{Z}) \to 0$$

for all i, the exactness for $i = 1$ shown in the proof of Corollary 2.5 and for $i = 2$ shown in the proof of Theorem 2.4. From this we see that if $H^2_c(X_b - Z, \mathbf{Z}) = \mathbf{Z}$, then $H^1(X_b, \mathbf{Z}) = H^2(X_b, \mathbf{Z}) = \mathbf{Z}$, while if $H^2_c(X_b - Z, \mathbf{Z}) = \mathbf{Z}^2$, then $H^1(X_b, \mathbf{Z}) = H^2(X_b, \mathbf{Z}) = \mathbf{Z}^2$, completing the proof in these cases.

Finally, suppose $H^2_c(X_b - Z, \mathbf{Z}) = 0$. Then $H^2_c(X_b, \mathbf{Z}) = 0$ and it follows as in the proof of Theorem 3.3 that $H^3_{X_b}(f^{-1}(U), \mathbf{Z}) = 0$, so one obtains from the relative cohomology sequence a surjection

$$H^2(f'^{-1}(U), \mathbf{Z}) \to H^2(f'^{-1}(U - \{b\}), \mathbf{Z}) \to 0.$$

The argument of the proof of Lemma 3.4 shows that $H^2(f'^{-1}(U - \{b\}), \mathbf{Z}) \to \Gamma(U - \{b\}, R^2 f'_* \mathbf{Z})$ is surjective, and hence so is $H^2(f'^{-1}(U), \mathbf{Z}) \to \Gamma(U - \{b\}, R^2 f'_* \mathbf{Z})$. Taking direct limits, one concludes that $H^2(X_b, \mathbf{Z}) = \ker(T - I)$ as desired, showing simplicity. But notice in fact this case can't occur, since as $H^2(X_b, \mathbf{Z}) = 0$, we also have $H^1(X_b, \mathbf{Z}) = \ker({}^tT - I) = 0$, contradicting $H^1(X_b, \mathbf{Z}) = H^1(Z, \mathbf{Z}) = \mathbf{Z}$. •

Remark 3.6. If $f : X \to B$ is a $\mathbf{Z}$-simple special Lagrangian T^3 fibration, we obtain some restrictions on the cohomology of a singular fibre X_b. Clearly $H^0(X_b, \mathbf{Z}) = H^3(X_b, \mathbf{Z}) = \mathbf{Z}$, so if $b_i = \operatorname{rank} H^i(X_b, \mathbf{Z})$, we will say for the duration of this remark that X_b is of type (b_1, b_2). Clearly $b_1, b_2 \le 3$, and if $b_2 = 3$, then X_b is non-singular. If $b_1 = 3$, then $b_2 = 3$, since $\bigwedge^2 H^1(X_b, \mathbf{Z}) \subseteq H^2(X_b, \mathbf{Z})$, and so X_b is non-singular. This also shows that if $b_1 = 2$ then $b_2 \ge 1$, and a similar argument shows that if $b_2 = 2$, then $b_1 \ge 1$. Thus the possible values for (b_1, b_2) are $(2,2)$, $(2,1)$, $(1,2)$, $(1,1)$, $(0,1)$, $(1,0)$ or $(0,0)$. We describe the probable topology of an integral singular fibre with each of the above possible cohomology groups.

$(2,2)$ Such cohomology is realised by a fibre of the form $I_1 \times S^1$, where I_1 denotes a Kodaira type I_1-fibre.

$(1,2)$ $S^1 \times T^2/\{pt\} \times T^2$. This was seen in the example in §1 of [14].

$(2,1)$ This was described in [14], Remark 1.4. Identify T^3 with the solid cube $[0,1]^3$ with opposite sides identified. Then take $T^3/\sim$, where $(x_1, x_2, x_3) \sim (x'_1, x'_2, x'_3)$ if and only if $(x_1, x_2, x_3) = (x'_1, x'_2, x'_3)$ or $(x_1, x_2) = (x'_1, x'_2) \in \partial([0,1])^2$. This fibre is singular along a figure eight.

$(1,1)$ There are two possibilities here. The fibre could be $II \times S^1$, where II denotes a Kodaira type II fibre, or it could be $T^3/\sim$, where $(x_1, x_2, x_3) \sim (x'_1, x'_2, x'_3)$ if and only if $(x_1, x_2, x_3) = (x'_1, x'_2, x'_3)$ or $(x_1, x_2) = (x'_1, x'_2) \in \{0,1\} \times [0,1]$ or $(x_1, x_2), (x'_1, x'_2) \in [0,1] \times \{0,1\}$. (This is the same as contracting one loop of the singular figure eight in the $(2,1)$ case to a point).

(0,1) $T^3/\sim$, where $(x_1,x_2,x_3) \sim (x_1',x_2',x_3')$ if and only if $(x_1,x_2,x_3) = (x_1',x_2',x_3')$ or $(x_1,x_2),(x_1',x_2') \in \partial[0,1]^2$. (This is equivalent to contracting the singular figure eight of the $(1,2)$ case to a point).

(1,0) $T^3/\sim$, where $(x_1,x_2,x_3) \sim (x_1',x_2',x_3')$ if and only if $(x_1,x_2,x_3) = (x_1',x_2',x_3')$ or $(x_1,x_2),(x_1',x_2') \in \partial[0,1]^2$ and $x_3 = x_3'$ or $x_3, x_3' \in \{0,1\}$.

(0,0) $T^3/\sim$, where $(x_1,x_2,x_3) \sim (x_1',x_2',x_3')$ if and only if $(x_1,x_2,x_3) = (x_1',x_2',x_3')$ or $(x_1,x_2,x_3),(x_1',x_2',x_3') \in \partial[0,1]^3$. (We are contracting the boundary of $[0,1]^3$ to a point. This is topologically a sphere).

One notes that for each fibre of type (m,n), there is a fibre of type (n,m) which should then be its dual. (In particular, the fibres of type $(2,2)$ and $(1,1)$ should be self-dual).

I cannot prove yet that this provides a complete classfication of integral three-dimensional singular fibres, but it seems to be a reasonable conjecture. In addition, these types of examples extend to higher dimensions, and one finds a much wider range of possible topologies, which nonetheless exhibit the desired duality.

Having proven **Z**-simplicity in some cases for special Lagrangian fibrations, we wish now to return to a more careful study of the Leray spectral sequence for special Lagrangian fibrations, with special consideration of the role torsion plays. First we make some observations on the Leray spectral sequence in any dimension.

Lemma 3.7. *If $f : X \to B$ is a $\mathbf{Z}$-simple special Lagrangian T^n-fibration with a section, then in the Leray spectral sequence for f, $E^2_{1,1} = E^\infty_{1,1}$ and $E^2_{1,n-1} = E^\infty_{1,n-1}$. In addition, the Leray filtration yields a surjection $H^n(X,\mathbf{Z})_{tors} \to (E^2_{1,n-1})_{tors}$.*

Proof. The only possible non-zero differential to or from $E^2_{1,1}$ is $d_2 : E^2_{1,1} \to E^2_{3,0} = H^3(B,\mathbf{Z})$. But since f has a section, the map $H^3(B,\mathbf{Z}) \to H^3(X,\mathbf{Z})$ is injective, and thus $d_2 = 0$. Thus $E^2_{1,1} = E^\infty_{1,1}$.

Next, $E^2_{1,n-1} = H^1(B, R^{n-1}f_*\mathbf{Z})$, and recall from [14] that that $H^1(B,R^{n-1}f_*\mathbf{Z})$ is the group of sections of f modulo homotopy, with a fixed section, say σ_0, the zero section. Then for any section σ, the cohomology class $[\sigma]-[\sigma_0] \in H^n(X,\mathbf{Z})$ and the element $[\sigma] \in H^1(B,R^{n-1}f_*\mathbf{Z})$ representing the section coincide up to sign in $E^\infty_{1,n-1}$ by [14], Theorem 4.1 (which holds with $\mathbf{Z}$ coefficients if f is $\mathbf{Z}$-simple). Thus $E^\infty_{1,n-1} = E^2_{1,n-1}$, and if $H^n(X,\mathbf{Z}) = F^0 \supseteq F^1 \supseteq \cdots$ is the Leray filtration on $H^n(X,\mathbf{Z})$, then $F^1/F^2 \cong E^2_{1,n-1}$. Since $F^0/F^1 \cong \mathbf{Z}$, $H^n(X,\mathbf{Z})_{tors} = F^1_{tors}$ and thus there is a map $H^n(X,\mathbf{Z})_{tors} \to (E^2_{1,n-1})_{tors}$. To see this map is surjective, suppose σ is a torsion section of f. Then σ must be disjoint from σ_0. Indeed, if $x \in \sigma \cap \sigma_0$, let $U \subseteq B$ be a small open neighborhood of $f(x) \in B$ in which σ is represented by a section $\tilde{\sigma} \in \Gamma(U, \mathcal{T}^*_B)$, such that $\tilde{\sigma}(f(x)) = 0$. Then if m is the order of the torsion section σ, $m\tilde{\sigma}$ is a non-zero section of $R^{n-1}_c f^\#_* \mathbf{Z}$ which is zero for at least one point, which is impossible. Thus σ and σ_0 are disjoint.

By [14], Theorem 4.1, $T^*_\sigma : H^*(X,\mathbf{Q}) \to H^*(X,\mathbf{Q})$ is a unipotent operator, but on the other hand $T^m_\sigma = I$ since σ is m-torsion. Thus $T^*_\sigma = I$, so $[\sigma] = [\sigma_0]$ in $H^n(X,\mathbf{Q})$ and $[\sigma]-[\sigma_0]$ is in fact a torsion element of $H^n(X,\mathbf{Z})$. This shows that the map $H^n(X,\mathbf{Z})_{tors} \to (E^2_{1,n-1})_{tors}$ is surjective. •

Even if in general the Leray spectral sequence for f does not degenerate, the above result might be sufficient for many applications; as we will see in later sections, $H^1(B, R^1 f_* \mathbf{R})$ and $H^1(B, R^{n-1} f_* \mathbf{R})$ play important roles in mirror symmetry.

Note that if $\mathbf{Z}$-simplicity fails because f has reducible fibres, we expect the second part of the above result to fail.

We now focus on the three dimensional case.

Proposition 3.8. *Let $f : X \to B$ and $\check{f} : \check{X} \to B$ be dual $\mathbf{Z}$-simple special Lagrangian T^3 fibrations, and suppose that $H^1(B, \mathbf{Z}) = 0$. Then $H^1(X, \mathbf{Z}) = 0$ if and only if $H^1(\check{X}, \mathbf{Z}) = 0$.*

Proof. Since $H^1(B, \mathbf{Z}) = 0$, $H^2(B, \mathbf{Z})$ is torsion, so

$$rank(H^1(X, \mathbf{Z})) = rank(H^0(B, R^1 f_* \mathbf{Z})).$$

But since X is Kähler, the first betti number of X is even, so if $H^1(X, \mathbf{Z}) \neq 0$ then $rank(H^0(B, R^1 f_* \mathbf{Z})) \geq 2$. The wedge of two independent sections of $R^1 f_* \mathbf{Z}$ yields a section of $R^2 f_* \mathbf{Z}$, and $R^2 f_* \mathbf{Z} \cong R^1 \check{f}_* \mathbf{Z}$, so $H^0(B, R^1 \check{f}_* \mathbf{Z}) \neq 0$, hence $H^1(\check{X}, \mathbf{Z}) \neq 0$. Repeating the same argument interchanging X and $\check{X}$ gives the result. •

Theorem 3.9. *Let $f : X \to B$, $\check{f} : \check{X} \to B$ be dual $\mathbf{Z}$-simple special Lagrangian T^3-fibrations with sections, and assume $H^1(X, \mathbf{Z}) = 0$. Then the Leray spectral sequences for f and $\check{f}$ with coefficients in $\mathbf{Z}$ degenerate at the E_2-term, and*

$$rank_{\mathbf{Z}} H^i(B, R^j f_* \mathbf{Z}) = rank_{\mathbf{Z}} H^{3-i}(B, R^{3-j} f_* \mathbf{Z}).$$

If in addition f and $\check{f}$ are $\mathbf{R}/\mathbf{Z}$-simple, then

$$Tors(H^i(B, R^j f_* \mathbf{Z})) \cong Tors(H^{4-i}(B, R^{3-j} f_* \mathbf{Z})).$$

Note that the additional assumption of $\mathbf{R}/\mathbf{Z}$-simplicity is not a particularly strong one. Indeed, we have seen that given suitable regularity hypotheses, we have only failed to show the G-simplicity condition for $p = 2$. But $\mathbf{Z}$-simplicity implies $R^2 f_* \mathbf{Z} \cong R^1 \check{f}_* \mathbf{Z}$ and $R^2 f_* \mathbf{R} \cong R^1 \check{f}_* \mathbf{R}$, whence $R^2 f_* \mathbf{R}/\mathbf{Z} \cong R^1 \check{f}_* \mathbf{R}/\mathbf{Z}$. Hence the existence of the dual fibration $\check{f}$ and the $\mathbf{R}/\mathbf{Z}$-simplicity condition for $p = 1$ implies it for $p = 2$.

Proof. The E_2-term of the Leray spectral sequence, by the arguments of [14], Lemma 2.4, looks like

$$\begin{array}{cccc} \mathbf{Z} & 0 & T^{2,3} & \mathbf{Z} \\ 0 & \mathbf{Z}^{h^{1,2}} \oplus T^{1,2} & \mathbf{Z}^{h^{1,1}} \oplus T^{2,2} & T^{3,2} \\ 0 & \mathbf{Z}^{h^{1,1}} \oplus T^{1,1} & \mathbf{Z}^{h^{1,2}} \oplus T^{2,1} & T^{3,1} \\ \mathbf{Z} & 0 & T^{2,0} & \mathbf{Z} \end{array} \tag{3.1}$$

with a similar diagram for $\check{f}$. Here $T^{i,j} = H^i(B, R^j f_* \mathbf{Z})_{tors}$. We are using $H^1(X, \mathbf{Z}) = H^1(\check{X}, \mathbf{Z}) = 0$ to obtain the zeroes on the left column and the top and bottom rows. Clearly the desired statement on ranks follows.

Recalling from [14], §2 that $\mathbf{Z}$-simplicity implies $R^if_*\mathbf{Z} \cong R^{3-i}\check{f}_*\mathbf{Z}$, it follows that if $\check{T}^{i,j} = H^i(B, R^j\check{f}_*\mathbf{Z})_{tors}$, then $\check{T}^{i,j} \cong T^{i,3-j}$.

Since f has a section, the argument of [14], Lemma 2.4, combined with the degeneration statement of Proposition 3.7, shows that the above spectral sequence degenerates. The same holds for $\check{f}$.

As for the statements about torsion, clearly $T^{2,3} \cong T^{2,0} \cong H^2(B,\mathbf{Z})$. To show the rest, we use Poincaré-Verdier duality (see for example [9]). In any dimension, applying duality to the map $s: B \to pt$, we obtain isomorphisms

$$\mathbf{R}\mathrm{Hom}(\mathbf{R}\Gamma(R^if_*\mathbf{Z}),\mathbf{Z}) \cong \mathbf{R}\Gamma\mathbf{R}\underline{\mathrm{Hom}}(R^if_*\mathbf{Z},\mathbf{Z}[n]).$$

Applying H^{-j} to both sides, we obtain

$$H^{-j}(\mathbf{R}\mathrm{Hom}(\mathbf{R}\Gamma(R^if_*\mathbf{Z}),\mathbf{Z})) \cong \mathrm{Ext}^{n-j}(R^if_*\mathbf{Z},\mathbf{Z}). \tag{3.2}$$

The left hand side is easily computed by choosing a complex of projective $\mathbf{Z}$-modules quasi-isomorphic to $\mathbf{R}\Gamma(R^if_*\mathbf{Z})$ and applying the algebraic universal coefficient theorem, which yields exact sequences

$$0 \to \mathrm{Ext}^1(H^{j+1}(B,R^if_*\mathbf{Z}),\mathbf{Z}) \to H^{-j}(\mathbf{R}\mathrm{Hom}(\mathbf{R}\Gamma(R^if_*\mathbf{Z}),\mathbf{Z})) \to \mathrm{Hom}(H^j(B,R^if_*\mathbf{Z}),\mathbf{Z}) \to 0.$$

The difficulty in applying Poincaré-Verdier duality for non-locally constant sheaves is the difficulty of comparing the Ext's and the cohomology groups. We will do this for $n=3, i=2$.

First on B_0 one has $\underline{\mathrm{Hom}}(R^2f_{0*}\mathbf{Z},\mathbf{Z}) \cong R^1f_{0*}\mathbf{Z}$ by Poincaré duality, and if $i: B_0 \hookrightarrow B$ is the inclusion, the natural map

$$i_*\underline{\mathrm{Hom}}(R^2f_{0*}\mathbf{Z},\mathbf{Z}) \to \underline{\mathrm{Hom}}(i_*R^2f_{0*}\mathbf{Z}, i_*\mathbf{Z})$$

is an isomorphism. Thus by $\mathbf{Z}$-simplicity, $\underline{\mathrm{Hom}}(R^2f_*\mathbf{Z},\mathbf{Z}) \cong R^1f_*\mathbf{Z}$. Also, $\mathbf{R}/\mathbf{Z}$-simplicity implies $\mathbf{Z}/m\mathbf{Z}$-simplicity for any m, and so a similar argument shows $\underline{\mathrm{Hom}}(R^2f_*\mathbf{Z},\mathbf{Z}/m\mathbf{Z}) \cong R^1f_*\mathbf{Z}/m\mathbf{Z}$.

So by the local-global Ext spectral sequence one has a five-term sequence

$$\begin{aligned} 0 &\to H^1(B,R^1f_*\mathbf{Z}) \to \mathrm{Ext}^1(R^2f_*\mathbf{Z},\mathbf{Z}) \to H^0(B,\underline{\mathrm{Ext}}^1(R^2f_*\mathbf{Z},\mathbf{Z})) \\ &\to H^2(B,R^1f_*\mathbf{Z}) \to \mathrm{Ext}^2(R^2f_*\mathbf{Z},\mathbf{Z}). \end{aligned} \tag{3.3}$$

I claim that $\underline{\mathrm{Ext}}^1(R^2f_*\mathbf{Z},\mathbf{Z})$ is a torsion-free sheaf. Indeed, apply $\underline{\mathrm{Hom}}(R^2f_*\mathbf{Z},\cdot)$ to the exact sequence

$$0 \to \mathbf{Z} \to \mathbf{Z} \to \mathbf{Z}/m\mathbf{Z} \to 0.$$

We obtain an exact sequence

$$0 \longrightarrow R^1f_*\mathbf{Z} \xrightarrow{\cdot m} R^1f_*\mathbf{Z} \longrightarrow R^1f_*\mathbf{Z}/m\mathbf{Z} \longrightarrow \underline{\mathrm{Ext}}^1(R^2f_*\mathbf{Z},\mathbf{Z}) \xrightarrow{\cdot m} \underline{\mathrm{Ext}}^1(R^2f_*\mathbf{Z},\mathbf{Z}).$$

But in fact $R^1f_*(\mathbf{Z}/m\mathbf{Z}) \cong R^1f_*\mathbf{Z}/mR^1f_*\mathbf{Z}$ since $R^2f_*\mathbf{Z}$ is torsion-free, so we see the multiplication by m map is injective on $\underline{\mathrm{Ext}}^1(R^2f_*\mathbf{Z},\mathbf{Z})$. Thus this sheaf is torsion free.

Now the left hand side of (3.2) is $\mathbf{Z}^{h^{1,2}} \oplus T^{2,2}$ for $j = 1$ and is $\mathbf{Z}^{h^{1,1}} \oplus T^{3,2}$ for $j = 2$. Thus by (3.2),

$$rank_{\mathbf{Z}}\mathrm{Ext}^1(R^2 f_*\mathbf{Z}, \mathbf{Z}) = h^{1,1} = rank_{\mathbf{Z}} H^1(B, R^1 f_*\mathbf{Z}),$$

so in (3.3) the fact that $H^0(B, \underline{\mathrm{Ext}}^1(R^2 f_*\mathbf{Z}, \mathbf{Z}))$ is torsion-free shows that the map

$$\mathrm{Ext}^1(R^2 f_*\mathbf{Z}, \mathbf{Z}) \to H^0(B, \underline{\mathrm{Ext}}^1(R^2 f_*\mathbf{Z}, \mathbf{Z}))$$

is zero. Thus we have

$$H^1(B, R^1 f_*\mathbf{Z}) \cong \mathrm{Ext}^1(R^2 f_*\mathbf{Z}, \mathbf{Z})$$

and

$$0 \to H^0(B, \underline{\mathrm{Ext}}^1(R^2 f_*\mathbf{Z}, \mathbf{Z})) \to H^2(B, R^1 f_*\mathbf{Z}) \to \mathrm{Ext}^2(R^2 f_*\mathbf{Z}, \mathbf{Z})$$

exact. Thus (3.2) implies $T^{1,1} \cong T^{3,2}$, and by using the same argument for $\check{f}$, $T^{1,2} \cong T^{3,1}$. In addition, $T^{2,1} = H^2(B, R^1 f_*\mathbf{Z})_{tors} \subseteq \mathrm{Ext}^2(R^2 f_*\mathbf{Z}, \mathbf{Z})_{tors} = T^{2,2}$. On the other hand, from (3.1) and Proposition 3.7 there are exact sequences (see the beginning of the proof of Theorem 3.10 for details of the first sequence)

$$0 \to T^{2,1} \to H^3(X, \mathbf{Z})_{tors} \to T^{1,2} \to 0,$$

$$0 \to T^{3,1} \to H^4(X, \mathbf{Z})_{tors} \to T^{2,2} \to 0,$$

and in addition for any oriented 6-manifold $H^3(X, \mathbf{Z})_{tors} \cong H^4(X, \mathbf{Z})_{tors}$ by Poincaré duality and the universal coefficient theorem. So $\#T^{2,2} = \#T^{2,1}$ and this implies $T^{2,2} \cong T^{2,1}$. •

Theorem 3.10. *Let $f : X \to B$, $\check{f} : \check{X} \to B$ be as in Theorem 3.9, and assume in addition that B is simply connected. Then there are non-canonical isomorphisms*

$$H^{even}(X, \mathbf{Z}[1/2]) \cong H^{odd}(\check{X}, \mathbf{Z}[1/2])$$
$$H^{odd}(X, \mathbf{Z}[1/2]) \cong H^{even}(\check{X}, \mathbf{Z}[1/2]).$$

In general, there are short exact sequences

$$0 \to H^2(B, R^1 f_*\mathbf{Z})_{tors} \to H^3(X, \mathbf{Z})_{tors} \to H^1(B, R^2 f_*\mathbf{Z})_{tors} \to 0$$

$$0 \to H^2(B, R^1 \check{f}_*\mathbf{Z})_{tors} \to H^3(\check{X}, \mathbf{Z})_{tors} \to H^1(B, R^2 \check{f}_*\mathbf{Z})_{tors} \to 0$$

and if they split, the above isomorphisms hold over $\mathbf{Z}$. This happens, for example, if both X and $\check{X}$ are simply connected. In any event,

$$\#H^{even}(X, \mathbf{Z})_{tors} = \#H^{odd}(\check{X}, \mathbf{Z})_{tors}$$

$$\#H^{odd}(X, \mathbf{Z})_{tors} = \#H^{even}(\check{X}, \mathbf{Z})_{tors}.$$

Proof. The Leray filtration on $H^3(X, \mathbf{Z})$ is

$$0 \subseteq F_0 = \mathbf{Z}[T^3] \subseteq F_1 \subseteq F_2 \subseteq F_3 = H^3(X, \mathbf{Z}).$$

Since the cohomology class $[T^3]$ is primitive in $H^3(X,\mathbf{Z})$, $F_0 \subseteq F_1$ is a primitive embedding and $(F_1)_{tors} = (F_1/F_0)_{tors} = T^{2,1}$ in the notation of the proof of Theorem 3.9. It then follows from Proposition 3.7 that there is an exact sequence

$$0 \to T^{2,1} \to H^3(X,\mathbf{Z})_{tors} \to T^{1,2} \to 0.$$

First assume this sequence for X and $\check{X}$ splits, so $H^3(X,\mathbf{Z})_{tors} = T^{2,1} \oplus T^{1,2}$. Now $H^3(X,\mathbf{Z})_{tors} \cong H^4(X,\mathbf{Z})_{tors}$, so $H^4(X,\mathbf{Z})_{tors} = T^{2,1} \oplus T^{1,2}$ also. Putting this together we see that

$$H^{even}(X,\mathbf{Z})_{tors} = T^{1,1} \oplus T^{2,1} \oplus T^{1,2}$$

and

$$\begin{aligned} H^{odd}(X,\mathbf{Z})_{tors} &= T^{3,2} \oplus T^{2,1} \oplus T^{1,2} \\ &= T^{1,1} \oplus T^{2,1} \oplus T^{1,2}. \end{aligned}$$

On the other hand

$$\begin{aligned} H^{odd}(\check{X},\mathbf{Z})_{tors} &= \check{T}^{1,1} \oplus \check{T}^{2,1} \oplus \check{T}^{1,2} \\ &= T^{1,2} \oplus T^{2,2} \oplus T^{1,1}. \\ &= H^{even}(\check{X},\mathbf{Z})_{tors} \end{aligned}$$

Since $T^{2,2} \cong T^{2,1}$ by Theorem 3.9, we are done.

Note that if X and $\check{X}$ are simply connected, $0 = T^{1,1} \cong \check{T}^{1,2}$ and $0 = \check{T}^{1,1} \cong T^{1,2}$, so the sequences trivially split. If the sequences don't split, then it is still clear that the numerical equalities hold.

Finally, we finish the proof of the theorem by showing the sequences do split over $\mathbf{Z}[1/2]$. We define a map $\phi : T^{1,2} \otimes_{\mathbf{Z}} \mathbf{Z}[1/2] \to H^3(X,\mathbf{Z}[1/2])_{tors}$. Indeed, for σ a torsion section of $f : X \to B$, given by $\sigma \in H^1(B, R^2 f_*\mathbf{Z})_{tors}$, we can set $\phi(\sigma) = (\log T_\sigma^*)([\sigma_0]) \in H^3(X,\mathbf{Z}[1/6])$ where $[\sigma_0] \in H^3(X,\mathbf{Z})$ is the cohomology class of the zero section. Here $\log T_\sigma^* = (T_\sigma^* - I) - \frac{1}{2}(T_\sigma^* - I)^2 + \frac{1}{3}(T_\sigma^* - I)^3$, as $(T_\sigma^* - I)^4 = 0$ by [14] Theorem 4.1. As observed in the proof of Theorem 3.7, $(T_\sigma^* - I)([\sigma_0]) = [\sigma] - [\sigma_0]$ is torsion and represents the class $\sigma \in H^1(B, R^2 f_*\mathbf{Z})_{tors}$. Thus $(T_\sigma - I)^3([\sigma_0]) \in H^3(B,\mathbf{Z}) \subseteq H^3(X,\mathbf{Z})$ must be zero as this element is also torsion. So $(\log T_\sigma^*)([\sigma_0]) = ((T_\sigma^* - I) - \frac{1}{2}(T_\sigma^* - I))([\sigma_0]) \in H^3(X,\mathbf{Z}[1/2])$. Furthermore,

$$\begin{aligned} \phi(\sigma+\tau) &= (\log T_{\sigma+\tau}^*)([\sigma_0]) \\ &= (\log T_\sigma^* \circ T_\tau^*)([\sigma_0]) \\ &= (\log T_\sigma^* + \log T_\tau^*)([\sigma_0]) \\ &= \phi(\sigma) + \phi(\tau), \end{aligned}$$

so ϕ is a group homomorphism. Thus ϕ gives the desired splitting over $\mathbf{Z}[1/2]$. •

The problem that arises with two-torsion in the above theorem seems at the moment to be unavoidable, and does not make the statement very aesthetically pleasing. The heart of this issue is the following: given a two-torsion element in $H^2(\check{X},\mathbf{Z})$, is the square of this element non-zero in $H^4(\check{X},\mathbf{Z})$? If it is non-zero, then it follows from [14], Theorem 4.1, that if σ is the corresponding torsion section of f, then $[\sigma] - [\sigma_0]$ is not two-torsion, making it unlikely that the exact sequences in Theorem 3.10 split over $\mathbf{Z}$.

Example 3.11. The only example personally known to me of a Calabi-Yau threefold X with $H^3(X, \mathbf{Z})_{tors}$ non-zero is the "Enriques threefold", obtained by dividing out $K3 \times E$ with the involution $(\iota, -1)$, where ι is the Enriques involution on the K3 surface. (See [4] for calculations of the cohomology of this threefold.) This possesses a special Lagrangian T^3 fibration f in much the same way as the examples in [16]. In fact $H^3(X, \mathbf{Z})_{tors} = \mathbf{Z}/2\mathbf{Z}$, and the fibration $f : X \to B$ is seen to have a torsion section, so $H^1(B, R^2 f_* \mathbf{Z})_{tors} = \mathbf{Z}/2\mathbf{Z}$.

Remark 3.12. If $f : X \to B$ does not have a section, then $\check{f} : \check{X} \to B$ does, as well as the Jacobian $j : J \to B$ of f. Then j and $\check{f}$ are dual, and Theorem 3.11 applies to this pair. In addition, $R^i f_* \mathbf{Z} = R^i j_* \mathbf{Z}$, so the E^2 term of the Leray spectral sequence for f is still given by (3.1), and $H^2(B, R^1 f_* \mathbf{Z})_{tors} \cong H^2(B, R^2 f_* \mathbf{Z})_{tors}$. However now the spectral sequence won't degenerate. Since there is no class in $H^3(X, \mathbf{Z})$ restricting to the generator of $H^3(X_b, \mathbf{Z})$, one of the differentials from $H^0(B, R^3 f_* \mathbf{Z})$ must be non-zero.

This now gives an explanation for the speculations of [5] of the role that $H^3(X, \mathbf{Z})_{tors}$ should play in mirror symmetry. There it was argued on physical grounds that the Kähler moduli space of a Calabi-yau threefold was in fact $H^2(X, \mathbf{C}/\mathbf{Z})$, so in fact it had one component for each element of $H^3(X, \mathbf{Z})_{tors}$. Thus the complex moduli space of the mirror $\check{X}$ should have a similar number of components. It was not clear what this meant. But in our current context it is clear: if $H^2(B, R^1 f_* \mathbf{Z})_{tors} \neq 0$, $f : X \to B$ will have many dual fibrations, only one of which will have a section. All the other duals are obtained by twisting the one with a section. The set of such dual fibrations is classified by $H^2(B, R^1 f_* \mathbf{Z})_{tors} = H^2(B, R^2 \check{f}_* \mathbf{Z})_{tors}$ and we have seen that these groups are related to $H^3(X, \mathbf{Z})_{tors}$. However, they do not necessarily coincide with $H^3(X, \mathbf{Z})_{tors}$, and this will lead us to modify the definition of B-field in Conjecture 6.6.

We end this section with some comments concerning the de Rham realisation of the Leray spectral sequence, which we will need later. In general, let $f : X \to S$ be a smooth map (i.e., f_* always surjective) of differentiable manifolds. Then one has an exact sequence of vector bundles

$$0 \to f^* \Omega^1_S \to \Omega^1_X \to \Omega^1_{X/S} \to 0.$$

This gives rise to a filtration $F^p \Omega^{\cdot}_X$ on the de Rham complex,

$$\Omega^{\cdot}_X = F^0 \Omega^{\cdot}_X \supseteq F^1 \Omega^{\cdot}_X \supseteq \cdots$$

such that

$$F^p \Omega^{p+q}_X / F^{p+1} \Omega^{p+q}_X = \bigwedge^p f^* \Omega^1_S \otimes \bigwedge^q \Omega^1_{X/S}.$$

This filtration gives rise to a spectral sequence with

$$E^0_{pq} = \Gamma(X, \bigwedge^p f^* \Omega^1_S \otimes \bigwedge^q \Omega^1_{X/S})$$

with differential d^0 being exterior differentiation along fibres of the map f. Then

$$E^1_{pq} = \Gamma(S, \Omega^p_S \otimes R^q f_* \mathbf{R}),$$

and d^1 is the Gauss-Manin connection ∇_{GM}. Here, given a form $\alpha \in F^p\Omega^{p+q}_X$ with $d^0\alpha = 0$, the element represented in E^1_{pq} is given as follows. For $v_1, \ldots, v_p \in \mathcal{T}_{B,b}$, the form $(\iota(v_1, \ldots, v_p)\alpha)|_{X_b}$ defines a well-defined cohomology class in $H^q(X_b, \mathbf{R})$. This yields an $R^q f_* \mathbf{R}$-valued p-form in E^1_{pq}. Next

$$E^2_{pq} = H^p(S, R^q f_* \mathbf{R}),$$

which coincides with the E^2-term of the Leray spectral sequence for f.

Now let us specialise to the case that $f : X \rightarrow B$ is a special Lagrangian T^n-fibration, $f_0 : X_0 \rightarrow B_0 = B - \Delta$ the smooth part of the fibration. On B_0, McLean's result gives a natural isomorphism

$$\mathcal{T}_{B_0} \cong R^1 f_{0*}\mathbf{R} \otimes C^\infty(B_0).$$

Mclean also defines an n-form on the base. This is given by

$$\Theta(v_1, \ldots, v_n) = \int_{X_b} (-\iota(v_1)\omega) \wedge \cdots \wedge (-\iota(v_n)\omega).$$

In canonical coordinates, this is

$$\Theta(\partial/\partial y_1, \ldots, \partial/\partial y_n) = \int_{X_b} dx_1 \wedge \cdots \wedge dx_n.$$

Here X_b is oriented canonically by Ω as it is special Lagrangian. Of course, this form goes to infinity at singular fibres. Another way to think about this form is via integration along fibres of $\omega^n/n!$. Since

$$\omega^n/n! = (-1)^{n(n+1)/2} dy_1 \wedge \cdots \wedge dy_n \wedge dx_1 \wedge \cdots \wedge dx_n,$$

$f_*(\omega^n/n!) = (-1)^{n(n+1)/2}\Theta$. Thus in particular,

$$\int_X \omega^n/n! = (-1)^{n(n+1)/2} \int_B \Theta.$$

We can identify $\bigwedge^{n-q} \mathcal{T}_{B_0}$ with $\Omega^q_{B_0}$ by contracting with Θ, and so obtain isomorphisms

$$\Omega^q_{B_0} \cong R^{n-q} f_{0*}\mathbf{R} \otimes C^\infty(B_0).$$

Thus the $E^1_{p,n-q}$ term of the de Rham realisation of the Leray spectral sequence for f_0 is $\Gamma(B_0, \Omega^p_{B_0} \otimes \Omega^q_{B_0})$.

Definition 3.13. *A cohomology class $[\alpha_0] \in H^1(B_0, R^{n-1} f_{0*}\mathbf{R})$ is symmetric if there is a representative $\alpha_0 \in E^1_{1,n-1} = \Gamma(B_0, \Omega^1_{B_0} \otimes \Omega^1_{B_0})$ of $[\alpha_0]$ which is invariant under the involution of $\Omega^1_{B_0} \otimes \Omega^1_{B_0}$ given by $a \otimes b \mapsto b \otimes a$. In other words, $\alpha_0 \in \Gamma(B_0, S^2\Omega^1_{B_0}) \subseteq \Gamma(B_0, \Omega^1_{B_0} \otimes \Omega^1_{B_0})$. A cohomology class $\alpha \in H^1(B, R^{n-1} f_* \mathbf{R})$ is said to be symmetric if its restriction to $H^1(B_0, R^{n-1} f_{0*}\mathbf{R})$ is symmetric.*

The following result, which identifies the symmetric cohomology classes, will be useful in §6 in studying the role of the B-field.

Theorem 3.14. *A cohomology class $[\alpha_0] \in H^1(B_0, R^{n-1}f_{0*}\mathbf{R})$ is symmetric if and only if $[\alpha_0] \wedge [\omega] \in H^2(B_0, R^n f_{0*}\mathbf{R})$ is zero. In particular, if $H^2(B, \mathbf{R}) = 0$, then all elements of $H^1(B, R^{n-1}f_*\mathbf{R})$ are symmetric.*

Proof. We work in action-angle coordinates, on a neighborhood U, so that the Gauss-Manin connection is the trivial connection. Thus if $y_1, \ldots, y_n$ are the action coordinates,

$$\nabla_{GM}(\sum f_{IJ}dy_I \otimes dy_J) = \sum d(f_{IJ}dy_I) \otimes dy_J.$$

In addition, in suitably ordered action-angle coordinates $\Theta = dy_1 \wedge \cdots \wedge dy_n$, from which one sees easily that the cohomology class of $(-1)^{i-1}dx_1 \wedge \cdots \wedge \hat{dx_i} \wedge \cdots \wedge dx_n$ on a fibre X_b is identified with $dy_i \in \Omega^1_{B,b}$.

Choose a representative $\alpha_0 \in \Gamma(B_0, \Omega^1_{B_0} \otimes \Omega^1_{B_0})$ for $[\alpha_0]$, so on U we write

$$\alpha_0 = \sum_{i,j} \alpha_{ij} dy_i \otimes dy_j, \tag{3.4}$$

with α_0 symmetric if and only if $\alpha_{ij} = \alpha_{ji}$. Locally, this class can also be represented by the n-form on $f^{-1}(U)$

$$\alpha_0 = \sum_{i,j} (-1)^{j-1}\alpha_{ij} dy_i \wedge dx_1 \wedge \cdots \wedge \hat{dx_j} \wedge \cdots \wedge dx_n,$$

from which we see that

$$\begin{aligned} \alpha_0 \wedge \omega &= \sum_{i,j} (-1)^{j-1}\alpha_{ij} dy_i \wedge dx_1 \wedge \cdots \wedge \hat{dx_j} \wedge \cdots \wedge dx_n \wedge dx_j \wedge dy_j \\ &= \sum_{i<j} (\alpha_{ji} - \alpha_{ij}) dy_i \wedge dy_j \wedge dx_1 \wedge \cdots \wedge dx_n. \end{aligned} \tag{3.5}$$

Thus we see α_0 is a symmetric representative for $[\alpha_0]$ if and only if $\alpha_0 \wedge \omega = 0$ in $\Gamma(B_0, \Omega^2_{B_0} \otimes \Omega^0_{B_0})$. Now $\alpha_0 \wedge \omega$ represents the cohomology class $[\alpha_0] \wedge [\omega] \in H^2(B_0, R^n f_{0*}\mathbf{R})$. This is zero if α_0 is symmetric. Conversely, suppose $[\alpha_0] \wedge [\omega]$ is the zero cohomology class. Then there exists a $\beta \in \Gamma(B_0, \Omega^1_{B_0} \otimes \Omega^0_{B_0})$ such that $\nabla_{GM}\beta = \alpha_0 \wedge \omega$. β also gives rise to an element $\beta' \in \Gamma(B_0, \Omega^0_{B_0} \otimes \Omega^1_{B_0})$ by using the map $a \otimes b \mapsto b \otimes a$. The following claim then proves the theorem.

Claim. $\alpha_0 + \nabla_{GM}\beta'$ is a symmetric representative for $[\alpha_0]$.

Proof. Locally write $\beta = \sum_i \beta_i dy_i \otimes 1$, so $\beta' = \sum_j \beta_j 1 \otimes dy_j$. Now $\nabla_{GM}(\beta) = \alpha_0 \wedge \omega$ means that

$$\frac{\partial \beta_j}{\partial y_i} - \frac{\partial \beta_i}{\partial y_j} = \alpha_{ji} - \alpha_{ij}.$$

Also,

$$\alpha_0 + \nabla_{GM}\beta' = \sum_{i,j} (\alpha_{ij} + \frac{\partial \beta_j}{\partial y_i}) dy_i \otimes dy_j.$$

But

$$\alpha_{ij} + \frac{\partial \beta_j}{\partial y_i} - (\alpha_{ji} + \frac{\partial \beta_i}{\partial y_j}) = 0,$$

so $\alpha_0 + \nabla_{GM}\beta'$ is a symmetric representative for $[\alpha_0]$. •

Remark 3.15. There is a related map which partly explains the interest in symmetric cohomology classes. If $f : X \to B$ is an integral, $\mathbf{Z}$-simple special Lagrangian fibration with a section, then there is a natural inclusion $R^{n-1}f_*\mathbf{R}/\mathbf{Z} \hookrightarrow \Lambda(X^\#)$. This is induced from the inclusion $R^{n-1}f_*\mathbf{Z} \hookrightarrow \Lambda(\mathcal{T}_B^*)$ and the inclusion obtained from this by tensoring with $\mathbf{R}$: $R^{n-1}f_*\mathbf{R} \hookrightarrow \Lambda(\mathcal{T}_B^*)$. Thus one obtains a natural map $H^1(B, R^{n-1}f_*\mathbf{R}/\mathbf{Z}) \to H^1(B, \Lambda(X^\#))$. To analyse this map, consider instead the map $H^1(B, R^{n-1}f_*\mathbf{R}) \to H^1(B, \Lambda(\mathcal{T}_B^*)) \cong H^2(B, \mathbf{R})$. This in fact coincides with the map $H^1(B, R^{n-1}f_*\mathbf{R}) \to H^2(B, R^n f_*\mathbf{R}) \cong H^2(B, \mathbf{R})$ obtained by cup-product with $-\omega$. It is easiest to see this over B_0: in this case, we have resolutions $0 \to R^{n-1}f_{0*}\mathbf{R} \to \Omega^{\cdot}_{B_0} \otimes \Omega^1_{B_0}$ and $0 \to \Lambda(\mathcal{T}_{B_0}^*) \to \Omega^{\cdot+1}_{B_0}$. The map $R^{n-1}f_{0*}\mathbf{R} \to \Lambda(\mathcal{T}_{B_0}^*)$ extends to a map of complexes $\Omega^{\cdot}_{B_0} \otimes \Omega^1_{B_0} \to \Omega^{\cdot+1}_{B_0}$ given by $\alpha \otimes \beta \mapsto \alpha \wedge \beta$. Thus α_0 as in (3.4) is mapped to $\sum_{i<j}(\alpha_{ij} - \alpha_{ji})dy_i \wedge dy_j$, which by (3.5) can be identified with $-\alpha_0 \wedge \omega$. Thus the map $H^1(B_0, R^{n-1}f_{0*}\mathbf{R}) \to H^1(B_0, \Lambda(\mathcal{T}_{B_0}^*))$ coincides with the map $\wedge(-\omega)$. It is not difficult to show this holds over B also, but we omit the cohomological argument.

Thus we see that the symmetric cohomology classes are those that map to zero in $H^1(B, \Lambda(X^\#))$.

§4. The symplectic form on D-brane moduli space.

Let B be a moduli space of special Lagrangian submanifolds in a Calabi-Yau manifold X of dimension n, along with a universal family

$$\begin{array}{ccc} \mathcal{U} & \subseteq & X \times B \\ \downarrow{\scriptstyle f} & & \\ B & & \end{array}$$

Let $p : \mathcal{U} \to X$ be the projection. For the moment we assume f is smooth, so that points in B are parametrizing only smooth special Lagrangian submanifolds. We do not assume these submanifolds are tori. Here $\dim B = \dim H^1(\mathcal{U}_b, \mathbf{R}) =: s$. The D-brane moduli space is the space of special Lagrangian submanifolds along with a choice of flat $U(1)$ connection modulo gauge equivalence, i.e. an element of $H^1(\mathcal{U}_b, \mathbf{R}/\mathbf{Z})$. Thus the D-brane moduli space $\mathcal{M}$ should be $R^1f_*(\mathbf{R}/\mathbf{Z})$. The prediction from string theory is that $\mathcal{M}$ should be a complex Kähler manifold, so we need to understand how to put these structures on $\mathcal{M}$. As long as there are no singular fibres to deal with, Hitchin has described how to put a complex and Kähler structure on $\mathcal{M}$. Here, we will describe a more coordinate independent way of describing the same Kähler form (i.e. a symplectic form in the absence of a complex structure). This both allows us to compute the cohomology class represented by this symplectic form and in principle should allow one to extend this construction to singular fibres.

In what follows, we assume the fibres of f are special Lagrangian with respect to the symplectic form ω and holomorphic n-form Ω, with the standard normalization $\omega^n/n! = (-1)^{n(n-1)/2}(i/2)^n\Omega\wedge\bar{\Omega}$. We also use a holomorphic n-form Ω_n normalised by

$$\Omega_n = \frac{\Omega}{\int_{\mathcal{U}_b} p^*\Omega}.$$

We use Ω_n instead of Ω for several reasons. First, we do not want the symplectic structure on $\mathcal{M}$ to depend on ω, but only on Ω. If we multiply ω by a constant, we must also rescale Ω. If we rescale Ω and use Ω instead of Ω_n in the construction below, then the symplectic form $\check{\omega}$ we construct on $\mathcal{M}$ also changes. Secondly, this normalization fits with the usual form of the mirror map as described in item (5) of the introduction.

To obtain a symplectic form on $\mathcal{M}$, we define a map

$$R^1 f_* \mathbf{Z} \to \mathcal{T}_B^*$$

in such a way that the canonical symplectic form on $\mathcal{T}_B^*$ descends to a symplectic form on $\mathcal{T}_B^*/R^1 f_*\mathbf{Z}$. We follow Hitchin's suggestion of computing the periods of $\operatorname{Im} p^*\Omega_n$. Now $(R^1 f_*\mathbf{Z})_b \cong H^1(\mathcal{U}_b, \mathbf{Z}) \cong H_{n-1}(\mathcal{U}_b, \mathbf{Z})$, so for $\gamma \in H_{n-1}(\mathcal{U}_b, \mathbf{Z})$, map γ to the differential

$$v \mapsto -\int_\gamma \iota(v) \operatorname{Im} p^*\Omega_n$$

where again we choose an arbitrary lifting of v to $\mathcal{U}$.

Lemma 4.1. *The image of $R^1 f_*\mathbf{Z}$ in $\mathcal{T}_B^*$ is Lagrangian with respect to the standard symplectic form on $\mathcal{T}_B^*$. Thus $\mathcal{M} = \mathcal{T}_B^*/R^1 f_*\mathbf{Z}$ inherits this symplectic form, which we will call $\check{\omega}$.*

Proof. See Hitchin's paper [18]. His proof is as follows: in a small open set of B, choose $\Gamma \subseteq \mathcal{U}$ a family of $n-1$-dimensional submanifolds representing a section of $R^1 f_*\mathbf{Z}$ over U, $\pi : \Gamma \to B$ the projection. Then the section of $\mathcal{T}_B^*$ obtained by taking periods with respect to Γ is just the 1-form $-\pi_*((p^* \operatorname{Im} \Omega)|_\Gamma)$. Since Ω is a closed form, so is this push-down, and hence $-\pi_*((p^* \operatorname{Im} \Omega|_\Gamma)$ is a Lagrangian section of $\mathcal{T}_B^*$ with respect to the standard symplectic form. •

We will now clarify what the cohomology class of $\check{\omega}$ is. To do so, we will compare the Leray spectral sequences for f and $\check{f} : \mathcal{M} \to B$, but will use the de Rham realisation of these spectral sequences discussed in §3, which we can do as f and $\check{f}$ are smooth. Our construction yields a canonical isomorphism $H_{n-1}(\mathcal{U}_b, \mathbf{Z}) \cong H_1(\mathcal{M}_b, \mathbf{Z})$ and hence a canonical isomorphism $H^{n-1}(\mathcal{U}_b, \mathbf{Z}) \cong H^1(\mathcal{M}_b, \mathbf{Z})$, which yields a canonical isomorphism

$$R^{n-1} f_*\mathbf{R} \cong R^1 \check{f}_*\mathbf{R}. \tag{4.1}$$

Since $\operatorname{Im} p^*\Omega_n$ restricts to zero on the fibres of f, $\operatorname{Im} p^*\Omega_n \in F^1\Omega^n_{\mathcal{U}}$, and since $d\Omega_n = 0$, $\operatorname{Im} p^*\Omega_n$ in particular gives rise to a class $[\operatorname{Im} p^*\Omega_n]$ in $E^1_{1,n-1} = \Gamma(B, \Omega^1_B \otimes R^{n-1} f_*\mathbf{R})$. Now on $\mathcal{M}$, $\check{\omega} \in F^1(\Omega^2_{\mathcal{M}})$, as $\check{\omega}$ restricts to zero on the fibres of $\check{f}$, and thus $\check{\omega}$ determines an element $[\check{\omega}]$ of $\check{E}^1_{1,1} = \Gamma(B, \Omega^1_B \otimes R^1 \check{f}_*\mathbf{R})$. Via the isomorphism (4.1) we can identify $E^1_{1,n-1}$ and $\check{E}^1_{1,1}$.

Proposition 4.2. *Under this identification, $[\check{\omega}] = [\operatorname{Im} p^*\Omega_n]$. Thus in particular, they represent the same class in $H^1(B, R^1 \check{f}_*\mathbf{R}) \cong H^1(B, R^{n-1} f_*\mathbf{R})$.*

Proof. A section of $\Gamma(B, \Omega^1_B \otimes R^{n-1} f_*\mathbf{R})$ associates, to any tangent vector $v \in \mathcal{T}_{B,b}$, an element of $H^{n-1}(\mathcal{U}_b, \mathbf{R})$. Specifically, $[\operatorname{Im} p^*\Omega_n]$ associates to a tangent vector $v \in \mathcal{T}_{B,b}$ the cohomology class represented by $\iota(v)(\mathrm{Im}\, p^*\Omega_n)$. To determine what cohomology class this is, we choose a basis $\gamma_1, \ldots, \gamma_s$ of $H_{n-1}(\mathcal{U}_b, \mathbf{Z})$ and calculate the periods

$$\int_{\gamma_i} \iota(v)\,\mathrm{Im}\, p^*\Omega_n.$$

On $\mathcal{M}$, $[\check{\omega}] \in \Gamma(B, \Omega^1_B \otimes R^1\check{f}_*\mathbf{R})$ is similarly represented by

$$v \mapsto \iota(v)\check{\omega},$$

and $\gamma_1, \ldots, \gamma_s$ also form a basis for $H_1(\mathcal{M}_b, \mathbf{Z})$ by construction. Recall that we embedded $H_{n-1}(\mathcal{U}_b, \mathbf{Z}) = H_1(\mathcal{M}_b, \mathbf{Z})$ in $\mathcal{T}^*_{B,b}$ by mapping γ_i to the differential

$$v \mapsto -\int_{\gamma_i} \iota(v)\,\mathrm{Im}\, p^*\Omega_n.$$

Now choosing local coordinates $y_1, \ldots, y_s$ on the base, $x_1, \ldots, x_s, y_1, \ldots, y_s$ canonical coordinates on $\mathcal{T}^*_B$,

$$\int_{\gamma_i} \iota(\partial/\partial y_j)\check{\omega} = -\int_{\gamma_i} dx_j = \int_{\gamma_i} \iota(\partial/\partial y_j)\,\mathrm{Im}\, p^*\Omega_n$$

by construction. Thus $[\check{\omega}] = [\mathrm{Im}\, p^*\Omega_n]$. •

We recall here for future use:

Observation 4.3. *(McLean [22]) Because $\mathcal{T}_{B,b}$ is naturally isomorphic to the space of harmonic 1-forms on $\mathcal{U}_b$, there is a metric h on $\mathcal{T}_B$ coming from the Hodge metric. Precisely, for $v \in \mathcal{T}_{B,b}$, $-(\iota(v)p^*\omega)|_{X_b}$ is the corresponding harmonic one-form, and $- * (\iota(v)p^*(\omega))|_{X_b} = (\iota(v)\,\mathrm{Im}\, p^*(\Omega))|_{X_b}$, and we define, for $v, w \in \mathcal{T}_{B,b}$,*

$$h(v,w) = -\int_{X_b} (\iota(v)p^*\omega) \wedge (\iota(v)\,\mathrm{Im}\, p^*\Omega).$$

This is a Riemannian metric on B.

Specialising down to the case that $f : X \to B$ is a special Lagrangian T^n-fibration with possible degenerate fibres, the above method gives us a way of constructing an open subset of the dual fibration along with a symplectic form on that open set. On $B_0 = B - \Delta$, we have defined an embedding $R^1 f_{0*}\mathbf{Z} \hookrightarrow \mathcal{T}^*_{B_0}$. This allows us to define $\check{X}_0$ via the exact sequence

$$0 \to R^1 f_{0*}\mathbf{Z} \to \mathcal{T}^*_{B_0} \to \check{X}_0 \to 0$$

and $\check{X}_0$ acquires a symplectic form $\check{\omega}$ inherited from the canonical symplectic form on $\mathcal{T}^*_{B_0}$.

Next we need to prove

Conjecture 4.4. *The embedding $R^1 f_{0*}\mathbf{Z} \hookrightarrow \mathcal{T}^*_{B_0}$ extends to an embedding*

$$R^1 f_* \mathbf{Z} \hookrightarrow \mathcal{T}^*_B.$$

If $\check{X}^\#$ is defined as $\mathcal{T}^*_B/R^1 f_*\mathbf{Z}$, then $\check{X}^\#$ is a manifold with symplectic form $\check{\omega}$ inherited from the standard symplectic form on $\mathcal{T}^*_B$. Furthermore, $\check{X}^\#$ can be compactified to a manifold $\check{X}$ with a map $\check{f} : \check{X} \to B$ extending $\check{f}^\#$, on which $\check{\omega}$ extends to a symplectic form on $\check{X}$.

This involves first understanding the asymptotic behaviour of the periods as one approaches Δ, as well as understanding the issue of compactification.

If this conjecture holds and f and $\check{f}$ are both $\mathbf{R}$-simple, then it is easy to see that $[\check{\omega}] \in H^1(B, R^1 \check{f}_* \mathbf{R})$ coincides with $[\mathrm{Im}\,\Omega_n] \in H^1(B, R^{n-1} f_* \mathbf{R})$. Indeed, by Proposition 4.2, these classes agree in $H^1(B_0, R^1 \check{f}_*\mathbf{R}) \cong H^1(B_0, R^{n-1} f_* \mathbf{R})$. However, since f and $\check{f}$ were assumed to be simple, $H^i_\Delta(B, R^j f_* \mathbf{R}) = 0$ for $i = 0$ and 1, and ditto for $\check{f}$. Thus there is an injection $H^1(B, R^i f_* \mathbf{R}) \hookrightarrow H^1(B - \Delta, R^i f_* \mathbf{R})$, and so the classes $[\check{\omega}]$ and $[\mathrm{Im}\,\Omega]$ agree also in $H^1(B, R^1 \check{f}_* \mathbf{R}) \cong H^1(B, R^{n-1} f_* \mathbf{R})$.

$\check{f} : \check{X} \to B$ is not the only possible Lagrangian fibration we might construct. This fibration possesses a Lagrangian section by construction. By Theorem 2.6, any element of $H^1(B, \Lambda(\check{X}^\#))$ gives rise to another, locally isomorphic Lagrangian fibration $\check{g} : \check{Y} \to B$. See also Remark 3.12 and Conjecture 6.6.

Remark 4.5. Having constructed a symplectic form $\check{\omega}$ on $\check{X}$, or on an open subset of $\check{X}$, $\check{\omega}^n/n!$ defines an orientation on $\check{X}$. Thus we can check that this agrees with the choice of orientation on $\check{X}$ made in [14], Convention 4.3. First, note that in having fixed Ω, we have fixed an orientation on the fibres of $f : X \to B$. If we have fixed canonical coordinates $y_1, \ldots, y_n, x_1, \ldots, x_n$, then $\Omega|_{X_b} = V dx_1 \wedge \cdots \wedge dx_n$ with V a real function and either $V > 0$ or $V < 0$. By changing the order of the variables y_i, we can ensure $V > 0$, and then $dy_1 \wedge \cdots \wedge dy_n$ yields a canonical orientation on B. We will always assume our coordinates are so oriented. Note that this orientation on B is the same as that induced by the n-form Θ on B_0.

Now let us check Convention 4.3 of [14] is correct. Recall that the convention of [14] for the cohomology class of a fibre, $[X_b]$, was that

$$\int_{X_b} \alpha = \int_X \alpha \wedge [X_b].$$

With $\alpha = [\Omega]$, we then have

$$0 < \int_{X_b} \Omega = \int_X \Omega \wedge [X_b].$$

We can take $[X_b]$ to be the pull-back of a nowhere zero n-form on B, locally $f dy_1 \wedge \cdots \wedge dy_n$. Then $\Omega \wedge [X_b] = V f dx_1 \wedge \cdots \wedge dx_n \wedge dy_1 \wedge \cdots \wedge dy_n$, while $\omega^n/n! = (-1)^{n(n-1)/2} dx_1 \wedge \cdots \wedge dx_n \wedge dy_1 \wedge \cdots \wedge dy_n$. Since $V > 0$, we need $sign(f) = (-1)^{n(n-1)/2}$. Thus we take $[X_b]$ locally to be of the form $(-1)^{n(n-1)/2}|f| dy_1 \wedge \cdots \wedge dy_n$. Now the dual class of $(-1)^n [X_b]$ in $H^n(B, R^n \check{f}_* \mathbf{R})$ will be locally represented by something like $(-1)^{n(n+1)/2} g dy_1 \wedge \cdots \wedge dy_n \wedge dx_1 \wedge \cdots \wedge dx_n$, $g > 0$. This is the same sign as $\check{\omega}^n/n! = (-1)^{n(n+1)/2} dy_1 \wedge \cdots \wedge dy_n \wedge dx_1 \wedge \cdots \wedge dx_n$, so the orientations agree.

In the case of torus fibrations, we now describe an alternative way of putting a symplectic form on $\check{X}_0$. We do this by providing an alternative description of the embedding $R^1 f_{0*}\mathbf{Z} \hookrightarrow \mathcal{T}^*_{B_0}$, using the Riemannian metric h on the base defined in Observation 4.3. Since we obtained in §2 an embedding $R^{n-1} f_{0*}\mathbf{Z} \to \mathcal{T}^*_{B_0}$, we obtain dually an embedding $R^1 f_{0*}\mathbf{Z} \to \mathcal{T}_{B_0}$. We will then use a normalised form of h to identify $\mathcal{T}_{B_0}$ and $\mathcal{T}^*_{B_0}$.

First, given that there is an identification between $H^1(X_b, \mathbf{R})$ and $\mathcal{T}_{B,b}$, via $v \in \mathcal{T}_{B,b} \mapsto -\iota(v)\omega$, we have also a canonical identification of $H_{n-1}(X_b, \mathbf{R})$ with $\mathcal{T}_{B,b}$, via Poincaré duality. In fact, write $X_b = V/\Lambda$, $V = \mathcal{T}^*_{B,b}$, $\Lambda = H_1(X_b, \mathbf{Z}) \subseteq V$, $V^\vee = \mathcal{T}_{B,b}$, $\Lambda^\vee = \{\varphi \in V^\vee | \varphi(\Lambda) \subseteq \mathbf{Z}\} \subseteq V^\vee$. There is a canonical identification of $\Lambda^\vee$ with $H^1(X_b, \mathbf{Z})$. On the other hand, the identification $\bigwedge^n \Lambda \cong \mathbf{Z}$ determined by the orientation on X_b gives us a natural identification of $H_{n-1}(X_b, \mathbf{Z}) = \bigwedge^{n-1} \Lambda$ with $\Lambda^\vee$ via the perfect pairing

$$\Lambda \times \bigwedge^{n-1} \Lambda \to \bigwedge^{n} \Lambda \xrightarrow{\cong} \mathbf{Z}.$$

(Note: Whenever we use Poincaré duality, there is an arbitrary choice of order in this pairing which may affect the signs of the isomorphisms. This was seen in [14], where certain conventions were chosen. Here we also make a choice, and keep in mind that we could just as well have chosen the pairing $\bigwedge^{n-1} \Lambda \times \Lambda \to \mathbf{Z}$.)

Proposition 4.6. *If $\alpha \in H^{n-1}(X_b, \mathbf{R})$, $\gamma \in H_{n-1}(X_b, \mathbf{Z}) \cong \Lambda^\vee \subseteq \mathcal{T}_{B,b}$ via the above identification, then*

$$\int_\gamma \alpha = -\int_{X_b} \iota(\gamma)\omega \wedge \alpha.$$

Proof. We compute both sides using local action-angle coordinates. Let $y_1, \cdots, y_n$ be action coordinates as in Remark 2.8. Then the lattice $\Lambda \subseteq V = \mathcal{T}^*_{B,b}$ is generated by $e_1, \ldots, e_n$, $e_i = dy_i$, and then $e_1^*, \ldots, e_n^*$ form a dual basis for $\Lambda^\vee \subseteq \mathcal{T}_{B,b}$, $e_i^* = \partial/\partial y_i$. Suppose $\alpha = dx_1 \wedge \cdots \wedge \hat{dx_j} \wedge \cdots \wedge dx_n$. Then

$$\begin{aligned} -\int_{X_b} \iota(e_i^*)\omega \wedge \alpha &= \int_{X_b} dx_i \wedge \alpha \\ &= (-1)^{j-1}\delta_{ij}. \end{aligned}$$

On the other hand, the isomorphism between $\Lambda^\vee$ and $\bigwedge^{n-1} \Lambda$ identifies e_i^* with $(-1)^{i-1} e_1 \wedge \cdots \wedge \hat{e}_i \wedge \cdots \wedge e_n$. The latter defines an oriented $n-1$-torus in X_b, namely the quotient of the subspace of V spanned by $e_1, \ldots, e_{i-1}, e_{i+1}, \ldots, e_n$ by the lattice generated by these vectors, and $\int_{e_i^*} \alpha$ is just the integral of α over this torus, which is clearly $(-1)^{j-1}\delta_{ij}$. Thus

$$\int_{e_i^*} \alpha = -\int_{X_b} \iota(e_i^*)\omega \wedge \alpha$$

and the result follows from linearity. •

This now gives us the opportunity to rephrase the embedding $H^1(X_b, \mathbf{Z}) \hookrightarrow \mathcal{T}^*_{B,b}$. Let h_n be the normalised metric on B_0 given by

$$h_n(v,w) = \frac{h(v,w)}{\int_{X_b} \Omega}.$$

Then we have

Proposition 4.7. *For $\gamma \in H^1(X_b, \mathbf{Z}) \cong H_{n-1}(X_b, \mathbf{Z}) = \Lambda^\vee \subseteq \mathcal{T}_{B,b}$, the 1-form*

$$v \mapsto -\int_\gamma \iota(v) \operatorname{Im} \Omega_n$$

coincides with the 1-form $-h_n(\gamma, \cdot)$. Thus the embedding $R^1 f_{0*}\mathbf{Z} \to \mathcal{T}^*_{B_0}$ previously defined coincides up to sign with the embedding $R^1 f_{0*}\mathbf{Z} \to \mathcal{T}_{B_0}$ composed with the isomorphism $\mathcal{T}_{B_0} \cong \mathcal{T}^*_{B_0}$ induced by the Riemannian metric h_n.

Proof. By Proposition 4.6,

$$\begin{aligned} -\int_\gamma \iota(v) \operatorname{Im} \Omega_n &= \int_{X_b} \iota(\gamma)\omega \wedge \iota(v) \operatorname{Im} \Omega_n \\ &= \frac{1}{\int_{X_b} \Omega} \int_{X_b} \iota(\gamma)\omega \wedge \iota(v) \operatorname{Im} \Omega \\ &= -\frac{h(\gamma, v)}{\int_{X_b} \Omega}. \bullet \end{aligned}$$

In fact, we see that h_n also describes the class $[\operatorname{Im} \Omega_n] \in E^1_{1,n-1} = \Gamma(B_0, \Omega^1_{B_0} \otimes R^{n-1} f_{0*}\mathbf{R})$:

Proposition 4.8. *Under the isomorphism*

$$E^1_{1,n-1} = \Gamma(B_0, \Omega^1_{B_0} \otimes R^{n-1} f_{0*}\mathbf{R}) \cong \Gamma(B_0, \Omega^1_{B_0} \otimes \Omega^1_{B_0})$$

given in §3, $[\operatorname{Im} \Omega_n]$ coincides with $h_n \in \Gamma(B_0, S^2\Omega^1_{B_0}) \subseteq \Gamma(B_0, \Omega^1_{B_0} \otimes \Omega^1_{B_0})$.

Proof. First note that for a point $b \in B$, $e_1^*, \ldots, e_n^*$ a basis of $H_{n-1}(X_b, \mathbf{Z}) \cong \Lambda^\vee \subseteq \mathcal{T}_{B,b}$, $[\operatorname{Im} \Omega_n]$ associates to a vector $v \in \mathcal{T}_{B,b}$ the class of $\iota(v) \operatorname{Im} \Omega_n \in H^{n-1}(X_b, \mathbf{R})$, and in terms of the periods,

$$\begin{aligned} \int_{e_i^*} \iota(v) \operatorname{Im} \Omega_n &= -\int_{X_b} \iota(e_i^*)\omega \wedge \iota(v) \operatorname{Im} \Omega_n \\ &= h_n(e_i^*, v). \end{aligned}$$

Thus, in action-angle coordinates as used in the proof of Theorem 3.14, $[\operatorname{Im} \Omega_n]$ corresponds to the element of $\Gamma(B_0, \Omega^1_{B_0} \otimes R^{n-1} f_{0*}\mathbf{R})$ given as

$$\sum_{i,j} h_n(\partial/\partial y_i, \partial/\partial y_j) dy_i \otimes (-1)^{j-1} dx_1 \wedge \cdots \wedge \widehat{dx_j} \wedge \cdots \wedge dx_n$$

which coincides with

$$\sum_{i,j} h_n(\partial/\partial y_i, \partial/\partial y_j) dy_i \otimes dy_j \in \Gamma(B_0, S^2\Omega^1_{B_0})$$

as desired. •

§5. Complex structures on special Lagrangian torus fibrations.

Recall from [18] the following: (for K3 surfaces, this was noticed in [31]; see also [30].)

Theorem 5.1. *Let X be a real $2n$-dimensional manifold. If Ω is a complex-valued C^∞ n-form on X satisfying the three properties*
(1) $d\Omega = 0$;
(2) Ω is locally decomposable (i.e. can be written locally as $\theta_1 \wedge \cdots \wedge \theta_n$ where $\theta_1, \ldots, \theta_n$ are 1-forms);
(3) $(-1)^{n(n-1)/2}(i/2)^n \Omega \wedge \bar\Omega > 0$ everywhere on X,

then Ω determines a complex structure on X for which Ω is a holomorphic n-form.

Theorem 5.2. *Let X be a real $2n$-dimensional manifold. Suppose ω is a symplectic form on X and Ω is a complex-valued n-form on X such that*
(1) Ω satisfies the conditions of Theorem 5.1;
(2) ω is a positive $(1,1)$ form in the complex structure of Theorem 5.1;
(3) $(-1)^{n(n-1)/2}(i/2)^n \Omega \wedge \bar\Omega = \omega^n/n!$.

Then Ω induces a complex structure on X such that ω is a Kähler form on X whose corresponding metric is Ricci-flat.

Now let $f : X \rightarrow B$ be an integral special Lagrangian fibration with a Lagrangian section, so that in local coordinates, ω takes the standard form. Let Ω be the holomorphic n-form on X normalised so that $\mathrm{Im}\,\Omega|_{X_b} = 0$ for all $b \in B$, $\omega^n/n! = (-1)^{n(n-1)/2}(i/2)^n \Omega \wedge \bar\Omega$, and $\int_{X_b} \Omega = Vol(X_b)$ with respect to the metric induced by the Kähler form ω. As before, we set $\Omega_n = \Omega / \int_{X_b} \Omega$.

Following the suggestion of [18], since Ω is locally decomposable, we can write, for local coordinates $y_1, \ldots, y_n$ on B as usual,

$$\Omega = V \bigwedge_i (dx_i + \sum_j \beta_{ij} dy_j),$$

where V is a real function of $y_1, \ldots, x_n$ and $V|_{X_b} dx_1 \wedge \cdots \wedge dx_n$ is the volume form on X_b, while β_{ij} is a complex valued function. Using Remark 4.5, we will always assume that $V > 0$. The forms of type $(1,0)$ are spanned by the 1-forms $\theta_i := dx_i + \sum_{i,j} \beta_{ij} dy_j$. Thus the entire complex structure is encoded in the matrix (β_{ij}). We now look to see how the conditions of Theorems 5.1 and 5.2 translate into conditions for the functions V and β_{ij}:

Calculation 5.3.

$$(-1)^{n(n-1)/2}(i/2)^n \Omega \wedge \bar\Omega = V^2 \det(\mathrm{Im}\,\beta) \omega^n / n!.$$

Proof. Write

$$\Omega = V\theta_1 \wedge \cdots \wedge \theta_n.$$

Now

$$\begin{aligned}\theta_i \wedge \bar\theta_i &= \sum_j (\bar\beta_{ij} - \beta_{ij}) dx_i \wedge dy_j + \cdots \\ &= -2i dx_i \wedge \left(\sum_j \operatorname{Im}\beta_{ij} dy_j\right) + \cdots.\end{aligned}$$

Thus

$$\Omega \wedge \bar\Omega = V^2(-2)^n i^n \det(\operatorname{Im}\beta) dx_1 \wedge dx_2 \wedge \ldots \wedge dy_n$$

so

$$(-1)^{n(n-1)/2}(i/2)^n \Omega \wedge \bar\Omega = (-1)^{n(n-1)/2} V^2 \det(\operatorname{Im}\beta) dx_1 \wedge dx_2 \wedge \ldots \wedge dy_n.$$

On the other hand,

$$\omega^n/n! = (-1)^{n(n-1)/2} dx_1 \wedge \cdots \wedge dy_n,$$

hence the result. •

Calculation 5.4. *ω is a positive form of type $(1,1)$ if and only if β is symmetric and $\operatorname{Im}\beta$ is positive definite.*

Proof. We first examine the condition that ω is of type $(1,1)$, i.e. we can write

$$\omega = \frac{i}{2} \sum_{i,j} h_{ij} \theta_i \wedge \bar\theta_j.$$

Now

$$\begin{aligned}\theta_i \wedge \bar\theta_j =& dx_i \wedge dx_j + dx_i \wedge \left(\sum_k \bar\beta_{jk} dy_k\right) - dx_j \wedge \left(\sum_k \beta_{ik} dy_k\right) \\ &+ \sum_{k,l} \beta_{ik}\bar\beta_{jl} dy_k \wedge dy_l\end{aligned}$$

and

$$\begin{aligned}\sum h_{ij}\theta_i \wedge \bar\theta_j =& \sum_{i<j} (h_{ij} - h_{ji}) dx_i \wedge dx_j + \sum_i dx_i \wedge \left(\sum_{j,k} h_{ij}\bar\beta_{jk} dy_k - \sum_{j,k} h_{ji}\beta_{jk} dy_k\right) \\ &+ \sum_{i,j,k,l} h_{ij}\beta_{ik}\bar\beta_{jl} dy_k \wedge dy_l.\end{aligned}$$

Thus in order for $\omega = \sum dx_i \wedge dy_i = \frac{i}{2}\sum h_{ij}\theta_i \wedge \bar\theta_j$, we must have, in particular, $h_{ij} = h_{ji}$, i.e. h is symmetric. On the other hand, since ω is real, $h_{ij} = \bar h_{ji}$, and thus the matrix h is real. Also,

$$\begin{aligned}I &= \frac{i}{2}(h\bar\beta - h\beta) \\ &= h \operatorname{Im}\beta.\end{aligned}$$

Thus we have $h = (\operatorname{Im}\beta)^{-1}$, and $\operatorname{Im}\beta$ is symmetric.

To ensure the last term vanishes, we need

$$\sum_{i,j} h_{ij}\beta_{ik}\bar{\beta}_{jl} = \sum_{i,j} h_{ij}\beta_{il}\bar{\beta}_{jk},$$

or equivalently, the matrix ${}^t\beta h\bar{\beta}$ is symmetric.

Now

$$\begin{aligned} {}^t\beta h\bar{\beta} &= (\operatorname{Re}{}^t\beta + i\operatorname{Im}\beta)(\operatorname{Im}\beta)^{-1}(\operatorname{Re}\beta - i\operatorname{Im}\beta) \\ &= (\operatorname{Re}{}^t\beta)(\operatorname{Im}\beta)^{-1}(\operatorname{Re}\beta) + i\operatorname{Re}\beta - i\operatorname{Re}{}^t\beta + \operatorname{Im}\beta, \end{aligned}$$

while

$$\begin{aligned} {}^t\bar{\beta} h\beta &= (\operatorname{Re}{}^t\beta - i\operatorname{Im}\beta)(\operatorname{Im}\beta)^{-1}(\operatorname{Re}\beta + i\operatorname{Im}\beta) \\ &= (\operatorname{Re}{}^t\beta)(\operatorname{Im}\beta)^{-1}(\operatorname{Re}\beta) + i\operatorname{Re}{}^t\beta - i\operatorname{Re}\beta + \operatorname{Im}\beta, \end{aligned}$$

so symmetry of ${}^t\beta h\bar{\beta}$ is equivalent to $\operatorname{Re}{}^t\beta = \operatorname{Re}\beta$.

Thus ω is of type $(1,1)$ if and only if β is symmetric. In addition, to ensure ω is a positive $(1,1)$ form, $h = (\operatorname{Im}\beta)^{-1}$ must be positive definite, so $\operatorname{Im}\beta$ must be positive definite. •

The real problem is understanding the condition $d\Omega = 0$. This is the heart of the difficulty, and we will return to this shortly.

We first connect Ω to the description of the choice of almost complex structure given in the introduction, namely as a choice of horizontal subspaces of an Ehresmann connection and the choice of a metric on the fibres.

Proposition 5.5. *The matrix* $(\operatorname{Im}\beta)^{-1}$ *is the matrix of the metric* (g_{ij}) *on the fibres of* f. *For a point* $x \in X_b^{\#}$, $J(\mathcal{T}_{X_b,x})$ *is spanned by the tangent vectors*

$$\{\partial/\partial y_j - \sum_i \operatorname{Re}\beta_{ij}\partial/\partial x_i | 1 \le j \le n\},$$

where $J : \mathcal{T}_{X^\#} \to \mathcal{T}_{X^\#}$ *is the almost complex structure induced by* Ω.

Proof. Since $\omega = \frac{i}{2}\sum_{i,j} h_{ij}\theta_i \wedge \bar{\theta}_j$ with $h = (\operatorname{Im}\beta)^{-1}$ is the Kähler form of the metric, the Kähler metric itself is $g = \sum_{i,j} h_{ij}\theta_i \otimes \bar{\theta}_j$. Thus $g_{ij} = g(\partial/\partial x_i, \partial/\partial x_j) = h_{ij}$, giving the interpretation of $\operatorname{Im}\beta$.

Next, let J be the almost complex structure on $\mathcal{T}_{X^\#}$ induced by Ω, and ${}^tJ$ the almost complex structure on $\mathcal{T}^*_{X^\#}$. Since the space spanned by $\theta_1, \ldots, \theta_n$ is the $+i$ eigenspace of ${}^tJ$ at a point $x \in X^\#$, the cotangent space $\mathcal{T}^*_{X,x}$ decomposes as $V_1 \oplus V_2$ with

$$\begin{aligned} V_1 &= span(\operatorname{Re}\theta_1, \ldots, \operatorname{Re}\theta_n) \\ V_2 &= span(\operatorname{Im}\theta_1, \ldots, \operatorname{Im}\theta_n), \end{aligned}$$

with ${}^tJ(V_1) = V_2$ and ${}^tJ(V_2) = V_1$. Note that $V_2 = span(dy_1, \ldots, dy_n)$ since $\operatorname{Im}\beta$ is invertible. Thus also $\mathcal{T}_{X,x} = V_1^\circ \oplus V_2^\circ$, V_i° the annihilator of V_i, with J interchanging

V_1° and V_2°. Now

$$V_2^\circ = span(\partial/\partial x_1, \ldots, \partial/\partial x_n)$$
$$V_1 = span(\{dx_i + \sum_j \mathrm{Re}\,\beta_{ij} dy_j\})$$
$$V_1^\circ = span(\{\partial/\partial y_j - \sum_i \mathrm{Re}\,\beta_{ij}\partial/\partial x_i\}).$$

Thus we see that $\mathrm{Re}\,\beta$ determines $J(\mathcal{T}_{X_b,x}) = V_1^\circ$ as claimed. •

Summarizing, we now have

Theorem 5.6. *Specifying an n-form Ω on $X^\#$ satisfying properties (2) and (3) of Theorem 5.1 and properties (2) and (3) of Theorem 5.2 and such that $f^\#$ is a special Lagrangian fibration with respect to Ω is equivalent to specifying*

(1) A metric (g_{ij}) on each fibre of $f^\# : X^\# \to B$.

(2) A splitting $\mathcal{T}_{X^\#} = \mathcal{T}_{X^\#/B} \oplus \mathcal{F}$, where $\mathcal{T}_{X^\#/B}$ is the subbundle of $\mathcal{T}_{X^\#}$ with $\mathcal{T}_{X^\#/B,x} = \mathcal{T}_{X_b^\#,x}$, and $\mathcal{F}$ is a Lagrangian subbundle of $\mathcal{T}_{X^\#}$.

Proof. Proposition 5.5 shows that Ω specifies the metric on the fibres and a splitting as desired with $\mathcal{F} = J(\mathcal{T}_{X^\#/B})$. This is clearly a Lagrangian subbundle since $\mathcal{T}_{X^\#/B}$ is.

Conversely, giving a splitting of the exact sequence

$$0 \longrightarrow \mathcal{T}_{X^\#/B} \longrightarrow \mathcal{T}_{X^\#} \stackrel{p}{\longrightarrow} \mathcal{F} \longrightarrow 0$$

determines $\mathrm{Re}\,\beta$ for us. Indeed, at a point $x \in X^\#$, with local coordinates as usual, such a splitting gives a map $s : \mathcal{F}_x \to \mathcal{T}_{X^\#,x}$, and there is a matrix (b_{ij}) such that

$$s(p(\partial/\partial y_j)) = \partial/\partial y_j - \sum b_{ij}\partial/\partial x_i.$$

We take $\mathrm{Re}\,\beta_{ij} = b_{ij}$. Note that the symmetry of the matrix b_{ij} is equivalent to $s(\mathcal{F}_x)$ being Lagrangian. Thus specifying (1) and (2) is equivalent, in local coordinates, to giving $\beta_{ij} = b_{ij} + ig^{ij}$, where g_{ij} is the metric on the fibre. Then we must have, with

$$\theta_i = dx_i + \sum_j \beta_{ij} dy_j,$$
$$\Omega = V\theta_1 \wedge \cdots \wedge \theta_n$$

for some real function V. Calculation 5.4 then tells us that ω is a positive $(1,1)$-form in the corresponding almost complex structure, since by construction β is symmetric and $\mathrm{Im}\,\beta$ is positive definite. Calculation 5.3 shows that $\omega^n/n! = (-1)^{n(n-1)/2}(i/2)^n\Omega\wedge\bar{\Omega}$ if and only if $V = \sqrt{\det(g_{ij})}$ since $(\sqrt{\det(g_{ij})})^2\det(g^{ij}) = 1$. •

We have now seen how the n-form Ω can be determined by choosing a matrix $\beta = (\beta_{ij}) = (b_{ij} + ig^{ij})$, so that

$$\Omega = \sqrt{\det(g_{ij})}\bigwedge_i (dx_i + \sum_j \beta_{ij} dy_j).$$

If β is chosen to be symmetric and $\operatorname{Im}\beta$ positive definite, then we have seen that $d\Omega = 0$ implies both the integrability of the almost complex structure induced by Ω and the Ricci-flatness of the metric induced by this complex structure and the standard symplectic form ω. At first sight, the condition $d\Omega = 0$ looks very complicated if one proceeds by brute force and tries to compute the exterior derivative of Ω. In fact, this condition can be simplified, and we wish to examine this here. We note that the calculation below of $d\Omega$ is quite similar to calculations carried out in [29] and [30] for an analagous situation in the study of deformations of complex structures on Calabi-Yau manifolds. However, we will introduce a formalism using differential operators to make the calculation easier.

To begin, first note that the integrability of the almost complex structure determined by Ω is a weaker condition than $d\Omega = 0$. Let us first understand this weaker condition.

Theorem 5.7. *The almost complex structure induced by Ω is integrable if and only if*

$$\sum_i \left(\frac{\partial \beta_{lk}}{\partial x_i}\beta_{ij} - \frac{\partial \beta_{lj}}{\partial x_i}\beta_{ik} \right) = \frac{\partial \beta_{lk}}{\partial y_j} - \frac{\partial \beta_{lj}}{\partial y_k}.$$

for all j, k and l.

Proof. Writing as before

$$\Omega = V\theta_1 \wedge \cdots \wedge \theta_n,$$

the almost complex structure is determined by the fact that $\theta_1, \ldots, \theta_n$ should span the space of $(1,0)$ forms. To show that the almost complex structure induced by $\theta_1, \ldots, \theta_n$ is integrable we need to show that $d\theta_i$ is of type $(2,0) + (1,1)$ for all i. Since

$$\theta_i = dx_i + \sum_{j=1}^{n} \beta_{ij} dy_j,$$

$\theta_1, \ldots, \theta_n, dy_1, \ldots, dy_n$ form a basis for the space of 1-forms, and thus we can write

$$d\theta_l = \sum_{i,j} \frac{1}{2} A^l_{ij}\theta_i \wedge \theta_j + \sum_{ij} B^l_{ij}\theta_i \wedge dy_j + \sum_{i,j} \frac{1}{2} C^l_{ij} dy_i \wedge dy_j.$$

The almost complex structure is integrable if and only if $C^l_{ij} = 0$ for all i, j and l. Here A^l and C^l are skew-symmetric matrices. Note

$$d\theta_l = \sum_{i,j} \frac{\partial \beta_{lj}}{\partial x_i} dx_i \wedge dy_j + \sum_{i,j} \frac{\partial \beta_{lj}}{\partial y_i} dy_i \wedge dy_j.$$

Since $d\theta_l$ contains no $dx_i \wedge dx_j$ terms, we must have $A^l_{ij} = 0$, and then $B^l_{ij} = \partial\beta_{lj}/\partial x_i$. Then the almost complex structure is integrable if and only if

$$\begin{aligned} d\theta_l &= \sum_{i,j} B^l_{ij}\theta_i \wedge dy_j \\ &= \sum_{i,j} \frac{\partial \beta_{lj}}{\partial x_i} dx_i \wedge dy_j + \sum_{i,j,k} \frac{\partial \beta_{lj}}{\partial x_i} \beta_{ik} dy_k \wedge dy_j \end{aligned}$$

which holds if and only

$$\sum_i \left(\frac{\partial \beta_{lk}}{\partial x_i}\beta_{ij} - \frac{\partial \beta_{lj}}{\partial x_i}\beta_{ik}\right) = \frac{\partial \beta_{lk}}{\partial y_j} - \frac{\partial \beta_{lj}}{\partial y_k}.$$

This is the desired condition. •

We wish to rephrase this condition. We are going work locally for the moment, fixing coordinates $y_1, \ldots, y_n$ on an open subset $U \subseteq B$, and consider all forms as living on $U \times \mathbf{R}^n$ with coordinates $y_1, \ldots, y_n, x_1, \ldots, x_n$. We then write

$$\beta = \sum_{i,j} \beta_{ij} dy_j \otimes \frac{\partial}{\partial x_i} \in \Gamma(U \times \mathbf{R}^n, f^*\Omega^1_B \otimes \mathcal{T}_{X/B}).$$

(This is not a coordinate independent expression. The correct coordinate independent expression would be

$$\sum_i dx_i \otimes \frac{\partial}{\partial x_i} + \beta \in \Gamma(f^{-1}(U), \Omega^1_X \otimes \mathcal{T}_{X/B}),$$

but for practical purposes it is more convenient to work with the above expression.)

Definition 5.8. *For expressions of the type $v = \sum_j dy_j \otimes v_j$, $w = \sum_j dy_j \otimes w_j$ with v_j, w_j vector fields, $v_j = \sum_i v_{ij}\partial/\partial x_i$, $w_j = \sum_i w_{ij}\partial/\partial x_i$, we define*

$$[v, w] = \sum_{l,m} [v_l, w_m] dy_l \wedge dy_m$$

where

$$[v_l, w_m] = \sum_{i,j} \left(v_{il}\frac{\partial w_{jm}}{\partial x_i} - w_{im}\frac{\partial v_{jl}}{\partial x_i}\right)\frac{\partial}{\partial x_j}$$

is the usual Lie bracket of vector fields. We also set

$$d_y v = \sum_{i,j,k} \frac{\partial v_{ij}}{\partial y_k} dy_k \wedge dy_j \otimes \frac{\partial}{\partial x_i}.$$

We then obtain

Theorem 5.9. *The almost complex structure induced by Ω is integrable if and only if*

$$d_y\beta - \frac{1}{2}[\beta, \beta] = 0.$$

Proof.

$$d_y\beta = \sum_{\substack{l\\ i<j}} \left(\frac{\partial \beta_{lj}}{\partial y_i} - \frac{\partial \beta_{li}}{\partial y_j}\right) dy_i \wedge dy_j \frac{\partial}{\partial x_l}$$

while

$$[\beta, \beta] = \sum_{\substack{i,j\\ l<m}} 2\left(\beta_{il}\frac{\partial \beta_{jm}}{\partial x_i} - \beta_{im}\frac{\partial \beta_{jl}}{\partial x_i}\right) dy_l \wedge dy_m \frac{\partial}{\partial x_j}.$$

Comparing with the formula of Theorem 5.7, we see that the two formulae are equivalent. •

Our next goal is to prove

Theorem 5.10. *$d\Omega = 0$ if and only if the almost complex structure induced by Ω is integrable and $d\Omega \subseteq F^2\Omega_X^{n+1}$.*

Thus, locally, one just needs to check that the coefficients of $dy_i \wedge dx_1 \wedge \cdots \wedge dx_n$, $1 \le i \le n$ in $d\Omega$ are zero, and also check the integrability condition.

To prove this theorem, we must introduce some additional algebraic structure to accomplish the calculation. We continue to use a choice of local coordinates, and work on the space $U \times \mathbf{R}^n$, $f : U \times \mathbf{R}^n \to U$ the projection. Let $\mathcal{T}_x$ be the subbundle of the tangent bundle of $U \times \mathbf{R}^n$ generated by $\partial/\partial x_1, \ldots, \partial/\partial x_n$, and let Ω_y be the subbundle of the cotangent bundle generated by $dy_1, \ldots, dy_n$. For $q \ge 0, p \le 0$, set

$$\Omega_U^{p,q} = \Gamma(\bigwedge^q \Omega_y \otimes \bigwedge^{-p} \mathcal{T}_x)$$

and set $\Omega_0 = dx_1 \wedge \cdots \wedge dx_n$. Then there is an isomorphism

$$\bigoplus_{i=0}^{p} \Omega_U^{p-n-i,i} \xrightarrow{\cong} \Gamma(\Omega^p_{U\times\mathbf{R}^n}) \tag{5.1}$$

where $\Omega^p_{U\times\mathbf{R}^n}$ denotes the sheaf of C^∞ p-forms on $U \times \mathbf{R}^n$. This isomorphism sends $\theta \otimes v$ to $\theta \wedge \iota(v)\Omega_0$. In particular, $\Gamma(\Omega^n_{U\times\mathbf{R}^n}) \cong \bigoplus_{p=0}^n \Omega_U^{-p,p}$.

For $\alpha \otimes \beta \in \Omega_U^{p,q}$, $\alpha' \otimes \beta' \in \Omega_U^{p',q'}$, we can define the product

$$(\alpha \otimes \beta) \cdot (\alpha' \otimes \beta') := (\alpha \wedge \alpha') \otimes (\beta \wedge \beta') \in \Omega_U^{p+p',q+q'}.$$

This satisfies the commutation relations

$$(\alpha \otimes \beta) \cdot (\alpha' \otimes \beta') = (-1)^{pp'+qq'}(\alpha' \otimes \beta') \cdot (\alpha \otimes \beta).$$

This gives us a bigraded ring structure on $\bigoplus \Omega_U^{p,q}$. Note that the subring $\bigoplus_{p=0}^n \Omega_U^{-p,p} \cong \Gamma(\Omega^n_{U\times\mathbf{R}^n})$ is in fact a commutative ring with 1, and $1 \in \Omega_U^{0,0}$ corresponds to Ω_0 under this isomorphism.

Lemma 5.11. *Under the isomorphism (5.1),*

$$\bigwedge_{i=1}^{n} \left(dx_i + \sum_j \beta_{ij} dy_j \right) = exp(\beta) := \sum_{p=0}^{\infty} \beta^p/p!$$

where

$$\beta = \sum_{i,j} \beta_{ij} dy_j \otimes \frac{\partial}{\partial x_i}$$

in the notation introduced above.

Proof. This is a straightforward though slightly tedious calculation. Here is where all the signs must be dealt with correctly. For this and subsequent calculations it is convenient to keep in mind that $\iota(\partial/\partial x_I)\Omega_0 = (-1)^M dx_{I^*}$, where $I^* = \{1, \ldots, n\} - I$ and $M = \#\{(i,j) | i \in I, j \in I^*, i > j\}$. •

The next step is to turn $\Omega_X^{\cdot,\cdot}$ into a double complex. We have exterior differentiation

$$d : \Gamma(\Omega^i_{U\times\mathbf{R}^n}) \to \Gamma(\Omega^{i+1}_{U\times\mathbf{R}^n}),$$

and under the isomorphism (5.1), it is clear that $d(\Omega_U^{p,q}) \subseteq \Omega_U^{p+1,q} \oplus \Omega_U^{p,q+1}$. Thus we can write $d = d_x + d_y$, so that $(\Omega_U^{\cdot,\cdot}, d_x, d_y)$ defines a bicomplex, with

$$d_x : \Omega_U^{p,q} \to \Omega_U^{p+1,q},$$

$$d_y : \Omega_U^{p,q} \to \Omega_U^{p,q+1}.$$

One checks that

$$d_x(dy_J \otimes \alpha) = (-1)^q \sum_{i=1}^n dy_J \otimes (\frac{\partial\alpha}{\partial x_i} \llcorner dx_i). \tag{5.2}$$

(Here, $(\partial/\partial x_I \wedge \partial/\partial x_i)\llcorner dx_i = \partial/\partial x_I$, and $\partial/\partial x_I \llcorner dx_i = 0$ if $i \notin I$.)

Let $D : \Omega_U \to \Omega_U$ be a graded endomorphism of $\Omega_U := \Omega_U^{\cdot,\cdot}$. We now recall what it means for D to be a differential operator of order $\leq r$. Put on $\Omega_U \otimes \Omega_U$ the anti-commutative algebra structure given by

$$(a \otimes b) \cdot (a' \otimes b') = (-1)^{(\deg a')\cdot(\deg b)} aa' \otimes bb'.$$

Here the dot is the standard dot product, keeping in mind Ω_U is bigraded. This turns $\Omega_U \otimes \Omega_U$ into a bigraded anti-commutative algebra. Let $\lambda : \Omega_U \to \Omega_U \otimes \Omega_U$ be given by $\lambda(a) = a \otimes 1 - 1 \otimes a$. We define

$$\Phi_D^r : \Omega_U^r \to \Omega_U$$

by

$$\Phi_D^r(a_1, \ldots, a_r) = m \circ (D \otimes id_{\Omega_U})(\prod_{i=1}^r \lambda(a_i)).$$

Here $m(a \otimes b) = ab$. We say D is a differential operator of order $\leq r$ if Φ_D^{r+1} is identically zero. Note that

$$\Phi_D^2(a,b) = D(ab) - D(a)b - (-1)^{(\deg a)\cdot(\deg b)} D(b)a + D(1)ab$$

and

$$\Phi_D^3(a,b,c) = \Phi_D^2(a,bc) - \Phi_D^2(a,b)c - (-1)^{(\deg b)\cdot(\deg c)}\Phi_D^2(a,c)b.$$

(See [20], §1.) As usual, the composition of differential operators of orders $\leq r$ and s is order $\leq r+s$.

Definition 5.12. *We set $d'_x : \Omega_U \to \Omega_U$ to be the operator acting on $\Omega_U^{p,q}$ by $(-1)^{p+q+1} d_x$.*

Lemma 5.13. *d_x is a differential operator of order ≤ 2 and d_y is a differential operator of order ≤ 1.*

Proof. That d_y is a differential operator of order ≤ 1 follows immediately from the definition, while (5.2) shows that d'_x can be written as a sum of a composition of two operators: differentiation in the direction $\partial/\partial x_i$ and $\alpha\otimes\beta \mapsto \alpha\otimes(-1)^{p+1}\beta\llcorner dx_i$. One checks easily that these are each first order operators, and hence d_x is second order. •

Now define, for $\alpha, \beta \in \Omega_X$,

$$[\alpha, \beta] := \Phi^2_{d'_x}(\alpha, \beta),$$

so

$$d'_x(\alpha\beta) = [\alpha, \beta] + d'_x(\alpha)\beta + (-1)^{(\deg\alpha)\cdot(\deg\beta)} d'_x(\beta)\alpha. \tag{5.3}$$

Since $\Phi^3_{d'_x} = 0$, we obtain

$$[\alpha, \beta\gamma] = [\alpha, \beta]\gamma + (-1)^{(\deg\beta)\cdot(\deg\gamma)}[\alpha, \gamma]\beta, \tag{5.4}$$

and from Lemma 5.13,

$$d_y(\alpha\beta) = d_y(\alpha)\beta + (-1)^{(\deg\alpha)\cdot(\deg\beta)} d_y(\beta)\alpha. \tag{5.5}$$

We note that this definition of bracket is an extension of Definition 5.8:

Proposition 5.14. *If $\alpha, \beta \in \Omega^{-1,1}$, then $[\alpha, \beta]$ as defined above agrees with the earlier definition.*

Proof. By linearity, we can assume that $\alpha = f dy_j \partial/\partial x_i$ and $\beta = g dy_l \partial/\partial x_k$. Then

$$\begin{aligned}
[\alpha, \beta] =& - d_x(\alpha\beta) + d_x(\alpha)\beta + d_x(\beta)\alpha \\
=& - d_x(fg dy_j \wedge dy_l \frac{\partial}{\partial x_i} \wedge \frac{\partial}{\partial x_k}) + d_x(f dy_j \frac{\partial}{\partial x_i}) g dy_l \frac{\partial}{\partial x_k} \\
&+ d_x(g dy_l \frac{\partial}{\partial x_k}) f dy_j \frac{\partial}{\partial x_i} \\
=& - \left(\frac{\partial f}{\partial x_k} g + f \frac{\partial g}{\partial x_k}\right) dy_j \wedge dy_l \frac{\partial}{\partial x_i} + \left(\frac{\partial f}{\partial x_i} g + f \frac{\partial g}{\partial x_i}\right) dy_j \wedge dy_l \frac{\partial}{\partial x_k} \\
&- g \frac{\partial f}{\partial x_i} dy_j \wedge dy_l \frac{\partial}{\partial x_k} + f \frac{\partial g}{\partial x_k} dy_j \wedge dy_l \frac{\partial}{\partial x_i} \\
=& f \frac{\partial g}{\partial x_i} dy_j \wedge dy_l \frac{\partial}{\partial x_k} - \frac{\partial f}{\partial x_k} g dy_j \wedge dy_l \frac{\partial}{\partial x_i},
\end{aligned}$$

proving the desired equality. •

Proof of Theorem 5.10. First suppose that $d\Omega = 0$. Then the almost complex structure is integrable by Theorem 5.1, and obviously $d\Omega \subseteq F^2\Omega_X^{n+1}$.

Conversely, suppose $d\Omega \subseteq F^2\Omega_X^{n+1}$ and the almost complex structure is integrable. We use the isomorphism (5.1) and Lemma 5.11 to write $\Omega = V\exp(\beta)$ for some $\beta \in \Omega^{-1,1}$, $V \in \Omega^{0,0}$. To show that $d\Omega = 0$, we need to show that each graded piece of $d\Omega$ is zero, i.e.

$$\frac{1}{n!}d_y(V\beta^n) + \frac{1}{(n+1)!}d_x(V\beta^{n+1}) = 0, \quad \text{for all } n \geq 0.$$

This is equivalent to

$$d_y(V\beta^n) = \frac{1}{n+1}d'_x(V\beta^{n+1}). \tag{5.6}$$

The fact that $d\Omega \subseteq F^2\Omega_X^{n+1}$ is equivalent to (5.6) for $n = 0$. Note that this states

$$d_y(V) = d'_x(V\beta) = [\beta, V] + d'_x(\beta)V \tag{5.7}$$

by (5.3), since $d'_x(V) = 0$. Observe that since $[\beta, \cdot]$ acts as an ordinary (non-graded) derivation on the commutative ring $\bigoplus_p \Omega^{-p,p}$, we have $[\beta, \beta^n] = n[\beta, \beta]\beta^{n-1}$.

We prove by induction that (5.7) implies

$$d'_x(V\beta^{n+1}) = (n+1)d_y(V)\beta^n + \frac{n(n+1)}{2}[\beta, \beta]V\beta^{n-1}. \tag{5.8}$$

Indeed this is true for $n = 0$, by (5.7). Then

$$\begin{aligned}
d'_x(V\beta^{n+1}) &= d'_x(\beta \cdot V\beta^n)\\
&= [\beta, V\beta^n] + d'_x(\beta)V\beta^n + d'_x(V\beta^n)\beta\\
&= [\beta, V]\beta^n + n[\beta, \beta]V\beta^{n-1} + d'_x(\beta)V\beta^n + d'_x(V\beta^n)\beta\\
&= d_y(V)\beta^n + n[\beta, \beta]V\beta^{n-1} + d'_x(V\beta^n)\beta
\end{aligned}$$

by (5.7), so by induction the desired result holds.

Next, we note using the integrability condition $d_y\beta = [\beta, \beta]/2$ and (5.8) that

$$\begin{aligned}
d_y(V\beta^n) &= d_y(V)\beta^n + d_y(\beta^n)V\\
&= d_y(V)\beta^n + nd_y(\beta)V\beta^{n-1}\\
&= d_y(V)\beta^n + \frac{n}{2}[\beta, \beta]V\beta^{n-1}\\
&= \frac{1}{n+1}d'_x(V\beta^{n+1}).
\end{aligned}$$

This proves (5.6), and hence the theorem. •

Next I would like to reinterpret the equations we've seen above so that they may look more natural. As we have seen earlier, $b = \operatorname{Re}\beta$ defines an Ehresmann connection whose horizontal subspaces, given by the subbundle $\mathcal{F}$, are Lagrangian. In local coordinates, this connection is determined by

$$b = \sum_{i,j} b_{ij}dy_j \otimes \frac{\partial}{\partial x_i} = \operatorname{Re}\beta.$$

It makes sense to define the covariant derivative with respect to this connection. This will be an operator

$$\nabla_b : \Gamma(f^*\Omega^q_B \otimes \mathcal{T}_{X^\#/B}) \to \Gamma(f^*\Omega^{q+1}_B \otimes \mathcal{T}_{X^\#/B})$$

defined by

$$\nabla_b \alpha := d_y\alpha - [b, \alpha].$$

It is easy to check that this definition is now independent of the choice of coordinates. The curvature tensor of ∇_b is then $F_b \in \Gamma(f^*\Omega^2_B \otimes \mathcal{T}_{X^\#/B})$ given by

$$F_b := d_y b - \frac{1}{2}[b, b].$$

It is easy to check that $F_b = 0$ if and only if the horizontal distribution $\mathcal{F}$ is integrable. Of course, an Ehresmann connection gives rise to parallel transport along a path contained in B_0; we say a family of p-forms on the fibres of $f_0 : X_0 \to B_0$ is *parallel* if it is invariant under parallel transport. If in local coordinates over $U \subseteq B_0$ this family of forms is written as $\alpha = \sum_I f_I dx_I$, f_I a function on $f^{-1}(U)$, α is parallel if

$$d_y\alpha - \mathcal{L}_b\alpha = 0;$$

by this we mean

$$\frac{\partial \alpha}{\partial y_j} - \mathcal{L}_{\sum_i b_{ij}\partial/\partial x_i}\alpha = 0$$

for each j. (For a similar treatment of Ehresmann connections, see [21]).

We can now rephrase the integrability conditions in a more invariant way.

Corollary 5.15. *Let $(\beta_{ij}) = (b_{ij} + ig^{ij})$, so we write $\beta = b + ig^{-1}$, $V = \sqrt{\det g}$. Then $d\Omega = 0$ if and only if*

$$F_b + \frac{1}{2}[g^{-1}, g^{-1}] = 0 \tag{5.9}$$

$$\nabla_b g^{-1} = 0 \tag{5.10}$$

(5.11) *$dx_1, \ldots, dx_n$ are harmonic forms on each fibre*

(5.12) *$V dx_1 \wedge \cdots \wedge dx_n$ is parallel.*

Proof. The first two equations are the real and imaginary parts of $d_y\beta - \frac{1}{2}[\beta, \beta] = 0$. The last two are (5.7) broken up again into its real and imaginary parts. •

We remark here that in the study of Ricci curvature in the context of Riemannian submersions, some similar structures arise. See [8], Chapter 9.

Corollary 5.16. *Suppose $d\Omega = 0$ and ∇_b is flat. Then the metric g is flat along the fibres.*

Proof. By (5.9) we have $[g^{-1}, g^{-1}] = 0$. We work on one fixed fibre X_b with coordinates $x_1, \ldots, x_n$. Let g^j be the vector field $\sum_i g^{ij}\partial/\partial x_i$, so $g^1, \ldots, g^n$ form a basis for $\mathcal{T}_{X_b}$ at each point of X_b, and $[g^i, g^j] = 0$. Let $\omega_1, \ldots, \omega_n$ be the dual basis of one-forms. Then

$$\begin{aligned} d\omega_i(g^j, g^k) &= g^j(\omega_i(g^k)) - g^k(\omega_i(g^j)) - \omega_i([g^j, g^k]) \\ &= 0 \end{aligned}$$

since $\omega_i(g^k) = \delta_{ik}$ is constant. Thus $\omega_1, \ldots, \omega_n$ are closed 1-forms, and so $\omega_i = d\varphi_i$ for a function φ_i on $\mathbf{R}^n$, the universal cover of X_b. φ_i will be a linear function plus a periodic function. Since $\omega_i = \sum_i g_{ij}dx_j$, we see that $g_{ij} = \partial\varphi_i/\partial x_j = g_{ji} = \partial\varphi_j/\partial x_i$, so there exists a function φ on $\mathbf{R}^n$ such that $\varphi_i = \partial\varphi/\partial x_i$, and $g_{ij} = \partial^2\varphi/\partial x_i\partial x_j$. φ satisfies the real inhomogeneous Monge-Ampère equation $\det(\partial^2\varphi/\partial x_i\partial x_j) = V^2$.

Next we prove V is constant. Note that $V^{-1}\omega_1\wedge\cdots\wedge\omega_n = V dx_1\wedge\cdots\wedge dx_n$, the volume form on X_b, so $*dx_i = \iota(g^i)(V^{-1}\omega_1\wedge\cdots\wedge\omega_n) = \pm V^{-1}\omega_1\wedge\cdots\wedge\hat{\omega}_i\wedge\cdots\wedge\omega_n$, and so $d(*dx_i) = \pm g^i(V^{-1})\omega_1\wedge\cdots\wedge\omega_n$. Thus, by (5.11), we must have $g^i(V^{-1}) = 0$ for all i so V^{-1}, hence V, is constant.

Thus φ is a solution to the equation

$$\det\left(\frac{\partial^2\varphi}{\partial x_i\partial x_j}\right) = C,$$

C a constant, and φ is a function on $\mathbf{R}^n$, with $\varphi = \varphi_{quad} + \varphi_{lin} + \varphi_{per}$, the decomposition into a quadratic, linear, and periodic part. We can of course assume that $\varphi_{lin} = 0$ and $\int_{X_b}\varphi_{per}dx_1\wedge\cdots\wedge dx_n = 0$. Then I claim $\varphi = \varphi_{quad}$, and so $g_{ij} = \partial^2\varphi/\partial x_i\partial x_j$ is constant. To see this, one applies a standard technique for non-linear elliptic partial differential equations. Let $\varphi_t = t\varphi + (1-t)\varphi_{quad}$. Let $m_{ij}(x_{kl})$ denote the ijth cofactor of the matrix (x_{kl}), so in particular

$$m_{ij}(x_{kl}) = \frac{\partial\det(x_{kl})}{\partial x_{ij}}.$$

Put

$$a_{ij} = \int_0^1 m_{ij}(\partial^2\varphi_t/\partial x_k\partial x_l)dt.$$

Now $(\partial^2\varphi/\partial x_i\partial x_j)$ is positive definite. So in fact is $(\partial^2\varphi_{quad}/\partial x_i\partial x_j)$. To see this, note that if we put $h_{ij} = h(\partial/\partial y_i, \partial/\partial y_j)$, we calculate

$$\begin{aligned} h_{ij} &= -\int_{X_b}\iota(\partial/\partial y_i)\omega\wedge\iota(\partial/\partial y_j)\,\mathrm{Im}\,\Omega \\ &= \int_{X_b} dx_i\wedge\left(V\sum_k(-1)^{k-1}g^{kj}dx_1\wedge\cdots\wedge\hat{dx_k}\wedge\cdots\wedge dx_n\right) \\ &= \int_{X_b} Vg^{ij}dx_1\wedge\cdots\wedge dx_n. \end{aligned}$$

Since V is constant, $h_{ij} = V\frac{\partial^2\varphi_{quad}}{\partial x_i\partial x_j}\int_{X_b} dx_1\wedge\cdots\wedge dx_n$. Thus since (h_{ij}) is positive definite, so is $(\partial^2\varphi_{quad}/\partial x_i\partial x_j)$. Thus $(\partial^2\varphi_t/\partial x_i\partial x_j)$ is positive definite for all $0 < t < 1$, so the matrix (a_{ij}) is also positive definite.

Claim:

$$\sum_{i,j} a_{ij}\frac{\partial^2(\varphi - \varphi_{quad})}{\partial x_i\partial x_j} = constant.$$

Proof.

$$\begin{aligned}\frac{\partial}{\partial t}\left(\det\left(\frac{\partial^2\varphi_t}{\partial x_k\partial x_l}\right)\right) &= \sum_{i,j} m_{ij}\left(\frac{\partial^2\varphi_t}{\partial x_k\partial x_l}\right)\frac{\partial^3\varphi_t}{\partial x_i\partial x_j\partial t}\\ &= \sum_{i,j} m_{ij}\left(\frac{\partial^2\varphi_t}{\partial x_k\partial x_l}\right)\frac{\partial^2(\varphi - \varphi_{quad})}{\partial x_i\partial x_j}.\end{aligned}$$

Integrating with respect to t gives the desired result. •

Now $\varphi - \varphi_{quad}$ is a periodic function, so applying the maximum principal (or minimum principal, depending on the sign of the constant), $\varphi - \varphi_{quad}$ is constant. Hence $\varphi = \varphi_{quad}$. •

This explains why in [18], the integrable complex structures constructed on torus fibrations had to have flat metric on the fibres given that $\operatorname{Re}\beta$ was taken to be zero in that paper.

§6. The complex structure on the mirror.

Having understood, at least to some extent, how one describes complex structures on torus fibrations, we now wish to explain how one should put a complex structure on the D-brane moduli space. We cannot solve this problem at present due to the complexity of the equation $d\Omega = 0$; however, here we will give guidelines as to where to look for the correct solutions.

We continue with an integral special Lagrangian torus fibration $f : X \to B$ with a Lagrangian section, along with forms ω and Ω. In §4, we have seen how to put a symplectic form $\check{\omega}$ on $\check{f} : \check{X}_0 \to B$ which has the property that $[\check{\omega}]$ and $[\operatorname{Im}\Omega_n]$ agree in $\Gamma(B_0, \Omega^1_{B_0} \otimes R^1\check{f}_{0*}\mathbf{R}) \cong \Gamma(B_0, \Omega^1_{B_0} \otimes R^{n-1}f_{0*}\mathbf{R})$. To specify the complex structure on $\check{X}$, we need to construct the form $\check{\Omega}$. This should, according to the appropriate conjectures, be determined by the B-field, i.e. something like an element $\mathbf{B} \in H^2(X, \mathbf{R}/\mathbf{Z})$, and ω the Kähler form on X. Certainly the first requirement for $\check{\Omega}$ should be that $[\omega]$ and $[\operatorname{Im}\check{\Omega}_n]$ should agree on $\Gamma(B_0, \Omega^1_{B_0} \otimes R^1f_{0*}\mathbf{R}) \cong \Gamma(B_0, \Omega^1_{B_0} \otimes R^{n-1}\check{f}_{0*}\mathbf{R})$, so that the double dual brings us back to X. The second requirement should involve $\operatorname{Re}\check{\Omega}_n$ and is much less precise at this point. We can only be guided by item (5) of the introduction, but will try to be more precise later.

Proposition 6.1. *If $\check{\Omega}_n$ is a normalised holomorphic n-form on $\check{X}_0$ making $\check{f}_0 : \check{X}_0 \to B$ special Lagrangian and if $\check{h}_n$ is the induced normalised Riemannian metric on the base, then $[\omega] = [\operatorname{Im}\check{\Omega}_n]$ in $\Gamma(B_0, \Omega^1_{B_0} \otimes R^1f_{0*}\mathbf{R}) \cong \Gamma(B_0, \Omega^1_{B_0} \otimes R^{n-1}\check{f}_{0*}\mathbf{R})$ if and only if $h_n = \check{h}_n$.*

Proof. Let $y_1, \ldots, y_n$ be action coordinates for f. (We now have two different special Lagrangian fibrations, f and $\check{f}$, hence two different sets of action coordinates.) Fixing $b \in B$, $X_b = \mathcal{T}^*_{B,b}/\Lambda$, Λ is generated by $e_1, \ldots, e_n$ with $e_i = dy_i$ and $\Lambda^\vee \subseteq \mathcal{T}_{B,b}$ is generated by $e_1^*, \ldots, e_n^*$, $e_i^* = \partial/\partial y_i$. Now $\check{X}_b$ is identified with $\mathcal{T}^*_{B,b}/\Lambda^\vee$, where e_i^* is identified with the 1-form $-h_n(e_i^*, \cdot) = -\sum_j (h_n)_{ij} dy_j$, and thus $e_i \in \Lambda$ is identified with $-\sum_j h_n^{ij} \partial/\partial y_j \in \mathcal{T}_{B,b}$. Thus

$$\begin{aligned}\check{h}_n(\sum_j h_n^{ij} \partial/\partial y_j, \partial/\partial y_k) &= -\int_{\check{X}_b} \iota(\sum_j h_n^{ij} \partial/\partial y_j)\check{\omega} \wedge \iota(\partial/\partial y_k) \operatorname{Im} \check{\Omega}_n \\ &= -\int_{e_i} \iota(\partial/\partial y_k) \operatorname{Im} \check{\Omega}_n\end{aligned}$$

by Proposition 4.6, where $e_i \in \Lambda = H_{n-1}(\check{X}_b, \mathbf{Z}) \cong H_1(X_b, \mathbf{Z})$. If $[\omega] = [\operatorname{Im} \check{\Omega}_n]$, this latter integral coincides with

$$-\int_{e_i} \iota(\partial/\partial y_k)\omega = \delta_{ik}.$$

Thus $\check{h}_n(\partial/\partial y_i, \partial/\partial y_j) = (h_n)_{ij}$, so $\check{h}_n = h_n$. The argument reverses to prove the converse. •

Moral 6.2. *$\check{\Omega}_n$ must be chosen on $\check{X}$ so that $\check{h}_n = h_n$.*

Remark 6.3. While this moral was deduced by beginning with a special Lagrangian torus fibration and applying the principal that double dualising should bring one back to the initial fibration, there is no reason this can't then be generalised to provide a guide for putting complex structures on more general D-brane moduli spaces. In the situation of §4, given a family $\mathcal{U} \rightarrow B$, one has the metric h_n on B. Then a holomorphic n-form should be chosen on $\mathcal{M}$ so that $\mathcal{M} \rightarrow B$ is special Lagrangian and the induced metric on B is h_n.

It is more difficult to say exactly what role the B-field plays. According to the conjecture originally stated in [16], and restated in the introduction, $\check{\Omega}_n$ should be chosen so that $[\check{\Omega}_n] - [\sigma_0] \in H^1(B, R^{n-1}\check{f}_*\mathbf{R})$ should coincide with the choice of the B-field $\mathbf{B} \in H^1(B, R^1 f_*\mathbf{R})$. This provides little guidance, but we will see an example of this below which may point in the correct direction for interpreting the B-field.

Example 6.4. (The Hitchin solution.) Hitchin [18] gives a choice of $\check{\Omega}_n$ which under certain assumptions about the metric h_n satisfies $d\check{\Omega}_n = 0$. He expresses it locally in terms of action-angle coordinates, but it can be written down in arbitrary coordinates on the base in a natural way. One takes for $\check{\nabla}_b$ the Gauss-Manin connection. This is a linear connection $\check{\nabla}_{GM}$ on $\mathcal{T}^*_{B_0} = (R^{n-1}\check{f}_{0*}\mathbf{R}) \otimes C^\infty(B_0)$, whose flat sections are the sections of $R^{n-1}\check{f}_{0*}\mathbf{R}$. Taking the horizontal subspaces of this connection, it is easy to see that these descend to give a flat Ehresmann connection on $\check{X}_0$, which one takes to define $\check{\nabla}_b$. Note that in action coordinates $v_1, \ldots, v_n$ for $\check{f}_0$, the Gauss-Manin connection is trivial, so in these coordinates one takes $b = 0$. Since now $\check{\nabla}_b$ is flat, we must have $\check{g}$ constant along fibres by Corollary 5.16. Thus

$$\begin{aligned} \check{h}_n(\partial/\partial y_i, \partial/\partial y_j) &= \int_{\check{X}_b} \check{V}\check{g}^{ij} dx_1 \wedge \cdots \wedge dx_n / \int_{\check{X}_b} \check{V} dx_1 \wedge \cdots \wedge dx_n \\ &= \check{g}^{ij}, \end{aligned}$$

and since we want $\check{h}_n = h_n$, we have no choice but to take $\check{g}^{ij} = (h_n)_{ij}$, giving rise to a choice of $\check{\Omega}_n$. To check to see if $d\check{\Omega}_n = 0$, one checks conditions (5.9)-(5.12). (5.9) and (5.11) are immediate, while (5.10) can be checked. (5.12) is then equivalent, in this case, to $\int_{\check{X}_b} \check{\Omega}_n$ being independent of b. But

$$\begin{aligned} \int_{\check{X}_b} \check{\Omega}_n &= \int_{\check{X}_b} \sqrt{\det \check{g}_{ij}} dx_1 \wedge \cdots \wedge dx_n \\ &= \frac{1}{\sqrt{\det(\check{h}_n)}} \int_{\check{X}_b} dx_1 \wedge \cdots \wedge dx_n. \end{aligned}$$

Hence this quantity must be independent of b for (5.12) to be satisfied. In particular, if $u_1, \ldots, u_n$ are action coordinates for f_0 and $v_1, \ldots, v_n$ are action coordinates on the same open subset for $\check{f}_0$, and $v_1, \ldots, v_n, x_1, \ldots, x_n$ are canonical coordinates, then $\int_{\check{X}_b} dx_1 \wedge \cdots \wedge dx_n$ is a constant independent of b, so in these coordinates, $d\check{\Omega}_n = 0$ is equivalent to

$$\det(\check{h}_n) = constant.$$

A simple calculation now shows that if this is the case, then $\check{\Omega}_n$ coincides with Hitchin's $\tilde{\Omega}^c$ (up to a constant factor) in [18], §6, where

$$\tilde{\Omega}^c = \bigwedge_i (dx_i + \sqrt{-1} du_i).$$

Thus one recovers [18], Proposition 5. Of course, $\check{\omega} = \sum dx_i \wedge dv_i$, and Hitchin views mirror symmetry as an exchanging of the roles of the two sets of action coordinates $\{u_i\}$ and $\{v_i\}$. By [18], Proposition 3, the condition $\det(\check{h}_n) = constant$ (in coordinates $v_1, \ldots, v_n$) is equivalent to the condition $\det(h_n) = constant$ (in coordinates $u_1, \ldots, u_n$). This holds in particular if the metric g_{ij} is constant on fibres.

Example 6.5. (The Hitchin solution twisted by the B-field.) Continuing with the above example, assume $d\check{\Omega}_n = 0$. Now choose a symmetric cohomology class $\mathbf{B} \in H^1(B, R^{n-1}\check{f}_*\mathbf{R})$, with a symmetric representative $b \in \Gamma(B_0, \Omega^1_{B_0} \otimes \Omega^1_{B_0})$; in action-angle coordinates for $\check{f}$, this will be of the form $\sum b_{ij} dv_i \otimes dv_j$. Now take, in this coordinate system, the n-form $\check{\Omega}_{n,b}$ to be given by the matrix $(\beta_{ij}) = (b_{ij} + \sqrt{-1}(h_n)_{ij})$. This in fact gives a well-defined n-form on all of $\check{X}_0$. Note that since (b_{ij}) is symmetric, so is β, and thus we just need to show that $d\check{\Omega}_{n,b} = 0$ in order to show that $\check{\Omega}_{n,b}$ induces a complex structure with a Ricci-flat metric. This closedness can be seen to be true as follows. If $U \subseteq B_0$ is a sufficiently small open set with action-angle coordinates $v_1, \ldots, v_n, x_1, \ldots, x_n$ for $\check{f}$, we can find an element $a \in \Gamma(U, \Omega^0_U \otimes \Omega^1_U)$ such that $\check{\nabla}_{GM}(a) = b$, since $\check{\nabla}_{GM}(b) = 0$. Here $a = \sum_i a_i 1 \otimes dv_i$, with $\partial a_i/\partial v_j = b_{ij}$. Thus by symmetry of b_{ij}, $(v_1, \ldots, v_n) \mapsto (v_1, \ldots, v_n, a_1, \ldots, a_n)$ gives a Lagrangian section σ of $\check{f}^{-1}(U) \to U$, and it is easy to see that

$$\check{\Omega}_{n,b} = T_\sigma^* \check{\Omega}_n,$$

where $\check{\Omega}_n$ is the Hitchin solution of Example 6.4. Since $d\check{\Omega}_n = 0$, $d\check{\Omega}_{n,b} = 0$ also.

Note also that if $b' = \sum b'_{ij} dy_i \otimes dy_j$ is a different symmetric representative for $\mathbf{B}$, then there exists an $a \in \Gamma(B_0, \Omega^0_{\check{B}_0} \otimes \Omega^1_{B_0})$ such that $\nabla_{GM}(a) = b' - b$. As before, a gives a Lagrangian section σ of $\check{f}_0$, and

$$T_\sigma^* \check{\Omega}_{n,b} = \check{\Omega}_{n,b'}.$$

Finally, one sees that $\check{\Omega}_{n,b} - \check{\Omega}_n$ represents the class $\mathbf{B} \in H^1(B, R^{n-1}\check{f}_*\mathbf{R})$, and $\check{\Omega}_{n,b}$ and $\check{\Omega}_n$ yield the same complex structure if $\mathbf{B} \in H^1(B, R^{n-1}\check{f}_*\mathbf{Z})$, for then there is a global section σ of $f : X \to B$ with $T_\sigma^* \check{\Omega}_n = \check{\Omega}_{n,b}$.

One can in fact go further. We have observed that for small open sets $U \subseteq B_0$, the complex and Kähler structures on $\check{X}_0$ induced by $\check{\Omega}_n$ and $\check{\Omega}_{n,b}$ on $\check{f}^{-1}(U)$ are isomorphic, so one should think of the Kähler structure induced by $\check{\Omega}_{n,b}$ as a torseur over that induced by $\check{\Omega}_n$. Specifically, fixing the complex structure $\check{\Omega}_n$ on $\check{X}_0$, note that the sheaf $\mathcal{A}$ on B_0 defined by

$$\mathcal{A}(U) = \{\text{sections } \sigma : U \to \check{f}^{-1}(U) \text{ such that } T_\sigma^* \check{\omega} = \check{\omega} \text{ and } T_\sigma^* \check{\Omega}_n = \check{\Omega}_n\}$$

coincides with $R^{n-1} f_* \mathbf{R}/\mathbf{Z}$. In fact, writing these conditions in the coordinates u_i and x_i of Example 6.4, one sees that the condition $T_\sigma^* \check{\Omega}_n = \check{\Omega}_n$ guarantees that the section σ is constant with respect to the Gauss-Manin connection. Thus the set of all special Lagrangian fibrations over B_0 obtained from $\check{f} : \check{X}_0 \to B_0$ by regluing using these translations is $H^1(B_0, R^{n-1}\check{f}_{0*}\mathbf{R}/\mathbf{Z})$. Because $\check{\omega}$ and $\check{\Omega}_n$ are preserved by these translations, they glue to give forms on the twisted fibrations. Thus each element $\mathbf{B} \in H^1(B_0, R^{n-1}\check{f}_{0*}\mathbf{R}/\mathbf{Z})$ gives rise to a fibration $\check{f}_{\mathbf{B}} : \check{X}_{0,\mathbf{B}} \to B_0$ with symplectic form $\check{\omega}_{\mathbf{B}}$ and holomorphic n-form $\check{\Omega}_{n,\mathbf{B}}$. This is a potentially wider class of examples than were constructed above using symmetric cohomology classes in $H^1(B, R^{n-1}\check{f}_*\mathbf{R})$; if $\mathbf{B}$ does not come from a symmetric class, then $\check{f}_{\mathbf{B}}$ will not possess a Lagrangian section, and may not even possess a topological section.

Note also that if f and $\check{f}$ are $\mathbf{R}/\mathbf{Z}$-simple, then

$$H^1(B, R^{n-1}\check{f}_*\mathbf{R}/\mathbf{Z}) \cong H^1(B, R^1 f_*\mathbf{R}/\mathbf{Z}).$$

This leads us to conjecture that the correct group for the B-field to live in is $H^1(B, R^1 f_*\mathbf{R}/\mathbf{Z})$. This new proposed definition for the B-field is dependent not just on X but on the topology of the fibration, and even in the threefold case does not necessarily coincide with $H^2(X, \mathbf{R}/\mathbf{Z})$, as we saw in Example 3.11. Nevertheless, I believe this is the correct interpretation of the B-field.

This now leads us to a refined mirror symmetry conjecture.

(1) *for each open set* $U \subseteq B_0 = B - \Delta$ *on which both* f *and* $\check{f}$ *have sections,* $f^{-1}(U) \to U$ *and* $\check{f}^{-1}(U) \to U$ *are topologically dual fibrations.*

(2) For $b \in B_0$, $\gamma \in H_1(\check{X}_b, \mathbf{Z}) \cong H_{n-1}(X_b, \mathbf{Z})$ and $v \in \mathcal{T}_{B,b}$,

$$\int_\gamma \iota(v)\check{\omega} = \int_\gamma \iota(v)\,\mathrm{Im}\,\Omega_n.$$

(3) For $\gamma \in H_{n-1}(\check{X}_b, \mathbf{Z}) \cong H_1(X_b, \mathbf{Z})$,

$$\int_\gamma \iota(v)\omega = \int_\gamma \iota(v)\,\mathrm{Im}\,\check{\Omega}_n.$$

(4) $\check{f}$ *possesses a topological section if*

$$\mathbf{B} \in H^1(B, R^1 f_*\mathbf{R})/H^1(B, R^1 f_*\mathbf{Z}) \subseteq H^1(B, R^1 f_*\mathbf{R}/\mathbf{Z}).$$

In this case, if $\check{\sigma}$ *is a topological section, then* $[\mathrm{Re}\,\check{\Omega}_n] - [\check{\sigma}]$ *defines a class in* $H^1(B, R^{n-1}\check{f}_*\mathbf{R})$ *which is well-defined modulo* $H^1(B, R^{n-1}\check{f}_*\mathbf{Z})$, *and agrees with* $\mathbf{B}$ *in*

$$H^1(B, R^1 f_*\mathbf{R})/H^1(B, R^1 f_*\mathbf{Z}).$$

(5) *If* $\check{J}$ *is the Jacobian of* $\check{X}$, *then* $\check{X}$ *is obtained from* $\check{J}$ *as a symplectic manifold via the image of* $\mathbf{B}$ *under the composed map* $H^1(B, R^1 f_*\mathbf{R}/\mathbf{Z}) \cong H^1(B, R^{n-1}\check{f}_*\mathbf{R}/\mathbf{Z}) \to H^1(B, \Lambda(\check{J}^\#))$ *of Remark 3.15.*

(6) *Once* $\check{X}$ *and* $\check{\omega}$ *are fixed,* $\check{\Omega}_n$ *is unique up to translation by a Lagrangian section of* $\check{J}$ *acting on* $\check{X}$.

I do not however suggest that $\check{f}: \check{X} \to B$ is obtained as a *Kähler* manifold as a torseur over some basic $\check{J} \to B$. While this occurred in Example 6.5, there is no reason to suspect this works when the metric on the fibres is not flat. We simply don't expect there to be isometries given by translation by a section. However, in some sense this might provide an initial approximation to the correct answer.

Remark 6.7. We can show a local form of the conjectured uniqueness. Suppose Ω_t is a family of holomorphic n-forms on X with respect to which a fixed symplectic form ω induces a Ricci-flat metric and $f: X \to B$ is special Lagrangian. In addition assume $[\Omega_t] \in H^n(X, \mathbf{C})$ is a fixed cohomology class. Then by local Torelli all Ω_t induce the same complex structure, so there exists diffeomorphisms $\phi_t : X \to X$ such that $\phi_t^*\Omega_t = \Omega_0$, $\phi_0 = id$. Now $\phi_t^*\omega$ is a symplectic form on X inducing a Ricci-flat metric in the complex structure induced by $\phi_t^*\Omega_t = \Omega_0$, and represents the same cohomology class as ω, so by uniqueness of Ricci-flat metrics, $\phi_t^*\omega = \omega$. Thus ϕ_t is a family of symplectomorphisms. Assuming $H^1(X, \mathbf{R}) = 0$, differentiating this family of diffeomorphisms at $t = 0$ yields a Hamiltonian vector field v induced by a Hamiltonian function H on X: $\iota(v)\omega = dH$. Then $\mathrm{Im}\,\Omega_t|_{X_b} = 0$ for all b implies that $(\mathcal{L}_v \mathrm{Im}\,\Omega_0)|_{X_b} = 0$ for all b. But $\mathcal{L}_v \mathrm{Im}\,\Omega_0 = d(\iota(v)\,\mathrm{Im}\,\Omega_0)$, and if Ω_0 is given as usual by a matrix $\beta = (\beta_{ij})$, $\beta_{ij} = b_{ij} + ig^{ij}$, then a simple calculation shows that

$$(d(\iota(v)\,\mathrm{Im}\,\Omega_0))|_{X_b} = -\sum_{i,j} V g^{ij} \frac{\partial^2 H}{\partial x_i \partial x_j} dx_1 \wedge \cdots \wedge dx_n.$$

Here we have used the fact that $\iota(\partial/\partial y_i)\,\mathrm{Im}\,\Omega_0$ is closed. Thus we see that on each fibre, H satisfies the second order elliptic partial differential equation

$$\sum_{i,j} g^{ij}\frac{\partial^2 H}{\partial x_i \partial x_j} = 0.$$

By the maximum principal, H cannot have a local maximum on each non-singular fibre unless H is constant on the fibre. Since the set of non-singular fibres is dense, we conclude that H is the pullback of a function on B. In particular, it follows from §2 that ϕ_t must be translation by a Lagrangian section.

Remark 6.8. One natural question is to determine the relationship between $Vol(X_b) = \int_{X_b}\Omega$ and $Vol(\check{X}_b) = \int_{\check{X}_b}\check{\Omega}$, since knowledge of the latter allows us to reconstruct $\check{\Omega}$ from $\check{\Omega}_n$. We can describe this relationship if the metric is constant along the fibres of $f : X \to B$ and $\check{f} : \check{X} \to B$. Let $y_1, \ldots, y_n$ be action coordinates for f. As in Example 6.4,

$$h_n^{ij} = g_{ij}.$$

Then

$$\begin{aligned} Vol(X_b) &= \int_{X_b} \sqrt{\det(g_{ij})}dx_1 \wedge \cdots \wedge dx_n \\ &= \sqrt{\det(h_n^{ij})}. \end{aligned}$$

On the other hand, for a fixed $b \in B$, $\check{X}_b$ can be written canonically as $\mathcal{T}^*_{B,b}/\Lambda^\vee$ as in Proposition 4.7, where $\Lambda^\vee$ is generated by the one-forms $h_n(\partial/\partial y_i, \cdot)$, i.e. the one-forms $\sum_j (h_n)_{ij}dy_j$. The metric on the fibre $\check{X}_b$ is still given by $\check{g}_{ij} = (h_n)^{ij}$, since $h_n = \check{h}_n$. Thus

$$\begin{aligned} Vol(\check{X}_b) &= \int_{\check{X}_b} \sqrt{\det(h_n^{ij})}dx_1 \wedge \cdots \wedge dx_n \\ &= \sqrt{\det(h_n^{ij})\det((h_n)_{ij})} \\ &= \frac{1}{Vol(X_b)}. \end{aligned}$$

This is the familiar "$R \mapsto 1/R$" relationship of T-duality. If the metric is not flat, we expect some corrections to the volume, and this may affect this relationship. However, as we shall see in §7, this relationship continues to hold for K3 surfaces.

We end this section with a brief discussion of the Yukawa coupling. Mirror symmetry instructs us that given a Calabi-Yau X, the Yukawa coupling on $\check{X}$ contains information about the genus 0 Gromov-Witten invariants on X, and in particular in the three dimensional case, these Gromov-Witten invariants can be completely recovered from the Yukawa coupling on $\check{X}$. How do we see the Yukawa coupling in the context of special Lagrangian fibrations?

Suppose that Conjecture 6.6 holds. Assume for simplicity that we only consider values of the B-field $\mathbf{B} \in H^1(B, R^{n-1}\check{f}_*\mathbf{R}/\mathbf{Z})$ which come from symmetric classes in $H^1(B, R^{n-1}\check{f}_*\mathbf{R})$, so that although $\mathbf{B}$ varies, we can fix the underlying manifold $\check{X}$

and symplectic form $\check{\omega}$ and simply let $\check{\Omega}_n$ vary; we denote the dependence on $\mathbf{B}$ by writing $\check{\Omega}_{n,\mathbf{B}}$. Of course, we will not have $\check{\Omega}_{n,\mathbf{B}+\alpha} = \check{\Omega}_{n,\mathbf{B}}$ for $\alpha \in H^1(B, R^{n-1}\check{f}_*\mathbf{Z})$, but merely expect that there exists a Lagrangian section σ of $\check{f} : \check{X} \to B$ with $T_\sigma^*\check{\Omega}_{n,\mathbf{B}} = \check{\Omega}_{n,\mathbf{B}+\alpha}$. The $(1, n-1)$-Yukawa coupling of interest is then, for tangent directions $\partial/\partial b_1, \ldots, \partial/\partial b_n \in H^1(B, R^{n-1}\check{f}_*\mathbf{R})$, the tangent space of the torus $H^1(B, R^{n-1}\check{f}_*\mathbf{R}/\mathbf{Z})$,

$$\left\langle \frac{\partial}{\partial b_1}, \ldots, \frac{\partial}{\partial b_n} \right\rangle = \int_{\check{X}} \check{\Omega}_{n,\mathbf{B}} \wedge \frac{\partial^n}{\partial b_1 \ldots \partial b_n} \check{\Omega}_{n,\mathbf{B}}.$$

In local coordinates, we write

$$\check{\Omega}_{n,\mathbf{B}} = V_{n,\mathbf{B}} \theta_1(\mathbf{B}) \wedge \cdots \wedge \theta_n(\mathbf{B}).$$

Note that in taking the n derivatives of $\check{\Omega}_{n,\mathbf{B}}$ by using the product rule, all terms still containing any undifferentiated θ_i will disappear after we wedge with $\check{\Omega}_{n,\mathbf{B}}$. Thus

$$\int_{\check{X}} \check{\Omega}_{n,\mathbf{B}} \wedge \frac{\partial^n}{\partial b_1 \ldots \partial b_n} \check{\Omega}_{n,\mathbf{B}} = \sum_{\sigma \in S_n} \int_{\check{X}} V_{n,\mathbf{B}}^2 \theta_1 \wedge \ldots \wedge \theta_n \wedge \frac{\partial \theta_1}{\partial b_{\sigma(1)}} \wedge \cdots \wedge \frac{\partial \theta_n}{\partial b_{\sigma(n)}}.$$

Writing $\theta_i = dx_i + \sum_j \check{\beta}_{ij}(\mathbf{B}) dy_j$,

$$\frac{\partial \theta_i}{\partial b_{\sigma(i)}} = \sum \frac{\partial \check{\beta}_{ij}}{\partial b_{\sigma(i)}} dy_j,$$

so the above integral is

$$\int_{\check{X}} V_{n,\mathbf{B}}^2 \sum_{\sigma \in S_n} \det \left(\frac{\partial \check{\beta}_{ij}}{\partial b_{\sigma(i)}} \right) dx_1 \wedge \cdots \wedge dx_n \wedge dy_1 \wedge \cdots \wedge dy_n.$$

So far, this is not particularly illuminating in the general case. However, for the twisted Hitchin solutions, it is an elementary calculation to show this integral can be evaluated in terms of the topological coupling on X, as expected. This comes from observing that if $\partial \theta_j / \partial b_i = \sum_k b_{jk}^i dy_k$, then in action-angle coordinates, where $V_n = 1$, the integrand above coincides with

$$(-1)^{n(n-1)/2} \bigwedge_{i=1}^{n} \left(\sum_{j,k} b_{jk}^i dx_j \wedge dy_k \right).$$

§7. K3 Surfaces.

Mirror symmetry for K3 surfaces has been completely understood using Torelli theorems for K3 surfaces. We will now show that the previous material of this paper gives us a differential geometric construction of mirror symmetry for K3 surfaces, and in doing so, we will show Conjecture 6.6 holds in two dimensions. In other words, given a special Lagrangian T^2-fibration on a K3 surface, and a choice of

B-field, we will construct the mirror in the sense made explicit in Conjecture 6.6. This will prove to be a variant of the mirror symmetry for K3 surfaces described in [6] and [11].

To begin, let S be a K3 surface with holomorphic 2-form Ω and Kähler form ω corresponding to a Ricci-flat metric. We insist on the usual normalisation and this implies in particular that $(\mathrm{Re}\,\Omega)^2 = (\mathrm{Im}\,\Omega)^2 = \omega^2 > 0$ and $(\mathrm{Re}\,\Omega) \wedge (\mathrm{Im}\,\Omega) = \omega \wedge (\mathrm{Re}\,\Omega) = \omega \wedge (\mathrm{Im}\,\Omega) = 0$. In order for S to possess a special Lagrangian fibration there must be a cohomology class $E \in H^2(S, \mathbf{Z})$ such that $E^2 = 0$ and $\omega.E = 0$. We assume such a class exists, and we fix it. We take E to be primitive. Now one constructs a special Lagrangian fibration on S by the usual hyperkähler trick, as originally suggested in [28]. First, multiply Ω by a phase $e^{i\theta}$ to ensure $\mathrm{Im}\,\Omega.E = 0$. Following the notation of [16], there is a complex structure K compatible with the Ricci-flat metric in which $\Omega_K = \mathrm{Im}\,\Omega + i\omega$ and $\omega_K = \mathrm{Re}\,\Omega$. Then special Lagrangian submanifolds on S are complex submanifolds in the K complex structure. Denote the K3 surface in the K complex structure by S_K. Then by construction $E.\Omega_K = 0$ so $E \in Pic(S_K)$. It is then standard that S_K possesses an elliptic fibration. However, the class of the fibre need not be E; E might be represented by a fibre plus a sum of -2 divisors. In any event, replace E by the class of the fibre, positively oriented. E will now remain fixed, and we have an elliptic fibration $f : S_K \to \mathbf{P}^1$, which we identify with a special Lagrangian T^2 fibration $f : S \to B = S^2$. We will assume f is integral, which is true if and only if there is no $\delta \in Pic(S_K)$ with $\delta^2 = -2$ and $\delta.E = 0$. This will certainly hold for general S.

We now consider the spectral sequence for f over $\mathbf{Z}$. Since there exists a class σ such that $\sigma.E = 1$, the sequence in fact degenerates and takes the form

$$\begin{array}{ccc} H^0(B,\mathbf{Z}) & 0 & H^2(B,\mathbf{Z}) \\ 0 & H^1(B, R^1 f_* \mathbf{Z}) & 0 \\ H^0(B,\mathbf{Z}) & 0 & H^2(B,\mathbf{Z}) \end{array}$$

It is also clear that $H^1(B, R^1 f_* \mathbf{Z})$ is canonically isomorphic to $E^\perp/E$.

We now wish to construct a mirror fibration given the data

$f : S \to B$ as above and a choice of B-field $\mathbf{B} \in E^\perp/E \otimes \mathbf{R}/\mathbf{Z} \cong H^1(B, R^1 f_* \mathbf{R}/\mathbf{Z})$.

We first recall some facts about elliptic fibrations. See [7] for general facts about the analytic theory of elliptic surfaces.

The fibration $f : S_K \to \mathbf{P}^1$ in general does not possess a holomorphic section; in fact for general choice of S, $\mathrm{Pic} S_K = \mathbf{Z}E$. However, there is a Jacobian fibration $j : J_K \to \mathbf{P}^1$ of f which is locally isomorphic to f, and which does possess a holomorphic section.

Proposition 7.1. *There is a diffeomorphism $\phi : J_K \to S_K$ over $\mathbf{P}^1$ which is holomorphic when restricted to each fibre. Furthermore, if $U \subseteq \mathbf{P}^1$ is an open subset on which there exists a biholomorphic map $\xi : j^{-1}(U) \to f^{-1}(U)$ over U, then $\phi^{-1} \circ \xi : j^{-1}(U) \to j^{-1}(U)$ is given by translation by a (not necessarily holomorphic) section of $j^{-1}(U)$.*

Proof. This follows easily from the fact that the "C^∞ Tate-Shafarevich group," i.e. the first cohomology group of the sheaf of C^∞ sections of $j : J_K^\# \to \mathbf{P}^1$, is zero.

Thus $f : S_K \to \mathbf{P}^1$ possesses a C^∞ section, and ϕ can be taken to identify this C^∞ section of f with a holomorphic section of j, such that ϕ is holomorphic on each fibre. •

We fix one holomorphic section σ_0 of $j : J \to \mathbf{P}^1$, and identify σ_0 with the topological section $\phi(\sigma_0)$ of $f : S \to B$. Having chosen this section, we can take it to be the zero section of $j : J_K \to \mathbf{P}^1$ and obtain a standard exact sequence

$$0 \to R^1 f_* \mathbf{Z} \to R^1 f_* \mathcal{O}_{S_K} \xrightarrow{\psi} J_K^{\#} \to 0$$

where $J_K^{\#}$ denotes the sheaf of holomorphic sections of $j : J_K \to \mathbf{P}^1$. Here $R^1 f_* \mathcal{O}_{S_K}$ can be identified with the normal bundle of the zero section, and the map ψ is just the fibre-wise exponential map. For K3 surfaces, $R^1 f_* \mathcal{O}_{S_K} \cong \omega_{\mathbf{P}^1}$. The underlying real bundle is $\mathcal{T}^*_{S^2}$. This also gives a map $\pi : \mathcal{T}^*_{S^2} \to S_K^{\#}$ with $\pi = \phi \circ \psi$.

Just as in the real case, the total space of $\omega_{\mathbf{P}^1}$, the holomorphic cotangent bundle of $\mathbf{P}^1$, has a canonical holomorphic symplectic form Ω_c. In local coordinates, if z is a coordinate on $\mathbf{P}^1$ and w the canonical coordinate on the cotangent bundle, then $\Omega_c = dw \wedge dz$. Furthermore, any holomorphic symplectic form on the cotangent bundle of $\mathbf{P}^1$ is proportional to Ω_c.

Proposition 7.2. *There is a map $\chi : \mathcal{T}^*_{S^2} \to \mathcal{T}^*_{S^2}$ given by fibrewise multiplication by a complex constant so that, for $\pi' = \pi \circ \chi$, $\pi'^*(\Omega_K) = \Omega_c + f^*\alpha$, where α is a 2-form on S^2.*

Proof. First note that there are two different complex structures on the total space of $\mathcal{T}^*_{S^2}$ in this picture: one is the standard complex structure coming from being the holomorphic line bundle $\mathcal{O}_{\mathbf{P}^1}(-2)$, while the other is induced by $\pi^*\Omega_K$. To distinguish between these two complex structures, let $\tilde{J}$ be the total space of $\mathcal{T}^*_{S^2}$ with the standard complex structure, and let $\tilde{S}$ denote the total space of $\mathcal{T}^*_{S^2}$ with the complex structure induced by $\pi^*\Omega_K$. Let $\tilde{j} : \tilde{J} \to S^2$, $\tilde{f} : \tilde{S} \to S^2$ be the projections.

Now let $U \subseteq S^2$ be a sufficiently small open set so that there exists a biholomorphic map $\xi : j^{-1}(U) \to f^{-1}(U)$ over U. By Proposition 7.1, $\phi^{-1} \circ \xi = T_\sigma$ for some section σ of $j^{-1}(U) \to U$, and if U is small enough, σ can be lifted to a section $\tilde{\sigma}$ of $\tilde{j}^{-1}(U) \to U$. We let $\tilde{\xi}$ denote translation by the section $\tilde{\sigma}$ (with the zero-section of $\mathcal{T}^*_{S^2}$ taken to be the origin). We then have a commutative diagram

$$\begin{array}{ccc} \tilde{j}^{-1}(U) & \xrightarrow{\tilde{\xi}} & \tilde{f}^{-1}(U) \\ \downarrow{\psi} & & \downarrow{\pi} \\ j^{-1}(U) & \xrightarrow{\xi} & f^{-1}(U) \end{array}$$

since $\pi \circ \tilde{\xi} = \phi \circ \psi \circ T_{\tilde{\sigma}} = \phi \circ T_\sigma \circ \psi = \xi \circ \psi$. In particular, since Ω_K is a holomorphic 2-form on $\tilde{f}^{-1}(U)$, $\tilde{\xi}^*\pi^*\Omega_K$ is a holomorphic 2-form on $\tilde{j}^{-1}(U)$, and hence can be written as $g dw \wedge dz$ in local coordinates, for some holomorphic function g. This function g must be constant along the fibres of $\tilde{j}$ since g must descend to a holomorphic function on the compact fibres of j. Thus on $\tilde{f}^{-1}(U)$,

$$\begin{aligned} \pi^*\Omega_K &= (\tilde{\xi}^{-1})^*(g dw \wedge dz) \\ &= T^*_{-\tilde{\sigma}}(g dw \wedge dz) \\ &= (g \circ T_{-\tilde{\sigma}}) dw \wedge dz + h dz \wedge d\bar{z} \end{aligned}$$

for some function h. Of course, $g \circ T_{-\tilde{\sigma}} = g$ since g is constant on fibres.

Let $i : \tilde{J} \to \tilde{S}$ be the identity map; this is of course non-holomorphic. The above equation shows that the $(2,0)$ part of $i^*\pi^*\Omega_K$, being locally of the form $gdz \wedge dw$, is in fact a holomorphic 2-form. In addition, this holomorphic 2-form is nowhere vanishing: if g vanishes then $\Omega_K \wedge \bar{\Omega}_K = 0$ at that point. Thus the $(2,0)$ part of $i^*\pi^*\Omega_K$ is proportional to Ω_c, say $C\Omega_c$. In addition, we then see from $d\Omega_K = 0$ that h must be constant along fibres and hence $\pi^*\Omega_K - C\Omega_c$ is the pullback of a $(1,1)$ form α on $\mathbf{P}^1$, i.e.

$$\pi^*\Omega_K = C\Omega_c + \tilde{f}^*\alpha.$$

Now let $\chi : \mathcal{T}^*_{S^2} \to \mathcal{T}^*_{S^2}$ be given by $w \mapsto C^{-1}w$. Then $\chi^*\pi^*\Omega_K = \Omega_c + \tilde{f}^*\alpha$ as desired. •

To sum up, replacing π by π', we now have a map $\pi : \mathcal{T}^*_{S^2} \to S^\#_K$ with kernel $R^1 f_*\mathbf{Z}$ and such that $\pi^*\Omega_K = \Omega_c + \tilde{f}^*\alpha$. This gives, identifying the underlying topological spaces $S^\#_K$ and $S^\#$, $\pi^*\omega = \mathrm{Im}(\Omega_c + \tilde{f}^*\alpha)$ and $\pi^*(\mathrm{Im}\,\Omega) = \mathrm{Re}(\Omega_c + \tilde{f}^*\alpha)$.

A local description of these forms are as follows: given complex canonical coordinates z, w on $\mathcal{T}^*_U$, z a coordinate on U, write $z = y_1 - iy_2$, $w = x_1 + ix_2$. The signs are chosen so that with real coordinates y_1, y_2 on the base, y_1, y_2, x_1, x_2 are canonical coordinates on $\mathcal{T}^*_{S^2}$. Then

$$\begin{aligned} \pi^*\omega &= \mathrm{Im}(dw \wedge dz + \tilde{f}^*\alpha) \\ &= dy_2 \wedge dx_1 + dx_2 \wedge dy_1 + \tilde{f}^*\,\mathrm{Im}\,\alpha \end{aligned} \tag{7.1}$$

and

$$\begin{aligned} \pi^*\,\mathrm{Im}\,\Omega &= \mathrm{Re}(dw \wedge dz + \tilde{f}^*\alpha) \\ &= dx_1 \wedge dy_1 + dx_2 \wedge dy_2 + \tilde{f}^*\,\mathrm{Re}\,\alpha. \end{aligned} \tag{7.2}$$

This completes our first goal of finding coordinates on $S^\#$ in which ω and $\mathrm{Im}\,\Omega$ have simple forms. Note that our map $\pi : \mathcal{T}^*_{S^2} \to S^\#$ is not the same as defined in §2 because the symplectic form is not the standard one. We will see why we have made this choice in the proof of the following theorem.

Theorem 7.3. *Conjecture 6.6 holds for the general integral special Lagrangian fibration $f : S \to B = S^2$.*

Proof. Choose a B-field $\mathbf{B} \in (E^\perp/E) \otimes \mathbf{R}/\mathbf{Z}$. Lift this to a representative $\mathbf{B} \in E^\perp/E \otimes \mathbf{R}$. At times, we will also further lift $\mathbf{B}$ to an element $\mathbf{B} \in E^\perp \otimes \mathbf{R}$ chosen so that $\mathbf{B}.[\sigma_0] = 0$, where σ_0 is the fixed topological section of $f : S \to B$ chosen previously. We are now trying to construct a special Lagrangian fibration $\check{f} : \check{S} \to B$ satisfying the properties of Conjecture 6.6. Because this fibration may not have a Lagrangian section, we first construct the Jacobian $\check{j} : \check{J} \to B$ (in the sense of §2) as a symplectic manifold. This Jacobian should be the dual fibration with a symplectic form $\check{\omega}_{\check{J}}$ as constructed in §4.

To do so, we reembed $R^1 f_*\mathbf{Z} \hookrightarrow \mathcal{T}^*_{S^2}$ using the periods given by $\mathrm{Im}\,\Omega_n$. Now $\Omega_n = \frac{1}{Vol(S_b)}\Omega$, and this embedding takes a cycle $\gamma \in H_1(S_b, \mathbf{Z}) \cong H^1(S_b, \mathbf{Z}) \subseteq \mathcal{T}^*_{S^2}$

to the one-form

$$v \mapsto -\int_\gamma \iota(v)\,\mathrm{Im}\,\Omega_n$$
$$= -\frac{1}{Vol(S_b)}\int_\gamma \iota(v)(dx_1\wedge dy_1 + dx_2\wedge dy_2)$$

which yields the 1-form $\gamma/Vol(S_b)$.

The moral is: *The dual lattice is simply the original lattice scaled by a factor of* $1/Vol(S_b)$.

We have in fact chosen the map π so that this would happen and so make it transparent that dualising does not change the topology of the fibration.

Instead of rescaling the lattice, it is easier to identify $\check{J}$ with S topologically, and rescale the symplectic form. Since in local coordinates we want $\check{\omega}_{\check{J}} = dx_1 \wedge dy_1 + dx_2 \wedge dy_2$, by (7.2) we take on $\check{J} = S$

$$\check{\omega}_{\check{J}} = (\mathrm{Im}\,\Omega - f^*\,\mathrm{Re}\,\alpha)/Vol(S_b).$$

How do we obtain $\check{S}$ as a symplectic manifold? Item (5) of Conjecture 6.6 instructs us to proceed as follows. $\mathbf{B} \in H^1(B, R^1f_*\mathbf{R}/\mathbf{Z}) \cong H^1(B, R^1\check{j}_*\mathbf{R}/\mathbf{Z})$ maps to an element of $H^1(B, \Lambda(\check{J}^\#))$. Having lifted $\mathbf{B}$ to an element of $H^1(B, R^1f_*\mathbf{R})$, we obtain in this way an element $-\mathbf{B}\wedge[\check{\omega}_{\check{J}}] \in H^2(B,\mathbf{R})$, via Remark 3.15, which then maps to the appropriate element of $H^1(B, \Lambda(\check{J}^\#))$. As remarked before Example 2.7, this means $\check{\omega} = \check{\omega}_{\check{J}} + \check{j}^*\alpha_1$, where α_1 is a form on B such that

$$\int_B \alpha_1 = -\mathbf{B}.[\check{\omega}_{\check{J}}].$$

Now $[\check{\omega}_{\check{J}}] = [\mathrm{Im}\,\Omega_n]$ by construction, so it would appear that we take $\int_B \alpha_1 = -\mathbf{B}.[\mathrm{Im}\,\Omega_n]$. There is a slight subtlety in this: here we are representing $\mathbf{B} \in H^1(B, R^1f_*\mathbf{R})$ as an element of $E^\perp/E \otimes \mathbf{R}$, but that does not mean that if we reinterpret $\mathbf{B}$ as a class in $H^1(B, R^1\check{j}_*\mathbf{R})$, this class will coincide with the original $\mathbf{B}$ in S under our identification of S with $\check{J}$. In fact, the correct choice is

$$\int_B \alpha_1 = \mathbf{B}.[\mathrm{Im}\,\Omega_n].$$

We will be vague about this here, but will see this sign change more explicitly shortly. We then set

$$\check{\omega} = \check{\omega}_{\check{J}} + \check{j}^*\alpha_1.$$

Note that the choice of α_1 is not important; any two choices representing the same cohomology class can be identified by translation by a section. Let $\check{S}$ be the symplectic manifold obtained with underlying manifold S and symplectic form $\check{\omega}$, and let $\check{f} : \check{S} \to B$ be the same map as $f : S \to B$. We note that the cohomology class of $\check{\omega}$ satisfies the relation

$$[\check{\omega}] = [\mathrm{Im}\,\Omega_n] + (\mathrm{Im}\,\Omega_n.(\mathbf{B} - [\sigma_0]))E.$$

Next we construct the form $\mathrm{Im}\,\check{\Omega}_n$. The first observation is that $\iota(v)\,\mathrm{Im}\,\check{\Omega}_n$ must be harmonic for any $v \in \mathcal{T}_{B,b}$. On the other hand, the same is true of $\iota(v)\check{\omega}$, and $\iota(v)\check{\omega} = adx_1 + bdx_2$ for a and b constant. Thus we already know, in the 2-dimensional case, the harmonic $n-1$ forms. This is a crucial point in dimension 2 which fails in higher dimensions. Applying item (3) of Conjecture 6.6, in the form given in Proposition 6.1, we can now determine $\check{\Omega}_n$ as follows. First

$$h_n(\partial/\partial y_i, \partial/\partial y_j) = \frac{\delta_{ij}}{Vol(S_b)} \int_{S_b} dx_1 \wedge dx_2$$

at a point $b \in B$, as is easily computed from the definition and (7.1), (7.2). Then in order for $\check{h}_n = h_n$, we must have

$$-\int_{S_b} \iota(\partial/\partial y_i)\check{\omega} \wedge \iota(\partial/\partial y_j)\,\mathrm{Im}\,\check{\Omega}_n = \frac{\delta_{ij}}{Vol(S_b)} \int_{S_b} dx_1 \wedge dx_2,$$

and a quick calculation shows this implies we locally can write

$$\mathrm{Im}\,\check{\Omega}_n = -\omega + h dy_1 \wedge dy_2,$$

where h is a function. But the condition that $d\,\mathrm{Im}\,\check{\Omega}_n = 0$ implies h is constant on fibres, so

$$\mathrm{Im}\,\check{\Omega}_n = -\omega + f^*\alpha_2$$

for some form α_2 on the base.

Here we see the sign reversal explicitly. It might be a bit surprising that we have obtained $-\omega$ instead of ω. But this is the fault of the identification we have chosen. We want $[\mathrm{Im}\,\check{\Omega}_n] = [\omega]$ as classes in $H^1(B, R^1\check{f}_*\mathbf{R}) \cong H^1(B, R^1 f_*\mathbf{R})$; it is only an accident of dimension that we have been able to identify S and $\check{S}$ as manifolds and then compare cohomology classes directly. In fact, this sign change *must* occur if we want to identify S and $\check{S}$ without changing the orientation of the fibres.

Condition (3) of Conjecture 6.6 does not tell us what α_2 must be; α_2 is not determined until one knows something about $\mathrm{Re}\,\check{\Omega}_n$. In fact, condition (4) tells us that we require

$$[\mathrm{Re}\,\check{\Omega}_n] = [\sigma_0] - \mathbf{B} \bmod E,$$

where again we are making use of the sign reversal observed above. Now our form $\mathrm{Im}\,\check{\Omega}_n$ constructed above satisfies

$$[\mathrm{Im}\,\check{\Omega}_n] = [-\omega] \bmod E,$$

which tells us that in order for $\check{\Omega}_n^2 = 0$, we must have $[\check{\Omega}_n]$ satisfying

$$[\check{\Omega}_n] = [\sigma_0] - (\mathbf{B} + i\omega) + (1 - (\mathbf{B} + i\omega)^2/2 + i(\omega.\sigma_0))E. \tag{7.3}$$

Here we have chosen a representative of $\mathbf{B} \in E^\perp$ such that $\mathbf{B}.\sigma_0 = 0$. Thus, in particular, we need to choose α_2 so that

$$\int_B \alpha_2 = -\mathbf{B}.\omega + \sigma_0.\omega.$$

Again, we need to ask how much freedom we have to choose α_2, given that we have fixed α_1. We had seen that α_1 could be chosen to be any representative of its cohomology class, as any choice could be obtained from any other by translating $\check{\omega}$ by a section of $\mathcal{T}^*_{S_2}$. Once we have fixed the form α_1, however, we can only translate by sections corresponding to 1-forms σ with $d\sigma = 0$. Let $\sigma = \sigma_1 dy_1 + \sigma_2 dy_2$. Then

$$T^*_\sigma(dy_2 \wedge dx_1 + dx_2 \wedge dy_1) = dy_2 \wedge dx_1 + dx_2 \wedge dy_1 - \left(\frac{\partial \sigma_1}{\partial y_1} + \frac{\partial \sigma_2}{\partial y_2}\right) dy_1 \wedge dy_2.$$

Thus $T^*_\sigma(\omega) - \omega = -f^*(d * \sigma)$, where $*$ denotes the Hodge $*$-operator in, say, the Fubini-Study metric on $B = \mathbf{P}^1$. Thus if α_2, α_2' are two 2-forms on B representing the same cohomology class, we just need to find a 1-form σ on B such that $d\sigma = 0$ and $d * \sigma = \alpha_2 - \alpha_2'$, and then $T^*_\sigma(\omega + f^*\alpha_2) = \omega + f^*\alpha_2'$. By the Hodge theorem, such a σ can always be found. Thus we have complete freedom to choose α_2, and any two choices are related by translation by a Lagrangian section.

This will be the last remaining choice in the construction which is not forced on us by any items in Conjecture 6.6; thus the uniqueness of item (6) of Conjecture 6.6 will hold, given that the lack of uniqueness in the lifting of $\mathbf{B}$ can be rectified by changing the choice of the zero section σ_0.

Finally, we set

$$\mathrm{Im}\,\check{\Omega} = (\mathrm{Im}\,\check{\Omega}_n)/Vol(S_b).$$

It now follows immediately from (7.1) and (7.2) that as forms,

$$(\mathrm{Im}\,\check{\Omega}) \wedge (\mathrm{Im}\,\check{\Omega}) = \check{\omega} \wedge \check{\omega} > 0,$$

and

$$(\mathrm{Im}\,\check{\Omega}) \wedge \check{\omega} = 0.$$

Thus $\check{\Omega}_K = \mathrm{Im}\,\check{\Omega} + i\check{\omega}$ is a 2-form which satisfies the conditions of Theorem 5.1 and hence determines a complex structure on $\check{S}$. We call $\check{S}$ with this complex structure $\check{S}_K$. As observed above, $[\check{\Omega}_n]$ must satisfy (7.3), so we need to look for a form $\mathrm{Re}\,\check{\Omega}$ such that

$$[\mathrm{Re}\,\check{\Omega}] = \frac{1}{Vol(S_b)}([\sigma_0] - \mathbf{B} - (\mathbf{B}^2 - \omega^2 - 2)E/2).$$

If $[\mathrm{Re}\,\check{\Omega}]$ is a Kähler class on $\check{S}_K$, then by Yau's theorem, there exists a unique Kähler form $\mathrm{Re}\,\check{\Omega}$ whose metric is Ricci-flat, so we only need to ensure $[\mathrm{Re}\,\check{\Omega}]$ is a Kähler class. We first note that $[\mathrm{Re}\,\check{\Omega}].[\mathrm{Im}\,\check{\Omega}] = [\mathrm{Re}\,\check{\Omega}].[\check{\omega}] = 0$, so $[\mathrm{Re}\,\check{\Omega}]$ is a $(1,1)$ class. Next we observe it is a positive class. Indeed, $\check{f} : \check{S}_K \rightarrow B$ is still a holomorphic elliptic fibration, and the complex structure on each fibre $\check{S}_{K,b}$ is the same as that of $S_{K,b}$. Since $[\mathrm{Re}\,\check{\Omega}].E > 0$, this shows then that $[\mathrm{Re}\,\check{\Omega}]$ is in fact positive on E. As long as $Pic(\check{S}_K)$ contains no -2 classes, this shows that $[\mathrm{Re}\,\check{\Omega}]$ is a Kähler class. Hence we obtain a Kähler form $\mathrm{Re}\,\check{\Omega}$ as desired, and set $\check{\Omega} = \mathrm{Re}\,\check{\Omega} + i\,\mathrm{Im}\,\check{\Omega}$. •

We make a few closing comments. First, in the above proof we can write

$$[\Omega] = Vol(S_b)([\sigma_0] + \check{\mathbf{B}} + i[\check{\omega}]) \bmod E$$

and

$$[\check{\Omega}] = \frac{1}{Vol(S_b)}([\sigma_0] - (\mathbf{B} + i[\omega])) \operatorname{mod} E.$$

Thus we see that $Vol(\check{S}_b) = 1/Vol(S_b)$, conforming with Remark 6.8. However this does not quite look like mirror symmetry: if we repeat the process we appear to get

$$[\check{\check{\Omega}}] = Vol(S_b)([\sigma_0] - (\check{\mathbf{B}} + i[\check{\omega}])) \operatorname{mod} E.$$

This again is the fault of our identifications, essentially having put in a 90° twist in $\mathcal{T}_B^*$ during each dualising. This is rectified on the double mirror by pulling back all forms by the fibrewise negation map on S. This acts trivially on $H^0(B, R^2 f_*\mathbf{R})$ and $H^2(B, f_*\mathbf{R})$, but by negation on $H^1(B, R^1 f_*\mathbf{R})$.

Finally we note that the construction of the mirror K3 surface given in this proof is of a different nature from previous constructions of mirror symmetry for K3 surfaces. Normally one appeals to the Torelli theorem to construct the mirror. Here, once we have produced a special Lagrangian fibration on S, we produce the mirror without an appeal to Torelli. Instead, we are essentially applying Yau's Theorem to solve the equations of Corollary 5.15. However we are still aided by some key points which don't hold in higher dimensions. These are:

(1) We know the harmonic $n-1$-forms on fibres, since $n-1=1$.

(2) We know the cohomology class of a holomorphic 2-form Ω if we know its class modulo E; this is completely determined by the requirement $[\Omega]^2 = 0$.

(3) We can use the hyperkähler trick.

§8. Postscript.

Since the initial version of this paper was prepared, there have been a number of new results which have greatly increased the evidence for the Strominger-Yau-Zaslow conjecture. The first is a result of W.-D. Ruan [25] constructing a Lagrangian T^3-fibration on the quintic threefold. The basic idea is to deform the Lagrangian T^3-fibration on the degenerate Calabi-Yau threefold $z_0z_1z_2z_3z_4 = 0$ in $\mathbf{P}^4$ induced by the moment map $\mu : \mathbf{P}^4 \to \Xi$ (where Ξ is the polytope corresponding to $\mathbf{P}^4$) as suggested in [16]. Such a procedure was also carried out in [32]. But Ruan introduces a new technique, using a symplectic flow, to move Lagrangian tori from the singular Calabi-Yau to the non-singular ones. Of course, this gives tori which are Lagrangian with respect to the Fubini-Study Kähler form, but by Moser's theorem, such a Calabi-Yau is symplectomorphic to the one whose Kähler form is given by the Ricci-flat metric.

Because this fibration constructed is not differentiable, the structure of Lagrangian fibrations studied in §2 does not apply, and in particular these fibrations look quite bad from the point of view of mirror symmetry. The discriminant locus is codimension one, which rules out the possibility of G-simplicity, and this construction has to be modified to obtain a fibration suitable for dualizing.

In my own paper [15], I addressed this issue in a completely different way. I worked in a completely topological setting, dealing with T^3-fibrations whose topological properties resemble those one expects of special Lagrangian fibrations as discussed in §1 and 2 here. I then showed how to use $\mathbf{Z}$ and $\mathbf{Z}/n\mathbf{Z}$-simplicity to classify monodromy near *semi-stable* fibres, i.e. fibres where the local monodromy group is unipotent. By construction of examples of such topological fibrations, one in fact proves this gives a complete classification of possible monodromy. In particular, using this classification, one accomplishes step (3) of the program outlined in the introduction for those topological T^3-fibrations with only semistable fibres. This gives one (but not necessarily the only!) solution to the compactification problem.

Next, I applied these ideas to an explicit example. In [25], Ruan had calculated the monodromy around his codimension one discriminant locus. Using this, one can explictly construct a torus fibre bundle with the same monodromy, and then glue in a fibration with suitable discriminant locus to match the same monodromy. This yields a fibration $f : X \to S^3$, and one can then prove that X is diffeomorphic to the quintic.

The best test of the Strominger-Yau-Zaslow conjecture is then to dualize this fibration using the general compactification method that we have developed, to obtain $\check{f} : \check{X} \to S^3$. One can then show that $\check{X}$ is indeed diffeomorphic to the mirror quintic, or rather to one particular minimal model of the mirror quintic. Different choices made for the fibration $f : X \to B$ yield different minimal models of the mirror. This now shows that at least on the topological level, the Strominger-Yau-Zaslow conjecture gives a satisfactory explanation of mirror symmetry.

More recently, Ruan has released another paper [26], announcing a construction of Lagrangian fibrations on the quintic and its mirror with the correct properties (in particular, these fibrations will be simple), and he shows that these fibrations are dual. The method is via a modification of his symplectic flow technique, though details of the construction are postponed until future papers. The fibrations constructed are, on the topological level, the same as those constructed in [26].

There still remains the hardest question: that of obtaining special Lagrangian fibrations. This remains a very difficult problem, despite this recent progress.

Bibliography

[1] Almgren, F. J., Jr. "Q Valued Functions Minimizing Dirichlet's Integral and the Regularity of Area Minimizing Rectifiable Currents up to Codimension Two," *Bull. Amer. Math. Soc. (N.S.)*, **8** (1983), 327–328.

[2] Almgren, F. J., Jr., "Q Valued Functions Minimizing Dirichlet's Integral and the Regularity of Area Minimizing Rectifiable Currents up to Codimension Two," preprint.

[3] Arnold, V. I., *Mathematical Methods of Classical Mechanics*, 2nd edition, Springer-Verlag, 1989.

[4] Aspinwall, P., "An $N = 2$ Dual Pair and a Phase Transition," *Nucl. Phys. B*, **460** (1996), 57–76.

[5] Aspinwall, P., and Morrison, D., "Stable Singularities in String Theory," with an appendix by M. Gross, *Comm. Math. Phys.* **178**, (1996) 115–134.

[6] Aspinwall, P., and Morrison, D., "String Theory on K3 surfaces," in *Essays on Mirror Manifolds II*, Greene, B.R., Yau, S.-T. (eds.) Hong Kong, International Press 1996, 703–716.

[7] Barth, W., Peters, C., and van de Ven, A., *Compact Complex Surfaces*, Springer-Verlag, 1984.

[8] Besse, A., *Einstein Manifolds*, Springer-Verlag, 1987.

[9] Borel, A. et al, *Intersection Cohomology*, Birkhäuser, 1984.

[10] Bredon, G., *Sheaf Theory*, 2nd edition, Springer-Verlag, 1997.

[11] Dolgachev, I., "Mirror Symmetry for Lattice Polarized K3 surfaces," Algebraic Geometry, **4**, *J. Math. Sci.* **81**, (1996) 2599–2630.

[12] Duistermaat, J., "On Global Action-Angle Coordinates," *Comm. Pure Appl. Math.*, **33** (1980) 687–706.

[13] Federer, H., *Geometric Measure Theory*, Springer-Verlag, 1969.

[14] Gross, M., "Special Lagrangian Fibrations I: Topology," to appear in the Proceedings of the Taniguchi Symposium on Integrable Systems and Algebraic Geometry.

[15] Gross, M., "Topological Mirror Symmetry," preprint, math.AG/9909015 (1999).

[16] Gross, M., and Wilson, P.M.H., "Mirror Symmetry via 3-tori for a Class of Calabi-Yau Threefolds," *Math. Ann.*, **309**, (1997) 505–531.

[17] Harvey, R., and Lawson, H.B. Jr., "Calibrated Geometries," *Acta Math.* **148**, 47-157 (1982).

[18] Hitchin, N., "The Moduli Space of Special Lagrangian Submanifolds," preprint, dg-ga/9711002.

[19] King, J., "The Currents Defined by Analytic Varieties," *Acta Math.*, **127** (1971), 185–220.

[20] Koszul, J.-L., "Crochet de Schouten-Nijenhuis et Cohomologie," in *The Mathematical Heritage of Elie Cartan (Lyon, 1984), Astérisque*, Numero Hors Serie, (1985) 257–271.

[21] Mangiarotti, L., and Modugno, M., "Graded Lie Algebras and Connections on a Fibered Space," *J. Math. Pures. et Appl.*, **63**, (1984), 111–120.

[22] McLean, R.C., " Deformations of Calibrated Submanifolds," Duke University Preprint, 1996.

[23] Morgan, F., *Geometric Measure Theory. A Beginner's Guide*, 2nd Edition, *Academic Press, Inc.*, 1995.

[24] Morrison, D.R., "The Geometry Underlying Mirror Symmetry," Preprint 1996.

[25] Ruan, W.-D., "Lagrangian Tori Fibration of Toric Calabi-Yau Manifold I," preprint, math.DG/9904012 (1999).

[26] Ruan, W.-D., "Lagrangian Tori Fibration of Toric Calabi-Yau Manifold III: Symplectic Topological SYZ Mirror Construction for General Quintics," preprint, math.DG/9909126 (1999).

[27] Simon, L., *Lectures on Geometric Measure Theory*, Proceedings of the Centre for Mathematical Analysis, Australian National University, **3**, *Australian National University, Centre for Mathematical Analysis, Canberra*, 1983.

[28] Strominger, A., Yau, S.-T., and Zaslow, E., "Mirror Symmetry is T-Duality," *Nucl. Phys.* **B479**, (1996) 243–259.

[29] Tian, G., "Smoothness of the Universal Deformation Space of Compact Calabi-Yau Manifolds and its Petersson-Weil Metric," in *Mathematical Aspects of String Theory, (San Diego, California, 1986)*, 629–646, *Adv. Ser. Math. Phys.* **1**, World Scientific Publishing.

[30] Todorov, A., "The Weil-Petersson Geometry of the Moduli Space of $SU(n \geq 3)$ (Calabi-Yau) Manifolds I," *Commun. Math. Phys* **126**, (1989) 325–346.

[31] Weil, A., "Final Report on Contract AF 18(603)-57," *Oeuvres Scientifiques*, Vol. II, 390–395.

[32] Zharkov, I., "Torus Fibrations of Calabi-Yau Hypersurfaces in Toric Varieties and Mirror Symmetry," preprint, alg-geom/9806091.

Mirror Principle I

Bong H. Lian,[1] Kefeng Liu,[2] and Shing-Tung Yau[3]

Abstract. We propose and study the following *Mirror Principle:* certain sequences of multiplicative equivariant characteristic classes on stable map moduli spaces can be computed in terms of certain hypergeometric type classes. As applications, we compute the equivariant Euler classes of obstruction bundles induced by any concavex bundles – including any direct sum of line bundles – on $\mathbf{P}^n$. This includes proving the formula of Candelas-de la Ossa-Green-Parkes for the instanton prepotential function for quintic in $\mathbf{P}^4$. We derive, among many other examples, the so-called multiple cover formula for GW invariants of $\mathbf{P}^1$. We also prove a formula for enumerating Euler classes which arise in the so-called local mirror symmetry for some noncompact Calabi-Yau manifolds. At the end we interprete an infinite dimensional transformation group, called the mirror group, acting on Euler data, as a certain *duality group* of the linear sigma model.

Appeared in Asian Journ. Math. Vol 1, No. 4 (1997).

---

[1]Department of Mathematics, Brandeis University, Waltham, MA 02154.

[2]Department of Mathematics, Stanford University, Stanford, CA 94305.

[3]Department of Mathematics, Harvard University, Cambridge, MA 02138.

Contents

1. Introduction 406
   1.1. The Mirror Principle 406
   1.2. Enumerative problems and the Mirror Conjecture . . 409
   1.3. Acknowledgements 412
2. Euler Data 413
   2.1. Preliminaries and notations 413
   2.2. Eulerity 415
   2.3. Concavex bundles 421
   2.4. Linked Euler data 428
   2.5. The Lagrange map and mirror transformations 431
3. Applications 437
   3.1. The first convex example: The Mirror Conjecture . . 440
   3.2. First concave example: multiple-cover formula 443
   3.3. Second concave example: $K_{\mathbf{P}^2}$ 444
   3.4. A concavex bundle on $\mathbf{P}^3$. 445
   3.5. A concavex bundle on $\mathbf{P}^4$. 446
   3.6. General concavex bundles 447
   3.7. Equivariant total Chern class 448
   3.8. Concluding remarks 449

1. Introduction

1.1. The Mirror Principle

In section 2, we develop a general theory of Euler data, and give many examples. In particular we introduce the notions of a convex and a concave bundles on $\mathbf{P}^n$, and show that they naturally give rise to Euler data. In section 3 we apply our method to compute the equivariant Euler classes, and their nonequivariant limits, of obstruction bundles induced by a convex or a concave bundle.

We briefly outline our approach for computing multiplicative equivariant characteristic classes on stable map moduli. This outline also fixes some notations used later. Our approach is partly inspired by the idea of mirror transformations (Candelas et al), the idea of the linear sigma model (Witten, Morrison-Plesser, Jinzenji-Nagura,..), and the idea of using the torus action and equivariant techniques (Ellinsrud-Stromme, Kontsevich, Givental,..). All these ideas are synthesized with what we call the *Mirror Principle*, which we now explain.

Let M be a projective manifold and $\beta \in H^2(M,\mathbf{Z})$. Let $\bar{\mathcal{M}}_{g,k}(\beta, M)$ be the stable map moduli space of degree β, arithmetic genus g, with k marked points. For a good introduction to stable maps, see [16]. Throughout this paper, we shall only deal with the case with $g = 0$.

We begin by analyzing two distinguished types of fixed points under an induced torus T action on $\bar{\mathcal{M}}_{0,0}(d,\mathbf{P}^n)$. Both types of fixed points reflect the structure of the stable map moduli space. A smooth fixed point we consider is a degree d cover of a T-invariant $\mathbf{P}^1$ joining two fixed points p_i, p_j in $\mathbf{P}^n$. A singular fixed point we consider is in the compactification divisor. It is given by gluing together two 1-pointed maps $(f_1, C_1, x_1) \in \bar{\mathcal{M}}_{0,1}(r,\mathbf{P}^n)$ and $(f_1, C_2, x_2) \in \bar{\mathcal{M}}_{0,1}(d-r,\mathbf{P}^n)$ at the marked points with $f_1(x_1) = f_2(x_2) = p_i \in \mathbf{P}^n$ (p_i being a T-fixed in $\mathbf{P}^n$), resulting in a degree d stable map (f, C). We consider two types of T-equivariant bundles V on $\mathbf{P}^n$, which we called convex and concave respectively (Definition 2.7). To be brief, we consider the convex case in this outline. A convex bundle $V \to \mathbf{P}^n$ induces on $\bar{\mathcal{M}}_{0,0}(d,\mathbf{P}^n)$ an obstruction bundle U_d whose fiber at (f, C) is the section space $H^0(C, f^*V)$. First we have the exact sequence over C

$$0 \to f^*V \to f_1^*V \oplus f_2^*V \to V|_{p_i} \to 0.$$

Passing to cohomology, we have

$$0 \to H^0(C, f^*V) \to H^0(C_1, f_1^*V) \oplus H^0(C_2, f_2^*V) \to V|_{p_i} \to 0.$$

Hence we obtain a similar exact sequence for the U_d restricted to a suitable fixed point set. Let b_T be any multiplicative T-equivariant characteristic class [23] for vector bundles. The exact sequence on the fixed point set above gives rise to the identity, which we call the *gluing identity*:

$$b_T(V) \cdot b_T(U_d) = b_T(U_r) \cdot b_T(U_{d-r}).$$

Let $M_d := \bar{\mathcal{M}}_{0,0}((1,d), \mathbf{P}^1 \times \mathbf{P}^n)$. This space has a $G = S^1 \times T$ action. There is a natural equivariant contracting map $\pi : M_d \to \bar{\mathcal{M}}_{0,0}(d,\mathbf{P}^n)$ given by

$$\pi : \ (f, C) \in M_d \mapsto (\pi_2 \circ f, C') \in \bar{\mathcal{M}}_{0,0}(d,\mathbf{P}^n)$$

where π_2 denotes the projection onto the second factor of $\mathbf{P}^1 \times \mathbf{P}^n$. Here $C' = C$ if $(\pi_2 \circ f, C)$ is still stable; and if $(\pi_2 \circ f, C)$ is unstable, this is the case when C is of the form $C = C_1 \cup C_0 \cup C_2$ with $\pi_2 \circ f(C_0)$

a point in $\mathbf{P}^n$, then C' is obtained from C by contracting the unstable component C_0.

Now via π, we pull back to M_d all the information obtained above on $\bar{\mathcal{M}}_{0,0}(d,\mathbf{P}^n)$. The reason is that there is an collapsing map $\varphi : M_d \to N_d = \mathbf{P}^{(n+1)d+n}$ which then allows us to perform computations on the linear object N_d. We call N_d *the linear sigma model and* M_d *the nonlinear sigma model.* There is a natural G action on N_d such that φ is G-equivariant. For example, to determine an equivariant cohomology class ω on N_d, we only need to know its restrictions $\iota^*_{p_{ir}}(\omega)$ to the $(n+1)(d+1)$ G-fixed points $\{p_{ir}\}$ in N_d. The corresponding weights of the G action on N_d are $\lambda_i + r\alpha$. Let Q_d be the push-forward of $b_T(V_d) = \pi^* b_T(U_d)$ into N_d, ie. $Q_d = \varphi_! \, b_T(V_d)$. Then the classes Q_d inherit the gluing identity (Theorem 2.8)

$$\iota^*_{p_{i,0}}(Q_0) \cdot \iota^*_{p_{ir}}(Q_d) = \iota^*_{p_{ir}}(Q_r) \cdot \iota^*_{p_{ir}}(Q_{d-r})$$

which is an identity in the ring $H^*_G(pt) = H^*(BG)$. The sequence Q_d is an example of what we call an *Euler data* (Definition 2.3). We summarize the various ingredients used, now and later, in our constructions:

$$\begin{array}{ccccccccc}
 & & V_d = \pi^* U_d & & U_d & & \rho^* U_d & & \\
 & & \downarrow & & \downarrow & & \downarrow & & \\
N_d & \xleftarrow{\varphi} & M_d & \xrightarrow{\pi} & \bar{\mathcal{M}}_{0,0}(d,\mathbf{P}^n) & \xleftarrow{\rho} & \bar{\mathcal{M}}_{0,1}(d,\mathbf{P}^n) & \xrightarrow{ev} & \mathbf{P}^n
\end{array}$$

where ρ forgets, and ev evaluates at, the marked point of a 1-pointed stable map. Also note that $U_d = \rho_!(ev^* V)$, ie. the direct image of $ev^* V$.

The gluing identity is not enough to determine all Q_d. In order to get further information we localize the Q_d to a fixed point in N_d whose inverse image in M_d is a *smooth* fixed point, and compute Q_d at the special values of $\alpha = (\lambda_i - \lambda_j)/d$. The property of $b_T(U_d)$ at a smooth fixed point comes into play here. For simplicity in this outline, let's consider for example the case when b_T is the equivariant Euler class e_T and $V = \mathcal{O}(l)$. In this case we find that, at $\alpha = (\lambda_i - \lambda_j)/d$,

$$\iota^*_{p_{i,0}}(Q_d) = \prod_{m=0}^{ld} (l\lambda_i - m(\lambda_i - \lambda_j)/d).$$

This immediately tells us that the Q_d should be compared with the sequence of classes

$$P: \;\; P_d = \prod_{m=0}^{ld} (l\kappa - m\alpha)$$

(Theorem 2.10), which has $\iota^*_{p_{i,0}}(P_d) = \iota^*_{p_{i,0}}(Q_d)$ at $\alpha = (\lambda_i - \lambda_j)/d$. We find that P is another example of an Euler data. This Euler data will then naturally give rise to a generating series of hypergeometric type, hence explains the very origin of these functions in enumerative problems on stable moduli! The same holds true for a large class of vector bundles V on $\mathbf{P}^n$.

Finally, under a suitable bound on $c_1(V)$, we can completely determine the Q_d, in terms of P_d by means of a mirror transformation argument (Theorem 3.9). We can in turn use Q_d to compute $e_T(U_d)$ and their nonequivariant limits (Theorem 3.2). Our approach, thus, makes the roles of three objects and their relationships quite transparent: certain fixed point sets, equivariant multiplicative characteristic classes, and series of hypergeometric type. Our method works well for many characteristic classes such as the Euler class and the total Chern class.

We summarize this *Mirror Principle*. Let b_T be a multiplicative equivariant characteristic class, V an equivariant vector bundle on $\mathbf{P}^n$ which induces a sequences of vector bundles $U_d \to \bar{\mathcal{M}}_{0,0}(d, \mathbf{P}^n)$.

1. (Euler Data) The behaviour of the U_d at a singular fixed point gives rise to the gluing identity. In turn this defines an Euler data Q_d on the linear sigma model N_d.
2. (Linking) The behaviour of the U_d at a smooth fixed point allows us to read off the restrictions of Q_d at certain fixed points of N_d for special values of weights. The restriction values determine a distinguished Euler data P_d to be compared with the Q_d.
3. (Mirror Transformations) Compute the Q_d and the $b_T(U_d)$ in terms of the P_d explicitly by means of a mirror transformation argument.

Our approach outlined here can be applied to a rather broad range of cases by replacing $\mathbf{P}^n$ by other manifolds. They include manifolds with torus action and their submanifolds. We will study the cases of toric varieties and homogeneous manifolds in a forth-coming paper [37]. On the other hand, we can even go beyond equivariant multiplicative characteristic classes. In a future paper, we will study a sequence of equivariant classes T_d of geometric origin on stable moduli satisfying our gluing identity.

1.2. Enumerative problems and the Mirror Conjecture

For the remarkable history of the Mirror Conjecture, see [14]. In 1990, Candelas, de la Ossa, Green, and Parkes [8] conjectured a formula for counting the number n_d of rational curves in every degree d on a general quintic in $\mathbf{P}^4$. The conjectured formula gives a generating function for the n_d explicitly in terms of certain hypergeometric functions. Their computation is based on the existence of "mirror manifolds", conjectured by a number of physicists including Dixon and Lerche, Vafa and Warner on the basis of physical intuition. Mirror symmetry took a dramatic turn upon the appearance of the papers of Greene-Plesser [21] and of Candelas et al [8]. In [21], they established the existence of mirror manifolds for a particular class of Calabi-Yau manifolds including the quintic.

On the mathematical side, it has been conjectured earlier by Clemens that the number of rational curves in every degree in a general quintic is finite. The conjectured formula agrees with a classical result in degree 1, an earlier computation by S. Katz in degree 2, and has been verified in degree 3 by Ellingsrud-Stromme. In 1994, using results of Uhlenbeck-Sacks, Gromov, Parker-Wolfson, and Witten, Ruan-Tian introduced the notions of J-holomorphic maps and symplectic Gromov-Witten (GW) invariants. Motivated by the same works, Kontsevich proposed an algebraic geometric analogue of GW invariants and stable maps. The latter is the algebraic geometric counterpart of J-holomorphic maps. Generalizations have been given by Li-Tian and Behrend-Fentachi. A recent paper of Li-Tian shows that the symplectic version and the algebraic geometric version of the GW theory are essentially the same in the projective category. Beautiful applications of ideas from quantum cohomology and GW theory have recently been done by Caporaso-Harris [10], Crauder-Miranda [11], DiFrancesco-Itzykson [12], Bryan-Leung [2], and others solving many important enumerative problems.

Closer to mirror symmetry, it is known that the degree k GW invariants for $\mathbf{P}^1$ (the so-called multiple-cover contribution) is given by k^{-3}. This was conjectured in [8], justified in [1][44] using a different compactification, and in [38] using the stable map compactification. According to [34], the number

$$K_d = \sum_{k|d} n_{d/k} k^{-3}$$

is the degree of the Euler class $c_{top}(U_d)$ for $U_d \to \bar{\mathcal{M}}_{0,0}(d, \mathbf{P}^4)$ induced by $\mathcal{O}(5) \to \mathbf{P}^4$.

By means of the torus action on $\mathbf{P}^4$, Ellingsrud-Stromme [15] and Kontsevich [34] have verified that n_3 and n_4 respectively agree with the conjectured formula. In some recent papers [18][19], Givental introduces some ideas which emphasize the use of equivariant version of quantum cohomology (see also Kim [27]). Unfortunately, the proof in [18]for the formula of Candelas et al, which has been read by many prominent experts, is incomplete. In light of our present paper, it is now clear that essentially all ingredients (equivariant cohomology theory, the Atiyah-Bott formula, equivariant Euler classes, obstruction bundles, fixed point structures, the linear sigma model, the graph construction, hypergeometric series, mirror transformations, etc) required to prove the formula of Candelas et al have all been well-known to experts since at least 1995. Thus the remaining nontrivial task, is to set up and put together in details all the pieces in a correct way to give a complete proof. In our present paper, we have introduced the substantial new ideas (beyond the known ingredients) needed to give the first complete proof. It is conceivable that our new ideas here can furnish the necessary parts missing in the paper [18]. More recently, fixed point method has also been applied by Graber-Pandharipande [20] to study GW invariants of $\mathbf{P}^n$.

We emphasize that the formula of Candelas et al, albeit conjectural, not only computes the K_d in terms of hypergeometric functions, but also provides a crucial guide for putting together various technical inputs in our proof. While applying the Atiyah-Bott localization formula on the stable moduli and on the linear sigma model yield important pieces of information about the cohomology classes in question, it is the formula of Candelas et al which provides the clues for how to put the pieces together – in terms of hypergeometric functions (which are periods of a holomorphic 3-form on a mirror manifold). Earlier attempts on the conjectured formula using quantum cohomology seems to have confused the issue considerably.

We believe that the machinery introduced in this paper will be useful for many other enumerative problems, aside from proving the formula of Candelas et al. In fact we have applied our machinery to problems in local mirror symmetry proposed by C.Vafa, S. Katz, and others. [4]

We now formulate one of our main theorems in this paper. Let

$$K_d = \int_{\bar{\mathcal{M}}_{0,0}(d,\mathbf{P}^4)} c_{top}(U_d), \qquad F(T) = \frac{5T^3}{6} + \sum_{d>0} K_d e^{dT}.$$

Consider the fourth order hypergeometric differential operator [8]:

$$L := (\frac{d}{dt})^4 - 5e^t(5\frac{d}{dt}+1)\cdots(5\frac{d}{dt}+4).$$

S. Katz has informed us that A. Elezi has also studied a similar problem.

By the Frobenius method, it is easy to show that

$$f_i := \frac{1}{i!}\left(\frac{d}{dH}\right)^i\Big|_{H=0} \sum_{d\geq 0} e^{d(t+H)} \frac{\prod_{m=1}^{5d}(5H+m)}{\prod_{m=1}^{d}(H+m)^5}, \quad i = 0, 1, 2, 3,$$

form a basis of solutions to the differential equation $L \cdot f = 0$. Let

$$T = \frac{f_1}{f_0}, \quad \mathcal{F}(T) = \frac{5}{2}\left(\frac{f_1}{f_0}\frac{f_2}{f_0} - \frac{f_3}{f_0}\right).$$

Theorem 1.1. *(The Mirror Conjecture)* $F(T) = \mathcal{F}(T)$.

The transformation on the functions f_i given by the normalization

$$f_i \mapsto \frac{f_i}{f_0}$$

and the change of variables

$$t \mapsto T(t) = \frac{f_1}{f_0}$$

are known as the *mirror transformation.* By the construction of Candelas et al, the functions $f_0, .., f_3$ are periods of a family of Calabi-Yau threefolds. By the theorem of Bogomolov-Tian-Todorov, these periods in fact determine the complex structure of the threefold.

A similar Mirror Conjecture formula holds true for a three dimensional Calabi-Yau complete intersection in a toric Fano manifold [37]. This will turn out to agree with the many beautiful mirror symmetry computations [40][36][5][9][4][30][31].

In this paper, to make the ideas clear we restrict ourselves to the simplest case, genus 0 stable moduli space for $\mathbf{P}^n$. In a forth-coming paper [37], we extend our discussions to toric varieties, homogeneous manifolds, and general projective manifolds. We also hope to eventually understand from our point of view the far reaching results for higher genus of [6], and the beautiful computations [17][13] for elliptic GW invariants.

1.3. Acknowledgements

We thank A. Todorov, A. Strominger, C. Vafa for helpful discussions. Our special thanks are due to J. Li who has been very helpful throughout our project.

2. Euler Data

One of the key ingredients in our approach is the linear sigma model, first introduced by Witten [45], and later used to study mirror symmetry by Morrison-Plesser [29], Jinzenji-Nagura [25], and others, resulting in new insights into the origin of hypergeometric series. In this paper, we consider the $S^1 \times T$-equivariant cohomology of the linear sigma model.

2.1. Preliminaries and notations

Let T be an r-dimensional real torus with a complex linear representation on $\mathbf{C}^{N+1}$. Let $\beta_0, .., \beta_N$ be the weights of this action. We consider the induced action of T on $\mathbf{P}^N$, and the T-equivariant cohomology with coefficients in $\mathbf{Q}$, which we shall denote by $H_T^*(-)$. Now $H_T^*(pt)$ is a polynomial algebra in r-variables, and β_i may be regarded as elements of $H_T^2(pt)$. *Throughout this paper, we shall follow the convention that such generators have degree 1.* It is known that the equivariant cohomology of $\mathbf{P}^N$ is given by [32]

$$H_T(\mathbf{P}^N) = H_T(pt)[\zeta]/\left(\prod_{i=0}^{N}(\zeta - \beta_i)\right).$$

Here ζ, which we shall call the equivariant hyperplane class, is a fixed lifting of the hyperplane class of $\mathbf{P}^N$. Each one-dimensional weight space in $\mathbf{C}^{N+1}$ becomes a fixed point p_i in $\mathbf{P}^N$. *We shall identify the rings $H_T^*(p_i)$ and $H_T^*(pt) = H^*(BT)$.* There are $N+1$ canonical restriction maps $\iota_{p_i}^* : H_T(\mathbf{P}^N) \to H_T(pt)$, given by $\zeta \mapsto \beta_i$, $i = 0, .., N$. There is also a push-forward map $H_T(\mathbf{P}^N) \to H_T(pt)$ given by integration over $\mathbf{P}^N$. By the localization formula, it is given by

$$\omega \mapsto \int_{\mathbf{P}^N} \omega = Res_\zeta \frac{\omega}{\prod_{i=0}^{N}(\zeta - \beta_i)}.$$

Two situations arise frequently in this paper. First consider the standard action of $T = (S^1)^{n+1}$ on $\mathbf{C}^{n+1}$, and let $\lambda = (\lambda_0, .., \lambda_n)$ denote the weights. On $\mathbf{P}^n$, there are $n+1$ isolated fixed points $p_0, .., p_n$. We shall denote the equivariant hyperplane class by p, the canonical restrictions by $\iota_{p_i}^* : \omega \mapsto \iota_{p_i}^*(\omega) = \omega(\lambda_i)$, and the push-forward by

$$pf : H_T(\mathbf{P}^n) \to H_T(pt) = \mathbf{Q}[\lambda].$$

We shall use the evaluation map $\lambda \mapsto 0$ on the ring $H_T^*(\mathbf{P}^n)$, and shall call this the *nonequivariant limit.* In this limit, p becomes the ordinary hyperplane class $H \in H^*(\mathbf{P}^n)$.

We now consider the second situation. For each $d = 0, 1, 2, ..$, consider the following complex linear action of the group $G := S^1 \times T$ on $\mathbf{C}^{(n+1)(d+1)}$. We let the group act on the (ir)-th coordinate line in $\mathbf{C}^{(n+1)(d+1)}$ by the weights $\lambda_i + r\alpha$ (λ_i being the weights of T as before, and α being a weight of S^1), $i = 0, .., n$, $r = 0, .., d$. Thus there are $(n+1)(d+1)$ isolated fixed points p_{ir} on the projective space $\mathbf{P}^{(n+1)d+n}$, given by those coordinate lines. In this case, we shall denote the equivariant hyperplane class by κ, the canonical restrictions by $\iota^*_{p_{ir}}$: $\omega \mapsto \iota^*_{p_{ir}}(\omega) = \omega(\lambda_i + r\alpha)$, and the push-forward by

$$pf_d : H_G(\mathbf{P}^{(n+1)d+n}) \to H_G(pt) = \mathbf{Q}[\alpha, \lambda].$$

Here we have abused the notation κ, using it to represent a class in $H_G(\mathbf{P}^{(n+1)d+n})$ for every d. But it should present no confusion in the context it arises.

Let N_d be the space of nonzero $(n+1)$-tuple of degree d homogeneous polynomials in two variables w_0, w_1, moduo scalar. There is a canonical way to identify N_d with $\mathbf{P}^{(n+1)d+n}$. Namely, a point $z \in \mathbf{P}^{(n+1)d+n}$ corresponds to the polynomial tuple $[\sum_r z_{0r} w_0^r w_1^{d-r}, .., \sum_r z_{nr} w_0^r w_1^{d-r}] \in N_d$. *This identification will be used throughout this paper.* The natural T-action on $(n+1)$-tuples together with the S^1-action on $[w_0, w_1] \in \mathbf{P}^1$ by weights $(\alpha, 0)$, induces a $S^1 \times T$-action on N_d given by $[f_0(w_0, w_1), .., f_n(w_0, w_1)] \mapsto [e^{\lambda_0} f_0(e^\alpha w_0, w_1), .., e^{\lambda_n} f_n(e^\alpha w_0, w_1)]$. This coincides with the $S^1 \times T$-action on $\mathbf{P}^{(n+1)d+n}$ described earlier.

Definition 2.1. *(Notations) We call the sequence of projective spaces $\{N_d\}$ the linear sigma model for $\mathbf{P}^n$. Here are some frequently used notations: $G := S^1 \times T$, $\mathcal{R} := \mathbf{Q}(\lambda)[\alpha]$, $\mathcal{R}^{-1} := \mathbf{Q}(\lambda, \alpha)$, $\mathcal{R}H^*_G(N_d) := H^*_G(N_d) \otimes_{\mathbf{Q}[\lambda,\alpha]} \mathcal{R}$, $\mathcal{R}^{-1}H^*_G(N_d) := H^*_G(N_d) \otimes_{\mathbf{Q}[\lambda,\alpha]} \mathcal{R}^{-1}$, and $deg_\alpha\ \omega$ means the degree in α of $\omega \in \mathcal{R}$.*

Obviously the maps $\iota^*_{p_{ir}}$, pf_d defined linearly over $\mathbf{Q}[\lambda, \alpha]$, can be extended $\mathcal{R}$- or $\mathcal{R}^{-1}$-linearly. There are two natural equivariant maps between the N_d, given by

$$I : N_{d-1} \to N_d, \quad [f_0, .., f_n] \mapsto [w_1 f_0, .., w_1 f_n]$$

$$\bar{\ } : N_d \to N_d, \quad [f_0(w_0, w_1), .., f_n(w_0, w_1)] \mapsto [f_0(w_1, w_0), .., f_n(w_1, w_0)].$$

The second map induces on equivariant cohomology $\mathcal{R}^{-1}H^*_G(N_d)$, $\bar{\kappa} = \kappa - d\alpha$, $\bar{\alpha} = -\alpha$, $\bar{\lambda}_i = \lambda_i$. In particular any $x \in \mathcal{R}$ has the form

$x = x_- + x_+$ with $\bar{x}_\pm = \pm x_\pm$. We also extend $\bar{\ }$ to the power series ring $\mathcal{R}[[e^t]]$ by leaving t invariant.

Composing a chain of d I's, we get a canonical inclusion $N_0 = \mathbf{P}^n \xrightarrow{I_d} N_d$. Note that the image of the fixed point p_i is $p_{i,0}$. For $\omega \in \mathcal{R}^{-1}H_G^*(N_d)$, we shall denote by $I_d^*(\omega) \in \mathcal{R}^{-1}H_G^*(N_0)$ the restriction of ω to N_0. Since S^1 acts trivially on $\mathbf{P}^n$, we can write

$$H_G^*(N_0) = H_T^*(\mathbf{P}^n)[\alpha].$$

In particular note that $H_T^*(\mathbf{P}^n)$ is invariant under $\bar{\ }$, and that $I_d^*(\kappa) = p$.

Obviously the set of classes $\omega \in H_T^*(\mathbf{P}^n)$ with $\iota_{p_i}^*(\omega) \neq 0$ for all i, is closed under multiplication. We localize the ring $H_T^*(\mathbf{P}^n)$ by allowing to invert such elements ω. We denote the resulting ring by $H_T^*(\mathbf{P}^n)^{-1}$.

Definition 2.2. *(Notations) The degree in α of a class $\omega \in H_T^*(\mathbf{P}^n)^{-1}[\alpha]$ will be denoted by $deg_\alpha\, \omega$. A class $\Omega \in H_T^*(\mathbf{P}^n)^{-1}$ with $\iota_{p_i}^*(\Omega) \neq 0$ for all i will be called invertible. Throughout this paper, Ω will denote a fixed but arbitrary invertible class. $\mathcal{S}$ denotes the set of sequences of cohomology classes Q : $Q_d \in \mathcal{R}^{-1}H_G^*(N_d)$, $d = 1, 2, \ldots$*

2.2. Eulerity

Definition 2.3. *We call a sequence Q : $Q_d \in \mathcal{R}H_G^*(N_d)$, $d = 1, 2, \ldots$, an Ω-Euler data if for all d, and $r = 0, .., d$, $i = 0, .., n$,*

$$(*) \quad \iota_{p_i}^*(\Omega)\ \iota_{p_{i,r}}^*(Q_d) = \overline{\iota_{p_{i,0}}^*(Q_r)}\ \iota_{p_{i,0}}^*(Q_{d-r}),$$

where $Q_0 := \Omega$. We denote by $\mathcal{A}^\Omega$ the set of Ω-Euler data.

When dealing only with one fixed class Ω at a time, we shall say Euler rather than Ω-Euler, and shall write $\mathcal{A}$ for the set of Euler data.

More explicitly, condition (*) can be written as $\Omega(\lambda_i)Q_d(\lambda_i + r\alpha) = \overline{Q_r(\lambda_i)}Q_{d-r}(\lambda_i)$. Applying this at $r = d$, we find that

$$Q_d(\lambda_i + d\alpha) = \overline{Q_d(\lambda_i)}. \tag{2.1}$$

Putting $\alpha = (\lambda_j - \lambda_i)/d$, we see that $Q_d(\lambda_j)$ at $\alpha = (\lambda_j - \lambda_i)/d$ coincides with $Q_d(\lambda_i)$ at $\alpha = (\lambda_i - \lambda_j)/d$. Applying both (*) and (2.1) at $\alpha = (\lambda_j - \lambda_i)/r$ (hence $\lambda_j = \lambda_i + r\alpha$), we get

$$\Omega(\lambda_i)Q_d(\lambda_j) = Q_r(\lambda_j)Q_{d-r}(\lambda_i) \ \ at\ \alpha = (\lambda_j - \lambda_i)/r.$$

Lemma 2.4. *(Reciprocity Lemma) If Q is an Euler data, then for $i, j = 0, .., n$, $r = 0, 1, .., d$, $d = 0, 1, 2, ..$, we have*
(i) $Q_d(\lambda_i + d\alpha) = \overline{Q_d(\lambda_i)}$.
(ii) $Q_d(\lambda_j)$ at $\alpha = (\lambda_j - \lambda_i)/d$ coincides with $Q_d(\lambda_i)$ at $\alpha = (\lambda_i - \lambda_j)/d$ for $d \neq 0$.
(iii) $\Omega(\lambda_i)Q_d(\lambda_j) = Q_r(\lambda_j)Q_{d-r}(\lambda_i)$ at $\alpha = (\lambda_j - \lambda_i)/r$ for $r \neq 0$.

Example 1. Let l be a positive integer. Put

$$P: \quad P_d = \prod_{m=0}^{ld} (l\kappa - m\alpha) \in H_G^*(N_d).$$

That this is an lp-Euler data follows from the identity:

$$l\lambda_i \prod_{m=0}^{ld} (l\lambda_i + (r-m)\alpha) = \prod_{m=0}^{lr} (l\lambda_i + m\alpha) \times \prod_{m=0}^{l(d-r)} (l\lambda_i - m\alpha).$$

This will arise naturally in the problem of computing the equivariant Euler classes of the obstruction bundles induced by $\mathcal{O}(l) \to \mathbf{P}^n$. (See below.)

Example 2. Put $\Omega = p^{-2}$, and

$$P: \quad P_d = \prod_{m=1}^{d-1} (\kappa - m\alpha)^2 \in H_G^*(N_d).$$

This Euler data will arise in the problem of computing the so-called multiple cover contributions, ie. the GW invariants for $\mathbf{P}^1$.

Example 3. Put $\Omega = (-3p)^{-1}$, and

$$P: \quad P_d = \prod_{m=1}^{3d-1} (-3\kappa + m\alpha) \in H_G^*(N_d).$$

This Euler data will arise in the problem of computing the equivariant Euler classes of the obstruction bundles induced by the canonical bundle of $\mathbf{P}^2$.

Example 4. Put $\Omega = -1$, and

$$P: \quad P_d = \prod_{m=0}^{2d} (2\kappa - m\alpha) \times \prod_{m=1}^{2d-1} (-2\kappa + m\alpha) \in H_G^*(N_d).$$

This Euler data will arise in the problem of computing the equivariant Euler classes of the obstruction bundles induced by $\mathcal{O}(2)\oplus\mathcal{O}(-2)$ on $\mathbf{P}^3$.

Example 5. It is easy to show that if Q is Ω-Euler and $a\in\mathbf{Q}(\lambda)$ is any nonzero element, then the data Q' : $Q'_d = a\cdot Q_d$ is $a\Omega$-Euler. Similarly, the data Q' : $Q'_d=(-1)^{d+1}Q_d$ is also $-\Omega$-Euler.

Example 6. We observe that the set of Euler data is a monoid, ie. it is closed under the product $Q_dQ'_d$, and has the unit given by $Q_d=1$ for all d. Hence the product of an Ω-Euler with an Ω'-Euler data is an $\Omega\Omega'$-Euler data. In the geometrically setting, this multiplicative property sometimes corresponds to taking intersection of two suitable projective manifolds. In this case, the class $\Omega\in H_T^*(\mathbf{P}^n)$ plays the role of the equivariant Thom class of the normal bundle of such a projective manifold.

Example 7. Let $Q_d=\kappa(\kappa-d\alpha)\in H_G^*(N_d)$. Then it is again trivial to check that Q is κ^2-Euler.

Example 8. Let Q be an Ω-Euler data, and Q' be an Ω'-Euler data. Suppose $Q_d/Q'_d\in\mathcal{R}H_G^*(N_d)$ for all $d>0$. Then it is immediate that they form a sequence, denoted by Q/Q', which is Ω/Ω'- Euler. As a special case, let $Q_d=l^2\kappa(\kappa-d\alpha)$ as in Example 7, and let P as in Example 1. Then P/Q is a $(lp)^{-1}$-Euler data. Example 2 is obtained by squaring this. Examples 3 and 4 can also be obtained in a similar way.

Example 9. Introduce a formal variable x. We can extend everything above by adjoining x, ie. by replacing the ground field $\mathbf{Q}$ by $\mathbf{Q}(x)$. For example, it is easy to show that

$$P:\quad P_d=\prod_{m=0}^{ld}(x+l\kappa-m\alpha)\in H_G^*(N_d)(x)$$

satisfies the gluing identity as in Example 1, thus is an Euler data in a more general sense. Such Euler data will appear in the computations of equivariant total Chern classes.

Example 10. Let $M_d^0:=\mathcal{M}_{0,0}((1,d),\mathbf{P}^1\times\mathbf{P}^n)$ be the moduli space of holomorphic maps $\mathbf{P}^1\to\mathbf{P}^1\times\mathbf{P}^n$ of bidegree $(1,d)$. Recall that M_d is the stable map compactification of M_d^0. Each map $f\in M_d^0$ can be represented by $f:[w_0,w_1]\mapsto[w_1,w_0]\times[f_0(w_0,w_1),..,f_n(w_0,w_1)]$, where f_i are degree d homogeneous polynomials. So there is an obvious map

$$\varphi: M_d^0\to N_d,\ f\mapsto[f_0,..,f_n]$$

which is $G=S^1\times T$ equivariant. For convenience we *define* $M_0:=N_0=\mathbf{P}^n$. With a bit of work (see below), it can be shown that the map φ has an equivariant regular extension to $\varphi : M_d \to N_d$. Let $(f, C) \in M_d$. Then C is an arithmetic genus 0 curve of the form $C = C_0 \cup C_1 \cup \cdots \cup C_N$ such that $\pi_1 \circ f : C_0 \tilde{\to} \mathbf{P}^1$, where π_1, π_2 are projections from $\mathbf{P}^1 \times \mathbf{P}^n$ to the first and second factors. Each C_j, $j > 0$, is glued to C_0 at some point $x_j \in C_0$. The map $\pi_2 \circ f : C_j \to \mathbf{P}^n$ is of degree d_j with $\sum_j d_j = d$, and $\pi_1 \circ f : C_j \to \mathbf{P}^1$ is constant map with image $\pi_1 \circ f(x_j) \in \mathbf{P}^1$. If we denote by $[\sigma_0, \cdots, \sigma_n]$ the degree d_0 polynomials representing $\pi_2 \circ f : C_0 \to \mathbf{P}^n$, then $\varphi : (f, C) \mapsto [\sigma_0 g, \cdots, \sigma_n g]$, where $g = \prod_j (a_j w_0 - b_j w_1)^{d_j}$ with $\pi_1 \circ f(x_j) = [a_j, b_j]$. Thus φ collapses all but one component of C.

The idea of using a collapsing map relating two moduli problems is not new. The map φ was known to Tian [43], and a similar map also appeared in [35] in which a collapsing map was used to relate two moduli spaces. The map φ was also used in [18]. Similar maps have also been studied in [24][26].

Let $V = \mathcal{O}(l)$ for $l > 0$, and consider the induced bundle $U_d \to \bar{\mathcal{M}}_{0,0}(d, \mathbf{P}^n)$. Pulling this back via the projection π, we get a bundle $V_d \to M_d$ (see Introduction). Let $e_T(V_d)$ be the equivariant Euler class of V_d, and let $\varphi_! e_T(V_d)$ be its push-forward via φ. The following theorem will be a special case of a general theorem proved in the next subsection.

Theorem 2.5. *The sequence $\varphi_! e_T(V_d) \in H_G^*(N_d)$ above is an lp-Euler data.*

We now return to the map φ. The reader who wishes to skip technical details can safely omit the proof.

Lemma 2.6. *There exists a morphism $\varphi : M_d \to N_d$. Moreover φ is equivariant with respect to the induced action of $S^1 \times T$.*

Proof: The following proof is given by J. Li. Let M_d be the moduli space of stable morphisms $f : C \to \mathbf{P}^1 \times \mathbf{P}^n$ from arithmetic genus 0 curves to $\mathbf{P}^1 \times \mathbf{P}^n$ of bi-degree $(1, d)$, and let N_d be the space of equivalence classes of $(n+1)$-tuples $(f_0, \ldots, f_n)$, where f_i are degree d homogeneous polynomials in two variables, and $(f_0, \ldots, f_n) \sim (f_0', \ldots, f_n')$ if there is a constant $c \neq 0$ such that $f_i = c \cdot f_i'$ for all i. We first define the morphism $\varphi : M_d \to N_d$. For convenience, we let $\mathcal{S}$ be the category of all schemes of finite type (over $\mathbf{C}$) and let

$$\mathcal{F} : \mathcal{S} \longrightarrow (\mathrm{Set})$$

be the the contra-variant functor that send any $S \in \mathcal{S}$ to the set of families of stable morphisms

$$F : \mathcal{X} \longrightarrow \mathbf{P}^1 \times \mathbf{P}^n \times S$$

over S, where $\mathcal{X}$ are families of connected arithmetic genus 0 curves, modulo the obvious equivalence relation. Note that $\mathcal{F}$ is represented by the moduli stack M_d. Hence to define φ it suffices to define a transformation

$$\Psi : \mathcal{F} \longrightarrow \mathrm{Mor}\,(-, N_d).$$

We now define such a transformation. Let $S \in \mathcal{S}$ and let $\xi \in \mathcal{F}(S)$ be represented by $(\mathcal{X}, F)$. We let p_i be the composite of F with the i-th projection of $\mathbf{P}^1 \times \mathbf{P}^n \times S$ and let p_{ij} be the composite of F with the projection from $\mathbf{P}^1 \times \mathbf{P}^n \times S$ to the product of its i-th and j-th components. We consider the sheaf $p_2^*\mathcal{O}_{\mathbf{P}^n}(1)$ on $\mathcal{X}$ and its direct image sheaf

$$\mathcal{L}_\xi = p_{13*}p_2^*\mathcal{O}_{\mathbf{P}^n}(1).$$

We claim that $\mathcal{L}_\xi$ is flat over S. Indeed, by argument in the proof of Theorem 9.9 in [22], it suffices to show that $\pi_{S*}(\mathcal{L}_\xi \otimes \pi_{\mathbf{P}^1}^*\mathcal{O}_{\mathbf{P}^1}(m))$ are locally free sheaves of $\mathcal{O}_S$-modules for $m \gg 0$, where $\pi_{\mathbf{P}^1}$ and π_S are the first and the second projections of $\mathbf{P}^1 \times S$. Clearly, this sheaf is isomorphic to $p_{3*}(p_2^*\mathcal{O}_{\mathbf{P}^n}(1) \otimes p_1^*\mathcal{O}_{\mathbf{P}^1}(m))$, which is locally free because

$$R^i p_{3*}(p_2^*\mathcal{O}_{\mathbf{P}^n}(1) \otimes p_1^*\mathcal{O}_{\mathbf{P}^1}(m)) = 0$$

for $i > 0$ and $m \gg 0$. For the same reasoning, the sheaves $\mathcal{L}_\xi$ satisfy the following base change property: let $\rho : T \to S$ be any base change and let $\rho^*(\xi) \in \mathcal{F}(T)$ be the pull back of ξ. Then there is a canonical isomorphism of sheaves of $\mathcal{O}_T$-modules

$$\mathcal{L}_{\rho^*(\xi)} \cong (\mathbf{1}_{\mathbf{P}^1} \times \rho)^*\mathcal{L}_\xi. \tag{2.2}$$

Since $\mathcal{L}_\xi$ is flat over S, we can define the determinant line bundle of $\mathcal{L}_\xi$, denoted by $\det(\mathcal{L}_\xi)$ [28]. The sheaf $\det(\mathcal{L}_\xi)$ is an invertible sheaf over $\mathbf{P}^1 \times S$. Using the Riemann-Roch theorem, one computes that its degree along fibers over S are d. Further, because $\mathcal{L}_\xi$ has rank one, there is a canonical homomorphism

$$\mathcal{L}_\xi \longrightarrow \det(\mathcal{L}_\xi), \tag{2.3}$$

so that its kernel is the torsion subsheaf of $\mathcal{L}_\xi$, denoted by $\mathrm{Tor}\,(\mathcal{L}_\xi)$. Let $w_0, \ldots, w_n$ be the homogeneous coordinate of $\mathbf{P}^n$ chosen before. $w_0, \ldots, w_n$ form a basis of $H^0(\mathbf{P}^n, \mathcal{O}_{\mathbf{P}^n}(1))$. Then their pull backs provide a collection of canonical sections of $\mathcal{L}_\xi$, and hence a collection of canonical sections

$$\sigma_{\xi,0}, \ldots, \sigma_{\xi,n} \in H^0(S, \pi_{S*}\det(\mathcal{L}_\xi)).$$

based on (2.3). Then after fixing an isomorphism

$$\det(\mathcal{L}_\xi) \cong \pi_S^*\mathcal{M} \otimes \pi_{\mathbf{P}^1}^* \mathcal{O}_{\mathbf{P}^1}(d) \tag{2.4}$$

for some invertible sheaf $\mathcal{M}$ of $\mathcal{O}_S$-modules, we obtain a section of

$$\pi_{S*}(\pi_{\mathbf{P}^1}^* \mathcal{O}_{\mathbf{P}^1}(d)) \otimes_{\mathcal{O}_S} \mathcal{M} \equiv H^0_{\mathbf{P}^1}(\mathcal{O}_{\mathbf{P}^1}(d)) \otimes_{\mathbf{C}} \mathcal{M}.$$

Finally, we let w_0, w_1 be the homogeneous coordinate of $\mathbf{P}^1$. Then the space $H^0_{\mathbf{P}^1}(\mathcal{O}_{\mathbf{P}^1}(d))$ is the space of degree d homogeneous polynomials in variables w_0 and w_1. This way, we obtain a morphism

$$\Psi(S) : S \longrightarrow N_d$$

that is independent of the isomorphisms (2.4). It follows from the base change property (2.2) that the collection $\Psi(S)$ defines a transformation

$$\Psi : \mathcal{F} \longrightarrow \mathrm{Mor}\,(-, N_d),$$

thus defines a morphism φ as desired.

It remains to check that for any $w \in S$, the sections

$$\sigma_{\xi,0}(w), \ldots, \sigma_{\xi,n}(w) \in H^0(\mathbf{P}^1, \det(\mathcal{L}_\xi) \otimes_{\mathcal{O}_S} k_w)$$

has the described vanishing property. Because of the base change property of $\mathcal{L}_\xi$, it suffices to check the statement when S is a point and $\xi \in \mathcal{F}_d(S)$ is the stable map $f : C \to \mathbf{P}^1 \times \mathbf{P}^n$. Let $x_1, \ldots, x_N$ be the set of points in $\mathbf{P}^1$ so that $p_1 : C \to \mathbf{P}^1$, where $p_1 = \pi_{\mathbf{P}^1} \circ f$, is not flat over these points. Let C_i be $p_1^{-1}(x_i)$ and let m_i be the degree of $f([C_i]) \in H_2(\mathbf{P}^n)$. Then $\mathcal{L}_\xi = p_{1*}p_2^*\mathcal{O}_{\mathbf{P}^n}(1)$ is locally free away from $x_1, \ldots, x_N$ and has torsion of length m_i at x_i. Then $\mathcal{L}_\xi/\mathrm{Tor}\,(\mathcal{L}_\xi)$ is locally free of degree $k - \sum m_i$. It is direct to check that the canonical inclusion

$$\mathcal{L}_\xi/\mathrm{Tor}\,(\mathcal{L}_\xi) \longrightarrow \det(\mathcal{L}_\xi) \cong \mathcal{O}_{\mathbf{P}^1}(d)$$

has cokernel supported on the union of $x_1, \ldots, x_N$ whose length at x_i is exactly m_i. The statement about the vanishing of $\sigma_{\xi,0}(w), \ldots, \sigma_{\xi,n}(w)$ follows immediately.

The fact that $\varphi : M_d \to N_d$ is $S^1 \times T$-equivariant as stated follows immediately from the fact that φ is induced by the transformation Ψ of functors. This completes the proof. □

2.3. Concavex bundles

Definition 2.7. *We call a T-equivariant vector bundle $V \to \mathbf{P}^n$ convex (resp. concave) if the T-equivariant Euler class $e_T(V)$ is invertible and if $H^1(C, f^*V) = 0$ (resp. $H^0(C, f^*V) = 0$) for every 0-pointed genus 0 stable map $f : C \to \mathbf{P}^n$. We call V a concavex bundle if it is a direct sum of a convex and a concave bundles. We denote by $V^{\pm}$ the convex and concave summands of V. By convention, we consider the zero bundle to be both convex and concave so that concavexity includes both convexity and concavity.*

The convexity of a bundle is analogous to the notion of convexity of a projective manifold introduced by Behrend-Manin.

For example $\mathcal{O}(l) \to \mathbf{P}^n$ is convex if $l > 0$, and concave if $l < 0$. Given any concavex bundle $V \to \mathbf{P}^n$, we have a sequence of induced bundles

$$U_d \to \bar{\mathcal{M}}_{0,0}(d, \mathbf{P}^n)$$

whose fiber at (f, C) is the space $H^0(C, f^*V^+) \oplus H^1(C, f^*V^-)$. Pulling back U_d via the contracting map $\pi : M_d \to \bar{\mathcal{M}}_{0,0}(d, \mathbf{P}^n)$, we get a sequence of bundles

$$V_d := \pi^* U_d \to M_d.$$

We denote by $e_T(V_d)$ the equivariant Euler class of V_d. We also introduce the notations:

$$\Omega^V := \frac{e_T(V^+)}{e_T(V^-)}$$

$$Q_d := \varphi_! e_T(V_d), \quad Q_0 := \Omega^V.$$

By *convention*, if V is the zero bundle, we set $e_T(V_d) = 1$, $e_T(V) = 1$, $\Omega^V = 1$.

Theorem 2.8. *The sequence $\varphi_! e_T(V_d) \in H_G^*(N_d)$ is an Ω^V-Euler data.*

Proof: We first discuss some preliminaries. Let M and N be two compact smooth manifolds with the action of a torus T, and $\varphi : M \to N$ be an equivariant map. Let F be one component of the fixed submanifold in N and i_F be the inclusion map F in N. Let $\phi_F = i_{F!}(1) \in H_T^*(N)$ denote the equivariant Thom class of the normal bundle of F in N. We then have, for any $\omega \in H_T^*(M)$

$$\int_M \omega\varphi^*(\phi_F) = \int_N \varphi_!(\omega)\phi_F = \int_F i_F^*(\varphi_!(\omega)).$$

On the other hand, let $\{P\}$ be the components of the fixed submanifold contained in $\varphi^{-1}(F)$. By the Atiyah-Bott [7][3] localization formula on M, we get

$$\sum_P \int_P i_P^*(\omega\varphi^*(\phi_F))/e_T(P/M) = \int_F i_F^*(\varphi_!(\omega)).$$

Here $e_T(P/M)$ denotes the equivariant Euler class of the normal bundle of P in M. The reason is that the contribution of the fixed point sets not contained in $\varphi^{-1}(F)$ is clearly zero. Actually if Q is a component not contained inside $\varphi^{-1}(F)$, its contribution to the localization is given by

$$\int_Q i_Q^*(\omega\varphi^*(\phi_F))/e_T(Q/M).$$

But by the naturality of the pull-backs, we have

$$i_Q^*\varphi^*(\phi_F) = \varphi_0^* i_E^*(\phi_F) = 0$$

where $E = \varphi(Q)$ is a fixed submanifold in N and φ_0 denotes the restriction of φ to Q. Note that if F is an isolated point, then $i_P^*\varphi^*(\phi_F)$ can be pulled out of the integral.

The above formula will be applied to the collapsing map $\varphi : M_d \to N_d$. All manifolds involved here are at worst orbifolds with finite quotient singularities, so the localization formula remains valid without any change as long as we consider the corresponding integrals in the orbifold sense.

We consider the S^1-action on the $\mathbf{P}^1$ factor in $\mathbf{P}^1 \times \mathbf{P}^n$ with weights $\alpha, 0$. Combining with the natural T-action on $\mathbf{P}^n$, we get the naturally induced $G = S^1 \times T$-actions on M_d and N_d, with respect to which the collapsing map φ is equivariant. As described in section 2.1, the G-fixed points in N_d are all of the form

$$p_{ir} = [0, \cdots, 0, w_0^r w_1^{d-r}, 0, \cdots, 0]$$

in which the only nonzero term is in the i-th position.

For each $r > 0$, let $\{F_r\} \subset \bar{\mathcal{M}}_{0,1}(r, \mathbf{P}^n)$ denote the T-fixed point components in $\bar{\mathcal{M}}_{0,1}(r, \mathbf{P}^n)$ with the marked point mapped onto the fixed point p_i in $\mathbf{P}^n$. Let $N(F_r) = N_{F_r/\bar{\mathcal{M}}_{0,1}(r,\mathbf{P}^n)}$ denote the normal bundle of F_r in $\bar{\mathcal{M}}_{0,1}(r, \mathbf{P}^n)$.

Let π_1, π_2 be the projections from $\mathbf{P}^1 \times \mathbf{P}^n$ onto the first and second factors. From the construction of φ, we see that the G-invariant submanifold that is mapped to p_{ir} consists of the following degree $(1, d)$ stable maps $f : C \to \mathbf{P}^1 \times \mathbf{P}^n$ with $C = C_1 \cup C_0 \cup C_2$. Here $C_0 \simeq \mathbf{P}^1$ and

$$\pi_2 \circ f(C_0) = [0, \cdots, 0, 1, 0 \cdots, 0] = p_i \in \mathbf{P}^n$$

where 1 is at the i-th position. The map $\pi_1 \circ f : C_0 \to \mathbf{P}^1$ is an isomorphism and maps $x_1 = C_0 \cap C_1$ and $x_2 = C_0 \cap C_2$ to 0 and ∞ respectively. Actually

$$\pi_1 \circ f(C_1) = 0, \quad \pi_1 \circ f(C_2) = \infty \ \ in \ \mathbf{P}^1,$$

ie. the curves C_1 and C_2 are respectively mapped to the points 0 and ∞ of $\mathbf{P}^1$.

The maps $\pi_2 \circ f$ restricted to C_j for $j = 1, 2$ are stable maps in $\bar{\mathcal{M}}_{0,1}(r, \mathbf{P}^n)$ and $\bar{\mathcal{M}}_{0,1}(d-r, \mathbf{P}^n)$ respectively. We consider $F_r \times F_{d-r}$ as a G-fixed submanifold of M_d by gluing each pair to C_0 at x_1 and x_2 respectively as above. It is easy to see that $\{F_r \times F_{d-r}\}$ are the G-fixed point sets in M_d whose image under φ is the fixed point p_{ir}.

We first consider a convex bundle V on $\mathbf{P}^n$, and the case $r \neq 0, d$. Then we have

$$Q_d(\lambda_i + r\alpha) = \int_{N_d} \phi_{p_{ir}} Q_d = \int_{M_d} \varphi^*(\phi_{p_{ir}})\, e_T(V_d). \tag{2.5}$$

Here $\phi_{p_{ir}}$ denote the equivariant Thom class of the G-fixed point p_{ir} in N_d. We will apply the localization formula to compute the right hand side of (2.5). First we need to know the normal bundle of the fixed points, which is, in the equivariant K-group of $F_r \times F_{d-r}$ [34][20],

$$\begin{aligned} N_{F_r \times F_{d-r}/M_d} &= N(F_r) + N(F_{d-r}) + [H^0(C_0, (\pi_1 \circ f)^* T\mathbf{P}^1)] \\ &\quad + [L_r \otimes T_{x_1} C_0] + [L_{d-r} \otimes T_{x_2} C_0] - [T_{p_i} P^n] - [A_{C_0}]. \end{aligned}$$

Here L_r denotes the line bundle on $\bar{\mathcal{M}}_{0,1}(r, \mathbf{P}^n)$ whose fiber at (f_1, C_1, x_1) is the tangent line at x_1. Likewise for L_{d-r}. Their contributions correspond to the deformation of the nodal points x_1 and x_2. The term $H^0(C_0, (\pi_1 \circ f)^* T\mathbf{P}^1)$ corresponds to the deformation of f restricted to C_0. The term $[A_{C_0}]$ is the bundle representing the infinitesimal automorphism of C_0 fixing the two points x_1, x_2. The term $-[T_{p_i} P^n]$ comes from gluing F_r and F_{d-r} onto C_0 and the property that $\pi_2 \circ f(C_0) = p_i$.

This gives the following formula for the corresponding equivariant Euler classes:

$$\begin{aligned} e_T(F_r \times F_{d-r}/M_d) &= e_T(N(F_r))e_T(N(F_{d-r}))e_T(L_r \otimes T_{x_1}C_0) \\ &\quad \times e_T(L_{d-r} \otimes T_{x_2}C_0)e_T(T_{p_i}\mathbf{P}^n)^{-1} \\ &\quad \times e_T(H^0(C_0,(\pi_1 \circ f)^*T\mathbf{P}^1))e_T^{-1}(A_{C_0}). \end{aligned}$$

Each term in this formula can be explicitly calculated. We clearly have $e_T(T_{p_i}\mathbf{P}^n) = \prod_{j\neq i}(\lambda_i - \lambda_j)$; the weights of $T_{x_1}C_0$ and $T_{x_2}C_0$ are α and $-\alpha$ respectively, therefore

$$e_T(L_r \otimes T_{x_1}C_0) = \alpha + c_1(L_r), \quad e_T(L_{d-r} \otimes T_{x_2}C_0) = -\alpha + c_1(L_{d-r})$$

where $c_1(L_r)$, $c_1(L_{d-r})$ are the restriction to F_r and F_{d-r} of the equivariant Chern classes of the line bundles L_r and L_{d-r} with respect to the induced T actions on $\bar{\mathcal{M}}_{0,1}(r,\mathbf{P}^n)$ and $\bar{\mathcal{M}}_{0,1}(d-r,\mathbf{P}^n)$. To compute $e_T(H^0(C_0,(\pi_1 \circ f)^*T\mathbf{P}^1))$ and $e_T(A_{C_0})$, first note that we have the standard exact sequence

$$0 \to O \to \mathcal{O}(1) \otimes \mathbf{C}^2 \to T\mathbf{P}^1 \to 0,$$

with O being the trivial bundle. From this we get

$$0 \to O \to H^0(C_0,\mathcal{O}(1)) \otimes \mathbf{C}^2 \to H^0(C_0,(\pi_1 \circ f)^*(T\mathbf{P}^1)) \to 0.$$

The weights of $H^0(C_0,\mathcal{O}(1))$ are $\alpha, 0$, the weights of $\mathbf{C}^2$ are $-\alpha, 0$ and the weight of O is 0. Therefore one finds that the weights of $H^0(C_0,(\pi_1 \circ f)^*(T\mathbf{P}^1))$ are α, $-\alpha$, 0.

For $[A_{C_0}]$, we have the exact sequence

$$0 \to A_{C_0} \to H^0(C_0,(\pi_1 \circ f)^*(T\mathbf{P}^1)) \to T_{x_1}C_0 \oplus T_{x_2}C_0 \to 0.$$

The weights of $T_{x_1}C_0$ and $T_{x_2}C_0$ are α and $-\alpha$ respectively. So $[A_{C_0}]$ contributes a 0 weight space which cancels with the 0 weight space of $[H^0(C_0,(\pi_1 \circ f)^*(T\mathbf{P}^1))]$. Therefore we will ignore the zero weights in our formulas and write as

$$e_T(H^0(C_0,(\pi_1 \circ f)^*(T\mathbf{P}^1))e_T^{-1}(A_{C_0}) = -\alpha^2.$$

When V_d is restricted to $F_r \times F_{d-r}$ considered as a fixed point set of M_d as before, we have the exact sequence:

$$0 \to V_d \to V_r|_{F_r} \oplus V_{d-r}|_{F_{d-r}} \to V|_{p_i} \to 0.$$

Note that $V_r|_{F_r}$ and $V_{d-r}|_{F_{d-r}}$ is the same as $\rho^*U_r|_{F_r}$ and $\rho^*U_{d-r}|_{F_{d-r}}$ which respectively are the restrictions to F_r and F_{d-r} of the pull-backs to $\bar{\mathcal{M}}_{0,1}(r, \mathbf{P}^n)$ and $\bar{\mathcal{M}}_{0,1}(d-r, \mathbf{P}^n)$ of the corresponding bundles on $\bar{\mathcal{M}}_{0,0}(r, \mathbf{P}^n)$ and $\bar{\mathcal{M}}_{0,0}(d-r, \mathbf{P}^n)$. Here $V|_{p_i}$ denotes the fiber of V at $p_i \in \mathbf{P}^n$.

Here comes the important point. The multiplicativity of equivariant Euler classes gives us

$$\Omega^V(\lambda_i) \cdot e_T(V_d) = e_T(V_r) \cdot e_T(V_{d-r}) = \rho^* e_T(U_r) \cdot \rho^* e_T(U_{d-r}),$$

when restricted to $F_r \times F_{d-r} \subset M_d$. Here $\rho : \bar{\mathcal{M}}_{0,1}(d, \mathbf{P}^n) \to \bar{\mathcal{M}}_{0,0}(d, \mathbf{P}^n)$ is the forgetting map (same notation for all d). Note that the above equality is just the pull-back via π from $\bar{\mathcal{M}}_{0,0}(d, \mathbf{P}^n)$ of the gluing identity discussed in the Introduction.

For the case of $r = 0$ or d, there is only one of the curves C_1 or C_2, ie. C is of the form $C_0 \cup C_2$ or $C_1 \cup C_0$. In this case we identify F_d as the fixed point set in M_d by gluing its marked point to C_0 at x_1 or x_2. The normal bundle in these two cases are respectively given by

$$N_{F_d/M_d} = N(F_d) + [H^0(C_0, (\pi_1 \circ f)^* T\mathbf{P}^1)] + [L_d \otimes T_{x_j} C_0] - [A_{C_0}]$$

in the K-group of F_d. Here L_d is the restriction to F_d of the line bundle on $\bar{\mathcal{M}}_{0,1}(d, \mathbf{P}^n)$ whose fiber is the tangent line at the marked point. For simplicity we write L as L_d in the following. For $j = 1, 2$, $T_{x_j} C_0$ is the tangent line of C_0 at the corresponding marked point x_j. In these two cases, one easily shows in the same way as above that the term $e_T(A_{C_0})$, except the 0 weight, contributes one nonzero weight $-\alpha$ or α respectively. Its 0 weight space still cancels with the 0 weight space of $[H^0(C_0, (\pi_1 \circ f)^* T\mathbf{P}^1)]$.

By putting all of the above computations together and combining with (2.5), we get, for $r \neq 0$, d,

$$\begin{aligned}\Omega^V(\lambda_i)Q_d(\lambda_i + r\alpha) &= \Omega^V(\lambda_i)\int_{M_d} \varphi^*(\phi_{p_{ir}})\, e_T(V_d) \\ &= -\alpha^{-2}\prod_{j\neq i}(\lambda_i - \lambda_j)e_T(p_{ir}/N_d) \\ &\times \sum_{F_r}\int_{F_r}\frac{\rho^* e_T(U_r)}{e_T(N(F_r))(\alpha + c_1(L_r))} \\ &\times \sum_{F_{d-r}}\int_{F_{d-r}}\frac{\rho^* e_T(U_{d-r})}{e_T(N(F_{d-r}))(-\alpha + c_1(L_{d-r}))}.\end{aligned} \tag{2.6}$$

Here $e_T(p_{ir}/N_d) = \iota^*_{p_{ir}}(\phi_{p_{ir}})$. Note that $\varphi^*(\phi_{p_{ir}})$ restricted to $F_r \times F_{d-r}$ is the same as $e_T(p_{ir}/N_d)$ which is a polynomial only in α and λ as given below. Similarly for $r = d$, we have

$$Q_d(\lambda_i + d\alpha) = \alpha^{-1}e_T(p_{id}/N_d)\sum_{F_d}\int_{F_d}\frac{\rho^* e_T(U_d)}{e_T(N(F_d))(\alpha + c_1(L))} \tag{2.7}$$

and for $r = 0$

$$Q_d(\lambda_i) = -\alpha^{-1}e_T(p_{i0}/N_d)\sum_{F_d}\int_{F_d}\frac{\rho^* e_T(U_d)}{e_T(N(F_d))(-\alpha + c_1(L))}. \tag{2.8}$$

We can easily compute $e_T(p_{ir}/N_d)$:

$$e_T(p_{ir}/N_d) = \prod_{j=0}^{n}\prod_{m=0}^{d}{}_{(j,m)\neq(i,r)}(\lambda_i - \lambda_j + (r - m)\alpha).$$

For $r = 0$ and d we have

$$\begin{aligned}e_T(p_{i0}/N_d) &= \prod_{j=0}^{n}\prod_{m=0}^{d}{}_{(j,m)\neq(i,0)}(\lambda_i - \lambda_j - m\alpha) \\ e_T(p_{id}/N_d) &= \prod_{j=0}^{n}\prod_{m=0}^{d}{}_{(j,m)\neq(i,d)}(\lambda_i - \lambda_j + (d - m)\alpha) \\ &= \prod_{j=0}^{n}\prod_{m=0}^{d}{}_{(j,m)\neq(i,0)}(\lambda_i - \lambda_j + m\alpha).\end{aligned}$$

The last two identities together with (2.7), (2.8) clearly gives us

$$Q_d(\lambda_i + d\alpha) = \overline{Q_d(\lambda_i)} \tag{2.9}$$

where $\bar{\alpha} = -\alpha$, $\bar{\lambda}_i = \lambda_i$. Finally our asserted quadratic relation:

$$\Omega^V(\lambda_i) Q_d(\lambda_i + r\alpha) = \overline{Q_r(\lambda_i)} Q_{d-r}(\lambda_i) \tag{2.10}$$

follows from (2.6), (2.7), (2.8), (2.9), and the following elementary identity:

$$\begin{aligned}
&\prod_{j=0_{j\neq i}}^{n} (\lambda_i - \lambda_j) \times \prod_{j=0}^{n} \prod_{m=0_{(j,m)\neq(i,r)}}^{d} (\lambda_i - \lambda_j + (r-m)\alpha) \\
&= \prod_{j=0}^{n} \prod_{m=0_{(j,m)\neq(i,0)}}^{r} (\lambda_i - \lambda_j + m\alpha) \\
&\quad \times \prod_{j=0}^{n} \prod_{m=0_{(j,m)\neq(i,0)}}^{d-r} (\lambda_i - \lambda_j - m\alpha)
\end{aligned}$$

Note that the last identity is just the interesting identity

$$e_T(T_{p_i}\mathbf{P}^n) \cdot e_T(p_{ir}/N_d) = e_T(p_{ir}/N_r) \cdot e_T(p_{i0}/N_{d-r}).$$

When V is concave, the fiber at $(f, C) \in \bar{\mathcal{M}}_{0,0}(d, \mathbf{P}^n)$ of the bundle U_d is $H^1(C, f^*V)$, and we need only one change in the above argument. The gluing exact sequence in the concave case is

$$0 \to V|_{p_i} \to V_d \to V_r|_{F_r} \oplus V_{d-r}|_{F_{d-r}} \to 0$$

instead. Therefore the gluing identity for equivariant Euler classes becomes

$$e_T(V_d) = e_T(V_r) \cdot e_T(V_{d-r})\ \iota_{p_i}^* e_T(V).$$

Since $\Omega^V = 1/e_T(V)$ for concave V, the quadratic relation (2.10) remains valid in this case.

The case when V is a direct sum of a convex and a concave bundle is also similar. $\square$

2.4. Linked Euler data

Definition 2.9. *Two sequences $P, Q \in \mathcal{S}$ are said to be linked if $\iota^*_{p_{i,0}}(P_d - Q_d) \in \mathcal{R}^{-1}$ vanish at $\alpha = (\lambda_i - \lambda_j)/d$ for all $j \neq i$, $d > 0$.*

Let V be a T-equivariant concavex bundle on $\mathbf{P}^n$, and $C \cong \mathbf{P}^1$ be any T-invariant line in $\mathbf{P}^n$. By Grothendieck's principle, we have the form

$$V|_C = \oplus_{a=1}^{N^+} \mathcal{O}(l_a) \oplus \oplus_{b=1}^{N^-} \mathcal{O}(-k_b)$$

for some positive integers l_a, k_b. (0 cannot occurs because $e_T(V)$ is invertible, by definitiion.) Assume that $\{l_a\}$ and $\{k_b\}$ are independent of C. This is the case, for example, if V is uniform [41]. *We call the numbers $(l_1, .., l_{N^+}; k_1, .., k_{N^-})$ the splitting type of V.* With this notations, we have

Theorem 2.10. *Let $Q_d = \varphi_! e_T(V_d)$ as before. At $\alpha = (\lambda_j - \lambda_i)/d$, $i \neq j$, we have*

$$\iota^*_{p_{j,0}}(Q_d) = \prod_a \prod_{m=0}^{l_a d} (l_a \lambda_j - m(\lambda_j - \lambda_i)/d) \times \prod_b \prod_{m=1}^{k_b d - 1} (-k_a \lambda_j + m(\lambda_j - \lambda_i)/d).$$

In particular the Euler data Q is linked to

$$P: \quad P_d = \prod_a \prod_{m=0}^{l_a d} (l_a \kappa - m\alpha) \times \prod_b \prod_{m=1}^{k_b d - 1} (-k_b \kappa + m\alpha).$$

Proof: Since we shall evaluate the class $\iota^*_{p_{j,0}}(Q_d)$ at $\alpha = (\lambda_j - \lambda_i)/d$, we introduce the notation $Q_d(\kappa, \alpha) = Q_d$, and denote the value of the class above by $Q_d(\lambda_j, (\lambda_j - \lambda_i)/d)$. First consider the case $V = \mathcal{O}(l)$ on $\mathbf{P}^n$. We consider a smooth point in $(f, C) \in M_d$ with $C = \mathbf{P}^1$ and in coordinates

$$f : C \to \mathbf{P}^1 \times \mathbf{P}^n, \quad [w_0, w_1] \mapsto [w_1, w_0] \times [0, \cdots, w_0^d, \cdots, w_1^d, \cdots, 0]$$

where in the last term, w_0^d is in the i-th position, w_1^d is in the j-th position, and all of the other components are 0. The image of (f, C) in N_d under φ is the smooth point

$$P_{ij} = [0, \cdots, w_0^d, \cdots, w_1^d, \cdots, 0].$$

It is easy to see that, if the weight of S^1 in the group $G = S^1 \times T$ is $\alpha = (\lambda_j - \lambda_i)/d$, then P_{ij} is fixed by the action of the subgroup of G with $\alpha = (\lambda_j - \lambda_i)/d$. So (f, C) is a smooth point in M_d fixed by the subgroup in $G = S^1 \times T$ with $\alpha = (\lambda_j - \lambda_i)/d$. The class $Q_d(\kappa, \alpha)$ restricted to P_{ij} is just $Q_d(\lambda_j, (\lambda_j - \lambda_i)/d)$.

At the points $(f, C) \in M_d$ and $P_{ij} \in N_d$, the map φ is a canonical identification. From definition, $Q_d(\kappa, \alpha)$ restricted to P_{ij} is the same as $e_T(V_d)$ restricted to $\varphi^{-1}(P_{ij}) = (f, C) \in M_d$, which by definition, is the same as $e_T(U_d)$ restricted to the T-fixed point $(\pi_2 \circ f, C)$ in $\bar{\mathcal{M}}_{0,0}(d, \mathbf{P}^n)$. Note that $(\pi_2 \circ f, C)$ is the degree d cover of the T-invariant line joining p_i and p_j in $\mathbf{P}^n$. Explicitly

$$\pi_2 \circ f : \ [w_0, w_1] \to [0, \cdots, w_0^d, \cdots, w_1^d, \cdots 0].$$

So the induced action of T on $C = \mathbf{P}^1$ has weights $\{\lambda_i/d, \lambda_j/d\}$.

Now let us compute $e_T(U_d)$ restricted to $(\pi_2 \circ f, C)$. When restricted $(\pi_2 \circ f, C) \in \bar{\mathcal{M}}_{0,0}(d, \mathbf{P}^n)$ the fiber of U_d is just the section space $H^0(C, (\pi_2 \circ f)^*\mathcal{O}(l)) = H^0(C, \mathcal{O}(ld))$. It has an explicit basis $\{w_0^k w_1^{ld-k}\}$ with $k = 0, \cdots, ld$. Since the T-weight of $w_0^k w_1^{ld-k}$ is $k\lambda_i/d + (ld - k)\lambda_j/d$, by multiplying them together we get

$$Q_d(\lambda_j, (\lambda_j - \lambda_i)/d) = \prod_{m=0}^{ld} (l\lambda_j - m(\lambda_j - \lambda_i)/d).$$

For a general convex vector bundle V on $\mathbf{P}^n$, the fiber of the induced bundle V_d restricted to $\varphi^{-1}(P_{ij}) = (f, C) \in M_d$ is $H^0(\mathbf{P}^1, (\pi_2 \circ f)^*V)$. By Grothendieck's principle, V restricted to the line spanning p_i, p_j splits into direct sum of line bundles $\{\mathcal{O}(l_a)\}$. Pulling them back to C via the degree d map $\pi_2 \circ f$, we get the direct sum of $\{\mathcal{O}(l_a d)\}$. Since V is convex, each $l_a > 0$. By applying the same computation to each summand, we get

$$Q_d(\lambda_j, (\lambda_j - \lambda_i)/d) = \prod_a \prod_{m=0}^{l_a d} (l_a\lambda_j - m(\lambda_j - \lambda_i)/d).$$

For a concave bundle V, we need only one minor change in the above argument. We leave to the reader as an exercise to check that for $V = \mathcal{O}(-k)$, $k > 0$,

$$Q_d(\lambda_j, (\lambda_j - \lambda_i)/d) = \prod_{m=1}^{kd-1} (-k\lambda_j + m(\lambda_j - \lambda_i)/d),$$

by using either the Atiyah-Bott fixed point formula or by writing down an explicit basis for $H^1(\mathbf{P}^1, \mathcal{O}(-kd))$. So for an arbitrary concave bundle V, we have the form

$$Q_d(\lambda_j, (\lambda_j - \lambda_i)/d) = \prod_b \prod_{m=1}^{k_b d-1} (-k_b\lambda_j + m(\lambda_j - \lambda_i)/d).$$

Similarly for an arbitrary concavex bundle V, we have the form

$$Q_d(\lambda_j, (\lambda_j - \lambda_i)/d) = \prod_a \prod_{m=0}^{l_a d} (l_a\lambda_j - m(\lambda_j - \lambda_i)/d)$$
$$\times \prod_b \prod_{m=1}^{k_b d-1} (-k_b\lambda_j + m(\lambda_j - \lambda_i)/d). \qquad \Box$$

Theorem 2.11. *Suppose P, Q are any linked Ω-Euler data. If*

$$deg_\alpha\ \iota^*_{p_{i,0}}(P_d - Q_d) \leq (n+1)d - 2$$

for all $i = 0, .., n$ and $d = 1, 2, ..,$ then $P = Q$.

Proof: By definition, $P_0 = Q_0 = \Omega$. We will show that $P_d = Q_d$, assuming that

$$P_r = Q_r, \quad r = 0, .., d-1. \tag{2.11}$$

Since the $\mathcal{R}$-valued pairing $pf_d(u \cdot v)$ on $\mathcal{R}H^*_G(N_d)$ is nondegenerate, it suffices to show that

$$L_s := pf_d(\kappa^s \cdot (P_d - Q_d))$$

is zero for all $s = 0, 1, 2, ...$ By the localization formula for pf_d, we get

$$L_s = \sum_{i=0}^{n} \sum_{r=0}^{d} (\lambda_i + r\alpha)^s \frac{\iota^*_{p_{ir}}(P_d - Q_d)}{\prod_{k=0}^{n} \prod_{m=0 (k,m)\neq(i,r)}^{d} (\lambda_i - \lambda_k + (r-m)\alpha)}.$$

Since P is an Euler data, it follows that $\iota^*_{p_{ir}}(P_d)$, for each $r = 1, .., d-1$, is expressible in terms of $P_1, .., P_{d-1}$. Likewise for Q. Thus by the inductive hypothesis (2.11), the sum over r above receives contributions only from the $r = 0, d$ terms. Applying the Reciprocity Lemma (i), we further simplify L_s to

$$L_s = \sum_{i=0}^{n} \left(\frac{\lambda_i^s\ A_i(\alpha)}{\alpha^d} + \frac{(\lambda_i + d\alpha)^s\ A_i(-\alpha)}{(-\alpha)^d} \right)$$
$$A_i(\alpha) := \frac{(-1)^d}{d! \prod_{k\neq i}(\lambda_i - \lambda_k)} \frac{\iota^*_{p_{i,0}}(P_d - Q_d)}{\prod_{k\neq i}\prod_{m=1}^{d}(\lambda_i - \lambda_k - m\alpha)}. \tag{2.12}$$

Since P, Q are linked Euler data, we have

$$\iota^*_{p_{i,0}}(P_d - Q_d) = 0 \tag{2.13}$$

at $\alpha = (\lambda_i - \lambda_k)/d$, $k \neq i$. By the inductive hypothesis (2.11) and the Reciprocity Lemma (iii), (2.13) holds at $\alpha = (\lambda_i - \lambda_k)/m$ for $m = 1, .., d$ as well. This shows that $A_i \in \mathcal{R} = \mathbf{Q}(\lambda)[\alpha]$ for all i. By assumption $deg_\alpha A_i < (n+1)d - 1 - nd = d - 1$. But since $L_s \in \mathcal{R}$ ie. polynomial in α, *for all* s, it follows easily that the A_i must be identically zero. □

2.5. The Lagrange map and mirror transformations

Throughout this subsection, we fix an invertible class Ω and shall denote by $\mathcal{A} = \mathcal{A}^\Omega$ the set of Ω-Euler data.

Definition 2.12. *An invertible map $\mu : \mathcal{A} \to \mathcal{A}$ is called a mirror transformation if for any $P \in \mathcal{A}$, $\mu(P)$ is linked to P. We call $\mu(P)$ a mirror transform of P.*

Definition 2.13. *(Notations) $\mathcal{S}_0$ denotes the set of sequences $B: \ B_d \in \mathcal{R}^{-1}H_G^*(N_0),\ \ d = 1, 2, \ldots$ We define the map $\mathcal{I} : \mathcal{S} \to \mathcal{S}_0$, $P \mapsto \mathcal{I}(P) = B$ where $B_d = I_d^*(P_d)$.*

Recall that any equivariant cohomology class $\omega \in \mathcal{R}^{-1}H_G^*(N_d)$ is determined by its restrictions $\iota^*_{p_{ir}}(\omega) \in \mathcal{R}^{-1}$, $i = 0, .., n$, $r = 0, .., d$. Conversely given any collection $\omega_{ir} \in \mathcal{R}^{-1}$, there exists a unique class $\omega \in \mathcal{R}^{-1}H_G^*(N_d)$ such that $\iota^*_{p_{ir}}(\omega) = \omega_{ir}$ for all i, r. In fact,

$$\omega = \sum_{i=0}^{n} \sum_{r=0}^{d} \omega_{ir} \prod_{(j,m)\neq(i,r)} \frac{\kappa - \lambda_j - m\alpha}{\lambda_i - \lambda_j - (m-r)\alpha}.$$

In particular given a sequence $B \in \mathcal{S}_0$, then for each d there is a unique class $P_d \in \mathcal{R}^{-1}H_G^*(N_d)$ such that

$$\iota_{p_{ir}}^*(P_d) = \iota_{p_i}^*(\Omega)^{-1}\,\overline{\iota_{p_i}^*(B_r)}\,\iota_{p_i}^*(B_{d-r}),\quad i=0,..,n,\quad r=0,..,d, \tag{2.14}$$

where we have set $B_0 := \Omega$. This defines a sequence $P \in \mathcal{S}$, hence a map

$$\mathcal{L}_\Omega : \mathcal{S}_0 \to \mathcal{S},\quad B \mapsto \mathcal{L}_\Omega(B) = P.$$

We shall call $\mathcal{L} = \mathcal{L}_\Omega$ the Lagrange map. By (2.14) at $r=0$, we get

$$\iota_{p_i}^*(B_d) = \iota_{p_{i,0}}^*(P_d) = \iota_{p_i}^* I_d^*(P_d),\quad i=0,..,n. \tag{2.15}$$

First, this implies that the two classes $B_d, I_d^*(P_d)$ on $N_0 = \mathbf{P}^n$ coincide for each d. Thus

$$B = \mathcal{I}(P) = \mathcal{I}\circ\mathcal{L}(B). \tag{2.16}$$

Thus $\mathcal{L} : \mathcal{S}_0 \to \mathcal{S}$ is a section of the onto map $\mathcal{I} : \mathcal{S} \to \mathcal{S}_0$. Second, substituting (2.15) into (2.14), we get

$$\iota_{p_i}^*(\Omega)\iota_{p_{ir}}^*(P_d) = \overline{\iota_{p_{i,0}}^*(P_r)}\;\iota_{p_{i,0}}^*(P_{d-r}). \tag{2.17}$$

If, furthermore, we have $P_d \in \mathcal{R}H_G^*(N_d)$ rather than in $\mathcal{R}^{-1}H_G^*(N_d)$, then eqn. (2.17) says that P is an Euler data.

(A) *The image $P = \mathcal{L}(B)$ of a given $B \in \mathcal{S}_0$ under the Lagrange map is an Euler data if $P_d \in \mathcal{R}H_G^*(N_d)$, $d > 0$.*

On the other hand, it is trivial to show that if $Q \in \mathcal{A} \subset \mathcal{S}$ then

$$Q = \mathcal{L}\circ\mathcal{I}(Q). \tag{2.18}$$

Now using $\mathcal{L}$ we can lift any map $\mu_0 : \mathcal{S}_0 \to \mathcal{S}_0$ to a map

$$\mu = \mathcal{L}\circ\mu_0\circ\mathcal{I} : \mathcal{S} \to \mathcal{S},$$

which *we shall call the Lagrange lift of μ_0*. Thus from eqns. (2.16) and (2.18), we have

(B) *Let $\mu_0 : \mathcal{S}_0 \to \mathcal{S}_0$ be invertible with inverse ν_0, and let μ, ν be their respective Lagrange lifts. Then $\mu\circ\nu = \nu\circ\mu = id_{\mathcal{A}}$ when restricted to Euler data.*

We now discuss the relationship between Euler data and series of hypergeometric type.

Definition 2.14. *Given any $B \in \mathcal{S}_0$, define*

$$HG[B](t) := e^{-pt/\alpha}\left(\Omega + \sum_{d>0} \frac{B_d\, e^{dt}}{\prod_{k=0}^{n}\prod_{m=1}^{d}(p-\lambda_k-m\alpha)}\right)$$

where $p \in H_G^*(N_0)$ is the equivariant hyperplane class of $N_0 = \mathbf{P}^n$.

Note that $HG[B](t)$ is a cohomology valued formal series. If $P: \ P_d = \prod_{m=0}^{(n+1)d}(l\kappa - m\alpha)$ as in Example 1, it is obvious that in the limit $\lambda \to 0$, we have

$$HG[\mathcal{I}(P)](t) = e^{-Ht/\alpha}\sum_{d\geq 0}\frac{\prod_{m=0}^{(n+1)d}((n+1)H-m\alpha)}{\prod_{m=1}^{d}(H-m\alpha)^{n+1}}e^{dt}$$

where $H \in H^*(\mathbf{P}^n)$ on the right hand side is the hyperplane class of $\mathbf{P}^n$. The coefficients of $(-\frac{H}{\alpha})^i$ for $i = 1,..,n$, are exactly solutions to a hypergeometric differential equation discussed in the Introduction.

We now consider a construction of mirror transformations. Let $B \in \mathcal{S}_0$, and set $B_0 := \Omega$. Given any power series $g \in e^t\mathcal{R}[[e^t]]$, there is a unique $\tilde{B} \in \mathcal{S}_0$ such that

$$HG[B](t+g) = HG[\tilde{B}](t).$$

In fact, since

$$HG[B](t+g) = e^{-pt/\alpha}e^{-pg/\alpha}\sum_{d\geq 0}\frac{B_d\, e^{dt}e^{dg}}{\prod_{k=0}^{n}\prod_{m=1}^{d}(p-\lambda_k-m\alpha)},$$

if we write $e^{dg} = \sum_{s\geq 0} g_{d,s}e^{st}$, $g_{d,s} \in \mathcal{R}$ and $e^{-pg/\alpha} = \sum_{s\geq 0} g'_s e^{st}$, $g'_s \in \mathcal{R}[p/\alpha]$, then it is straightforward to find that

$$\begin{aligned}\tilde{B}_d &= B'_d + \sum_{r=0}^{d-1} g'_{d-r}B'_r \prod_{j=0}^{n}\prod_{m=r+1}^{d}(p-\lambda_j-m\alpha)\\ B'_d &:= B_d + \sum_{r=0}^{d-1} g_{r,d-r}B_r \prod_{j=0}^{n}\prod_{m=r+1}^{d}(p-\lambda_j-m\alpha).\end{aligned} \tag{2.19}$$

Thus we have an invertible transformation $\mu_0 : \mathcal{S}_0 \to \mathcal{S}_0$, $B \mapsto \tilde{B}$. Similarly, given any power series $f \in e^t\mathcal{R}[[e^t]]$ we have an invertible transformation $\mu_0 : \mathcal{S}_0 \to \mathcal{S}_0$, $B \mapsto \tilde{B}$, such that

$$e^{f/\alpha}\, HG[B](t) = HG[\tilde{B}](t).$$

Again if we write $e^{f/\alpha} = \sum_{s\geq 0} f_s e^{st}$, $f_s \in \mathcal{R}[\alpha^{-1}]$, then

$$\tilde{B}_d = B_d + \sum_{r=0}^{d-1} f_{d-r} B_r \prod_{j=0}^{n} \prod_{m=r+1}^{d} (p - \lambda_j - m\alpha). \tag{2.20}$$

We now make an important observation about the transformation μ_0 in each case above. For $d > 0$ the class $\prod_{j=0}^{n}(p - \lambda_j - d\alpha)$ always vanishes when restricted to the fixed points $p_i \in \mathbf{P}^n$, at $\alpha = (\lambda_i - \lambda_j)/d$. It follows immediately from (2.19) and (2.20) that $\iota_{p_i}^*(B_d)$, $\iota_{p_i}^*(\tilde{B}_d)$ always agree (whenever defined for all d) at $\alpha = (\lambda_i - \lambda_j)/d$, for $j \neq i$. To summarize:

(C) *Given $g, f \in e^t\mathcal{R}[[e^t]]$, let $\mu_0 : \mathcal{S}_0 \to \mathcal{S}_0$, $B \mapsto \tilde{B}$, be the invertible transformation defined by*

$$e^{f/\alpha}\, HG[B](t+g) = HG[\tilde{B}](t).$$

Suppose B is such that all values of $\iota_{p_i}^*(B_d)$ are well-defined at $\alpha = (\lambda_i - \lambda_j)/d$, $j \neq i$. Then these values are preserved under μ_0.

Obviously if P is an Euler data and $B = \mathcal{I}(P)$, then the restrictions $\iota_{p_i}^*(B_d) = \iota_{p_{i,0}}^*(P_d) \in \mathcal{R}$ are polynomial in α. Hence they are always well-defined at $\alpha = (\lambda_i - \lambda_j)/d$, $j \neq i$.

Lemma 2.15. *Let μ be the Lagrange lift of the above transformation $\mu_0 : \mathcal{S}_0 \to \mathcal{S}_0$, $B \mapsto \tilde{B}$. Then μ is a mirror transformation. In particular, if P is an Euler data, then $\tilde{P} = \mu(P)$ is an Euler data with*

$$e^{f/\alpha}\, HG[\mathcal{I}(P)](t+g) = HG[\mathcal{I}(\tilde{P})](t).$$

Proof: The second assertion follows from the first assertion and the fact that $\mathcal{I} \circ \mu = \mathcal{I} \circ \mathcal{L} \circ \mu_0 \circ \mathcal{I} = \mu_0 \circ \mathcal{I}$.

It suffices to consider the two cases $f = 0$, $g = 0$, separately. In each case we let P be an Euler data, and denote

$$\tilde{P} = \mu(P), \quad B = \mathcal{I}(P), \quad \tilde{B} = \mu_0(B).$$

Since $\mu := \mathcal{L} \circ \mu_0 \circ \mathcal{I}$, we have $\tilde{P} = \mathcal{L}(\tilde{B})$. In each case *we will show that $\tilde{P}$ is an Euler data.* We claim that this suffices. First, by statement

(B) above, μ is invertible as a transformation on the set $\mathcal{A}$ of Euler data. Second, by statement (C) above, the restrictions $\iota_{p_i}^*(B_d) = \iota_{p_{i,0}}^*(P_d)$ and $\iota_{p_i}^*(\tilde{B}_d) = \iota_{p_{i,0}}^*(\tilde{P}_d)$ agree at $\alpha = (\lambda_i - \lambda_j)/d$, $j \neq i$. Thus $\tilde{P}$ is linked to P. So, by definition, μ is a mirror transformation.

We now proceed to checking Eulerity of $\tilde{P}$. Since $P = \mathcal{L}(B)$, (2.14) holds. Multiply both sides of eqn. (2.14) by the respective sides of the following identity:

$$e^{d\tau} \frac{e^{(\lambda_i + r\alpha)(t-\tau)/\alpha}}{\prod_{j=0}^n \prod_{m=0\,(j,m)\neq(i,r)}^d (\lambda_i + r\alpha - \lambda_j - m\alpha)}$$

$$= \frac{1}{\prod_{j\neq i}(\lambda_i - \lambda_j)} \times e^{\lambda_i t/\alpha} \frac{e^{rt}}{\prod_{j=0}^n \prod_{m=1}^r (\lambda_i - \lambda_j + m\alpha)}$$

$$\times e^{-\lambda_i \tau/\alpha} \frac{e^{(d-r)\tau}}{\prod_{j=0}^n \prod_{m=1}^{d-r} (\lambda_i - \lambda_j - m\alpha)},$$

and then sum over $i = 0, .., n$, and $r = 0, .., d$. The result is

$$e^{d\tau}\, pf_d(P_d\, e^{\kappa(t-\tau)/\alpha})$$

$$= \sum_{r=0}^d pf\left(\Omega^{-1}\, \overline{[e^{-pt/\alpha} \frac{B_r\, e^{rt}}{\prod_{j=0}^n \prod_{m=1}^r (p - \lambda_j - m\alpha)}]}\right.$$

$$\left. \times [e^{-p\tau/\alpha} \frac{B_{d-r}\, e^{(d-r)\tau}}{\prod_{j=0}^n \prod_{m=1}^{d-r} (p - \lambda_j - m\alpha)}]\right.$$

Now summing this over $d = 0, 1, 2, ..$, we get:

$$\sum_{d\geq 0} e^{d\tau}\, pf_d(P_d\, e^{\kappa(t-\tau)/\alpha}) = pf\left(\Omega^{-1}\, \overline{HG[B](t)}\, HG[B](\tau)\right). \quad (2.21)$$

Likewise, of course, for $\tilde{P}$ and $\tilde{B}$.

First case:

$$HG[\tilde{B}](t) = HG[B](t + g(e^t)). \quad (2.22)$$

By (2.21), we have

$$pf\left(\Omega^{-1}\, \overline{HG[\tilde{B}](t)}\, HG[\tilde{B}](\tau)\right) = \sum_{d\geq 0} e^{d\tau}\, pf_d(\tilde{P}_d\, e^{\kappa(t-\tau)/\alpha})$$

$$pf\left(\Omega^{-1}\, \overline{HG[B](t + g(e^t))}\, HG[B](\tau + g(e^\tau))\right)$$

$$= \sum_{d\geq 0} e^{d(\tau + g(e^\tau))}\, pf_d\left(P_d\, e^{\kappa(t + \bar{g}(e^t) - \tau - g(e^\tau))/\alpha}\right).$$

By (2.22), we can equate the two right hand sides above. Setting $q = e^\tau$, $\zeta = (t-\tau)/\alpha$, we get

$$\sum_{d\geq 0} q^d \, pf_d(\tilde{P}_d \, e^{\zeta\kappa}) = \sum_{d\geq 0} q^d e^{dg(q)} \, pf_d \tag{2.23}$$

$$\times \left(P_d \, e^{\kappa\zeta} \, e^{\kappa(g_+(qe^{\zeta\alpha})-g_+(q))/\alpha} \, e^{-\kappa(g_-(qe^{\zeta\alpha})+g_-(q))/\alpha}\right)$$

where $g = g_+ + g_-$ with $\bar{g}_\pm = \pm g_\pm$. Obviously for any $g(q) \in \mathcal{R}[[q]]$, $g_+(qe^{\zeta\alpha}) - g_+(q) \in \alpha \cdot \mathcal{R}[[q,\zeta]]$. Since the involution $\omega \mapsto \bar{\omega}$ on $\mathcal{R}$ simply changes the sign of α, the fact that g_- is odd shows that $g_-(q) \in \alpha \cdot \mathcal{R}[[q]]$. Likewise for $g_-(qe^{\zeta\alpha})$. We know that $P_d \in \mathcal{R}H_G^*(N_d)$ (since P is Euler), and that pf_d maps $\mathcal{R}H_G^*(N_d)$ to $\mathcal{R}$. So the right hand side of (2.23) now clearly lies in $\mathcal{R}[[q,\zeta]]$. So, likewise for the left hand side of (2.23). It follows that

$$pf_d(\tilde{P}_d \, \kappa^s) \in \mathcal{R}, \quad s = 0,1,2,... \tag{2.24}$$

A priori $\tilde{P}_d \in \mathcal{R}^{-1}H_G^*(N_d)$ has the form

$$\tilde{P}_d = a_N\kappa^N + \cdots + a_0, \quad a_i \in \mathcal{R}^{-1}, \quad N = (n+1)d + n.$$

Since $pf_d(\kappa^N) = 1$, it follows from (2.24) that $a_N, .., a_0 \in \mathcal{R}$. Hence $\tilde{P}_d \in \mathcal{R}H_G^*(N_d)$ (rather than in $\mathcal{R}^{-1}H_G^*(N_d)$). By statement (A) above, $\tilde{P}$ is an Euler data.

Second case:

$$HG[\tilde{B}](t) = e^{f/\alpha} \, HG[B](t).$$

Again applying (2.21) and writing $f \in e^t\mathcal{R}[[e^t]]$ as $f = f_+ + f_-$ with $\bar{f}_\pm = \pm f_\pm$, we get

$$\sum_{d\geq 0} q^d \, pf_d(\tilde{P}_d \, e^{\zeta\kappa}) = e^{-\bar{f}(e^t)/\alpha} \, e^{f(e^\tau)/\alpha} \, pf\left(\Omega^{-1} \, \overline{HG[B](t)} \, HG[B](\tau)\right)$$

$$= e^{-(f_+(qe^{\zeta\kappa})-f_+(q))/\alpha} \, e^{(f_-(qe^{\zeta\kappa})+f_-(q))/\alpha}$$

$$\times \sum_{d\geq 0} q^d \, pf_d(P_d \, e^{\zeta\kappa}).$$

The right hand side lies in $\mathcal{R}[[q,\zeta]]$ as before, implying that $\tilde{P}$ is Euler. □

3. Applications

Definition 3.1. *A concavex bundle V on $\mathbf{P}^n$ is called a critical bundle if the induced bundle $U_d \to \bar{\mathcal{M}}_{0,0}(d,\mathbf{P}^n)$ has rank dim $\bar{\mathcal{M}}_{0,0}(d,\mathbf{P}^n) = (n+1)d+n-3$. We denote the nonequivariant Euler class by $c_{top}(U_d)$.*

Recall the notation that given a concavex bundle $V = V^+ \oplus V^-$, we have

$$\Omega^V := e_T(V^+)/e_T(V^-)$$
$$Q: \quad Q_d = \varphi_! e_T(V_d), \quad d > 0.$$

By Theorem 2.8 the sequence Q is an Ω^V-Euler data. If V is a critical bundle, introduce

$$K_d := \int_{\bar{\mathcal{M}}_{0,0}(d,\mathbf{P}^n)} c_{top}(U_d), \qquad \Phi := \sum_{d>0} K_d e^{dt}.$$

Theorem 3.2. *Let V be a concavex bundle on $\mathbf{P}^n$.*
(i) The restrictions $I_d^*(Q_d) \in H_G^*(\mathbf{P}^n)$ has $deg_\alpha I_d^*(Q_d) \le (n+1)d-2$.
(ii) If V is critical, then in the nonequivariant limit $\lambda \to 0$,

$$\int_{\mathbf{P}^n} e^{-Ht/\alpha} \frac{I_d^*(Q_d)}{\prod_{m=1}^d (H - m\alpha)^{n+1}} = \alpha^{-3}(2 - d\, t)K_d$$
$$\int_{\mathbf{P}^n} \left(HG[\mathcal{I}(Q)](t) - e^{-Ht/\alpha}\Omega^V\right) = \alpha^{-3}(2\Phi - t\Phi').$$

Proof: The second equality in (ii) follows trivially from the first equality.

By eqn. (2.8) in the proof of Theorem 2.8, we have

$$Q_d(\lambda_i) = \phi_{p_i}(\lambda_i)\phi_{\mathbf{P}^n}(\lambda_i) \sum_{F_d} \int_{F_d} \frac{\rho^* e_T(U_d)}{e_T(N(F_d))[\alpha(\alpha - c_1(L))]} \tag{3.1}$$

where $\phi_{p_i} := \prod_{j\neq i}(p - \lambda_j)$, $\phi_{\mathbf{P}^n} := \prod_{j=0}^n \prod_{m=1}^d (p - \lambda_j - m\alpha) \in H_T^*(\mathbf{P}^n)$. From the localization formula, we deduce that

$$\frac{\rho^* e_T(U_d)}{\alpha(\alpha-c_1(L))ev^*\phi_{p_i}=\int_{\mathbf{P}^n} ev_!\left(\frac{\rho^* e_T(U_d)}{\alpha(\alpha-c_1(L))}\right)\phi_{p_i}=\iota_{p_i}^* ev_!\left(\frac{\rho^* e_T(U_d)}{\alpha(\alpha-c_1(L))}\right).} Thus (3.1) canbe$$

$\iota_{p_i}^*[\phi_{\mathbf{P}^n}\ ev_!\left(\frac{\rho^* e_T(U_d)}{\alpha(\alpha-c_1(L))}\right)], \quad i = 0,..,n. It follows that I_d^*(Q_d) = \phi_{\mathbf{P}^n}\ ev_!\left(\frac{\rho^* e_T(U_d)}{\alpha(\alpha-c_1(L))}\right).(3.2) This shows that \deg_\alpha I_d^*(Q_d) \leq deg_\alpha \phi_{\mathbf{P}^n} - 2 = (n+1)d-2$, proving (i).

Since $Q_d = \varphi_! e_T(V_d) \in H_G^*(N_d)$, their nonequivariant limit $\lambda \to 0$ exist. In this limit (3.2) gives

$$\begin{aligned}
A &:= \int_{\mathbf{P}^n} e^{-Ht/\alpha} \frac{I_d^*(Q_d)}{\prod_{m=1}^d (H-m\alpha)^{n+1}} \\
&= \int_{\bar{\mathcal{M}}_{0,1}(d,\mathbf{P}^n)} e^{-ev^*Ht/\alpha} \frac{\rho^* c_{top}(U_d)}{\alpha(\alpha-c_1(L))} \\
&= \int_{\bar{\mathcal{M}}_{0,0}(d,\mathbf{P}^n)} c_{top}(U_d)\ \rho_! \left(\frac{e^{-ev^*Ht/\alpha}}{\alpha(\alpha-c_1(L))}\right).
\end{aligned}$$

Now $c_{top}(U_d)$ has degree the same as the dimension of $\bar{\mathcal{M}}_{0,0}(d,\mathbf{P}^n)$. The second factor in the last integrand contributes a scalar factor given by integration over a generic fiber E (which is a $\mathbf{P}^1$) of ρ. So we pick out the degree 1 term in $\frac{e^{-ev^*Ht/\alpha}}{\alpha(\alpha-c_1(L))}$, which is just $\frac{-ev^*Ht}{\alpha^3} + \frac{c_1(L)}{\alpha^3}$. Restricting to the generic fiber E, say over $(f,C) \in \bar{\mathcal{M}}_{0,0}(d,\mathbf{P}^n)$, the evaluation map ev is equal to f, which is a degree d map $E \cong \mathbf{P}^1 \to \mathbf{P}^n$. It follows that

$$\int_E ev^*H = d.$$

Moreover, since $c_1(L)$ restricted to E is just the first Chern class of the tangent bundle to E, it follows that

$$\int_E c_1(L) = 2.$$

So we have

$$A = (-\frac{d\,t}{\alpha^3} + \frac{2}{\alpha^3})K_d. \quad \square$$

It is easy to work out the complete list of *critical* concavex vector bundles V on $\mathbf{P}^n$ which are direct sums of line bundles. Such a V is of the form

$$\begin{aligned}
V &= V^+ \oplus V^- \\
V^+ &= \oplus_{a=1}^{N^+} \mathcal{O}(l_a) \\
V^- &= \oplus_{b=1}^{N^-} \mathcal{O}(-k_b)
\end{aligned}$$

where $l_1, .., l_{N^+}$, $k_1, .., k_{N^-}$ are positive integers. By Riemann-Roch, the bundles U_d that V induces on $\bar{\mathcal{M}}_{0,0}(d, \mathbf{P}^n)$ has *rank* $U_d = d(\sum l_a + \sum k_b) + N^+ - N^-$, which must be $(n+1)d + n - 3$ for all d if V is critical. Thus we must have

$$\begin{aligned} \sum l_a + \sum k_b &= n+1 \\ N^+ - N^- &= n-3. \end{aligned} \tag{3.3}$$

The complete list of critical bundles that are also direct sums of line bundles is:

$$\begin{aligned}
\mathbf{P}^1 &: \ \mathcal{O}(-1) \oplus \mathcal{O}(-1) \\
\mathbf{P}^2 &: \ \mathcal{O}(-3) \\
\mathbf{P}^3 &: \ \mathcal{O}(2) \oplus \mathcal{O}(-2), \mathcal{O}(3) \oplus \mathcal{O}(-1) \\
\mathbf{P}^4 &: \ \mathcal{O}(5), \quad \mathcal{O}(2) \oplus \mathcal{O}(2) \oplus \mathcal{O}(-1) \\
\mathbf{P}^5 &: \ \mathcal{O}(2) \oplus \mathcal{O}(4), \quad \mathcal{O}(3) \oplus \mathcal{O}(3) \\
\mathbf{P}^6 &: \ \mathcal{O}(2) \oplus \mathcal{O}(2) \oplus \mathcal{O}(3) \\
\mathbf{P}^7 &: \ \mathcal{O}(2) \oplus \mathcal{O}(2) \oplus \mathcal{O}(2) \oplus \mathcal{O}(2).
\end{aligned} \tag{3.4}$$

Note that we have excluded the critical bundles in which the hyperplane bundle $\mathcal{O}(1)$ occurs because in the nonequivariant limit it only reduces a given case of $\mathbf{P}^n$ to $\mathbf{P}^{n-1}$. For example, even though the bundle $\mathcal{O}(1) \oplus \mathcal{O}(-1) \oplus \mathcal{O}(-1)$ on $\mathbf{P}^2$ is certainly critical, computing the K_d for the induced bundles is equivalent to doing the same with $\mathcal{O}(-1) \oplus \mathcal{O}(-1)$ on $\mathbf{P}^1$. It is curious to note that the numerical conditions (3.3) is rather similar to the condition for having a projective complete intersection Calabi-Yau threefold. In fact five of the examples on $\mathbf{P}^4$ through $\mathbf{P}^7$ above involving only positive bundles are exactly the cases in which each critical bundle cuts out a complete intersection Calabi-Yau threefold. We also note that, with the exception of the $\mathbf{P}^1$ case, the three examples which involve negative bundles in fact correspond to *noncompact Calabi-Yau threefolds*. The total space of $\mathcal{O}(-3) \to \mathbf{P}^2$, the total space of $\psi^*\mathcal{O}(-2) \to X$, where $\psi : X \hookrightarrow \mathbf{P}^3$ is a quadric, and the total space of $\psi^*\mathcal{O}(-1) \to \mathbf{P}^4$ where $\psi : X \hookrightarrow \mathbf{P}^4$ is the intersection of two quadrics, all three are noncompact Calabi-Yau. These examples arise in the so-called *local mirror symmetry*. In the next subsection, we shall compute the Euler classes of the induced bundles for the list above.

3.1. The first convex example: The Mirror Conjecture

Throughout this subsection, we set $l = n+1$, consider the convex bundle $V = \mathcal{O}(l)$ on $\mathbf{P}^n$, and fix $\Omega^V = lp$. P, Q shall denote the following two linked Euler data (cf. Theorems 2.8, 2.10.):

$$P: \ P_d = \prod_{m=0}^{ld} (l\kappa - m\alpha)$$

$$Q: \ Q_d = \varphi_! e_T(V_d).$$

Consider the hypergeometric differential equation

$$\left((\frac{d}{dt})^n - le^t(l\frac{d}{dt}+1)\cdots(l\frac{d}{dt}+n) \right) h(t) = 0.$$

We have seen that a basis f_i, $i = 0, .., n-1$, of solutions can be read off from the hypergeometric series (cf. Introduction) in the limit $\lambda \to 0$:

$$\begin{aligned} HG[\mathcal{I}(P)](t) &= e^{-Ht/\alpha} \sum_{d\geq 0} \frac{\prod_{m=0}^{ld}(lH - m\alpha)}{\prod_{m=1}^{d}(H - m\alpha)^{n+1}} e^{dt} \\ &= lH \left(f_0 - f_1 \frac{H}{\alpha} + f_2 \frac{H^2}{\alpha^2} - \cdots \right). \end{aligned}$$

Recall that

$$T(t) := \frac{f_1}{f_0} = t + \frac{g_1}{f_0} \tag{3.5}$$

is the mirror map of Candelas et al, where

$$f_0 := \sum_{d\geq 0} \frac{(ld)!}{(d!)^{n+1}} e^{dt}, \qquad g_1 := \sum_{d\geq 1} \frac{(ld)!}{(d!)^l} \sum_{m=d+1}^{ld} \frac{l}{m} e^{dt}.$$

Lemma 3.3. *In the limit $\lambda \to 0$, we have $HG[\mathcal{I}(Q)](T(t)) = \frac{1}{f_0} HG[\mathcal{I}(P)](t)$.*

Proof: By expanding in powers of α^{-1} and using the assumption that $l = n+1$, we get

$$\begin{aligned} HG[\mathcal{I}(P)](t) &= e^{-pt/\alpha} \sum_{d\geq 0} \frac{\prod_{m=0}^{ld}(lp - m\alpha)}{\prod_{k=0}^{n}\prod_{m=1}^{d}(p - \lambda_k - m\alpha)} e^{dt} \\ &= lp \left[f_0 - \alpha^{-1}(p\, f_1 + g_2 \sum_{k=0}^{n} \lambda_k) + \cdots \right] \end{aligned} \tag{3.6}$$

where $g_2 = \sum_{d\geq 1} \frac{(ld)!}{(d!)^l} \sum_{m=1}^{d} \frac{1}{m} e^{dt}$.

Put $f := (log\ f_0)\alpha + \frac{g_2}{f_0} \sum_{k=0}^{n} \lambda_k \in e^t \mathcal{R}[[e^t]]$. By Lemma 2.15, we have a mirror transformation μ such that

$$HG[\mathcal{I}(\tilde{P})](t) = e^{f/\alpha}\ HG[\mathcal{I}(P)](t) \tag{3.7}$$

where $\tilde{P} = \mu(P)$. Substituting (3.6) into (3.7), we get

$$\begin{aligned} HG[\mathcal{I}(\tilde{P})](t) &= (1 + \alpha^{-1} \frac{g_2}{f_0} \sum_{k=0}^{n} \lambda_k + \cdots) \\ &\times\ f_0^{-1}\ lp \left[f_0 - \alpha^{-1}(p\ f_1 + g_2 \sum_{k=0}^{n} \lambda_k) + \cdots \right] \\ &= lp - \alpha^{-1} lp^2 \frac{f_1}{f_0} + \cdots . \end{aligned} \tag{3.8}$$

By Lemma 2.15 again, we have a mirror transformation ν such that

$$HG[\mathcal{I}(\tilde{Q})](t) = HG[\mathcal{I}(Q)](t + \frac{g_1}{f_0})$$

where $\tilde{Q} = \nu(Q)$. Since, by Theorem 3.2 (i), $I_d^*(Q_d) = O(\alpha^{(n+1)d-2})$, it is straightforward to find that

$$HG[\mathcal{I}(\tilde{Q})](t) = e^{-p(t+\frac{g_1}{f_0})/\alpha}(lp+\cdots) = lp - \alpha^{-1} lp^2 (t + \frac{g_1}{f_0}) + \cdots . \tag{3.9}$$

From (3.8) and (3.9), we conclude that for $d > 0$,

$$\frac{I_d^*(\tilde{P}_d - \tilde{Q}_d)}{\prod_{k=0}^{n} \prod_{m=1}^{d} (p - \lambda_k - m\alpha)} \equiv 0$$

modulo order α^{-2}, and hence

$$deg_\alpha\ \iota_{p_{i},0}^{*}(\tilde{P}_d - \tilde{Q}_d) \leq (n+1)d - 2.$$

But $\tilde{P} = \mu(P)$ is linked to P, and $\tilde{Q} = \nu(Q)$ is linked to Q. Since P and Q are linked, it follows that $\tilde{P}$ and $\tilde{Q}$ are also linked. By Theorem 2.11, we have $\tilde{P} = \tilde{Q}$. In particular, we have

$$HG[\mathcal{I}(Q)](T(t)) = HG[\mathcal{I}(\tilde{Q})](t) = HG[\mathcal{I}(\tilde{P})](t) = e^{f/\alpha}\ HG[\mathcal{I}(P)](t).$$

Since both P_d, Q_d lie in $H_G^*(N_d)$, their nonequivariant limit exist. Taking $\lambda \to 0$ yields our assertion. □

Throughout the rest of this subsection, we set $l = n + 1 = 5$ and consider the critical bundle $\mathcal{O}(5) \to \mathbf{P}^4$. We assume that we have taken the nonequivariant limit $\lambda \to 0$. Recall that

$$F(T) := \frac{5T^3}{6} + \sum_{d>0} K_d e^{dT}, \quad \mathcal{F}(T) := \frac{5}{2}\left(\frac{f_1}{f_0}\frac{f_2}{f_0} - \frac{f_3}{f_0}\right).$$

Theorem 3.4. *(The Mirror Conjecture) $F = \mathcal{F}$.*

Proof: Since

$$HG[\mathcal{I}(P)](t) = 5H\left(f_0 - f_1\frac{H}{\alpha} + f_2\frac{H^2}{\alpha^2} - f_3\frac{H^3}{\alpha^3}\right), \tag{3.10}$$

we will prove that (cf. [8])

$$HG[\mathcal{I}(Q)](T) = 5H\left(1 - T\frac{H}{\alpha} + \frac{F'}{5}\frac{H^2}{\alpha^2} - \frac{TF' - 2F}{5}\frac{H^3}{\alpha^3}\right). \tag{3.11}$$

Eqns. (3.10), (3.11) and the preceding lemma imply $F = \mathcal{F}$. Denote the right hand side of (3.11) by R. Then

$$e^{HT/\alpha}R = 5H\left(1 + \frac{\Phi'}{5}\frac{H^2}{\alpha^2} + \frac{2\Phi}{5}\frac{H^3}{\alpha^3}\right) = 5H + O(e^T)$$

where $\Phi := F - \frac{5T^3}{6} = O(e^T)$. Similarly $e^{HT/\alpha}HG[\mathcal{I}(Q)](T) = 5H + O(e^T)$, which also has no *polynomial dependence* on T. So (3.11) is equivalent to

$$\int_{\mathbf{P}^4} e^{-HT/\alpha}\left(e^{HT/\alpha}HG[\mathcal{I}(Q)]\right) = \int_{\mathbf{P}^4} e^{-HT/\alpha}\left(e^{HT/\alpha}R\right).$$

By Theorem 3.2(ii), this left hand side is

$$\int_{\mathbf{P}^4} HG[\mathcal{I}(Q)] = \alpha^{-3}(2\Phi - T\Phi') + \int_{\mathbf{P}^4} e^{-HT/\alpha}5H = \alpha^{-3}(2F - TF'),$$

which coincides with $\int_{\mathbf{P}^4} R$. □

It is straightforward to generalize Theorem 3.4 to all other critical *convex* bundles V in the list (3.4). In each case, Ω^V becomes $\prod_a l_a p$ and the Euler data P to be linked with $Q: \ Q_d := \varphi_! e_T(V_d)$ is given by $P_d = \prod_a \prod_{m=0}^{l_a d}(l_a\kappa - m\alpha)$. In the nonequivariant limit, the hypergeometric series $HG[\mathcal{I}(P)](t)$ will produce some hypergeometric functions $f_0, .., f_3$ defining the function $\mathcal{F}$. The generating function F for the K_d is modified by simply replacing the term $\frac{5T^3}{6}$ by $\frac{T^3}{6}\int_X \psi^* H^3$, where $\psi : X \to \mathbf{P}^n$ is the Calabi-Yau cut out by V. With these minor modifications in each case, Theorem 3.4 holds. We leave the details as an exercise for the reader.

3.2. First concave example: multiple-cover formula

Let V be the bundle $\mathcal{O}(-1)\oplus\mathcal{O}(-1)$ on $\mathbf{P}^1$. For $d > 1$, V induces a rank $2d-2$ bundle $U_d \to \mathcal{M}_{0,0}(d, \mathbf{P}^1)$ whose fiber at (f, C) is the space $H^1(C, f^*V)$, thus V is a critical concave bundle on $\mathbf{P}^1$. We set $\Omega^V = 1/e_T(V) = p^{-2}$. We shall compute the equivariant classes $Q_d := \varphi_! e_T(V_d)$, and the numbers K_d for this critical bundle. Note that by definition $Q_1 = 1$ and $K_1 = 1$. As a consequence of Theorem 2.10,

$$\iota_{p_{i,0}}^*(Q_d) = \prod_{m=1}^{d-1}(\lambda_i - m(\lambda_i - \lambda_j)/d)^2$$

at $\alpha = (\lambda_i - \lambda_j)/d$, $j \neq i$. Thus Q is linked to

$$P: \ P_d := \prod_{m=1}^{d-1}(\kappa - m\alpha)^2,$$

which is a p^{-2}-Euler data. Obviously $deg_\alpha I_d^*(P_d) = 2d-2$. It follows from Theorem 3.2 (i) that $deg_\alpha I_d^*(P_d - Q_d) \le 2d-2$, implying $Q = P$ by Theorem 2.11.

Corollary 3.5. $K_d = d^{-3}$.

Proof: By Theorem 3.2(ii) in the limit $\lambda \to 0$, we have

$$\int_{\mathbf{P}^1} e^{-Ht/\alpha}\frac{I_d^*(Q_d)}{\prod_{m=1}^d (H - m\alpha)^2} = \alpha^{-3}(2 - d\,t)K_d.$$

Since $Q = P$, we have $I_d^*(Q_d) = I_d^*(P_d) = \prod_{m=1}^{d-1}(H - m\alpha)^2$, giving

$$\int_{\mathbf{P}^1} e^{-Ht/\alpha}\frac{I_d^*(Q_d)}{\prod_{m=1}^d (H - m\alpha)^2} = \alpha^{-3}d^{-3}(2 - d\,t). \qquad \square$$

3.3. Second concave example: $K_{\mathbf{P}^2}$

Let V be the canonical bundle $\mathcal{O}(-3) \to \mathbf{P}^2$. For $d > 0$, this bundle induces a rank $3d-1$ bundle $U_d \to \bar{\mathcal{M}}_{0,0}(d, \mathbf{P}^2)$. Thus V is a critical concave bundle. We set $\Omega^V = 1/e_T(V) = (-3p)^{-1}$. We shall compute the equivariant classes $Q_d := \varphi_! e_T(V_d)$, and the numbers K_d for this critical bundle. As a consequence of Theorem 2.10, we have

$$\iota^*_{p_{i,0}}(Q_d) = \prod_{m=1}^{3d-1} (-3\lambda_i + m(\lambda_i - \lambda_j)/d).$$

at $\alpha = (\lambda_i - \lambda_j)/d$, $j \neq i$. Thus Q is linked to

$$P: \quad P_d := \prod_{m=1}^{3d-1} (-3\kappa + m\alpha).$$

which is a $(-3p)^{-1}$-Euler data.

Corollary 3.6. *$HG[\mathcal{I}(Q)](t+g) = HG[\mathcal{I}(P)](t)$ where $g := \sum_{d>0} \frac{(-1)^d}{d} \frac{(3d)!}{d!^3} e^{dt}$.*

Proof: By expanding in powers of α^{-1}, we get

$$\begin{aligned} HG[\mathcal{I}(P)](t) &= e^{-pt/\alpha} \\ &\times \left((-3p)^{-1} + \sum_{d>0} \frac{\prod_{m=1}^{3d-1}(-3p + m\alpha)}{\prod_{k=0}^{2} \prod_{m=1}^{d} (p - \lambda_k - m\alpha)} e^{dt} \right) \\ &= (-3p)^{-1} + \alpha^{-1} \frac{t+g}{3} + \cdots . \end{aligned} \tag{3.12}$$

As before, it is now straightforward to show that

$$HG[\mathcal{I}(Q)](t+g) \equiv HG[\mathcal{I}(P)](t)$$

modulo order α^{-2}. Once again by Theorem 2.11, the two sides are equal identically. □

Using Theorem 3.2(ii) and the preceding corollary, we obtain the

K_d, for $d = 1, .., 10$:

d	K_d
1	3
2	$-\frac{45}{8}$
3	$\frac{244}{9}$
4	$-\frac{12333}{64}$
5	$\frac{211878}{125}$
6	$-\frac{102365}{6}$
7	$\frac{64639725}{343}$
8	$-\frac{1140830253}{512}$
9	$\frac{6742982701}{243}$
10	$-\frac{36001193817}{100}$

3.4. A concavex bundle on $\mathbf{P}^3$.

Let $V = \mathcal{O}(2) \oplus \mathcal{O}(-2)$ on $\mathbf{P}^3$. This is a direct sum of a convex and a concave bundle. The induced bundle $U_d \to \bar{\mathcal{M}}_{0,0}(d, \mathbf{P}^3)$, with fiber at (f, C) being $H^0(C, f^*\mathcal{O}(2)) \oplus H^1(C, f^*\mathcal{O}(-2))$, has rank $4d$. We set $\Omega^V = e_T(\mathcal{O}(2))/e_T(\mathcal{O}(-2)) = -1$. We shall compute the equivariant classes $Q_d := \varphi_! e_T(V_d)$, and the numbers K_d for this critical bundle. As a consequence of Theorem 2.10, we have

$$\iota^*_{p_{i,0}}(Q_d) = \prod_{m=0}^{2d} (2\lambda_i - m(\lambda_i - \lambda_j)/d) \times \prod_{m=1}^{2d-1} (-2\lambda_i + m(\lambda_i - \lambda_j)/d).$$

at $\alpha = (\lambda_i - \lambda_j)/d$, $j \neq i$. Thus Q is linked to

$$P: \quad P_d := \prod_{m=0}^{2d} (2\kappa - m\alpha) \times \prod_{m=1}^{2d-1} (-2\kappa + m\alpha).$$

which is a -1-Euler data.

Corollary 3.7. $HG[\mathcal{I}(Q)](t + g) = HG[\mathcal{I}(P)](t)$ *where* $g :=$

$\sum_{d>0} \frac{1}{d} \frac{(2d)!^2}{d!^4} e^{dt}$.

Proof: By expanding in powers of α^{-1}, we get

$$\begin{aligned} HG[\mathcal{I}(P)](t) &= e^{-pt/\alpha} \\ &\times \left(-1 + \sum_{d>0} \frac{\prod_{m=0}^{2d}(2p - m\alpha) \times \prod_{m=1}^{2d-1}(-2p + m\alpha)}{\prod_{k=0}^{3} \prod_{m=1}^{d}(p - \lambda_k - m\alpha)} e^{dt} \right) \\ &= -1 + \alpha^{-1} p(t+g) + \cdots, \end{aligned} \tag{3.13}$$

which, as in the previous examples, agrees with $HG[\mathcal{I}(Q)](t+g)$ up to order α^{-2}. Hence Theorem 2.11 yields our assertion. □

Using Theorem 3.2(ii) and the preceding corollary, we obtain the K_d, for $d = 1, .., 10$:

d	K_d
1	-4
2	$-\frac{9}{2}$
3	$-\frac{328}{27}$
4	$-\frac{777}{16}$
5	$-\frac{30004}{125}$
6	$-\frac{4073}{3}$
7	$-\frac{2890808}{343}$
8	$-\frac{7168777}{128}$
9	$-\frac{285797488}{729}$
10	$-\frac{714787509}{250}$

3.5. A concavex bundle on $\mathbf{P}^4$.

Consider now the critical bundle $V = \mathcal{O}(2) \oplus \mathcal{O}(2) \oplus \mathcal{O}(-1)$ on $\mathbf{P}^4$. The induced bundle $U_d \to \bar{\mathcal{M}}_{0,0}(d, \mathbf{P}^3)$, has rank $5d+1$. We set $\Omega^V = e_T(\mathcal{O}(2))^2 / e_T(\mathcal{O}(-1)) = -4p$. We shall compute the equivariant classes $Q_d := \varphi_! e_T(V_d)$, and the numbers K_d for this critical bundle. As a consequence of Theorem 2.10, we have

$$\iota_{p_{i,0}}^*(Q_d) = \prod_{m=0}^{2d} (2\lambda_i - m(\lambda_i - \lambda_j)/d)^2 \times \prod_{m=1}^{d-1} (-\lambda_i + m(\lambda_i - \lambda_j)/d).$$

at $\alpha = (\lambda_i - \lambda_j)/d$, $j \neq i$. Thus Q is linked to

$$P: \ P_d := \prod_{m=0}^{2d} (2\kappa - m\alpha)^2 \times \prod_{m=1}^{d-1} (-\kappa + m\alpha).$$

which is a $-4p$-Euler data.

Corollary 3.8. *$HG[\mathcal{I}(Q)](t+g) = HG[\mathcal{I}(P)](t)$ where $g := \sum_{d>0} \frac{(-1)^d}{d} \frac{(2d)!^2}{d!^4} e^{dt}$.*

Proof: By expanding in powers of α^{-1}, we get

$$\begin{aligned} HG[\mathcal{I}(P)](t) &= e^{-pt/\alpha} \left(-4p + \sum_{d>0} \frac{\prod_{m=0}^{2d}(2p - m\alpha)^2 \times \prod_{m=1}^{d-1}(-p + m\alpha)}{\prod_{k=0}^{4} \prod_{m=1}^{d} (p - \lambda_k - m\alpha)} e^{dt} \right) \\ &= -4p + \alpha^{-1} 4p^2 (t+g) + \cdots, \end{aligned} \tag{3.14}$$

which, as in the previous examples, agrees with $HG[\mathcal{I}(Q)](t+g)$ up to order α^{-2}. Hence Theorem 2.11 yields our assertion. □

We can work out the K_d here as we did before. The K_d here can be obtained by taking K_d from the preceding example on $\mathbf{P}^3$, and multiply it by $4(-1)^d$. This is so because in the nonequivariant limit, the hypergeometric series $HG[\mathcal{I}(P)](t)$ (cf. (3.13) and (3.14)) in this example on $\mathbf{P}^4$ and the preceding example on $\mathbf{P}^3$ are related by first a multiplication of $4p$ followed by a change of variable $e^{dt} \mapsto (-1)^d e^{dt}$.

3.6. General concavex bundles

In fact the examples above are representative of the most general concavex bundle. Let $V = V^+ \oplus V^-$ be a concavex bundle on $\mathbf{P}^n$, and let Q, P be as defined in Theorem 2.10, and assume that V has splitting type $(l_1, .., l_{N^+}; k_1, .., k_{N^-})$. Note that $\sum l_a + \sum k_b$ is the value of the class $c_1(V^+) - c_1(V^-)$ on a T-invariant $\mathbf{P}^1$ in $\mathbf{P}^n$.

Theorem 3.9. *If $d(\sum l_a + \sum k_b) - N^- \leq (n+1)d - 2$ for all $d > 0$, then $Q = P$. If $d(\sum l_a + \sum k_b) - N^- \leq (n+1)d$ for all $d > 0$, then there exists a mirror transformation μ, depending only on the l_a, k_b, such that $Q = \mu(P)$.*

Proof: By definition of P in Theorem 2.10,

$$deg_\alpha I_d^*(P_d) = d\left(\sum l_a + \sum k_b\right) - N^-.$$

Consider the first case, where this is bounded above by $(n+1)d-2$ for all d. Then by Theorem 3.2,

$$deg_\alpha(I_d^*(P_d) - I_d^*(Q_d)) \leq (n+1)d - 2,$$

implying $Q = P$ by Theorem 2.11.

Consider now the second case. Obviously our assumption implies that $\sum l_a + \sum k_b \leq (n+1)$. It is trivial to show that the only possibilities not covered by the first case are: (1) $N^- = 0$ and $\sum l_a = n+1$; (2) $N^- = 1$ and $\sum l_a + k_1 = n+1$. (3) $N^- = 0$ and $\sum l_a = n$; In each of these cases, a mirror transformation can be constructed by immitating the previous examples in a straightforward way. Case (1) immitates the example $\mathcal{O}(5) \to \mathbf{P}^4$, while cases (2), (3) immitate the example $\mathcal{O}(-3) \to \mathbf{P}^2$. It is obvious that in each case, the mirror transformation depends only on the data l_a, k_b. □

Corollary 3.10. *Under the same hypotheses as in the preceding theorem, the Euler data Q : $Q_d = \varphi_! e_T(V_d)$ depends only on the splitting type, ie. the numbers l_a, k_b, of the concavex bundle V on $\mathbf{P}^n$.*

Note that not every concavex bundle on $\mathbf{P}^n$ is a direct sum of line bundles. For example the tangent bundle is convex, but is not a direct sum of line bundles.

3.7. Equivariant total Chern class

For simplicity, we restrict to convex bundles. Let V be a rank r convex bundle on $\mathbf{P}^n$, and let

$$c_T(V) = x^r + x^{r-1}c_1(V) + \cdots + c_r(V)$$

be the T-equivariant Chern polynomial of V. Similarly we denote by $c_T(U_d)$ the equivariant Chern polynomial for U_d. As explained in Example 9, we can extend the notion of Euler data Q allowing Q_d to depend on x polynomially, simply by replacing the ground field $\mathbf{Q}$ by $\mathbf{Q}(x)$. Then a similar argument as in Theorem 2.8 shows that the sequence $Q_d := \varphi_!(\pi^* c_T(U_d))$ is also an Euler data in the generalized sense. Moreover, the analogue of Theorem 2.10 holds, ie. at $\alpha = (\lambda_j - \lambda_i)/d$, we have the form

$$\iota_{p_{j,0}}^*(Q_d) = \prod_a \prod_{m=0}^{l_a d} (x + l_a\lambda_j - m(\lambda_j - \lambda_i)/d).$$

Hence Q is linked to the Euler data

$$P: \quad P_d = \prod_a \prod_{m=0}^{l_a d} (x + l_a \kappa - m\alpha).$$

Again, under a suitable bound on $c_1(V)$, one can easily relate P, Q by a generalized (depending on x) mirror transformation.

For example, by taking the bundle $\mathcal{O}(4)$ on $\mathbf{P}^3$, and applying the above result, we can compute the nonequivariant limits of all $\int_{\bar{\mathcal{M}}_{0,0}(d,\mathbf{P}^3)} c_{4d}(U_d)$. They are expected to count rational curves in a $\mathbf{P}^1$-family of K3 hypersurfaces in $\mathbf{P}^3$. Similarly we can take $\mathcal{O}(3)$ on $\mathbf{P}^2$ and compute $\int_{\bar{\mathcal{M}}_{0,0}(d,\mathbf{P}^2)} c_{3d-1}(U_d)$ which should count the number of rarional curves in a $\mathbf{P}^2$-family of elliptic curves in $\mathbf{P}^2$. Details will be reported in full in our forth-coming paper [37].

3.8. Concluding remarks

The most important result we establised in this paper is the Mirror Principle. For simplicity we have restricted our examples in this paper, to studying only Euler classes and total Chern classes. As mentioned in the Introduction, the Mirror Principle works well for any multiplicative equivariant characteristic classes. We shall study in details more examples of the total Chern class in our forthcoming paper [37]. Generalization to manifolds with torus action will also be dealt with in details there.

Finally, we make a tantalizing observation which might be of both physical and mathematical significance. As we have seen, the set of linked Euler data has an infinite dimensional transformation group - the mirror group. For suitable concavex bundle $V \to \mathbf{P}^n$, two special linked Euler data (cf. Theorem 2.10) $Q: \; Q_d = \varphi_! e_T(V_d)$ arising from the nonlinear sigma model (the stable map moduli), and P the corresponding Euler data of hypergeometric type, are related by a mirror transformation. Since the mirror group is so big, there are many other Euler data which are linked to P and can be obtained simply by acting on P by the mirror group. From the physical point of view, P arises from type IIB string theory while Q arises from type IIA string theory, and mirror symmetry is a duality between the two. This relationship manifests itself on the linear sigma model as a duality transformation. This suggests that some other Euler data linked to P may arise from some other string theories which are dual to type IIA and IIB, via more general mirror transformations. From the point of view of moduli theory, P is associated to the linear sigma model compatification N_d of the moduli space M_d^0 we discussed in

Example 10. Whereas Q is associated to the nonlinear sigma model M_d, which is the stable map compactification of M_d^0. This suggests that some other Euler data linked to P may correspond to other compactifications of M_d^0. If true, we will have an association between string theories, linked Euler data, and compactifications of moduli space of maps, all in the same picture, whereby there is a duality in each kind which one sees in the linear sigma model. It would be interesting to understand this duality more precisely.

References

[1] P. Aspinwall and D. Morrison, *Topological field theory and rational curves*, Commun. Math. Phys. 151 (1993), 245-262.

[2] J. Bryan and N. Leung, *The enumerative geometry of K3 surfaces and modular forms*, alg-geom/9711031.

[3] M. Atiyah and R. Bott, *The moment map and equivariant cohomology*, Topology 23 (1984) 1-28.

[4] V. Batyrev and L. Borisov, *On Calabi-Yau complete intersections in toric varieties*, alg-geom/9412017.

[5] V. Batyrev and D. van Straten, *Generalized hypergeometric functions and rational curves on Calabi-Yau complete intersections in toric varieties*, Comm. Math. Phys. 168 (1995) 495-533.

[6] M. Bershadsky, S. Cecotti, H. Ooguri, C. Vafa, *Kodaira-Spencer theory of gravity and exact results for quantum string amplitude*, Commun. Math. Phys. 165 (1994) 311-428.

[7] R. Bott, *A residue formula for holomorphic vector fields*, J. Diff. Geom. 1 (1967) 311-330.

[8] P. Candelas, X. de la Ossa, P. Green, and L. Parkes, *A pair of Calabi-Yau manifolds as an exactly soluble superconformal theory*, Nucl. Phys. B359 (1991) 21-74.

[9] P. Candelas, X. de la Ossa, A. Font, S. Katz and D. Morrison, *Mirror symmetry for two parameter models I*, hep-th/9308083.

[10] L. Caporaso and J. Harris, *Parameter spaces for curves on surfaces and enumeration of rational curves*, alg-geom/9608024.

[11] B. Crauder and R. Miranda, *Quantum cohomology of rational surfaces*, Progress in Mathematics 129, Birkhauser (1995) 33.

[12] P. DiFrancesco and C. Itzykson, *Quantum intersection rings*, Progress in Mathematics 129, Birkhauser (1995) 81.

[13] R. Dijkgraaf, *Mirror symmetry and elliptic curves*, Progress in Mathematics 129, Birkhauser (1995) 149.

[14] *Essays on Mirror Manifolds*, ed. S.T. Yau, International Press.

[15] G. Ellingsrud and S.A. Stromme, *Bott's formula and enumerative geometry*, JAMS 9 (1996), no. 1, 175-193.

[16] W. Fulton and R. Pandharipande, *Notes on stable maps and quantum cohomology*, alg-geom/9608011.

[17] E. Getzler, *The elliptic Gromov-Witten invariants on CP^3*, alg-geom/9612009.

[18] A. Givental, *Equivariant Gromov-Witten invariants*, alg-geom/9603021

[19] A. Givental, *A mirror theorem for toric complete intersections*, alg-geom/9701016.

[20] T. Graber and R. Pandharipande, *Localization of Virtual classes*, alg-geom/9708001.

[21] B. Greene and R. Plesser, *Duality in Calabi-Yau moduli space*, Nucl. Phys. B338 (1990) 15-37.

[22] R. Hartshorne, *Algebraic Geometry.* Graduate Texts in Mathematics 52. Berlin, Heidelberg, New York Springer 1977

[23] F. Hirzebruch, *Topological methods in algebraic geometry*, Springer-Verlag, Berlin 1995, 3rd Ed.

[24] Y. Hu, *Moduli spaces of stable polygons and symplectic structures on on $\bar{\mathcal{M}}_{0,n}$*, Comp. Math. (to appear), alg-geom/9701011.

[25] M. Jinzenji and M. Nagura, *Mirror symmetry and exact calculation of $N-2$ point correlation function on Calabi-Yau manifold embedded in CP^{N-1}*, hep-th/9409029.

[26] M. Kapranov, *Chow quotients of Grassmannian*, I.M. Gel'fand Seminar, AMS (1993).

[27] B. Kim, *On equivariant quantum cohomology*, IMRN 17 (1996) 841.

[28] F. Knudsen, and D. Mumford, *The projectivity of the moduli space of stable curves I*, preliminaries on 'det' and 'div'. Math. Scand. 39 (1976) 19-55.

[29] D. Morrison and R. Plesser, *Summing the instantons: quantum cohomology and mirror symmetry in toric varieties*, alg-geom/9412236.

[30] S. Hosono, A. Klemm, S. Theisen and S.T. Yau, *Mirror symmetry, mirror map and applications to complete intersection Calabi-Yau spaces*, hep-th/9406055.

[31] S. Hosono, B.H. Lian and S.T. Yau, *GKZ-generalized hypergeometric systems and mirror symmetry of Calabi-Yau hypersurfaces*, alg-geom/9511001, Commun. Math. Phys. 182 (1996) 535-578.

[32] W.Y. Hsiang, *On characteristic classes and the topological Schur lemma from the topological transformation groups viewpoint*, Proc. Symp. Pure Math. XXII. (1971) 105-112.

[33] S. Katz, *On the finiteness of rational curves on quintic threefolds*, Comp. Math. 60 (1986) 151-162.

[34] M. Kontsevich, *Enumeration of rational curves via torus actions.* In: The Moduli Space of Curves, ed. by R.Dijkgraaf, C.Faber, G.van der Geer, Progress in Math.vol.129, Birkhäuser, 1995, 335–368.

[35] J. Li, *Algebraic geometric interpretation of Donaldson's polynomial invariants*, Journ. Diff. Geom. 37 (1993) 417-466.

[36] A. Libgober and J. Teitelboim, Duke Math. Journ., Int. Res. Notices 1 (1993) 29.

[37] B. Lian, K. Liu, and S.T. Yau, *Mirror Principle II*, in preparation.

[38] Yu.I. Manin, *Generating functions in algebraic geometry and sums over trees.* In: The Moduli Space of Curves, ed. by R.Dijkgraaf, C.Faber, G.van der Geer, Progress in Math.vol.129, Birkhäuser, 1995, 401–418.

[39] Yu.I. Manin, *Frobenius Manifolds, quantum cohomology, and moduli spaces (Chapter I,II,III).*, Max-Planck Inst. preprint MPI 96-113.

[40] D. Morrison, *Picard-Fuchs Equations and Mirror Maps for Hypersurfaces*, in [14].

[41] C. Okonek, M. Schneider and H. Spindler *Vector bundles on complex projective spaces*, Progress in Math., Birkhauser.

[42] Y.B. Ruan and G. Tian, *A mathematical theory of quantum cohomology*, Journ. Diff. Geom. Vol. 42, No. 2 (1995) 259-367.

[43] G. Tian, *private communication*, November 1995.

[44] C. Voisin, *A mathematical proof of a formula of Aspinwall-Morrison*, Comp. Math. 104 (1996) 135-151.

[45] E. Witten, *Phases of N=2 theories in two dimension*, hep-th/9301042.

Mirror Principle II

Bong H. Lian,[1] Kefeng Liu,[2] and Shing-Tung Yau[3]

Dedicated to Professor Michael Atiyah.

Abstract.

We generalize our theorems in *Mirror Principle I* to a class of balloon manifolds. Many of the results are proved for convex projective manifolds. In a subsequent paper, Mirror Principle III, we will extend the results to projective manifolds without the convexity assumption.

First appeared in Asian Journ. Math. Vol 3, No 1, (March, 1999)

---

[1] Department of Mathematics, Brandeis University, Waltham, MA 02154.

[2] Department of Mathematics, Stanford University, Stanford, CA 94305.

[3] Department of Mathematics, Harvard University, Cambridge, MA 02138.

Contents

1. Introduction . . . 456
1.1. Main Ideas . . . 457
2. Set-up . . . 460
2.1. Equivariant localization . . . 460
2.2. Functorial localization formula . . . 461
2.3. Balloon manifolds . . . 463
2.4. Sigma models . . . 466
2.5. Regularity of the collapsing map . . . 471
3. The Gluing Identity . . . 474
4. Euler Data . . . 479
4.1. An algebraic property . . . 481
5. Linking and Uniqueness . . . 482
6. Mirror Transformations . . . 486
7. From stable map moduli to Euler data . . . 491
7.1. The Euler data Q . . . 491
7.2. Linking theorem for A . . . 493
8. Applications . . . 495
8.1. Toric manifolds . . . 495
8.2. Chern polynomials for mixed bundles . . . 496
8.3. Convex bundle . . . 499
8.4. A complete intersection in $\mathbf{P}^1 \times \mathbf{P}^2 \times \mathbf{P}^2$ . . . 500
8.5. $V = \mathcal{O}_1(-2) \otimes \mathcal{O}_2(-2)$ on $\mathbf{P}^1 \times \mathbf{P}^1$ . . . 501
9. Generalizations and Concluding Remarks . . . 502
9.1. A weighted projective space . . . 502
9.2. General projective balloon manifolds . . . 502
9.3. A General Mirror Formula . . . 503
9.4. Formulas without T-action . . . 504

1. Introduction

For the long history of mirror symmetry, consult [17]. For a brief description of more recent development, see the introduction in [37][38]. The present paper is a sequel to *Mirror Principle I* [37]. Here, we generalize all the results there to a class of T-manifolds which we call balloon manifolds. These results were announced in [38].

Let X be a projective n-fold, and $d \in H^2(X, \mathbf{Z})$. Let $M_{0,k}(d, X)$ denote the moduli space of k-pointed, genus 0, degree d, stable maps $(C, f, x_1, .., x_k)$ on X [32]. Note that our notation is without the bar. By the work of [34](cf. [7][19]), each nonempty

$M_{0,k}(d,X)$ admits a cycle class $LT_{0,k}(d,X)$ in the Chow group of degree $dim\ X + \langle c_1(X), d\rangle + n - 3$. This cycle plays the role of the fundamental class in topology, hence $LT_{0,k}(d,X)$ is called the virtual fundamental class.

Let V be a convex vector bundle on X. (ie. $H^1(\mathbf{P}^1, f^*V) = 0$ for every holomorphic map $f : \mathbf{P}^1 \to X$.) Then V induces on each $M_{0,k}(d,X)$ a vector bundle V_d, with fiber at $(C, f, x_1, .., x_k)$ given by the section space $H^0(C, f^*V)$. Let b be any multiplicative characteristic class [26]. (ie. if $0 \to E' \to E \to E'' \to 0$ is an exact sequence of vector bundles, then $b(E) = b(E')b(E'')$.) The problem we study here is to compute the characteristic numbers

$$K_d := \int_{LT_{0,0}(d,X)} b(V_d)$$

and their generating function:

$$\Phi(t) := \sum K_d\ e^{d\cdot t}.$$

There is a similar and equally important problem if one starts from a concave vector bundle V [37]. (ie. $H^0(\mathbf{P}^1, f^*V) = 0$ for every holomorphic map $f : \mathbf{P}^1 \to X$.) More generally, V can be a direct sum of a convex and a concave bundle. Important progress made on these problems has come from mirror symmetry. All of it seems to point toward the following general phenomenon [12], which we call *the Mirror Principle.* Roughly, it says that the function $\Phi(t)$ can be computed by a change of variables in terms of certain explicit special functions, loosely called generalized hypergeometric functions.

When X is a toric manifold with $c_1(X) \geq 0$, b is the Euler class, and V is a sum of line bundles, there is a general formula derived in [29] from mirror symmetry. This formula was later studied in [21] based on a series of axioms.

Acknowledgements. We thank B. Cui, J. Li, T.-J. Li, and G. Tian, for numerous helpful discussions with us during the course of this project. B.H.L.'s research is supported by NSF grant DMS-9619884. K.L.'s research is supported by NSF grant DMS-9803234 and the Terman fellowship. S.T.Y.'s research is supported by DOE grant DE-FG02-88ER25065 and NSF grant DMS-9803347.

1.1. Main Ideas

We now sketch our main ideas for computing the classes $b(V_d)$.

Step 1. Localization on the linear sigma model. Consider the moduli spaces $M_d(X) := M_{0,0}((1,d), \mathbf{P}^1 \times X)$. The projection $\mathbf{P}^1 \times$

$X \to X$ induces a map $\pi : M_d(X) \to M_{0,0}(d, X)$. Moreover, the standard action of S^1 on $\mathbf{P}^1$ induces an S^1 action on $M_d(X)$. We first study a slightly different problem. Namely consider the classes $\pi^* b(V_d)$ on $M_d(X)$, instead of $b(V_d)$ on $M_{0,0}(d, X)$. First, there is a canonical way to embed fiber products (see below)

$$F_r = M_{0,1}(r, X) \times_X M_{0,1}(d - r, X)$$

each as an S^1 fixed point component into $M_d(X)$. Let $i_r : F_r \to M_d(X)$ be the inclusion map. Second, there is an evaluation map $e : F_r \to X$ for each r. Third, suppose that there is a projective manifold W_d with S^1 action, that there is an equivariant map $\varphi : M_d(X) \to W_d$, and embeddings $j_r : X \to W_d$, such that the diagram

$$\begin{array}{ccc} F_r & \xrightarrow{i_r} & M_d(X) \\ e\downarrow & & \downarrow\varphi \\ X & \xrightarrow{j_r} & W_d \end{array}$$

commutes. Let α denotes the weight of the standard S^1 action on $\mathbf{P}^1$. Then applying the localization formula [3], this diagram allows us to recast our problem to one of studying the S^1-equivariant classes

$$Q_d := \varphi_! \pi^* b(V_d)$$

defined on W_d. Moreover we can expand the class

$$A_d := \frac{j_0^* Q_d}{e_{S^1}(X_0/W_d)}$$

on X in powers of α^{-1}, and find that it is of order α^{-2}.

The spaces W_d in the commutative diagram above are called the linear sigma model of X. They have been introduced in [39] following [45] when X is a toric manifold,

Step 2. Gluing identity. Consider the vector bundle $\mathcal{U}_d := \pi^* V_d \to M_d(X)$, restricted to the fixed point components F_r. A point in (C, f) in F_r is a pair $(C_1, f_1, x_1) \times (C_2, f_2, x_2)$ of 1-pointed stable maps glued together at the marked points, ie. $f_1(x_1) = f_2(x_2)$. From this, we get an exact sequence of bundles on F_r:

$$0 \to i_r^* \mathcal{U}_d \to U_r' \oplus U_{d-r}' \to e^* V \to 0.$$

Here $i_r^* \mathcal{U}_d$ is the restriction to F_r of the bundle $\mathcal{U}_d \to M_d(X)$. And U_r' is the pullback of the bundle $U_r \to M_{0,1}(d, X)$ induced by V, and similarly for U'_{d-r}. Taking the multiplicative characteristic class b, we get the identity on F_r:

$$e^*b(V)b(i_r^*\mathcal{U}_d) = b(U'_r)b(U'_{d-r}).$$

This is what we call the *gluing identity.* This may be translated to a similar quadratic identity, via Step 1, for Q_d in the equivariant cohomology groups $H^*_{S^1}(W_d)$. The new identity is called the Euler data identity.

Step 3. Linking theorem. The construction above is functorial, so that if X comes equipped with a torus T action, then the entire construction becomes $G = S^1 \times T$ equivariant and not just S^1 equivariant. In particular, the Euler data identity is an identity of G-equivariant classes on W_d. Our problem is to first compute the G-equivariant classes Q_d on W_d satisfying the Euler data identity, and with the property that $A_d \sim \alpha^{-2}$. Note that the restrictions $Q_d|_p$ to the T fixed points p in $X_0 \subset W_d$ are polynomial functions on the Lie algebra of G. Suppose that X is a balloon manifold. Then it can be shown that (with a nondegeneracy assumption on $e_G(X_0/W_d)$) the classes Q_d are uniquely determined by the values of the $Q_d|_p$, when α is some scalar multiple of a weight on the tangent space T_pX. These values of $Q_d|_p$ can be computed explicitly by exploiting the structure of a balloon manifold.

Once these values are known, it is often easy to manufacture explicit G-equivariant classes $\tilde{Q}_d$ with the restrictions $\tilde{Q}_d|_p$ having the above same values, and satisfying the Euler data identity. In this case, we say that the data $\tilde{Q}_d$ are linked to the data Q_d. By a suitable change of variables, one can also arrange that $\frac{j_0^*\tilde{Q}_d}{e_{S^1}(X_0/W_d)} \sim \alpha^{-2}$. By the preceding discussion, we get $Q_d = \tilde{Q}_d$.

Step 4. Computing $\Phi(t)$. Once the classes $Q_d = \varphi_! \pi^* b(V_d)$ are determined, we can unwind the many maps used in Step 1. The preceding computations can be done simply in the form of power series. This finally computes the generating function $\Phi(t)$.

The answer for $\Phi(t)$ is given in the form of Conjecture 9.1. In this paper, for clarity, we restrict ourselves to the case when the tangent bundle of X is convex. We prove that Conjecture 9.1 holds whenever X is a balloon manifold having a linear sigma model W_d such that $e_G(X_0/W_d)$ satisfies a nondegeneracy condition.

In the nonconvex case, we must replace $M_{0,k}(d, X)$ by Li-Tian's virtual fundamental cycle [34] for the purpose of localization and integration. The sequel, Mirror Principle III, to this paper will be devoted entirely to dealing with the added technicality arising from this replacement. All the results in this paper will generalize with only slight modifications as a result of this replacement, but with no change to the overall conceptual framework.

By the equivalence, established in [35], of symplectic GW theory and algebraic GW theory for projective manifolds, we also expect that the results in this paper can be readily generalized to the symplectic case [43][44].

2. Set-up

2.1. Equivariant localization

We first discuss some basic facts about localization. The key technique of our proof is the equivariant localization formula, due to Atiyah-Bott [2][10][3], and Berline-Vergne [9]. For an orbifold version of the localization formula, see [31]. The spirit of the localization we'll use is closer to the Bott residue formula. We first explain this formula.

Let X and Y be two spaces, by which we mean compact manifolds or orbifolds, with a torus T-action. When an orbifold is involved, the integral and localization formulas should be taken in the orbifold sense. Let $\{F\}$ be the components of the fixed point set. Let $H_T^*(\cdot)$ denote the equivariant cohomology group with complex coefficient, and $i_F : F \to X$ the inclusion map. We say that equivariant localization holds on X, if the two maps

$$i_F^* : \; H_T^*(X) \to H_T^*(F), \quad i_{F!} : \; H_T^*(F) \to \; H_T^*(X).$$

which are respectively the pull-back, and the Gysin map, are such that the following formulas holds: given any equivariant cohomology class ω on X, we have

$$\omega = \sum_F i_{F!} \left(\frac{i_F^* \omega}{e_T(F/X)} \right).$$

This formula is equivalent to the integral version of the localization formula

$$\int_X \omega = \sum_F \int_F \frac{i_F^* \omega}{e_T(F/X)}.$$

An important fact about equivariant theory is that, if V is an equivariant vector bundle on an orbifold X, then any characteristic class of V has an equivariant extension. Let $T = S^1$ for simplicity. If c_{2k} is a characteristic class of degree $2k$, then its equivariant extension can be represented by the form

$$c = c_{2k} + c_{2k-2} + \cdots + c_0$$

in the equivariant cohomology of X.

Here is a way to calculate the terms in the localization formula. Assume that each fixed point component F is smooth. If c_{2k} is a Chern class, then by using splitting principle, it can be expressed as a symmetric function of the Chern roots: $P(x_1, \cdots, x_l)$ where $l = rank\ V$. When V can be decomposed into a direct sum of line bundles on F:

$$V|_F = L_1 \oplus \cdots \oplus L_l$$

with the T-action on L_j given by, say the character $e^{2\pi\sqrt{-1}n_j t}$, then the restriction of its equivariant counterpart c to F is

$$i_F^* c = P(c_1(L_1) + n_1 t, \cdots, c_1(L_l) + n_l t).$$

The computation of the equivariant Euler class of F in X is similar. When the restriction of TX to F has a decompostion into line bundles

$$TX|_F = E_1 \oplus \cdots \oplus E_n$$

where T acts on E_j by the character $e^{2\pi\sqrt{-1}m_j t}$, then

$$e_T(F/X) = \prod_j (c_1(E_j) + m_j t).$$

In the above, the n_j's and m_j's are integers for if X is a manifold, and are ratonal numbers if X is an orbifold.

2.2. Functorial localization formula

In this subsection, we derive two formulas which are often used in our work. Let X, Y be two T-spaces, and

$$f : \ X \to Y$$

be an equivariant map. Let E be a fixed point component in Y, and $F := f^{-1}(E)$ be a fixed component in X. Let g be the restriction of f to F, and $j_E : E \to Y$, $i_F : F \to X$ be the inclusion maps. Thus we have the commutative diagram:

$$\begin{array}{ccc} F & \xrightarrow{i_F} & X \\ g \downarrow & & \downarrow f \\ E & \xrightarrow{j_E} & Y. \end{array}$$

Lemma 2.1. *Given any class $\omega \in H_T^*(X)$, we have the equality on E:*

$$\frac{j_E^* f_!(\omega)}{e_T(E/Y)} = g_! \left(\frac{i_F^* \omega}{e_T(F/X)} \right)$$

Proof: Let us consider localization of $\omega f^* j_{E!}(1)$ on X,

$$\omega f^* j_{E!}(1) = i_{F!} \left(\frac{i_F^*(\omega f^* j_{E!}(1))}{e_T(F/X)} \right).$$

Note the contributions from fixed components other than F vanish. Applying the push-forward $f_!$ to both sides, we get

$$f_!(\omega) j_{E!}(1) = f_! i_{F!} \left(\frac{i_F^*(\omega f^* j_{E!}(1))}{e_T(F/X)} \right).$$

Now $f \circ i_F = j_E \circ g$ which, implies

$$f_! i_{F!} = j_{E!} g_!, \quad i_F^* f^* = g^* j_E^*.$$

Thus we get

$$f_!(\omega) j_{E!}(1) = j_{E!} g_! \left(\frac{i_F^*(\omega)\, g^* e_T(E/Y)}{e_T(F/X)} \right).$$

Applying j_E^* to both sides, we then arrive at

$$\begin{aligned} j_E^* f_!(\omega)\, e_T(E/Y) &= e_T(E/Y)\, g_! \left(\frac{i_F^*(\omega)\, g^* e_T(E/Y)}{e_T(F/X)} \right) \\ &= e_T(E/Y)^2\, g_! \left(\frac{i_F^*(\omega)}{e_T(F/X)} \right). \end{aligned}$$

Since $e_T(E/Y)$ is invertible, our assertion follows. □

The same argument applies to the case when E and F are T-invariant subspaces. A slightly different argument for the proof of the above lemma will be given in our subsequent paper. We will also need the following formula, which is actually a special case of [18], Theorem 6.2. Here we include a proof for the convenience of the reader. The spaces involved are T-spaces, that is, T-manifolds or orbifolds.

Lemma 2.2. *Suppose we have a T-equivariant commutative diagram*

$$\begin{array}{ccc} V & \xrightarrow{i} & W \\ g \downarrow & & \downarrow f \\ Z & \xrightarrow{j} & Y \end{array}$$

such that $f^*j_!(1) = i_!(1)$. *Then for any class* ω *on* $H_T^*(W)$, *we have the following equality on* Z*:*

$$j^*f_!(\omega) = g_!i^*(\omega).$$

Proof: By assumption, we have

$$\omega\ f^*j_!(1) = \omega\ i_!(1) = i_!i^*(\omega).$$

Applying $j^*f_!$ to both sides, on the one hand we get

$$j^*f_!(\omega\ f^*j_!(1)) = j^*(f_!(\omega)\ j_!(1)) = j^*f_!(\omega)\ e_T(Z/Y).$$

On the other hand, we get

$$j^*f_!(i_!i^*(\omega)) = j^*j_!g_!i^*(\omega) = g_!i^*(\omega)\ e_T(Z/Y).$$

Thus our assertion follows. □

The case we will use in this paper is when Z and $V = f^{-1}(Z)$ are both T invariant submanifolds of same codimension, in which the condition in the Lemma clearly holds.

2.3. Balloon manifolds

By a *balloon manifold*, we mean a complex projective T-manifold X with the following properties. There are only finite number of T-fixed points. At each fixed point p, the T-weights on the isotropic representation T_pX are pairwise linearly independent. This class of manifolds were introduced by Goresky-Kottwitz-MacPherson [22]. (We refer the reader to [24] for an excellent exposition.) Throughout this paper, we assume that X *is convex*, ie. $H^1(\mathbf{P}^1, f^*TX) = 0$ for any holomorphic map $f : \mathbf{P}^1 \to X$.

One important property of a balloon n-fold is that at each fixed point p, there are exactly n balloons, ie. T-invariant $\mathbf{P}^1$, each balloon connecting p to one other fixed point q. The induced action on each balloon is the standard rotation with two fixed points p and q. (see [22][25]). We *denote by* pq *the balloon connecting the fixed points* p, q. Toric manifolds, complex C-spaces and spherical manifolds are examples of balloon manifolds.

We fix a T equivariant embedding of X into the product of projective spaces

$$\mathbf{P}(n) := \mathbf{P}^{n_1} \times \cdots \times \mathbf{P}^{n_m}$$

such that the pull-backs of the hyperplane classes $H = (H_1, \cdots, H_m)$ generate $H^2(X, \mathbf{Q})$. We use the same notations for the corresponding equivariant classes of the H's, and their restrictions to X. For $\omega \in H^2(X)$ and $d \in H_2(X)$, we denote their pairing by $\langle \omega, d \rangle$.

For convenience, we introduce the following notations:

$$H = (H_1, .., H_m)$$
$$H \cdot \zeta = H_\zeta = H_1\zeta_1 + \cdots + H_m\zeta_m$$
$$H(p) = (H_1(p), .., H_m(p))$$
$$H_\zeta(p) = H_1(p)\zeta_1 + \cdots + H_m(p)\zeta_m.$$

Here $\zeta = (\zeta_1, .., \zeta_m)$ are formal variables. We denote by $K^\vee \subset H_2(X)$ the set of points in $H_2(X, \mathbf{Z})_{free}$ in the dual of the closure of the Kähler cone of X. Since $K^\vee$ is a semigroup in $H_2(X)$, it defines a partial ordering $\succ$ on the lattice $H_2(X, \mathbf{Z})_{free}$. That is, $d \succeq r$ iff $d - r \in K^\vee$. Let $\{H_j^\vee\}$ be the basis dual to the $\{H_j\}$ in $H_2(X)$. If $d \succeq r$ for two classes $d, r \in H_2(X)$, then $d - r = d_1 H_1^\vee + \cdots + d_m H_m^\vee$ for nonnegative integers $d_1, \cdots, d_m$.

We also consider a balloon manifold as a symplectic manifold with a symplectic structure given by $\omega = H_\zeta$ for some generic ζ. By the convexity theorem of Atiyah [2] and Guillemin-Sternberg [23], the image of the moment map μ_ζ in the dual Lie algebra $\mathcal{T}^*$ is a convex polytope, known as the moment polytope. When X is a toric manifold, the moment polytope is known as a Delzant polytope [15]. In this case, it is well-known that the normal fan of this polytope is the defining fan of X.

We say X a multiplicity-free manifold, if for each point p in $\mathcal{T}^*$, the inverse image $\mu_\zeta^{-1}(p)$ is connected.

Lemma 2.3. *Let X be a multiplicity-free balloon manifold, then $H(p) \neq H(q)$ for any two distinct fixed points p and q in X.*

Proof: Let μ_ζ denote the moment map of the T-action on X with respect to the symplectic form $H_\zeta = \langle H, \zeta \rangle$ for a generic choice of $\zeta \in \mathbf{C}^k$. Then the image of $\mu_\zeta : X \to \mathcal{T}^*$ is a convex polytope whose vertices are given the images of the fixed points $\{p\}$. The weights of H_ζ at the fixed points, up to an over all translation, are the same as $\{\mu_\zeta(p)\}$ which are all different. Since X is multiplicity-free, the inverse image of each vertex contains only one fixed point in X. Since ζ is generic, this implies that the $H(p)$'s are distinct at different fixed point. □

We shall assume throughout this paper that $H(p) \neq H(q)$ *for all distinct fixed points* p, q *in* X. Equivalently, if $c(p) = c(q)$ for all $c \in H_T^2(X)$, then $p = q$. This condition is also equivalent to the statement that the moment map with respect to $\omega = H_\zeta$ and the T

action is injective to the set of vertices of the moment polytope, when restricted to the fixed points X^T. By the above lemma we know that toric manifolds and compact homogeneous manifolds all satisfy this condition.

When X is a toric n-fold, we have $N = m + n$ T-invariant divisors in X. Let $D_a = c_1(L_a)$, $a = 1,..,N$, be the equivariant first Chern classes of the corresponding equivariant line bundles. These T divisors correspond 1-1 with the one-cones of the defining fan of X [40]. Moreover the fixed points correspond 1-1 with the n-cones. Labelling the n-cones by $p \in X^T$, we have a balloon [40] pq in X iff the n-cones p, q intersect in a codimension one subcone. Since X is smooth, hence the n-cones are regular, there are exactly n balloons pq for each fixed q. One can give a dual description of all these by using the Delzant polytope.

Returning to the general case, suppose that X is a balloon manifold, and that we have equivariant classes $\{D_a\}$ in $H_T^2(X)$ with the following property. At every fixed point p, $D_a(p)$ is either zero or it is a weight on T_pX. Let pq be a balloon in X. The induced T-action on pq is the standard rotation with fixed points p, q. By applying the localization formula on $pq \simeq \mathbf{P}^1$ and the integral $\langle c, [pq]\rangle$, we have

$$c(q) = c(p) + \langle c, [pq]\rangle\lambda$$

for all $c \in H_T^*(X)$, where λ is the weight on the tangent line $T_q(pq)$. Let $\lambda = D_a(q)$. Specializing to $c = D_a$, we get $D_a(q) = D_a(p) + \langle D_a, [pq]\rangle D_a(q)$. This shows that $\langle D_a, [pq]\rangle \neq 0$. For otherwise we would have $D_a(q) = D_a(p) \neq 0$, and this would mean that $D_a(p)$ is a weight on $T_p(po)$ for some edge po running *in the direction* of $D_a(q) = D_a(p)$ from p to o. So we had three vertices lying joined in a line from q to p to o in the moment graph. This would mean that there is a pair of linearly dependent weights on the tangent space T_pX, which can't happen in a balloon manifold. A similar argument shows that $\langle D_a, [pq]\rangle = 1$.

Lemma 2.4. *Let $\omega = H_\zeta$ and $p, q \in X^T$, $r \succ 0$ and λ be a weight on T_qX. If $\omega(q) = \omega(p) + \langle\omega, r\rangle\lambda$ for generic ζ, then p, q are joined by a balloon, $r = [pq]$, and λ is the weight on the tangent line $T_q(pq)$.*

Proof: It suffices to prove that p, q are joined by a balloon. The last two conclusions then follow immediately. Then under the corresponding moment map μ, p, q are mapped to $\omega(p), \omega(q)$ (up to an overall affine transformation), which are distinct because

$$\omega(q) - \omega(p) = \langle\omega, r\rangle\lambda \neq 0. \tag{2.1}$$

Since λ is a weight on T_qX, there is an edge emanating from the point $\omega(q)$ in the direction of λ, ending at some other vertex $\omega(o)$, where qo is a balloon in X. If $\omega(p) \neq \omega(o)$, we would have three distinct vertices of the moment polytope lying on a single line. Thus $\omega(p) = \omega(o)$, which implies that $p = o$. $\square$

Lemma 2.5. *The zero class $\omega = 0$ is the only class in $H_T^*(X)$ with the property that*

$$\int_X \omega\ e^{H_\zeta} = 0$$

for all generic $\zeta \in \mathbf{C}$.

Proof: Suppose

$$\int_X \omega\ e^{H_\zeta} = 0.$$

By localization, we have

$$\sum_{p \in X^T} \frac{\omega(p)}{e_T(p/X)} e^{H_\zeta(p)} = 0.$$

But since the vectors $H(p)$ are all distinct, those exponential functions in ζ are linearly independent over the field $\mathbf{Q}(\mathcal{T}^*)$, implying that $\omega(p) = 0$ for all p. Thus $\omega = 0$. $\square$

2.4. Sigma models

Let X be balloon manifold with a fixed T-equivariant embedding $X \to \mathbf{P}(n)$, as discussed above. We write

$$M_d(X) := M_{0,0}((1,d), \mathbf{P}^1 \times X).$$

Since X is assumed to be convex, $M_d(X)$ is an orbifold. The standard S^1 action on $\mathbf{P}^1$ together with the T action on X induce a $G = S^1 \times T$ action on $M_d(X)$.

Here is a description of some S^1 fixed point components F_r, labelled by $0 \preceq r \preceq d$, inside of $M_d(X)$. Let F_r be the fiber product

$$F_r := M_{0,1}(r, X) \times_X M_{0,1}(d-r, X)$$

More precisely, consider the map

$$ev_r \times ev_{d-r} : M_{0,1}(r, X) \times M_{0,1}(d-r, X) \to X \times X$$

given by evaluations at the marked points; and

$$\Delta:\ X\to X\times X$$

the diagonal map. Then

$$F_r=(ev_r\times ev_{d-r})^{-1}\Delta(X).$$

Note that $F_d=M_{0,1}(d,X)$ by convention. The set F_r can be identified with an S^1 fixed point component of $M_d(X)$ as follows. Consider the case $r\neq 0,d$ first. Given a point $(C_1,f_1,x_1)\times(C_2,f_2,x_2)$ in F_r, we get a new curve C by gluing C_1,C_2 to $\mathbf{P}^1$ with x_1,x_2 glued to $0,\infty\in\mathbf{P}^1$ respectively. The new curve C is mapped into $\mathbf{P}^1\times X$ as follows. Map $\mathbf{P}^1\subset C$ identically onto $\mathbf{P}^1$, and collapse C_1,C_2 to $0,\infty$ respectively; then map C_1,C_2 into X with f_1,f_2 respectively, and collapse the $\mathbf{P}^1$ to $f(x_1)=f(x_2)$. This defines a point (C,f) in $M_d(X)$. For $r=0$, we glue (C_1,f_1,x_1) to $\mathbf{P}^1$ at x_1 and 0. For $r=d$, we glue (C_2,f_2,x_2) to $\mathbf{P}^1$ at x_2 and ∞. We will identify F_r as a subset of $M_d(X)$ as above, and let

$$i_r:F_r\to M_d(X)$$

denotes the inclusion map. Clearly, we also have an evaluation map

$$e_r:F_r\to X$$

which sends a pair in F_r to the value at the marked point. In the following, we will simply write e_r as e without causing any coonfusion.

We call a compact manifold or orbifold W_d with $G=S^1\times T$ action a *linear sigma model* of degree d for X, if the following conditions are satisfied:

1. The S^1 action on W_d has fixed point components given by X_r, labelled by $0\preceq r\preceq d$, and each X_r is T-equivariantly isomorphic to X.
2. There is a G-equivariant birational map φ from $M_d(X)$ to W_d, such that $\varphi|_{F_r}=e$, and $\varphi^{-1}(X_r)=F_r$.
3. All equivariant cohomology classes in $H_G^2(W_d)$ are lifted from $H_T^2(X)$, and the lift $\hat{D}\in H_G^2(W_d)$ of $D\in H_T^2(X)$ restricts to $D+\langle D,r\rangle\alpha$ on X_r.
4. The G-equivariant Euler class of the normal bundle of X_0 in W_d has the form

$$e_G(X_0/W_d)=\prod_a\prod_{m_a}(D_a-m_a\alpha)$$

where the m_a's are positive integers and the D_a's are classes in $H_T^2(X)$, such that at a given T fixed point p in X, the nonzero $D_a(p)$'s are multiples of distinct weights of T_pX.

Here a birational map, in algebraic geometry language, is a regular morphism which is an isomorphism when restricted to a Zariski open set in $M_d(X)$.

Note W_d need not be unique. We identify X_r with X by assumption 1, and denote by

$$j_r : X_r \to W_d$$

the inclusion map.

We call a balloon manifold X *admissible* if it has a linear sigma model W_d for each d, and that $H_\zeta(p) \neq H_\zeta(q)$ for any two distinct fixed points p, q in X. The main result in this paper is to show that the mirror principle holds for any admissible balloon manifold.

Remark 2.6. *Condition 4 is actually assuming more than what we need. This condition can be replaced by the following weaker, but more technical condition. For each fixed point p and for any d, as a function of α, $e_G(X_0/W_d)|_p$ has possible zero only at either 0 or a multiple of a weight λ on T_pX. In addition if $[pq]$ is a balloon and $d = \delta[pq]$, then λ/δ is at worse a simple zero. For example, the following form would meet this criterion:*

$$e_G(X_0/W_d) = \frac{\prod_a \prod_{m_a}(D_a - m_a\alpha)}{\prod_b \prod_{n_b}(D_b - n_b\alpha)}$$

where the m_a, n_b are nonzero scalars.

Example 1: Projective space $\mathbf{P}^n$ with $W_d = \mathbf{P}^{(n+1)d+n}$ is admissible. The existence of φ was proved in [37], which is the so-called Li-Tian map. The lifted hyperplane class κ has the required property that

$$j_r^*\kappa = H + \langle H, r\rangle\alpha = H + r\alpha.$$

The equivariant Euler class

$$e_G(\mathbf{P}^n_0/W_d) = \prod_{i=0}^{n}\prod_{m=1}^{d}(H - \lambda_i - m\alpha)$$

where λ_i's denote the weights of the torus T action on $\mathbf{P}^n$. Clearly the equivariant classes $\{H - \lambda_i\}$ has the required property.

Example 2: More generally for $\mathbf{P}(n)$, we can take $N_d(\mathbf{P}(n))$ to be W_d. In fact the S^1 fixed point components on $N_{k,l}$ are exactly $k+1$ copies $\mathbf{P}^l_r$, $r = 0, .., k$, of $\mathbf{P}^l$. Each $\mathbf{P}^l_r$ consists of $l+1$ tuples of monomials, each being a scalar multiple of $w_0^r w_1^{k-r}$. Similarly the S^1 fixed point components on $N_d(\mathbf{P}(n))$ are copies $\mathbf{P}(n)_r$, $0 \preceq r \preceq d$, of

$\mathbf{P}(n)$. All equivariant cohomology classes in $H^2_G(N_d(\mathbf{P}(n))$ are lifted from $H^2_T(\mathbf{P}^n)$ (cf [37]). Let κ_i be the lift of the hyperplane class H_i, of the ith factor $\mathbf{P}^{n_i}$. Then

$$j_r^* \kappa_i = H_i + \langle H_i, r \rangle \alpha$$

where j_r denotes the inclusion of $\mathbf{P}(n)_r$ in $N_d(\mathbf{P}(n))$. By using the formula in [37], it is easy to show the equivariant Euler class $e_G(\mathbf{P}(n)_0/N_d(\mathbf{P}(n)))$, which is a product of e_G's in last example, has the required property.

Example 3: Let $N_{k,l}$ be the space of $l+1$ tuples $[f_0, .., f_l]$ of degree k polynomials $f_i(w_0, w_1)$, modulo scalar. Thus $N_{k,l} \cong \mathbf{P}^{(l+1)k+l}$. It is called the linear sigma model for $\mathbf{P}^l$. (See [37].) Let

$$N_d(\mathbf{P}(n)) := N_{d_1,n_1} \times \cdots \times N_{d_m,n_m}.$$

Recall that we have a collapsing map $\varphi : M_k(\mathbf{P}^l) \to N_{k,l}$, which is $G := S^1 \times T$ equivariant. By taking composite with the projection from $M_d(\mathbf{P}(n))$ to each $M_{d_j}(\mathbf{P}^{n_j})$, we obtain a G-equivariant map

$$M_d(\mathbf{P}(n)) \to N_d(\mathbf{P}(n))$$

which we also denote by φ. Note that $M_d(X)$ can be viewed as a cycle in $M_d(\mathbf{P}(n))$. We denote the image cycle $\varphi_!(M_d(X))$ in $N_d(\mathbf{P}(n))$ φ by $N_d(X)$.

If $N_d(X)$ is a manifold or an orbifold, then Properties 1-3 are automatically satisfied, if furthermore $e_G(X_0/W_d)$ has property 4, then we can simply take $W_d = N_d(X)$ as the linear sigma model.

Example 4: Convex toric varieties. In this case W_d is a toric n-fold, as introduced by Witten [45]and used first by Morrison-Plesser [39]to study quantum cohomology. Recall that a toric manifold X can be realized as the GIT quotient $\mathbf{C}^N//T^m_{\mathbf{C}}$ where $T^m_{\mathbf{C}}$ is a m-dimensional complex torus acting on $\mathbf{C}^N$. Here $m = rank\ H^2(X, \mathbf{Z})$, $N = n + m$. Let $[z_1, \cdots, z_N]$ denote the coordinates on $\mathbf{C}^N$. Then each z_j can be viewed as a section of a line bundle L_j on X [14][39]. Modulo the induced action by T^m_C from $\mathbf{C}^N$, a map from $\mathbf{P}^1$ into X is uniquely represented by an N-tuple of polynomials

$$[f_1(w_0, w_1), \cdots, f_N(w_0, w_1)]$$

where f_j is a section of the line bundle $O(l_j)$ over $\mathbf{P}^1$ with $l_j = \langle c_1(L_j), d \rangle$. Let $\mathbf{C}^N(d)$ be the vector space of N-tuple of polynomials of degree $(d_1, \cdots, d_N)$ as above. Then as described in [39], W_d is the GIT quotient by the induced action of T_C^m on it:

$$W_d = \mathbf{C}^N(d)//T_C^m.$$

Let $M_d^o(X)$ denote the set of points (f, C) in $M_d(X)$ such that $C \simeq \mathbf{P}^1$. We call $M_d^o(X)$ the smooth part of $M_d(X)$. We can define a map φ_o from $M_d^o(X)$ to W_d in the following way: each (f, C) gives a map from $\mathbf{P}^1$ to X, and modulo the induced T_C^m action, uniquely determines N-tuple of polynomials as above, therefore gives a point in W_d, which we define to be the image of (f, C) under φ_o. This is clearly a canonical identification.

It is not difficult to see that the S^1-fixed components in W_d can be described as GIT quotient,

$$X_r \simeq \{[a_1 w_0^{\langle D_0, r\rangle} w_1^{\langle D_0, d-r\rangle}, \cdots, a_N w_0^{\langle D_N, r\rangle} w_1^{\langle D_N, d-r\rangle}] | a \in \mathbf{C}^N\}//T_C^m.$$

The equivariant Euler class of its normal bundle in W_d is

$$e_G(X_r/W_d) = \prod_{a=1}^{N} \prod_{k=0, k\neq \langle D_a, r\rangle}^{\langle D_a, d\rangle} (D_a + \langle D_a, r\rangle\alpha - k\alpha).$$

Here $D_a = c_1(L_a)$ is the equivariant first Chern class of the line bundle L_a corresponding to the ath component in the coordinates of X. The lift $\hat{D}_a$ of D_a to W_d clearly has the property

$$j_r^* \hat{D}_a = D_a + \langle D_a, r\rangle\alpha.$$

As pointed out in [39], the cohomology of W_d are generated by the $\hat{D}_a$. Thus W_d has properties 1, 3, and 4. In the next subsection, we establish that the φ_o extends to a regular G-equivariant map φ from $M_d(X)$ to W_d, with property 2. So, for a toric X we can take its linear sigma model to be W_d as constructed above. It follows that X is an admissible balloon manifold.

Example 5: Our method works well even for certain singular manifolds. We take weighted projective space as example to illustrate the ideas. Let $\mathbf{P}_a^n$ with $a = (a_0, \cdots, a_n)$ be a weighted projective space. Let $[z_0, \cdots, z_n]$ be the coordinates for $\mathbf{P}_a^n$, then z_j can be considered as a section of the line bundle $O(a_j)$. Then the linear sigma model for this weighted projective space is the induced weighted quotient by S^1 on the space of $n+1$ tuple of polynomials $[f_0(w_0, w_1), \cdots, f_1(w_0, w_1)]$ where f_j is a section of the line bundle $O(da_j)$ on $\mathbf{P}^1$.

It is known that $\mathbf{P}_a^n$ is equivalent to $\mathbf{P}^n/Z_a$ where Z_a is a finite group. The space $M_d(\mathbf{P}_a^n)$ for $\mathbf{P}_a^n$ is equal to $M_d(\mathbf{P}^n)/Z_a$. On the other hand we can also take $N_d(\mathbf{P}^n)/Z_a$ as the linear sigma model W_d. In this case W_d is an orbifold, a weighted projective space. Since the action of Z_a commutes with the action of torus T, we see that the induced collapsing map

$$\varphi : \quad M_d(\mathbf{P}_a^n) \to N_d(\mathbf{P}_a^n) = W_d$$

is clearly a regular map. The corresponding equivariant Euler class has the expression:

$$e_G(\mathbf{P}_a^n/W_d) = \prod_{j=0}^{n} \prod_{m=1}^{da_j} (a_j H - \lambda_j - m\alpha)$$

with H the T-equivariant hyperplane class and λ_j's the weights of the T-action.

Examples of singular toric varieties will be discussed again in our subsequent paper, in which resolution of singularities will be used to reduce to the smooth case. The above example was motivated by a question of Mazur, who suggested that the situation of counting rational curves in orbifolds is similar to certain Diophantine problem in number theory.

Example 6: For a general projective manifold X embedded in $\mathbf{P}(n)$, assume it is defined by a system of polynomial equations $P(z^1, \cdots, z^n) = 0$ where $z^j = (z_1^j, \cdots, z_{n_j}^j)$ denotes the coordinate of $\mathbf{P}^{n_j}$. Assume the variety defined by the induced equation $P(f^1, \cdots, f^n) = 0$ in $N_d(\mathbf{P}(n))$ where $f^j = (f_1^j(w_0, w_1), \cdots, f_{n_j}^j(w_0, w_1))$ is the tuple of polynomials, the coordinates for the linear sigma model $N_d(\mathbf{P}^{n_j})$, is an orbifold. Then we can take it to be our linear sigma model W_d. Note that $N_d(X)$ in Example 3 is embedded inside this W_d. Very likely they are the same.

Though we don't know whether this variety is an orbifold or not, it is clear that the fixed point compoenents in the above variety are given by X_r's. In fact, we only need to assume that the localization formula holds on it. This is the case if the fixed point components embedded into W_d as local complete intersection subvarieties. We conjecture that this is the case for any convex projective manifold. Later, we will state a general conjectural Mirror Formula in terms of this W_d.

2.5. Regularity of the collapsing map

For a toric manifold X, the following lemma show that W_d described in Example 4 is a linear sigma model of X.

Lemma 2.7. *For toric manifold X, there is a regular extension*

$$\varphi: \ M_d(X) \to W_d$$

of the map φ_o in Example 4 above.

Proof: We simply follow the argument in [37], together with the construction in [14]. We will define a morphism $\varphi : M_d(X) \to W_d$. Let $\mathcal{S}$ be the category of all schemes of finite type (over $\mathbf{C}$) and let

$$\mathcal{F} : \mathcal{S} \longrightarrow (\mathrm{Set})$$

be the the contra-variant functor that send any $S \in \mathcal{S}$ to the set of families of stable morphisms

$$F : \mathcal{X} \longrightarrow \mathbf{P}^1 \times X \times S$$

over S, where $\mathcal{X}$ are families of connected arithmetic genus 0 curves, modulo the obvious equivalence relation. Note that $\mathcal{F}$ is represented by the moduli stack $M_d(X)$. Hence to define the morphism φ, it suffices to define a transformation

$$\Psi : \mathcal{F} \longrightarrow \mathrm{Mor}\,(-, W_d).$$

We now define such a transformation. Let $S \in \mathcal{S}$ and let $\xi \in \mathcal{F}(S)$ be represented by $(\mathcal{X}, F)$. We let p_i be the composite of F with the i-th projection of $\mathbf{P}^1 \times X \times S$ and let p_{ij} be the composite of F with the projection from $\mathbf{P}^1 \times X \times S$ to the product of its i-th and j-th components. We consider the sheaf $p_2^* \mathcal{O}_X(L_j)$ on $\mathcal{X}$ and its direct image sheaf

$$\mathcal{L}_{j,\xi} = p_{13*} p_2^* \mathcal{O}_X(L_j).$$

Here the L_j are the line bundles on X, as defined in Example 4. As in [37], one can show that $\mathcal{L}_{j,\xi}$ is flat in a standard way.

For the same reasoning, the sheaves $\mathcal{L}_{j,\xi}$ satisfy the following base change property: let $\rho : T \to S$ be any base change and let $\rho^*(\xi) \in \mathcal{F}(T)$ be the pull back of ξ. Then there is a canonical isomorphism of sheaves of $\mathcal{O}_T$-modules

$$\mathcal{L}_{j,\rho^*(\xi)} \cong (\mathbf{1}_{\mathbf{P}^1} \times \rho)^* \mathcal{L}_{j,\xi}. \tag{2.2}$$

Since $\mathcal{L}_{j,\xi}$ is flat over S, we can define the determinant line bundle of $\mathcal{L}_{j,\xi}$, denoted by $\det(\mathcal{L}_{j,\xi})$ which is an invertible sheaf over $\mathbf{P}^1 \times S$. Using the Riemann-Roch theorem, one finds that its degree along fibers over S is $l_j = \langle c_1(L_j), d\rangle$. Furthermore, because $\mathcal{L}_{j,\xi}$ has rank one, there is a canonical homomorphism

$$\mathcal{L}_{j,\xi} \longrightarrow \det(\mathcal{L}_{j,\xi}), \tag{2.3}$$

so that its kernel is the torsion subsheaf of $\mathcal{L}_{j,\xi}$.

Let z_j be the j-th homogeneous coordinate of X (Example 4). Then z_j is a section in $H^0(X, L_j)$. Its pull-back is a section of $\mathcal{L}_{j,\xi}$, which induces a section

$$\sigma_{j,\xi} \in H^0(S, \pi_{S*}\det(\mathcal{L}_{j,\xi})).$$

based on (2.3). Then after fixing an isomorphism

$$\det(\mathcal{L}_{j,\xi}) \cong \pi_S^*\mathcal{M} \otimes \pi_{\mathbf{P}^1}^*\mathcal{O}_{\mathbf{P}^1}(l_j) \tag{2.4}$$

for some invertible sheaf $\mathcal{M}$ of $\mathcal{O}_S$-modules, where $l_j = \langle c_1(L_j), d\rangle$. We then obtain a section in

$$\pi_{S*}(\pi_{\mathbf{P}^1}^*\mathcal{O}_{\mathbf{P}^1}(l_j)) \otimes_{\mathcal{O}_S} \mathcal{M} \equiv H^0_{\mathbf{P}^1}(\mathcal{O}_{\mathbf{P}^1}(l_j)) \otimes_{\mathbf{C}} \mathcal{M}.$$

So $\sigma_{j,\xi}$ is determined up to certain constant λ_j coming from $\mathcal{M}$.

Now apply the above argument to each L_j, $j = 1, \cdots, N$, we get N sections $[\sigma_{1,\xi}, \cdots, \sigma_{N,\xi}]$. Let w_0, w_1 be the homogeneous coordinate of $\mathbf{P}^1$, we will write $\sigma_{j,\xi} = f_j(w_0, w_1)$ as a homogeneous polynomial of degree l_j. In this way we get a point in $\mathbf{C}^N(d)$. The constants λ_j from $\mathcal{M}$ in choosing $\sigma_{j,\xi}$ must satisfy the relation $\prod_j \lambda_j^{\langle m, n_j\rangle} = 1$. Here the n_j are vectors in an integral lattice, which generate the 1-cones in the defining fan of X, and m is any element in the dual lattice. (See [14].)

For such λ_j's we can then find an element g in T_C^m such that $[\lambda_1 f_1, \cdots, \lambda_N f_N]$ is transformed to $[f_1, \cdots, f_N]$ by g, therefore they represent the same point in W_d. In this way, after taking GIT quotient by T_C^m, the induced action on $[f_1(w_0, w_1), \cdots, f_N(w_0, w_1)]$ from the action on $\mathbf{C}^N$, for each map $(f, C) \in M_d(X)$, we have obtained canonically a point in W_d, therefore a morphism

$$\Psi(S) : S \longrightarrow W_d$$

that is independent of the isomorphisms (2.4). It follows from the base change property (2.2) that the collection $\Psi(S)$ defines a transformation

$$\mathbf{\Psi} : \mathcal{F} \longrightarrow \mathrm{Mor}\,(-, W_d),$$

thus defines the morphism φ as desired.

The fact that $\varphi : M_d(X) \to W_d$ is $S^1 \times T^N$-equivariant follows immediately from the fact that φ is induced by the transformation Ψ of functors. This completes the proof. □

For another proof of the above lemma we can proceed as follows. We use the notations as in the above Example 3. We show that the regularity of the collapsing map for $\mathbf{P}(n)$ induces the regularity of the collapsing map for X. For this we only need to prove that $N_d(X)$, the image $\varphi(M_d(X))$ in $N_d(\mathbf{P}(n))$ of the collapsing map for $\mathbf{P}(n)$, lies in W_d.

First, we show that W_d lies in $N_d(\mathbf{P}(n))$. Note that both W_d and $N_d(\mathbf{P}(n))$ are toric manifolds, and a Zariski open subset W_d^o in W_d is embedded G-equivariantly in $N_d(\mathbf{P}(n))$. Also the G-fixed points in W_d are all in X_r, therefore in $\mathbf{P}(n)_r$ and $N_d(P(n))$. Any point in W_d is in the closure of a generic $G_{\mathbf{C}}$ orbit in W_d passing through two G fixed points in X_r's. By the equivariance, this $G_{\mathbf{C}}$ orbit is also in $N_d(\mathbf{P}(n))$, therefore the closure of this orbit lies in $N_d(\mathbf{P}(n))$.

Second, we show that $N_d(X)$ lies in W_d. For this we note that φ_o extends to $M_d(X)$, since it is actually the restriction of the corresponding map on $M_d(\mathbf{P}(n))$. Now by taking closure of the inclusion $\varphi(M_d^o(X)) \subseteq W_d$, which is induced from the canonical identification, we get

$$\varphi(M_d(X)) = \overline{\varphi(M_d^o(X))} \subseteq \overline{W_d} = W_d,$$

since W_d is itself closed.

3. The Gluing Identity

Returning to the general case, we let X be an admissible balloon manifold from now on. In this section, we apply the functorial localization formula to the linear sigma model. The argument used here is modelled on the one used in [37], except that the T-action is *not* used here. Thus all the results in this section hold for manifolds without T action. We will have more to say about the mirror principle without T action later.

Recall that we have the commutative diagram:

$$\begin{array}{ccc} F_r & \xrightarrow{i_r} & M_d(X) \\ e\downarrow & & \downarrow\varphi \\ X_r & \xrightarrow{j_r} & W_d. \end{array}$$

We also have the natural forgetting map $\rho : M_{0,1}(d,X) \to M_{0,0}(d,X)$, and the projection map $\pi : M_d(X) \to M_{0,0}(d,X)$. Note that we have a commutative diagram

$$\begin{array}{ccc} M_d(X) & & \\ \pi \downarrow & \nwarrow i_0 & \\ M_{0,0}(d,X) & \overset{\rho}{\longleftarrow} & M_{0,1}(d,X). \end{array}$$

Let $\varphi : M_d(X) \to W_d$, $e : F_r \to X_r$ play the respective roles of $f : X \to Y$, $g : F \to E$ in the functorial localization formula. Then it follows that

Lemma 3.1. *Given any G-equivariant cohomology class ω on $M_d(X)$, we have the following equality on X_r for $0 \preceq r \preceq d$:*

$$\frac{j_r^* \varphi_!(\omega)}{e_G(X_r/W_d)} = e_! \left(\frac{i_r^*(\omega)}{e_G(F_r/M_d(X))} \right).$$

Actually this lemma may be viewed as an equivariant version of the so-called excess intersection formula of [18], Theorem 6.3.

Let L_r denote the line bundle on $M_{0,1}(r,X)$ whose fiber at $(f,C;x)$ is the tangent line at the marked point $x \in C$. Let π_1 denote the projection from $\mathbf{P}^1 \times X$ to $\mathbf{P}^1$.

The normal bundle of F_r in $M_d(X)$ can be computed just as in [37]. For $r \neq 0, d$, we have

$$N(F_r/M_d(X)) = H^0(C_0, (\pi_1 \circ f)^* T\mathbf{P}^1) + T_{x_1}C_0 \otimes L_r + T_{x_2}C_0 \otimes L_{d-r} - A_{C_0}.$$

Here we have used the notations as in [37]: a point (f_1, C_1, x_1) in $M_{0,1}(r,X)$ and a point (f_2, C_2, x_2) in $M_{0,1}(d-r,X)$ is glued to $C_0 \simeq \mathbf{P}^1$ at 0 and ∞ respectively to get the point (f,C) in $M_d(X)$ with $C \simeq C_1 \cup C_0 \cup C_2$. Since x_1 and x_2 are mapped to the same point in X under the projection $\pi_2 : \mathbf{P}^1 \times X \to X$, so this point can be considered as a point in F_r by gluing together (f_1, C_1, x_1) and (f_2, C_2, x_2) at the marked points. Similarly, for $r = 0, d$, we have

$$N(F_0/M_d(X)) = H^0(C_0, (\pi_1 \circ f)^* T\mathbf{P}^1) + T_{x_1}C_0 \otimes L_d - A_{C_0}$$

and

$$N(F_d/M_d(X)) = H^0(C_0, (\pi_1 \circ f)^* T\mathbf{P}^1) + T_{x_2}C_0 \otimes L_d - A_{C_0}.$$

In the above $H^0(C_0, (\pi_1 \circ f)^* T\mathbf{P}^1)$ corresponds to the deformation of C_0; $T_{x_1}C_0 \otimes L_r$ and $T_{x_2}C_0 \otimes L_{d-r}$ correspond respectively to the deformations of the nodal points x_1 and x_2; A_{C_0} denotes the automorphism group to be quotiented out.

The equivariant Euler classes of the normal bundles above are computed as in [37], to which we refer the readers for details. For $r \neq 0, d$, the equivariant Euler classes are:

$$e_G(F_r/M_d(X)) = -\alpha(-\alpha + c_1(L_{d-r})) \cdot \alpha(\alpha + c_1(L_r))$$

where the two factors on the right hand side are pullbacked to F_r from $M_{0,1}(d-r, X)$, $M_{0,1}(r, X)$ respectively. For $r = 0, d$, we have

$$e_G(F_0/M_d(X)) = -\alpha(-\alpha + c_1(L_d)),\ e_G(F_d/M_d(X)) = \alpha(\alpha + c_1(L_d))$$

respectively. Combining this with the preceding lemma, we get the following equality on $X = X_0$:

$$\frac{j_0^* \varphi_!(\omega)}{e_G(X_0/W_d)} = ev_! \left(\frac{i_0^*(\omega)}{\alpha(\alpha - c_1(L_d))} \right).$$

Here we have dropped the subscript from ev_d. In particular, if ψ is a class on $M_{0,0}(d, X)$, then for $\omega = \pi^* \psi$, we get $i_0^*(\omega) = i_0^*(\pi^* \psi) = \rho^* \psi$. This yields

Lemma 3.2. *Given any T-equivariant cohomology class ψ on $M_{0,0}(d, X)$, we have the following equality on X:*

$$\frac{j_0^* \varphi_!(\pi^* \psi)}{e_G(X_0/W_d)} = ev_! \left(\frac{\rho^* \psi}{\alpha(\alpha - c_1(L_d))} \right).$$

Lemma 3.3. *For $0 \preceq r \preceq d$, we have the following equality on X:*

$$e_G(X_r/W_d) = \overline{e_G(X_0/W_r)} e_G(X_0/W_{d-r}).$$

In particular, we have

$$e_G(X_d/W_d) = \overline{e_G(X_0/W_d)}.$$

Proof: Consider the commutative diagram

$$\begin{array}{ccc} F_r & \xrightarrow{\Delta_0} & M_{0,1}(r, X) \times M_{0,1}(d-r, X) \\ e \downarrow & & \downarrow ev_r \times ev_{d-r} \\ X & \xrightarrow{\Delta} & X \times X \end{array} \tag{3.1}$$

where Δ is the diagonal map, and Δ_0 is the inclusion induced by Δ. In particular, by definition we have $(ev_r \times ev_{d-r})^* \Delta_!(1) = (\Delta_0)_!(1)$. So we have

$$\Delta^*(ev_r \times ev_{d-r})_!(\omega) = e_! \Delta_0^*(\omega) \tag{3.2}$$

for any class ω on $M_{0,1}(r, X) \times M_{0,1}(d-r, X)$. Now put $\omega = \frac{1}{\alpha(\alpha+c_1(L_r))} \times \frac{1}{\alpha(\alpha-c_1(L_{d-r}))}$. Then (3.2) becomes

$$(ev_r)_! \frac{1}{e_G(F_0/M_r(X))} \cdot (ev_{d-r})_! \frac{1}{e_G(F_0/M_{d-r}(X))} = e_! \frac{1}{e_G(F_r/M_d(X))}.$$

Since $\varphi : M_d(X) \to W_d$ is an isomorphism on a Zariski open set, we see that $\varphi_!(1) = 1$. In fact, by Prop. (5.3.3) [1], $\varphi_!$ preserves degree because since $M_d(X)$ and W_d have the same dimension. So $\varphi_!(1) \in H_G^0(W_d)$ is a constant. By restricting to the Zariski open set on which φ is an isomorphism, we find $\varphi_!(1) = 1$.

By taking $\psi = 1$ in the preceding lemma, we get

$$(ev_r)_! \frac{1}{e_G(F_0/M_r(X))} = \frac{1}{e_G(X_0/W_r)},$$
$$(ev_{d-r})_! \frac{1}{e_G(F_0/M_{d-r}(X))} = \frac{1}{e_G(X_0/W_{d-r})}.$$

By taking $\omega = 1$ in Lemma 3.1, we get

$$e_! \frac{1}{e_G(F_r/M_d(X))} = \frac{1}{e_G(X_r/W_d)}.$$

Combining the last four equations yields our assertion. $\square$

Fix a T-equivariant multiplicative class b_T. Fix a T-equivariant bundle of the form $V = V^+ \oplus V^-$, where $V^\pm$ are respectively the convex/concave bundles. (cf. [37].) We call such a V a mixed bundle. We assume that

$$\Omega := \frac{b_T(V^+)}{b_T(V^-)}$$

is a well-defined invertible class on X. By convention, if $V = V^\pm$ is purely convex/concave, then $\Omega = b_T(V^\pm)^{\pm 1}$. Recall that the bundle $V \to X$ induces the bundles

$$V_d \to M_{0,0}(d, X), \quad U_d \to M_{0,1}(d, X), \quad \mathcal{U}_d \to M_d(X).$$

Moreover, they are related by $U_d = \rho^* V_d$, $\mathcal{U}_d = \pi^* V_d$, Throughout this section, we denote

$$Q: \quad Q_d := \varphi_!(\pi^* b_T(V_d)).$$

If ω is a class on W_d, we write

$$i_r^*\omega^v := \frac{j_r^*\omega}{e_G(X_r/W_d)}$$

which is a class on $X = X_r$.

Lemma 3.4. *For* $0 \preceq r \preceq d$,

$$\Omega\ i_r^* Q_d^v = \overline{i_0^* Q_r^v}\ i_0^* Q_{d-r}^v.$$

Proof: For simplicity, let's consider the case $V = V^+$. The general case is entirely analogous.

Recall that a point (f, C) in $F_r \subset M_d$ comes from gluing together a pair of stable maps $(f_1, C_1, x_1), (f_2, C_2, x_2)$ with $f_1(x_1) = f_2(x_2) = p \in X$. From this, we get an exact sequence over C:

$$0 \to f^*V \to f_1^*V \oplus f_2^*V \to V|_p \to 0.$$

Passing to cohomology, we have

$$0 \to H^0(C, f^*V) \to H^0(C_1, f_1^*V) \oplus H^0(C_2, f_2^*V) \to V|_p \to 0.$$

Hence we obtain an exact sequence of bundles on F_r:

$$0 \to i_r^*\mathcal{U}_d \to U_r' \oplus U_{d-r}' \to e^*V \to 0.$$

Here $i_r^*\mathcal{U}_d$ is the restriction to F_r of the bundle $\mathcal{U}_d \to M_d(X)$. And U_r' is the pullback of the bundle $U_r \to M_{0,1}(d, X)$, and similarly for U_{d-r}'. Taking the multiplicative characteristic class b_T, we get the identity on F_r:

$$e^*b_T(V)b_T(i_r^*\mathcal{U}_d) = b_T(U_r')b_T(U_{d-r}').$$

This is what we call the *gluing identity.*

Now put

$$\omega = \frac{b_T(U_r)}{e_G(F_r/M_r(X))} \times \frac{b_T(U_{d-r})}{e_G(F_0/M_{d-r}(X))}.$$

From the commutative diagram (3.1), we have the identity:

$$\Delta^*(ev_r \times ev_{d-r})_!(\omega) = e_!\Delta_0^*(\omega).$$

On one hand is

$$\begin{aligned}\Delta^*(ev_r \times ev_{d-r})_!(\omega) &= (ev_r)_! \frac{b_T(U_r)}{e_G(F_r/M_r(X))} \cdot (ev_{d-r})_! \frac{b_T(U_{d-r})}{e_G(F_0/M_{d-r}(X))} \\ &= (ev_r)_! \frac{\rho^* b_T(V_r)}{e_G(F_r/M_r(X))} \cdot (ev_{d-r})_! \frac{\rho^* b_T(V_{d-r})}{e_G(F_0/M_{d-r}(X))} \\ &= \overline{i_0^* Q_r^v}\; i_0^* Q_{d-r}^v,\end{aligned}$$

the last equality being a consequence of Lemma 3.2. On the other hand, applying the gluing identity, we have

$$\begin{aligned}e_! \Delta_0^*(\omega) &= e_! \left(\frac{b_T(U_r')}{\alpha(\alpha + c_1(L_r))} \frac{b_T(U_{d-r}')}{\alpha(\alpha - c_1(L_{d-r}))} \right) \\ &= e_! \frac{e^* b_T(V) i_r^* b_T(\mathcal{U}_d)}{e_G(F_r/M_d(X))} \\ &= b_T(V)\; e_! \frac{i_r^* b_T(\mathcal{U}_d)}{e_G(F_r/M_d(X))} \\ &= b_T(V) i_r^* Q_d^v,\end{aligned}$$

the last equality being a consequence of Lemma 3.1. This proves our assertion. □

Remark 3.5.

(a) If we take V to be the trivial line bundle, and b_T to be the total Chern class, then the preceding lemma reduces to Lemma 3.3.

(b) All the lemmas in this section, in fact, holds for a general projective manifold X without T-action, provided that we still have the S^1-equivariant map $\varphi : M_d(X) \to W_d$, with properties 1.-2. stated in section 2. All G-equivariant classes above are then replaced by their S^1-equivariant counterparts.

4. Euler Data

Notations: We denote by κ_i the G-equivariant class on W_d with the property that $j_r^* \kappa_i = H_i + \langle H_i, r\rangle\alpha$. By the localization theorem, κ_i is determined by these restriction conditions, and is a class in the localized equivariant cohomology of W_d. More generally a class $\phi \in H_T^2(X)$ has a G-equivariant lift $\hat{\phi} \in H_G^2(W_d)$ determined by $j_r^* \hat{\phi} = \phi + \langle \phi, r\rangle\alpha$. We denote by $\langle H_T^2(X)\rangle$ the ring generated by $H_T^2(X)$, and by R_d the ring generated by their lifts $\hat{\phi}$. We put $\mathcal{R} = \mathbf{Q}(\mathcal{T}^*)[\alpha]$, where $\mathbf{Q}(\mathcal{T}^*)$ is the rational function field on the Lie algebra of T. For convenience, we introduce the notations

$$\kappa \cdot \zeta = \kappa_\zeta := \kappa_1\zeta_1 + \cdots + \kappa_m\zeta_m$$
$$i_r^*\omega^v := \frac{j_r^*\omega}{e_G(X_r/W_d)}$$

where ω is a class on W_d.

It is often necessary to work over a larger field than $\mathbf{C}$ for coefficients of cohomology groups. For example when we consider the case of the equivariant Chern polynomial c_T, a formal variable x is introduced. In this case we replace everywhere the scalars $\mathbf{C}$ by $\mathbf{C}(x)$. This will be implicit in all of the discussion below.

Recall the localization formula:

$$\int_{W_d} \omega = \sum_{0 \preceq r \preceq d} \int_X \frac{j_r^*(\omega)}{e_G(X_r/W_d)}.$$

We shall often apply the following version:

$$\int_{W_d} \omega\, e^{\kappa_\zeta} = \sum_{0 \preceq r \preceq d} \int_X i_r^*\omega^v\; e^{H_\zeta + \langle H_\zeta, r\rangle\alpha}.$$

Definition 4.1. *Fix an invertible class* $\Omega \in H_T^*(X)^{-1}$. *A list* $P:\ P_d \in H_G^*(W_d)^{-1}$, $d \succ 0$, *is a* Ω*-Euler data if on* X,

$$\Omega\; i_r^*P_d^v = \overline{i_0^*P_r^v}\; i_0^*P_{d-r}^v$$

(called Euler data identity) for all $r \preceq d$, *and the* $\int_{W_d} P_d \cdot \omega$ *are polynomial in* α *for all* $\omega \in R_d$. *By convention we set* $P_0 = \Omega$.

Example 0. In the last section we have proved, using the *gluing identity*, that the data $Q:\quad Q_d = \varphi_!(\pi^* b_T(V_d))$ associated with a mixed bundle V and a multiplicative class b_T satisfies the Euler data identity. This indicates that the gluing identity is really the geometric origin of Euler data. This is what motivates our definition of Euler data. Note that since Q_d is the equivariant push-forward of a class in $H_G^*(M_d(X))$, the polynomial condition on Q is automatic. This condition will be needed when mirror transformation is discussed.

Example 1. Let L be any equivariant line bundle with $c_1(L) \geq 0$. Let $\hat{L}$ be the G-equivariant lift of $c_1(L)$.

$$P_d = \prod_{k=0}^{\langle c_1(L), d\rangle} (\hat{L} - k\alpha)$$

is an Ω-Euler data where $\Omega = c_1(L)$.

Example 2. Let L be any equivariant line bundle with $c_1(L) < 0$. Let $\hat{L}$ be the G-equivariant lift of $c_1(L)$.

$$P_d = \prod_{k=1}^{-\langle c_1(L),d\rangle -1} (\hat{L} + k\alpha)$$

is an Ω-Euler data where $\Omega = c_1(L)^{-1}$.

Example 3. If P, P' are Ω, Ω'-Euler data respectively, then $P \cdot P'$ is a $\Omega\Omega'$-Euler data as shown in [37].

Example 4. Let L be as in Example 1, and x be a formal variable. Then

$$P_d = \prod_{k=0}^{\langle c_1(L),d\rangle} (x + \hat{L} - k\alpha)$$

is an Ω-Euler data where $\Omega = c_T(L)$ denotes the Chern polynomial.

In each of Examples 1-4 above, the Euler data identity follows immediately from the algebraic identity $\Omega\ j_r^* P_d = \overline{j_0^* P_r}\ j_0^* P_{d-r}$, and Lemma 3.3.

Strictly speaking, in the examples above, we must require that $c_1(L)$ be an invertible class. This requirement can be easily met by twisting L by a trivial line bundle on which T acts by a suitable weight. In the end, we will only be interested in the nonequivariant limit of an Euler data. Thus the choice of twisting is of no consequence at the end. Alternatively, we can consider the Chern polynomial or the total Chern class (which is automatically invertible) instead of the first Chern class.

4.1. An algebraic property

Let $\mathcal{S}$ denotes the set of sequences $B: \quad B_d \in H_G^*(X)^{-1}, \quad d \succeq 0$. By convention, we set $B_0 = \Omega$.

Definition 4.2. *Given any $B \in \mathcal{S}$, define the formal series*

$$HG[B](t) := e^{-H\cdot t/\alpha} \left(\Omega + \sum_{d \succ 0} B_d\ e^{d\cdot t} \right).$$

Note that $e^{H\cdot t/\alpha} HG[B](t)$ takes value in the ring $H_G(X)^{-1}[[K^\vee]]$. (Notations: if R is a ring, hen $\mathcal{R}[[K^\vee]] := \{\sum_{d\in\Lambda} a_d e^{d\cdot t} | a_d \in \mathcal{R}\}$. We use the notations $e^{d\cdot t} = e^{\langle H_t, d\rangle}$ interchangeably.)

Let P be an Euler data, and let B be the list with $B_d := i_0^* P_d^v$. By the localization formula and the Euler data identity, we have

$$\int_{W_d} P_d\, e^{\kappa\zeta} = \sum_{r \preceq d} \int_X i_r^* P_d^v\, e^{H_\zeta + \langle H_\zeta, r\rangle \alpha}$$
$$= \sum_{r \preceq d} e^{-d\cdot\tau} \int_X \Omega^{-1} \overline{\left[e^{-H_t/\alpha} i_0^* P_r^v\, e^{r\cdot t}\right]} \left[e^{-H_\tau/\alpha} i_0^* P_{d-r}^v\, e^{(d-r)\cdot\tau}\right].$$

Here $t = \zeta\alpha + \tau$. Note that $\bar{\zeta} = -\zeta$, $\bar{\alpha} = -\alpha$, and all other variables are invariant under the "bar" operation. Now multiply both sides by $e^{d\cdot\tau}$ and sum over $d \in K^\vee$, we get the formula:

$$\sum_d e^{d\cdot\tau} \int_{W_d} P_d\, e^{\kappa\zeta} = \int_X \Omega^{-1}\, \overline{HG[B](\zeta\alpha+\tau)}\; HG[B](\tau). \qquad (4.1)$$

By definition, the coefficient of $e^{d\cdot\tau}$ on the right hand side is a power series in ζ with coefficients which are polynomial in α, ie. the series lies in $\mathcal{R}[[e^\tau, \zeta]]$.

Conversely, given $B \in \mathcal{S}$ such that

$$\int_X \Omega^{-1}\, \overline{HG[B](\zeta\alpha+\tau)}\; HG[B](\tau) \in \mathcal{R}[[e^\tau, \zeta]],$$

there exists a unique Euler data $P: P_d$ satisfying (4.1). Namely, P_d is defined by the conditions

$$j_r^* P_d = \Omega^{-1} e_G(X_r/W_d)\, \overline{B_r}\, B_{d-r}.$$

Thus an Euler data P gives rise to a list $B \in \mathcal{S}$ in a canonical way. Abusing the terminology, *we shall call such a B an Euler data.*

5. Linking and Uniqueness

Lemma 5.1. *Let $\omega \in H_T^*(X)^{-1}(\alpha)$. Suppose that*

(a) for each $q \in X^T$, $\omega_q(\alpha) := \omega(\alpha)|_q$ is a Laurent polynomial in α with $deg_\alpha \omega(\alpha) \le -2$;

(b) the power series in ζ: $\int_X \left(\omega(\alpha) e^{H_\zeta} + \omega(-\alpha) e^{H_\zeta + \langle H_\zeta, d\rangle\, \alpha}\right)$ has coefficients which are polynomial in α.

Then $\omega = 0$.

Proof: Suppose $\omega \neq 0$, and we will get a contradiction. By assumption (a), we can write

$$\frac{\omega_q(\alpha)}{e_T(q/X)} = a_q\alpha^{-k} + b_q\alpha^{-k+1} + \cdots$$

which is a *finite* sum, with a_q independent of α and $k \geq 2$. By supposition, not all the a_q are zero. By localization, we get

$$\begin{aligned}&\int_X \left(\omega(\alpha)e^{H_\zeta} + \omega(-\alpha)e^{H_\zeta + \langle H_\zeta, d\rangle\ \alpha}\right)\\ &= \sum_q \Big((a_q\alpha^{-k} + b_q\alpha^{-k+1} + \cdots)e^{H_\zeta(q)}\\ &\qquad + (a_q(-\alpha)^{-k} + b_q(-\alpha)^{-k+1} + \cdots)e^{H_\zeta(q) + \langle H_\zeta, d\rangle\alpha}.\end{aligned}$$

By assumption (b), order by order in ζ, this expression is polynomial in α. Since $k \geq 2$, the polar term with α^{-k} must vanish, and so

$$\sum_q a_q e^{H_\zeta(q)}(1 + (-1)^k) = 0.$$

Since not all a_q are zero and the functions $e^{H_\zeta(q)}$ are linearly independent, it follows that k is odd. Now the coefficient of α^{-k+1} becomes

$$\sum_q e^{H_\zeta(q)}(2b_q - a_q\langle H_\zeta, d\rangle) = 0.$$

Again by linear independence of the exponential functions, it follows that $a_q = 0 = b_q$ for all q, which is a contradiction. □

Lemma 5.2. *Suppose A, B are Euler data with $A_r = B_r$ for all $r \prec d$. Suppose that the $(A_d - B_d)|_q$, $q \in X^T$, are Laurent polynomial in α. Suppose also that $deg_\alpha(A_d - B_d) \leq -2$. Then $A_d = B_d$.*

Proof: It suffices to show that

$$\omega(\alpha) := A_d - B_d$$

has property (b) of the preceding lemma.

Let A'_d be the coefficient of $e^{d\cdot\tau}$ in the series

$$\int_X \Omega^{-1}\overline{HG[A](\zeta\alpha+\tau)}HG[A](\tau).$$

Likewise for the B'_d. Since A, B are Euler data, A'_d, B'_d are power series in ζ with coefficients which are polynomial in α. Explicitly,

$$A'_d = \sum_{r \preceq d} \int_X \Omega^{-1} e^{H_\zeta} A_r e^{r\cdot\zeta\alpha}\ A_{d-r}$$

and likewise for the B'_d. Using that $A_r = B_r$, $r \prec d$, and that $A_0 = B_0 = \Omega$, we see that $A'_d - B'_d$ is a sum over r with only two surviving terms, corresponding to $r = 0, d$. That is,

$$A'_d - B'_d = \int_X \left(\omega(\alpha)e^{H_\zeta} + \omega(-\alpha)e^{H_\zeta + \langle H_\zeta, d\rangle\ \alpha}\right).$$

Since both A'_d, B'_d have coefficients which are polynomial in α, this shows that the class $\omega(\alpha)$ has property (b) of the preceding lemma. □

Definition 5.3. *Two Euler data A, B are linked if for every balloon pq in X and every $d = \delta[pq] \succ 0$,*

$$(A_d - B_d)|_q$$

is regular at $\alpha = \frac{\lambda}{\delta}$ where λ is the weight on the tangent line $T_q(pq)$.

Suppose A, B both come from Euler data Q, P respectively, ie. $A_d = i_0^* Q_d^v$ and $B_d = i_0^* P_d^v$. Suppose also that

$$j_0^*(P_d)|_q = j_0^*(Q_d)|_q \quad at\ \alpha = \lambda/\delta. \tag{5.1}$$

whenever $d = \delta[pq] \succ 0$ as above. Recall that $\alpha = \lambda/\delta$ is at worst a simple pole of $1/e_G(X_0/W_d)|_q$. It follows that $(A_d - B_d)|_q$ is regular at this value. This shows that the conditions (5.1) guarantee that A, B are linked.

Theorem 5.4. *Suppose A, B are linked Euler data satisfying the following properties: for $d \succ 0$,*

(i) If $q \in X^T$, the only possible poles of $(A_d - B_d)|_q$ are scalar multiples of a weight on T_qX.

(ii) $deg_\alpha(A_d - B_d) \le -2$.

Then $A = B$.

Proof: We will prove, by induction, the assertion that $A_d = B_d$ for all d. If $d = 0$, there is nothing to prove. Suppose the assertion holds for all $r \prec d$. Set $\omega_q(\alpha) := (A_d - B_d)|_q$ as before. We will show, under assumption (i), that the $\omega_q(\alpha)$ are Laurent polynomial in α. It follows then, from the preceding lemma and assumption (ii), that $A_d = B_d$.

Let $\lambda \in \mathcal{T}^* - 0$. We will show that each $\omega_q(\alpha)$ is regular at $\alpha = \lambda$. Recall the power series in ζ: A'_d, B'_d, with coefficients polynomial in α as in the preceding proof. Thus for any integers $k, l \geq 0$,

$$Res_{\alpha=\lambda}\left((\alpha-\lambda)^k(\alpha+\lambda)^l(A'_d - B'_d)\right) = 0.$$

Also recall that

$$\begin{aligned} A'_d - B'_d &= \int_X \left(\omega(\alpha)e^{H_\zeta} + \omega(-\alpha)e^{H_\zeta + \langle H_\zeta, d\rangle\ \alpha}\right) \\ &= \sum_{q\in X^T} \frac{1}{e_T(q/X)}\left(\omega_q(\alpha)\ e^{H_\zeta(q)} + \omega_q(-\alpha)\ e^{H_\zeta(q)+\langle H_\zeta, d\rangle\alpha}\right). \end{aligned}$$

From the preceding two equations, we get

$$\begin{aligned} 0 = \sum_{q\in X^T} \frac{1}{e_T(q/X)} &\left(e^{H_\zeta(q)}\ Res_{\alpha=\lambda}(\alpha-\lambda)^k(\alpha+\lambda)^l\omega_q(\alpha)\right. \\ &\left. + e^{H_\zeta(q)+\langle H_\zeta, d\rangle\lambda} Res_{\alpha=\lambda}(\alpha-\lambda)^k(\alpha+\lambda)^l\omega_q(-\alpha)\right). \end{aligned} \tag{5.2}$$

If $Res_{\alpha=\lambda}(\alpha-\lambda)^k(\alpha+\lambda)^l\omega_q(-\alpha) = 0$ for all q, then the preceding equation shows that $Res_{\alpha=\lambda}(\alpha-\lambda)^k(\alpha+\lambda)^l\omega_q(\alpha) = 0$ for all q, because the vectors $H_\zeta(q)$ are distinct. Similarly if $Res_{\alpha=\lambda}(\alpha-\lambda)^k(\alpha+\lambda)^l\omega_q(\alpha) = 0$ for all q, then we have $Res_{\alpha=\lambda}(\alpha-\lambda)^k(\alpha+\lambda)^l\omega_q(-\alpha) = 0$ for all q. In either case, we conclude that each $\omega_q(\alpha)$ is regular at $\alpha = \lambda$. So if $\alpha = \lambda$ is a pole of a $\omega_q(\alpha)$, then we necessarily have $Res_{\alpha=\lambda}(\alpha-\lambda)^k(\alpha+\lambda)^l\omega_q(\alpha) \neq 0$ *and* $Res_{\alpha=\lambda}(\alpha-\lambda)^k(\alpha+\lambda)^l\omega_p(-\alpha) \neq 0$ for some p, q and some l, k, such that

$$H_\zeta(q) = H_\zeta(p) + \langle H_\zeta, d\rangle\lambda,$$

to ensure cancellation of the exponential functions in (5.2). Note that since $d \succ 0$ and $\lambda \neq 0$, we have $p \neq q$. By our assumption (i), the pole $\alpha = \lambda$ of $\omega_q(\alpha)$ must be of the form $\lambda = \frac{\lambda'}{\delta} \neq 0$ for some weight λ' on T_qX, and some scalar $\delta \neq 0$. By Lemma 2.4, p, q must be joined by a balloon, $d = \delta[pq]$, and λ' is the weight on the tangent line $T_q(pq)$.

Thus if d is not a multiple of $[pq]$, then we have shown that $\omega_q(\alpha)$ is regular away from $\alpha = 0$.

Now suppose that $d = \delta[pq]$, and consider the only possible pole of $\omega_q(\alpha)$ at $\alpha = \frac{\lambda'}{\delta} \neq 0$, as above. By hypothesis, A, B are linked. But this means that $\omega_q(\alpha)$ is regular at $\alpha = \frac{\lambda'}{\delta} \neq 0$. □

Remark 5.5. *In our applications later, the situation is better then the conditions (i)-(ii) demand. We will have two Euler data A, B such that A_d, B_d separately, rather than just $A_d - B_d$, will satisfy both conditions (i)-(ii) at the outset. In this situation, to prove that $A = B$, it suffices to prove that they are linked.*

6. Mirror Transformations

Throughout this section, we fix an invertible class Ω on X, and will denote by $\mathcal{A}$ the set of Ω-Euler data.

Definition 6.1. *A map $\mu : \mathcal{A} \to \mathcal{A}$ is called a mirror transformation if it preserves linking. In other words, $\mu(A)$ and A are linked for any $A \in \mathcal{A}$. We call $\mu(A)$ a mirror transform of A.*

We now consider a construction of mirror transformations, as motivated by the classic example of [12]. Consider a transformation $\mu : \mathcal{S} \to \mathcal{S}$, $B \to \tilde{B}$, of the type

$$\tilde{B}_d = B_d + \sum_{r \prec d} a_{d,r} B_r \tag{6.1}$$

where the $a_{d,r} \in H_G^*(X)^{-1}$ are a given set of coefficients. This transformation is obviously invertible, and preserves $B_0 = \Omega$.

Lemma 6.2. *Suppose that $B, \tilde{B}$ are both Euler data. Let $d = \delta[pq] \succ 0$ for some balloon pq in X. Suppose that the coefficients in (6.1) are such that their restrictions $a_{d,r}(q)$, $r \prec d$, to the fixed point q are regular at $\alpha = \lambda/\delta$ where λ is the weight on $T_q(pq)$. Then $(\tilde{B}_d - B_d)|_q$ is regular at $\alpha = \lambda/\delta$.*

Proof: From (6.1), it suffices to show that the functions $B_r|_q$, $0 \prec r \prec d$, are regular at $\alpha = \lambda/\delta$. Suppose the contrary that some $B_r|_q = 0$ has a pole of order $k+1$ there. Since B is a Euler data, we know that

$$B'_r := \sum_{s \preceq r} \int_X \Omega^{-1} e^{H_\zeta} B_s e^{\langle H_\zeta, s\rangle \alpha} B_{r-s}$$

is a power series in ζ with coefficients polynomial in α.

By the localization formula,

$$B'_r = \sum_{s \preceq r} \sum_{o \in X^T} \frac{1}{e_T(o/X)} e^{H_\zeta(o) + \langle H_\zeta, s\rangle \alpha} \Omega(o)^{-1} B_s(o) B_{r-s}(o).$$

Now multiply both sides by $(\alpha - \lambda/\delta)^k$ and take residue at $\alpha = \lambda/\delta$. We get

$$0 = \sum_{s \preceq r} \sum_{o \in X^T}$$

$$\frac{1}{e_T(o/X)} e^{H_\zeta(o) + \langle H_\zeta, s\rangle \lambda/\delta} Res_{\alpha=\lambda/\delta} (\alpha - \lambda/\delta)^k \Omega(o)^{-1} B_s(o) B_{r-s}(o).$$

By assumption, the summand above with $s = 0, o = q$ is nonzero. Observe that this term has an exponential factor $e^{H_\zeta(q)}$. Thus in order to cancel this term, any other term contributing to this cancellation must have an identical exponential factor. This means that

$$H_\zeta(q) = H_\zeta(o) + \langle H_\zeta, s\rangle \lambda/\delta$$

for some s with $s \preceq r$, and some $o \in X^T$. By Lemma 2.4, this implies that $s = \delta[pq]$, contradicting that $s \preceq r \prec d$. □

Definition 6.3. *The transformation (6.1) is said to have the regularity property if for every balloon pq in X and $d = \delta[pq]$, the coefficients are such that their restrictions $a_{d,r}(q)$, $r \prec d$, are regular at $\alpha = \lambda/\delta$ where λ is the weight on $T_q(pq)$.*

Thus the preceding lemma says that transformation (6.1) having the regularity property preserves linking.

Again, motivated by [12] and [28], we consider the following special types of transformations. Given a power series $f \in \mathcal{R}[[K^\vee]]$ with no constant term, we have an invertible transformation $\mu_f : \mathcal{S} \to \mathcal{S}$, $B \mapsto \tilde{B}$, such that

$$e^{f/\alpha} HG[B](t) = HG[\tilde{B}](t).$$

In fact, we have

$$\tilde{B}_d = B_d + \sum_{r \prec d} f_{d-r} B_r$$

where $e^{f/\alpha} = \sum_{s \succeq 0} f_s e^{s \cdot t}$, $f_s \in \mathcal{R}[\alpha^{-1}]$. This is clearly a transformation of type (6.1) having the regularity property. (In fact, all the coefficients f_{d-r} are regular away from $\alpha = 0$.)

Given power series $g = (g_1, .., g_m)$, $g_j \in \mathcal{R}[[K^\vee]]$ with no constant term, we have an invertible transformation $\nu_g : \mathcal{S} \to \mathcal{S}$, $B \mapsto \tilde{B}$, such that

$$HG[B](t+g) = HG[\tilde{B}](t).$$

In fact since

$$HG[B](t+g) = e^{-H\cdot t/\alpha} e^{-H\cdot g/\alpha} \sum_{d \succeq 0} B_d \, e^{d\cdot t} e^{d\cdot g},$$

if we write $e^{d\cdot g} = \sum_{s\succeq 0} g_{d,s} e^{s\cdot t}$, $g_{d,s} \in \mathcal{R}$ and $e^{-H\cdot g/\alpha} = \sum_{s\succeq 0} \hat{g}_s e^{s\cdot t}$, $\hat{g}_s \in \mathcal{R}[H/\alpha]$, then

$$\tilde{B}_d = B_d + \sum_{r \prec d} a_{d,r} B_r$$

where the $a_{d,r} \in H_G^*(X)^{-1}$ are quadratic expressions in the $g, \hat{g}$. Thus we obtain another transformation $\mathcal{S} \to \mathcal{S}$ of type (6.1), again having the regularity property.

Theorem 6.4. *The transformations $\mu_f, \nu_g : B \mapsto \tilde{B}$ above each defines a mirror transformation. That is, if B is a Euler data then $\mu_f(B)$ and $\nu_g(B)$ are both Euler data linked to B.*

Proof: Let B be a given Euler data. We have seen that the preceding lemma guarantees that $\mu_f(B)$, $\nu_g(B)$ are linked to B. So it suffices to show that they are Euler data.

First case: set $\tilde{B} = \nu_g(B)$, ie.

$$HG[\tilde{B}](t) = HG[B](t + g(e^t)). \tag{6.2}$$

(Here e^t means the variables $(e^{t_1}, .., e^{t_m})$.) Set $t = \zeta\alpha + \tau$, $q_i = e^{\tau_i}$. On the one hand, we have

$$\int_X \Omega^{-1} \, \overline{HG[B](t)} \; HG[B](\tau) = \sum_{d,m} q^d \, \zeta^m \, B'_{d,m}(\alpha) \tag{6.3}$$

for some $B'_{d,m} \in \mathcal{R}$. Now compare

$$(*) \quad \int_X \Omega^{-1} \overline{HG[B](t)}\, HG[B](\tau) = \int_X \Omega^{-1} e^{H\zeta} \sum \bar{B}_d e^{d\cdot\tau} e^{d\cdot\zeta\alpha} \times \sum B_d e^{d\cdot\tau}$$

$$(**) \quad \int_X \Omega^{-1} \overline{HG[B](t+g(e^t))}\, HG[B](\tau + g(e^\tau))$$
$$= \int_X \Omega^{-1} e^{H\zeta} e^{H\cdot(\bar{g}(qe^{\zeta\alpha})-g(q))/\alpha} \times \sum \bar{B}_d e^{d\cdot(\tau+g(q))} e^{d\cdot(\bar{g}(qe^{\zeta\alpha})-g(q))+d\cdot\zeta\alpha} \times \sum B_d e^{d\cdot(\tau+g(q))}.$$

This shows that the series (**) can be obtained from (*) by the replacements $\tau \mapsto \tau + g(q)$, $\zeta \mapsto \zeta + (\bar{g}(qe^{\zeta\alpha}) - g(q))/\alpha$. Thus combining (6.2) and (6.3), we get

$$\int_X \Omega^{-1} \overline{HG[\tilde{B}](t)}\, HG[\tilde{B}](\tau) = \sum_{d,m} q^d e^{d\cdot g(q)} \left(\zeta + (\bar{g}(qe^{\zeta\alpha}) - g(q))/\alpha\right)^m B'_{d,m}(\alpha). \tag{6.4}$$

Now write $g = g_+ + g_-$ with $\bar{g}_\pm = \pm g_\pm$. Obviously for any $g(q) \in \mathcal{R}[[q]]$, $g_+(qe^{\zeta\alpha}) - g_+(q) \in \alpha \cdot \mathcal{R}[[q,\zeta]]$. Since the involution $\omega \mapsto \bar{\omega}$ on $\mathcal{R}$ simply changes the sign of α, the fact that g_- is odd shows that $g_-(q) \in \alpha \cdot \mathcal{R}[[q]]$. Likewise for $g_-(qe^{\zeta\alpha})$. This shows that (6.4) lies in $\mathcal{R}[[q,\zeta]]$. This completes our proof in this case.

Second case: set $\tilde{B} = \mu_f(B)$, ie.

$$HG[\tilde{B}](t) = e^{f/\alpha}\, HG[B](t).$$

Again writing $f \in \mathcal{R}[[e^t]]$ as $f = f_+ + f_-$ with $\bar{f}_\pm = \pm f_\pm$, we get

$$\int_X \Omega^{-1} \overline{HG[\tilde{B}](t)}\, HG[\tilde{B}](\tau) = e^{-\bar{f}(e^t)/\alpha}\, e^{f(e^\tau)/\alpha} \times \int_X \Omega^{-1} \overline{HG[B](t)}\, HG[B](\tau)$$
$$= e^{-(f_+(qe^{\zeta\alpha})-f_+(q))/\alpha}\, e^{(f_-(qe^{\zeta\alpha})+f_-(q))/\alpha} \times \int_X \Omega^{-1} \overline{HG[B](t)}\, HG[B](\tau).$$

The right hand side lies in $\mathcal{R}[[q,\zeta]]$ as before. □

Remark 6.5. *All mirror transformations we will use later will be of the type μ_f, ν_g as above. Moreover, all Euler data we will encounter will have property (i) of Theorem 5.4. The transformations μ_f, ν_g clearly preserve this property.*

Theorem 6.6. *Suppose that A, B have property (i) of Theorem 5.4, and that A, B are linked. Suppose that A is an Euler data with $deg_\alpha A_d \leq -2$ for all $d \prec 0$, and that there exists power series $f \in \mathcal{R}[[K^\vee]]$, $g = (g_1, .., g_m)$, $g_j \in \mathcal{R}[[K^\vee]]$, all without constant term, such that*

$$e^{f/\alpha} HG[B](t) = \Omega - \Omega \frac{H \cdot (t+g)}{\alpha} + O(\alpha^{-2}) \tag{6.5}$$

when expanded in powers of α^{-1}. Then

$$HG[A](t+g) = e^{f/\alpha}\, HG[B](t).$$

Proof: By Theorem 6.4, f, g define two mirror transformations μ_f, ν_g, with

$$\begin{aligned} HG[\tilde{B}](t) &= e^{f/\alpha} HG[B](t) \\ HG[\tilde{A}](t) &= HG[A](t+g) \end{aligned} \tag{6.6}$$

where $\tilde{B} = \mu_f(B)$, $\tilde{A} = \nu_g(A)$. Now both $\tilde{B}, \tilde{A}$ have property (i) of Theorem 5.4. (See remark after Theorem 6.4.)

Since $deg_\alpha A_d \leq -2$, $HG[\tilde{A}](t)$ has the same asymtotic form as $HG[\tilde{B}](t)$ in eqn. (6.5) *mod* $O(\alpha^{-2})$. It follows that

$$e^{H \cdot t/\alpha}\, HG[\tilde{A} - \tilde{B}](t) \equiv O(\alpha^{-2}),$$

or equivalently $deg_\alpha(\tilde{A}_d - \tilde{B}_d) \leq -2$. Thus $\tilde{A}, \tilde{B}$ satisfy condition (ii) of Theorem 5.4. Since A is linked to B, it follows that $\tilde{A}$ is linked to $\tilde{B}$. By Theorem 5.4, we conclude that $\tilde{A} = \tilde{B}$. Now our assertion follows from eqns. (6.6). □

Remark 6.7. *The preceding theorem says that one way to compute A (or Q) is by first finding an explicit Euler data B linked to A, and then relate A and B via mirror transformations.*

7. From stable map moduli to Euler data

Fix an admissible balloon manifold with $c_1(X) \geq 0$. Fix a T-equivariant multiplicative class b_T. Its nonequivariant limit is denoted by b. Fix a T-equivariant bundle of the form $V = V^+ \oplus V^-$, where $V^\pm$ are respectively the convex/concave bundles. As before, we write

$$\Omega = \frac{b_T(V^+)}{b_T(V^-)}.$$

Let V_d be the bundle induced by V on the 0-pointed degree d stable map moduli of X. Throughout this section, we denote

$$\begin{aligned} Q: \quad & Q_d := \varphi_!(\pi^* b_T(V_d)) \\ & K_d := \int_{M_{0,0}(d,X)} b(V_d) \\ & \Phi := \sum K_d \, e^{d\cdot t} \\ A: \quad & A_d := i_0^* Q_d^v. \end{aligned}$$

Note that all these objects depend on the choice of b_T and V, though the notations do not reflect this.

7.1. *The Euler data Q*

Theorem 7.1. *(i)* $deg_\alpha A_d \leq -2$.
(ii) If for each d the class $b_T(V_d)$ has homogeneous degree the same as the degree of $M_{0,0}(d,X)$, then in the nonequivariant limit we have

$$\int_X e^{-H\cdot t/\alpha} A_d = \alpha^{-3}(2 - d\cdot t) K_d$$

$$\int_X \left(HG[A](t) - e^{-H\cdot t/\alpha}\Omega \right) = \alpha^{-3}(2\Phi - \sum t_i \frac{\partial \Phi}{\partial t_i}).$$

Proof: Earlier we have proved that

$$A_d = i_0^* Q_d^v = ev_! \left(\frac{\rho^* b_T(V_d)}{\alpha(\alpha - c_1(L))} \right),$$

where $L = L_d$ is the line bundle on $\mathcal{M}_{0,1}(d,X)$ whose fiber at a point $(f, C; x)$ is the tangent line at x.

Assertion (i) now follows immediately from this formula. The second equality in assertion (ii) follows from the first equality. By the above formula again,

$$\begin{aligned} I &:= \int_X e^{-H\cdot t/\alpha} A_d \\ &= \int_{M_{0,1}(d,X)} e^{-ev^*H\cdot t/\alpha} \frac{\rho^* b(V_d)}{\alpha(\alpha - c_1(L))} \\ &= \int_{M_{0,0}(d,X)} b(V_d)\ \rho_! \left(\frac{e^{-ev^*H\cdot t/\alpha}}{\alpha(\alpha - c_1(L))} \right). \end{aligned}$$

Now $b(V_d)$ has homogeneous degree the same as the dimension $M_{0,0}(d,X)$. The second factor in the last integrand contributes a scalar factor given by integration over a generic fiber E (which is a $\mathbf{P}^1$) of ρ. So we pick out the degree 1 term in $\frac{e^{-ev^*H\cdot t/\alpha}}{\alpha(\alpha-c_1(L))}$, which is just $\frac{-ev^*H\cdot t}{\alpha^3} + \frac{c_1(L)}{\alpha^3}$. Restricting to the generic fiber E, say over $(f,C) \in M_{0,0}(d,X)$, the evaluation map ev is equal to f, which is a degree d map $E \cong \mathbf{P}^1 \to X$. It follows that

$$\int_E ev^* H = d.$$

Moreover, since $c_1(L)$ restricted to E is just the first Chern class of the tangent bundle to E, it follows that

$$\int_E c_1(L) = 2.$$

So we have

$$I = (-\frac{d\cdot t}{\alpha^3} + \frac{2}{\alpha^3})K_d. \quad \square$$

Theorem 7.2. *More generally suppose b_T is an equivariant multiplicative class of the form*

$$b_T(V) = x^r + x^{r-1}b_1(V) + \cdots + b_r(V), \quad rk\ V = r$$

where x is a formal variable, b_i is a characteristic class of degree i. Suppose $s := rk\ V_d - dim\ M_{0,0}(d,X) \geq 0$ is independent of $d \succ 0$.

Then

$$\frac{1}{s!}\left(\frac{d}{dx}\right)^s |_{x=0} \int_X e^{-H\cdot t/\alpha} A_d = \alpha^{-3} x^{-s}(2 - d\cdot t)K_d$$

$$\frac{1}{s!}\left(\frac{d}{dx}\right)^s |_{x=0} \int_X (HG[A](t) - e^{-H} \cdot t/\alpha\Omega$$

$$= \alpha^{-3}x^{-s}(2\Phi - \sum t_i \frac{\partial \Phi}{\partial t_i}).$$

Proof: The proof is entirely analogous to (ii) above. □

7.2. Linking theorem for A

Now consider a mixed bundle $V = V^+ \oplus V^-$ on X. Fixed a choice of equivariant multiplicative class b_T. We assume that V has the following property: there exists nontrivial T-equivariant line bundles $L_1^+, .., L_{N^+}^+; L_1^-, .., L_{N^-}^-$ on X with $c_1(L_i^+) \geq 0$ and $c_1(L_j^-) < 0$, such that for any balloon $pq \cong \mathbf{P}^1$ in X we have

$$V^\pm|_{pq} = \oplus_{i=1}^{N^\pm} L_i^\pm|_{pq}.$$

Note that $N^\pm = rk\ V^\pm$. We also require that

$$b_T(V^+)/b_T(V^-) = \prod_i b_T(L_i^+)/\prod_j b_T(L_j^-).$$

In this case we call the list $(L_1^+, .., L_{N^+}^+; L_1^-, .., L_{N^-}^-)$ the splitting type of V. Note that V is not assumed to split over X. Given such a bundle V and a choice of multiplicative class b_T, we obtain an Euler data $Q: Q_d = \varphi_!(\pi^* b_T(V_d))$ (or A) as before.

Theorem 7.3. *Let $b_T = e_T$ be the equivariant Euler class. Let pq be a balloon, $d = \delta[pq] \succ 0$, and λ be the weight on the tangent line $T_q(pq)$. Then at $\alpha = \lambda/\delta$, we have*

$$j_0^*(Q_d)|_q = \prod_i \prod_{k=0}^{\langle c_1(L_i^+),d\rangle} (c_1(L_i^+)|_q - k\lambda/\delta)$$

$$\times \prod_j \prod_{k=1}^{-\langle c_1(L_j^-),d\rangle - 1} \left(c_1(L_j^-)|_q + k\lambda/\delta\right).$$

In particular Q is linked to

$$P: \quad P_d = \prod_i \prod_{k=0}^{\langle c_1(L_i^+), d\rangle} (\hat{L}_i^+ - k\alpha) \times \prod_j \prod_{k=1}^{-\langle c_1(L_j^-), d\rangle - 1} (\hat{L}_j^- + k\alpha).$$

Proof: We first consider one positive line bundle L. As in [37], we consider a point $(f, C) \in M_d(X)$ where f is δ-cover from $C = \mathbf{P}^1$ to the balloon $pq \simeq \mathbf{P}^1$. For $\alpha = \lambda/\delta$, this map can be written as

$$f: \ C \to \mathbf{P}^1 \times pq \subset \mathbf{P}^1 \times X$$

where the second map is the inclusion. In terms of coordinates we can write the first map as

$$f: \ [w_0, w_1] \to [w_1, w_0] \times [w_0^\delta, w_1^\delta].$$

Note that the T-action induces standard rotation on $pq \simeq \mathbf{P}^1$ with the weights λ_1, λ_2 and $\lambda = \lambda_1 - \lambda_2$. It is now easy to see that this point (f, C) is fixed by the subgroup of G with $\alpha = \lambda/\delta$. On the other hand as argued in [37], $(\pi_2 \circ f, C)$ is then a smooth fixed point in $M_{0,0}(d, X)$ under the T-action. The restriction $j_0^* Q_d|_p$ with $\alpha = \lambda/\delta$ is equal to the value of $e_T(\mathcal{U}_d)$ at (f, C). This, in turn, is equal to the restriction of $e_T(V_d)$ at $(\pi_2 \circ f, C)$ in $M_{0,0}(d, X)$.

Assume the restriction of L to $pq \simeq \mathbf{P}^1$ is $\mathcal{O}(l)$ with $l = \langle c_1(L), [pq]\rangle$. We compute that the equivariant Euler class restricted to this point $(\pi_2 \circ f, C)$. As in [37], we get

$$e_T(U_d) = \prod_{m=0}^{l\delta} (l\lambda_1 - m\frac{\lambda}{\delta}).$$

Also note that $c_1(L)(p) = l\lambda_1$ and $d = \delta[pq]$, this implies that $Q_d = \varphi_!(\pi^* e_T(V_d))$ is linked to

$$P_d = \prod_{m=0}^{\langle c_1(L), d\rangle} (c_1(\hat{L}) - m\alpha).$$

Similarly for a concave line bundle L, if its restriction to the balloon pq is $\mathcal{O}(-l)$ with $-l = \langle c_1(L), [pq]\rangle$, then

$$e_T(U_d) = \prod_{m=-}^{l\delta - 1} (-l\lambda_1 + m\frac{\lambda}{\delta})$$

which implies the formula that in this case Q_d is linked to

$$P_d = \prod_{m=1}^{-\langle c_1(L),d\rangle -1} (c_1(\hat{L}) + m\alpha).$$

The general case is just a product of these cases. □

Similarly we can prove the following formula for the Chern polynomial.

Theorem 7.4. *Let $b_T = c_T$ be the equivariant Chern polynomial. Let pq be a balloon, $d = \delta[pq] \succ 0$, and λ be the weight on the tangent line $T_q(pq)$. Then at $\alpha = \lambda/\delta$, we have*

$$j_0^*(Q_d)|_q = \prod_i \prod_{k=0}^{\langle c_1(L_i^+),d\rangle} \left(x + c_1(L_i^+)|_q - k\lambda/\delta\right)$$
$$\times \prod_j \prod_{k=1}^{-\langle c_1(L_j^-),d\rangle -1} \left(x + c_1(L_j^-)|_q + k\lambda/\delta\right).$$

In particular Q is linked to

$$P: \ P_d = \prod_i \prod_{k=0}^{\langle c_1(L_i^+),d\rangle} (x + \hat{L}_i^+ - k\alpha) \times \prod_j \prod_{k=1}^{-\langle c_1(L_j^-),d\rangle -1} (x + \hat{L}_j^- + k\alpha).$$

8. Applications

8.1. Toric manifolds

We call a toric manifold X *reflexive* if its defining fan satisfies the following combinatorial condition: the convex hull of the primitive generators of the 1-cones in the fan is a reflexive polytope. It has been shown [4][41] that a pair of polar reflexive polytopes gives rise to a pair of mirror (in the sense of Hodge numbers) Calabi-Yau varieties, by taking anti-canonical hypersurfaces in the corresponding reflexive toric manifolds. It has been conjectured that [5] a similar statement holds for complete intersections in toric manifolds. It is known that [29] a toric manifold X is reflexive iff $c_1(X) \geq 0$. We shall assume that X is reflexive. Recall that for a (convex) toric manifold X, we have

$$e_G(X_0/W_d) = \prod_a \prod_{k=1}^{\langle D_a, d\rangle} (D_a - k\alpha)$$

where each D_a is the T-equivariant first Chern classes of the line bundles corresponding to a T-invariant hypersurfaces in X.

8.2. Chern polynomials for mixed bundles

To proceed, we make two further choices: let b_T be the T-equivariant Chern polynomial c_T, and let $V = V^+ \oplus V^-$ be a mixed bundle with splitting type $(L_1^+, .., L_{N^+}^+; L_1^-, .., L_{N^-}^-)$. Here the L's are T-equivariant line bundles on X with

$$\begin{aligned}
&c_1(L_i^+) \geq 0, \quad c_1(L_j^-) < 0, \\
&\Omega := c_T(V^+)/c_T(V^-) = \prod_i (x + c_1(L_i^+)) / \prod_j (x + c_1(L_j^-)) \\
&\sum_i c_1(L_i^+) - \sum_j c_1(L_j^-) = c_1(X).
\end{aligned}$$

From this, we get an Ω-Euler data $Q: \; Q_d = \varphi_!(\pi^* c_T(V_d))$ as before. By the Linking Theorem, Q is linked to the Euler data

$$P: \; P_d = \prod_i \prod_{k=0}^{\langle c_1(L_i^+), d\rangle} (x + \hat{L}_i^+ - k\alpha) \times \prod_j \prod_{k=1}^{-\langle c_1(L_j^-), d\rangle - 1} (x + \hat{L}_j^- + k\alpha).$$

As before, we set

$$B: \; B_d = i_0^* P_d^v, \qquad A: \; A_d = i_0^* Q_d^v.$$

We consider three separate cases. We will be using the elementary formula

$$\prod_{k=1}^{M} (\frac{\omega}{\alpha} - k) \equiv (-1)^M M! (1 - \frac{\omega}{\alpha} \sum_{k=1}^{M} \frac{1}{k}) \tag{8.1}$$

where " $\equiv$" here means equal *mod* $O(\alpha^{-2})$, to compute the leading terms of

$$\begin{aligned}B_d = &\prod_i \prod_{k=0}^{\langle c_1(L_i^+),d\rangle} (x + c_1(L_i^+) - k\alpha) \\ &\times \prod_j \prod_{k=1}^{-\langle c_1(L_i^-),d\rangle - 1} (x + c_1(L_i^-) + k\alpha) \\ &\times \frac{1}{\prod_a \prod_{k=1}^{\langle D_a,d\rangle} (D_a - k\alpha)} \\ &= \Omega\ c_T(V^-)\ \alpha^{-N^-} \prod_i \prod_{k=1}^{\langle c_1(L_i^+),d\rangle} (\frac{x + c_1(L_i^+)}{\alpha} - k) \\ &\times \prod_j \prod_{k=1}^{-\langle c_1(L_i^-),d\rangle - 1} (\frac{x + c_1(L_i^-)}{\alpha} + k) \\ &\times \frac{1}{\prod_a \prod_{k=1}^{\langle D_a,d\rangle} (\frac{D_a}{\alpha} - k)}.\end{aligned} \tag{8.2}$$

First suppose that $rk\ V^- = N^- \geq 2$. In this case we have

$$deg_\alpha B_d = -rk\ V^- \leq -2$$

and hence

$$HG[B](t) \equiv \Omega - \Omega \frac{H \cdot t}{\alpha}.$$

By Theorem 6.6, we conclude that $A = B$ and $Q = P$. This completes the computation of A and Q in this case.

Now consider the case $rk\ V^- = N^- = 1$, hence V^- is a line bundle. In this case we have

$$\begin{aligned}B_d &\equiv \alpha^{-1}\ \Omega\ (x + c_1(V^-))\ (-1)^{\langle c_1(V^-),d\rangle} \\ &\qquad \times (-\langle c_1(V^-), d\rangle - 1)! \frac{\prod_i \langle c_1(L_i^+), d\rangle!}{\prod_a \langle D_a, d\rangle!} \\ &=: \alpha^{-1}\ \Omega\ (\sum_i H_i \phi_{d,i} + \psi_d)\end{aligned}$$

where the $\phi_{d,i} \in \mathbf{Q}$, $\psi_d \in \mathbf{Q}[\mathcal{T}^*, x]$, are determined uniquely by the writing $c_1(V^-) \in H_T^2(X)$ in the last equality, according to the decomposition $H_T^2(X) = \oplus_{i=1}^m \mathbf{Q} H_i \oplus \mathcal{T}^*$. Hence we get

$$e^{-H \cdot t/\alpha}\ B_d \equiv \Omega\ (\alpha^{-1} H \cdot \phi_d + \alpha^{-1} \psi_d).$$

Summing over $d \in K^\vee$, we get

$$HG[B](t) \equiv \Omega(1 - \alpha^{-1}H \cdot (t + F) + \alpha^{-1}G)$$
$$F := -\sum \phi_d \, e^{d \cdot t}$$
$$G := \sum \psi_d \, e^{d \cdot t}.$$

From this we get

$$e^{-G/\alpha} \; HG[B](t) \equiv \Omega - \Omega \frac{H \cdot (t + F)}{\alpha}$$

By Theorem 6.6, we conclude that

$$e^{-G/\alpha} HG[B](t) = HG[A](t + F). \tag{8.3}$$

This completes our computation of A and Q in this case.

Recall that

$$\begin{aligned} dim \; M_{0,0}(d, X) &= \langle c_1(X), d\rangle + n - 3 \\ rk \; V_d &= \sum_i \langle c_1(L_i^+), d\rangle - \sum_j \langle c_1(L_j^-), d\rangle + N^+ - N_- \\ &= \langle c_1(X), d\rangle + rk \; V^+ - rk \; V^-. \end{aligned}$$

To applied Theorem 7.2, we assume that $rk\; V^+ - rk\; V^- \geq n-3$, and we determine all K_d immediately. Explicitly (in the nonequivariant limit $\mathcal{T}^* \to 0$):

$$\frac{1}{s!}\left(\frac{d}{dx}\right)^s |_{x=0} \int_X \left(e^{-G/\alpha} HG[B](t) - e^{-H \cdot \tilde{t}/\alpha}\Omega\right) \tag{8.4}$$

$$= \alpha^{-3} x^{-s} (2\Phi(\tilde{t}) - \sum_i \tilde{t}_i \frac{\partial \Phi(\tilde{t})}{\partial \tilde{t}_i}).$$

where $s := rk\; V^+ - rk\; V^- - (n-3)$, $\tilde{t} := t + F(t)$. Note that this same formula applies also when $rk\; V^- \geq 2$, whereby we put $G, F = 0$.

We now consider the case when V is purely convex: $N^- = 0$.

8.3. Convex bundle

We will denote the L_i^+ simply by L_i. Using formulas (8.1) and (8.2), we get

$$B_d \equiv \Omega \frac{\prod_i \langle c_1(L_i), d\rangle !}{\prod_a \langle D_a, d\rangle !}$$
$$\times \left(1 + \alpha^{-1} \sum_a D_a \sum_{k=1}^{\langle D_a, d\rangle} \frac{1}{k} - \alpha^{-1} \sum_i (x + c_1(L_i)) \sum_{k=1}^{\langle c_1(L_i), d\rangle} \frac{1}{k}\right)$$
$$=: \Omega\lambda_d + \alpha^{-1} \sum_i H_i \phi_{d,i} + \alpha^{-1} \psi_d$$

Here the $\lambda_d, \phi_{d,i} \in \mathbf{Q}$, $\psi_d \in \mathbf{Q}[\mathcal{T}^*, x]$ are determined uniquely by the writing each $D_a, c_1(L_i) \in H^2_{\mathcal{T}}(X)$ in the last equality, according to the decomposition $H^2_{\mathcal{T}}(X) = \oplus_{i=1}^m \mathbf{Q}H_i \oplus \mathcal{T}^*$. Since $e^{-H\cdot t/\alpha} \equiv 1 - \alpha^{-1} H \cdot t$, we get

$$e^{-H\cdot t/\alpha}\, B_d \equiv \Omega(\lambda_d - \alpha^{-1} H \cdot (\lambda_d t - \phi_d) + \alpha^{-1} \psi_d).$$

Summing over $d \in K^\vee$, we get

$$HG[B](t) \equiv \Omega\left(F_0 - \alpha^{-1} H \cdot (F_0 t + F) + \alpha^{-1} G\right)$$
$$F_0 := 1 + \sum \lambda_d\, e^{d\cdot t}$$
$$F := -\sum \phi_d\, e^{d\cdot t}$$
$$G := \sum \psi_d\, e^{d\cdot t}.$$

Put $f := \alpha\ log\ F_0 - \frac{G}{F_0}$. Then we get

$$e^{f/\alpha}\, HG[B](t) \equiv \Omega - \Omega \frac{H \cdot (t + \frac{F}{F_0})}{\alpha}$$

By Theorem 6.6, we conclude that

$$e^{f/\alpha}\, HG[B](t) = HG[A](t + \frac{F}{F_0}). \tag{8.5}$$

This completes our computation of A and Q in this case.

Again to apply Theorem 7.2, we assume that $rk\ V \geq n-3$, and determine all K_d immediately. Explicitly:

$$\frac{1}{s!}\left(\frac{d}{dx}\right)^s |_{x=0} \int_X \left(e^{f/\alpha}\, HG[B](t) - e^{-H\cdot\tilde{t}/\alpha}\Omega\right) \tag{8.6}$$
$$= \alpha^{-3}x^{-s}(2\Phi(\tilde{t}) - \sum_i \tilde{t}_i \frac{\partial\Phi(\tilde{t})}{\partial\tilde{t}_i}).$$

where $s := rk\ V - (n-3)$, $\tilde{t} := t + \frac{F(t)}{F_0(t)}$.

Let us now specialize to the case $rk\ V = n - 3$ (ie. $s = 0$), and $V = \oplus_i L_i$. We can then set $x = 0$, so that $b_T = c_T$ becomes the equivariant Euler class e_T, and the K_d is just the intersection numbers for $e(V_d)$. Then the formula (8.6) yields the general formula derived in [28] and in [27], on the basis of the conjectural mirror correspondence. Note that

$$F_0 = \sum \frac{\prod_i \langle c_1(L_i), d\rangle!}{\prod_a \langle D_a, d\rangle!} e^{d\cdot t}$$

is an example of a hypergeometric function [20]. It has been proved in [29] that F_0 is the unique holomorphic period of Calabi-Yau hypersurfaces near the so-called large radius limit. For the purpose of comparison, we should mention that the definition of Φ here differs from the prepotential in [28][27] by a degree three polynomial in $\tilde{t}$, and the definition of the hypergeometric series $HG[B](t)$ here differs from that denoted by $w_0(x,\rho)$ in [28][27] by an irrelevant overall constant factor.

Precursors to the above general formula have been many examples [27][6][13][8]. We now specialize to a few numerical examples which have been frequently studied by both physicists and mathematicians alike.

8.4. A complete intersection in $\mathbf{P}^1 \times \mathbf{P}^2 \times \mathbf{P}^2$

The complete intersection of degrees $(1,3,0)$, $(1,0,3)$ in this 5-dimensional toric manifold X has been studied in [27] using mirror symmetry, and in [30] computing some of the intersection numbers K_d for the Euler class $b = e$ in terms of modular forms.

From our point of view, that complete intersection correspond to the following choice of convex bundle:

$$V = \mathcal{O}_1(1) \otimes \mathcal{O}_2(3) \oplus \mathcal{O}_1(1) \otimes \mathcal{O}_3(3)$$

where $\mathcal{O}_i(l)$ denotes the pullback of $\mathcal{O}(l)$ from the ith factor. The Kähler cone of X is abviously generated by the hyperplanes

H_1, H_2, H_3 from the three factors of X, and hence $K^\vee$ can be identified with the set of $d = (d_1, d_2, d_3) \in \mathbf{Z}^3_{\geq 0}$. We consider intersection numbers K_d for the Euler class $b = e$ as before. Thus we set $\Omega = e(V) = (H_1 + 3H_2)(H_1 + 3H_3)$. The Euler data P we need in eqn. (8.6) is given by

$$P_d = \prod_{k=0}^{d_1+3d_2} (\kappa_1 + 3\kappa_2 - k\alpha) \times \prod_{k=0}^{d_1+3d_3} (\kappa_1 + 3\kappa_3 - k\alpha)$$

$$j_0^*(P_d) = \prod_{k=0}^{d_1+3d_2} (H_1 + 3H_2 - k\alpha) \times \prod_{k=0}^{d_1+3d_3} (H_1 + 3H_3 - k\alpha).$$

The linear sigma model is $W_d = N_d(\mathbf{P}(n)) = N_{d_1,1} \times N_{d_2,2} \times N_{d_3,2}$. The equivariant Euler class, after taking nonequivariant limit with respect to the T action, is given by

$$e_G(X_0/W_d) = \prod_{m=1}^{d_1} (H_1 - m\alpha)^2 \prod_{m=1}^{d_2} (H_2 - m\alpha)^3 \prod_{m=1}^{d_3} (H_3 - m\alpha)^3.$$

Now we can easily write down the hypregeometric series and all the K_d can be computed by our formula (8.6) at once using the obvious intersection form on X, given by the relations:

$$\int_X H_1 H_2^2 H_3^2 = 1, \;\; H_1^2 = 1, \;\; H_2^3 = 1, \;\; H_3^3 = 1.$$

Once we have the hypergeometric series, the corresponding Picard-Fuchs equation can be easily written down as given in [27].

8.5. $V = \mathcal{O}_1(-2) \otimes \mathcal{O}_2(-2)$ *on* $\mathbf{P}^1 \times \mathbf{P}^1$

Here we denote by $\mathcal{O}_i(l)$ the pullback of $\mathcal{O}(l)$ from the ith factor of $X = \mathbf{P}^1 \times \mathbf{P}^1$. Our bundle V has $rk\ V^+ - rk\ V^- = n - 3 = -1$. Thus we can apply our formula (8.4) with $x = 0$. We put $\Omega = \frac{1}{H_1 H_2}$. The Euler data P in eqn. (8.4) that compute the K_d is now given by:

$$P_d = \prod_{k=1}^{2d_1-1} (-2\kappa_1 + k\alpha) \times \prod_{k=1}^{2d_2-1} (-2\kappa_2 + k\alpha).$$

The corresponding equivariant Euler class, after taking the nonequivariant limit with respext to the T-action is

$$e_G(X_0/W_d) = \prod_{m=1}^{d_1}(H_1 - m\alpha)^2 \prod_{m=1}^{d_2}(H_2 - m\alpha)^2.$$

Again one can immediately write down the hypergeometric series as well as the corresponding Picard-Fuchs equation by using our mirror principle.

9. Generalizations and Concluding Remarks

9.1. A weighted projective space

Consider the following example: the concave bundle $V = O(-6)$ over $\mathbf{P}_{3,2,1}$, $\Omega = \frac{1}{6H}$. This example will be studied in our subsequent paper by using resolution of singularities. This is an example of "local mirror symmetry" studied in physics [33]. The mirror formula there can derived as a special case of our general result. In fact, the Euler data which computes the K_d in this case is determined by

$$j_0^* P_d = \prod_{m=1}^{6d-1}(-6H + m\alpha).$$

The corresponding equivariant Euler class, after taking nonequivariant limit with respect to the T action, is:

$$e_G(X_0/W_d) = \prod_{m=1}^{d}(H - m\alpha) \prod_{m=1}^{2d}(2H - m\alpha) \prod_{m=1}^{3d}(3H - m\alpha).$$

The corresponding hypergeometric series and Picard-Fuchs equation can be immediately written down. It turns out that the hypergeometric series gives the periods of a meromorphic 1-form for a family of elliptic curves [33].

9.2. General projective balloon manifolds

Let X be a projective manifold embedded in $\mathbf{P}(n)$, with a system of homogeneous polynomial defining equations $P(z^1, \cdots, z^n) = 0$, where $z^j = (z_1^j, \cdots, z_{n_j}^j)$. For each P, by taking the coefficients of each monomial $w_0^a w_1^b$ in $P(f^1, \cdots, f^n) = 0$, where $f^j = [f_1^j(w_0, w_1), \cdots, f_{n_j}^j(w_0, w_1)]$ for $j = 1, \cdots, k$ is the tuple of polynomials that define the coordinates of $N_d(\mathbf{P}(n))$, we get several equations of the same degree as P. These equations together define a projective variety, which we denote by $N_d(X)$, in $N_d(\mathbf{P}(n))$.

As discussed earlier, we see that the S^1 fixed point components in $N_d(X)$ are given by the X_r's which are copies of X. We do not know whether the localization formula holds on $N_d(X)$. The localization formula holds if the fixed point components embedded into W_d as local complete intersection subvarieties. It is likely that this is the case for any convex projective manifold. If this is true, then we can take $N_d(X)$ to be the linear sigma model W_d for X. Then our mirror principle may apply readily to compute multiplicative characteristic numbers on $M_{0,0}(d,X)$ in terms of the hypergeometric series.

9.3. A General Mirror Formula

Many of our results so far are proved for projective manifolds without T-action. Here we first discuss a formula for computing the numbers

$$K_d = \int_{M_{0,0}(d,X)} b(V_d)$$

for a general convex projective n-fold X *without T-action.* For simplicity, let's focus on the case when the multiplicative class b is the Chern polynomial c, and V is a direct sum of line bundles on X. There is a similar formulation in the general case. We fix a projective embedding $X \to \mathbf{P}(n)$, as before. Note that the map $\varphi : M_d(X) \to N_d(\mathbf{P}(n))$ is now only S^1-equivariant. Recall that the subvariety $W_d := \varphi(M_d(X)) \subset N_d(\mathbf{P}(n))$ contains as S^1 fixed point components copies of X: X_r, $0 \preceq r \preceq d$. We assume that the localization formula holds on it.

We denote by $e_{S^1}(X_0/W_d)$ the equivariant Euler class of the normal bundle of X_0 in W_d. Let

$$V = V^+ \oplus V^-, \quad V^+ := \oplus L_i^+, \quad V^- := \oplus L_j^-$$

satisfying $c_1(V^+) - c_1(V^-) = c_1(X)$ and $rk(V^+) - rk(V^-) - (n-3) \geq 0$, where the $L_i^\pm$ are respectively convex/concave line bundles on X.

Let

$$\Omega = B_0 := c(V^+)/c(V^-) = \prod_i (x + c_1(L_i^+))/\prod_j (x + c_1(L_j^-))$$

$$B_d := \frac{1}{e_{S^1}(X_0/W_d)} \times \prod_i \prod_{k=0}^{\langle c_1(L_i^+),d\rangle} (x + c_1(L_i^+) - k\alpha)$$

$$\times \prod_j \prod_{k=1}^{-\langle c_1(L_j^-),d\rangle-1} (x + c_1(L_j^-) + k\alpha).$$

$$HG[B](t) := \sum B_d e^{d\cdot t}$$

$$\Phi(t) := \sum K_d e^{d\cdot t}.$$

Conjecture 9.1. *There exist unique power series $G(t), F(t)$ such that the following formula holds:*

$$\frac{1}{s!}\left(\frac{d}{dx}\right)^s|_{x=0} \int_X \left(e^{-G/\alpha} HG[B](t) - e^{-H\cdot\tilde{t}/\alpha}\Omega\right)$$

$$= \alpha^{-3} x^{-s} (2\Phi(\tilde{t}) - \sum_i \tilde{t}_i \frac{\partial \Phi(\tilde{t})}{\partial \tilde{t}_i}).$$

where $s := rk\ V^+ - rk\ V^- - (n-3)$, $\tilde{t} := t + F(t)$. Moreover G, F are determined by the condition that the integrand on the left hand side is of order $O(\alpha^{-2})$.

9.4. Formulas without T-action

One of our key ingredient, the functorial localization formula plays an important role in relating the data on $M_d(X)$ and those on W_d. It turns out that similar formula holds in K-theory. It holds even when X has no group action. This indicates that our method may be extended to compute K-theory multiplicative type characteristic classes on $M_d(X)$ (and ultimately on $M_{0,0}(d, X)$), in terms of certain q-hypergeometric series, even for projective manifolds without group action.

We now write down the relevant localization formulas for convex X without torus action, both in cohomology and in K-theory. The notations and proofs are basically the same as before. Given a manifold X, let's assume that there is a linear sigma model W_d.

Lemma 9.2. *For any equivariant cohomology class ω on $M_d(X)$, the following equality holds on X_r for any $0 \preceq r \preceq d$:*

$$\frac{j_r^*\varphi_!(\omega)}{e_{S^1}(X_r/W_d)} = e_![\frac{i_r^*\omega}{e_{S^1}(F_r/M_d(X))}].$$

Here $e_{S^1}(\cdot)$ denotes the S^1-equivariant Euler class. As in the cases we have studied earlier, the left hand side of the above formula indicates that when $V = L$ is a line bundle, we should compare the Euler data $Q_d = \varphi_!\pi^*e(V_d)$, to the Euler data given by

$$P_d = \prod_{m=0}^{\langle c_1(L),d\rangle} (c_1(L) - m\alpha).$$

What is left is to develope uniqueness and mirror transformations, which we are unable to achieve at this moment, though they can be easily axiomized.

Now let us look at K-theory formula, which can be proved by using equivariant localization in K-theory. First following the same idea, we get the explicit formula as follows: given any equivariant element V in $K(W_d)$, we have

$$V = \sum_r j_{r!}\frac{j_r^*V}{E_G(X_r/W_d)}$$

where $E_G(X_r/W_d)$ is the equivariant Euler class of the normal bundle of X_r in $N_d(X)$. Here the push-forward and pull-backs by j_r denote the corresponding operations in K-theory. By taking $V = 1$, we get

$$e_![\frac{1}{E_G(F_r/M_d(X))}] = \frac{j_r^*\varphi_!(1)}{E_G(X_r/W_d)}.$$

Second, we have the following lemma; If $\varphi_!(1) = 1$, which is the case if $X = \mathbf{P}^n$, this formula gives explicit formulas for some K-theoretic characteristic numbers of the moduli spaces.

Lemma 9.3. *Given any equivariant element V in $K_G(M_d(X))$, then we have formula*

$$e_![\frac{i_r^*V}{E_G(F_r/M_d(X))}] = \frac{j_r^*(\varphi_!V)}{E_G(X_r/W_d)}$$

where $E_G(\cdot)$ denotes the equivariant Euler class of the corresponding normal bundle in K-group.

In particular we have explicit expressions from the decomposition of the normal bundles:

$$E_G(F_r/M_d(X)) = (1-e^{\alpha})(1-e^{-\alpha})(1-e^{\alpha}L_r)(1-e^{-\alpha}L_{d-r})$$

and similarly

$$E_G(F_0/M_d(X)) = (1-e^{\alpha})(1-e^{\alpha}L_d),\ E_G(F_d/M_d(X)) = (1-e^{-\alpha})(1-e^{-\alpha}L_d).$$

For a toric manifold X, we also have the explicit class in K-group,

$$E_G(X_0/W_d) = \prod_a \prod_{m=1}^{\langle D_a,d\rangle} (1-e^{m\alpha}[D_a])$$

where $[D_a]$ are the equivariant line bundle corresponding to the T divisors D_a.

If V is a multipicative type K-theory characteristic class, then we can develope a similar theory of Euler data and uniqueness. These result can also be extended to the nonconvex case without a hitch.

References

[1] C. Allday, V. Puppe, *Cohomology methods in transformation groups*, Cambridge University Press, 1993.

[2] M. Atiyah, *Convexity and commuting Hamiltonians*, Bull. London Math. Soc. 14 (1982) 1-15.

[3] M. Atiyah and R. Bott, *The moment map and equivariant cohomology*, Topology 23 (1984) 1-28.

[4] V. Batyrev, *Dual polyhedra and mirror symmetry for Calabi-Yau hypersurfaces in toric varieties*, J. Alg. Geom. 3 (1994) 493-535.

[5] V. Batyrev and L. Borisov, *On Calabi-Yau complete intersections in toric varieties*, alg-geom/9412017.

[6] V. Batyrev and D. van Straten, *Generalized hypergeometric functions and rational curves on Calabi-Yau complete intersections in toric varieties*, Comm. Math. Phys. 168 (1995) 495-533.

[7] K. Behrend and B. Fantechi, *The intrinsic normal cone*, Invent. Math. 128 (1997) 45-88.

[8] P. Berglund, S. Katz and A. Klemm, *Mirror Symmetry and the Moduli Space for Generic Hypersurfaces in Toric Varieties*, hep-th/9506091, Nucl. Phys. B456 (1995) 153.

[9] N. Berline and M. Vergne, *Classes caracteéristiques equivariantes, Formula de localisation en cohomologie equivariante.* 1982.

[10] R. Bott, *A residue formula for holomorphic vector fields*, J. Diff. Geom. 1 (1967) 311-330.

[11] M. Brion, *Equivariant cohomology and equivariant intersection theory*, math.AG/9802063.

[12] P. Candelas, X. de la Ossa, P. Green, and L. Parkes, *A pair of Calabi-Yau manifolds as an exactly soluble superconformal theory*, Nucl. Phys. B359 (1991) 21-74.

[13] P. Candelas, X. de la Ossa, A. Font, S. Katz and D. Morrison, *Mirror symmetry for two parameter models I*, hep-th/9308083, Nucl. Phys. B416 (1994) 481-538.

[14] D. Cox, *The functor of a smooth toric variety*, alg-geom/9312001.

[15] T. Delzant, *Hamiltoniens périodiques et image convex de l'application moment*, Bull. Soc. Math. France 116 (1988) 315-339.

[16] D. Edidin and W. Graham,*Equivariant intersection theory*, alg-geom/9609018.

[17] S.T. Yau, ed.,*Essays on Mirror Manifolds I*, International Press, Hong Kong 1992.

[18] W. Fulton, *Intersection Theory*, Springer-Verlag, 2nd Ed., 1998.

[19] K. Fukaya and K. Ono, *Arnold conjecture and Gromov-Witten invariants,* preprint, 1996.

[20] I. Gel'fand, M. Kapranov and A. Zelevinsky, *Hypergeometric functions and toral manifolds*, Funct. Anal. Appl. 23 (1989) 94-106.

[21] A. Givental, *A mirror theorem for toric complete intersections*, alg-geom/9701016.

[22] M. Goresky, R. Kottwitz and R. MacPherson, *Equivariant cohomology, Koszul duality and the localization theorem*, Invent. Math. 131 (1998) 25-83.

[23] V. Guillemin and S. Sternberg, *Convexity properties of the moment mapping*, Invent. Math. 67 (1982) 491-513.

[24] V. Guillemin and S. Sternberg, *Supersymmetry and equivariant de Rham theory*, preprint.

[25] V. Guillemin and C. Zara, *Equivariant de Rham Theory and Graphs*, math.DG/9808135.

[26] F. Hirzebruch, *Topological methods in algebraic geometry*, Springer-Verlag, Berlin 1995, 3rd Ed.

[27] S. Hosono, A. Klemm, S. Theisen and S.T. Yau, *Mirror symmetry, mirror map and applications to complete intersection Calabi-Yau spaces*, hep-th/9406055, Nucl. Phys. B433 (1995) 501-554.

[28] S. Hosono, B.H. Lian and S.T. Yau, *GKZ-generalized hypergeometric systems and mirror symmetry of Calabi-Yau hypersurfaces*, alg-geom/9511001, Commun. Math. Phys. 182 (1996) 535-578.

[29] S. Hosono, B.H. Lian and S.T. Yau, *Maximal Degeneracy Points of GKZ Systems*, alg-geom/9603014, Journ. Am. Math. Soc., Vol. 10, No. 2 (1997) 427-443.

[30] S. Hosono, M. Saito and J. Stienstra, *On mirror conjecture for Schoen's Calabi-Yau 3-folds*, preprint 1997.

[31] A. Kresch, *Cycle groups for Artin Stacks,* math. AG/9810166.

[32] M. Kontsevich, *Enumeration of rational curves via torus actions.* In: The Moduli Space of Curves, ed. by R.Dijkgraaf, C.Faber, G.van der Geer, Progress in Math.vol.129, Birkhäuser, 1995, 335–368.

[33] W. Lerche, P. Mayr, N. Warner, *Noncritical Strings, Del Pezzo Singularities and Seiberg-Witten Curves,* hep-th/9612085.

[34] J. Li and G. Tian, *Virtual moduli cycle and Gromov-Witten invariants of algebraic varieties,* J. of Amer. math. Soc. 11, no. 1, (1998) 119-174.

[35] J. Li and G. Tian, *Virtual moduli cycle and Gromov-Witten invariants of general symplectic manifolds. Topics in symplectic 4-manifolds (Irvine, CA, 1996), 47-83, First Int. Press Lect. Ser. I,* Internat. Press, Cambridge, MA, 1998.

[36] J. Li and G. Tian, *Comparison of algebraic and symplectic GW-invariants,* To appear in Asian Journal of Mathematics, 1998.

[37] B. Lian, K. Liu and S.T. Yau, *Mirror Principle I,* Asian J. Math. Vol. 1, No. 4 (1997) 729-763.

[38] B. Lian, K. Liu and S.T. Yau, *Mirror Principle, A Survey,* to appear.

[39] D. Morrison and R. Plesser, *Summing the instantons: quantum cohomology and mirror symmetry in toric varieties,* alg-geom/9412236.

[40] T. Oda, *Convex Bodies and Algebraic Geometry,* Series of Modern Surveys in Mathematics 15, Springer Verlag (1985).

[41] S.-S. Roan, *The mirror of Calabi-Yau orbifolds,* Intern. J. Math. 2 (1991), no. 4, 439-455.

[42] Y.B. Ruan, *Virtual neighborhoods and pseudo-holomorphic curves,* math.AG/9611021.

[43] Y.B. Ruan and G. Tian, *A mathematical theory of quantum cohomology,* Journ. Diff. Geom. Vol. 42, No. 2 (1995) 259-367.

[44] B. Siebert, *Gromov-Witten invariants for general symplectic manifolds,* alg-geom/9608005.

[45] E. Witten, *Phases of N=2 theories in two dimension,* hep-th/9301042.

Differential Equations from Mirror Symmetry

Bong H. Lian,[1] and Shing-Tung Yau[2]

Abstract: We discuss a method for deriving differential equations for the prepotential and the mirror map arising from a pair of Calabi-Yau manifold. Examples with one Kähler modulus are given. Here we find that the differential equations we derive almost entirely characterize the prepotential and the mirror map in question. Some of the results here have been announced previously in [3].

July 99

[1] Department of Mathematics, Brandeis University, Waltham, MA 02154.

[2] Department of Mathematics, Harvard University, Cambridge, MA 02138.

1. Introduction

We construct a new type of differential equations which govern the B-model prepotential and the mirror map which arise from a pair of mirror manifolds. By this we mean a pair of Calabi-Yau threefolds whereby the B-model prepotential of one threefold agrees with the A-model prepotential of the other threefold. The existence of those equations are in fact closedly related to the special geometry of the moduli spaces of Calabi-Yau manifolds.

In this paper we will restrict ourselves to the case where the mirror pair X, Y have $h^{1,1}(Y) = 1 = h^{2,1}(X) = 1$. Our construction can be generalized to the multi-moduli cases as well.

2. The mirror map and the prepotential

Let $\pi : \mathcal{X} \to \mathbf{P}^1$ be a smooth algebraic family over $\mathbf{P}^1$. We assume that the fibers $X_z := \pi^{-1}(z)$, except at finitely many points, are Calabi-Yau threefolds with $h^{2,1}(X_z) = 1$, and that the singular locus is a divisor with normal crossing. In a neighborhood of a smooth fiber X_z, we have a map from this into a universal family of Calabi-Yau threefolds, which by the theorem of Bogomolov-Tian-Todorov, is a smooth family over a disk in $H^{2,1}(X_z)$. We assume that this map is locally 1-1. We shall fix a smooth base point $z_0 \in \mathbf{P}^1$. Away from the singularity, there is a natural line bundle over the base whose fiber over z is $H^{3,0}(X_z)$. There is a section Ω whose value $\Omega(z)$ at z is a nowhere vanishing holomorphic 3-form on X_z. It is clear that Ω is unique up to a choice of a holomorphic function. For a given symplectic base $\alpha_1, , \alpha_2, \beta^1, \beta^2$ of $H^3(X_{z_0})$ with $\langle \alpha_i, \beta^j \rangle = \delta_i^j$, we can consider the integrals of $\Omega(z)$ over them near z_0. Put together, they form a 4-vector of (multivalued) holomorphic function which we call a period vector $\vec{\pi}$. It can be shown that (see [2][4]) that the period vector takes the form $(\xi_0, \xi_1, \frac{\partial G}{\partial \xi_0}, \frac{\partial G}{\partial \xi_1})$ where G is a homogeneous function of degree 2 of ξ_0, ξ_1. By a change of variable, $t = \xi_1/\xi_0$, it follows that $F = \xi_0^{-2} G$ is a function of t only, and we can write

$$\vec{\pi} = \xi_0(1, t, F', 2F - tF'). \tag{2.1}$$

It is well-known that the components of the period vectors are flat section of a bundle equipped with a natural flat connection. Concretely, the normalization of Ω can be choosen so that the components of the period vectors are solutions to a fourth order ordinary differential equations on $\mathbf{P}^1$ with regular singularities. It is known as a

Picard-Fuchs equation. In particular it is an ODE with polynomial coefficients.

Under suitable conditions, Mirror Symmetry interpretes the function F as a holomorphic function defined on the tube domain $\mathcal{T} = H^{1,1}(Y, \mathbf{R}) + \sqrt{-1}K(Y)$, where Y is another Calabi-Yau threefold and $K(Y)$ is its Kähler cone. The change of variable $t = \xi_1/\xi_0$ is interpreted as a mapping from $\mathbf{P}^1$ into $\mathcal{T}$. Note that as given, this mapping is only defined near a based point z_0. Inverting the mapping locally, we get a holomorphic function $z(t)$ which we call a *mirror map*. In Physics, F''' is known as a Yukawa coupling.

Note that both $z(t)$ and $F(t)$ are both defined at least as functions from a domain in $\mathbf{C}$, independent of the above Mirror Symmetry interpretation. In the context of mirror symmetry, many physicists have posted the following question (we thank M. Bershadsky and C. Vafa for communicating this question to us): *Is the function F governed by a differential equation?* In this paper, we answer this question in the affirmative. We shall construct a (rather complicated) 7th order polynomial differential equation for the function

$$K := F'''$$

with constant coefficients. More generally, we have

Theorem 2.1. *Let L be a 4th order linear differential operator on $\mathbf{P}^1$ with polynomial coefficients. Suppose $\xi_0, \xi_1, \xi_0 F'(\frac{\xi_1}{\xi_0}), \xi_0(2F(\frac{\xi_1}{\xi_0}) - \frac{\xi_1}{\xi_0}F'(\frac{\xi_1}{\xi_0}))$ are linearly independent solutions to $Lf = 0$ near 0 for some holomorphic function F on a domain. Then $K(t) := F'''(t)$ satisfies a 7th order polynomial differential equation with constant coefficients. Likewise for $z(t)$ as defined above.*

The result is really local in nature and can be formulated entirely in a disk. The details of the construction of the 7th order differential equation for $z(t)$ was done in [3] and shall not be repeated here. We shall first prove the theorem, and then return to examples in the last section. The hypotheses of the theorem will be in force throughout this paper.

The basic idea of the proof is this: ODEs with solutions above can be viewed as the differential algebraic analogue of polynomial equations in two variables. The functions F and z play the roles of the two variables. In the algebraic case, one can use eliminating theory to construct algebraic equations for each variables by "separating" the variables. The strategy is to try to do the same in the differential algebraic case. We achieve this by developing and using some elementary tools in differential graded algebra.

Acknowledgements. We thank A. Klemm and S.S. Roan for participating in the initial stage of this project a few years ago. Special thanks to C. Doran for proofreading this paper. B.H.L. is supported by NSF grant DMS-9619884. S.T.Y.is supported by DOE grant DE-FG02-88ER25065 and NSF grant DMS-9803347.

3. Change of variables

Note that the functions $z(t)$ and $F(t)$ depend only on the ratio of two solutions ξ_1, ξ_0, hence remain unchanged if we perform a change of variables $f(z) = A(z)g(z)$ on the equation $Lf = 0$. We can choose A so that the new equation reads:

$$g''''(z) + a_2(z)g''(z) + a_1(z)g'(z) + a_0(z)g(z) = 0 \qquad (3.1)$$

where the a_i are rational functions.

Lemma 3.1. $a_1(z) = a_2'(z)$.

For the proof of this, see [2]. Under a change of variable $z \mapsto t$, (3.1) becomes

$$\tilde{L}g := g''''(z(t)) + c_2 g''(z(t)) + c_2' g'(z(t)) + c_0 g(z(t)) = 0 \qquad (3.2)$$

where

$$\begin{aligned} c_2 =& a_2 z'^2 - \frac{15}{2}(\frac{z''}{z'})^2 + 5\frac{z^{(3)}}{z'} = a_2 z'^2 + 5\{z(t), t\} \\ c_0 =& a_0 z'^4 + \frac{3}{2}\frac{da_2}{dz} z'^2 z'' - \frac{3}{4} a_2 z''^2 - \frac{135 z''^4}{16 z'^4} \\ &+ \frac{3}{2} a_2 z' z^{(3)} + \frac{75 z''^2 z^{(3)}}{4 z'^3} - \frac{15 {z^{(3)}}^2}{4 z'^2} - \frac{15 z'' z^{(4)}}{2 z'^2} + \frac{3 z^{(5)}}{2 z'} \end{aligned} \qquad (3.3)$$

The prime here means $\frac{d}{dt}$. Note that c_0, c_2 are differential rational functions of $z(t)$, ie. a rational function in $z, z', z'', \ldots$ Now (3.2)has four linear independent solutions of the form $u(1, t, F', 2F - tF')$ where $u = \xi_0/A$. We may now view (3.2) together with its four special solutions as a system of polynomial ODEs in $z(t), F(t), u(t)$. Our polynomial ODEs for $z(t), F(t)$ with constant coefficients will be obtained by separating the variables.

First we claim that

$$K = const.\ u^{-2} = const.\ \frac{A^2}{\xi_0^2}. \tag{3.4}$$

Substituting the 4 special solutions above in (3.2), we get

$$\begin{aligned} u^{(4)} + c_2 u'' + c_2' u' + c_0 u &= 0 \\ 4u''' + 2c_2 u' + c_2' u &= 0 \\ K(6u'' + c_2 u) + 4K'u' + K''u &= 0 \\ 2Ku' + K'u &= 0. \end{aligned} \tag{3.5}$$

Our claim follows from solving the 4th equation above.

Note that given this K, the 2nd equation in (3.5)can be obtained by differentiating the 3th equation. From now on we want to view the system (3.5)of equations as a system of ODEs in the two dependent variables $K(t), z(t)$ and one independent variable t. The system (3.5)can be written as

$$\begin{aligned} c_2(z;t) &= r_2(K;t) \\ c_0(z;t) &= r_0(K;t) \end{aligned} \tag{3.6}$$

where c_2, c_0 are rational function of $z, z', ..$ defined in (3.3). The r_2, r_0 are rational function of $K, K', ..$ given by

$$\begin{aligned} r_2 &= \frac{2K''}{K} - \frac{5}{2}\left(\frac{K'}{K}\right)^2 \\ r_0 &= \frac{-35\,K'^4}{16\,K^4} + \frac{5\,K'^2\,K''}{K^3} - \frac{5\,K''^2}{4\,K^2} - \frac{2\,K'\,K^{(3)}}{K^2} + \frac{K^{(4)}}{2\,K}. \end{aligned} \tag{3.7}$$

Note that the system (3.6)of equations are derived with no special assumption about the coordinate t other than that it is the ratio of two linearly independent solutions: $t = \xi_1/\xi_0$. Thus the *form* of the system (3.6)must remain the same if we use another ratio $\tilde{t}$. Thus the new equations will read:

$$\begin{aligned} c_2(z(\tilde{t});\tilde{t}) &= r_2(K(\tilde{t});\tilde{t}) \\ c_0(z(\tilde{t});\tilde{t}) &= r_0(K(\tilde{t});\tilde{t}). \end{aligned} \tag{3.8}$$

4. Some differential algebra

If we assign the weight 1 to each t differentiation (ie each prime), then both c_2, r_2 in (3.6)are expressions of weight 2. Similarly both c_0, r_0 are expression of weight 4. This motivates the following considerations. Let $x, x', .., y, y', ..$ be formal variables. We assume that $wt\ x^{(k)} = wt\ y^{(k)} = k$. We may also freely use some other formal variables $\alpha, \alpha', .., \beta, \beta', ..$ which will bear the analogous meaning. A rational function $f(x, x', .., y, y', ..)$ in these variables will be called a differential rational function. We usually denote it as $f[x, y]$. If it depends only on the x, we write $f[x]$. Given two such functions, say $f[x], g[y]$, it makes sense to compose them. Namely we do so in an obvious way $f[g[x]]$ according to the usual rule of differentiation. Let $X[x], Y[x], A[y], B[y]$ be differential rational functions, homogeneuos with positive weights $|X| = |A|$, $|Y| = |B|$. We would like to study the system of ODEs:

$$\begin{aligned} X[x] &= A[y] \\ Y[x] &= B[y]. \end{aligned} \tag{4.1}$$

We write

$$I(A[y], B[y]) := \{P[a, b] | P[A[y], B[y]] \equiv 0\}. \tag{4.2}$$

This is obviously a differential ideal in the ring of differential polynomials in two variables α, β. If α, β are assign the weight $|A|, |B|$ respectively, then $I(A[y], B[y])$ is in fact a graded ideal. Note that if $x(t), y(t)$ are solutions to the system (4.1), then every element $P[a, b] \in I(A[y], B[y])$ gives a differential equation

$$P[X[x(t)], Y[x(t)]] = 0 \tag{4.3}$$

for the function $x(t)$. This equation can of course be trivial.

We shall construct a nontrivial differential equation for the function $K(t)$ by first studying the differential graded ideal $I(c_2, c_0)$ where c_2, c_0 are given in (3.3). We will prove that there exists an element of minimal weight in this ideal. In this sense, this will be the simplest polynomial differential equation for $K(t)$.

Let's consider the effect under the formal substitution law $t \mapsto \tilde{t} = \frac{at+b}{ct+d}$, $x \mapsto \tilde{x}$, $y \mapsto \tilde{y}$ on a homogeneous differential rational function of the form $P[X[x], Y[x]]$ of weight l. It is called covariant if it transforms like $P[X[x], Y[x]] \mapsto \gamma^{2l} P[X[\tilde{x}], Y[\tilde{x}]]$. Suppose the differential rational functions $X[x], Y[x]$ transform as follows:

$$\begin{aligned} X[z] &\mapsto \sum_{j=0}^{n} \gamma^{2n-j} c^j Q_j[X[\tilde{z}], Y[\tilde{z}]] \\ Y[z] &\mapsto \sum_{j=0}^{m} \gamma^{2m-j} c^j R_j[X[\tilde{z}], Y[\tilde{z}]] \end{aligned} \tag{4.4}$$

where $\gamma = ct + d$, n, m are the respective weights of X, Y, and the Q_j, R_j are some fixed differential polynomials of respective weights $n - j, m - j$. We also assume the same transformation law for the differential polynomials $A[y], B[y]$, but only with $x, \tilde{x}$ replaced by $y, \tilde{y}$ respectively. Note that the leading term must be given by $Q_0[X, Y] = X$ and $R_0[X, Y] = Y$. The assumption above simply says that *the differential polynomial* $\mathbf{C}(t)$*-algebra generated by* X, Y *is closed under the transformations from* $SL(2, \mathbf{C})$.

Lemma 4.1. *Suppose that* $P[\alpha, \beta]$ *is an element in* $I(A[y], B[y])$ *of minimal weight. Then* $P[X[x], Y[x]]$ *is covariant.*

Proof: It is enough to show that $P[\alpha, \beta] \mapsto \gamma^{2l} P[\alpha, \beta]$ under formal substitution:

$$\begin{aligned} \alpha &\mapsto \sum_{j=0}^{n} \gamma^{2n-j} c^j Q_j[\alpha, \beta] \\ \beta &\mapsto \sum_{j=0}^{m} \gamma^{2m-j} c^j R_j[\alpha, \beta]. \end{aligned} \tag{4.5}$$

Obvioiusly under (4.5), we have

$$P[\alpha, \beta] \mapsto \sum_{j=0}^{l} \gamma^{2l-j} c^j S_j[\alpha, \beta] \tag{4.6}$$

for some differential polynomials S_j of weights $l - j$. We must show that $S_j = 0$ for all $j > 0$. By assumption A, B have similar transformation law (4.5). More precisely we have

$$P[A[y], B[y]] \mapsto \sum_{j=0}^{l} \gamma^{2l-j} c^j S_j[A[\tilde{y}], B[\tilde{y}]] \tag{4.7}$$

Since $P[\alpha, \beta]$ is an element of the ideal $I(A[y], B[y])$, it follows that $P[A[y], B[y]]$ is identically zero. Hence the right hand side of (4.7)must also be identically zero. Thus we have $S_j[A[\tilde{y}], B[\tilde{y}]] = 0$ for all j. Since $\tilde{y}$ is just a formal variable, the same holds true with $\tilde{y}$ replaced by y. This means that the $S_j[\alpha,\beta]$ are in the ideal $I(A[y], B[y])$. But S_j has weight $l-j$ and so by the minimality of $P[\alpha,\beta]$, we have $S_j[\alpha,\beta] = 0$ for all $j > 0$. Thus (4.6)now says exactly that P is covariant of weight l. This completes the proof. □

4.1. SL(2) action on solution set

We now return to our original system of equations (3.6), and diagress slightly to consider its general solutions. Given a complex parameter t, let $\tilde{t} = \frac{at+b}{ct+d}$ where $ad - bc = 1$. Viewing $t \to \tilde{t}$ as a change of coordinates, we have the formal relations:

$$\begin{aligned} c_2(z(\tilde{t});\tilde{t}) &= \gamma^4 c_2(z(\tilde{t});t) \\ c_0(z(\tilde{t});\tilde{t}) &= \gamma^8 c_0(z(\tilde{t});t) + 3\gamma^7 c c_2'(z(\tilde{t});t) + 6\gamma^6 c^2 c_2(z(\tilde{t});t). \end{aligned} \tag{4.8}$$

Proposition 4.2. *Let $z(t), K(t)$ be solutions to (3.6). Then so are*

$$\begin{aligned} \tilde{z}(t) &:= z(\frac{at+b}{ct+d}) \\ \tilde{K}(t) &:= C\gamma^{-4} K(\frac{at+b}{ct+d}) \end{aligned} \tag{4.9}$$

for any constant C.

Proof: It follows immediately from (3.6)that the equations are invariant under the scaling of K by a constant. So without loss of generality, we can set $C = 1$. By direct computation, we find that

$$\begin{aligned} r_2(K(\tilde{t});\tilde{t}) &= \gamma^4 r_2(\gamma^{-4}K(\tilde{t});t) \\ r_0(K(\tilde{t});\tilde{t}) &= \gamma^8 r_0(\gamma^{-4}K(\tilde{t});t) + 3\gamma^7 c r_2'(\gamma^{-4}K(\tilde{t});t) + 6\gamma^6 c^2 r_2(\gamma^{-4}K(\tilde{t});t). \end{aligned} \tag{4.10}$$

Applying (3.6), (4.8) and (4.10), we get

$$\begin{aligned} c_2(z(\tilde{t});t) &= \gamma^{-4} c_2(z(\tilde{t});\tilde{t}) = \gamma^{-4} r_2(K(\tilde{t});\tilde{t}) = r_2(\gamma^{-4}K(\tilde{t});t) \\ c_0(z(\tilde{t});t) &= \gamma^{-8} c_0(z(\tilde{t});\tilde{t}) - 3\gamma^{-1} c c_2'(z(\tilde{t});t) - 6\gamma^{-2} c^2 c_2(z(\tilde{t});t) \\ &= r_0(\gamma^{-4}K(\tilde{t});t) \qquad \square \end{aligned} \tag{4.11}$$

It is interesting to see how the new solution given is *constructed.* Fix a solution $(z(t), K(t))$ to the system (3.6). And consider the same system with z, K replaced by $\tilde{z}, \tilde{K}$. Set $\tilde{z}(t) = z(\frac{at+b}{ct+d})$ where $z(t)$ is assumed to be a given solution. We would like to solve for $\tilde{K}(t)$ in terms of the given $K(t)$. We shall see that the general solution is precisely $\tilde{K}(t) = C\gamma^{-4}K(\frac{at+b}{ct+d})$. We must solve

$$\begin{aligned} r_2(K(\tilde{t});\tilde{t}) &= \gamma^4 r_2(\tilde{K}(t);t) \\ r_0(K(\tilde{t});\tilde{t}) &= \gamma^8 r_0(\tilde{K}(t);t) + 3\gamma^7 c r_2'(\tilde{K}(t);t) + 6\gamma^6 c^2 r_2(\tilde{K}(t);t). \end{aligned} \tag{4.12}$$

Set $y(t) = Log\ K(t)$. Then we get

$$\begin{aligned} r_2(K(t);t) &= 2y'' - \frac{y'^2}{2} \\ r_0(K(t);t) &= \frac{y^{(4)}}{2} + \frac{y''^2}{4} - \frac{y''y'^2}{2} + \frac{y'^4}{16}. \end{aligned} \tag{4.13}$$

Thus the first equation in (4.12)becomes

$$\frac{d^2y(\tilde{t})}{d\tilde{t}^2} - \frac{1}{4}(\frac{dy(\tilde{t})}{d\tilde{t}})^2 = \gamma^4(\frac{d^2\tilde{y}(t)}{dt^2} - \frac{1}{4}(\frac{d\tilde{y}(t)}{dt})^2). \tag{4.14}$$

Since $y(t)$ is given, this is a Ricatti equation in the variable $\frac{d\tilde{y}(t)}{dt}$. It's easy to verify that its general solution is

$$\tilde{y}(t) := y(\frac{at+b}{ct+d}) - 4Log\ \gamma. \tag{4.15}$$

By the standard classical method, we find that the general solution to (4.14) is

$$y_{gen}(t) = \tilde{y}(t) - \int \frac{4C exp(\frac{1}{2}\tilde{y}(t))}{1 + C\int exp(\frac{1}{2}\tilde{y}(t))} \tag{4.16}$$

where C is an arbitrary constant. Writing it in terms of $K(t)$, we get

$$\tilde{K}(t) = \frac{\gamma^{-4}K(\frac{at+b}{ct+d})}{\left(B + C\int \gamma^{-2}K(\frac{at+b}{ct+d})^{1/2}\right)^4}. \tag{4.17}$$

Now this satisfy the second equation in (4.12) iff $C = 0$.

5. ODEs for the Yukawa Coupling – Existence

Our problem now is to construct an element $P[\alpha,\beta]$ of $I(c_2,c_0)$, which will give a nontrivial differential equation for $K(t)$. The last section indicates that we should look for a *covariant* element $P[\alpha,\beta]$ such that

$$P[c_2,c_0]\equiv 0 \tag{5.1}$$

identically. We shall do this in three steps. Let α,β be formal variable of weight 2,4 respectively. We shall call a differential polynomial $P[\alpha,\beta]$ a *simple covariant* of weight k if

$$P[c_2,c_0]=p(z)z'^k \tag{5.2}$$

for some rational function $p(z)$. It is obvious that $P[\alpha,\beta]$ is automatically a covariant. First we find two simple covariants $P_1[\alpha,\beta], Q_1[\alpha,\beta]$ of lowest weights. Our original system (3.6) then becomes

$$\begin{aligned} P_1[r_2[K],r_0[K]] =&p(z)z'^{|P_1|}\\ Q_1[r_2[K],r_0[K]] =&q(z)z'^{|Q_1|} \end{aligned} \tag{5.3}$$

for some rational functions $p(z),q(z)$. We then eliminate z from this pair of equations. Finally we prove that the resulting equation governing K is nontrivial.

Proposition 5.1. *Let*

$$\begin{aligned} \rho[\alpha,\beta] =&100\beta-9\alpha^2-30\alpha''\\ \chi[\alpha,\beta] =&-32\rho[\alpha,\beta]^2\alpha-45(\frac{d}{dt}\rho[\alpha,\beta])^2+40\rho[\alpha,\beta]\frac{d^2}{dt^2}\rho[\alpha,\beta]. \end{aligned} \tag{5.4}$$

They are the unique simple covariants of weights 4,10 respectively. Moreover any other simple covariant not proportional to either of them has higher weight.

Note that these two simple covariants are in fact universal: they are independent of the original 4th order differential operator. Thus they are independent of the initial data (a_2,a_0).

Proof: We prove this by solving exactly the condition (5.2) in the differential graded algebra generated by α,β up to weight 10. Since α,β have weights 2,4, the only graded pieces we have to look are of weights between 2,10. It is easy to work out a basis in terms of $\alpha, \alpha', .., \beta, \beta'$ for each graded pieces. Solving the above condition yields the desired result. □

It is interesting to note that the equations (5.4)defines an isomorphism

$$\iota : (\alpha, \beta) \mapsto (\rho[\alpha, \beta], \chi[\alpha, \beta]) \tag{5.5}$$

between the two fields of differential rational functions: one is generated by ρ, χ, the other by α, β.

We also note that

$$\begin{aligned} \rho[c_2, c_0] &= \rho[a_2(z), a_0(z)]z'^4 \\ \chi[c_2, c_0] &= \chi[a_2(z), a_0(z)]z'^{10}. \end{aligned} \tag{5.6}$$

Here we have slightly abused the notation: $\rho[a_2(z), a_0(z)]$ really means $100a_0(z) - 9a_2(z)^2 - 30a_2''(z)$, and similarly for $\chi[a_2(z), a_0(z)]$. Thus both are just rational functions in the variable z.

If $\rho[a_2(z), a_0(z)] = 0$ (or $\chi[a_2(z), a_0(z)] = 0$), then it's easy to show that $\rho[r_2[K], r_0[K]] = 0$ (or $\chi[r_2[K], r_0[K]] = 0$) is a nontrivial ODE for K. Thus from now on, we assume that neither is zero. Set

$$\begin{aligned} p(z) &= \rho[a_2(z), a_0(z)] \\ q(z) &= \chi[a_2(z), a_0(z)] \end{aligned} \tag{5.7}$$

Then we have

$$5\chi[c_2, c_0]\rho[c_2, c_0]' - 2\chi[c_2, c_0]'\rho[c_2, c_0] = (5q\frac{dp}{dz} - 2p\frac{dq}{dz})z'^{15}. \tag{5.8}$$

It follows that

$$\begin{aligned} \frac{\rho[c_2, c_0]^5}{\chi[c_2, c_0]^2} &= \frac{p^5}{q^2} \\ \frac{(5\chi[c_2, c_0]\rho[c_2, c_0]' - 2\chi[c_2, c_0]'\rho[c_2, c_0])^2}{\chi[c_2, c_0]^3} &= \frac{(5q\frac{dp}{dz} - 2p\frac{dq}{dz})^2}{q^3}. \end{aligned} \tag{5.9}$$

Lemma 5.2. *For any $f(z), g(z) \in \mathbf{C}(z)$, there is a nonzero polynomial $Q(a, b)$ such that $Q(f(z), g(z)) \equiv 0$ identically. If we write $f = f_1/f_2$, $g = g_1/g_2$, then such a Q may be choosen so* *that deg_a Q and deg_b Q are both bounded from above by* $N_{max} := max(max(deg\ f_1, deg\ f_2) + max(deg\ g_1, deg\ g_2) - 2, 1)$.

Proof: We consider the system of homogeneous linear equations resulting from the coefficients of the powers of z in

$$\sum_{i,j=0}^{N} a_{ij} f(z)^i g(z)^j = 0. \tag{5.10}$$

It is easy to verify that for $N = N_{max}$, the linear system has more variables than equations. □

In particular if we now take f, g to be the two right hand sides of (5.9), then it follows that we have a nonzero polynomial Q with the above vanishing property. Fix such a Q and define a new differential polynomial in the variables ξ, η of weights 4,10 respectively:

$$R[\xi, \eta] := \eta^{5N} Q(\frac{\xi^5}{\eta^2}, \frac{(5\eta\xi' - 2\eta'\xi)^2}{\eta^3}). \tag{5.11}$$

It is obvious that $R[\xi, \eta]$ is nonzero. Also by construction we have $R[\rho[c_2, c_0], \chi[c_2, c_0]] \equiv 0$ identically.

Proposition 5.3. $R[\rho[r_2[K], r_0[K]], \chi[c_2[K], c_0[K]]] = 0$ *is a non-*

trivial ODE for K.

Proof: By direct computation, we find

$$\rho[r_2[K], r_0[K]] =$$
$$175\, K'^4 - 280\, K \frac{K'^2\, K'' + 49\, K^2\, K''^2 + 70\, K^2\, K'\, K^{(3)} - 10\, K^3\, K^{(4)}}{K^4}$$

$$\begin{aligned}
\chi[r_2[K], r_0[K]] =4 \Big(&1225000\, K'^{10} - 4900000\, K\, K'^8\, K'' \\
&+ 6737500\, K^2\, K'^6\, K''^2 r - 3626000\, K^3\, K'^4\, K''^3 \\
&+ 689675\, K^4\, K'^2\, K''^4 - 360836\, K^5\, K''^5 \\
&+ 1225000\, K^2\, K'^7\, K^{(3)} - 2940000\, K^3\, K'^5\, K''\, K^{(3)} \\
&+ 1690500\, K^4\, K'^3\, K''^2\, K^{(3)} \\
&+ 695800\, K^5\, K'\, K''^3\, K^{(3)} \\
&+ 367500\, K^4\, K'^4\, {K^{(3)}}^2 - 705600\, K^5\, K'^2\, K''\, {K^{(3)}}^2 \\
&- 235200\, K^6\, K''^2\, {K^{(3)}}^2 + 117600\, K^6\, K'\, {K^{(3)}}^3 \\
&- 175000\, K^3\, K'^6\, K^{(4)} + 420000\, K^4\, K'^4\, K''\, K^{(4)} \\
&- 562800\, K^5\, K'^2\, K''^2\, K^{(4)} + 203000\, K^6\, K''^3\, K^{(4)} \\
&+ 42000\, K^5\, K'^3\, K^{(3)}\, K^{(4)} + 142800\, K^6\, K'\, K''\, K^{(3)}\, K^{(4)} \\
&- 16800\, K^7\, {K^{(3)}}^2\, K^{(4)} - 12000\, K^6\, K'^2\, {K^{(4)}}^2 \\
&- 26400\, K^7\, K''\, {K^{(4)}}^2 + 94500\, K^5\, K'^3\, K''\, K^{(5)} \\
&- 103950\, K^6\, K'\, K''^2\, K^{(5)} - 31500\, K^6\, K'^2\, K^{(3)}\, K^{(5)} \\
&+ 37800\, K^7\, K''\, K^{(3)}\, K^{(5)} + 9000\, K^7\, K'\, K^{(4)}\, K^{(5)} \\
&- 1125\, K^8\, {K^{(5)}}^2 - 17500\, K^5\, K'^4\, K^{(6)} \\
&+ 28000\, K^6\, K'^2\, K''\, K^{(6)} - 4900\, K^7\, K''^2\, K^{(6)} \\
&- 7000\, K^7\, K'\, K^{(3)}\, K^{(6)} + 1000\, K^8\, K^{(4)}\, K^{(6)}\Big) / K^{10}
\end{aligned} \tag{5.12}$$

Observe that $R[\xi, \eta]$ has either a nontrivial leading power in η', or else it has a leading power in η. In the first case, we see

$R[\rho[r_2[K], r_0[K]], \chi[c_2[K], c_0[K]]]$ has a nontrivial leading term containing the 7th derivative $K^{(7)}$. In the second case we have a nontrivial leading term containing $K^{(6)}$. In either case, the resulting ODE is nontrivial. □

This completes the proof of our main theorem.

6. Remarks and examples

The proof above indicates also that we can put a bound on the minimal weight of our ODE for K. Since Q obviously has weight zero, the differential polynomial R above has weight $5N$. Thus the minimal weight is no more than $5N_{max}$.

We now return to the discussion of the Picard-Fuchs equation of a family of Calabi-Yau 3-folds. In the case of the mirror quintic, we have the famous Picard-Fuchs equation

$$\left(\Theta^4 - 5z(5\Theta+4)(5\Theta+3)(5\Theta+2)(5\Theta+1)\right) f(z)) = 0 \qquad (6.1)$$

where $\Theta := z\frac{d}{dz}$. In this case one finds that $N_{max} = 1400$. This turns out to be far from optimum. Using the computer, we find an ODE for K of weight 360.

In practice, we can simplify the computation for an ODE significantly as follows. If we regard $5\eta\xi' - 2\eta'\xi$ as another variable, then the construction in the last section guarantees that there have a polynomial in $\xi, \eta, \delta := 5\eta\xi' - 2\eta'\xi$ which gives a nontrivial ODE for K.

Thus it is enough to look in the polynomial ring in ξ, η, δ where the weights of the variables are respectively $4, 10, 15$. Our construction guarantees that any zero polynomial quasi-homogeneous polynomial $P(\xi, \eta, \delta)$ satisfying

$$P(\rho[c_2, c_0], \chi[c_2, c_0], 5\rho[c_2, c_0]'\chi[c_2, c_0] - 2\rho[c_2, c_0]\chi[c_2, c_0]') \equiv 0 \quad (*) \qquad (6.2)$$

identically gives a a nontrivial ODE for K. We want to find one with the minimum weight.

Lemma 6.1. *A necessary and sufficient condition for a nonzero polynomial $P(\xi, \eta, \delta)$ satisfying (*) to be a minimum of weight n is that every polynomial $T(\xi, \eta, \delta)$ satisfying (*) of weight $n-1, n-2, n-3$, or $n-4$ is zero.*

Proof: The necessity of the condition is clear. To show sufficiency, note that if $T(\xi, \eta, \delta)$ is a polynomial of weight lower that $n-4$ and satisfies (*), then we can multiply it by a suitable power – say ξ^k – so that $\xi^k T(\xi, \eta, \delta)$ has weight $n-1, n-2, n-3$, or $n-4$ because ξ has weight 4. Now it must be zero, implying that T itself is zero. □

When we try to find a polynomial $P(\xi, \eta, \delta)$ satisfying (*) and of minimal weight, say n, the lemma saves us from checking the the lower weight polynomials – all we have to check is that there is none in weights $n-4$ to $n-1$. For example, in the case of the mirror quintic. We verify by computer that there are no nonzero polynomial $T(\xi, \eta, \delta)$ satisfying (*) and has weight between 359 and 356. Moreover there are exactly 127 monomials of weight 360 in the three variables. Up to multiple, exactly one linear combination of them satisfies (*).

Here we shall give the simplest example known to us: here we consider the family of Calabi-Yau 3-folds X mirror to the complete intersection of 4 quadrics in $\mathbf{P}^7$. In this case, the minimal polynomial satisfying (*) has weight 180:

$$
\begin{aligned}
&P(\delta, \chi, \rho) = \\
&3783403212890625\,\chi^{18} + 52967644980468750\,\chi^{15}\,\delta^{2} \\
&+ 292835408677734375\,\chi^{12}\,\delta^{4} + 833559395864062500\,\chi^{9}\,\delta^{6} \\
&+ 1301823644717109375\,\chi^{6}\,\delta^{8} + 1064406315612768750\,\chi^{3}\,\delta^{10} \\
&+ 357449882108765625\,\delta^{12} + 9097175898878906250\,\chi^{16}\,\rho^{5} \\
&+ 75543680906950781250\,\chi^{13}\,\delta^{2}\,\rho^{5} \\
&- 55168781762820937500\,\chi^{10}\,\delta^{4}\,\rho^{5} \\
&- 1235933279927738437500\,\chi^{7}\,\delta^{6}\,\rho^{5} \\
&- 2628328829388247068750\,\chi^{4}\,\delta^{8}\,\rho^{5} \\
&- 1669442421173622243750\,\chi\,\delta^{10}\,\rho^{5} \\
&- 3163959222224629737093 75\,\chi^{14}\,\rho^{10} \\
&- 10413036935813864040750 00\,\chi^{11}\,\delta^{2}\,\rho^{10} \\
&+ 1397061241390545045311250\,\chi^{8}\,\delta^{4}\,\rho^{10} \\
&- 9780717526289292060000\,\chi^{5}\,\delta^{6}\,\rho^{10} \\
&+ 3088961515882945520173125\,\chi^{2}\,\delta^{8}\,\rho^{10} \\
&+ 726375263921582813504122500\,\chi^{12}\,\rho^{15} \\
&- 2401967567306257982918892000\,\chi^{9}\,\delta^{2}\,\rho^{15} \\
&+ 2931906039367569842399977800\,\chi^{6}\,\delta^{4}\,\rho^{15} \\
&- 2592007729730548310729752000\,\chi^{3}\,\delta^{6}\,\rho^{15} \\
&+ 264772114318563252921325000\,\delta^{8}\,\rho^{15} \\
&+ 7317735278685046995613249290 24\,\chi^{10}\,\rho^{20} \\
&- 1384453886791545382987331665920\,\chi^{7}\,\delta^{2}\,\rho^{20}
\end{aligned}
$$

$$
\begin{aligned}
&+ 1003786188392583028918031769600\,\chi^4\,\delta^4\,\rho^{20}\\
&- 9265029998433113889422540800\,0\,\chi\,\delta^6\,\rho^{20}\\
&+ 264379950716374035480557566033920\,\chi^8\,\rho^{25}\\
&- 323653884996678359415539902709760\,\chi^5\,\delta^2\,\rho^{25}\\
&+ 105122101152057682020817226956800\,\chi^2\,\delta^4\,\rho^{25}\\
&+ 4885370016741424964003892343865344 0\,\chi^6\,\rho^{30}\\
&- 3315342376066498968351383146725376 0\,\chi^3\,\delta^2\,\rho^{30}\\
&+ 528120679253988321156369324441600\,\delta^4\,\rho^{30}\\
&+ 49655388960105132238220106179962470 40\,\chi^4\,\rho^{35}\\
&- 12389340807480736990290861242921779 20\,\chi\,\delta^2\,\rho^{35}\\
&+ 2650217711622663559007619458167681843 20\,\chi^2\,\rho^{40}\\
&+ 582240682567099819640139229658876372582 4\,\rho^{45}
\end{aligned}
$$

(6.3)

Note that since δ, χ, ρ are of weights 15,10,4 respectively, P is a quasi-homogeneous polynomial of weight 180. Each of the 37 terms in this polynomial corresponds to a partition of 180 by 15,10,4.

In the second example, we consider the complete intersection of two cubic in $\mathbf{P}^5$. In this case, we find that the ODE for K is given by a polynomial of weight 330. There are 108 partitions of 330 by 15,10,4. Thus this polynomial has a maximum length of 108.

We have verified that in all cases above, our differential equation determines the Yukawa coupling $K(t)$ up to the first two Fourier coefficients.

References

[1] P. Candelas, X. de la Ossa, P. Green, and L. Parkes, *A pair of Calabi-Yau manifolds as an exactly soluble superconformal theory*, Nucl. Phys. B359 (1991) 21-74.

[2] S. Ferrara, C. Kounnas, D. Lüst, and F. Swirner, *Automorphic Functions and Special Kähler Geometry*, In *Essays on Mirror Manifolds*, Ed. S.T. Yau, International Press 1992.

[3] A. Klemm, B.H. Lian, S.S. Roan, and S.T. Yau, *A Note on ODEs from Mirror Symmetry*, Conference Proceeding in honor of I. Gel'fand (1995).

[4] A. Strominger, *Commun. Math. Phys. 133 (1990) 163.*

Surveys in Differential Geometry, vol. 5

Heat Kernels, Symplectic Geometry, Moduli Spaces and Finite Groups

Kefeng Liu

1. Introduction

In this note we want to discuss some applications of heat kernels in symplectic geometry, moduli spaces and finite groups. More precisely we will prove the nonabelian localization formula in symplectic geometry, derive formulas for the symplectic volume and intersection numbers of the moduli space of flat connections on a Riemann surface, and obtain several quite general formulas for the numbers of solutions of equations in finite groups. Several new formulas for the push-forward measures by various maps between Lie groups are also obtained. In solving these problems, we will use the corresponding heat kernels on Euclidean spaces, on Lie groups and on finite groups.

The discussions of some aspects of the first two applications have appeared in [Liu], [Liu1] and [Liu2], here we will only sketch the main ideas. The purpose to include them here is to unify the discussions by using heat kernels. The third and the fourth applications were worked out through many discussions with P. Diaconis. Several results for compact Lie groups in Section 4 were also motivated by his conjectures and his results for finite groups

The main idea we use is very simple and goes as follows. We consider a map between two spaces $f : M \to N$. In many problems we are interested in understanding the inverse image $f^{-1}(x_0)$ for some point $x_0 \in N$. Assume that there is a heat kernel $H(t, x, x_0)$ on N. We consider the integral of the pull-back of $H(t, x, x_0)$ by f:

$$I(t) = \int_M H(t, f(y), x_0) dy$$

with respect to certain measure dy on M. We then perform the computations of this integral in two different ways, similar to the heat kernel proofs of the Atiyah-Singer index formula and the Atiyah-Bott fixed point formula. On one hand, when t goes to 0, $I(t)$ will localize to an integral over a neighborhood of $f^{-1}(x_0) \subset M$, on the other hand, by using the special properties of symplectic manifolds or representation theory of the corresponding groups, we can explicitly compute $I(t)$ globally in our cases. In this way, we obtain formulas about the geometry and topology of $f^{-1}(x_0)$

in terms of certain global information on M and N. We will give several examples here, to all of which we can apply this simple idea.

(1). The moment map $\mu: \ M \to g^*$, where M is a symplectic manifold, and g^* is the dual of the Lie algebra of the group G acting on M. In this case we are interested in the symplectic reduction $M_G = \mu^{-1}(0)/G$.

(2). The holonomy map $f: \ G^{2g} \to G$ where G is a compact Lie group and f is the product of the commutators

$$f(x_1, y_1, \cdots x_g, y_g) = \prod_{j=1}^{g} [x_j, y_j].$$

In this case we are interested in the space $f^{-1}(c)/Z_c$, which is the moduli space of flat connections on a flat G-bundle on a Riemann surface. Here $c \in G$ and Z_c denotes the stabilizer of c. More general examples were studied in [Liu2].

(3). Equivalently we can consider the map

$$f: \ G^{2g} \times \prod_{j=1}^{n} O_{c_j} \to G$$

where O_{c_j} is the conjugacy class in G passing through $c_j \in G$.

Another interesting map is the n-commutator map $f^{(n)}: \ G^n \to G$ with

$$f^{(n)}(x_1, x_2, \cdots, x_n) = [x_1, [x_2, \cdots [x_{n-1}, x_n]]].$$

In this case, an inductive formula for the push-forward by $f^{(n)}$ of the Riemannian measure on G^n can be obtained.

(4). For a finite group G, we can consider the maps similar to those in (2) and (3) or more generally we can consider

$$f: \ G^n \times \prod_{j=1}^{n} H_j \to G$$

where $H_1, \cdots, H_n$ are subgroups of G and

$$f(x_1, \cdots, x_n; z_1, \cdots, z_n) = \prod_{j=1}^{n} x_j z_j x_j^{-1}.$$

In this case the integral to define $I(t)$ is replaced by the sum over $G^n \times \prod_j H_j$, and the limit as t goes to 0 gives the number of solutions to the equation in $G^n \times \prod_{j=1}^{n} H_j$:

$$\prod_{j=1}^{n} x_j z_j x_j^{-1} = e.$$

More general examples than the above cases will also be considered in this note.

The interesting point here is that, the heat kernel method supplies a unified way to deal with several seemingly unrelated problems in geometry, topology and finite group theory. In Section 2 we derive a nonabelian localization formula in symplectic geometry from the heat kernel point of view. Here we use the heat kernel of Euclidean space. Note that, different treatment of the result has been discussed first by Witten [W], then in [Wu], [JK] and [Liu]. In Section 3 we obtain the formulas for the symplectic volume and some intersection numbers of the moduli space by using the heat kernel of the Lie group G. Some detailed discussion in this section has appeared in [Liu1], [Liu2], and in [BL]. In Section 4 we derive various formulas for the push-forward by those maps in (2) and (3) of the Riemaniann measures. In Section 5 we derive several formulas for counting the numbers of solutions of certain polynomial equations in finite groups. All of the results in this section grew out of discussions with P. Diaconis who has quite different proofs for these formulas by using combinatorics methods. One of the formulas in Section 5 has been proved and used as the main tools in [St] to solve the long-standing Brauer p-block conjecture.

I want to dedicate this paper to the memory of Prof. Qi-Ming Wang who passed away ten years ago. In 1988, after reading my master thesis, Prof. Wang encouraged me to follow Prof. S.-T. Yau to study geometry. This is the turning point of my career.

Finally I would like to thank P. Diaconis for the many stimulating and enlightening discussions.

2. Symplectic reduction

Let M be a compact symplectic manifold with a symplectic form ω. Assume that the compact Lie group G acts on M. Let g denote the Lie algebra of G and g^* be its dual. Assume that the G-action has a moment map

$$\mu : \quad M \to g^*.$$

With the metric induced from the Killing form on G, we can identify g^* to the Euclidean space $\mathbf{R}^n$. Let us denote the metric on g^* by $< \cdot, \cdot >$. Let

$$H(t, x, x_0) = \frac{1}{(4\pi t)^{\frac{n}{2}}} \exp(-\frac{\|x - x_0\|^2}{4t})$$

be the heat kernel on g^*. Assume $0 \in g^*$ is a regular value of μ. We are interested in studying the symplectic reduction $M_G = \mu^{-1}(0)/G$. We consider the integral of the pull-back of the heat kernel against the symplectic volume on M :

$$I(t) = \int_M H(t, \mu(y), 0) e^{\omega}.$$

As in the derivation of the index formula, we will compute this integral in two different ways: local and global.

(i) Local computation. Let t go to 0, then the integral $I(t)$ localizes to a neighborhood of $\mu^{-1}(0)$, which can be identified as $\mu^{-1}(0) \times B_\delta$ where B_δ denotes a small ball of radius δ in g^*. More precisely we can write the above integral as

$$I(t) = \int_{\mu^{-1}(0) \times B_\delta} H(t, \mu(y), 0) e^{\omega} + O(e^{-\delta^2/4t}).$$

When restricted to $\mu^{-1}(0) \times B_\delta$, the symplectic structure ω has a canonical expression in terms of the induced symplectic form ω_0 on the symplectic reduction $M_G = \mu^{-1}(0)/G$:

$$\omega = \pi^*\omega_0 + d(\alpha, \theta)$$

where $\pi : \mu^{-1}(0) \to M_G$ denotes the quotient map which gives a principal G-bundle, and θ is a connection form of this bundle. Also α denotes the coordinate on g^* and (α, θ) denotes the obvious paring. And in this local model μ is just the simple projection map: $\mu : \mu^{-1}(0) \times g^* \to g^*$ defined by $\mu(x, \alpha) = \alpha$.

By substituting the local expression of ω into the integrand we obtain

$$I(t) = \int_{\mu^{-1}(0)\times g^*} H(t, \alpha, 0)e^{\pi^*\omega_0 + d(\alpha,\theta)} + O(e^{-\delta^2/4t}).$$

We now use the equalities:

$$d(\alpha, \theta) = (d\alpha, \theta) + (\alpha, d\theta), \text{ and } F = d\theta - \theta \wedge \theta$$

where F denotes the curvature of θ. From these we derive

$$I(t) = \int_{\mu^{-1}(0)} e^{\pi^*\omega_0} \wedge \theta \int_{g^*} H(t, \alpha, 0)e^{(\alpha, F)} D\alpha + O(e^{-\delta^2/4t})$$

where $D\alpha$ denotes the volume form of g^* and $\wedge\theta = \theta_1 \wedge \cdots \wedge \theta_n$. Here we have used the fact that, by a degree count, in $e^{(d\alpha,\theta)}$ only the term $D\alpha \wedge \theta$ will contribute to the integral. Note that $\wedge\theta$ is actually a volume form of G.

By performing a simple Gaussian integral on $g^* \simeq \mathbf{R}^n$, we get

$$I(t) = |G| \int_{M_G} e^{\omega_0 - t<F,F>} + O(e^{-\delta^2/4t})$$

where $< \cdot, \cdot >$ also denotes the inner product on g induced from the Killing form, and $|G|$ denotes the corresponding volume of G.

(ii) Global computation. On the other hand we can rewrite $I(t)$ in terms of equivariant cohomology class.

$$\begin{aligned} I(t) &= \int_g \int_M e^{-t<\varphi,\varphi>} e^{\omega + i(\mu,\varphi)} d\varphi \\ &= \int_g e^{-t<\varphi,\varphi>} \int_M e^{\omega + i(\mu,\varphi)} d\varphi. \end{aligned}$$

Here $< \varphi, \varphi >$ denotes the inner product induced on g by the Killing form. Note that the first identity actually corresponds to the Fourier transform of $H(t, x, 0)$. By comparing the computations in (i) and (ii), we have the following equality of Witten which we summarize as a proposition.

Proposition 1: *We have the following formula:*

$$|G| \int_{M_G} e^{\omega_0 - t<F,F>} = \lim_{t\to 0} \int_g \int_M e^{-t<\varphi,\varphi>} e^{\omega + i(\mu,\varphi)} d\varphi$$

This formula expresses the geometric and topological information on M_G in terms of the information on M. This is the spirit of the nonabelian localization of Witten.

By using an idea of Wu [Wu], we can actually go a little bit further. Let T denote the maximal torus of G and $\{P\}$ be its fixed point components. By using Atiyah-Bott localization formula, we can express the integral

$$\int_M e^{\omega+i(\mu,\varphi)}$$

as the sum of the integrals over the fixed point components $\{P\}$. Here note that $\omega + i(\mu, \varphi)$ is a G-equivariant cohomology class. The interested reader can also interprete the derivation of the general formula in [Liu] by using the idea in this section.

3. Moduli spaces of flat connections

In this section we will derive the symplectic volume formula and some results about the intersection numbers of the moduli space of flat connections on a principal flat G-bundle on a Riemann surface. Some of the discussions about moduli spaces in this section are basically contained in [Liu1] and [Liu2] where we considered maps like

$$f: \ G^{2g} \times G^n \to G$$

with

$$f(x_1, y_1, \cdots, x_g, y_g; z_1, \cdots z_n) = \prod_{j=1}^{g} [x_j, y_j] \prod_{i=1}^{n} z_i c_i z_i^{-1}$$

for certain fixed generic points $c_1, \cdots, c_n \in G$. Here G is a semisimple simply connected compact Lie group. We refer the reader to [Liu1] and [Liu2] for the details of the study of these maps by using heat kernels. Here let us consider an equivalent map given by

$$f: \ G^{2g} \times \prod_{i=1}^{n} O_{c_i} \to G$$

with

$$f(x_1, \cdots, y_g; z_1, \cdots, z_n) = \prod_{j=1}^{g} [x_j, y_j] \prod_{i=1}^{n} z_i$$

which is more commonly used to describe the moduli spaces. Here x_j, $y_j \in G$ and $z_i \in O_{c_i}$. Recall that O_{c_i} is the conjugacy class of c_i in G.

Now let us start to derive the formulas. We equip G with the bi-invariant metric induced by the Killing form, then the explicit expression of the heat kernel on G is given by

$$H(t, x, y) = \frac{1}{|G|} \sum_{\lambda \in P_+} d_\lambda \cdot \chi_\lambda(xy^{-1}) e^{-tp_c(\lambda)}$$

where $|G|$ denotes the volume of G and P_+ denotes all irreducible representations of G, which can be identified as a lattice in t^*, the dual of the Lie algebra of the maximal torus T. Also d_λ and χ_λ denote the dimension and respectively the character of the representation λ, and $p_c(\lambda) = |\lambda + \rho|^2 - |\rho|^2$ with

$$\rho = \frac{1}{2} \sum_{\alpha \in \Delta +} \alpha$$

the half sum of the positive roots.

Recall that the moduli space we are interested in is just

$$M_c = M_{c_1, \cdots, c_n} = f^{-1}(e)/G$$

where G acts on $G^{2g} \times \prod_{i=1}^n O_{c_i}$ by the conjugation γ:

$$\gamma : G \to G^{2g} \times \prod_i O_{c_i}$$

with

$$\gamma(w)(x_1, \cdots, y_g; z_1, \cdots, z_n) = (wx_1w^{-1}, \cdots, wy_gw^{-1}; wz_1w^{-1}, \cdots, wz_nw^{-1}).$$

Similarly we consider the integral

$$I(t) = \int_{h \in G^{2g} \times \prod_j O_{c_j}} H(t, f(h), e) dh$$

where dh denotes the volume form of the bi-invariant metric on $G^{2g} \times \prod_j O_{c_j}$ induced from the Killing form. We will again perform the computation of $I(t)$ in two different ways: local and global.

(i) Local computation. Let $Z(G)$ denote the center of G and $|Z(G)|$ denote the number of elements in $Z(G)$. As t goes to 0, a computation of the Gaussian integral as in [Liu1] gives us

$$I(t) = \frac{|G|}{|Z(G)|} \int_{M_c} d\nu_c + O(e^{-\delta^2/4t})$$

where $d\nu_c$ is the Reidemeister torsion $\tau(C_c')$ of the complex

$$C_c' : \qquad 0 \to g \overset{d\gamma}{\to} g^{2g} \times \prod_j Z_{c_j}^{\perp} \overset{df}{\to} g \to 0$$

where $Z_{c_j}^{\perp}$ denotes the tangent space to O_{c_j} at c_j. It is clear that $Z_{c_j}^{\perp}$ can be viewed as the image of the map

$$(I - \mathrm{Ad}(c_j)) : \ g \to g.$$

By using the Poincare duality of the cochain complexes of the Riemann surface, one can show that the torsion $\tau(C_c')$ is related to the L^2-volume of M_c in the following way:

$$\tau(C_c')^2 = j(c)^2 \|\det TM_c\|_{L^2}^2$$

where

$$j(c)^2 = \prod_j |\det(I - \mathrm{Ad}(c_j))|$$

can be considered as the torsion of the boundary of the Riemann surface.

But from the definition we know that, up to a normalization by a factor of 2π, the L^2-volume is exactly the symplectic volume:

$$(2\pi)^{2N_c}\frac{\omega_c^{N_c}}{N_c!} = \|\det TM_c\|_{L^2}$$

for the moduli space M_c. Here N_c denotes the complex dimension of M_c, and ω_c denotes the canonical symplectic form on M_c induced from the Poincare duality. So we have

$$\tau(C_c') = d\nu_c = (2\pi)^{2N_c}|j(c)|\frac{\omega^{N_c}}{N_c!}$$

where

$$|j(c)| = \prod_j |\det(I - \mathrm{Ad}(c_j))|^{\frac{1}{2}}.$$

We refer the reader to [W], [BL], [Liu1], and [Liu2] for a proof of this relation.

(ii) Global computation. The global computation is achieved by using the character relations:

$$\int_G \chi_\lambda(wyzy^{-1}z^{-1})dz = \frac{|G|}{d_\lambda}\chi_\lambda(wy)\chi_\lambda(y^{-1}),$$

$$\int_G \chi_\lambda(wy)\chi_\lambda(y^{-1})dy = \frac{|G|}{d_\lambda}\chi_\lambda(w)$$

and the formula

$$\int_{O_{c_j}} h(g)dv_g = \frac{|J(c_j)|}{|Z_{c_j}|}\int_G h(gc_jg^{-1})dg$$

for any continuous function h on O_{c_j}. Here Z_{c_j} denotes the stabilizer of c_j, $|Z_{c_j}|$ its induced Riemannian volume, and dv_g is the induced volume form on O_{c_j}. And note that we have used the notation:

$$J(c_j) = \det(I - \mathrm{Ad}(c_j)).$$

By applying these formulas inductively we get

$$\int_{G^{2g}\times\prod_i O_{c_i}} \chi_\lambda\Big(\prod_j [x_j, y_j]\prod_i z_i\Big)\prod_j dx_j dy_j \prod_i dz_i$$

$$= |G|^{2g+n}\frac{j(c)^2}{\prod_i |Z_{c_i}|}\frac{\prod_i \chi_\lambda(c_i)}{d_\lambda^{2g+n-1}}.$$

By putting the above computations together, we obtain the following formula which we also summarize as a proposition.

Proposition 2: *We have the following formula for the symplectic volume of M_c:*

$$\int_{M_c} e^{\omega_c} = |Z(G)| \frac{|G|^{2g+n-2}|j(c)|}{(2\pi)^{2N_c}\prod_j |Z_{c_j}|} \sum_{\lambda \in P_+} \frac{\prod_j \chi_\lambda(c_j)}{d_\lambda^{2g+n-2}} e^{-tp_c(\lambda)} + O(e^{-\delta^2/4t}).$$

As in [Liu1] or [Liu2], we can then take derivatives with respect to the c_j's on both sides of the above identity to get intersection numbers on the moduli space M_c or M_u for u an element in the center $Z(G)$. We refer the reader to [Liu1] and [Liu2] for the details of the derivation. Here we only mention an interesting vanishing theorem that we can easily obtain from the above formula. For simplicity let us take $n = 1$. The general case is the same.

In fact let us introduce a function on the dual Lie algebra t of the maximal torus T, $\pi(\lambda) = \prod_{\alpha \in \Delta^+} < \alpha, \lambda >$, from which we construct a differential operator $\pi(\partial)$ such that, for $C \in t$ with $u \exp C = c$,

$$\pi(\partial)e^{(\lambda,\, C)} = \pi(\lambda)e^{(\lambda,\, C)}$$

where (λ, C) denotes the natural pairing. By applying $\pi(\partial)^{2g}$ to $I(t)$, we get, up to a constant,

$$\pi(\partial)^{2g}I(t) = |Z(G)| \frac{|G|^{2g-1}|j(c)|}{2\pi^{N_c}|Z_c|} \sum_{\lambda \in P_+} d_\lambda \chi_\lambda(c) e^{-tp_c(\lambda)} + O(e^{-\delta^2/4t}).$$

Note that, if $c \neq e$ where e denotes the identity element of G, then the sum on the right hand side, which is just the heat kernel $H(t, c, e)$, has limit 0 as t goes to 0. This in particular implies that $\lim_{t\to 0} I(t)$ is a piecewise polynomial in C of degree at most $2g|\Delta^+|$. From this one easily deduces certain new vanishing theorems for the intersection numbers of the moduli spaces. See [Liu1] and [Liu2] for the details on the explicit formula for certain intersection numbers.

We remark that, to get more complete information about integrals on the moduli space M_c, we can consider the integral

$$I(t) = \int_{G^{2g} \times \prod_j O_{c_j}} F(h)H(t, f(h), e)dh$$

where $F(h)$ is some G-invariant function on $G^{2g} \times \prod_j O_{c_j}$.

On one hand, as t goes to 0, $I(t)$ has limit given by

$$\frac{|G|}{|Z(G)|} \int_{M_c} \bar{F} e^{\omega_c}$$

where $\bar{F}$ denotes the function on M_c induced by $F(h)$. This is the result of the local computation. Note that in principle the integral of any cohomology class on M_c can be written in the above form.

On the other hand, by Peter-Weyl theorem, we know that $F(h)$ can be expressed as the combinations of the characters of G. By using the product formula for characters

$$\chi_\mu \cdot \chi_\lambda = \sum_\nu C^\nu_{\mu\lambda} \cdot \chi_\nu$$

where $C^{\nu}_{\mu\lambda}$ is the Clebsch-Gordon coefficients, we can perform the global computation to get an explicit infinite sum over P_+. As an example the interested reader may try to derive the formula by taking $F(h) = \prod_{j=1}^{k} \chi_{\mu_j}(x_j^{-1})$ for $k \leq g-1$.

4. The push-forward of measures

Now motivated by a conjecture of Diaconis, we consider the push-forward measure by the following map:

$$f: \quad G^{2g} \to G, \; f(x_1, \cdots, y_g) = \prod_j [x_j, y_j].$$

Let dh denote the Riemannian volume of G^{2g}, then Diaconis conjectures that the push-forward measure is given by the following formula:

$$f_* dh(x) = |G|^{2g-1} \sum_{\lambda \in P_+} \frac{\chi_\lambda(x^{-1})}{d_\lambda^{2g-1}} dx$$

if the right hand side converges. Otherwise we should use the normalized limit

$$f_* dh(x) = |G|^{2g-1} \lim_{t \to 0} \sum_{\lambda \in P_+} \frac{\chi_\lambda(x^{-1})}{d_\lambda^{2g-1}} e^{-tp_c(\lambda)} dx.$$

This conjecture can be proved by using the above heat kernel idea. Assume $f_* dh(x) = F(x)dx$ as forms on G, then we only need to show that

$$F(x) = |G|^{2g-1} \sum_{\lambda \in P_+} \frac{\chi_\lambda(x^{-1})}{d_\lambda^{2g-1}}.$$

It is clear that we have the identity

$$I(t) = \int_{G^{2g}} H(t, f(h), x) dh = \int_G H(t, y, x) F(y) dy.$$

Note that, as t goes to 0, the right hand side has limit given by $F(x)$, and the right hand side can be calculated by using the character relations. This immediately gives the above conjectured equality.

In fact we can prove similar formula for the push-forward of the Riemannian measure by the more general map

$$f: \quad G^{2g} \times \prod_j O_{c_j} \to G, \quad f(x_1, \cdots, y_g; z_1, \cdots, z_n) = \prod_j [x_j, y_j] \prod_j z_j,$$

from which we get the following formula.

Proposition 3: *Let dh denotes the Riemannian measure on $G^{2g} \times \prod_j O_{c_j}$. Then the following formula holds on G*

$$f_* dh(x) = |G|^{2g+n-2} \frac{j(c)^2}{\prod_j |Z_{c_j}|} \sum_{\lambda \in P_+} \frac{\prod_j \chi_\lambda(c_j)}{d_\lambda^{2g+n-1}} \chi_\lambda(x^{-1}) dx.$$

Here we assume the right hand side converges, otherwise we should consider the normalized limit as above.

Now we want to consider a slightly different situation. Let H_j, $j = 1, \cdots, n$, be subgroups of G, we consider the following map

$$f: \ G^n \times \prod_j H_j \to G, \ \ f(x_1, \cdots, x_n; u_1, \cdots, u_n) = \prod_j x_j u_j x_j^{-1}.$$

The same argument of using heat kernel will give us an interesting formula for the push-forward measure. Indeed we consider the integral

$$I(t) = \int_{h \in G^n \times \prod_j H_j} H(t, f(h), x) dh = \int_G H(t, y, x) F(y) dy.$$

As t goes to 0, the right hand side is $F(x)$, and the left hand side can be calculated by using the character relations, from which we get

Proposition 4: *Let dh denote the Riemannian measure on $G^n \times \prod_j H_j$. Then the following formula holds:*

$$f_* dh(x) = |G|^{n-1} \sum_{\lambda \in P_+} \frac{\prod_j \int_{H_j} \chi_\lambda(u_j) du_j}{d_\lambda^{n-1}} \chi_\lambda(x^(- 1) dx.$$

One may also consider the map

$$f: \ G^{2g} \times G \to G$$

with

$$f(x_1, \cdots, y_g; z) = \prod_j [x_j, y_j] z^2,$$

as well as the map

$$f: \ G^{2g} \times G \times G \to G$$

with

$$f(x_1, \cdots, y_g; w; z) = \prod_j [x_j y_j] w z w^{-1} z.$$

We leave these as exercises to the interested reader. In these cases one needs the formula

$$\int_G \chi_\lambda(x^2) dx = |G| f_\lambda$$

where $f_\lambda = 1$ if λ has real structure, -1 if λ has a quartenionic structure, 0 otherwise.

More generally, we can consider the problem of solving equations

$$f_j(x_1, \cdots, x_m) = c_j, \ j = 1, \cdots, n$$

in Lie groups. To understand the measure or number of solutions of these equations, we may apply the heat kernel on the corresponding group G to the map

$$f: \quad G^m \to G^n$$

with $f = (f_1, \cdots, f_n)$, and consider the integral

$$I(t) = \int_{G^n} \prod_{j=1}^{n} H(t, f_j(h), c_j) dh.$$

In many cases we can derive various interesting formulas. As an example, let us consider the map

$$f^{(n)}: \quad G^n \to G, \quad f^{(n)}(x_1, \cdots, x_n) = [x_1, [x_2, [\cdots, x_n]]].$$

This is the n-commutator map. We will derive an inductive formula for the push-forward measure, Let us define $Q_n(x)$ by the identity :

$$f_*^{(n)} dh(x) = Q_n(x) dx.$$

Then we will prove

Proposition 5: *The following formula holds:*

$$Q_n(w) = \lim_{t \to 0} \sum_{\lambda \in P_+} e^{-tp_c(\lambda)} \int_G \chi_\lambda(g) \chi_\lambda(g^{-1} w^{-1}) Q_{n-1}(g) dg.$$

To prove this formula, we consider the integral

$$I(t) = \int_{G^n} H(t, f^{(n)}(h), x) dh = \int_G H(t, y, x) Q_n(y) dy.$$

As t goes to 0, the right hand side is nothing but $Q_n(x)$, while the left hand side is reduced to the computation of the integrals like

$$\int_{G^n} \chi_\lambda([x_1, [\cdots, x_n]] x^{-1}) \prod_{j=1}^{n} dx_j$$

which, after integration with respect to x_1 by using the character relation, is equal to

$$\frac{|G|}{d_\lambda} \int_{G^{n-1}} \chi_\lambda([x_2, [\cdots, x_n]]) \chi_\lambda([x_2, [\cdots, x_n]]^{-1} x^{-1}) \prod_{j=2}^{n} dx_j$$

$$= \frac{|G|}{d_\lambda} \int_G \chi_\lambda(w) \chi_\lambda(w^{-1} x^{-1}) f_*^{(n-1)} [\prod_{j=2}^{n} dx_j](w)$$

$$= \frac{|G|}{d_\lambda} \int_G \chi_\lambda(w) \chi_\lambda(w^{-1} x^{-1}) Q_{n-1}(w) dw.$$

Here the second identity is from the change of variable

$$w = [x_2, [\cdots, x_n]].$$

By putting all of the above formulas together, we get the wanted formula.

5. Counting solutions in finite groups

It turns out that all of the results for compact Lie groups in the above section have analogues for finite groups. Even the proofs are basically the same, if we replace the integrals in the last section by sums.

As pointed to me by P. Diaconis, the heat kernel for a finite group G is given by a similar expression:

$$H(t,x,y) = \frac{1}{|G|}\sum_{\lambda\in P_+} d_\lambda \cdot \chi_\lambda(xy^{-1})e^{-tp_c(\lambda)}$$

where $|G|$ denotes the number of elements in G, P_+ still denotes all of the irreducible representations which is a finite set, and $p_c(\lambda)$ is a function on P_+. In fact the only property we need for this function is that its limit, as t goes to 0, is the delta function:

$$\delta(xy^{-1}) = \frac{1}{|G|}\sum_{\lambda\in P_+} d_\lambda \cdot \chi_\lambda(xy^{-1})$$

where $\delta(xy^{-1}) = 1$ if $x = y$ and 0 otherwise.

Let O_{c_j} with $c_1,\cdots,c_n \in G$ denote the conjugacy class of c_j in G. Let $S_{g,n}$ denote the number of solutions in $G^{2g} \times \prod_j O_{c_j}$ of the equation

$$\prod_{j=1}^{g}[x_j,y_j]\prod_{j=1}^{n} z_j = e$$

where $x_j, y_j \in G$ and $z_j \in O_{c_j}$.

This is related to the Chern-Simons theory with finite gauge group. Let S be a compact Riemann surface. For $n = 0$, $S_{g,n}$ counts the number of elements in the set $\mathrm{Hom}\,(\pi_1(S), G)$. We will derive a general formula for $S_{g,n}$. The special case when $n = 0$ was proved in [FQ] by using topological field theory. As pointed out in [FQ], such formula was actually known to Serre.

It is interesting to consider some variations of the above equation. Given subgroups $H_1,\cdots,H_n$ in G, we may consider the equation in $G^n \times \prod_j H_j$:

$$\prod_{j=1}^{n} x_j w_j x_j^{-1} = e$$

where $x_j \in G$ and $w_j \in H_j$.

We can also consider the n-commutator equation

$$[x_1,[x_2,\cdots,x_n]] = e.$$

In all of these cases we will give general formulas. Certainly one may try to find more general equations by combining the above equations together, or try to figure out equations of other types.

As in the compact Lie group cases, all of the formulas will follow naturally from the heat kernel trick. For simplicity and compatibility with last section, let us introduce a notation: for a finite set G and any function f on it, let us write

$$\int_G f(g)dg = \sum_{g\in G} f(g).$$

We start from the first problem. We consider the map

$$f: \quad G^{2g} \times \prod_j O_{c_j} \to G$$

with

$$f(x_1, y_1, \cdots, x_g, y_g; z_1, \cdots, z_n) = \prod_{j=1}^{n} [x_j, y_j] \prod_{j=1}^{n} z_j,$$

and introduce the integral, more precisely the sum,

$$I(t) = \int_{G^{2g}\times\prod_j O_{c_j}} H(t, f(h), e)dh.$$

(i) Local computation. As t goes to 0, the delta-function property of the heat kernel tells us that the limit is exactly the number $S_{g,n}$.

(ii) Global computation. On the other hand we can explicitly calculate the integral by using the formulas

$$\int_G \chi_\lambda(wyzyz^{-1})dz = \frac{|G|}{d_\lambda}\chi_\lambda(wy)\chi_\lambda(y^{-1}),$$

$$\int_G \chi_\lambda(wy)\chi_\lambda(y^{-1})dy = \frac{|G|}{d_\lambda}\chi_\lambda(w),$$

and if Z_{c_j} denotes the stabilizer of c_j,

$$\int_{O_{c_j}} h(g)dv_g = \frac{1}{|Z_{c_j}|}\int_G h(gc_jg^{-1})dg.$$

Here as our convention, the integral means taking sums over G, and $|G|$, $|Z_{c_j}|$ denote the number of elements in G and Z_{c_j} respectively.

By putting the above two computations together, we get the following:

Proposition 6: *We have the following formula:*

$$S_{g,n} = \frac{|G|^{2g+n-1}}{\prod_{j=1}^{n}|Z_{c_j}|}\sum_{\lambda\in P_+}\frac{\prod_{j=1}^{n}\chi_\lambda(c_j)}{d_\lambda^{2g+n-2}}.$$

For the second equation, let us define $Q_n(w)$ to be the number of solutions in G^n for the n-commutator equation, then we will derive an induction formula which was first derived by Diaconis by using combinatorics:

Proposition 7: *The following formula holds:*

$$Q_n(w) = \sum_{\lambda\in P_+}\sum_{g\in G}\chi_\lambda(g)\chi_\lambda(g^{-1}w^{-1})Q_{n-1}(g).$$

To prove this we consider the map

$$f^{(n)}: \quad G^n \to G, \quad f(x_1, \cdots, x_n) = [x_1, [x_2, \cdots, x_n]]$$

and the function:

$$I(t) = \int_{G^n} H(t, f^{(n)}(h), w)dh = \int_G H(t, y, w)Q_n(y)dy$$

(i) Local computation. As t goes to 0, the limit is clearly the function $Q_n(w)$.

(ii) Global computation. By using the character formulas, similar to the compact Lie group case, we get

$$\int_{G^n} \chi_\lambda(f(h)w^{-1})dh = \frac{|G|}{d_\lambda} \int_G \chi_\lambda(g)\chi_\lambda(g^{-1}w^{-1})Q_{n-1}(g)dg.$$

We then obtain the wanted formula by identifying the above two computations.

For the third equation, let R_n be the number of solutions in $G^n \times \prod_{j=1}^n H_j$ such that, for $x_j \in G$ and $z_j \in H_j$,

$$\prod_{j=1}^n x_j z_j x_j^{-1} = e.$$

Let us consider the map

$$f: \quad G^n \times \prod_{j=1}^n H_j \to G$$

with

$$f(x_1, \cdots, x_n; z_1, \cdots, z_n) = \prod_{j=1}^n x_j z_j x_j^{-1}.$$

Again we consider the function

$$I(t) = \int_{G^n \times \prod_{j=1}^n H_j} H(t, f(h), e)dh,$$

and perform the local and global computations.

(i) Local computation. As t goes to 0, the limit is given by the R_n.

(ii) Global computation. Still by using the character formulas we get the expression:

$$\int_{G^n \times \prod_i H_i} \chi_\lambda(\prod_{j=1}^n x_j z_j x_j^{-1}) \prod_j dx_j dz_j = |G|^n \sum_{\lambda \in P_+} \frac{\prod_j \int_{H_j} \chi_\lambda(z_j)dz_j}{d_\lambda^{n-1}}.$$

So we have obtained our next proposition:

Proposition 8: *We have the following expression for R_n:*

$$R_n = |G|^{n-1} \sum_{\lambda \in P_+} \frac{\prod_j \sum_{z_j \in H_j} \chi_\lambda(z_j)}{d_\lambda^{n-2}}.$$

More generally, we can find the number of solutions in finite groups of some general equations like

$$f_j(x_1, \cdots, x_n) = c_j, \quad c,\ x_j \in G,\ j = 1, \cdots, m$$

by considering the map

$$f: \ G^n \to G^m,\ f(h) = (f_1(h), \cdots, f_m(h))$$

and the function from the heat kernel:

$$I(t) = \int_{G^n} \prod_{j=1}^{n} H(t, f_j(h), c_j^{-1}) dh.$$

Then similarly we can perform the global and local computations as above to get explicit formulas.

We leave to the reader as an exercise to find the numbers of solutions in the equations like

$$\prod_{j=1}^{g} [x_j, y_j] z^2 = e, \text{ in } G^{2g} \times G$$

and

$$\prod_{j=1}^{g} [x_j, y_j] w z w^{-1} z = e, \text{ in } G^{2g} \times G \times G.$$

Another interesting problem is to consider the integrals, or sums, like

$$I(t) = \int_{G^{2g}} F(h) H(t, f(h), e) dh$$

for some G-invariant function F on G^{2g}. As $t \to 0$, the limit of $I(t)$ compute

$$\sum_{h \in M} \bar{F}(h)$$

where M denotes the set of solutions of one of the equations in this section. By expressing $F(h)$ in terms of the characters of G, we can obtain certain explicit expression of this sum in terms of the representation data of G.

Note that the numbers of solutions of some of the above equations correspond to the Hurwitz numbers, the numbers of coverings of a given Riemann surface. In [St], the formula in Proposition 6 was used to derive the general case of the Brauer p-block conjecture.

6. Concluding remarks

There are some other interesting cases to which one may consider to apply the above heat kernel method. In [Liu2] we tried to find invariants for knots and 3-manifolds. It may be interesting to combine this method with the work in [Hu].

As one can see that, symplectic structure has played the important role in our computations, such as in the derivation of the nonabelian localization formula and the intersection number formulas for moduli spaces. Recently P. Xu has explained to me some very general constructions of moment maps to Poisson manifolds. It should be interesting to apply our method to such general situation. An interesting example is to consider the group-valued moment map as introduced in [AMM], one may consider the integral of the pull-back of the heat kernel on the Lie group G by this map against the measure introduced in [AMM]. This should give us some quite interesting formulas relating the symplectic reduction and the representations of the Lie group. More precisely this should give us formulas similar to Witten's nonabelian localization formula. I thank Prof. S. Wu for the discussions concerning this point.

References

[AMM] A. Alekseev, A. Malkin, E. Meinrenken, *Lie group valued moment maps*, preprint 1997.

[BL] J.-M. Bismut and F.Labourie, *Symplectic geometry and the Verlinde formula*, prerint 1999.

[FQ] D. Freed, F. Quinn, *Chern-Simons theory for finite group*, Commun. Math. Phys. 156 (1993), 435-472.

[Hu] J. Huebschmann, *Extended moduli spaces, the Kan construction , and lattice gauge theory*, Topology 35 (1999), 555-596.

[JK] L. Jeffrey, F. Kirwan, *Localizations for nonabelian group actions*, Topology 34 (1995), 291-328.

[Liu] K. Liu, *Remarks on nonabelian localization*, IMRN Int. Math. Res. Notices, 13 (1995),683-691.

[Liu1] K. Liu, *Heat kernel and moduli space*, MRL Math. Res. Letter, 3 (1996), 743-762.

[Liu2] K. Liu, *Heat kernel and moduli space II*, MRL Math. Res. Letter, 4 (1997), 569-588.

[St] S. P. Strunkov, *On equations in finite groups and invariants of representations for their subgroups*, J. Pure and App. Alg. 107 (1996) 303-307.

[W] E. Witten, *On quantum gauge theory in two dimensions*, Comm. Math. Phys. 141(1991), 153-209.

[Wu] S. Wu : *An integral formula for the square of moment maps of circle actions*, Lett. Math. Phys. 29 (1993), 311-328.

Surveys in Differential Geometry, vol. 5

A Brief Tour of GW Invariants

Jun Li
Department of Mathematics
Stanford University
Stanford, CA 94305
and
Gang Tian
Department of Mathematics
Massachusetts Institute of Technology
Cambridge, MA 02139

The machinery of virtual moduli cycle was initiated by authors in 1995-96 to attack a class of topological invariants, especially Gromov-Witten invariant. The development of this machinery went through several stages during which several people made key contribution to its development. This machinery is essential to the study of mathematical problems in super-string theory. It also played key role in symplectic topology and enumerative problems in algebraic geometry. In this article, we will lead a brief tour though the development of this machinery. We will explain the basic idea to its construction and the means to its application. We hope that this will provide a manageable account on how to apply this machinery for non-specialists.

The era of using non-linear analysis to define topological invariants began with Donaldson's celebrated work of his invariants of smooth 4-manifolds. This invariant is a counting invariant that counts ASD-connections of principle G-bundles over 4-manifolds. The prototype of a similar invariant on counting rational curves in a symplectic manifold was proposed by Ruan in early 90's [**Ru1**], inspired by Gromov's work on pseudo-holomorphic maps [**Gr**] and Witten's work on non-linear σ-models in string theory [**Wi1, Wi2**]. This counting began to attract attention in part because of the notion of the Quantum cohomology and the Mirror symmetry conjecture. However, its mathematical foundation was not developed until the end of 1993 when Ruan and the second author constructed the symplectic invariants and proved their basic property for the class of semi-positive symplectic manifolds. This class of invariants are now referred to as Gromov-Witten invariants. Overtime, people began to ponder if similar invariants by counting curves can be defined for any symplectic manifolds and algebraic varieties. The notion of stable morphisms was introduced by Kontsevich [**Ko**] to construct a compactified moduli spaces for this purpose. In 1995, the authors succeeded in constructing such invariants of smooth projective varieties. They reported their construction in the Santa Cruz conference in algebraic geometry that year [**LT3**]. Their method is to construct the virtual moduli cycle of the moduli space of stable morphisms and define Gromov-Witten invariants as the integration of the tautological topological classes against this cycle [**LT1**]. They constructed such cycles by constructing virtual normal cones using Kuranishi maps of the moduli space. In early 1996, Behrend-Fantechi [**BF**] gave an alternative construction of such cycles. They constructed similar cones as Artin stacks over the moduli stack. In 1996, Fukaya-Ono [**FO**] and the authors [**LT2**] independently constructed the Gromov-Witten invariants for general symplectic manifolds, after constructing the virtual moduli cycle in analytic category. Different constructions were achieved later by Siebert [**Si1**] and Ruan [**Ru2**]. Finally, the authors in 1997 [**LT4**] and Siebert in 1998 [**Si2**] proved that the algebraic construction and the analytic construction of Gromov-Witten invariants coincide in case the underlining manifold is a smooth projective variety. All these work concluded the task of constructing Gromov-Witten invariants.

Research on Gromov-Witten invariants over the years has contributed, directly or indirectly, to the progress in other mathematical research. Here we can only single out a few such areas. Among them is the Mirror Symmetry Conjecture for quintic Calabi-Yau manifold, which relates the *virtual* counting of rational curves to variations of Hodge structure of its mirror. After work of many, including Candelas et al [**Ca**], Kontsevich [**Ko**] and Givental [**Gi**], this conjecture was finally proved rigorously by Lian-Liu-Yau [**LLY**] in 1997. Because of the limitation of time and space, we will not be able to cover this and others in any details, suffices to say that they include the development of Floer-homology for symplectic manifolds and the proof of the Arnold conjecture, the localization formula for equivalent virtual moduli cycles, the research for relative GW invariants and the research for Frobenius manifold structure.

We now describe the composition of this paper. The main body of this survey paper consists of three parts, in order they discuss the algebraic construction of the the GW invariants, the symplectic construction of the GW invariants and the equivalence of these two constructions. In the first part, we will describe the geometric motivation of our original construction of virtual moduli cycles. We will explain why the *obstruction theory* of the moduli space is an essential ingredient for this construction and how *Kuranishi maps* and *normal cones* appear naturally in this construction. We will describe briefly the notions that appear in this construction and state the main theorem. We will refer the details of this construction and other related issues to our original paper [**LT1**].

For the construction of symplectic GW invariants, we will present a slightly different approach as to the one presented in [**LT2, LT4**]. The purpose for doing this is to present a construction that requires as little smoothness as possible. We will introduce a class of weakly Fredholm V-bundle and show that they admit Euler classes, just like the usual case. The new ingredient is that to a weakly Fredholm V-bundle $E \to X$, among other things, the base space X is only assumed to have a locally closed partition (called a stratification) into smooth Banach orbifolds, and the bundle E is only assumed to be smooth along each stratum. The benefit of this, as is clear for the GW invariants, is that many moduli spaces admit geometric stratifications of which their strata are easily seen to be smooth. It is to study the normal structures of these strata in the total spaces where the technical difficulty arises. Our approach thus provides a more topological construction that bypass many of the tedious, if not impossible, technical study of singularities. We hope this will be proved to be useful in the study of similar problems in the future.

In the last part, we will briefly mention the equivalence of the above two constructions.

1. GW invariant as counting invariant

We begin with a brief account of GW invariants and its first construction by Ruan and the second author for semi-positive manifolds. Let (X,ω) be a smooth symplectic manifold and J an almost complex structure tame to ω. For non-negative integers g, n and homology class $A \in H_*(X,\mathbb{Z})$ we form the moduli space $\mathfrak{M}_{g,k}(X,A)$ of the equivalence classes of pseudo-holomorphic maps $f:C \to X$ such that C is an n-pointed genus g smooth Riemann surface[1], f is pseudo-holomorphic and $f_*([\Sigma]) = A$. In this article, we will use $f:C \to X$ to denote such a map with marked points on C implicitly understood. In case we want to stress the marked points, we will use $f:D \subset C \to X$ with $D \subset C$ the divisor of the marked points. Here we say that $f: C \to X$ is equivalent to $f': C' \to X$ if there is an isomorphism $\rho: C \to C'$ as pointed Riemann surfaces so that $f' \circ \rho = f$. By choosing ω and J generic, we can make $\mathfrak{M}_{g,k}(X,A)$ smooth (indeed with at most orbifold singularities) of complex dimension

$$r = (3 - \dim_{\mathbb{C}} X)(g-1) + k + A \cdot c_1(X). \tag{1.1}$$

It has an obvious evaluation map

$$\mathrm{ev}: \mathfrak{M}_{g,k}(X,A) \longrightarrow X^k$$

that sends any map $(f, C, x_1, \cdots, x_k)$ to $(f(x_1), \cdots, f(x_k))$. Then for any class $[\alpha] \in H_*(X)$, we choose its cycle representative $\alpha \subset X$ and investigate the set

$$\mathrm{ev}^{-1}(\alpha) \subset \mathfrak{M}_{g,k}(X,A). \tag{1.2}$$

If we choose $[\alpha]$ to be a degree $2r$ class, the above set will be a finite set if various transversality conditions are satisfied. By incorporating the almost complex structure of X, it is standard to show that it is an oriented set. Thus we can algebraically count this set. This counting is the GW invariant of (X,ω).

For the moment, this counting still depends on the choice of the symplectic form ω, the almost complex structure J and the cycle representative α. These combined form the defining data of the set $\mathrm{ev}^{-1}(\alpha)$. To simplify the notation, we will use Λ to denote this set of defining data and use

$$\Phi(\Lambda) \subset \mathfrak{M}_{g,k}(X,A)_\Lambda$$

to denote the pair in (1.2) defined based Λ. We say Λ is generic if various transversality conditions hold in constructing the set $\Phi(\Lambda)$.

We now explain why the algebraic count $\#\Phi(\Lambda)$ is a topological invariant. Let Λ be any generic defining data. Then $\#\Phi(\Lambda)$ is an invariant if it remains constant after perturbation of Λ. In [**RT1**], Ruan and the second author showed that in case $g = 0$ and (X,ω) is semi-positive[2], then any two generic defining data Λ_0 and Λ_1

[1] We defer the definition to Chapter 3.

[2] A symplectic manifold (X,ω) is semi-positive if for any pseudo-holomorphic map $f: C \to X$ we have $f_*([C]) \cdot c_1(X) \geq 0$.

in the same homotopy class can be connected by a generic path Λ_t such that

$$\Phi(\Lambda_{[0,1]}) = \bigcup_{t\in[0,1]} \Phi(\Lambda_t)$$

form an oriented cobordism between $\Phi(\Lambda_0)$ and $\Phi(\Lambda_1)$. To accomplish this, they first showed that $\Phi(\Lambda_{[0,1]})$ is a smooth manifold with boundary $\Phi(\Lambda_1) - \Phi(\Lambda_0)$. The critical step is to show that it is also closed. For this, they compactified the moduli space $\mathfrak{M}_{0,k}(X,A)_{\Lambda_t}$ and studied the closure of $\Phi(\Lambda_{[0,1]})$ in the union of these compactifications. Relying on semi-positivity condition, they showed that $\Phi(\Lambda_{[0,1]})$ is indeed closed, thus proving the counting $\#\Phi(\Lambda)$ does not depend on the variation of Λ. This defines the GW invariant

$$\gamma^X_{A,0,n} : H^*(X^k, \mathbb{Q}) \longrightarrow \mathbb{Q}$$

that counts rational curves in X. We note that the class of semi-positive manifolds includes all Calabi-Yau manifolds. Later they generalized their construction to cover the GW invariants of all genus for semi-positive manifolds [**RT2**]. However, their approach relied on semi-positivity condition in part because in general the compactification of $\mathfrak{M}_{g,k}(X,A)$ is too large to prove the independence of the counting $\#\Phi(\lambda)$.

2. GW-invariants of algebraic varieties

2.1. Why virtual cycle and what is a virtual cycle. In this section, we will discuss the algebraic construction of virtual moduli cycle and its application to defining GW invariants of smooth projective varieties. In the following, we will use morphisms, schemes and other commonly used terminology in algebraic geometry. For those unfamiliar to such terminologies, they can simply replace morphisms by holomorphic maps and schemes by complex varieties, possibly singular.

The first step to the construction of GW invariants for general varieties (or symplectic manifolds) is to think of it as the outcome of integrating certain tautological topological classes on the moduli space over its fundamental class. For this, we need to work with the compactified moduli space.

Let X be a smooth projective variety and $\mathfrak{M}_{g,k}(X,A)$ the moduli space of stable morphisms[3] $f: D\subset C \to X$ from n-pointed (arithmetic) genus g nodal curve $D \subset C$ to X of degree $f_*([C]) = A$. The notion of stable morphism was introduced by Kontsevich [**Ko**]. The moduli space $\mathfrak{M}_{g,k}(X,A)$ admits a natural evaluation map ev. When the moduli space $\mathfrak{M}_{g,k}(X,A)$ has pure dimension r (in (1.1)) and the subset of maps with smooth domains is dense in $\mathfrak{M}_{g,k}(X,A)$, then the GW invariant is

$$\gamma^X_{A,g,k}(\alpha) = \int_{[\mathfrak{M}_{g,k}(X,A)]} \mathrm{ev}^*(\alpha), \qquad \alpha \in H^*(X^k, \mathbb{Q}). \tag{2.1}$$

The equivalence of this with the one based on counting is apparent from Poincare duality.

This approach has the advantage that even in case the moduli $\mathfrak{M}_{g,k}(X,A)$ has bigger than expected dimension, we can still define the integral (1.1) if we can replace $[\mathfrak{M}_{g,k}(X,A)]$ with an cycle that "imitate" the fundamental class of the

[3] A morphism $f: D\subset C \to X$ is stable if $D \subset C$ is a pre-stable n-pointed nodal curve and if there is no non-trivial vector field of C vanishing at the marked points and the nodal points of C so that $f_*(v) = 0$. An equivalent definition can be found in section 3.5.

moduli space. One way to visualize this cycle is this: We deform X to a family of varieties, say X_t with $X_0 = X$. Then $A \in H_2(X,\mathbb{Z})$ induces $A_t \in H_2(X_t,\mathbb{Z})$, at least for small t (in analytic topology). Then $\mathfrak{M}_{g,n}(X_t, A_t)$ form a family of moduli spaces not necessarily flat. Now imagine for $t \neq 0$ we can find X_t so that for all $t \neq 0$ the spaces $\mathfrak{M}_{g,n}(X_t, A_t)$ have expected dimensions. Then the limit $\lim[\mathfrak{M}_{g,n}(X_t, A_t)]$ when $t \to 0$ will be a cycle in $H_{2r}(\mathfrak{M}_{g,k}(X, A))$. This limit cycle will be the virtual moduli cycle. It is to construct this cycle the machinery of virtual cycles was developed.

2.2. Algebraic construction of virtual cycle. We now discuss the algebraic construction of virtual cycle. We begin with an example that will illustrate this construction. Let

$$\begin{array}{c} E \\ s \uparrow\downarrow \\ W \end{array} \tag{2.2}$$

be a section of a vector bundle over a smooth projective variety W. Let $X = s^{-1}(0)$ be the subscheme of W. We think of W as the ambient space and s the defining equation of X. Let $n = \dim W$ and $m = \operatorname{rank} E$. Since $X \subset W$ was cut out by m equations, the expected dimension

$$\text{exp.}\dim X = n - m.$$

In case X has pure dimension $n-m$, we will take its fundamental class $[X]$ its virtual cycle $[X]^{\text{vir}}$. Note that this is the Euler class of E, viewed as a class in $A_{n-m}W$.[4] In general, we demand $[X]^{\text{vir}} \in A_*X$ to be the class so that it will remain constant if we vary X by varying the section s. This leaves us the only option that $[X]^{\text{vir}}$ is the Euler class $e(E)$. To make it a class in A_*X, we will take the localized Euler class

$$e_{\text{loc}}(E, s) \in A_{n-m}s^{-1}(0) = A_{n-m}X$$

thanks to the section s. The localized Euler class is constructed as follows [**Fu**]: Let $N_{X/W}$ be the normal cone to X in W. It is canonically a subcone of $E|_X$. Let $\eta_E : X \to E|_X$ be the zero section and $\eta_E^* : A_*E|_X \to A_*X$ the Gysin map. Then

$$e_{\text{loc}}(E, s) = \eta_E^*[N_{X/W}] \in A_*X.$$

Analytically, $[N_{X/W}]$ can be viewed as the limit current (cycle) of Γ_{ts} when $t \to 0$, where Γ_s is the graph of s in E and $\eta_E^*[N_{X/W}]$ can be viewed as a class in $H_*(X)$ resulting from intersecting $N_{X/W}$ with a generic section of $E|_X$. This interpretation shows that $\eta_E^*[N_{X/W}]$ is the usual Euler class when viewed as an element in $H_*(W)$.

It is this localized Euler class which we define to be the virtual cycle $[X]^{\text{vir}}$.

To push this construction to a general moduli space, say $\mathbf{M}$, we face the improbable task of embedding it *naturally* into some ambient space $\mathbf{W}$ as the vanishing locus of a section $\mathbf{s}$ of a vector bundle $\mathbf{E}$ over $\mathbf{W}$. Here we stress the naturality of such choice since otherwise the virtual moduli cycle will not be a naturally defined object. For instance, if $\mathbf{M} \subset \mathbf{W}$ is the zero locus of a section $\mathbf{s}$ of $\mathbf{E}$, a section of $\mathbf{E} \oplus \mathbf{C}_\mathbf{W}$, then $\mathbf{M}$ will be the zero locus of $\mathbf{s} \oplus \mathbf{0}$, where $\mathbf{0}$ is the zero section of the trivial line bundle $\mathbf{C}_\mathbf{W}$. However, $e(\mathbf{E}) \neq e(\mathbf{E} \oplus \mathbf{C}_\mathbf{W})$. To satisfies this "naturality" condition, we ask ourselves *what is the canonical defining equation(s) of a moduli*

[4] Z_*W is the group of algebraic cycles and A_*X is Z_*X modulo the rational equivalence relation.

space. The key to this question rests on the notion of *the obstruction theory and its Kuranishi map.*

We recall the notion of an obstruction theory to deformation of a point p in a scheme X.

DEFINITION 2.1. *(X,p) is said to have an obstruction theory with coefficients in V, where V is a finite dimensional $\mathbb{C}$-linear space, if the following holds: For any triple (A,I,φ), where A is an Artin ring, $I \subset A$ is an ideal annihilated by the maximal ideal $\mathfrak{m} \subset A$ and $\varphi: \operatorname{Spec} A/I \to X$ is a morphism such that $\varphi(0) = p$, there is a function $\operatorname{Ob}(A,I,\varphi) \in V \otimes_{\mathbb{C}} I$ such that (1) $\operatorname{Ob}(A,I,\varphi) = 0$ if and only if there is an extension $\tilde{\varphi}: \operatorname{Spec} A \to X$ of φ and (2) this function $\operatorname{Ob}(A,I,\varphi)$ satisfies base change property.*

We skip the exact wording of the base change property since it does not affect our discussion later. It can be found in [**LT1**].

The following example illuminates the notion of obstruction theory.

EXAMPLE 2.2. *Let W be a smooth scheme and $X \subset W$ be a subscheme defined by the vanishing of a section s of a vector bundle E. Let $p \in X$ and*

$$V = \operatorname{Coker}\{ds(p) : T_pW \longrightarrow E|_p\}. \tag{2.3}$$

Then s induces an obstruction theory of (X,p) with coefficients in V.

There is an obvious way to construct this obstruction theory. Since W is smooth, $\varphi: \operatorname{Spec} A/I \to X$ extends to $\psi: \operatorname{Spec} A \to W$. Since $s \circ \psi(\operatorname{Spec} A/I) = 0$ and $\mathfrak{m} \cdot I = 0$, $s \circ \psi$ induces an element in $E|_p \otimes_{\mathbb{C}} I$. We denote by

$$\operatorname{Ob}(A,I,\varphi) \in V \otimes_{\mathbb{C}} I$$

its image in $V \otimes_{\mathbb{C}} I$ under the quotient map $E|_p \otimes_{\mathbb{C}} I \to V \otimes_{\mathbb{C}} I$. It is straightforward to show that this assignment satisfies all requirements of an obstruction theory, hence is an obstruction theory to deformation of p in X.

Now let (X,p) be any scheme endowed with an obstruction theory with coefficients in V. Let $\hat{X}$ be the formal completion of X along p. ($\hat{X}$ sometimes is called the germ of X at p.) Assume $n = \dim T_pX$ and $m = \dim V$. We pick a dual basis $(z) = (z_1, \cdots, z_n)$ of T_pX. Following Kuranishi's construction of his map, there is a pair (ι, Φ), where

$$\iota: \hat{X} \hookrightarrow \operatorname{Spec} \mathbb{C}[[z]] \quad \text{and} \quad \Phi \in V \otimes \mathbb{C}[[z]] \tag{2.4}$$

so that $\hat{X}$ is the vanishing locus of Φ and the obstruction theory induced by Φ coincide with the obstruction theory given. We call (ι, Φ) a Kuranishi map of the obstruction theory.

It is instructive to work out an explicit example. Let $s(z) = 2z^3 + z^4$. It defines the subscheme $X = \{z^3 = 0\}$ in $\mathbb{C}$. For triple $A = \mathbb{C}[t]/(t^3)$, $I = (t^2)$ and $\varphi_2(t) = z : \operatorname{Spec} A/I$ to X, the obstruction class $\operatorname{Ob}(A,I,\varphi) = 0$. Obviously φ_2 extends to $\varphi_3(t) = z : \operatorname{Spec} A \to X$. Now let $A' = \mathbb{C}[t]/(t^4)$ and $I' = (t^3)$. We compute $\operatorname{Ob}(A', I', \varphi_3) = 2$, the coefficient of the leading non-zero term in s. This reveals the fact that the leading non-trivial terms in a Kuranishi map is determined by the obstruction theory, and the *non-uniqueness* of Kuranishi maps comes from the freedom of higher order terms. In this case, any function $\Phi(z) = 2z^3 + O(z^4)$ is a Kuranishi map.

So far we have settled the question of finding natural defining equation of a moduli space: *The moduli space should come with an obstruction theory and the associated Kuranishi maps will serve as its defining equation.* The strategy to construct the virtual cycle is clear now: *We will use the Kuranishi maps to construct a virtual cone in a vector bundle E and use the Gysin map of its zero section to define the virtual cycle.* There remain two technical difficulties: One is that the Kuranishi maps are not unique; The second is that they are local. They only exist as formal functions.

We are still in the set up of (2.2). According to the plan, we intend to reconstruct the normal cone $N_{X/W}$ near p by the Kuranishi maps of the obstruction theory. Let (ι, Φ) be a Kuranishi map of the obstruction theory of $(\hat{X}, p)$, as in (2.2). We consider the normal cone

$$N_\Phi := N_{\hat{X}/\operatorname{Spec}\mathbb{C}[[z]]} \subset V \times \hat{X} \tag{2.5}$$

which is canonically a subcone of $V \times \hat{X}$. The cone N_Φ is closely related to the normal cone $N_{X/W}$ at p.

LEMMA 2.3. *Let the notation be as before and let $\Phi_1, \Phi_2 \in V \otimes \mathbb{C}[[z]]$ be any two Kuranishi maps of $(\hat{X}, p)$. Then there is a vector bundle automorphism ϕ of $V \times \hat{X}$ (as bundle over $\hat{X}$) that restricts to the identity homomorphism at the closed fiber $V \times \{p\}$ such that $\phi^{-1}(N_{\Phi_1}) = N_{\Phi_2}$.*

The upshot of this Lemma is that though Kuranishi maps are not unique, the normal cones they defined are almost unique. The only ambiguity comes from those automorphisms whose restrictions to the closed fiber are the identity automorphisms.

We now investigate the choice of the vector bundle E. It is understood now that the key that makes E the vector bundle of our choice is that the sheaf

$$\mathcal{T} = \operatorname{Coker}\{\Omega_W^\vee|_X \to \mathcal{O}_X(E)\}, \tag{2.6}$$

which has the property that at each point $p \in X$ the space $\mathcal{T} \otimes_{\mathcal{O}_X} \mathbb{C}_p$ is the obstruction space to deformation of p in X, is a quotient sheaf of $\mathcal{O}_X(E)$. We call $\mathcal{T}$ the obstruction sheaf. We remark that not all obstruction theories has this property. Those that do will be called *perfect obstruction theories*.

DEFINITION 2.4. *A scheme Y is said to have a perfect obstruction theory if there is a perfect complex $\mathcal{T}^\bullet = [\mathcal{T}_1 \to \mathcal{T}_2]$ in the derived category $\mathcal{D}^b(X)$ of which the following holds: For any triple (A, I, φ), where A is a $\mathbb{C}$-algebra, $I \subset A$ an ideal so that $I^2 = 0$ and $\varphi : \operatorname{Spec} A/I \to X$ is a morphism, there is a function $\operatorname{Ob}(A, I, \varphi) \in \mathfrak{h}^2(\mathcal{T}^\bullet \otimes_{\mathcal{O}_X} I)$ such that*
(1) $\operatorname{Ob}(A, I, \varphi) = 0$ if and only if there is an extension $\tilde{\varphi}: \operatorname{Spec} A \to X$ of φ.
(2) If $\operatorname{Ob}(A, I, \varphi) = 0$ then the space of all extensions $\tilde{\varphi}$ are parameterized by $\mathfrak{h}^1(\mathcal{T}^\bullet \otimes_{\mathcal{O}_X} I)$.
(3) the function $\operatorname{Ob}(\cdot, \cdot, \cdot)$ satisfies base change property.

This definition is almost the same as Definition (2.1) except that now A can be any $\mathbb{C}$-algebra. Thus this covers deformation of *all* points in the sense of Grothendieck. For those who are unfamiliar with the derived category, they can replace $[\mathcal{T}_1 \to \mathcal{T}_2]$ with a complex of locally free sheaves of $\mathcal{O}_X$ modules $\mathcal{E}^\bullet = [\mathcal{E}_1 \to \mathcal{E}_2]$, and $\mathfrak{h}^j(\mathcal{T}^\bullet \otimes_{\mathcal{O}_X} I)$ by the kernel and the cokernel of $\mathcal{E}_1 \otimes_{\mathcal{O}_X} I \to \mathcal{E}_2 \otimes_{\mathcal{O}_X} I$ for $i = 1$ and 2 respectively.

Let Y be any projective scheme endowed with a perfect obstruction theory with associated complex $\mathcal{T}^\bullet$. Let E be a vector bundle on Y so that the $\mathfrak{h}^2(\mathcal{T}^\bullet)$ is a quotient sheaf of $\mathcal{O}_X(E)$. For any $p \in Y$, we denote by $\hat{Y}$ the formal completion of Y along p and V_p the vector space $\mathfrak{h}^2(\mathcal{T}^\bullet \otimes \mathbb{C}_p)$, the obstruction space to deformation of p in $\hat{Y}$.

BASIC CONSTRUCTION 2.5. *Let the notation be as before. There is a cone scheme N so that for each $p \in Y$ there is a vector bundle homomorphism*

$$\phi : E|_{\hat{Y}} \longrightarrow V_p \times \hat{Y}$$

that extends the given homomorphism $E|_p \to V_p$ so that

$$N|_{\hat{Y}} = \phi^{-1}(N_\Phi). \tag{2.7}$$

In example (2.2), the cone so constructed is exactly the normal cone $N_{X/W}$ when using the obstruction theory induced by the section s with the associated complex $\Omega_W^\vee|X \to \mathcal{O}_X(E)$ and the vector bundle E.

DEFINITION-THEOREM 2.6. *Let X be any projective scheme endowed with a perfect obstruction theory with associated complex $\mathcal{T}^\bullet = [\mathcal{T}_1 \to \mathcal{T}_2]$. Then X has a canonical virtual cycle $[X]^{\mathrm{vir}}$ which is constructed according to the following recipe:*
(1) We pick a vector bundle E over X so that the sheaf cohomology $\mathfrak{h}^2(\mathcal{T}^\bullet)$ is a quotient sheaf of $\mathcal{O}_X(E)$.
(2) We form a cone $N \subset E$ according to the basic construction (2.5).
(3) We define the virtual cycle $[X]^{\mathrm{vir}} = \eta_E^*[N]$.

Here is a few remarks. First, since $\mathcal{T}^\bullet$ is a perfect complex with two terms, its cokernel is a well-defined sheaf of $\mathcal{O}_X$-modules. Because X is projective, the vector bundle E always exists; Second, the existence of the cone N satisfying rule (2.7) does not follow from the existence of the Kuranish maps. The existence is proved by constructing a relative Kuranishi family. Once such cone exists, then it will by unique, following (2.7); Thirdly, the cycle $[X]^{\mathrm{vir}}$ so defined is independent of the choice of E. One caution, the virtual cycle only exists as a class in A_*X. It does not have a canonical cycle representative, just like the Euler class has not canonical submanifold representative.

2.3. GW Invariants. we now show that this machinery can be applied to construct the GW invariants of all smooth projective varieties. Following the blueprint of the construction, we need to describe the canonical obstruction theory of the moduli space $\mathfrak{M}_{g,k}(X, A)$ of stable morphisms, to workout a global vector bundle that makes the obstruction sheaf its quotient sheaf and then use the Gysin map to define the virtual cycle.

Let $f : D \subset C \to X$ be a stable morphism. It associates a complex $\mathcal{A}_f = [f^*\Omega_X \to \Omega_C(D)]$ (in the derived category) of degree -1 and 0. The obstruction theory of f is given by the following Lemma.

LEMMA 2.7. *The space of first order deformation of f is naturally parameterized by the extension group $\mathrm{Ext}^1_C(\mathcal{A}_f, \mathcal{O}_C)$. The deformation functor $\mathfrak{Mor}_f$ admits a canonical obstruction theory with coefficients in the extension group $\mathrm{Ext}^2_C(\mathcal{A}_f, \mathcal{O}_C)$.*

The proof of this Lemma is a standard deformation-obstruction computation, and can be found in [**LT1, Rn**]. Now we let (z) be a multi-variable dual to the first extension group $\mathrm{Ext}^1(\mathcal{A}_f, \mathcal{O}_C)$ and let J be the ideal generated by the components of the Kuranishi map of its obstruction theory. Then over $\operatorname{Spec}\mathbb{C}[[z]]/J$ there is a universal family to deformation of f. Sometimes the scheme $\operatorname{Spec}\mathbb{C}[[z]]/J$ is called the universal family to deformation of f.

The obstruction space fits into the exact sequence of cohomology groups

$$\longrightarrow \operatorname{Ext}^1(\Omega_C(D), \mathcal{O}_C) \longrightarrow H^1(f^*\Omega_X^\vee) \longrightarrow \operatorname{Ext}^2(\mathcal{A}_f, \mathcal{O}_C) \longrightarrow 0.$$

An easy consequence of this is the following smoothness result for convex manifolds[5].

COROLLARY 2.8. *Let X be a smooth convex variety and $f : D \subset C \to X$ be a stable morphism of which C has genus 0. Then there is no obstruction to deformation of f.*

It is known that the germ of $\mathfrak{M}_{g,k}(X, A)$ at f is a quotient of the universal family to deformation of f.

LEMMA 2.9. *Let* $\operatorname{Aut}(f)$ *be the automorphism group of f. Then G acts naturally on* $\operatorname{Ext}^1(\mathcal{A}_f, \mathcal{O}_C)$ *and on the universal family* $\operatorname{Spec}\mathbb{C}[[z]]/J$ *to the deformations of f. Further, the germ of $\mathfrak{M}_{g,k}(X, A)$ at f is canonically isomorphic to the quotient* $(\operatorname{Spec}\mathbb{C}[[z]]/J)/G$.

Note that it is possible that the local universal family to deformation of f is smooth while the the moduli space is singular at f. In general there are two sources for the singularities of a moduli space: One is from the non-trivial automorphism group of an object and the other is the non-trivial obstruction to deformations.

One essential precondition for constructing virtual cycle is the existence of a perfect obstruction theory. When $S \subset \mathfrak{M}_{g,k}(X, A)$ is an open subset over which a universal family exists, say $f : D \subset C \to X$, then S has a perfect obstruction theory with the associated complex

$$\mathcal{E}xt^1_{C/S}(\mathcal{A}_f, \mathcal{O}_C) \xrightarrow{\times 0} \mathcal{E}xt^2_{C/S}(\mathcal{A}_f, \mathcal{O}_C). \tag{2.8}$$

Because there are maps $f \in \mathfrak{M}_{g,k}(X, A)$ that has non-trivial automorphism groups, the moduli space *does not* admit universally families. Nevertheless, since all such automorphism groups are finite, we have the following

LEMMA 2.10. *For any points $\xi \in \mathfrak{M}_{g,k}(X, A)$ we can find a scheme S acted on by $G =$* $\operatorname{Aut}(f)$ *and a G-equivariant family of stable morphisms over S so that the tautological morphism $S/G \to \mathfrak{M}_{g,k}(X, A)$ induced by this family is an open neighborhood (i.e. open embedding) of $\xi \in \mathfrak{M}_{g,k}(X, A)$. Let $f : D \subset C \to X$ be such family over S. Then S admits a canonical perfect obstruction theory with associated complex (2.8).*

The Lemma suggests that we should work with pairs (S, G), called charts of $\mathfrak{M}_{g,k}(X, A)$, and their tautological families. When all charts S are smooth, this is the classical notion of orbifold. For us, we need to deal with singular charts as well. This leads us to the notion of Mumford-Deligne stacks or $\mathbb{Q}$-schemes. (We will call it stack since we will only work with Delign-Mumford stacks.)

The notion of stack is widely accepted nowadays among algebraic geometers working with moduli problems. It is less so for people working in other areas. To completely develope the notion of stacks from scratch is technically demanding.

---

[5]A complex manifold is said to be convex if for any morphism $f : \mathbf{P}^1 \to X$ we have $H^1(f^*T_X) = 0$.

However, the underlining idea behind it is quite simple. It was developed exactly to study a moduli space like $\mathfrak{M}_{g,k}(X,A)$: *it has a collection* (S_α, G_α), *where* G_α *are finite group acts on* S_α, *so that each* S_α/G_α *is an open subset of* $\mathfrak{M}_{g,k}(X,A)$, *and they together form an open covering of* $\mathfrak{M}_{g,k}(X,A)$. Each (S_α, G_α) in this collection is called a chart of the stack. The informed readers will now suspect that there should be compatibility condition of (S_α, G_α) and (S_β, G_β) in case $S_\alpha \cap S_\beta \neq \emptyset$,[6] etc.. This is certainly the case. For brevity we will not state them here but instead say they satisfy the obvious compatibility condition. The upshot of using charts (S_α, G_α) instead of open subsets S_α/G_α is that the former carry certain important data of which the later does not. For instance, an ordinary orbifold has smooth charts but not smooth neighborhood. Our moduli space $\mathfrak{M}_{g,k}(X,A)$ has charts carrying tautological families.

We are now ready to construct the virtual moduli cycle of $\mathfrak{M}_{g,k}(X,A)$. We pick a collection of charts of $\mathfrak{M}_{g,k}(X,A)$, say $\{(S_\alpha, G_\alpha)\}$, with $f_\alpha : D_\alpha \subset C_\alpha \to X$ the tautological family over S_α. The associated complex of the obstruction theory of f_α is $\mathcal{E}xt^\bullet_{C_\alpha/S_\alpha}(\mathcal{A}_\alpha, \mathcal{O}_{S_\alpha})$, where $\mathcal{A}_\alpha = [f_\alpha^* X \to \Omega_{C_\alpha}(D_\alpha)]$. Following the recipe, it remains to find a global vector bundle over $\mathfrak{M}_{g,k}(X,A)$ that makes the obstruction sheaves its quotient sheaves. Working with stack, it amounts to find a collection of vector bundles E_α over S_α, each is G_α-equivariant, and G_α-equivariant quotient homomorphisms

$$\mathcal{O}_{S_\alpha}(E_\alpha) \longrightarrow \mathcal{E}xt^2_{C_\alpha/S_\alpha}(\mathcal{A}_\alpha, \mathcal{O}_{C_\alpha}),$$

that are compatible over the intersections $S_\alpha \cap S_\beta$. In our case, such a vector bundle can be constructed explicitly [**LT1**]. We still denote it by $\{E_\alpha\}$. Then from the Basic Construction 2.5, we can construct (virtual normal) cone cycle $N_\alpha \subset E_\alpha$. Since such construction is unique, $N_\alpha \subset E_\alpha$ are G_α-equivariant and are compatible.

In the end, we need to intersect this collection of cone-cycles $[N_\alpha] \in Z_* E_\alpha$ with the collection of zero sections of E_α. One way to do this is to descend E_α to a V-vector bundle on $\mathfrak{M}_{g,k}(X,A)$ which contains the descendents of $[N_\alpha]$ as a subcone cycle. The other is to view the collection E_α as a vector bundle over the stack $\mathfrak{M}_{g,k}(X,A)$. We denote $\mathbf{E}$ the collection $\{E_\alpha\}$ either as a V-vector bundle or a vector bundle over a stack. We denote by $[\mathbf{N}] \subset \mathbf{E}$ the corresponding virtual normal cone cycle. As before, if we let $\eta_{\mathbf{E}}$ be the zero section of $\mathbf{E}$, then the virtual moduli cycle should be

$$[\mathfrak{M}_{g,k}(X,A)]^{\mathrm{vir}} = \eta^*_{\mathbf{E}}([\mathbf{N}]) \in A_* \mathfrak{M}_{g,k}(X,A),$$

where $\eta^*_{\mathbf{E}}$ is an appropriately defined Gysin homomorphism of the zero section $\eta_{\mathbf{E}}$. Such Gysin homomorphism is known to exists [**Vi**]. The result of this Gysin homomorphism takes values in the group of cycles with rational coefficient.

Once we have the virtual moduli cycle, we define the GW invariant to be

$$\gamma^X_{A,g,k}(\alpha) = < [\mathfrak{M}_{g,k}(X,A)]^{\mathrm{vir}}, \mathrm{ev}^*(\alpha) > \in \mathbb{Q}, \qquad \alpha \in A^* X^k.$$

This completes the task of constructing the GW invariants of smooth projective varieties.

3. Symplectic GW invariants

The main purpose of this chapter is to construct the virtual moduli cycles for general symplectic manifolds. Accordingly, in this chapter we will work with

[6]The intersection is $S_\alpha \times_{\mathfrak{M}_{g,k}(X,A)} S_\beta$ in Grothendieck's topology.

analytic category. For instance, we will view the moduli space of k-pointed genus g stable curves $\mathfrak{M}_{g,k}$, $2g+k \geq 3$, as a compact orbifold.

Let X be a smooth symplectic manifold with a given symplectic form ω of complex dimension n, and let $A \in H_2(X,\mathbb{Z})$. We state the main theorem of this chapter.

THEOREM 3.1. *Let (X,ω) be a compact symplectic manifold of complex dimension n. Then for each g, k and A, there is a virtual fundamental class*

$$e_{A,g,k}(X) \in H_r(\mathfrak{M}_{g,k} \times X^k, \mathbb{Q}),$$

where $r = 2c_1(X)(A) + 2(n-3)(1-g) + 2k$. Moreover, this $e_{A,g,k}(X)$ is a symplectic invariant.

As an application, let us define the GW-invariants now. Let $2g+k \geq 3$. We define

$$\psi^X_{A,g,k} : H^*(\mathfrak{M}_{g,k}, \mathbb{Q}) \times H^*(X^k, \mathbb{Q}) \mapsto \mathbb{Q}, \tag{3.1}$$

to be the integrals

$$\psi^X_{A,g,k}(\beta, \alpha) = \int_{e_{A,g,k}(X)} \mathrm{pr}_1^* \beta \wedge \mathrm{pr}_2^* \alpha \tag{3.2}$$

where $\beta \in H^*(\mathfrak{M}_{g,k}, \mathbb{Q})$, $\alpha \in H^*(X^k, \mathbb{Q})$ and pr_i is the projection of $\mathfrak{M}_{g,k} \times X^k$ to its component. All $\gamma^X_{A,g,k}$ are symplectic invariants of (X,ω).

To construct such cycles, we will introduce the notion of smoothly stratified orbispaces, weakly pseudocycles and weakly Fredholm V-bundles over any smoothly stratified orbispaces. This is what we will do in the first part of this chapter. After this we will sketch the construction of the virtual moduli cycles for weakly Fredholm V-bundles. Finally, we will apply this construction to stable maps and consequently prove the above theorem.

3.1. Smoothly stratified orbispaces. In this section, we introduce a class of topological spaces that admit a class of stratified structures. In this note, all topological spaces are Hausdorff.

We first introduce the notion of *smoothly stratified spaces.*

DEFINITION 3.2. *By a smoothly stratified space we mean a topological space X with a locally finite*[7] *partition $X = \bigcup_{\alpha \in I} X_\alpha$, called a stratification, by locally closed subsets X_α, called strata, such that each X_α is a smooth Banach manifold.*

Note that our definition differs from the usual one in that we do not require knowledge of the normal cone structure along each stratum X_α. Let X and Y be smoothly stratified spaces. We will denote by X^{top} the topological space X without the stratification. A map $f : X \to Y$ between two smoothly stratified spaces is a map $f : X^{\mathrm{top}} \to Y^{\mathrm{top}}$ such that it maps strata of X into strata of Y. f is smooth if $f|_{X_\alpha}$ are smooth. A smooth manifold is viewed as a smoothly stratified space whose stratification consists of a single stratum X. In the following we will call smoothly stratified space simply stratified space if no confusion will arise.

Next we introduce the notion of *smoothly stratified orbispaces* and their *local uniformization charts.* We let X be a topological space with a locally finite, locally closed stratification $X = \bigcup_{\alpha \in I} X_\alpha$.

[7]By this we mean for each $x \in X$ the set $\{\alpha \in I \mid a \in X_\alpha\}$ is finite.

DEFINITION 3.3. *Let X be as before. A local uniformization chart of X consists of a smoothly stratified space $\tilde{\mathbf{U}} = \cup_{\alpha\in I}\tilde{\mathbf{U}}_\alpha$, a finite group $G_{\tilde{\mathbf{U}}}$ acting effectively on $\tilde{\mathbf{U}}$ by smooth maps and a $G_{\tilde{\mathbf{U}}}$-invariant continuous map $\pi_{\mathbf{U}} : \tilde{\mathbf{U}}^{\mathrm{top}} \to X^{\mathrm{top}}$ such that*
(1) $\mathbf{U} = \pi_{\mathbf{U}}(\tilde{\mathbf{U}})$ is an open neighborhood of X;
(2) each stratum $\mathbf{U}_\alpha$ of $\tilde{\mathbf{U}}$ is invariant under $G_{\tilde{\mathbf{U}}}$;
(3) $\pi_{\mathbf{U}}$ induces a covering map from $\tilde{\mathbf{U}}/G_{\tilde{\mathbf{U}}}$ onto $\mathbf{U}$ as stratified spaces that respect the induced stratum $\tilde{\mathbf{U}}_\alpha/G_{\tilde{\mathbf{U}}}$ of $\tilde{\mathbf{U}}/G_{\tilde{\mathbf{U}}}$ and $\tilde{\mathbf{U}} \cap X_\alpha$ of $\mathbf{U}$.

Again, if there is no confusion we will call $(\tilde{\mathbf{U}}, G_{\tilde{\mathbf{U}}})$ a chart and call $\mathbf{U}$ the quotient of the chart $(\tilde{\mathbf{U}}, G_{\tilde{\mathbf{U}}})$. Let $\mathbf{V} \subset \mathbf{U}$ be an open subset, then $(\pi_{\tilde{\mathbf{U}}}^{-1}(\mathbf{V}), G_{\tilde{\mathbf{U}}})$ defines a new chart of X, called the restriction of $(\tilde{\mathbf{U}}, G_{\tilde{\mathbf{U}}})$ to $\mathbf{V}$. For simplicity, we will often denote it by $(\tilde{\mathbf{U}}, G_{\tilde{\mathbf{U}}})|_{\mathbf{V}}$. When $p \in \mathbf{U}$, we sometimes call $(\tilde{\mathbf{U}}, G_{\tilde{\mathbf{U}}})$ a chart of p. Now let X be as before and let $(\tilde{\mathbf{U}}, G_{\tilde{\mathbf{U}}})$ and $(\tilde{\mathbf{V}}, G_{\tilde{\mathbf{V}}})$ be two charts of X. We say the later is finer than the former if there is a homomorphism $G_{\tilde{\mathbf{V}}} \to G_{\tilde{\mathbf{U}}}$ and an equivariant smooth covering map $\varphi_{\tilde{\mathbf{U}}\tilde{\mathbf{V}}}$ belongs to the square

$$\begin{array}{ccc} \pi_{\tilde{\mathbf{V}}}^{-1}(\mathbf{U}\cap\mathbf{V}) & \xrightarrow{\varphi_{\tilde{\mathbf{U}}\tilde{\mathbf{V}}}} & \pi_{\tilde{\mathbf{U}}}^{-1}(\mathbf{U}\cap\mathbf{V}) \\ {\scriptstyle \pi_{\mathbf{V}}}\downarrow & & {\scriptstyle \pi_{\mathbf{U}}}\downarrow \\ \mathbf{U}\cap\mathbf{V} & = & \mathbf{U}\cap\mathbf{V} \end{array} \tag{3.3}$$

Note that if $\mathbf{U} \cap \mathbf{V} = \emptyset$, then $\tilde{\mathbf{U}}$ is automatically finer than $\tilde{\mathbf{V}}$.

We now define the notion of smoothly stratified orbispace.

DEFINITION 3.4. *We say that X is a smoothly stratified orbispace if it can be covered by a collection of charts as in the Definition 3.3 such that for any two charts $(\tilde{\mathbf{U}}_i, G_{\tilde{\mathbf{U}}_i})$ and $(\tilde{\mathbf{U}}_j, G_{\tilde{\mathbf{U}}_j})$ and $x \in \mathbf{U}_i \cap \mathbf{U}_j$, there is a chart $(\tilde{\mathbf{V}}, G_{\tilde{\mathbf{V}}})$ of $x \in X$ such that $(\tilde{\mathbf{V}}, G_{\tilde{\mathbf{V}}})$ is finer than both $(\tilde{\mathbf{U}}_i, G_{\tilde{\mathbf{U}}_i})$ and $(\tilde{\mathbf{U}}_j, G_{\tilde{\mathbf{U}}_j})$.*

Later, we will abbreviate smoothly stratified orbispaces to SS-orbispaces. Clearly, each smoothly stratified space is a SS-orbispace. A smoothly stratified space is smooth if all its charts, namely the $\tilde{\mathbf{U}}_i$ in the Definition 3.4, are smooth. In particular, a finite dimensional smooth smoothly stratified space is an orbifold. Note that according to this definition, each stratum X_α of X is a Banach orbifold.

If X and Y are two SS-orbispaces, a map $f : X \to Y$ is a map $f^{\mathrm{top}} : X^{\mathrm{top}} \to Y^{\mathrm{top}}$ such that for each $x \in X$ there is a chart $(\tilde{\mathbf{U}}, G_{\tilde{\mathbf{U}}})$ of $x \in X$, a chart $(\tilde{\mathbf{V}}, G_{\tilde{\mathbf{V}}})$ of $f(x) \in Y$, a homomorphism $G_{\tilde{\mathbf{U}}} \to G_{\tilde{\mathbf{V}}}$ and a $G_{\tilde{\mathbf{U}}}$-equivariant map $\tilde{f} : \tilde{\mathbf{U}} \to \tilde{\mathbf{V}}$ such that it descends to $f^{\mathrm{top}}|_{\mathbf{U}} : \mathbf{U} \to \mathbf{V} \subset X$. We call such $\tilde{f}$ a local representative of f. We say f is smooth if all its local representatives are smooth. A smooth map $f : X \to Y$ with smooth inverse f^{-1} is called an isomorphism.

3.2. Weakly pseudocycles. In this subsection, we will work with a continuous map $f : X \to Y$ between two topological spaces X and Y. All homology theories are with rational coefficients.

We begin with pseudomanifold. An oriented pseudomanifold S of dimension d in X is a pair (ρ, M), where M is an oriented, smooth manifold of dimension d with boundary ∂M with corner[8] and $\rho : M \to X$ is a continuous map. For a

[8] A manifold of dimension d with boundary with corner is is a manifold with boundary except that the neighborhoods of a boundary point are isomorphic to the neighborhoods of a boundary

pseudomanifold $S = (f, M)$ we define ∂S to be $(f, \partial M)$. We define $-S$ to be $(f, \overline{M})$, where $\overline{M}$ is M with the reversed orientation. We define the sum of two pseudomanifolds (f, M) and (g, N) to be $(f \coprod g, M \cup N)$. We then can define the sum $\sum n_i \alpha_i$, where $n_i \in \mathbb{Z}$, of pseudomanifolds inductively. For any (f, M), we say it is 0 if there are two disjoint open subsets U_1 and U_2 of M so that the complement of their union in M is an at most $d-1$ dimensional set[9], such that there is an orientation reversing isomorphism $\rho : U_1 \to U_2$ so that $f|_{U_1} \equiv f \circ \rho$. We can talk about rational pseudomanifolds, which are formal sum of oriented pseudomanifolds with rational coefficients. A rational pseudomanifold is zero if an integer multiple of it is integral and is zero.

Let $f : X \to Y$ be a continuous map between topological spaces. We now define weakly f-pseudocycles.

DEFINITION 3.5. *A rational weakly d-dim'l f-pseudocycle is a triple (ρ, M, K), where (ρ, M) is a rational pseudomanifold in X, K is a closed subset in Y of homological dimension*[10] *no more than $d-2$ such that the closure of $M - (f \circ \rho)^{-1}(K)$ is compact in M and the rational $(d-1)$-dimensional pseudomanifold $(\rho, \partial M - (f \circ \rho)^{-1}(K))$ is 0.*

Two such cycles (ρ_1, M_1, K_1) and (ρ_2, M_2, K_2) are quasi-cobordant if there is a $(d+1)$-dimensional pseudomanifold (ρ, M) in X, a closed subset $K \subset Y^{\text{top}}$ of homological dimension no more than $d-1$ such that $M - (f \circ \rho)^{-1}(K)$ is compact in M, K_1 and K_2 are contained in K and $(\rho_1, M_1) - (\rho_2, M_2) = (\rho, \partial M)$. Two such cycles S and S' are equivalent if there is a chain of weakly f-pseudocycle $S_0, \cdots, S_l$ so that $S = S_0$, S_i is quasi-cobordant to S_{i+1} and $S_l = S'$.

The usefulness of this definition is illustrated by the following Proposition.

PROPOSITION 3.6. *Let (ρ, M, K) be a weakly d-dim'l f-pseudocycle. Then $f \circ \rho : M \to Y$ defines a unique element in $H_d(Y^{\text{top}}, K) = H_d(Y^{\text{top}})$. Further, two equivalent such cycles define identical elements in $H_d(Y)$.*

PROOF. First, the two homology groups are canonically isomorphic since K has homological dimension $\leq d-2$. Now we show that $f \circ \rho$ defines a cycle in $H_d(Y^{\text{top}}, K)$. Since M is smooth and $M - (f \circ \rho)^{-1}(K)$ is pre-compact in M, we can give M a triangulation so that the boundary of $f \circ \rho(M)$ is contained in K, thus it defines an element in $H_d(Y, K)$. For the same reason, two quasi-cobordant and thus two equivalent such cycles define identical elements in $H_d(Y)$. □

3.3. Weakly Fredholm V-bundles. In this section, we introduce the notion of weakly Fredholm V-bundles over a SS-orbispace $X = \cup_{\alpha \in I} X_\alpha$. We assume throughout this chapter that $0 \in I$ and the stratum X_0 is dense in X. Also, if $\tilde{\mathbf{U}}$ is a chart of X, then $\tilde{\mathbf{U}}$ has a stratification $\cup_{\alpha \in I} \tilde{\mathbf{U}}_\alpha$ so indexed that $\tilde{\mathbf{U}}_\alpha$ corresponds to the stratum $\mathbf{U} \cap X_\alpha$.

DEFINITION 3.7. *Let X be an SS-orbispace. A V-bundle over X is a pair $(\mathbf{E}, \mathbf{P})$, where $\mathbf{E}$ is an SS-orbispace and $\mathbf{P} : \mathbf{E} \to X$ is a projection (of SS-orbispaces) such that X (resp. $\mathbf{E}$) is covered by a set of charts $(\tilde{\mathbf{U}}_i, G_{\tilde{\mathbf{U}}_i})$ (resp. $(\mathbf{E}_{\tilde{\mathbf{U}}_i}, G_{\tilde{\mathbf{U}}_i})$)*

point of $L = \{x_1 \geq 0, \cdots, x_k \geq 0\}$. The boundary of such manifold is also a manifold with boundary with corner, following the convention that the boundary of L is the disjoint union of k manifolds $L \cap \{x_i = 0\}$.

[9]By this we mean it is the image of $d-1$ dimensional manifold.

[10]By this we mean $\max\{k; H_k(K, \mathbb{Q}) \neq 0\}$.

of which the following hold:
(1) If $\tilde{\mathbf{U}}_i$ *is a chart over* $\mathbf{U}_i \subset X$ *then* $\mathbf{E}_{\tilde{\mathbf{U}}_i}$ *is a chart over* $\mathbf{P}^{-1}(\mathbf{U}_i)$ *and there is a projection* $\mathbf{P}_{\tilde{\mathbf{U}}_i}:\mathbf{E}_{\tilde{\mathbf{U}}_i} \to \tilde{\mathbf{U}}_i$ *representing* $\mathbf{P}$ *over* $\mathbf{U}_i$.
(2) Let $\tilde{\mathbf{U}}_\alpha$ *be any stratum of* $\tilde{\mathbf{U}}_i$, *then restricting to* $\tilde{\mathbf{U}}_\alpha$ *the bundle* $\mathbf{E}_{\tilde{\mathbf{U}}_i}|_{\tilde{\mathbf{U}}_\alpha}$ *is a smooth Banach vector bundle over* $\tilde{\mathbf{U}}_\alpha$ *with a* $G_{\tilde{\mathbf{U}}_i}$*-linear action. Further the stratum* $\mathbf{E}_{\tilde{\mathbf{U}}_i}$ *are pull-back of the stratum of* $\tilde{\mathbf{U}}_i$.
(3) For any two charts $(\tilde{\mathbf{U}}_i, G_{\tilde{\mathbf{U}}_i})$ *and* $(\tilde{\mathbf{U}}_j, G_{\tilde{\mathbf{U}}_j})$ *of* X *(over* $\mathbf{U}_i$ *and* $\mathbf{U}_j$ *respectively) and* $x \in \mathbf{U}_i \cap \mathbf{U}_j$, *there is a chart* $(\tilde{\mathbf{U}}_k, G_{\tilde{\mathbf{U}}_k})$ *of* x *finer than both* $(\tilde{\mathbf{U}}_i, G_{\tilde{\mathbf{U}}_i})$ *and* $(\tilde{\mathbf{U}}_j, G_{\tilde{\mathbf{U}}_j})$ *such that there are* $G_{\mathbf{U}}$*-equivariant isomorphisms*

$$\mathbf{E}_{\tilde{\mathbf{U}}_k}|_{\mathbf{U}_i\cap\mathbf{U}_k} \cong \varphi^*_{\tilde{\mathbf{U}}_k\tilde{\mathbf{U}}_i}(\mathbf{E}_{\tilde{\mathbf{U}}_i}) \quad \text{and} \quad \mathbf{E}_{\tilde{\mathbf{U}}_k}|_{\mathbf{U}_j\cap\mathbf{U}_k} \cong \varphi^*_{\tilde{\mathbf{U}}_k\tilde{\mathbf{U}}_j}(\mathbf{E}_{\tilde{\mathbf{U}}_j})$$

preserving the Banach bundle structures along each stratum, where $\varphi_{\tilde{\mathbf{U}}_k\tilde{\mathbf{U}}_i}$ *is the map given in (3.3).*
(4) Let $\mathbf{O}_{\mathbf{E}_{\tilde{\mathbf{U}}_i}}$ *be the zero sections of* $\mathbf{E}_{\tilde{\mathbf{U}}_i}$, *then they define a SS-orbi-subspace* $\mathbf{O}_{\mathbf{E}}$ *of* $\mathbf{E}$. *We require that it is isomorphic to* X *under the projection* $\mathbf{P}$.

Now let $\mathbf{E}$ be a V-fiber bundle over X. A section of $\mathbf{E}$ is a map $\Phi : X \to \mathbf{E}$ as SS-orbispaces so that $\mathbf{P}\circ\Phi : X \to X$ is an isomorphism. We say Φ has compact support if $\Phi^{-1}(O_{\mathbf{E}})$ as a subspace in X^{top} is compact.

We fix a V-bundle over X with a section Φ with compact support. We now describe the weakly Fredholm structure of this bundle with section in terms of its local finite approximations.

DEFINITION 3.8. *A local finite approximation of* $(\mathbf{E},\mathbf{P},\Phi)$ *consists of a chart* $(\tilde{\mathbf{U}}, G_{\tilde{\mathbf{U}}})$ *of* X, *a chart* $(\mathbf{E}_{\tilde{\mathbf{U}}}, G_{\tilde{\mathbf{U}}})$ *of* $\mathbf{E}$, *a representative* $\Phi_{\tilde{\mathbf{U}}} : \tilde{\mathbf{U}} \to \mathbf{E}_{\tilde{\mathbf{U}}}$ *of* Φ *and a finite rank equi-rank* $G_{\tilde{\mathbf{U}}}$*-linearized vector bundle* $\mathbf{F}$ *over* $\tilde{\mathbf{U}}$ *such that:*
(1) $\mathbf{F}$ *is a* $G_{\tilde{\mathbf{U}}}$*-equivariant SS-subspace of* $\mathbf{E}_{\tilde{\mathbf{U}}}$. *Further restricting to each stratum* $\tilde{\mathbf{U}}_\alpha \subset \tilde{\mathbf{U}}$ *the inclusion* $\mathbf{F}|_{\tilde{\mathbf{U}}_\alpha} \subset \mathbf{E}_{\tilde{\mathbf{U}}}|_{\tilde{\mathbf{U}}_\alpha}$ *is a smooth sub-vector bundle.*
(3) $U := \Phi^{-1}_{\tilde{\mathbf{U}}}(\mathbf{F}) \subset \tilde{\mathbf{U}}$ *is a finite dimension equi-dimension SS-orbispace (with strata* $U = \cup_{\alpha\in I} U_\alpha$*). Further we require* $U_0 = U \cap \tilde{\mathbf{U}}_0$ *is dense in* U *and its complement has codimension at least 2 in* U;
(4) $F := \mathbf{F}|_U$ *is a continuous vector bundle. Further its restriction to each stratum* U_α *it is a smooth vector bundle and the map* $\phi_F|_{U_\alpha}$, *where* $\phi_F = \Phi_{\tilde{\mathbf{U}}}|_U : U \to F$, *is a smooth section;*
(5) At each $w \in \Phi^{-1}(0) \cap \tilde{\mathbf{U}}_0$, *the differential*

$$d\Phi(w) : T_w\tilde{\mathbf{U}}_0 \longrightarrow \mathbf{E}_{\tilde{\mathbf{U}}_0}|_w \,/\, \mathbf{F}|_w$$

is surjective.

Such local finite approximation will be denoted by $(\tilde{\mathbf{U}}, G_{\tilde{\mathbf{U}}}, \mathbf{E}_{\tilde{\mathbf{U}}}, \mathbf{F})$.

An orientation of $(\tilde{\mathbf{U}}, G_{\tilde{\mathbf{U}}}, \mathbf{E}_{\tilde{\mathbf{U}}}, \mathbf{F})$ is a $G_{\tilde{\mathbf{U}}}$-invariant orientation of the real line bundle $\wedge^{\text{top}}(TU)\otimes\wedge^{\text{top}}(F)^{-1}$ over U_0. Such orientations are given by $G_{\tilde{\mathbf{U}}}$-invariant non-vanishing sections of $\wedge^{\text{top}}(TU)\otimes\wedge^{\text{top}}(V)^{-1}$ over U_0.

We call $\operatorname{rank} F - \dim U$ the index of $(\tilde{\mathbf{U}}, G_{\tilde{\mathbf{U}}}, \mathbf{E}_{\tilde{\mathbf{U}}}, \mathbf{F})$.

Now assume that $(\tilde{\mathbf{U}}', G_{\tilde{\mathbf{U}}'}, \mathbf{E}_{\tilde{\mathbf{U}}'}, \mathbf{F}')$ is another local finite approximation of identical index over $\mathbf{U}' \subset X$.

DEFINITION 3.9. *We say that* $(\tilde{\mathbf{U}}', G_{\tilde{\mathbf{U}}'}, \mathbf{E}_{\tilde{\mathbf{U}}'}, \mathbf{F}')$ *is finer than* $(\tilde{\mathbf{U}}, G_{\tilde{\mathbf{U}}}, \mathbf{E}_{\tilde{\mathbf{U}}}, \mathbf{F})$ *if we have the following:*

(1) $(\tilde{\mathbf{U}}', G_{\tilde{\mathbf{U}}'})$ *is finer than* $(\tilde{\mathbf{U}}, G_{\tilde{\mathbf{U}}})$.
(2) Let $\varphi : \pi_{\tilde{\mathbf{U}}'}^{-1}(\mathbf{U} \cap \mathbf{U}') \to \pi_{\tilde{\mathbf{U}}}^{-1}(\mathbf{U} \cap \mathbf{U}')$ *(which is* $\varphi_{\mathbf{U}\mathbf{U}'}$ *in (3.3)) be the covering map, then* $\varphi^*\mathbf{F} \subset \varphi^*\mathbf{E}_{\tilde{\mathbf{U}}} \equiv \mathbf{E}_{\tilde{\mathbf{U}}'}|_{\mathbf{U}' \cap \mathbf{U}}$ *is a smoothly stratified subbundle of* $\mathbf{F}'$.
(3) Let (U, F) *and* (U', F') *be those defined in the Definition 3.8 for* $\tilde{\mathbf{U}}$ *and* $\tilde{\mathbf{U}}'$ *respectively, then for any* w *in* $U_0' \cap \varphi^{-1}(U)$*, the natural homomorphism*

$$T_w U' / T_{\varphi(w)} U \longrightarrow (F'/\varphi^* F)|_w$$

is an isomorphism.
(4) In case both $(\tilde{\mathbf{U}}, G_{\tilde{\mathbf{U}}}, \mathbf{E}_{\tilde{\mathbf{U}}}, \mathbf{F})$ *and* $(\tilde{\mathbf{U}}', G_{\tilde{\mathbf{U}}'}, \mathbf{E}_{\tilde{\mathbf{U}}'}, \mathbf{F}')$ *are oriented, then we require that the orientations of* $(\tilde{\mathbf{U}}, G_{\tilde{\mathbf{U}}}, \mathbf{E}_{\tilde{\mathbf{U}}}, \mathbf{F})$ *and* $(\tilde{\mathbf{U}}', G_{\tilde{\mathbf{U}}'}, \mathbf{E}_{\tilde{\mathbf{U}}'}, \mathbf{F}')$ *are isomorphic through the isomorphism*

$$\wedge^{\mathrm{top}}(T_w U') \otimes \wedge^{\mathrm{top}}(F'|_w)^{-1} \cong \wedge^{\mathrm{top}}(T_{\varphi(w)} U) \otimes \wedge^{\mathrm{top}}(F|_{\varphi(w)})^{-1},$$

induced by the isomorphism in (3), where w *is any point in* $U_0' \cap \varphi^{-1}(U_0)$.

Note that from (5) of the Definition 3.8, $U_0' \cap \varphi^{-1}(U)$ is a locally closed smooth submanifold of codimension $\operatorname{rank} F' - \operatorname{rank} F$ in U'. Hence the homomorphism (3) in the above Definition is well-defined.

Now let $\mathfrak{A} = \{(\tilde{\mathbf{U}}_i, G_{\tilde{\mathbf{U}}_i}, \mathbf{E}_{\tilde{\mathbf{U}}_i}, \mathbf{F}_i)\}_{i \in \mathcal{K}}$ be a collection of oriented finite approximations of $(X, \mathbf{E}, \Phi)$. In the following, we will denote by $\mathbf{U}_i$ the open subsets of X such that $\Lambda_i = (\tilde{\mathbf{U}}_i, G_{\tilde{\mathbf{U}}_i})$ is a uniformization chart over $\mathbf{U}_i$. We say $\mathfrak{A}$ covers $\Phi^{-1}(0)$ if $\Phi^{-1}(0)$ is contained in the union of $\mathbf{U}_i$ in X.

DEFINITION 3.10. *An index* r *oriented pre-weakly Fredholm structure of the triple* $(\mathbf{B}, \mathbf{E}, \Phi)$ *is a collection* $\mathfrak{A} = \{(\tilde{\mathbf{U}}_i, G_{\tilde{\mathbf{U}}_i}, \mathbf{E}_{\tilde{\mathbf{U}}_i}, \mathbf{F}_i)\}_{i \in \mathcal{K}}$ *of index* r *oriented local finite approximations such that* $\mathfrak{A}$ *covers* $\Phi^{-1}(0)$ *and that for any* $(\tilde{\mathbf{U}}_i, G_{\tilde{\mathbf{U}}_i}, \mathbf{E}_{\tilde{\mathbf{U}}_i}, \mathbf{F}_i)$ *and* $(\tilde{\mathbf{U}}_j, G_{\tilde{\mathbf{U}}_j}, \mathbf{E}_{\tilde{\mathbf{U}}_j}, \mathbf{F}_j)$ *in* $\mathfrak{A}$ *with* $p \in \mathbf{U}_i \cap \mathbf{U}_j$*, there is a local finite approximation* $(\tilde{\mathbf{U}}, G_{\tilde{\mathbf{U}}}, \mathbf{E}_{\tilde{\mathbf{U}}}, \mathbf{F}) \in \mathfrak{A}$ *such that* $p \in \mathbf{U}$ *and* $(\tilde{\mathbf{U}}, G_{\tilde{\mathbf{U}}}, \mathbf{E}_{\tilde{\mathbf{U}}}, \mathbf{F})$ *is finer than both* $(\tilde{\mathbf{U}}_i, G_{\tilde{\mathbf{U}}_i}, \mathbf{E}_{\tilde{\mathbf{U}}_i}, \mathbf{F}_i)$ *and* $(\tilde{\mathbf{U}}_j, G_{\tilde{\mathbf{U}}_j}, \mathbf{E}_{\tilde{\mathbf{U}}_j}, \mathbf{F}_j)$.

It follows easily from this definition that for any number of local finite approximations $\{(\tilde{\mathbf{U}}_i, G_i, \mathbf{E}_i, \mathbf{F}_i)\}_{1 \le i \le l}$ in $\mathfrak{A}$ and $x \in \cap_{i=1}^{l} \mathbf{U}_i$, there is a local finite approximation $(\tilde{\mathbf{U}}, G_{\tilde{\mathbf{U}}}, \mathbf{E}_{\tilde{\mathbf{U}}}, \mathbf{F})$ which contains x and is finer than all $(\tilde{\mathbf{U}}_i, G_i, \mathbf{E}_i, \mathbf{F}_i)$.

Let $\mathfrak{A}'$ be another index r oriented pre-weakly Fredholm structure of $(X, \mathbf{E}, \Phi)$. We say that $\mathfrak{A}'$ is finer than $\mathfrak{A}$ if for any $(\tilde{\mathbf{U}}, G_{\tilde{\mathbf{U}}}, \mathbf{E}_{\tilde{\mathbf{U}}}, \mathbf{F}) \in \mathfrak{A}$ over $\mathbf{U} \subset X$ and $p \in \mathbf{U} \cap \Phi^{-1}(0)$, there is a $(\tilde{\mathbf{U}}', G_{\tilde{\mathbf{U}}'}, \mathbf{E}_{\tilde{\mathbf{U}}'}, \mathbf{F}') \in \mathfrak{A}'$ over $\mathbf{U}'$ such that $p \in \mathbf{U}'$ and $(\tilde{\mathbf{U}}', G_{\tilde{\mathbf{U}}'}, \mathbf{E}_{\tilde{\mathbf{U}}'}, \mathbf{F}')$ is finer than $(\tilde{\mathbf{U}}, G_{\tilde{\mathbf{U}}}, \mathbf{E}_{\tilde{\mathbf{U}}}, \mathbf{F})$. We say that these two structures $\mathfrak{A}_1$ and $\mathfrak{A}_2$ are equivalent if $\mathfrak{A}_1$ is finer than $\mathfrak{A}_2$ and vice verse.

DEFINITION 3.11. *A* V*-vector bundle with section* $(X, \mathbf{E}, \Phi)$ *is weakly Fredholm* V*-bundle if it admits an oriented pre-weakly Fredholm structure such that* $\Phi^{-1}(0)$ *is compact and is contained in finitely many strata of* X.

We will abbreviate a weakly Fredholm V-bundle to WFV-bundle.

DEFINITION 3.12. *Two WFV-bundles* $(X, \mathbf{E}, \Phi)$ *and* $(X, \mathbf{E}', \Phi')$ *are homotopic if there is a WFV-bundle* $(X \times [0,1], \overline{\mathbf{E}}, \Psi)$ *such that it restricts to* $(X, \mathbf{E}, \Phi)$ *over* $X \times \{0\}$ *and to* $(X', \mathbf{E}', \Phi')$ *over* $X \times \{1\}$.

3.4. Construction of virtual moduli cycles. The following is the main result on constructing virtual moduli cycles.

Let X be any SS-orbispace, let Y be a smooth orbifold and $f: X^{\text{top}} \to Y^{\text{top}}$ be a smooth map. We index the stratification $X = \cup_{\alpha\in I} X_\alpha$ and $Y = \cup_{\alpha\in I} Y_\alpha$ so that $f(X_\alpha) \subset Y_\alpha$.

THEOREM 3.13. *Let $(X, \mathbf{E}, \Phi)$ be a WFV-bundle of relative index r. Let $\mathfrak{A}$ be the collection in the definition of WFV-bundle. We assume further that for any $(\tilde{\mathbf{U}}, G_{\tilde{\mathbf{U}}}, \mathbf{E}_{\tilde{\mathbf{U}}}, \mathbf{F}_{\tilde{\mathbf{U}}})$ in $\mathfrak{A}$ and $U = \cup_{\alpha\in I} U_\alpha$ the stratification, the representative $f_{\tilde{\mathbf{U}},\alpha}: U_\alpha \to Y_\alpha$ is a submersion. Then we can assign to $(X, \mathbf{E}, \Phi)$ an equivalence class of weakly f-pseudocycle $e(X, \mathbf{E}, \Phi)$, called the Euler class of $(X, \mathbf{E}, \Phi)$. It depends only on the homotopy class of $(X, \mathbf{E}, \Phi)$. Further, the image of this class under f_* is a well-defined homology class in $H_r(Y, \mathbb{Q})$.*

The class $e(X, \mathbf{E}, \Phi)$ or its image under f_* is the virtual cycle we set to construct.

Sometimes it is desirable to include the case where Y is a more general topological space. For instance if we want the Euler class to be a class in $H_r(X^{\text{top}})$, then we can take $f = \text{id}_X : X \to X$ and hence $Y = X$ will be an infinite dimensional space. The proof of Theorem 3.13 can be generalized to such case if $f : X \to Y$ satisfies certain topological condition. However, when Y is infinite dimensional, we have not found a satisfactory condition that is general enough to include many interesting examples. We will leave this to our future investigation.

In the remainder of this subsection, we will briefly sketch how to use the WFV structure of $(X, \mathbf{E}, \Phi)$ to construct the required Euler class. The complete account of this approach is a modification of the construction in [**LT2, LT4**] and will appear elsewhere [**LT5**].

The idea of the construction is as follows: Given a local finite approximation $(\tilde{\mathbf{U}}, G_{\tilde{\mathbf{U}}}, \mathbf{E}_{\tilde{\mathbf{U}}}, \mathbf{F})$ over $\mathbf{U} \subset X$, we can associate to it a finite dimensional model consisting of a finite dimensional SS-orbispace U, a smoothly stratified finite rank V-bundle F over U and a smooth section $\phi : U \to F$. Following the topological construction of Euler class, we like to perturb ϕ to a new section $\tilde{\phi}: U \to F$ so that $\tilde{\phi}$ is as transversal to the zero section of F as possible. Here the difficulty with this is that F is only smooth along, not near, strata of U. Thus the notion of transversality is ill-defined near points where F is not smooth. Here is our solution: We find a homological dimension $r - 2$ set $K \subset U$ so that we can perturb ϕ to $\tilde{\phi}$ so that $\tilde{\phi}$ is transversal to the zero section of F away from K. Since the dimension of the zero locus of $\tilde{\phi}$ is r, the set K will not affect the construction of an r dimensional cycle. In the case we work with many local finite approximations, we will choose these perturbations so that their vanishing locus patch together to form a well-defined weakly f-pseudocycle in X. The Euler class of $(X, \mathbf{E}, \Phi)$ is the equivalent class it represents.

As is clear from this description, the main difficulty of this construction is to make sure that these perturbed vanishing locus patch together. We will do perturbation one chart at a time, and in the mean time make sure that the perturbation on the current chart agrees with the perturbations chosen in the preceding charts. One difficulty arises in that the intersection of charts are not open. However if the current chart is finer than all the preceding charts, then the intersection will be an open subset in the preceding charts and a locally closed subset in the current chart. Hence our first order is to pick a good set of charts to begin with. Let

$\mathfrak{A} = \{(\tilde{\mathbf{U}}_i, G_i, \mathbf{E}_i, \mathbf{F}_i)\}_{i\in\mathcal{K}}$ be the weakly smooth structure of $(X, \mathbf{E}, \Phi)$. For simplicity, we will denote the projection $\pi_{\mathbf{U}_i}: \tilde{\mathbf{U}}_i \to \mathbf{U}_i$ by π_i, denote $\Phi_i^{-1}(\mathbf{F}_i)$ by U_i, denote $\mathbf{F}_i|_{U_i}$ by F_i and denote the restriction $\Phi_i|_{U_i}: U_i \to \tilde{\mathbf{E}}_i|_{U_i}$ by $\phi_i: U_i \to F_i$. Without loss of generality, we can assume that for any approximation $(\tilde{\mathbf{U}}_i, G_i, \mathbf{E}_i, \mathbf{F}_i) \in \mathfrak{A}$ over $\mathbf{U}_i$ and any $\mathbf{U}' \subset \mathbf{U}_i$, the restriction $(\tilde{\mathbf{U}}_i, G_i, \mathbf{E}_i, \mathbf{F}_i)|_{\mathbf{U}'}$ is also a member in $\mathfrak{A}$.

The following lemma enables us to pick a good set of charts.

LEMMA 3.14. *There is a finite collection $\mathcal{L} \subset \mathcal{K}$ and a total ordering of $\mathcal{L}$ of which the following holds:*
(1) The set $\Phi^{-1}(0)$ is contained in the union $\cup\{U_i \mid i \in \mathcal{L}\}$;
(2) For any pair $i < j \in \mathcal{L}$, the approximation $(\tilde{\mathbf{U}}_j, G_j, \mathbf{E}_j, \mathbf{F}_j)$ is finer than the approximation $(\tilde{\mathbf{U}}_i, G_i, \mathbf{E}_i, \mathbf{F}_i)$.

PROOF. We now outline the proof, which is elementary.

We will use the stratified structure of $(X, \mathbf{E}, \Phi)$ as we did in [LT2]. Let X_α, $\alpha \in I_0$ be all the strata of X that intersect $\Phi^{-1}(0)$. We give an order to I_0 so that for each $\alpha \in I_0$, $\Phi^{-1}(0) \cap \bigcup_{\beta\geq\alpha} X_\beta$ is compact. For convenience, we assume $I_0 = \{1, \cdots, k\}$.

To prove the lemma, we suffice to construct subsets $\mathcal{L}_\alpha^l \subset \mathcal{K}$, where $\alpha \in I_0$ and $0 \leq l \leq l_\alpha$, such that

$\mathbf{A}_1$: For any distinct $i, j \in \mathcal{L}_\alpha^l$, $\mathbf{U}_i \cap \mathbf{U}_j = \emptyset$;

$\mathbf{A}_2$: For each $\alpha \in I_0$ and $l \leq l_\alpha$, the set

$$\mathbf{Z}_\alpha^l = \Phi^{-1}(0) \cap X_\alpha - \cup_{j\geq l} \cup_{\beta\in\mathcal{L}_\alpha^j} \mathbf{U}_\beta$$

is contained in the image of finitely many manifolds of dimension less than l in X_α;

$\mathbf{A}_3$: For any pair of distinct $(i, j) \in \mathcal{L}_\alpha^l \times \mathcal{L}_{\alpha'}^{l'}$ with $(\alpha, l) < (\alpha', l')$, i.e. $\alpha \leq \alpha'$ or $\alpha = \alpha'$ and $l \leq l'$, the chart $(\tilde{\mathbf{U}}_i, G_i, \mathbf{E}_i, \mathbf{F}_i)$ is finer than $(\tilde{\mathbf{U}}_j, G_j, \mathbf{E}_j, \mathbf{F}_j)$.

In particular, all $\mathbf{U}_i$ with $i \in \bigcup_{0\leq l\leq l_\alpha} \mathcal{L}_\alpha^l$ cover $\Phi^{-1}(0) \cap X_\alpha$.

We will construct $\mathcal{L}_\alpha^l$ inductively, starting from the largest $\alpha \in I_0$. Note that $\mathcal{L}_\alpha^l$ may be empty. We now assume that for some $\alpha \in I_0$ we have constructed $\mathcal{L}_{\alpha+1}^l, \mathcal{L}_{\alpha+2}^l, \cdots$ for all l. We now construct $\mathcal{L}_\alpha^l$. We first pick a finite $\mathcal{L}' \subset \mathcal{K}$ such that $\{\mathbf{U}_i | i \in \mathcal{L}'\}$ covers $\mathbf{Z}_{\alpha+1}^0$. This is possible since it is compact. Let l_α be the maximum of $\{\dim(U_i \cap X_\alpha) | i \in \mathcal{L}'\}$. By assumption, $l_\alpha > 0$. Since $\tilde{\mathbf{U}}_i$ is a SS-space with strata $\pi_i^{-1}(\mathbf{U}_i \cap X_\alpha)$, we can find open subsets $\mathbf{U}_i' \subset \mathbf{U}_i$ such that (1) let $\tilde{\mathbf{U}}_i' = \pi_i^{-1}(\mathbf{U}_i)$ then $\tilde{\mathbf{U}}_i'^{-} \cap \pi_i^{-1}(X_\alpha)$ is the same as the closure of $\tilde{\mathbf{U}}_i' \cap \pi_i^{-1}(X_\alpha)$ in $\pi_i^{-1}(X_\alpha)$; (2) Let $R = \tilde{\mathbf{U}}_i' \cap U_i \cap \pi_i^{-1}(X_\alpha)$, then $R^- - R$ (in $U_i \cap \pi_i^{-1}(X_\alpha)$) is a smooth submanifold in $U_i \cap \pi_i^{-1}(X_\alpha)$ of dimension less than l_α. After fixing an ordering on $\mathcal{L}'$, we define $\mathcal{L}_\alpha^{l_\alpha}$ to be the collection of charts

$$(\tilde{\mathbf{U}}_i, G_i, \mathbf{E}_i, \mathbf{F}_i)|_{\mathbf{U}_i' \setminus \bigcup_{j<i} \mathbf{U}_j'^{-}}, \qquad i \in \mathcal{L}'. \tag{3.4}$$

Clearly, $\mathcal{L}_\alpha^{l_\alpha}$ satisfies $\mathbf{A}_1 - \mathbf{A}_3$. Now we assume that $\mathcal{L}_\alpha^{l_\alpha}, \cdots, \mathcal{L}_\alpha^l$ have been constructed. We want to construct $\mathcal{L}_\alpha^{l-1}$. For each $x \in \mathbf{Z}_\alpha^l$, we can find a finite approximation $(\tilde{\mathbf{U}}_i, G_i, \mathbf{E}_i, \mathbf{F}_i)$ such that it contains x and it is finer than all approximations in $\cup_{l'\geq l} \mathcal{L}_\alpha^{l'}$. We denote this collection by $\mathcal{L}'$ again. Since $\mathbf{Z}_\alpha^l$ is compact, we can pick a finite subcollection, still denote by $\mathcal{L}'$, such that $\{\mathbf{U}_i | i \in \mathcal{L}'\}$ covers $\mathbf{Z}_\alpha^l$. By our choice of the charts, $\mathbf{Z}_\alpha^l$ is contained in the image of finitely many submanifolds in X_α of dimension less than l. Next we choose $\mathbf{U}_i' \subset \mathbf{U}_i$ such that $\pi_i^{-1}(\mathbf{U}_i')$, $i \in \mathcal{L}'$, still cover $\mathbf{Z}_\alpha^l$ and that if we let $R = \pi_i^{-1}(\mathbf{U}_i') \cap \mathbf{Z}_\alpha^l$, then $R^- - R$ is a finite union of smooth submanifolds of dimensions $\leq l-1$. Now we fix an ordering on $\mathcal{L}'$, as before, and define $\mathcal{L}_\alpha^{l-1}$ to be the collection (3.4). By continuing this procedure for $l-2,\cdots,0$, we obtain the desired $\mathcal{L}_\alpha^l$ for $0 \leq l \leq l_\alpha$. This completes the proof of the Lemma. □

We now briefly sketch the strategy of the proof of the Theorem 3.13. Let $(\tilde{\mathbf{U}}_i, G_i, \mathbf{E}_i, \mathbf{F}_i)_{\alpha\in\mathcal{L}}$ be the collection of local finite approximations given by Lemma 3.14. We denote by $\phi_i : U_i \to F_i$ the corresponding finite models. As stratified set $U_i = \cup_{\alpha\in I_0} U_{i,\alpha}$, where I_0 is the ordered index set in the proof of Lemma 3.14. Since $\Phi^{-1}(0)$ is compact, we can find open $\tilde{\mathbf{W}}_i \subset \tilde{\mathbf{U}}_i$ with $\pi_i^{-1}(\pi_i(\tilde{\mathbf{W}}_i)) = \tilde{\mathbf{W}}_i$ such that $\cup\{\pi_i(\tilde{\mathbf{W}}) \mid i \in \mathcal{L}\}$ covers $\Phi^{-1}(0)$ and $\tilde{\mathbf{W}}_i \cap U_i$ is precompact in U_i.[11] Note that by assumption, we can assume that for each $\alpha \in I_0$, $f|_{U_{i,\alpha}} : U_{i,\alpha} \to Y_\alpha$ is a submersion.

Before we perturb the sections ϕ_i, we need to clarify by which we mean a perturbation is generic. We begin with the general situation. Let $p: U \to V$ be a smooth map between two open smooth manifolds. Let $K \subset U$ be a pre-compact subset and $s \in \Gamma_U(E)$ is a smooth section of a smooth vector bundle over U. We say t is a p-generic perturbation of s relative to K with compact support if the following holds. (1) The support of $s-t$ is pre-compact in U and (2) there is a precompact open neighborhood K^{nbfd} of $K^- \subset U$ so that the graph of t is transversal to the zero section in K^{nbhd} and $p: K^{\mathrm{nbhd}} \cap t^{-1}(0) \to V$ is a strong immersion. Here by a strong immersion $p: A \to B$ between open manifolds we mean p is an immersion and further for any $x, y \in A$ so that $p(x) = p(y)$ then $p_* T_x A$ is transversal to $p_* T_y A$.

LEMMA 3.15. *Let the situation be as before. Suppose dim $U-$ rank $E <$ dim V and $f: U \to V$ is a submersion. Then for any pre-compact $K \subset U$ and $s \in \Gamma_U(E)$, the space of f-generic perturbations of s relative to K with compact support, denoted B_0, is dense and open in the space $B = \{h \in \Gamma_U(E) \mid h - s$ is compact$\}$ with Whitney C^∞ topology.*

PROOF. The proof is an easy application of the ordinary transversality theorem. It is clear that $B_0 \subset B$ is open. It remains to show it is dense. First, by the usual transversality theorem we can find a small perturbation t of s with compact support $t-s$ such that the graph of t is transversal to the zero section in a neighborhood of K^-. We then can find a diffeomorphism $\varphi: U \to U$ sufficiently close to Id_U with $\{x \in U \mid \varphi(x) \neq x\}$ precompact in U so that $p: \varphi(t^{-1}(0)) \to V$ is a strong submersion in a neighborhood of K^-. We pick an isomorphism $\phi: E \cong \varphi^* E$ so that it is close to $E \equiv E$ and away from a compact set $L \supset K$ of U the isomorphism ϕ reduces to $E \equiv E$. Here of course we assume $\varphi|_{U-L} = \mathrm{Id}_{U-L}$. Then the corresponding section $\phi^*(t)$ is a small perturbation of t and is in B_0. This proves the Lemma. □

We now show how to construct the perturbations of ϕ_i one at a time. Without loss of generality, we assume $r \leq \; dim Y$, since otherwise $H_r(Y) = 0$. We begin with the first member in $\mathcal{L}$, which is 1. To make the notation easy to follow, we denote the finite model $\phi_1 : U_1 \to F_1$ by $\phi: U \to F$, $\tilde{\mathbf{W}}_1 \subset \tilde{\mathbf{U}}_1$ by $\tilde{\mathbf{W}}$ and $U_{1,\alpha}$ by U_α. We first look at the largest $\alpha \in \mathcal{I}$ so that $U_\alpha \neq \emptyset$. Assume it is not dense in U. Recall $f: X \to Y$ is the map given and $f: U_\alpha \to Y_\alpha$ is a submersion. We then pick a small f-generic perturbation of $\phi|_{U_\alpha}$ relative to $U_\alpha \cap \tilde{\mathbf{W}}^-$ of compact

[11] By this we mean its closure is compact.

support, denoted by $\phi_\alpha : U_\alpha \to Y_\alpha$. Since U_α is not dense in U, by assumption, it has codimension at least 2. Thus

$$\dim f(\phi_\alpha^{-1}(0) \cap \tilde{\mathbf{W}}^-) \le r - 2.$$

Since Y is a smooth orbifold, we can find a closed neighborhood $L_\alpha \subset Y$ of $f(\phi_\alpha^{-1}(0) \cap \tilde{\mathbf{W}}^-)$ such that it has homological dimension $\le r-2$ and its boundary is a smooth orbifold in Y. Let $M_\alpha = f^{-1}(L_\alpha)$.

Let $\beta \in \mathcal{L}$ is the next largest element so that $U_\beta \ne \emptyset$. We next extend ϕ_α to U_β. Since the vanishing locus of ϕ_α in $\tilde{\mathbf{W}}^- \cap U_\beta$ is contained in the interior point of $M_\alpha \cap U_\beta$, we can extend ϕ_α to a neighborhood of $U_\alpha \subset U_\alpha \cup U_\beta$ so that its zero locus in $\tilde{\mathbf{W}}$ is still in $M_\alpha \cap (U_\alpha \cup U_\beta)$. Hence, we can find a small perturbation of $\phi|_{U_\alpha \cup U_\beta}$ so that it is an extension of ϕ_α, it is an f-generic perturbation of $\phi|_{U_\alpha \cup U_\beta}$ relative to $\tilde{\mathbf{W}} - M_\alpha$ with compact support. Let the new section be $\phi_\beta \in \Gamma_{U_\alpha \cup U_\beta}(F)$. Hence $f : \phi_\beta^{-1}(0) \cap \tilde{\mathbf{W}}_\alpha^- \to Y$ is a strong immersion away from L_α. Without loss of generality, we can assume that $f(\phi_\beta^{-1}(0))$ intersects the boundary of L_α transversely, in the sense of orbifold. Since it has dimension $\le r - 2$, assuming U_β is not dense either, we can find a closed strong retraction neighborhood L'_β of $f(\phi_\beta^{-1}(0) \cap \tilde{\mathbf{W}}^-) - L_\alpha$ in Y. Hence $L_\alpha \cup L'_\beta$ will have homological dimension $\le r-2$ as well. We let $L_\beta = L_\alpha \cup L'_\beta$ and $M_\beta = f^{-1}(L_\beta)$. Continue this procedure, we can find a closed neighborhood $L \subset Y$ of homological dimension $\le r - 2$ and an f-generic perturbation of ϕ relative to $\tilde{\mathbf{W}} \cap U - f^{-1}(L)$ of compact support. Let the perturbed section be $\tilde{\phi}_1$. Let $U_1^n \subset U = U_1$ be a pre-compact neighborhood containing $U \cap \tilde{\mathbf{W}}^-$ so that $\tilde{\phi}$ is transversal to the zero section in $U_1^n - L$. Let $n_1 =$ the product of the number of the sheets of the covering $\tilde{\mathbf{U}}_1/G_2 \to \mathbf{U}_1$ with the order of G_1. We define $\Delta_1 = (1/n_1)[\pi_1(\tilde{\phi}_1^{-1}(0))]$, considered as a rational pseudomanifold.

Next, we continue this procedure to U_2. Let U_{12} be $\varphi_{12}^{-1}(U_1)$, where φ_{12} is defined in (3.3). Then $\varphi_{12}^{-1}(U_1 \cap U_1^n - L_1)$ is a smooth submanifold of U_2. Hence, we can extend $\varphi_{12}^*(\tilde{\phi})$ to a neighborhood of R of $U_{12} \subset U_2$, say ϕ', so that it is a small f-generic perturbation of ϕ_2 relative to $R - f^{-1}(L)$ with compact support. We then extend this small perturbation to whole U_2, say $\tilde{\phi}_2$, so that there is a subset $L_2 \subset Y$ of homological dimension $\le r - 2$ so that $\tilde{\phi}_2$ is an extension of $\varphi_{12}^*(\tilde{\phi}_1)$, it is an f-generic perturbation of ϕ_2 relative to $\tilde{\mathbf{W}}_2^- - f_2^{-1}(L_2)$ with compact support. Let $\Delta = (1/n_2)[\pi_2(\tilde{\phi}_2^{-1}(0))]$. Clearly, $\Delta_1 = \Delta_2$ when restricted to $\mathbf{U}_1 \cap \mathbf{U}_2 - f^{-1}(L_2)$.

Continue this procedure until we reach the largest $m \in \mathcal{L}$. We let $L_m \subset Y$ be the corresponding closed neighborhood of homological dimension $\le r-2$. Then by our construction, for each $i, j \in \mathcal{L}$,

$$\Delta_i|_{\mathbf{U}_i \cap \mathbf{U}_j - f^{-1}(L_m)} = \Delta_j|_{\mathbf{U}_i \cap \mathbf{U}_j - f^{-1}(L_m)}$$

as pseudomanifolds. We let Δ be the patch together of these Δ_i. Since $\{\tilde{\mathbf{W}}_i \mid i \in \mathcal{L}\}$ covers $\Phi^{-1}(0)$, if we choose the perturbations $\tilde{\phi}_i$ sufficiently close to ϕ_i for all i, then the pair (Δ, L_m) will be a weakly f-pseudocycle. This cycle is the Euler class of $(X, \mathbf{E}, \Phi)$.

3.5. Stable maps. In this section, we review the definition of stable maps.

Let X be a smooth symplectic manifold with a given symplectic form ω, and let $A \in H_2(X, \mathbb{Z})$ and let $g, n \in \mathbb{Z}$ be fixed once and for all. In the following, by a k-pointed prestable curve, we mean a connected complex curve with k marked points and only normal crossing singularity.

We recall the notion of stable C^2-maps [**LT2**, Definition 2.1].

DEFINITION 3.16. *A k-pointed stable map is a collection $(f, \Sigma, x_1 \ldots, x_k)$, where $(\Sigma, x_1, \ldots, x_k)$ is a k-pointed connected prestable curve and $f: \Sigma \to X$ is a continuous map such that*
(1) the composite $f \circ \pi$ is C^2-smooth, where $\pi: \tilde{\Sigma} \to \Sigma$ is the normalization of Σ and
(2) any component $R \subset \tilde{\Sigma}$ satisfying $(f \circ \pi)_*([R]) = 0 \in H_2(X, \mathbb{Z})$ must have $3 \geq 2g(R)+$ the number of distinguished points on R, where the set of distinguished points on $\tilde{\Sigma}$ consists of all preimages of the marked points and the nodal points of Σ.

For convenience, we will abbreviate $(f, \Sigma, x_1, \ldots, x_k)$ to $(f, \Sigma, (x_i))$ or to (f, Σ) or simply to f, if no confusion will arise. We call Σ the domain of f, with marked points understood. Two stable maps $(f, \Sigma, (x_i))$ and $(f', \Sigma', (x_i'))$ are said to be equivalent if there is an isomorphism $\rho: \Sigma \to \Sigma'$ such that $f' \circ \rho = f$ and $x_i' = \rho(x_i)$. When $(f, \Sigma) \equiv (f', \Sigma')$, such a ρ is called an automorphism of (f, Σ). We will denote by Aut_f the group of automorphisms of (f, Σ).

We let $\mathbf{B}^X_{A,g,k}$ be the space of equivalence classes $[f, \Sigma]$ of C^2-stable maps (f, Σ) such that the arithmetic genus of Σ is g and $f_*([\Sigma]) = A \in H_2(X; \mathbb{Z})$. With X, A, g and k understood, we will simply write $\mathbf{B}^X_{A,g,k}$ as $\mathbf{B}$. There is a natural topology on $\mathbf{B}$, which we will define at Section 3.7. However, $\mathbf{B}$ is not a smooth Banach manifold as one hopes. It does admit a stratification with smooth strata, which we now describe.

Given any almost complex structure J compatible with ω, one can define a generalized bundle $\mathbf{E}$ over $\mathbf{B}$ as follows. Let (f, Σ) be any stable map and let $\tilde{f}: \tilde{\Sigma} \to X$ be the composite of f with $\pi: \tilde{\Sigma} \to \Sigma$. We define $\Lambda^{0,1}_f$ to be the space of all C^1-smooth sections of $(0,1)$-forms of $\tilde{\Sigma}$ with values in $\tilde{f}^* TX$. Here, by a $\tilde{f}^* TX$-valued $(0,1)$-form we mean a section ν of $T\tilde{\Sigma} \otimes \tilde{f}^* TX$ over $\tilde{\Sigma}$ such that $J \cdot \nu = -\nu \cdot \mathbf{j}$, where $\mathbf{j}$ denotes the complex structure of $\tilde{\Sigma}$. Assume that (f, Σ) and (f', Σ') are two equivalent stable maps with the associated isomorphism $\rho: \Sigma \to \Sigma'$, then there is a canonical isomorphism $\Lambda^{0,1}_{f'} \cong \Lambda^{0,1}_f$. It follows that the automorphism group Aut_f acts on $\Lambda^{0,1}_f$ and the quotient $\Lambda^{0,1}_f / \mathrm{Aut}_f$ is independent of the choice of the representative in the equivalence class of (f, Σ). So we can define $\Lambda^{0,1}_{[f]}$ to be $\Lambda^{0,1}_f / \mathrm{Aut}_f$. Put

$$\mathbf{E} = \bigcup_{[f] \in \mathbf{B}} \Lambda^{0,1}_{[f]}.$$

It has an obvious projection $\mathbf{P}: \mathbf{E} \to \mathbf{B}$ whose fibers are finite quotients of infinite dimensional linear spaces.

There is a natural map

$$\Phi_J : \mathbf{B} \longrightarrow \mathbf{E} \quad \text{satisfying} \quad \mathbf{P} \circ \Phi_J = \mathrm{Id}_{\mathbf{B}},$$

defined as follows: For any stable map f, we define $\Phi_J(f)$ to be the image of $d\tilde{f} + J \cdot d\tilde{f} \cdot \mathbf{j} \in \Lambda^{0,1}_f$ in $\Lambda^{0,1}_{[f]}$. Obviously, we have $\Phi_J(f) = \Phi_J(f')$ if f and f' are equivalent. Thus Φ_J descends to a map $\mathbf{B} \to \mathbf{E}$, which we still denote by Φ_J.

We will denote by $\mathfrak{M}^X_{A,g,k}$ the moduli space of J-holomorphic stable maps with homology class A, having k-marked points and whose domains have arithmetic genus g. Clearly, $\mathfrak{M}^X_{A,g,k} = \Phi_J^{-1}(0)$.

3.6. Stratifying the space of stable maps. In this section, we give a natural stratification of $\mathbf{B} = \mathbf{B}^X_{A,g,n}$ and study its basic properties.

Given a stable map (f, Σ), we can associate to it a dual graph Γ_f as follows: Each irreducible component Σ_α of Σ corresponds to a vertex v_α in Γ_f together with a marking (g_α, a_α), where g_α is the geometric genus of Σ_α and a_α is the homology class $f_\alpha([\Sigma_\alpha])$ in X; For each marked point x_i of Σ_α we attach a leg to v_α; For each intersection point of distinct components Σ_α and Σ_β we attach an edge joining v_α and v_β and for each self-intersection point of Σ_α we attach a loop to the vertex v_α. In the following, we will denote by $\mathrm{Ver}(\Gamma)$ the set of all vertices of Γ and $\mathrm{Ed}(\Gamma)$ the set of all edges of Γ.

Clearly, the dual graph Γ_f is independent of the representatives in $[f]$, so we can simply denote the dual graph of $[f]$ by $\Gamma_{[f]} = \Gamma_f$. Moreover, the genus g of $[f]$ is the same as the genus of $\Gamma_{[f]}$, which is defined to be the sum of g_α for all $\alpha \in \mathrm{Ver}(\Gamma)$ and the number of holes in the graph $\Gamma_{[f]}$. Also, the homology class a of $f_*([\Sigma])$ is the sum of all a_α, which is defined to be the homology class of the graph $\Gamma_{[f]}$.

Given any graph Γ with genus g and homology class a, we let $\mathbf{B}(\Gamma)$ be the space of all equivalence classes $[f]$ of stable maps in $\mathbf{B}$ with $\Gamma_{[f]} = \Gamma$. Clearly, $\mathbf{B}$ is a disjoint union of all $\mathbf{B}(\Gamma)$.

The main result of this section is the following.

PROPOSITION 3.17. *For each dual graph Γ the space $\mathbf{B}(\Gamma)$ is a smooth Banach orbifold and the restriction of $\mathbf{E}$ to $\mathbf{B}(\Gamma)$ is a smooth orbifold bundle. Further the restriction of Φ_J to $\mathbf{B}(\Gamma)$ is a smooth orbifold section.*

PROOF. We first prove that $\mathbf{B}(\Gamma)$ is a Banach orbifold.

It suffices to construct the neighborhoods of $[f]$ in $\mathbf{B}(\Gamma)$. Let $f = (f, \Sigma, (x_i))$ and let $\tilde{\Sigma}$ be the normalization of Σ. Then

$$\tilde{\Sigma} = \bigcup \left\{ \tilde{\Sigma}_\alpha \mid \alpha \in \mathrm{Ver}(\Gamma) \right\}, \tag{3.5}$$

where $\tilde{\Sigma}_\alpha$ is the normalization of the component Σ_α corresponding to α. Recall that the distinguished points of $\tilde{\Sigma}_\alpha$ are preimages of marked points and the nodal points of Σ. Note that each edge $e \in Ed(\Gamma)$ corresponds to a node in Σ, thus has two distinguished points in $\tilde{\Sigma}$. We denote them by z_{e1} and z_{e2}.

To avoid any confusion, we denote by $\tilde{\Sigma}^{\mathrm{top}}$ the underlining real 2-dimensional manifold of $\tilde{\Sigma}$. A complex structure on $\tilde{\Sigma}^{\mathrm{top}}$ is given by an almost complex structure which is a homomorphism $\mathbf{j} : T\tilde{\Sigma}^{\mathrm{top}} \mapsto T\tilde{\Sigma}^{\mathrm{top}}$ with $\mathbf{j}^2 = -\mathrm{Id}$. Two almost complex structures $\mathbf{j}$ and $\mathbf{j}'$ give rise to the same complex structure if and only if there is a diffeomorphism ϕ of $\tilde{\Sigma}^{\mathrm{top}}$ such that $\mathbf{j}' = d\phi \cdot \mathbf{j} \cdot (d\phi)^{-1}$. Let $\mathbf{j}_t$ be a family of almost complex structures such that $\mathbf{j}_0$ is the given complex structure of $\tilde{\Sigma}$. Then the derivative $\mathbf{v}(\mathbf{j}_0) := (d/dt\, \mathbf{j}_t)_{t=0}$ is a $T\tilde{\Sigma}$-valued (0,1)-form. If $\mathbf{j}'_t$ is another family of almost complex structures such that each $\mathbf{j}'_t$ induces the same complex structure as $\mathbf{j}_t$ does, then the corresponding (0,1)-form $\mathbf{v}(\mathbf{j}'_0)$ can be written as $\mathbf{v}(\mathbf{j}_0) + \bar{\partial} u$ for some section u of $T\tilde{\Sigma}$. This shows that local complex deformations of $\tilde{\Sigma}$ are parameterized by an open subset of $H^1_{\tilde{\Sigma}}(T\tilde{\Sigma})$. Thus a local universal family $\tilde{\mathcal{U}}$

of marked curve $(\tilde{\Sigma}, (\tilde{x}_i), (z_{e1}.z_{e2}))$, where $\tilde{x}_i \in \tilde{\Sigma}$ is the preimage of $x_i \in \Sigma$ and $e \in \mathrm{Ed}(\Gamma)$, is of the form

$$Q \times \prod_i P_i \times \prod_e P_{e1} \times P_{e2}, \tag{3.6}$$

where Q, P_i, P_{e1} and P_{e2} are small neighborhoods of 0 in $H^1_{\tilde{\Sigma}}(T\tilde{\Sigma})$ and small neighborhoods of $\tilde{x}_i$, z_{e1} and z_{e2} in $\tilde{\Sigma}$, respectively. Here as usual $(\tilde{x}_i)$ is the k-marked points and (z_{e1}, z_{e2}) denotes $2 \cdot \#\mathrm{Ed}(\Gamma)$-marked points, after fixing an ordering of $\mathrm{Ed}(\Gamma)$.

By the Serre duality, we have $H^1_{\tilde{\Sigma}}(T\tilde{\Sigma}) = H^0_{\tilde{\Sigma}}(T^*\tilde{\Sigma}^{\otimes 2})$. The latter is the space of holomorphic quadratic differential forms.

Let $\mathcal{V}$ be a small neighborhood of 0 in the space of $\tilde{f}^*TX$-valued vector fields along $\tilde{\Sigma}$ in the C^2-topology. Here $\tilde{f}: \tilde{\Sigma} \to X$. For each $v \in \mathcal{V}$, we can associate to it a unique C^2-map $\exp_{\tilde{f}}(v)$ from $\tilde{\Sigma}$ into X, where exp is the exponential map of a fixed metric on X.

We denote by G_f the automorphism group of f. It is a finite group and acts naturally on $H^1_{\tilde{\Sigma}}(T\tilde{\Sigma})$. This is because each $\sigma \in G_f$ lifts naturally to an automorphism of $\tilde{\Sigma}$, so it induces an action on $H^1_{\tilde{\Sigma}}(T\tilde{\Sigma})$. Without loss of generality, we can assume that both Q and $\prod_i P_i \times \prod_e P_{e1} \times P_{e2}$ are G_f-invariant. Now we define a natural G_f-action on $\tilde{\mathcal{U}} \times \mathcal{V}$: Any point in $\tilde{\mathcal{U}} \times \mathcal{V}$ is of the form

$$(q, (x_i'), (z_{e1}', z_{e2}'), v), \quad q \in Q, x_i' \in P_i, (z_{e1}', z_{e2}') \in P_{e1} \times P_{e2} \text{ and } v \in \mathcal{V}.$$

Given any $\sigma \in G_f$, we define

$$\sigma(q, (x_i'), (z_{e1}', z_{e2}'), v) = (\sigma_*(q), (\sigma(x_i')), (\sigma(z_{e1}'), \sigma(z_{e2}')), v \cdot \sigma^{-1}).$$

Next we construct a product manifold X_Γ as follows: for each edge or loop e of Γ, we associate to it two copies of X, say X_{e1}, X_{e2}. Define $X_\Gamma = \prod_e X_{e1} \times X_{e2}$. We also define $\Delta_\Gamma = \prod_e \Delta_e$, where Δ_e is the diagonal of $X_{e1} \times X_{e2}$. We can define a smooth evaluation map $\pi: \tilde{\mathcal{U}} \times \mathcal{V} \to X_\Gamma$ by

$$\pi(q, (x_i'), (z_{e1}', z_{e2}'), v) = \prod_e \left(\exp_{\tilde{f}}(v)(z_{e1}), \exp_{\tilde{f}}(v)(z_{e2})\right).$$

One can easily show that π is transversal to Δ_Γ. So $\tilde{\mathbf{W}} = \pi^{-1}(\Delta_\Gamma)$ is smooth. The finite group G_f acts on $\tilde{\mathbf{W}}$ smoothly, thus $\tilde{\mathbf{W}}/G_f$ is a neighborhood of f in $\mathbf{B}(\Gamma)$. This proves the first part of the Proposition.

We now prove the remainder part of the Proposition. Note that each point in $\tilde{\mathbf{W}}$ is represented by a stable map $(f, \Sigma, (x_i))$ and $(\tilde{\Sigma}, (x_i), (z_{e1}, z_{e2})) \in \tilde{\mathcal{U}}$. Let $\tilde{\mathbf{E}}_f$ be the set of all C^1-smooth, $\tilde{f}^*TX$-valued (0,1)-forms over $\tilde{\Sigma}$. Put

$$\tilde{\mathbf{E}}_{\tilde{\mathbf{W}}} = \bigcup_{f \in \tilde{\mathbf{W}}} \tilde{\mathbf{E}}_f.$$

Clearly, $\tilde{\mathbf{E}}_{\tilde{\mathbf{W}}}$ is a smooth Banach bundle, and G_f, which acts on $\tilde{\mathbf{W}}$, lifts to a linear action on $\tilde{\mathbf{E}}_{\tilde{\mathbf{W}}}$ such that the natural projection $\tilde{\mathbf{E}}_{\tilde{\mathbf{W}}} \to \tilde{\mathbf{W}}$ is G_f-equivariant. Moreover, $\mathbf{E}|_{\mathbf{W}} = \tilde{\mathbf{E}}_{\tilde{\mathbf{W}}}/G_f$, so $\mathbf{E}$ is an orbifold bundle over $\mathbf{B}$.

Finally, the section Φ_J lifts to a section $\tilde{\Phi}_{\mathbf{W}}$ over $\tilde{\mathbf{W}}$ defined by

$$\tilde{\Phi}_{\mathbf{W}}(f) = d\tilde{f} + J \cdot d\tilde{f} \cdot \mathrm{j}_\Sigma \in \tilde{\mathbf{E}}_f.$$

Obviously, it is smooth. This proves the Proposition. □

3.7. Topology of the space of stable maps. Now it is time to describe the topology of **B**. Let $[f]$ be any stable map in **B** represented by a stable map $(f, \Sigma, (x_i))$. We want to construct the neighborhoods of $[f]$ in **B**. Let Γ be the dual graph of f. For each $\alpha \in \mathrm{Ver}(\Gamma)$, we define k_α to be the number of edges and legs attached to it. Here we count a loop to α as two edges. As before we still denote by Σ_α the corresponding component and $\tilde{\Sigma}_\alpha$ its normalization. We define $\mathrm{Ver}_u(\Gamma) \subset \mathrm{Ver}(\Gamma)$ to be $\{\alpha \in \mathrm{Ver}(\Gamma) \mid k_\alpha < 3 \text{ and } g_\alpha = 0\}$ and define $\mathrm{Ver}_s(\Gamma)$ the complement of $\mathrm{Ver}_u(\Gamma)$. When $\alpha \in \mathrm{Ver}_u(\Gamma)$, then $\tilde{\Sigma}_\alpha$ contains one or two distinguished points. We add two or one marked point(s) to Σ_α according to whether $\tilde{\Sigma}_\alpha$ contains one or two distinguished point. We also require that at these added points the curve Σ is smooth and the differential df is injective. Note that this is always possible since $f_*([\Sigma_\alpha]) \neq 0$, by the stability of f. We denote by $(y_j)_{1 \leq j \leq l}$ the set of all added points to Σ. Then $(\Sigma, (x_i, y_j))$ is a Deligne-Mumford stable curve with $k + l$ marked points.

As a stable curve with $k+l$ marked points, local deformations of $(\Sigma, (x_i, y_j))$ are parameterized by admissible quadratic differentials on Σ. An admissible quadratic differential q is a meromorphic quadratic differential with at most simple poles at x_i or y_j and double poles at nodes satisfying: if w_1, w_2 are local coordinates of Σ near a node, i.e., Σ is defined by $w_1 w_2 = 0$ in $\mathbb{C}^2$ near such a node, then

$$\lim_{w_1 \to 0, w_2 = 0} w_1^2 \frac{q}{dw_1^2} = \lim_{w_2 \to 0, w_1 = 0} w_2^2 \frac{q}{dw_2^2}.$$

Neighborhoods of $(\Sigma, (x_i, y_j))$ in $\overline{\mathfrak{M}}_{g,k+l}$ can be constructed as follows: let $\tilde{G}_f$ be the automorphism group of $(\Sigma, (x_i, y_j))$, then a neighborhood $\mathcal{U}$ of $(\Sigma, (x_i, y_j))$ in $\overline{\mathfrak{M}}_{g,k+l}$ is of the form $\tilde{\mathcal{U}}/\tilde{G}_f$, where $\tilde{\mathcal{U}}$ is a small neighborhood of the origin in the space of admissible quadratic differentials.

For each y_i, we choose a codimension two submanifold $H_j \subset X$ such that H_j intersects $f(\Sigma)$ uniquely and transversely at $f(y_j)$. We orient H_j so that it has positive intersection with $f(\Sigma)$.

We fix a compact set $K \subset \Sigma \backslash Sing(\Sigma)$ containing all marked points (x_i, y_j). We may assume that K is G_f-invariant. Let $\mathcal{CU}$ be the universal curve over $\tilde{\mathcal{U}}$. We then choose a diffeomorphism ϕ from a neighborhood of $K \times \tilde{\mathcal{U}}$ into $\mathcal{CU}$ such that F preserves fibers over $\tilde{\mathcal{U}}$ and restricts to the identity map on $K \times \{(\Sigma, (x_i, y_j))\}$. We also fix a $\delta > 0$.

To each collection of $\mathcal{U}$, H_j, K, δ and ϕ given as above, we can associate to it a neighborhood $\mathbf{U} = \mathbf{U}(\mathcal{U}, H_j, K, \delta, \phi)$ as follows: define $\tilde{\mathbf{U}} = \tilde{\mathbf{U}}(\mathcal{U}, H_j, K, \delta, \phi)$ to be the set of all tuples $(f', \Sigma', (x_i', y_j'))$ satisfying: (1) $(\Sigma', (x_i', y_j'))$ is in $\tilde{\mathcal{U}}$; (2) f' is a continuous map from Σ' into X with $f'(y_j) \in H_j$; (3) f' lifts to a C^2-map $\tilde{\Sigma}' \to X$; (4) $\|f' \cdot \phi - f\|_{C^2(K)} < \delta$; (5) $d_X(f(\Sigma), f'(\Sigma')) < \delta$, where d_X denotes the distance function of a fixed Riemannian metric on V. Note that the topology of Σ' may be different from that of Σ. We define $\tilde{\mathbf{U}}(\mathcal{U}, H_j, K, \delta, \phi)$ to be a local uniformization chart of **B**.

Given any tuple $(f', \Sigma', (x_i', y_j'))$ in $\tilde{\mathbf{U}}$, we call $\mathbf{f}' = (f', \Sigma', (x_i'))$ its descendant. One can show that $\mathbf{f}'$ is a stable map and gives rise to a point $[\mathbf{f}']$ in **B**. Let **U** be the set of all equivalence classes of stable maps descended from tuples in $\tilde{\mathbf{U}}$. Then **U** is a neighborhood of $[f]$.

The topology of **B** is generated by all such neighborhoods $\mathbf{U}(\mathcal{U}, H_j, K, \delta, \phi)$.

Now assume that δ and $\tilde{\mathcal{U}}$ are sufficiently small. Then there is a natural action of G_f on $\tilde{\mathbf{U}}$: let $\sigma \in G_f$ and $\tilde{\mathbf{f}}' = (f', \Sigma', (x_i', y_j')) \in \tilde{\mathbf{U}}$, then σ acts on $\tilde{\mathcal{U}}$ by sending Σ' to $\sigma(\Sigma')$ and x_i to $\sigma(x_i)$. We define

$$\sigma(\tilde{\mathbf{f}}') = (f' \cdot \sigma^{-1}, \sigma(\Sigma'), (\sigma(x_i'), y_j'')),$$

where for each j, y_j'' is the unique point in $\sigma(\Sigma')$ near y_j such that $f'(\sigma^{-1}(y_j'')) \in H_j$. The existence and uniqueness of such y_j'' is assured by the assumption that H_j are transversal to $f(\Sigma)$. Note that y_j'' may not be y_j'. Clearly, $\tilde{\mathbf{f}}'$ and $\sigma(\tilde{\mathbf{f}}')$ descend to the identical $[\mathbf{f}']$ in $\mathbf{B}$. Conversely, if $\tilde{\mathbf{f}}' = (f', \Sigma', (x_i', y_j'))$ and $\tilde{\mathbf{f}}'' = (f'', \Sigma'', (x_i'', y_j''))$ descend to the same $[\mathbf{f}']$ in $\mathbf{U}$, then there is a biholomorphic map $\tau : \Sigma' \mapsto \Sigma''$ such that $f' = f'' \circ \tau$ and $\tau(x_i') = x_i''$. When δ and $\tilde{\mathcal{U}}$ are sufficiently small, we may assume that τ induces a biholomorphic map, denoted by σ, of $(\Sigma, (x_i))$. This σ acts on $\tilde{\mathbf{U}}$ as defined above. Thus one can easily see that $\sigma(\mathbf{f}') = \mathbf{f}''$. It follows that $\mathbf{U}$ is of the form $\tilde{\mathbf{U}}/G_f$. Therefore, the neighborhoods $\mathbf{U}(\mathcal{U}, H_j, K, \delta, \phi)$ are the quotients of local uniformization charts $\tilde{\mathbf{U}}(\mathcal{U}, H_j, K, \delta, \phi)$. This proves

THEOREM 3.18. *Equiped with the topology described above, $\mathbf{B}$ is a SS-orbispace.*

Moreover, we have

THEOREM 3.19. *Assume that (X, ω) is a compact manifold with a compatible almost complex structure J. Then the moduli space $\mathfrak{M}^X_{A,g,k}$ is compact in $\mathbf{B}$ in the above topology.*

The compactness theorem of this sort first appeared in the work of Gromov [Gr], further studied and extended by Parker and Wolfson, Pansu and Ye. A proof was also given in [RT1] for holomorphic maps of any genus, following Sacks and Uhlenbeck on harmonic maps in early 80's.

Given a local uniformization chart $\tilde{\mathbf{U}}$, we define

$$\mathbf{E}_{\tilde{\mathbf{U}}} = \bigcup_{(f', \Sigma', (x_i', y_j')) \in \tilde{\mathbf{U}}} \tilde{\mathbf{E}}_{(f', \Sigma', (x_i', y_j'))},$$

where $\tilde{\mathbf{E}}_{(f', \Sigma', (x_i', y_j'))}$ consists of all C^1-smooth, f'^*TX-valued (0,1)-forms over the normalization of Σ'. All such $\mathbf{E}_{\tilde{\mathbf{U}}}$'s form charts of $\mathbf{E}$. One can show that the conditions in Definition 3.7 are all satisfied. Therefore, we have proved

THEOREM 3.20. *$(\mathbf{B}, \mathbf{E}, \Phi_J)$ is a V-vector bundle.*

4. Proof of the main theorem

By Theorem 3.13, in order to prove our main theorem, we need to show that $(\mathbf{B}, \mathbf{E}, \Phi_J)$ is a WFV-bundle. By the results in the last section, we suffice to construct local finite approximations of $(\mathbf{B}, \mathbf{E}, \Phi_J)$.

We continue to use the notations developed so far. Let $\tilde{\mathbf{U}}$ be a chart of $\mathbf{B}$ with the corresponding group G, and let $\mathbf{E}_{\tilde{\mathbf{U}}}$ be the corresponding chart of $\mathbf{E}$ over $\tilde{\mathbf{U}}$. We know that $\tilde{\mathbf{U}}$ is of the form $\tilde{\mathbf{U}}(\mathcal{U}, H_j, K, \delta, \phi)$. Let $\tilde{\mathcal{U}}$ be the local uniformization of $\mathcal{U}$ and $C\mathcal{U}$ be the universal curve over $\tilde{\mathcal{U}}$. We may assume that $\mathcal{U}$ is sufficiently small.

Recall that a TX-valued (0,1)-form on $\mathcal{CU} \times X$ is an endomorphism $v : T\mathcal{CU} \mapsto TX$ such that $J \cdot v = -v \cdot \mathbf{j}_{\mathcal{CU}}$, where $\mathbf{j}_{\mathcal{CU}}$ is the complex structure on $\mathcal{CU}$. Let $\Lambda^{0,1}(\mathcal{CU}, TX)_0$ be the space of all C^∞-smooth TX-valued (0,1)-forms on $\mathcal{CU} \times X$ which vanish near Sing($\mathcal{CU}$). Here Sing($\mathcal{CU}$) denotes the set of the singularities in the fibers of $\mathcal{CU}$ over $\tilde{\mathcal{U}}$.

Given each $v \in \Lambda^{0,1}(\mathcal{CU}, TX)_0$, we can associate a section of $\mathbf{E}_{\tilde{\mathbf{U}}}$ as follows: For each $\mathbf{f} = (f, \Sigma, (x_i, y_j)) \in \tilde{\mathbf{U}}$, we define $v|_{\mathbf{f}}$ by

$$v|_{\mathbf{f}}(x) = v(x, f(x)), \; x \in \Sigma.$$

Clearly, $v|_{\mathbf{f}}$ is a section on the fiber of $\mathbf{E}_{\tilde{\mathbf{U}}}$ over $\mathbf{f}$. This way we obtain a section $\mathbf{f} \mapsto v|_{\mathbf{f}}$ over $\tilde{\mathbf{U}}$. To avoid introducing new notations, we still denote this section by v. Now assume G_f is non-trivial. Then for $\sigma \in G_f$ the pull-back $\sigma^*(v_i)$ is a section over $\tilde{\mathbf{U}}$. Without loss of generality, we can assume the $l \cdot |G_f|$ sections

$$\{\sigma^*(v_i) \mid 1 \le i \le l, \sigma \in G_f\}$$

of $\mathbf{E}_{\tilde{\mathbf{U}}}$ is linearly independent everywhere. We define $\mathbf{F} = \mathbf{F}(v_1, \cdots, v_l)$ to be the subbundle in $\mathbf{E}_{\tilde{\mathbf{U}}}$ generated by the above $l \cdot |G_f|$ sections. $\mathbf{F}$ is a trivial vector bundle and is a G_f-equivariant subbundle of $\mathbf{E}_{\tilde{\mathbf{U}}}$.

LEMMA 4.1. *Suppose* $\mathbf{f} = (f, \Sigma, (x_i, y_j)) \in \tilde{\mathbf{U}}$ *and* $\mathrm{Sp}(v_1, \cdots, v_l)|_{\mathbf{f}}$ *are transverse to* L_f*, where* L_f *is the linearization of the Cauchy-Riemann equation at* f*. We further assume that in the definition of* $\tilde{\mathbf{U}}$*,* δ *is sufficiently small and* K *is sufficiently big. Then* $\Phi_J^{-1}(\mathbf{F})$ *is a smooth manifold of dimension* $r + l$*, where* r *is the index of* L_f *which can be computed in terms of* $c_1(X)$*, the homology class of* $f(\Sigma)$*, the genus of* Σ *and the number of marked points.*

Using this lemma, one can easily show that $(\tilde{\mathbf{U}}, G_f, \mathbf{E}_{\tilde{\mathbf{U}}}, \mathbf{F})$ is a local finite approximation. Furthermore, one can assign it a natural orientation by the canonical orientation on the determinant line $\det(L_f)$ of L_f. Note that L_f is different from a J-invariant $\bar\partial$-operator $\bar\partial_f$ by an zero-order operator, so $\det(L_f) = \det(\bar\partial_f)$, hence, the canonical orientation is given by the canonical one on $\det(\bar\partial_f)$. Lastly, it is easy to see that strata of the finite model U of $(\tilde{\mathbf{U}}, \mathbf{F})$ mapped submersively to strata of $\mathfrak{M}_{g,k} \times^k$, if we choose enough sections $v_1, \cdots, v_l$. This completes the proof of the main theorem.

5. Final remark

We now make a few remark to tie up the lose end of our presentation.

The first is the equivalence of the algebraic and the symplectic definition of GW invariants. This was generaly believed to be true and was proved in [**LT4, Si2**].

THEOREM 5.1. *Let* X *be any smooth projective variety with a Kähler form* ω*. Then the algebraically constructed GW-invariants of* X *coincide with the analytically constructed GW-invariants of the symplectic manifold* $(X^{\mathrm{top}}, \omega)$.

The proof relied on the fact that we can find holomorphic finite approximation $\phi: U \to F$, in the symplectic construction of GW invariants, so that the obstruction theory induced by ϕ is identical to the canonical obstruction of $\mathfrak{M}_{g,k}(X, A)$. We refer the details to our paper above.

The second is some basic properties of the GW invariants. One property of virtual moduli cycle we stressed earlier is that if X_T is a smooth family of peojective varieties with smooth base and if A_t extends to a global constant family in $\cup_t A_* X_t$, it exists after a base change of T if necessary, then the GW invariant $\gamma^{X_t}_{A_t,g,k}$ should be a local constant. This is indeed the case. It is proved by showing that the virtual cycle $[\mathfrak{M}_{g,k}(X_t,A_t)]^{\mathrm{vir}}$ is the image of the Gysin map $\eta_t^*[\mathfrak{M}_{g,k}(X_T,A_T)]^{\mathrm{vir}}$, where η is defined by the square

$$\begin{array}{ccc} \mathfrak{M}_{g,k}(X_t,A_t) & \xrightarrow{\eta} & \mathfrak{M}_{g,k}(X_T,A_t) \\ \downarrow & & \downarrow \\ \{t\} & \longrightarrow & T \end{array}$$

The second is that the GW invariants are expected to satisfy several important relations, among them the associativity (or composition) law. This again can be proved by studying some geometric (co)cycle Δ in the moduli of curves $\mathfrak{M}_{g,k}$ and the virtual cycle of

$$\mathfrak{M}_{g,k}(X,A) \times_{\mathfrak{M}_{g,k}} \Delta.$$

We should also mention that the alternative construction of Behrend-Fantechi [**BF**] used the notion of cotangent complex to represent the obstruction theory of the moduli stack $\mathfrak{M}_{g,k}$, and then constructed a cone, called virtual normal cone, as an Artin stack over the D-M stack $\mathfrak{M}_{g,k}(X,A)$. In the end, they use a vector bundle E over $\mathfrak{M}_{g,k}(X,A)$, similar to the one mentioned earlier, to obtain a cone in E and hence obtain the virtual cycle using the Gysin map of the zero section of E. Recently this last step of using vector bundle E was relaxed by the work of Kresch [**Kr**] after he developed the intersection theory on Artin stack. It is interesting to note our symplectic construction of GW invariants *did not* use the existence of such global vector bundle, thus it can serve as a hint that the global bundle E is not essential to the construction of virtual cycle. We note that the symplectic constructions by Siebert [**Si1**] and Ruan [**Ru2**] both relied on such global vector bundles.

References

[Be] K. Behrend, *Gromov-Witten invariants in alebraic geometry*, Invent. Math. **127**(1997), no. 3, 601-617.

[BF] K. Behrend and B. Fantechi, *The intrinsic normal cone*, Invent. Math. **128**(1997), no. 1, 45-88.

[Ca] P. Candelas, X. Ossa, P. Green and L. parkes, *A pair of Calabi-yau manifolds as an exactly solvable superconformal theory*, Nucl. Phys. B359 (1991), 21-74.

[DM] P. Deligne and D. Mumford, *The irreducibility of the space of curves of given genus*, Publ. I.H.E.S. **45**(1969), 101-145.

[FO] K. Fukaya and K. Ono, *Arnold conjecture and Gromov-Witten invariants*, preprint, 1996.

[Fu] W. Fulton *Intersection theory,* Ergebnisse der Math. und ihrer Grenzgebiete 3. Folge Band 2, Springer 1984.

[Gi] A. Giventál, *Equivariant Gromov-Witten invariants*, Internat. Math. Res. Notices (1996), no. 13, 613–663.

[Gr] M. Gromov, *Pseudoholomorphic curves in symplectic manifolds*, Invent. Math. **82**(1985), no. 2, 307-347.

[KM] M. Kontsevich and Y. Manin, *Gromov-Witten classes, quantum cohomology, and enumerative geometry*, Comm. Math. Phys. **164**(1994), no. 3, 525-562.

[Ko] M. Kontsevich, *Enumeration of rational curves via torus actions*, The Moduli Space of Curves, edited by R.Dijkgraaf, C.Faber, G. van der Geer, Progress in Mathematics vol. 129, Birkhauser, 1995.

[Kr] A. Kresch, *Cycle groups for Artin stacks*, preprint math.AG/9819166.

[LT1] J. Li and G. Tian, *Virtual moduli cycles and Gromov-Witten invariants of algebraic varieties*, J. Amer. Math. **11**(1998), no. 1, 119-174.

[LT2] J. Li and G. Tian, *Virtual moduli cycles and Gromov-Witten invariants of general symplectic manifolds*, Topics in symplectic 4-manifolds (Irvine, CA, 1996), 47–83, First Int. Press Lect. Ser., I, Internat. Press, Cambridge, MA, 199.

[LT3] J. Li and G. Tian, *Algebraic and symplectic geometry of virtual moduli cycles*, Algebraic geometry—Santa Cruz 1995, 143–170, Proc. Sympos. Pure Math., 62, Part 2, Amer. Math. Soc., Providence, RI, 1997.

[LT4] J. Li and G. Tian, *Comparison of algebraic and symplectic Gromov-Witten invariants*, to appear in Asian J. Math.

[LT5] J. Li and G. Tian, in preparation.

[LLY] K-F. Liu, B. Lian and S-T. Yau, *Mirror principle I*, Asian J. math. **1**(1997), no. 4, 729-763.

[PW] T. Parker and J. Wolfson, *Pseudo-holomorphic maps and bubnle trees*, J. Geom. Anal. **3**(1993), no. 1, 63-98.

[Rn] Z. Ran, *Deformations of maps*, Algebraic curves and projective geometry, E. Ballico, C. Ciliberto, eds, Lecture notes in Math. 1389.

[Ru1] Y. Ruan, *Topological Sigma model and Donaldson type invariants in Gromov theory*, Duke Math. J. **83**(1996), no. 2, 461–500.

[Ru2] Y. Ruan, *Virtual neighborhoods and pseudo-holomorphic curves*, preprint, 1996.

[RT1] Y.B. Ruan and G. Tian, *A mathematical theory of quantum cohomology*, J. Diff. Geom. **42**(1995), no. 2, 259-367.

[RT2] Y.B. Ruan and G. Tian, *Higher genus symplectic invariants and sigma model coupled with gravity*, Invent. Math. **130**(1997), no. 3, 455–516.

[Si1] B. Siebert, *Gromov-Witten invariants for general symplectic manifolds*, preprint, 1996, (AG/9608005).

[Si2] B. Siebert, *Algebraic and symplectic Gromov-Witten invariants coincide*, preprint, 1998, (AG/9804108).

[Vi] A. Vistoli, *Intersection theory on algebraic stacks and on their moduli spaces*, Invent. Math. **97**(1989), no. 3, 613-670.

[Wi1] E. Witten, *Topological sigma models*, Comm. Math. Phys. **118**(1988), no. 3, 411-449.

[Wi2] E. Witten, *Two dimensional gravity and intersection theory on moduli space*, Surveys in differential geometry (Cambridge, MA, 1990), 243–310, Lehigh Univ., Bethlehem, PA, 1991.

Made in United States
North Haven, CT
05 November 2026

21764580R00317